BIFURCATION THEORY
AND APPLICATIONS
IN SCIENTIFIC DISCIPLINES

ANNALS OF THE NEW YORK ACADEMY OF SCIENCES
Volume 316

BIFURCATION THEORY
AND APPLICATIONS
IN SCIENTIFIC DISCIPLINES

Edited by Okan Gurel and Otto E. Rössler

The New York Academy of Sciences
New York, New York
1979

Library of Congress Cataloging in Publication Data

Main entry under title:

Bifurcation theory and applications in scientific disciplines.

(Annals of the New York Academy of Sciences; v. 316)
Papers from a conference held Oct. 31–Nov. 4, 1977, sponsored by the New York Academy of Sciences and the University of Tübingen.
Includes bibliographical references and indexes.
1. Bifurcation theory—Congresses. I. Gurel, Okan.
II. Rössler, Otto E. III. New York Academy of Sciences. IV. Tübingen. Universität. V. Series: New York Academy of Sciences. Annals; v. 316.
Q11.N5 vo. 316 [QA370] 508'.ls [501'.515] 78-32114
ISBN 0-89766-000-5

SP
Printed in the United States of America
ISBN 0-89766-000-5

ANNALS OF THE NEW YORK ACADEMY OF SCIENCES
VOLUME 316
February 28, 1979

BIFURCATION THEORY AND APPLICATIONS IN SCIENTIFIC DISCIPLINES*

Editors and Conference Chairmen
Okan Gurel and Otto E. Rössler

CONTENTS

*This series of papers is the result of a conference entitled Bifurcation Theory and Applications in Scientific Disciplines, held on October 31, November 1, 2, 3, and 4, 1977 and sponsored by The New York Academy of Sciences and the University of Tübingen.

Financial assistance was received from:

- DEUTSCHE FORSCHUNGSGEMEINSCHAFT
- INTERNATIONAL BUSINESS MACHINES CORPORATION
- SHELL DEVELOPMENT COMPANY
- UNION CARBIDE CORPORATION
- U.S. ARMY RESEARCH OFFICE

PREFACE

Okan Gurel

IBM Corporation
White Plains, New York 10604

In recognition of the stimulating inspiration that was provided by the work of Professor Eberhard Hopf, this volume of the *Annals* is dedicated to him on his 75th year.

The following is a brief history of the development of the concept of bifurcation, a preface for this collection of papers presented at The New York Academy of Sciences Conference on Bifurcation Theory and Applications in Scientific Disciplines in October 1977.

There are a few individuals whose pioneering ideas and efforts open up many avenues for understanding problems of theoretical and practical importance. J. H. Poincaré (1854–1912) introduced many such ideas, most of which led to new developments in both mathematics and scientific disciplines in general. The concept of bifurcation belongs to one of Poincaré's pioneering studies. In 1881, Poincaré introduced the basic elements entering the theory of bifurcation. In an 1885 paper,[1] he defined bifurcation, discussed the elements of the theory, and described the phenomenon to an extremely detailed extent.

The basic elements of the theory may be grouped into three areas; critical solutions, dynamic stability, and structural (in)stability. The advances made in these three areas form the body of the work done in the bifurcation field. The concept basically implies that system behavior goes through some fundamental changes, *qualitative* in nature. However, one of the conditions of bifurcation phenomena is a change in the number of critical solutions, a quantitative aspect. Because of this change in critical solutions, the Poincaré bifurcation concept is extended over some buckling problems of engineering origin. In fact, Leonard Euler (1707–1783), who discussed the problem of the buckling of columns in his 1774 study "De curvis elasticis," may be named as the eminent scientist of the "pre-bifurcation" period. Many problems in physics, such as phase transitions, or in engineering, such as buckling of structures, are in a sense extensions of Euler's concept of buckling (branching).

An important observation here is that in both Euler's approach and Poincaré's bifurcation concept the study is based on analyzing how the *parameters* effect the qualitative variations in the solution space.

The theoretical formulation of the bifurcation concept is certainly due to Poincaré. In fact, because of this formalization, three fundamental sections of bifurcation theory have subsequently inspired researchers to study either each section separately or to advance these sections without making a distinct separation between them. The earliest significant work is that of Poincaré's contemporary, A. M. Liapunov (1857–1918). In 1892, Liapunov published his treatise on the stability of critical solutions.[2] An important aspect of this work is that it is also qualitative.

It is interesting that during the period of half a century after the introduction of the bifurcation concept there was no other activity except for the work of

1

0077-8923/78/0316-0001 $1.75/1 © 1979, NYAS

Liapunov. The first attempt after this long pause is again by a Russian mathematician, A. A. Andronov (1901–1952), who in 1937 together with L. Pontriagin published his article on coarse systems,[3] which was the beginning of a series of studies on bifurcation problems by Andronov and his colleagues. The significance of Andronov's work comes from the fact that this is the first paper where the concept of structural (in)stability of Poincaré is elaborated. Shortly after, E. Hopf (1902–) published his paper in 1942,[4] where he elaborated on the bifurcation of limit cycles from singular points by giving his now celebrated theorem of bifurcation.

Although Poincaré's collected works came out in 1951,[5] unfortunately it was in the original language, French, making it not easily accessible for the English-speaking community.

Because of the leadership of S. Lefschetz (1884–1972), many mathematical studies in Russian, particularly those of the Andronov school were introduced to Western mathematicians. In fact Lefschetz in his 1957 book coined the word *structural stability*,[6] replacing the Russian term *coarse system* (earlier translated into French as *systèmes grossières*). Shortly after, M. M. Peixoto in 1959,[7] and S. Smale in 1960,[8] published their studies on the concept of structural stability. It is interesting to note that Lefschetz also introduced Liapunov's work to the English-speaking world, thus playing another important role in the field of bifurcations.

After many papers and studies, Andronov's book specifically on bifurcations on a plane,[9] was published by his colleagues and students posthumously in 1967. This was translated into English in 1971.

The modern version of dynamic concepts started by Poincaré so many years earlier appeared in S. Smale's 1967 article.[10] This paper became a significant reference in dynamic systems in general. About the same time, in 1968, R. Thom published his article on structural stability where the *catastrophe theory* was introduced.[11] It is of course clear that in bifurcations, as Poincaré stated, structural (in)stability is important.

Independent from the main emphasis of bifurcations, E. Lorenz published his article in 1963 concerning turbulence.[12] The solution corresponding to turbulence belongs to critical solutions, discussed almost a three quarters of a century prior to this work by Poincaré. The significance of the Lorenz paper is that together with Hopf's paper they started a new flow of research both theoretical and applied in the field of critical solutions.

In 1973, the term *peeling* was proposed to replace the term bifurcation because the concept of peeling is more general than *bi-furcation* and also *branching* used in connection with Euler problems.[13]

On the side of applications, in engineering there are the earliest examples of Euler branching. Applications in the physical sciences are phase transitions and some hydrodynamic problems. In chemical fields, because of oscillations as formulated in the models of some chemical kinetics, the bifurcation phenomena, if they enter into these problems, become more of the Poincaré type than the Euler type. In fact, many interesting developments in the recent past originated from the studies in chemical kinetics such as those by E. E. Sel'kov. In biological models, more complicated solutions may be expected as in the abstract models of O. E. Rossler. It should be added that an important interdisciplinary field based on the similarities of "qualitative changes of the systems," with bifurcations playing a central role, is *Synergetics* introduced by H. Haken.[14]

The theory and applications of the bifurcation concept have been expanding in the recent past and there have been a few conferences, with proceedings appearing as an outcome of these symposia. That of J. B. Keller and S. Antman (1969)[15] was published based on seminar presentations on Euler-type branching in engineering problems. A more recent book by P. H. Rabinowitz is the proceedings of another conference on applications in physical and engineering problems with an emphasis on Euler-type branching.[16] An earlier treatise along these lines is by M. M. Vainberg and V. A. Trenogin[17] where the Liapunov-Schmidt technique for branching is discussed at length. In a book edited by P. Bernard and T. Ratiu[18] are reported the presentations made at the Turbulence Seminar organized by A. Chronin, J. Marsden, and S. Smale. The interest is still growing and a platform to discuss what has been accomplished and how each individual effort fits into the field needed to be discussed. The New York Academy of Sciences Conference on Bifurcation Theory and Applications in Scientific Disciplines became the first bifurcation conference addressing all the areas of interest in the field of bifurcations. In order to include all the possible developments and unify the efforts in many different directions, researchers from various disciplines gathered in New York City. E. Hopf, S. Smale, E. Lorenz, D. Ruelle, J. M. T. Thompson, G. W. Hunt, D. H. Sattinger, R. Noyes, J. E. Marsden, and P. H. Rabinowitz, to name a few, and many others active in the field were the participants at the Conference.

In addition to the contributions discussed above, various papers and books outlining further studies in theoretical as well as in application-related topics exist. The New York Academy of Sciences Conference provided a unified platform where efforts in many different directions were discussed and related to each other. This included not only the sections on basic and applied sciences where a number of contributions are well known, but also disciplines such as ecology and economics where the role of bifurcation is still debated. There are two interacting directions in which the field will expand: critical solutions and their application fields. Critical solutions characterizing the behavior of a system are becoming extremely interesting and intricate. The hydrodynamic phenomenon of turbulence and the chaotic behavior of interacting elements are identified as critical solutions. Mathematicians are proving theorems stating necessary and sufficient conditions for the existence of such solutions. The impact of models from the applied fields on these new critical solutions is obvious. In turn, the new discoveries in critical solutions will open up possibilities for modeling complex systems with an increasing exactness. Applied scientists are incorporating these new mathematical findings in their analyses of applications to practical problems. In both theory and applications, the need for ingenious techniques adapted from applied mathematics will be undoubtedly felt; moreover, the use of computers as a tool to identify, classify, and apply new findings will be immensely beneficial.

To summarize an extensive conference such as this one is almost impossible. One can only guide each reader to the area of interest. With this in mind, to have a structure, the papers are grouped under subsections, where the reader will notice a certain amount of overlapping.

The organization of this conference is the result of the close cooperation of dedicated individuals. We would like to thank every member of the Conference Committee, Dr. Srinivasan, Chairman, and in particular, the subcommittee members Drs. Aissen and Chi for their continuous interest and support. Mrs. Ellen A. Marks and Miss Renee Wilkerson, who carried the burden of preparation for the conference from the very first day to the end efficiently and smoothly, are the key elements in the success of the conference. Miss Ann Collins expertly fulfilled the publicity task in the most pleasant fashion. Mr. Bill Boland, Mr. Bill Klein, and Mr. Richard Malloy whose expertise made the conference proceedings available to the scientific world in a remarkably short time without sacrificing in superb quality, were always most accommodating and full of enthusiasm. We all owe so much to these experts, and fall short in expressing our gratitude.

We reiterate our gratitudes once again to all the sponsoring institutions and to The New York Academy of Sciences.

References

1. POINCARÉ, H. 1885. Sur l'équilibre d'une masse fluide animeé d'un movement de rotation. Acta Math. **7**: 259–380.
2. LIAPUNOV, A. M. 1892. Problème général de la stabilité du mouvement. (1907 French translation of 1892 Russian memoires.) Reprinted 1949 by Princeton University Press. Princeton, N. J.
3. ANDRONOV, A. & L. PONTRIAGIN. 1937. Grubie Sistemi. Dokl. Akad. Nauk SSSR. Tom **14**(5): 247–251.
4. HOPF, E. 1942. Abzweigung einer periodischen Lösung von einer stationaren Lösung eines Differentialsystems, Berichten der Math.-Phys. Klass der Sachlischen Akademie der Wissenschaften zu Leipzig. **94**: 3–22.
5. POINCARÉ, H. 1951. Oeuvres. Gauthier-Villars. Paris.
6. LEFSCHETZ, S. 1957. Differential Equations: Geometric Theory. Interscience Publishers. New York, N.Y.
7. PEIXOTO, M. M. 1959. On structural stability. Ann. Math. **69**(2): 199–222.
8. SMALE, S. 1960. On dynamical systems. Bol. Soc. Mat. Mexicana. 195–198.
9. ANDRONOV, A. A., E. A. LEONTOVICH, I. I. GORDON & A. G. MAIER. 1973. Theory of Bifurcations of Dynamics Systems on a Plane. John Wiley and Sons. New York, N.Y. (Translation of the Russian 1967 edition.)
10. SMALE, S. 1967. Differentiable dynamical systems. Bull. Amer. Math. Soc. **73**: 747–817.
11. THOM, R. 1968. Topological methods in biology. Topology **8**: 313–385.
12. LORENZ, E. 1963. Deterministic nonperiodic flow. J. Atmos. Sci. **20**: 130–141.
13. GUREL, O. 1973; 1975. Peeling and nestling of a striated singular point. Not. Amer. Math. Soc. **20**: A380; Collect. Phenom. **2**: 89–97.
14. HAKEN, H. & R. GRAHAM. 1971. Synergetik-die Lehre vom Zusammenwirken. Umschau **6**: 191–195.
15. KELLER, J. B. & S. ANTMAN. 1969. Bifurcation Theory and Nonlinear Eigenvalue Problems. W. A. Benjamin. New York, N.Y. (1969).
16. RABINOWITZ, P. H. 1977. Applications of Bifurcation Theory. Academic Press. New York, N.Y.
17. VAINBERG, M. M. & V. A. TRENOGIN. 1974. The Theory of Branching of Solutions of Non-linear Equations. Noordhoff International Publishers. Leiden, Netherlands. (Translated from the 1969 Russian edition).
18. BERNARD, P. & T. RATIU, Eds. 1977. Turbulence Seminar, Berkeley 1976/1977. Organized by A. Chorin, J. Marsden & S. Smale. Lecture Notes in Mathematics, No. 615. Springer-Verlag. New York, N.Y.

POINCARÉ'S BIFURCATION ANALYSIS

Okan Gurel

IBM Corporation
White Plains, New York 10604

In his 1885 paper, Poincaré studied the equilibrium of a fluid in rotational motion.[1] The two questions he considered are: *What are the* equilibrium figures? and, *What are the conditions of stability of this equilibrium?*

If $F(x_1, \ldots, x_n)$ is a function of forces, the equilibrium lies where all the derivatives of this function vanish; i.e., where,

$$\frac{dF}{dx_1} = \frac{dF}{dx_2} = \cdots = \frac{dF}{dx_n} = 0 \tag{1}$$

where dF/dx_i is a partial derivative. Poincaré calls (1) *equations of equilibrium,* and their roots, *equilibrium forms.* If the system depends on a parameter y, the roots of the equilibrium equations can be expressed as continuous functions of y; thus,

$$x_1 = \phi_{i1}(y), x_2 = \phi_{i2}(y), \ldots, x_n = \phi_{in}(y), \quad i = 1, \ldots, n \tag{2}$$

Poincaré calls these a *linear sequence of roots.* For the same values of y, two or more roots may coincide; thus $\phi_{ij}(y) = \phi_{kj}(y), i \neq k, j = 1, \ldots, n$.

The Hessian Δ is the functional determinant of the n derivatives dF/dx_1, $\ldots, dF/dx_n$ with respect to n variables $x_1, \ldots, x_n$. Poincaré states that:

THEOREM 1. The necessary and sufficient condition for two or more roots to coincide is that Δ vanish[1] (p.262).

It is assumed that at $y = y_0$ the function $\Delta(y)$ changes its sign. Thus the series $\phi_1(y_0), \phi_2(y_0), \ldots, \phi_n(y_0)$, consists of real roots. It is also possible that as y varies between $y_0 - \epsilon$ and $y_0 + \epsilon$, Δ may either go from negative to positive values, or have the same sign on either side of y_0.

As y varies, an equilibrium form Equation 2 may vary to provide diverse equilibrium forms such that one equilibrium form belongs to two or more of the linear sequence mentioned above. Then, this is called a *bifurcation form.* In fact, for a given value of y infinitely close to the y_0 value corresponding to the bifurcation form, *two* equilibrium forms that differ by an infinitely small amount from the bifurcation form can be found.

Also, as y varies, the two linear sequence of real equilibrium forms come close to overlapping, but then disappear because the roots of the equilibrium equation become imaginary. The corresponding equilibrium form is called a *limit form.* Poincaré[1] states: *An equilibrium form may be a bifurcation form or a limit form only when* Δ *becomes zero* (p.270). If Δ passes zero and changes its sign, the corresponding equilibrium point may not be a limit form because the nearby equilibrium forms are supposed to be real. Therefore, this is always a *bifurcation form.*

5

0077–8923/78/0316–0005 $1.75/1 © 1979, NYAS

Jules Henri Poincaré (1854–1912). (From Slosson.[12])

If there is no change of sign, it is still not a limit form; moreover, it may be a bifurcation form, but not always.

The stability of an equilibrium form is defined in terms of *stability coefficients*. These are,

$$\frac{d^2F}{dx_1^2}, \frac{d^2F}{dx_2^2}, \ldots, \frac{d^2F}{dx_n^2} \tag{3}$$

This is based on the assumption that $d^2F/dx_i dx_j = 0$, $x_i \neq x_j$. For Δ to vanish, one or more of these terms should also vanish. Among the remaining terms, those with negative values indicate stability. For stability it is necessary and sufficient that all the terms are negative. As y varies, if $y > y_0$, stability exist, but if $y < y_0$, instability exists. Then this indicates an *exchange of stabilities*. Poincaré concludes:

THEOREM 2. For an equilibrium form corresponding to a real linear sequence to be a bifurcation form, it is sufficient not only that Δ changes sign, but also that some of the stability coefficients also change sign (p.276).

This theorem specifically points out that in order to have bifurcation there must be (1) a sign change in stability coefficients, and (2) a change in the number of equilibrium forms.

Following his discussion of *bifurcation equilibrium* (p.261) and the *change of stability* (p.270), Poincaré continues with a discussion on the *stability of relative equilibrium* (p.293). He states:

It is very easy to find the condition of stability of absolute equilibrium of a physical system with respect to fixed axes. For such an equilibrium to be stable, it is necessary and sufficient that the function of forces be maximum. However, the problem of the stability of the *relative equilibrium* of a physical system with respect to moving axes is extremely complicated.

In recognizing this complexity, Poincaré makes the distinction between *secular stability* and *ordinary stability*. Suppose the position of the system under consideration with respect to moving axes is defined by n variables $x_1, \ldots, x_n$, chosen such that equilibrium is at $x_1 = x_2 = \ldots - x_n = 0$. If the system is perturbed by a very small amount from its equilibrium position, the values of $x_1, \ldots, x_n$ become very small, and if the quadratic terms of these quantities are neglected, the differential equations that define the variations become linear. The solutions of these linear equations are expressed in terms of the exponential of λ_m, $m = 1, 2, \ldots, 2n$, where the λ_m are given by an algebraic equation of degree $2n$ such that to find the conditions of stability, it is sufficient to discuss this equation of λ. Poincaré, of course, refers to characteristic equation and the characteristic values as roots of this equation. If all of the λ_m are pure imaginary, stability exists. If their real parts are zero or negative, stability still exists. However any λ_m has its real part positive, we have instability.

Assuming that the potential energy of a system vanishes at the equilibrium point, Poincaré states: In the case of the absolute stability, *that is to say, when there is no motion, for stability it is sufficient that the quadratic form of potential energy is positive definite* (p.295). This condition is also sufficient, but not neces-

Aleksandr Mikhaylovich Liapunov (1857–1918). (From Ishlinskogo.[13])

sary, if there is motion. If the stability coefficients are *all* negative, stability will exist even if there is motion. This is the fundamental theorem of Liapunov.[2]

When *relative equilibrium* is stable, we have *ordinary stability,* however if equilibrium remains stable even after introducing, say, viscosity terms, we have *secular stability.* Without secular stability, it is possible to have ordinary stability, such as in the case of a very small viscosity. For secular stability to exist, it is necessary and sufficient that the potential energy be positive definite (Liapunov stability); that is, *all of the stability coefficients are negative* (p.295).

The conditions for the ordinary stability are much more complicated. We should add that this stability *can never take place, if the number of positive stability coefficients is odd* (p.295). Referring to Laplace's *Livre* IV, Chapter II, No. 14, Poincaré gives a theorem of Laplace: *If potential energy is not positive definite there can be no stability* (p.298). From this Poincaré concludes:

If a system is in a state of relative equilibrium, *a figure of revolution, this equilibrium does not have ordinary stability.*

It can be seen that, when the system becomes unstable only one (odd number) of the λ's becomes null, thus the conclusion is proved. An important point is stated as follows:

Stability can still exist if there is only a small difference between the exterior and primitive figures of a system; however, the divers parallels tend to turn with different velocities (p.299).

In conclusion it can be seen that THEOREMS 1 and 2 above remain almost unaltered:

(Theorem 1) If one of the stability coefficients in a linear sequence of the equilibrium figure changes its sign, it is true that the corresponding form is a bifurcation.

(Theorem 2) The exchange of stabilities so far as secular stability is concerned also holds true where conditions are not changed very much by additional forces. The same holds true for ordinary stability, however, only if the stability coefficients vanish one at a time.

In fact, we have seen that we may have ordinary stability when only one coefficient of stability, or, in general, an odd number of these coefficients, is positive and all the others are negative.

It should be remarked here that the work of Andronov and Pontryagin[3] on the definition of structurally stable systems as applied by Andronov *et al.*[4] to systems on a plane is an extension of Poincaré's work. Poincaré's conditions for bifurcation state that $\Delta = 0$ (structurally unstable), while the Andronov-Pontryagin conditions for structural stability require $\Delta \neq 0$ (Ref. 4, p.50).

Discussions of periodic solutions and the appearance or disappearance of various periodic solutions are given in Poincaré's treatise on celestial mechanics.[5] In particular, the connection between bifurcation[1] and periodic solutions[5] can be made via the concept of stability. For the differential equations in Hamiltonian, or canonical, forms, an integral F of the system of differential equations,

$$\frac{dx_i}{dt} = X_i(x_1, \ldots, x_n), \quad i = 1, \ldots, n \tag{4}$$

can be viewed as a function of force given in Equation 1 (Ref. 5a, p.8). A solution

Aleksandr Aleksandrovich Andronov (1901–1952). (From Reference 14.)

of Equation 4 is defined as,

$$x_i = \phi_i(t), \quad i = 1, \ldots, n \tag{5}$$

A periodic solution is defined as,

$$\phi_i(t) = \phi_i(t + T) \tag{6}$$

where T is the period[5] (p.79). We can see that if $x_i = b_i$ $(i = 1, \ldots, n)$ corresponds to the initial values of x_i, then after period T,

$$\phi_i = b_i + f_i, \quad i = 1, \ldots, n \tag{7}$$

The f_i may be expanded in a power series about x_i and m, where m is a parameter of the X_i.

The characteristic equation of Equation 4 may be written as,

$$\begin{vmatrix} \dfrac{dX_1}{dx_1} - s & \dfrac{dX_1}{dx_2} & \cdots \\ \vdots & \vdots & \vdots \\ \dfrac{dX_n}{dx_1} & \cdots & \dfrac{dX_n}{dx_n} - s \end{vmatrix} = 0 \tag{8}$$

which yields n roots $s_1, \ldots, s_n$ (Ref. 5a, p.157).

It can be verified that the functional determinant of f with respect to b, for $m = 0$ and $b_i = 0$, is equal to,

$$\Delta = (e^{s_1 T} - 1) \ldots (e^{s_n T} - 1) \tag{9}$$

To have a periodic solution with a period T it is sufficient that

$$f_i = 0, \quad i = 1, \ldots, n \tag{10}$$

If the determinant Δ in (9) vanishes there can be one or more periodic solutions that vanish as $m \to 0$, satisfying (10). For Δ in (9) to vanish, it is sufficient that one $e^{s_i T} = 1$. This implies that one of the roots of (8) becomes null (Ref. 5a, p.159). Therefore if $\Delta = \partial(f_i)/\partial(b_j) \gtrless 0$, Equation 4 must have a periodic solution for small values of m (p.180).

In studying stability, Poincaré refers to the characteristic exponents and the stability of the periodic solutions (5b, p.343). Since Equation 4 depends on a parameter m, as m approaches a critical value, Poincaré states: *We see that two periodic solutions approach each other, merge, and then disappear* (Ref. 5b, p.346). He further states: *We have, therefore, two analytical sequences of periodic solutions, for* m = 0, *they merge, at this moment the two sequences exchange stabilities* (p.347). He concludes: *Therefore, if, for a certain value of* m, *a periodic solution gains or loses stability (in such a way that the exponent* a *becomes null), it merges with another periodic solution with which it exchanges its stability* (p.349).

Poincaré's comment that a change of sign in the real parts of imaginary roots implies stability changes necessary for bifurcations to periodic solutions is discussed by Hopf.[6] His notions have been highly influential in recent studies on bifurcations to periodic solutions.

Without further elaboration, various examples may be studied using the original definitions of bifurcation given by Poincaré in the following examples.

EXAMPLE 1 (Van der Pol equation).
  Differential equations: $\dot{x} = y$, $\dot{y} = -m(x^2 - 1)y - x$
  Equilibrium equations:

$$\frac{dF}{dx} = y = 0$$

$$\frac{dF}{dy} = -m(x^2 - 1)y - x = 0$$

Equilibrium forms: $x = 0, y = 0$

Hessian:

$$\frac{d^2F}{dx^2} = 0, \frac{d^2F}{dxdy} = 1, \frac{d^2F}{dydx} = -1 - my2x, \frac{d^2F}{dy^2} = -m(x^2 - 1),$$

$$\Delta = \begin{vmatrix} 0 & 1 \\ -1 - my2x & -m(x^2 - 1) \end{vmatrix} = 1 + 2mxy$$

Stability: At $x = y = 0$, linearized equations are,

$$\dot{x} = y$$

$$\dot{y} = -x + my$$

The characteristic equation:

$$\begin{vmatrix} -s & 1 \\ -1 & m - s \end{vmatrix} = s^2 - ms + 1 = 0$$

$$s = \frac{m}{2} \pm \sqrt{\frac{m^2}{4} - 1}$$

Therefore the origin is stable for $m < 0$, and unstable for $m > 0$. Furthermore, for $m > 2$ it is a focus; for $m < 2$ it is a node. It is, of course, well known that the Van der Pol equation has not only a singular point at the origin, but also a limit cycle surrounding it with a proper stability corresponding to the stability of $(0, 0)$.
  Bifurcation: For bifurcation Δ must be zero. However, at $x = y = 0$, $\Delta = 1$. For any value of m, the origin cannot be a bifurcation form, although the stability changes as m crosses the zero value.

EXAMPLE 2 (Truncated equations of a harmonic oscillator).
  Differential equations: $\dot{x} = y, \dot{y} = -ax(1 - b^2x^2)$.
  Equilibrium equations:

$$\frac{dF}{dx} = y = 0$$

$$\frac{dF}{dy} = -ax(1 - b^2x^2) = 0$$

Equilibrium forms:

$$x = y = 0, \quad \text{and} \quad y = 0, \ x = \pm \frac{1}{b}.$$

Hessian:

$$\frac{d^2F}{dx^2} = 0, \ \frac{d^2F}{dxdy} = 1, \ \frac{d^2F}{dydx} = -a + ab^23x^2, \ \frac{d^2F}{dy^2} = 0.$$

$$\Delta = \begin{vmatrix} 0 & 1 \\ -a + ab^23x^2 & 0 \end{vmatrix} = a - ab^23x^2$$

At $x = y = 0$, $\Delta = a$, therefore at $a = 0$, Δ vanishes.

Stability: At $x = y = 0$, the linearized equations are $\dot{x} = y$, $\dot{y} = -ax$. The characteristic equation is,

$$\begin{vmatrix} -s & 1 \\ -a & -s \end{vmatrix} = s^2 + a = 0$$

$$s_1 = -(-a)^{1/2}, \ s_2 = (-a)^{1/2}$$

For $a > 0$, $(0, 0)$ is a center and for $a < 0$, $(0, 0)$ is a saddle point. By the definition of Poincaré there is a stability change.

Bifurcation: Both $\Delta = 0$ and the stability change are satisfied at $a = 0$. However, one more condition that the number of equilibrium forms changes is not satisfied, therefore $(0,0)$ is *not* a bifurcation form.

For the equilibrium forms $x = \pm 1/b$, $y = 0$, substituting the x value in Δ, it is found that for $a = 0$, Δ vanishes. The characteristic equation becomes $s^2 - 2a = 0$ which yields $s_{1,2} = \pm(-2a)^{1/2}$. For $a < 0$, these equilibrium forms are saddle points and for $a > 0$ they are centers. No change in the number of equilibrium forms takes place, therefore, $(+1/b, 0)$ is not a bifurcation form.

EXAMPLE 3.

Differential equations.

$$\dot{x} = y$$

$$\dot{y} = -p(x + x^3)$$

Equilibrium equations:

$$\frac{dF}{dx} = y = 0, \ \frac{dF}{dy} = -p(x + x^3) = 0$$

Equilibrium forms: $x = y = 0$ (the only one)

Hessian:

$$\frac{d^2F}{dx^2} = 0, \ \frac{d^2F}{dxdy} = 1, \ \frac{d^2F}{dydx} = -p(1 + 3x^2), \ d^2F/dy^2 = 0$$

$$\Delta = \begin{vmatrix} 0 & 1 \\ -p(1 + 3x^2) & 0 \end{vmatrix} = p(1 + 3x^2)$$

At $x = y = 0$, $\Delta = p$, therefore for $p = 0$ it vanishes.

Stability: The linearized equations at $(0,0)$ are $\dot{x} = y$, $\dot{y} = -px$. The characteristic equation is

$$\begin{vmatrix} -s & 1 \\ -p & -s \end{vmatrix} = s^2 + p = 0$$

$$s_1 = (-p)^{1/2}, \quad s_2 = -(-p)^{1/2}$$

If $p < 0$, $(0,0)$ is a saddle point and if $p > 0$, it is a center. By the definition given by Poincaré, it shows a change in stability.

Bifurcation: At $p = 0$, Δ vanishes. Stability also changes. The number of equilibrium forms remain the same, therefore $(0,0)$ is *not* a bifurcation form.

EXAMPLE 4. Lorenz equations.[7]

Differential equations:

$$\dot{x} = -mx + my$$

$$\dot{y} = -x(z - r) - y$$

$$\dot{z} = xy - bz$$

Equilibrium equations:

$$-mx + my = 0$$

$$-x(z - r) - y = 0$$

$$xy - bz = 0$$

Equilibrium forms:

$$x = y = z = 0.$$

$$z = r - 1, \, x = y = \sqrt{(r - 1)b}$$

$$z = r - 1, \, x = y = -\sqrt{(r - 1)b}$$

Hessian:

$$\Delta = \begin{vmatrix} -m & m & 0 \\ -(z - r) & -1 & -x \\ y & x & -b \end{vmatrix} = -m[b(z - r + 1) + x(x + y)]$$

For $m = 0$, Δ is always zero. At $x = y = z = 0$, $b = 0$ or $r = 1$ would make Δ zero. At the second and third equilibrium points, $b = 0$, $r = 1$ are the parameter values where Δ becomes zero.

Stability in the neighborhood of $(0,0,0)$: At $(0,0,0)$ the linearized equations are such that the characteristic equation becomes,

$$\begin{vmatrix} -m - s & m & 0 \\ r & -1 - s & 0 \\ 0 & 0 & -b - s \end{vmatrix} = (b + s)[-s^2 - (m + 1)s + rm - m] = 0$$

The characteristic roots are,

$$s_1 = -b$$

$$s_{2,3} = -\frac{(1 + m)}{2} \pm \frac{\sqrt{(m + 1)^2 + 4m(r - 1)}}{2}$$

The values shown in TABLE 1 divide the parameter space into regions with different stability properties.

TABLE 1

r	$m < -1$	$m = -1$	$-1 < m < 0$	$m = 0$	$0 < m$	
$r < 0$	$s_2 > 0$	$s_2 > 0$	$s_2 > 0$	$s_2 = 0$	$r\uparrow$ ∂B $\begin{cases} s_2 < 0 \\ s_3 < 0 \end{cases}$	
	$s_3 < 0$	$s_3 < 0$	$s_3 < 0$	$s_3 < 0$	$\boxed{r = \partial B \begin{cases} s_2 = 0 \\ s_3 = 0 \end{cases}}$	
					$r\downarrow$ ∂B Re < 0	
$0 < r < 1$	$s_2 > 0$	$s_2 > 0$	$s_2 > 0$	$s_2 = 0$	$s_2 < 0$	
	$s_3 < 0$	$s_3 < 0$	$s_3 < 0$	$s_3 < 0$	$s_3 < 0$	
$r = 1$	$s_2 = 0$	$s_2 = 0$	$s_2 = 0$	$s_2 = 0$	$s_2 = 0$	
	$s_3 < 0$	$s_3 = 0$	$s_3 < 0$	$s_3 < 0$	$s_3 < 0$	
$1 < r$	$r\downarrow$ ∂A $s_2 = 0$	Re $= 0$	$r\downarrow$ ∂A $s_2 = 0$		
	$s_3 < 0$	s_2, s_3	$s_3 < 0$	$s_2 = 0$	$s_2 > 0$	
		(pure Im)				
	$\boxed{r = \partial A \; s_2 = 0 \atop s_3 = 0}$		$\boxed{r = \partial A \; s_2 = 0 \atop s_3 = 0}$	$s_3 < 0$	$s_3 < 0$	
	$r\uparrow$ ∂A Re > 0		$r\uparrow$ ∂A Re < 0			

For positive values of b, s_1 is always negative; for the negative values of b, s_1 is always positive; for $b = 0$, $s_1 = 0$. Therefore, at $b = 0$, one of the characteristic values vanishes, thus the stability also changes through $b = 0$. On the other hand, s_2 and s_3 are completely analyzed on the m,r-plane by setting the relations shown above for s_2 and s_3 equal to zero (FIGURE 1).

The Number of Equilibrium Forms: The preceding stability analysis is applicable in the neighborhood of the origin $(0, 0, 0)$. In order to have bifurcation at r, m, and b values, where (1) the Hessian becomes zero, and (2) the stability changes, we must also find a change in the number of equilibrium forms. Ignoring the occurrence of periodic solutions, such as limit cycles, the possibilities of singular points can be studied by examining the equations given for the second and third singular points as summarized in TABLE 2.

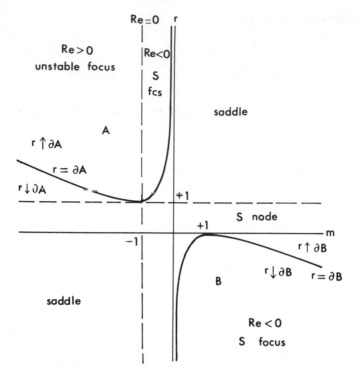

FIGURE 1. Regions of different stability properties on the m,r-plane of the parameter space.

For all the values of m TABLE 2 remains the same. One can easily see that if r crosses the value 1, the number of singular points can change from one to three. For $r = 1$ the unique singular point is a triple value point. For $r \neq 1$, there is one real point or three distinct singular points. At $b = 0$, the entire z-axis becomes an equilibrium form.

   Bifurcation: It should be noted that bifurcations of $(0, 0, 0)$ take place first. That is, the system has only one singular point at the origin (the "1" in the square

TABLE 2

r	b		
	<0	$=0$	>0
<1	3	∞ z-axis	1
$=1$	1	∞ z-axis	1
>1	1	∞ z-axis	3

in TABLE 2). The system has two states: one exhibits a stable singular point ($b > 0$, $r < 1$) and the other, an unstable singular point ($b < 0$, $r > 1$). As proper variations are followed along b and r, *multiple* points appear. However, only those variations along r lead to bifurcations. The parameters m and b are not bifurcation parameters. For $(0, 0, 0)$, the possibilities of bifurcation are listed in TABLE 3.

Stability in the Neighborhood of the Second and Third Equilibrium Forms: Let us now assume that the bifurcation of the origin has taken place. At the second and third equilibrium forms (singular points), linearized equations are such that the characteristic equation becomes,

$$\begin{vmatrix} -m - s & m & 0 \\ 1 & -1 - s & -\sqrt{b(r - 1)} \\ \sqrt{b(r - 1)} & \sqrt{b(r - 1)} & -b - s \end{vmatrix} = 0$$

Note that this is for the second equilibrium form. For the third, the signs of the terms with the square roots change. Therefore, there should be some sign changes

TABLE 3

$\Delta = 0$	Stability Change	Number of Equilibrium Forms	Type of Equilibrium Form
$b = 0$	None (Saddle point remains)	Changes	No bifurcation
$r = 1$	$b \neq 0$	Changes	Bifurcation
$m = 0$	$r \neq 1$	No change	No bifurcation

in forming the determinant. However, since the 0 in the first row is a multiplier in the terms which change sign, the same determinant is obtained for both the second and third equilibrium forms. The explicit form of the characteristic equation becomes,

$$-s^3 - (b + m + 1)s^2 - (bm + br)s + 2mb(1 - r) = 0$$

Let the three roots be s_1, s_2, and s_3. Therefore, this equation can be written as

$$(s - s_1)(s - s_2)(s - s_3) = 0$$

Comparing the coefficients of the terms with equal powers of s for the two equations, we obtain,

$$s_1 + s_2 + s_3 = -(b + m + 1)$$

$$s_1 s_2 + s_1 s_3 + s_2 s_3 = b(m + r)$$

$$s_1 s_2 s_3 = -2mb(r - 1)$$

From the third equation, $s_2 s_3$ can be expressed in terms of s_1 and substituted into the second equation. From the first equation, $s_2 + s_3$ can also be expressed in terms of s_1 and substituted into the second equation. The resulting equation is a polynomial in s_1,

$$s_1{}^3 + s_1{}^2(m + b + 1) + s_1 b(m + r) + 2mb(r - 1) = 0$$

Since $m, b > 0$ and $r > 1$, the only possible solution for the above equation is for $s_1 < 0$. The possibilities for s_2 and s_3 are listed below:

1. $s_2, s_3 < 0$ implies the second and third points are stable nodes.
2. $s_2, s_3 > 0$ implies the second and third points are both saddle points.
3. s_2 and s_3 are complex conjugates with Re < 0, stable focus.
4. s_2 and s_3 are complex conjugates with Re > 0, unstable focus with stable manifold.

For $s_2 = $ Re $+ i$Im, and $s_3 = $ Re $- i$Im, the above equations become,

$$s_1 + 2\text{Re} = -(b + m + 1)$$

$$s_1 2\text{Re} + (\text{Re}^2 + \text{Im}^2) = b(m + r)$$

$$s_1(\text{Re}^2 + \text{Im}^2) = -2mb(r - 1)$$

Eliminating first $(\text{Re}^2 + \text{Im}^2)$ and then s_1 from the above equations a relation between r and Re can be obtained as follows,

$$r = \frac{[2\text{Re}(A + 2\text{Re})^2 + B(A + 2 + 2\text{Re})]}{[2B - b(A + 2\text{Re})]}$$

where $A = m + b + 1$ and $B = mb$. When Re becomes zero the characteristic complex roots may change sign, thus by setting Re $= 0$ we find that,

$$r = \frac{B(A + 2)}{(2B - bA)} = \frac{m(A + 2)}{(2m - A)}.$$

Therefore this value of r corresponds to the point where a stability change takes place.

The Number of Equilibrium Forms: As discussed earlier when the Hessian becomes zero at $r = 1$, the only change in the number of equilibrium forms occurs; namely, the splitting into three singular points. As one moves along r and reaches the critical value just found, there can be no other singular point branching off the second and the third points.

Bifurcation: Along the bifurcation parameter r, the first bifurcation takes place when $r = 1$. Here, not only does the Hessian become zero, but the number of equilibrium points increases as a stability change takes place. As one moves along r at another point, the stability property of the origin remains the same, however that of the second and third singular points changes. Moreover, we have established that in the domain of the parameter space, (where b is positive) on which the second change in stability calculations were based, the system must remain stable as long as the $b = 0$ border is not crossed. There is no change in the stability property of the origin while the second and third points have manifolds spiraling out. This indicates instability at these points. Then these spirals must

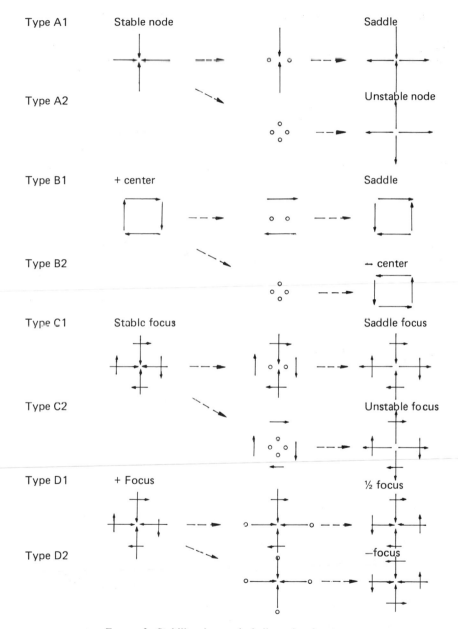

FIGURE 2. Stability changes in 2-dimensional systems.

lead to a stable equilibrium form that cannot be another point, but a limiting object. In fact the limiting object has the following properties.

PROPERTY i.   As r passes the r value found above by setting Re = 0, a stable form appears around the two bifurcated points behaving like limit cycles. This equilibrium form becomes larger as the r values increase, and at one point they combine to form a stable surface surrounding both singular points.

PROPERTY ii.   Since the expression relating r to Re is such that for Re = 0 there is only one possible value of r for any particular set of parameters m and b, both positive for the stable system under consideration, this bifurcation value is the last possible one for the equations discussed. That is to say, after the origin bifurcates to yield two more singular points and then those points bifurcating to form a stable limiting object, there can be no other bifurcation. All other possible phenomena are, if they exist, of a nonbifurcation type.

PEELING STUDIES

In Gurel[8] it is shown that singular points may be classified according to the flow pattern in the neighborhood of a singular point and that various levels can be attached to them. Following a similar argument, several possible stability changes may be listed. The stability changes for 2-dimensional systems are shown in FIGURE 2.

We should mention in passing that although transitions C1 and D1 may not be obtained algebraically, flow patterns for the saddle focus of C1 and the $\frac{1}{2}$ focus of D1 are illustrated in FIGURE 3.

The characteristic roots of the characteristic equation may not reveal the nature of the stability change at the critical value of the bifurcation parameter. The first three examples discussed earlier, all 2-dimensional, have the following

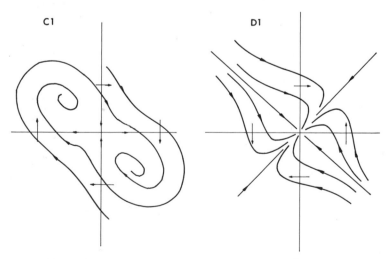

FIGURE 3. Flow patterns for the saddle focus C_1 and $\frac{1}{2}$ focus D_1.

Example 1

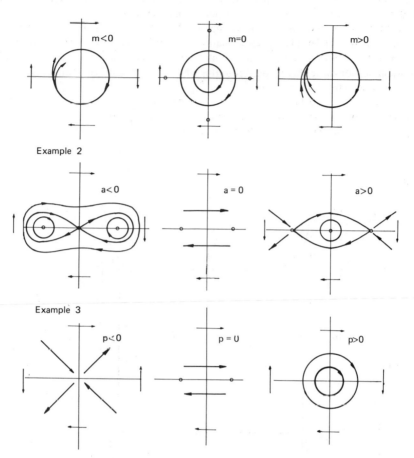

Example 2

Example 3

FIGURE 4. Elements of stability changes given in EXAMPLES 1, 2, and 3.

stability changes: In Example 1, at $m = 0$, the characteristic values at $x = y = 0$ are $\pm i$; this happens in the case of Type C2 (FIGURE 2). In Examples 2 and 3, for $a = 0$ and $p = 0$, respectively, at the origin the characteristic roots are double roots and are zero; this happens for A2 and B2, however both examples are of Type B1.

In short, these first three examples illustrate various types of stability changes; they are summarized as follows:

Example 1. For $m = 0, x = y = 0$, Type C2.

Example 2. For $a = 0, x = y = 0$, Type B1.

Example 3. For $p = 0, x = y = 0$, Type B1.

The fourth example is 3-dimensional, so it is not covered here, however it can be similarly analyzed as will be shown further on in this section.

In terms of the definition of partial peeling given in Gurel[9], the stability changes A1, B1, C1, and D1, if accompanied by a bifurcation, or peeling, in more general terms, correspond to stability changes in the case of partial peeling.[10] Phase planes for these three examples after and at the "stability change" value of the parameters are shown in FIGURE 4.

The equilibrium equations of Poincaré, sometime referred to as *isoclines,* are also significant in connection with PROPERTY i. Determining the location of singular points, ii. Deciding on the behavior of the singular points as the stability

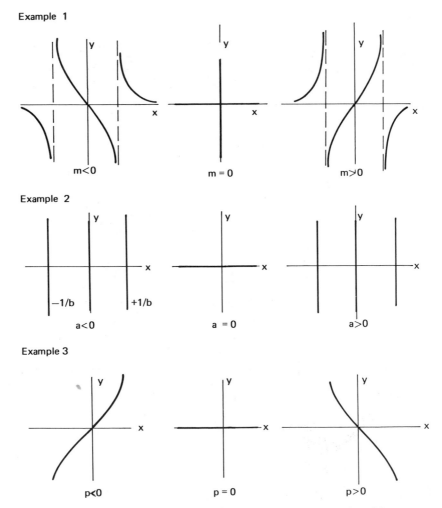

FIGURE 5. Poincaré's equilibrium equations for EXAMPLES 1, 2, and 3.

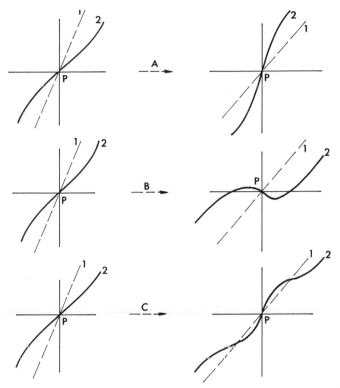

FIGURE 6. A = stability change; B = bifurcation to a periodic solution; C = bifurcation to multiple solutions.

changes. Equilibrium equations for the examples discussed earlier are sketched in FIGURE 5.

Stability changes are reflected in the relationship between the isoclines. In Figure 6, various possibilities are illustrated for 2-dimensional systems. In this figure, the following variations can be seen.

        A: Stability change only
        B: Stability change and bifurcation to a periodic solution
        C: Stability change and bifurcation to multiple solutions

Poincaré defines B and C as *bifurcations* (Poincaré bifurcations) and B^{-1} and C^{-1} as *limits*. Using the terminology in Gurel,[10]

        B: Peeling
        C: Degenerate peeling

We shall now consider the 3-dimensional example of Lorenz. TABLE 4 shows stability changes similar to those given in FIGURE 2 for 2-dimensional systems. The various possibilities can be easily listed by referring to Types A through D

TABLE 4

STABILITY CHANGES FOR THE 3-DIMENSIONAL CASE

A^{3s}	stable 3-dimensional node	becomes or or	$A1^{1u,2s}$ $A2^{1s,2u}$ $A3^{3u}$
$B^{1s,+c}$	stable + center	becomes or or or	$B1^{1u,2s} = A1^{1u,2s}$ $B2^{1s,-c}$ $B3^{1u,-c}$ $B4^{1s,2u} = A2^{1s,2u}$
$C^{1s,2\bar{s}}$	stable focus	becomes or or or	$C1^{1s,1\bar{s},1\bar{u}}$ $C2^{1s,2\bar{u}}$ $C3^{1u,1\bar{s},1\bar{u}}$ $C4^{1u,-2\bar{u}}$
$D^{1s,+2\bar{s}}$	stable + focus	becomes or or or	$D1^{1s,+1/2s,-1/2s}$ $D2^{1s,-2\bar{s}}$ $D3^{1u,+1/2s,-1/2s}$ $D4^{1u,-2\bar{s}}$

At $x = y = z = 0$

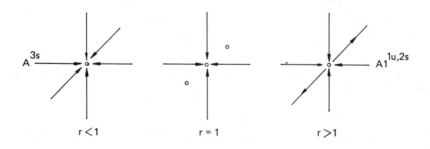

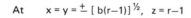

At $x = y = \pm\,[\,b(r-1)\,]^{1/2}, \quad z = r-1$

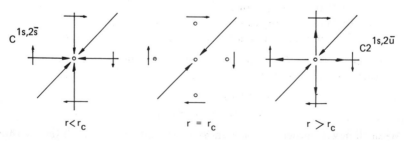

FIGURE 7. Peeling at the singular points of Lorenz equations, and types of stability changes at the peeling values of the parameter r.

of FIGURE 2. We shall use the following notation:

$$A^{3s}, \quad B^{1s,+c}, \quad C^{1s,2\bar{s}}, \quad D^{1s,+2\bar{s}}$$

where, e.g., 3s = stable in three dimensions, +c = center in one direction, $2\bar{s}$ = stable focus in two dimensions (u = unstable in TABLE 4.)

A table similar to TABLE 4 can be made for A^{3u}, $B^{1u,+c}$, etc. For the Lorenz system, we can refer to TABLE 4 and analyze peeling phenomena at the peeling values of the parameters. For the case of $b, m > 0$, there are two values of r where peeling takes place.

For $r = 1$, at $x = y = z = 0$, $A1^{1u,2s}$ takes place (FIGURE 7).

For

$$r = \frac{m(A + 2)}{(2m - A)}$$

at

$$x = y = \pm \sqrt{b(r - 1)}, z = r - 1,$$

$C2^{1s,2u}$ takes place (FIGURE 7).

For parameter values of $b < 0$, stability changes can easily be found. In this case, the generating singular point at the origin is Type A^{3u} an unstable node that peels into three singular points by $A1^{1s,2u}$ and, similarly, peeling at the points away from the origin $C2^{1u,2s}$ takes place. As a result of this type of stability change, the flow pattern around these points is quite different from the closed limiting object obtained for the positive b case. In fact it is a cylindrical pattern extending to infinity. Various trajectories corresponding to these cases are reported elsewhere.[11]

REFERENCES

1   POINCARÉ, H. Sur l'équilibre d'une masse fluide animée d'un mouvement de rotation. Acta Mathematica (16 September 1885). 7: 259–380. See also Oeuvres. 1952. Gauthier-Villars, Paris. 7: 40–140.

2.   LIAPUNOV, A. M. 1949. Problème Générale de la Stabilité du Mouvement (1907 French translation of 1892 Russian mémoires). Reprinted by Princeton Univ. Press, Princeton, N. J. See also O. Gurel & L. Lapidus. 1968. Stability via Liapunov's second method. Ind. Eng. Chem. 60: 12–26, and O. Gurel & L. Lapidus. 1969. A guide to the generation of Liapunov functions. Ind. Eng. Chem. 61: 30–41.

3.   ANDRONOV, A. & L. PONTRYAGIN. 1937. Grubye Sistemy. Dokl. Akad. Nauk SSSR 14(5):247–251. See also ANDRONOV, A. 1956. Sobranie Trudov. Akademiya Nauk Soyuza SSR. Izdatek'stvi A.N. SSSR:183–187.

4.   ANDRONOV, A. A., E. A. LEONTOVICH, I. I. GORDON & A. G. MAIER. 1973. Theory of Bifurcations of Dynamic Systems on a Plane. John Wiley, New York.

5.   POINCARÉ, H. 1892. Les Méthodes Nouvelles de la Méchanique Céleste. a) Vol. I & b) Vol. III. Gauthier-Villars, Paris.

6.   HOPF, E. 1942. Abzweigung einer periodischen Lösung von einer stationären Lösung eines Differentialsystems, Berichten der Math.-Phys. Klass der Sächlischen Akademie der Wissenschaften zu Leipzig. XCIV: 3–22. See English translation by L. N. Howard & N. Kopell. 1976. In The Hopf Bifurcation and Its Applications. J. E. Marsden & M. McCracken, Eds. Springer-Verlag, New York. 163–193.

7. LORENZ, E. N. 1963. Deterministic nonperiodic flow. J. Atmos. Sci. **20:** 130–141.
8. GUREL, O. 1973. A classification of singularities of (X, f). Math. Syst. Theory. **7:** 154–163.
9. GUREL, O. 1975. Partial peeling. *In* Dynamical Systems. L. Cesari, J. Hale & J. LaSalle, Eds. Academic Press, New York. 255–259.
10. GUREL, O. 1975. Peeling and nestling of a striated singular point. Collect. Phenom. **2:** 89–97.
11. GUREL, O. Peeling studies of complex dynamic systems. *In* Simulation of Systems. L. Dekker, Ed. Proc. 8th AICA Congr. Delft, the Netherlands. August 23–28, 1976. North-Holland Publ. Co., Amsterdam. 53–58.
12. SLOSSON, E. E. 1914. Major Prophets of To-Day. Reprinted 1968. Books for Libraries Press. Freeport, N.Y.
13. ISHLINSKOGO, A. YU. 1967. Razvitie Mekhaniki v SSSR. Sovetskaya Nauka i Tekhnika za 50 Let (1917–1967). :80. Izd. Nauka. Moscow.
14. RAZVITIE FIZIKI V SSSR. 1967. Kniga Pervaya. Sovetskaya Nauka i Tekhnika za 50 Let (1917–1967). :221. Izd. Nauka. Moscow.

INTRODUCTION

Eberhard Hopf

Indiana University
Bloomington, Indiana

All of us have been looking forward to a conference on bifurcations for some time, and we are deeply grateful to The New York Academy of Sciences and the University of Tübingen for their enormous help in preparing this Conference on Bifurcation Theory and Applications in Scientific Disciplines. I am very grateful to have been invited to this Conference.

Thirty-five years ago I wrote a paper on what is now called the *Hopf bifurcation*. At that time I expected an extension of this theory from ordinary differential systems to partial differential systems would come, but I did not have the slightest idea that this generalized theory would find such a large number of applications in so many different disciplines. In this present session we may expect very interesting, purely mathematical contributions to the theory of bifurcations.

A GENERAL BIFURCATION THEOREM*

Jane Cronin

*Department of Mathematics
Rutgers University
New Brunswick, New Jersey 08903*

INTRODUCTION

The term *bifurcation* refers here simply to the behavior of solutions of a given equation with a parameter ϵ for values of ϵ near 0. It is assumed that fairly complete information is provided about solutions of the equation at $\epsilon = 0$. A technique will be described for studying bifurcation problems of the form,

$$f(x, \epsilon) = 0 \tag{1}$$

where f is a continuous mapping from $R^n \times R$ into R^n (i.e., $x \in R^n$ and $\epsilon \in R$). Then this technique will be applied to the study of functional equations on which a Lyapunov-Schmidt reduction can be used.

We assume that Equation 1 has the solution $x = 0$, $\epsilon = 0$ and ask whether (1) has solutions x (especially solutions near $x = 0$) if ϵ is near but distinct from zero. We are concerned only with a qualitative analysis of the question (i.e., only existence questions will be considered).

The elementary approach to the study of Equation 1 is to impose hypotheses so that the implicit function theorem can be applied (i.e., to assume that the matrix $f_x(0, 0)$ is nonsingular). In this case Equation 1 has, for $|\epsilon|$ sufficiently small, a unique solution $x(\epsilon)$, which is a continuous function of ϵ.

A somewhat weaker hypothesis is the assumption that there exists an open neighborhood N of 0 in R^n such that the topological degree, $\deg[f, \bar{N}, 0]$, is nonzero. It then follows from the basic properties of topological degree that if $|\epsilon|$ is sufficiently small, Equation 1 has a solution $x(\epsilon) \in N$. However the solution $x(\epsilon)$ is not, in general, unique, and we can draw no conclusions as to whether the solution varies continuously with ϵ.

A third approach is to assume that $f(0, \epsilon) = 0$, 0 is an eigenvalue of $f_x(0,0)$ of odd multiplicity, and there exists $\epsilon_0 > 0$ such that if $0 < |\epsilon| \le \epsilon_0$ then 0 is not an eigenvalue of $f_x(0, \epsilon)$.

A simple sufficient condition for the latter assumption is that $\det[f_{x\epsilon}(0,0)] \ne 0$ (in the first two approaches, no condition on how f depends on ϵ was imposed). The condition $\det[f_{x\epsilon}(0,0)] \cong 0$ is the first such condition. The technique that will be described later depends heavily on investigating how f depends on ϵ.

By a simple topological degree argument, it follows that there exist $R > 0$ and $\bar{\epsilon} > 0$ such that if $0 < \epsilon_1 < \bar{\epsilon}$ and $0 < r < R$, then there exists $\tilde{\epsilon}$ such that $-\epsilon_1 < \tilde{\epsilon} < \epsilon_1$, and $f(x, \tilde{\epsilon}) = 0$ has a solution $x = \tilde{x}$, where $|\tilde{x}| = r$. This state-

*The research in this paper was supported by the U.S. Army Research Office (Durham), Grant No. DAHCO4-75-G0148.

0077-8923/78/0316-0028 $1.75/1 © 1979, NYAS

ment is just a finite-dimensional version of the classical theorem of Krasnosel'skii (Ref. 5, p. 196).

There are obvious difficulties with the first two approaches. First, the hypothesis that the matrix $f_x[0,0]$ is nonsingular is often not satisfied in concrete problems (i.e., equations of the form (1), which stem from functional bifurcation problems). For example, the hypothesis is not satisfied for resonance problems in ordinary differential equations.

Second, the hypothesis that $\deg[f, \overline{N}, 0]$ is nonzero is weaker, but this may be difficult to verify (it must be shown that the degree is defined and is nonzero).

A less obvious, but equally serious, difficulty with these two methods is the conclusion that if $|\epsilon|$ is sufficiently small, Equation 1 has a solution. This conclusion is actually too strong for many important cases. In general, it cannot be expected that (1) has solutions for *both* positive and negative values of ϵ, as the simple scalar equation, $x^2 - \epsilon = 0$, shows.

It may be worth noting that in the conclusion of the Hopf bifurcation theorem, periodic solutions are obtained only for $\epsilon \geq 0$ or $\epsilon \leq 0$. The theorem can be proved by using the implicit function theorem, but the proof is not direct since ϵ is not regarded as the independent variable. Instead, a strategically chosen new independent variable is introduced.

The third approach does not have this difficulty. All the solutions may correspond either to *positive* or *negative* values of ϵ. But this method has two difficulties: first, the hypothesis that 0 is an eigenvalue of odd multiplicity is a demanding one that cannot, in general, be satisfied in concrete problems; second, the information obtained about solutions is limited. For example, the conclusion does not tell us whether there is a continuum of values of ϵ for which (1) has a solution. Also there is no information as to whether the solutions obtained depend continuously on ϵ.

In the section headed "Bifurcation Theorem" a method in the form of a theorem is proposed for studying Equation 1 which avoids most of these difficulties. This method is based on the use of topological degree; however the theorem can be stated in elementary terms without reference to topological degree. Basically, the idea of the method is to investigate how the topological degree changes as ϵ changes. It is shown that if a simple hypothesis is imposed on how the function f changes with respect to ϵ at $(0,0)$, and if a reasonable set S is chosen, then one of the two numbers, $\deg[f(\cdot, \epsilon), S, 0]$, where $\epsilon > 0$, and $\deg[f(\cdot, \epsilon), S, 0]$, where $\epsilon < 0$, must be nonzero; hence, $f(x, \epsilon) = 0$ has a solution in the interior of S. The hypotheses of the theorem can be expressed as nonzero conditions on simple, practically computable expressions that depend on f. Although the proof of the theorem depends almost entirely on the use of topological degree, neither the hypothesis nor the conclusion of the theorem requires its use for their statements.

In the section headed "Bifurcation in Functional Equations" applications of the bifurcation theorem to functional nonlinear and quasilinear equations are described. The functional equations are abstract equations in Banach space. There are many concrete examples of such equations including integral, integro-differential, and nonlinear elliptic equations.

The bifurcation results for functional equations that are obtained can be regarded as extensions of the classical Krasnosel'skii bifurcation theorem. However they are different from known extensions due to Ize[4] and Dancer[3] (see also Nirenberg, Ref. 6, p. 88ff). in a number of ways.

First, we obtain a stronger conclusion (i.e., solutions for each sufficiently small positive ϵ or for each negative ϵ with $|\epsilon|$ sufficiently small); second, the conditions imposed on how f varies with respect to ϵ are different from those described by Krasnosel'skii. Also we are not restricted to Fredholm operators of index zero. As will be pointed out later, the technique is applicable if the index is positive.

Finally, considerations for quasilinear equations are not limited to local studies.

BIFURCATION THEOREM

BIFURCATION THEOREM. Suppose U is a bounded open set in R^n and I is the interval $(-\epsilon_0, \epsilon_0)$, where $\epsilon_0 > 0$, on the real line. Let G be an open set in R^n such that $\overline{U} \subset G$ and let $f(x, \epsilon)$ be a differentiable mapping from $G \times I$ into R^n. Assume that the following conditions are satisfied:

1. There exists $p_0 \in \partial U$ such that $f(p_0, 0) = 0$.
2. If $p \in \partial U - \{p_0\}$, then $f(p, 0) \neq 0$.
3. There exists a neighborhood N of p_0 in R^n such that $f[N \cap \partial U, \epsilon]$ is a surface Σ_ϵ which has a tangent hyperplane H_ϵ at $f(p_0, \epsilon)$ and H_ϵ has normal N_ϵ which is a continuous function of ϵ.
4. The vector $(\partial f/\partial \epsilon)(p_0, 0)$ is nonzero, and $[(\partial f/\partial \epsilon)(p_0, 0)] \cdot N_0 \neq 0$.

Conclusion: There exists $\epsilon_1 > 0$ such that if $0 < |\epsilon| < \epsilon_1$, then the Brouwer degree

$$\deg[f(\cdot, \epsilon), \overline{U}, 0] \tag{2}$$

is defined. For all $\epsilon \in (0, \epsilon_1)$, Expression 2 has the same value [denote it by $d(+)$] and for all $\epsilon \in (-\epsilon_1, 0)$, Expression 2 has the same value [denote it by $d(-)$] and

$$|d(+) - d(-)| = 1 \tag{3}$$

Before proving the theorem, we make a few remarks about its significance:

i. From (3), it follows that if $-\epsilon_1 < \bar{\epsilon} < 0$ and $0 < \tilde{\epsilon} < \epsilon_1$, then

$$\deg[f(\cdot, \bar{\epsilon}), \overline{U}, 0] \neq \deg[f[\cdot, \tilde{\epsilon}), \overline{U}, 0].$$

Hence either $\deg[f(\cdot, \epsilon), \overline{U}, 0]$ is nonzero for all $\epsilon \in (-\epsilon_1, 0)$ or $\deg[f(\cdot, \epsilon), \overline{U}, 0]$ is nonzero for all $\epsilon \in (0, \epsilon_1)$. Therefore either $f(x, \epsilon) = 0$ (Equation 1) has a solution $x \in \overline{U}$ for all $\epsilon \in (-\epsilon_1, 0)$ or for all $\epsilon \in (0, \epsilon_1)$. Thus the conclusion of the theorem, and hence the entire theorem, can be stated without referring to topological degree.

ii. The hypotheses of the theorem can be verified by showing that certain explicit expressions, which are often easily computed, are nonzero. For example, consider Condition 3. Suppose that $N \cap \partial U$ is described by the equations:

$$x_1 = u_1$$
$$\ldots\ldots$$
$$x_{n-1} = u_{n-1}$$
$$x_n = g(u_1, \ldots, u_{n-1})$$

where g is a differentiable function. Then the surface $f[N \cap \partial U, \epsilon]$ is described by

$$f[x_1, \ldots, x_{n-1}, g(x_1, \ldots, x_{n-1}), \epsilon]$$

Suppose $p_0 = 0$. If f_i denotes the ith component of f and if

$$h_i(x_1, \ldots, x_{n-1}) = f_i[x_1, \ldots, x_{n-1}, g(x_1, \ldots, x_{n-1}), \epsilon]$$

for $i = 1, \ldots, n$, then a sufficient condition that the tangent hyperplane H of Condition 3 exists is that the matrix,

$$\mathfrak{M}_\epsilon = \begin{bmatrix} \dfrac{\partial h_1}{\partial x_1} & \dfrac{\partial h_1}{\partial x_2} & \cdots & \dfrac{\partial h_1}{\partial x_{n-1}} \\[2mm] \dfrac{\partial h_2}{\partial x_1} & \dfrac{\partial h_2}{\partial x_2} & \cdots & \dfrac{\partial h_2}{\partial x_{n-1}} \\[2mm] \cdot\cdot\cdot\cdot\cdot\cdot\cdot\cdot\cdot\cdot \\[1mm] \dfrac{\partial h_n}{\partial x_1} & \dfrac{\partial h_n}{\partial x_2} & \cdots & \dfrac{\partial h_n}{\partial x_{n-1}} \end{bmatrix} \tag{4}$$

where each entry, evaluated at $x_1 = x_2 = \cdots x_{n-1} = 0$, has rank $n - 1$. Also the ith component of the normal N_ϵ is

$$\det \begin{bmatrix} \dfrac{\partial h_1}{\partial x_1} & \dfrac{\partial h_1}{\partial x_2} & \cdots & \dfrac{\partial h_1}{\partial x_{n-1}} \\[2mm] \cdot\cdot\cdot\cdot\cdot\cdot\cdot\cdot\cdot\cdot \\[1mm] \dfrac{\partial h_{i-1}}{\partial x_1} & \dfrac{\partial h_{i-1}}{\partial x_2} & \cdots & \dfrac{\partial h_{i-1}}{\partial x_{n-1}} \\[2mm] \dfrac{\partial h_{i+1}}{\partial x_1} & \dfrac{\partial h_{i+1}}{\partial x_2} & \cdots & \dfrac{\partial h_{i+1}}{\partial x_{n-1}} \\[2mm] \cdot\cdot\cdot\cdot\cdot\cdot\cdot\cdot\cdot\cdot \\[1mm] \dfrac{\partial h_n}{\partial x_1} & \dfrac{\partial h_n}{\partial x_2} & \cdots & \dfrac{\partial h_n}{\partial x_{n-1}} \end{bmatrix}$$

Hence N is nonzero. The condition that matrix $\mathfrak{M}$ (Equation 4) have rank $n - 1$ is sufficient but not necessary.

iii. Condition 3 is not simply a consequence of the differentiability. This is shown by the following example. Let $f(\cdot, 0)$ be the mapping of R^2 into R^2 (using complex variable notation) $z \rightarrow z^2$, and suppose

$$U = \{(x, y) \mid 0 \le x^2 + y^2 \le r^2, y \ge 0\}$$

where r is a positive number. Then

$$N \cap \partial U = \{(x, y) \mid -\eta \le x \le \eta; y = 0\}$$

where η is a positive number and

$$f[N \cap \partial U] = \{(x,y) \mid 0 \leq x \leq \eta^2; y = 0\}$$

Thus Σ is not a smooth surface. Actually in this case the surface $f[N \cap \partial U, 0]$ is described by

$$h_1(x) = x^2$$
$$h_2(x) = 0$$

and matrix $\mathfrak{M}$ is

$$\begin{bmatrix} 2x & 0 \\ 0 & 0 \end{bmatrix}$$

which clearly has rank 0 at $x = 0$ (i.e., each entry is zero at $x = 0$).

This example shows that in applying the theorem, the choice of the set U is crucial. In a corollary to the theorem, a constructive way to choose U will be described. We merely observe here that U need not be "small" although in some applications it will be so.

iv. From the proof of the theorem, it will be clear that the hypotheses of the theorem can be considerably reduced. The "smoothness" of Condition 3 is convenient, but stronger than actually required. After the theorem is proved, it will be indicated explicitly how the hypotheses can be weakened.

v. For certain applications, it is important to know that the solution $x \in U$ of Equation 1 is distinct from p_0. But this follows at once from the finite Taylor expansion for $f(x, \epsilon)$ around $f(p_0, 0)$, if $|\epsilon|$ is sufficiently small, because $f_\epsilon(0,0) \neq 0$.

Proof of the theorem: For the proof of the theorem, we use the following two lemmas.

LEMMA 1. There exists $r_1 > 0$ such that if $0 < r < r_1$ and N_r is a ball neighborhood of radius r of p_0 in R^n, then $\overline{N}_r \subset N$ (the neighborhood in Condition 3, and if L is a ray in R^n starting from the point $f(p_0, 0) = 0$, and L is a ray in R^n starting from the point $f(p_0, 0) = 0$, and L is parallel to and has the same direction as the vector $(\partial f/\partial \epsilon)(p_0, 0)$ then for all sufficiently small nonzero $|\epsilon|$, L has intersection number $+1$ or -1 with $f[N_r \cap \partial u, \epsilon]$, if $\epsilon > 0$ and L has a null intersection (therefore intersection number zero) with $f[N_r \cap \partial U, \epsilon]$, if $\epsilon < 0$ (see FIGURE 1).

Proof: If r_1 is sufficiently small and if $|\epsilon|$ is sufficiently small and nonzero, then it follows from Conditions 3 and 4 that $0 \notin f[N_r \cap \partial U, \epsilon]$. Also $f[N_r \cap \partial U, \epsilon]$ is sufficiently well approximated by its tangent hyperplane H_ϵ at $f(p_0, \epsilon)$ so that the intersection number of L and $f[N_r \cap \partial U, \epsilon]$ is equal to the intersection number of L and H_ϵ, if the latter intersection number is defined—the equality follows from the definition of the intersection number.[1,2] But if ϵ is sufficiently small and positive L is not parallel to $H_{\hat\epsilon}$ (by Condition 4), and hence the intersection number is defined. Moreover, since L has the same direction as $(\partial f/\partial \epsilon) \cdot (p_0, 0)$, the intersection number is $+1$ or -1. If ϵ is negative and $|\epsilon|$ and r_1 are both sufficiently small, then $-L[H_\epsilon N_r] = \phi$, and the intersection number is 0. $\square$

LEMMA 2. If $K = \partial U - N_r$, L is as in Lemma 1, and $|\epsilon|$ is sufficiently small, than the intersection number I of L and $f(K, \epsilon)$ is independent of ϵ.

Proof: This follows from the basic property of the intersection number.[1] $\square$

With LEMMAS 1 and 2, we prove the theorem as follows: If $\bar{\epsilon} > 0$ and $\bar{\epsilon}$ is sufficiently small, then

$$\deg[f(\cdot, \bar{\epsilon}), \overline{U}, 0] = \pm 1 + I$$

If $\bar{\epsilon} < 0$ and $|\bar{\epsilon}|$ is sufficiently small, then $\deg[f(\cdot, \bar{\epsilon}), \overline{U}, 0] = I$. $\square$

Inspection of the proof of theorem shows clearly how the hypotheses of the theorem can be weakened. For example, Conditions 3 and 4 can be replaced by conditions 3a and 4a:

3a. There exists a half-cone with vertex 0; i.e., a set C which, if the proper coordinates are used, can be described as:

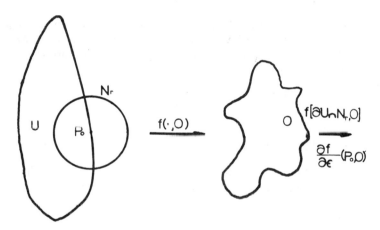

FIGURE 1.

$$C = \{x_1, \ldots, x_n) \mid 0 \leq \alpha_2 x_2^2 + \cdots + \alpha_n x_n^2 \leq x_1^2 \leq r\}$$

where $\alpha_2, \ldots, \alpha_n$, r are positive numbers and $x_j \geq 0$ for $j = 1, \ldots, n$ such that $C - \{0\} \subset f[U, 0]$.

4a. The vector $(\partial f / \partial \epsilon)(p_0, 0)$ is nonzero and if its initial point is at 0, then this vector is contained in the half-cone:

$$\tilde{C} = \{(x_1, \ldots, x_n) \mid 0 \leq \alpha_1 x_1^2 + \alpha_2 x_2^2 + \cdots + \alpha_n x_n^2; \quad x_j \leq 0, \quad j = 1, \ldots, n\}$$

However the verification of Conditions 3 and 4 seems considerably easier than trying to determine if 3a and 4a hold.

As mentioned earlier, in application of the bifurcation theorem, the choice of the set U is crucial. A method for selecting U which is computable in many cases will now be described.

Let $Jf(x, \epsilon)$ denote the Jacobian of f with respect to x at (x, ϵ), i.e.,

$$Jf(x, \epsilon) = \det \begin{bmatrix} \dfrac{\partial f_1}{\partial x_1}(x, \epsilon) & \cdots & \dfrac{\partial f_1}{\partial x_n}(x, \epsilon) \\ \dfrac{\partial f_n}{\partial x_1}(x, \epsilon) & \cdots & \dfrac{\partial f_n}{\partial x_n}(x, \epsilon) \end{bmatrix}$$

Assume that $Jf(0, 0) = 0$; i.e., the implicit function theorem cannot be applied to $f(x, \epsilon) = 0$ at the initial solution $x = 0$ and $\epsilon = 0$. Now suppose for all (x, ϵ) in some neighborhood of $(0, 0)$, it is true that $(x, \epsilon) \neq (0, 0)$ implies $Jf(x, \epsilon) \neq 0$. It follows then that, if N is a ball neighborhood of $x = 0$ in R^n, N is of sufficiently small radius, and $0 \in f(\partial N, 0)$, then $\deg[f(\cdot, 0), \overline{N}, 0] \neq 0$. Hence for all ϵ with $|\epsilon|$ sufficiently small, $f(x, \epsilon) = 0$ has a solution $x \in N$ (we need not use the bifurcation theorem).

Assume, instead, that $\quad \dfrac{\partial}{\partial x_1} Jf(0, 0) \neq 0$

Then, by the implicit function theorem, $Jf(x_1, \ldots, x_n, 0) = 0$ can be solved uniquely for x_1 as a function of $x_2, \ldots, x_n$ in a neighborhood of 0. Write the solution as $x_1 = j(x_2, \ldots, x_n)$ and let A be the vector which is normal to the surface

$$\{\Sigma : (x_2, \ldots, x_n) \rightarrow f[j(x_2, \ldots, x_n), x_2, \ldots, x_n]\}$$

That is, assume that the matrix,

$$M = \left[\frac{\partial f_i}{\partial x_k} [j(x_2, \ldots, x_n), x_2, \ldots, x_n] \right]$$

$(i = 1, \ldots, n; k = 2, \ldots, n)$ has rank $n - 1$ at $x_2 = x_3 = \cdots = x_n = 0$ (so that Σ is a smooth surface) and let A be the n-vector whose qth component is

$$\det \left[\frac{\partial f_i}{\partial x_k} [j(x_2, \ldots, x_n), x_2, \ldots, x_n] \right]_q$$

where $\qquad \left[\dfrac{\partial f_i}{\partial x_k} [j(x_2, \ldots, x_n), x_2, \ldots, x_n] \right]_q$

is the $(n - 1) \times (n - 1)$ matrix obtained by letting $i = 1, \ldots, q - 1$, $q + 1, \ldots, n$, $k = 2, \ldots, n$, and $x_2 = x_3 = \cdots = x_n = 0$ (since M has rank $n - 1$, then A is nonzero.) Notice that it is not necessary to know $j(x_2, \ldots, x_n)$ explicitly in order to determine the elements of M.

For example, take $i = 1$, $k = 2$. Then, by the chain rule and the implicit function theorem, it follows that

$$\frac{\partial f_1}{\partial x_2} [j(x_2, \ldots, x_n), x_2, \ldots, x_n]$$

$$= \frac{\partial f_1}{\partial x_1} \frac{\partial j}{\partial x_2} + \frac{\partial f_1}{\partial x_2}$$

$$= \frac{\partial f_1}{\partial x_1} \frac{\partial (Jf)}{\partial x_2} \frac{\partial (Jf)^{-1}}{\partial x_1} + \frac{\partial f_1}{\partial x_2}$$

COROLLARY. If $(\partial/\partial x_1)Jf(0,0) \neq 0$, matrix M has rank $n - 1$, and $A \cdot (\partial f/\partial \epsilon)(0,0) \neq 0$, then the conclusion of the bifurcation theorem holds.

The bifurcation theorem has two serious drawbacks. First, the conclusion of the theorem tells us that the equation $f(x, \epsilon) = 0$ has at least one solution x for each ϵ such that $0 < \epsilon < \epsilon_1$ or that the equation has at least one solution x for each ϵ such that $-\epsilon_1 < \epsilon < 0$. But no method is suggested for determining whether solutions exist for positive ϵ or negative ϵ. Secondly, no information is obtained concerning the (possible) uniqueness of x as a solution for fixed ϵ and no information about whether the solution x is a continuous function of ϵ is obtained. These gaps are typical of conclusions obtained by the application of topological methods. A sharpening of this theorem that yields results about whether solutions exist for for positive or negative ϵ and unpublished results about uniqueness and continuity of solutions have been obtained by Stephen Hoyle, a doctoral student at Rutgers.

BIFURCATION IN FUNCTIONAL EQUATIONS

We consider two functional equations: a nonlinear equation, for which a local study will be made, and a quasilinear equation.

Nonlinear Equation

Let X and Y be Banach spaces and $L: X \to Y$ be a Fredholm operator. Also let G be a continuous mapping from $X \times R$ into Y with the following properties:

a. $G(0,0) = 0$.
b. There exists a neighborhood N of 0 in X such that if $x, \bar{x} \in N$ and $\epsilon \in R$, then

$$\| G(x, \lambda) - G(\bar{x}, \lambda) \| \leq M(x + \bar{x}, \lambda) \| x - \bar{x} \| \tag{5}$$

where M is a continuous mapping from $X \times R$ into R such that $M(0,0) = 0$, and if $(x, \lambda) \neq (0,0)$, then $M(x, \lambda) > 0$.

We consider the equation,

$$Lx + G(x, \lambda) = 0 \tag{6}$$

and study the problem of determining solutions x of (6), where $x \in N$, for $|\lambda|$ sufficiently small. This is an abstract version of a classical nonlinear problem: the local study of solutions of nonlinear functional equations. A detailed study of many classes of such equations is given in Nirenberg[6] and Vainberg and Trenogin.[7] Notice that although the parameter λ is a real number, we can use this approach to deal with a more general class of equations other than (6). That is, suppose $G = G(x, y)$, where y is an element of a Banach space. Then if we take $y = \lambda y_0$, where y_0 is fixed and λ is a real parameter, the equation takes the form of (6).

If there exists L^{-1}, then if $|\lambda|$ and $\|x\|$ are sufficiently small, the mapping $L^{-1}G(x, \lambda)$, for fixed λ, is a contraction mapping, and Equation 6 has the unique solution $x = 0$ for $|\lambda|$ sufficiently small. So we are really interested only in the case in which the null space of L is nontrivial. To deal with this case, we apply the Lyapunov-Schmidt reduction technique to Equation 6.

Let X_0 be the finite-dimensional null space of L and let X^1 be the closed subspace of X that is complementary to X_0, i.e., $X = X_0 + X^1$ (direct sum). Then $x \in X$ implies that x can be represented uniquely as a sum $x = x_0 + x^1$, where $x_0 \in X_0$ and $x^1 \in X^1$. We use $x = (x^1, x_0)$ in referring to this decomposition.

Next, let $Y^1 = L(X)$. Since L is Fredholm, then $Y = Y_0 + Y^1$ (direct sum), where Y_0 is a finite-dimensional subspace of Y.

Finally let E^1, E_0 be the continuous projections of Y onto Y^1, Y_0, respectively. Now we may rewrite (6) as

$$L(x^1, x_0) + G[(x^1, x_0), \lambda] = 0$$

or, since $L(x_0) = 0$,

$$L(x^1, 0) + G[(x^1, x_0), \lambda] = 0 \tag{7}$$

Applying E^1, E_0 to (7), we obtain

$$E^1 L(x^1, 0) + E^1 G[(x^1, x_0), \lambda] = 0 \tag{8}$$

and

$$E_0 G[(x^1, x_0), \lambda] = 0 \tag{9}$$

Since E^1 is the projection of Y onto $L(X)$, then (8) may be rewritten,

$$L(x^1, 0) + E^1 G[(x^1, x_0), \lambda] = 0 \tag{10}$$

Let $K = [L/X^1]^{-1}$. Applying K to (10), we obtain

$$(x^1, 0) + K E^1 G[(x^1, x_0), \lambda] = 0 \tag{11}$$

Let $K E^1 G$ be denoted by H. Then (11) may be rewritten as,

$$x^1 + H[(x^1, x_0), \lambda] = 0 \tag{12}$$

For fixed x_0 and λ such that $\|x_0\|$ and $|\lambda|$ are sufficiently small, it follows, by Condition b on G, that if $\|x^1\|$ is sufficiently small, the mapping H is a contraction; hence Equation 12 can be solved uniquely for x^1 as a continuous function of x_0 and λ; thus,

$$x^1 = x^1(x_0, \lambda) \tag{13}$$

Substituting from (13) into (9), we obtain

$$E_0 G[x^1(x_0, \lambda), x_0), \lambda] = 0, \tag{14}$$

and the original problem of solving Equation 6 is now reduced to the problem of solving Equation 14 for x_0 as a function of λ.

If the index of L is zero (i.e., if $m = \dim X_0 = \dim Y_0$) then (14) is an example of Equation 1, and the bifurcation theorem can be applied to Equation 14 with $\epsilon = \lambda$.

If the index of L is positive (i.e., if $m = \dim X_0 > \dim Y_0 = k$) the bifurcation theorem can also be applied by fixing $m - k$ of the components of x_0 in the left side of (14). The remaining expression is a mapping from R^k into R^k with a parameter λ, and, again, the bifurcation theorem can be applied. It is easy to show that in this case the topological degree is zero, therefore some technique like the bifurcation theorem must be used. Notice also that since all our considerations here are local, then in studying the mapping

$$E_0 G[(x^1(x_0, \lambda), x_0), \lambda]$$

it is sufficient to study the lowest-order terms.

Note that since G is a "higher order" in x, uniformly in λ (property b of G), then the corollary to the bifurcation theorem has only very limited value. It would be applicable only in the case where G had strictly second-order terms and m or k had the value 2. The kind of weakened sufficient condition mentioned at the end of Remark ii (p. 31) would be useful in this case.

Quasilinear Equation

Now consider the quasilinear equation

$$Lx + \lambda N(x, \lambda) = 0 \qquad (15)$$

where, as before, L denotes a Fredholm operator from Banach space X into Banach space Y and N denotes a differentiable operator from $X \times R$ into Y which satisfies a Lipschitz condition in x uniformly in λ. We wish to determine solutions x of (15) when $|\lambda|$ is sufficiently small.

The essential difference between the studies of Equations 6 and 15 is that the study of (6) is entirely local, i.e., we search for solutions (x, λ) in a neighborhood of $(0, 0)$, whereas the study of (15) is not entirely local. In the notation of the discussion of Equation 6, we search for solutions (x, λ) of (15), where $x = (x^1, x_0)$, and x^1 and λ are both near 0, but there is no restriction on the size of $|x_0|$.

Using the nonlinear equation notation, we can rewrite (15) as:

$$L(x^1, 0) + \lambda N[(x^1, x_0), \lambda] = 0 \qquad (16)$$

Applying E^1 and E_0 to (16), we obtain

$$L(x^1) + \lambda E^1 N[(x^1, x_0), \lambda] = 0 \qquad (17)$$

and
$$\lambda E_0 N[(x^1, x_0), \lambda] = 0 \qquad (18)$$

Applying K to (18) yields,

$$x^1 + \lambda L E^1 N[x^1, x_0), \lambda] = 0 \qquad (19)$$

This equation clearly has the initial solution $x^1 = 0$, $\lambda = 0$, and x_0 arbitrary but fixed. Also if $|\lambda|$ is sufficiently small then, since N satisfies a Lipschitz condition in x uniformly in λ, it follows that (19) can be solved uniquely for x^1 as a continuous function of x_0 and λ, where both x^1 and λ are in a neighborhood of 0; that is, we obtain the solution,

$$x^1 = x^1(x_0, \lambda) \qquad (20)$$

Substituting from (20) into (18) and dividing by λ, we obtain:

$$E_0 N[(x^1(x_0, \lambda), x_0), \lambda] = 0 \qquad (21)$$

As before, we can apply the bifurcation theorem to (21). Also, as before, we must consider separately the two cases: index $L = 0$ and $L > 0$. Since the mapping N is not necessarily of a higher order, the corollary has a wider application than in the case of the nonlinear equation.

Let
$$g(x_0, \lambda) = E_0 N[(x^1(x_0, \lambda), x_0), \lambda]$$

If $N(0, 0) = 0$, then a straightforward computation shows that

$$\frac{\partial g}{\partial \lambda}(0, 0) = \left(\frac{\partial}{\partial \lambda} E_0 N\right)(0, 0)$$

where $E_0 N$ denotes $E_0 N(x, \lambda)$. Also $g(x_0, 0)$, which describes the surface S (when appropriate restrictions are placed on x_0), is given by $g(x_0, 0) = E_0 N[x_0, 0]$. Thus verification of the conditions in the bifurcation theorem involves only fairly straightforward computations provided the projection operator E_0 is explicitly known.

REFERENCES

1. ALEXANDROFF, P. & H. HOPF. 1945. Topologie. Edwards Bros., Ann Arbor.
2. CRONIN, J. 1964. Fixed Points and Topological Degree in Nonlinear Analysis. Am. Math. Soc.
3. DANCER, E. N. 1971. Bifurcation theory in real Banach Spaces. J. London Math. Soc. (3) **23**: 699–734.
4. IZE, J. 1976. Bifurcation theory for Fredholm operators. Am. Math. Soc., Providence (Memoirs of the Am. Math. Soc. **174**).
5. KRASNOSEL'SKII, M. A. 1964. Topological Methods in the Theory of Nonlinear Integral Equations. Macmillan, New York.
6. NIRENBERG, L. 1974. Topics in Nonlinear Functional Analysis. Courant Inst. Math. Sci.
7. VAINBERG, M. M. & V. A. TRENOGIN. 1974. Theory of Branching of Solutions of Nonlinear Equations. Noordhoff International Publ., Leyden.

NECESSARY AND SUFFICIENT CONDITIONS FOR COOPERATIVE PEELING OF MULTIPLE GENERATING SINGULAR POINTS

Okan Gurel

IBM Corporation
White Plains, New York 10604

The solution space $X \subseteq R^n$ and a parameter space $P \subseteq R_m$ are given. The vector field $g:(X, P) \rightarrow (X, P)$ maps (X, P) onto itself. A point in the product space (X, P) is given with coordinates $x \in X$ and $p \in P$. For a dynamical system determined by vector differential equation $dx/dt = f(X, P)$, the Hessian Δ is formed as functional determinant of the matrix whose elements are derivatives of f with respect to n coordinates $x_1, \ldots, x_n$ of X locally.

One of the Poincaré conditions for bifurcation at a singular point x^0 is that Δ evaluated at x^0 vanishes at the p^0 value[1,2]; thus, the Poincaré condition: $\Delta(x^0, p^0) = 0$. This yields p^0 bifurcation values of the bifurcation parameters.

Matrix $M(x, p)$ is an $n \times n$ matrix whose determinant is $\Delta(x, p)$. We can write $M(x, p)$ in a canonical form which we shall denote by $M_c(x, p)$. Evaluated at x^0, the generating singular point $\Delta_c(x^0, p)$, where x^0 is the only root of $f(x, p) = 0$, has the form,

$$\Delta_c(x^0, p) = m_{11}(x^0, p)m_{22}(x^0, p) \ldots m_{nn}(x^0, p)$$

When $\Delta_c = 0$, and assuming that the other two Poincaré conditions are satisfied,[1] the p values found become the bifurcation values of the parameters. Not only the bifurcation values p^0 but also the parameters playing a role in bifurcation p_p are thus determined: $\Delta_c(x^0, p_p{}^0) = 0$. This is possible for $m_{ii}(x^0, p_p{}^0) = 0$, for any one i, or any combination of indices $i = 1, \ldots, n$. For brevity, $\Delta_c{}^0$ will be used to denote this determinant. Each vanishing m_{ii} corresponds to a vanishing eigenvalue λ_i, thus a possible stability alteration is expected.

If at $p = p_p{}^0$, all the $m_{ii} = 0$, then we have peeling.[3] For at least one $m_{ii} \neq 0$, while some others vanish, this is a partial peeling.[4] In such latter cases usually the critical limiting set $CLS[X]^{3,5}$ contains multiple singular points and/or limiting objects including the generating singular points. Let us assume that, in addition to the generating singular point, there are k more singular points in $CLS[X]$, thus $x^j \in CLS[X], j = 0, 1, \ldots, k$.

At each new x^j, we could evaluate the matrix M. These are put into a canonical form with the diagonal elements $m_{ii}{}^j = m_{ii}(x^j, p)$, and denoted by,

$$M_c{}^0, M_c{}^1, \ldots, M_c{}^k$$

Taking diagonals of $M_c{}^j, j = 0, 1, \ldots, k$ as columns of a new matrix $\overline{M}$, we define $\overline{M}$ which combines all the singular points,

0077-8923/78/0316-0039 $1.75/1 © 1979, NYAS

. . .

$$M_c^{\;j} = \begin{bmatrix} m_{11}^{\;j} & & & \mathbf{0} \\ & m_{22}^{\;j} & & \\ & & \ddots & \\ \mathbf{0} & & & m_{nn}^{\;j} \end{bmatrix} \quad \text{and} \quad \overline{M} = \begin{bmatrix} m_{11}^{\;0} & m_{11}^{\;1} & \cdots & m_{11}^{\;k} \\ m_{22}^{\;0} & m_{22}^{\;2} & \cdots & m_{22}^{\;k} \\ \vdots & \vdots & & \vdots \\ m_{nn}^{\;0} & m_{nn}^{\;1} & \cdots & m_{nn}^{\;k} \end{bmatrix}$$

where $m_{ii}^{\;j} = m_{ii}(x^j, p)$—see FIGURE 1.

Should $n \neq k$, $\overline{M}$ may be put into a square matrix by augmenting, i.e., adding $(n - k)$ dummy columns, if $k < n$ and $(k - n)$ dummy rows, if $k > n$. Each element of a new dummy row or column may be taken equal to a constant as α_s or β_s, $s - 1, 2, \quad$, they may be taken equal to 1 (see Gurel[6]).

REMARK 1. $\Delta_c^{\;j} = 0$ implies that one or more $m_{ii}^{\;j}$ vanish. Thus by writing char acteristic equations,

$$\prod_{i=1}^{n} (m_{ii}^{\;j} - \lambda_i) = 0$$

we find one or more $\lambda_i = m_{ii}^{\;j}$ vanishing. We conjecture that the determinant of $\overline{M}$, denoted by $\bar{\Delta}$, would correspond to an index related to combined peeling of the CLS[X].

REMARK 2. It is clear that if (complete) peeling takes place at any of the singular points, all λ's through 0, thus $M_c^{\;j} = 0$ is a null matrix. This implies that peeling of a singular point in CLS[X] results in $\bar{\Delta} = 0$.

COOPERATIVE PEELING (necessary) CONDITION I. While $\Delta c^j \neq 0$ for all j, $\bar{\Delta} = 0$ implies "Cooperative peeling," and results in determining peeling parameters and their values at points in the parameter space where combined peeling takes place. The other two Poincaré conditions[2]: (a) change of stability at the generating singular point and (b) a change in the number of characteristic solutions (i.e., singular points, periodic solutions, etc.) can be replaced by the following single condition.

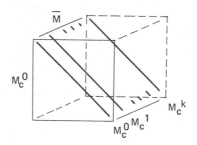

$M_c^{\;j}$ = canonical matrix at point x^j,

$\overline{M}$ = <u>Combined matrix</u>

FIGURE 1. Construction of the combined matrix from canonical matrices at points x^j, $j = 0, 1, \ldots, k$.

COOPERATIVE PEELING (sufficient) CONDITION II. For the combined peeling to take place, in addition to Condition I, it is necessary that CLS[X] is replaced by CLS'[X].

REMARK 3. It is clear that CLS[X] by definition possesses the same stability property as that of the generating singular point. Since CLS[X] → CLS'[X] implies topological a change, a change in number is included. Therefore, Condition II covers both (a) and (b) for multiple singular points.

EXAMPLE. For the Lorenz equations[2,5] at the generating singular point diag $M_c^0[-m, (r-1), -b]$, thus $\Delta_c^0 = mb(r-1) = 0$ results in $m = 0$, $b = 0$, or $r = 1$ as parameter values. However, only r is the peeling parameter. After $r = 1$ value the generating singular point $x_0 = y_0 = z_0$ partially peels to form three singular points.[4] The set CLS[X] consists of these three points, a saddle, and two stable foci.

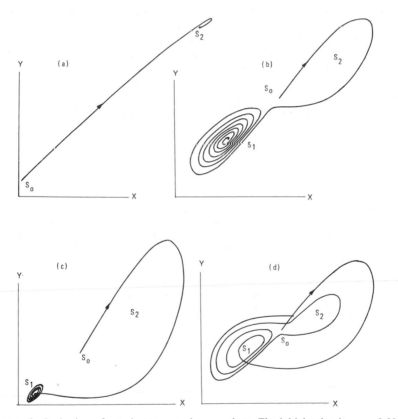

FIGURE 2. Projection of a trajectory on the x, y-plane. The initial value is $x_0 = 0.3037$, $y_0 = 0.3027$, and $z_0 = 0.03536$, the final time is 10 units, and the parameter values are $m = 10$, $b = \frac{8}{3}$, while: (a) $r = 2$ (greater than 1) approaching S_2; (b) $r = 15$ (greater than 1) approaching S_1; (c) $r = 30$ (greater than 24.374) first moving to S_1 and then (slowly) approaching the Lorenz attractor from the S_1-side; (d) $r = 45$ (greater than 24.374) (rapidly) approaching the Lorenz attractor.

At the newly created foci, diag $M_c^1 = [-m, -2, -A^2]$ and diag $M_c^2 = [-m, -2, -A^2]$, where $A = [b(r - 1)]^{1/2}$. Note that Δ_c^1 and Δ_c^2 vanish for $m = 0$, $b = 0$, and $r = 1$; these are the same values for $\Delta_c^0 = 0$. Therefore we do not have new information as to the possible peeling at these new singular points.

The study of the stability changes at the singular points S_1 and S_2 reveals that at $r = m(A + 2)/(2m - A)$ the two stable singular points become unstable.[4] For example, for $m = 10$, $b = \frac{8}{3}$, the value for r is computed as $r = 24.374$.

The combined matrix, however, is formed by taking M_c^0, M_c^1 and M_c^2 as,

$$\overline{M} = \begin{bmatrix} -m & -m & -m \\ (r - 1) & -2 & -2 \\ -b & -A^2 & -A^2 \end{bmatrix}$$

Since the two last columns are equal, this is a singular matrix, and the determinant vanishes for all combinations of m, b, and r. Therefore the values of the parameters, where a stability change takes place (e.g., $r = 24.374$), are candidates for peeling. In fact, at this value of r not only does the stability change take place, but CLS[X] also changes, and it contains only the Lorenz attractor. Thus, a change in the number of critical solutions also appears.

A set of simulation results for various parameter values are given in FIGURE 2, where (a) and (b) correspond to the cases in which CLS[X] consists of three singular points, while (c) and (d) correspond to the cases in which CLS[X] is the Lorenz attractor.

REFERENCES

1. POINCARÉ, H. 1885. Sur l'equilibre d'une masse fluide animée d'un mouvement de Rotation. Acta Mathematica. **7**: 259–380.
2. GUREL, O. Poincaré's Bifurcation analysis. *In* Bifurcation Theory and Applications in Scientific Disciplines. O. Gurel & O. E. Rossler, Eds. Ann. N. Y. Acad. Sci. In this volume.
3. GUREL, O. 1975. Peeling and nestling of a striated singular point. Collect. Phenom. **2**: 89–97.
4. GUREL, O. 1976. Partial Peeling. *In* Dynamical Systems II. L. Cesari, et al., Eds. Academic Press, Inc., New York 255–259.
5. GUREL, O. 1976. Peeling studies of complex dynamic systems. *In* Simulation of Systems. L. Dekker, Ed. North-Holland Publ. Co., the Netherlands 53–58.
6. GUREL, O. 1977. Decomposed partial peeling and limit bundles. Phys. Lett. **61A**: 219–223.

ON STABILITY OF STATIONARY POINTS OF
TRANSFORMATION GROUPS

Morris W. Hirsch

Department of Mathematics
University of California
Berkeley, California 94720

INTRODUCTION

Let $F: \mathbb{R}^n \to \mathbb{R}^n$ be a C^1 (= continuously differentiable) vector field on Euclidean n-space, and consider the associated differential equation

$$\frac{dx}{dt} = F(x). \tag{1}$$

For each $y \in \mathbb{R}^n$ let $t \to \phi(t, y)$ be the solution to (1) which passes through y when $t = 0$, so that

$$\phi(0, y) = y, \quad \frac{\partial \phi}{\partial t}\bigg|_{t=0} = F.$$

Assume for simplicity that solutions are defined for all t. Thus we have a C^1 map $\phi: \mathbb{R} \times \mathbb{R}^n \to \mathbb{R}^n$. For each $t \in \mathbb{R}$ we obtain a C^1 map

$$\phi_t: \mathbb{R}^n \to \mathbb{R}^n$$

$$y \to \phi(t, y).$$

Uniqueness of solutions implies that

$$\phi_{s+t} = \phi_s \circ \phi_t,$$

the composition of ϕ_s and ϕ_t, and clearly ϕ_0 is the identity map of $\mathbb{R}^n$. In this way we obtain a homomorphism $t \to \phi_t$ from the additive group of real numbers to the group of homeomorphisms of $\mathbb{R}^n$. Since in fact each ϕ_t is C^1 with a C^1 universe ϕ_{-t}, we actually have a homomorphism into the group of C^1 diffeomorphisms,

$$\alpha: \mathbb{R} \longrightarrow \text{Diff } \mathbb{R}^n,$$

$$\alpha(t) = \phi_t.$$

We call α the *flow* defined by (1).

Many questions in dynamical systems can be profitably studies in terms of flows. Some flows have a *section*, that is, a hypersurface $S \subset \mathbb{R}^n$ transverse to the vector field, which orbits repeatedly intersect. In this way one is led to consider the "first return" map $f: S \to S$. Such an f generates (under suitable hypotheses) a homomorphism from the group $\mathbb{Z}$ of integers into Diff S.

A homomorphism of a group G into the group of homeomorphisms of a space X is called an *action* of G on X. In the examples above we have actions of $\mathbb{R}$ and

43

0077-8923/78/0316-0043 $1.75/1 © 1979, NYAS

of $\mathbb{Z}$. Actions of other groups also occur. For example, let $F, E: \mathbb{R}^n \to \mathbb{R}^n$ be two vector fields of the type considered above, such that their Lie bracket $[E, F] = 0$. Then their corresponding flows commute and we obtain an action of $\mathbb{R} \times \mathbb{R}$ on $\mathbb{R}^n$. The presence of a group G of symmetries of a dynamical system often leads to an action of $G \times \mathbb{R}$.

Two central themes in dynamical systems are *recurrence* and *stability*. The book Topological Dynamics[2] of Gottschalk and Hedlund was the first systematic investigation of recurrence in the context of group actions. A great deal of work in this field is currently being done, much of it in ergodic theory and foliation theory. Questions of stability, however, have barely been touched for groups other than $\mathbb{Z}$ or $\mathbb{R}$.[3,4]

The simplest stability problem is that of stability of a stationary point under perturbation of a group action. The rest of this article discusses sufficient conditions for such stability.

The Problem

Consider a topological group G acting differentiably on a manifold M by a continuous homomorphism $\alpha: G \to \text{Diff } M$. Let $p \in M$ be a *stationary point* of the action, that is,

$$\alpha(g)p = p \quad \text{for all} \quad g \in G.$$

Call p *stable* if nearby actions have nearby stationary points (precise definitions are given below). The question is: *under what conditions is a stationary point stable under perturbations of the action?*

For simplicity I consider only the case $M = \mathbb{R}^n$, and G is always assumed to be generated by some compact subset $K \subset G$. For example, G might be a finitely generated discrete group, or a connected Lie group.

Now some precise definitions.

The topology on $\text{Diff } \mathbb{R}^n$ is that of uniform convergence of diffeomorphisms and their first derivatives on compact sets. There is a complete metric for this topology.

An *action* α of G on $\mathbb{R}^n$ is a continuous homomorphism $\alpha: G \to \text{Diff } \mathbb{R}^n$. The set of all such actions is denoted $A(G, \mathbb{R}^n)$. A *neighborhood* $\mathfrak{N} \subset A(G, \mathbb{R}^n)$ of α is a set of the form $\mathfrak{N} = \mathfrak{N}(\alpha, s, \delta)$ where $s > 0$, $\delta > 0$, and $\beta \in \mathfrak{N}$ if and only if the following condition holds: for all $g \in K$ and all $x \in \mathbb{R}^n$ with $|x| \leq s$,

$$|\alpha(g)x - \beta(g)x| < \delta$$

and

$$|| D\alpha(g)(x) - D\beta(g)(x) || < \epsilon.$$

Here $D\alpha(g)(x)$ is the linear operator on $\mathbb{R}^n$ which is the derivative of $\alpha(g)$ at x, and $|| \ ||$ denotes the usual operator norm.

We can now define what it means for a stationary point $p \in \mathbb{R}^n$ to be *stable* for α: for every $\epsilon > 0$ there is a neighborhood $\mathfrak{N}$ of α such that if $\beta \in \mathfrak{N}$ then β has a stationary point q with $|p - q| < \epsilon$.

EXAMPLES AND THEOREMS

Example 1. Let $G = \mathbb{Z}$. Let $\alpha : G \to$ Diff $\mathbb{R}^n$ be generated by the diffeomorphism $\alpha(1) = f$. The stationary points of α are the fixed points of f. It is well known that a fixed point p is stable if 1 is not an eigenvalue of $Df(p)$. This condition is equivalent to the graph of f being transverse to the diagonal in $\mathbb{R}^n \times \mathbb{R}^n$ at (p,p). It is by no means *necessary* for stability. For example $0 \in \mathbb{R}$ is a stable fixed point of $f(x) = x + x^3$.

It is important to notice a certain uniqueness property of the stationary point p in Example 1: There is a neighborhood $U \subset \mathbb{R}^n$ of p such that p is the only stationary point of α in U. The same is true for perturbations of α: it follows easily from the inverse function theorem that neighborhoods U of p and $\mathfrak{N}$ of α can be chosen so that every $\beta \in \mathfrak{N}$ has a unique stationary point in U. A stable stationary point with this property is called *uniquely stable.*

Example 2. Let $\phi_t : \mathbb{R}^n \to \mathbb{R}^n$ be a flow defined by a C^1 vector field F, so that

$$F(x) = \frac{\mathrm{d}}{\mathrm{d}t}\bigg|_{t=0} \phi_t(x).$$

Then $p \in \mathbb{R}^n$ is stationary for the action $\phi : \mathbb{R} \to$ Diff $\mathbb{R}^n$ if and only if $F(p) = 0$. Suppose that p is stationary and $DF(p)$ is invertible. By the inverse function theorem any C^1 vectorfield sufficiently near F must have a zero near p. Thus p is stable for perturbations of the vector field.

The topology on the space of actions is not defined in terms of the vector field F, but directly in terms of the flow. In fact it is not true that every C^1 flow has a C^1 vector field, although it does have a continuous vector field by a result of Chernoff and Marsden.[1] In any case we can easily replace the condition that $DF(p)$ is nonsingular by a condition expressed purely in terms of the flow. It is equivalent to saying: *for some $t \in \mathbb{R}$, $D\phi_t(p)$ does not have 1 for an eigenvalue.* The equivalence follows easily from the well-known formula

$$D\phi_t(p) = \exp [tDF(p)].$$

The equivalent condition makes sense for actions of groups other than $\mathbb{R}$, and leads to wide generalization of Example 2.

THEOREM 1. Let $\alpha : G \to$ Diff $\mathbb{R}^n$ be an action of an Abelian group G. A stationary point $p \in \mathbb{R}^n$ is stable provided there exists $g_0 \in G$ with $I - D\alpha(g_0)_p$ invertible.

Proof. Put $f = \alpha(g_0) \in$ Diff $\mathbb{R}^n$. Then p is a uniquely stable fixed point for f, by Example 1 and the remark following it. Therefore there exists a neighborhood U of p, and a neighborhood $\mathfrak{N}$ of α, such that for every $\beta \in \mathfrak{N}$, $\beta(g_0)$ has a unique stationary point $q_\beta \in U$. Moreover $q_\beta \to p$ as $\beta \to \alpha$. Let $\mathfrak{N}_1 \subset \mathfrak{N}$ be a smaller neighborhood such that $\beta \in \mathfrak{N}_1$ implies $\beta(g)q_\beta \in U$ for all $g \in K$ where $K \subset G$ is a compact generating set. Such an $\mathfrak{N}_1$ exists because $\beta(g)q_\beta \to p$ as $\beta \to \alpha$, uniformly for $g \in K$.

I claim q_β is stationary for β if $\beta \in \mathfrak{N}_1$. It suffices to prove $\beta(g)q_\beta = q\beta$ for all $g \in K$. Now

$$\beta(g)q_\beta = \beta(g)\beta(g_0)q_\beta$$

$$= \beta(g_0)\beta(g)q_\beta.$$

Therefore $\beta(g)q_\beta$ *is a fixed point of* $\beta(g_0)$, *and it lies* in U. By uniqueness of such a fixed point, $\beta(g)q = q_\beta$. Thus q_β is stationary for β because K generates G.

Example 3. Let $\alpha : \mathbb{Z} \oplus \mathbb{Z} \to \text{Diff } \mathbb{R}^2$ be the linear action generated by

$$\alpha(1, 0) = \begin{bmatrix} 1 & 0 \\ 0 & -1 \end{bmatrix} = f_1,$$

$$\alpha(0, 1) = \begin{bmatrix} -1 & 0 \\ 0 & 1 \end{bmatrix} = f_2.$$

Then the stationary point $0 \in \mathbb{R}^2$ is stable. What must be shown is that if g_1, $g_2 \in \text{Diff } \mathbb{R}^2$ commute and are sufficiently C^1 close to f_1, f_2, then they have a common fixed point near 0. I do not know how to prove this directly. But the proof is easy if we use the fact that $\mathbb{Z} \oplus \mathbb{Z}$ is an Abelian group. For we apply Theorem 1 to $g_0 = (1, 1)$:

$$\alpha(g_0) = \begin{bmatrix} -1 & 0 \\ 0 & -1 \end{bmatrix}.$$

Thus $I - D\alpha(g_0)(0)$ is invertible, so 0 is a stable stationary point.

The method of proof used in Theorem 1 extends to more general situations. All that was really needed was the *normality* of the subgroup generated by g_0. A more general result is:

THEOREM 2. Let $\alpha : G \to \text{Diff}(\mathbb{R}^n)$ be an action and $p \in \mathbb{R}^n$ a stationary point. Suppose $H \subset G$ is a subgroup such that p is uniquely stable for the restricted action $\alpha \mid H : H \to \text{Diff}(\mathbb{R}_n)$. If H is a normal subgroup then p is uniquely stable for α.

Proof. Let $U \subset \mathbb{R}^n$ be a neighborhood of p for which there is a neighborhood $\mathfrak{N}_0 \subset A(H, \mathbb{R}^n)$ of $\alpha \mid H$ such that if $\gamma \in \mathfrak{N}_0$ then γ has a unique stationary point $q(\gamma) \in U$, and $q(\gamma) \to p$ has $\gamma \to \alpha \mid H$. Let $\mathfrak{N} \subset A(G, \mathbb{R}^n)$ be a neighborhood of α so small that if $\beta \in \mathfrak{N}$ then $\gamma = \beta \mid H \in \mathfrak{N}_0$, and also $\beta(k)q(\beta \mid H) \in U$ for all $k \in K$ (= a compact generating set for G). Let $\beta \in \mathfrak{N}$. I claim $q(\beta \mid H) = q$ is fixed for all $\beta(k), k \in K$. For if $k \in K$ then

$$\beta(k)q = \beta(k)\beta(H)q \quad \text{(as sets)}$$

$$= \beta(H)B(k)q \quad \text{(by normality)}.$$

Hence $\beta(k)q$ is a stationary point in U for $\alpha \mid H$. By uniqueness $\beta(k)q = q$. Q.E.D.

A nilpotent group G may be defined as one having the property that for any subgroup $H_0 \subset G$ there is a finite chain of subgroups $H_0 \subset \ldots \subset H_n = G$ where each H_{i-1} is normal in H_i. The group of upper triangular $m \times m$ matrices with values of 1 on the diagonal is nilpotent. One can use Theorem 2 to prove:

THEOREM 3. Let G be a nilpotent group and p a stationary point for an action α of G on $\mathbb{R}^n$. Then p is uniquely stable provided

$$I - D\alpha(g)(p)$$

is invertible for some $g \in G$.

Example 4. Let $G = SL(n)$ be the group of $m \times m$ real matrices of determinant 1, acting as usual on $\mathbb{R}^n$. Then the origin $0 \in \mathbb{R}^n$ is stable if n is even. For let $H = \{\pm I\}$. Then H is Abelian and the matrix $-I$ does not have 1 as an eigenvalue. By Theorem 1, 0 is stable for the action of H. Since H is normal, 0 is stable for $SL(m)$ by Theorem 2. This result is also true for odd m, but needs a different proof; see Example 6.

Example 5. Let S be the group of 2×2 real matrices

$$A_{a,b} = \begin{bmatrix} a & b \\ 0 & 1 \end{bmatrix}, \, a > 0.$$

S is solvable but not nilpotent. For every real number t an action $\alpha_t : S \rightarrow \text{Diff}(\mathbb{R})$ is defined by

$$\alpha_t(A_{a,b})(x) = ax + tb.$$

The action α_0 has $0 \in \mathbb{R}$ as a stationary point; 0 is *unstable* because α_t has no stationary point for $t \neq 0$, and $\alpha_t \rightarrow \alpha_0$ as $t \rightarrow 0$. This example also shows that Theorem 2 fails for solvable groups.

I mention without proof two other sufficient conditions for stability, Theorems 4 and 5.

THEOREM 4 (Palais[5]). If G is compact every stationary point is stable.

Example 6. The origin in $\mathbb{R}^n$ is stable for the usual action α of $SL(n)$. The case n even was covered in Example 4. Since $SO(n)$ is compact, 0 is stable for it by Palais' Theorem. If β is a small perturbation of α then $\beta \mid SO(n)$ has a stationary point near 0. The isotropy group of q is a closed subgroup Γ, and $SO(n) \subset \Gamma \subset SL(n)$. The orbit of q under $\beta(SL(n))$ is the image of an immersion of $SL(n)/\Gamma$, so that

$$\dim SL(n) - \dim \Gamma \leq n.$$

It follows that $\Gamma \neq SO(n)$ if $n \geq 3$. But $SO(n)$ is a maximal proper closed subgroup; hence $\Gamma = SL(n)$.

THEOREM 5. Let G be a Lie group; $H \subset G$ a closed subgroup; α an action of G on $\mathbb{R}^n$; $p \in \mathbb{R}^n$ a stationary point of $\alpha \mid H$. Let E be the quotient of the Lie algebra of G by that of H, and for $h \in H$ let $A(h): E \rightarrow E$ be the linear operator induced by the adjoint representation of G. Then p is a stable stationary point for α provided there exists $h \in H$ such that $D\alpha(h)(p)$ and $A(h)$ have no eigenvalues in common.

Example 7. Let S be as in Example 5. For every $s \in \mathbb{R}$ define an action

$$\gamma_s : S \rightarrow \text{Diff } \mathbb{R},$$

$$\gamma_s(A_{a,b})x = a^s x.$$

Then 0 is unstable for $s = 1$ (Example 6) and for $s = 0$ (Exercise!). For $s \neq 0, 1$ it is not hard to use Theorem 3 to prove that 0 is stable: take $H = \{A_{a,b}: b = 0\}$.

Theorem 5 can be used to prove that many representations of semisimple groups have a stable stationary point at the origin.

I end with a conjecture and an unsolved problem.

Conjecture. Every stationary point of every semisimple Lie group is stable.

Problem. Is $0 \in \mathbb{R}^3$ stable for the action of $\mathbb{Z} \oplus \mathbb{Z} \oplus \mathbb{Z}$ generated by 180° rotations about the three coordinate axes?

REFERENCES

1. CHERNOFF, P. & J. MARSDEN. 1970. On continuity and smoothness of group actions. Bull. Amer. Math. Soc., **75**: 1044–1049.
2. GOTTSCHALK, W. & G. HEDLUND. 1955. Topological Dynamics. American Mathematical Society Colloquium Publications. Vol. 36. American Mathematical Society. Providence, R.I.
3. HIRSCH, A. 1970. Foliations and noncompact transformation groups. Bull. Amer. Math. Soc. **76**: 1020–1023.
4. PUGH, C. & M. SHUB. 1975. Axiom A actions. Inventiones Math. **29**: 7–38.
5. PALAIS, R. 1961. Equivalence of nearby differentiable actions of a compact group. Bull. Amer. Math. Soc. **67**: 362–364.

SPONTANEOUS SYMMETRY BREAKING
IN NONLINEAR PROBLEMS

D. H. Sattinger

School of Mathematics
University of Minnesota
Minneapolis, Minnesota 55455

BIFURCATION AT MULTIPLE EIGENVALUES

Suppose we want to investigate the bifurcation of solutions of a system of equations in a Banach space given by,

$$G(\lambda, u) = 0 \tag{1}$$

in the neighborhood of a known solution (λ_0, u_0). If $L_0 = G_u(\lambda_0, u_0)$ is a Fredholm operator of index 0 and $n = \dim \ker G_u(\lambda_0, u_0) \geq 1$, then, by the Liapunov-Schmidt method, the bifurcation problem can be reduced to that of solving a system of algebraic equations,

$$F_i(\lambda, z_1, \ldots, z_n) = 0, \quad i = 1, \ldots, n \tag{2}$$

If $G(\lambda, u)$ is an analytic operator, then the F_i are also analytic.

In practice the computation of even the lowest-order terms of the F_i is a nontrivial matter, especially if Equation 1 is a particularly complicated system of equations, as in elasticity of fluid mechanics. Moreover, given n equations in n unknown, virtually anything can happen in the way of solution sets and their stabilities, and the algebraic problem can be quite complex.

In many problems of physical interest, however, the multiplicity of the branch point can be traced to an underlying symmetry of the problem. This phenomenon is well known in quantum mechanics, where the invariance of the Hamiltonian under a symmetry group leads to a degeneracy of the energy levels. Group representation theory is an important tool in analyzing the splitting of the energy levels of a Hamiltonian under symmetry-destroying perturbations (e.g., the Stark effect); but these methods also apply in a natural and elegant way to the nonlinear problems of bifurcation theory. The applicability of group representation theory rests on the tensor character of the bifurcation equations on the one hand, and the theory of tensor products of group representations on the other.[24]

Moreover, it is very often the case in physical applications, especially in the area of mechanics, that the systems of equations, even though nonlinear, are covariant with respect to a transformation group. For example, the Hamiltonian equations of celestial mechanics or the partial differential equations governing the dynamics of a homogeneous continuum are covariant with respect to the Euclidean group of rigid motions. Let us assume therefore, that the mapping G is *covariant* with respect to a transformation group $\mathcal{G}$. That is, let T_g be a representation of $\mathcal{G}: T_{g_1 g_2} = T_{g_1} T_{g_2}$, and assume that,

$$H_1: T_g G(\lambda, u) = G(\lambda, T_g u) \tag{3}$$

49

0077-8923/78/0316-0049 $1.75/1 © 1979, NYAS

This is a natural assumption in physical theories and is a mathematical expression of the axiom that the equations of mathematical physics are independent of the observer.

From Equation 3 it follows that $T_g G_u(\lambda, u) = G_u(\lambda, T_g u) T_g$, so if u_0 is a solution which happens to be invariant under the entire group $\mathcal{G}$, $T_g u_0 = u_0$, we have $T_g L_0 = L_0 T_g$, where $L_0 = G_u(\lambda_0, u_0)$. Therefore $N_0 = \ker L_0$ is invariant under T_g, and $T_g \mid N_0$ is a finite-dimensional representation of $\mathcal{G}$. Let us write the bifurcation equations (**2**) in the form $F(\lambda, v) = 0$, where $v \in N_0$ and $F: C \times N_0 \to N_0$.

THEOREM 1. If $G(\lambda, u)$ is covariant and $T_g u_0 = u_0$ then the bifurcation equations are also covariant: $T_g F(\lambda, v) = F(\lambda, T_g v)$.[24]

Let us now expand F in a power series in V.

$$F(\lambda, v) = A(\lambda)v + B_2(\lambda, v, v) + B_3(\lambda, v, v, v) + \cdots$$

Then we must have,

$$T_g A(\lambda) = A(\lambda) T_g \qquad (4)$$

$$T_g B_2(\lambda, v, w) = B_2(\lambda, T_g v, T_g w) \qquad (5)$$
$$\vdots$$

Throughout this paper assume that $H_2: N_0$ is irreducible under T_g; then from Equation 4 and H_2 it follows (by Schur's lemma[31]) that $A(\lambda) = \sigma(\lambda)I$, where I is the identity. Suppose for convenience $\lambda_0 = 0$ and $\sigma(\lambda) = C_1 \lambda + C_2 \lambda^2 + \cdots$. Then, by various scaling arguments, the bifurcation problem can be reduced to an analysis of the equations,

$$\lambda w = B_k(w) \qquad (6)$$

where B_k is the first nonvanishing term in F, homogeneous of degree k. Call (**6**) *reduced bifurcation equations*. It can be shown that the stability of the bifurcating solutions can be determined to lowest-order from an analysis of the Jacobian of (**6**) at a solution (Ref. 25, Theorem 7.2).

The group theoretic approach, then, is to compute the lowest nonvanishing terms B_k, find all solutions of (**6**), and determine their stability in the neighborhood of the branch point. This attack not only allows us to bypass the numerical difficulties inherent in the Liapunov-Schmidt procedure; but it also provides us with a systematic approach to bifurcation at multiple eigenvalues and with a way of classifying multiple eigenvalue bifurcation points.

We would argue that the group theoretic approach is the natural one to take in treating bifurcation at multiple eigenvalues. These arguments give much more precise information about the bifurcation point than, say, topological degree arguments or Ljusternik-Schnirelman theory. Moreover, group theoretic approaches have proved extremely fruitful in modern physics, for example, in the physics of elementary particles. These ideas have also been important in classical physics as well, for example in determining the possible forms of constitutive laws in continuum mechanics from the hypothesis that they are frame indifferent (i.e., covariant). For a survey of recent work in group theory and spontaneous symmetry breaking in physics, see Birman[3], Michel[6], and Haken.[11]

Let us return to Equation 6. It turns out that, depending on the representation $T_g \mid N_0$, there may be more than one covariant term of lowest degree k. In that case we arrive at a system of reduced equations of the form,

$$\lambda w = \mathcal{C}_1 B_k^{(1)}(w) + \mathcal{C}_2 B_k^{(2)}(w) + \cdots + \mathcal{C}_\ell B_k^{(\ell)}(w) \qquad (6a)$$

where the coefficients $\mathcal{C}_1 \ldots \mathcal{C}_\ell$ are parameters that depend on the original parameters of the systems. The multiplicity ℓ of covariant terms of degree k can be computed directly from a knowledge of the representation $T_g \mid N_0$, and does not depend on the particular structure of the equations at hand (see Jarić and Birman[4,5] for a general method of computing the multiplicities).

When there are multiple covariant tensors, as in Equation 6a the possibility of *selection mechanisms* arises. The stability of the various bifurcating solutions depends on the relative sizes of the parameters $\mathcal{C}_1, \ldots, \mathcal{C}_\ell$. In the Bénard problem, for example, $\ell = 2$ when $k = 3$, and there occurs a selection mechanism for the stability of rolls and hexagons. We shall discuss the Bénard problem in greater detail later.

In general, the computation of the exact dependence of the parameters $\mathcal{C}_1, \ldots, \mathcal{C}_\ell$ on the original parameters of the problem is a difficult matter—equivalent to the calculation of the full set of bifurcation equations (2). Such calculations could be (and have been) carried out numerically in specific cases,[6] but it is not clear to me that such a direct numerical approach is the best one. What we are really after is a classification of the types of transitions that can take place, and for that purpose it is sufficient to consider the parameters $\mathcal{C}_1, \ldots, \mathcal{C}_\ell$ as free parameters, more or less like the *control parameters* in Thom's classification of the elementary singularities.[10,29] In that way we can classify the spontaneous symmetry breaking transitions that a physical system may undergo purely on algebraic and geometric grounds.

There are some differences between the group theoretic classification of singularities and Thom's classification. Thom's approach has been to classify singularities of structurally stable mappings or of their structurally stable unfoldings. In order to obtain a finite set of cases some constraints must be placed on the problem, otherwise we face an infinite number of possibilities. Thom has observed that if the case is restricted to stable unfoldings of a scalar function with four or fewer parameters, a finite classification scheme is obtained—the now famous "seven elementary catastrophes."

What is being proposed here is somewhat different. In the first place, the covariant mappings are not structurally stable. Simple examples show that the zero set of a mapping is altered drastically when the symmetry of the mappings is destroyed. The assumption of symmetry of a physical system must therefore be considered a mathematical idealization of the problem whose sole justification is that it considerably simplifies the analysis of the problem, and yet still gives physically meaningful results.

What is being proposed here, then, is a classification of the possible symmetry-breaking transitions which may take place within the constraint of an overall symmetry of the governing system of equations. A deeper analysis, based on the symmetry-destroying imperfections of the problem and their effect on the structure of the solution set (i.e., the structure of the singularity), would be extremely interesting, both mathematically and physically.

In the following two sections we shall describe some details of this approach in two classical problems that have attracted widespread interest—the Bénard problem and the buckling of spherical shells.

The Bénard Problem

If a layer of fluid is heated from below, convective instabilities set in when the temperature drop exceeds a certain critical value, and the convective motions that evolve often display a striking cellular structure.[12,15,28] The mathematical model most often analyzed is that of the Boussinesq equations in an infinite plane layer of fluid. The infinite plane layer model has the advantage that it simplifies the mathematical analysis and gives good quantitative predictions for the onset of instability and of the value of the critical wave number. It has the disadvantage, however, of introducing a number of mathematical degeneracies into the problem, one of which is the multiplicity of solutions that may bifurcate at the onset of instability. The question then arises as to what mechanisms, if any, govern the selection of a particular pattern.

The Boussinesq equations consist of a nonlinear elliptic system of partial differential equations for a vector-valued function $u = \text{col}(u_1(\underline{x}), u_2(\underline{x}), u_3(\underline{x}), \theta(\underline{x}), p(\underline{x}))$, where u_1, u_2, u_3 are the Cartesian components of the velocity, θ is the temperature, and p is the pressure. The covariance of the Boussinesq equations is easily established along the same lines as that of the Navier Stokes equations.[13] The following analysis applies to *any* problem that is covariant with respect to the group of rigid motions. For simplicity we restrict ourselves to scalar values functions: the algebra is the same as in the Boussinesq equations.

The appearance of cellular solutions can be described mathematically as the bifurcation of doubly periodic solutions. Let Λ be a lattice of vectors in the plane: If $\underline{\omega}$, $\underline{\omega}' \in \Lambda$ then $m\underline{\omega} + n\underline{\omega}' \in \Lambda$, for any integers m and n. A function u is doubly periodic if $u(\underline{x} + \underline{\omega}) = u(\underline{x})$, for all $\underline{\omega} \in \Lambda$. We shall say that u is Λ-periodic if we wish to specify a particular lattice. We denote by $\mathcal{E}(\Lambda)$ the subgroup of rigid motions that leaves the class of Λ-periodic functions invariant.

Consider the bifurcation of nontrivial doubly periodic solutions of (1) from the rest state $u = 0$. Let the parameter λ be chosen so that the critical parameter value is $\lambda_0 = 0$. We denote by $\eta(\Lambda)$ the kernel of L_0 in the subspace of Λ-periodic functions. Since $\eta(\Lambda)$ is invariant under the full group of translations, we may choose a basis for $\eta(\Lambda)$ of the form $\{e^{i<\underline{\omega}_j, \underline{x}>}, \ j = 1, 2, \ldots\}$, where the vectors $\underline{\omega}_1, \underline{\omega}_2, \ldots$ lie in the plane and $<\ ,\ >$ is the Euclidean inner product. The functions $e^{i<\underline{\omega}_k, \underline{x}>}$ are the one-dimensional irreducible representations of the translation group and are known as the *Bloch* functions in solid state physics. Since $e^{i<\underline{\omega}_j, \underline{x}>}$ is Λ-periodic we must have $e^{i<\underline{\omega}_j, \underline{x}+\underline{\omega}>} = e^{i<\underline{\omega}_j, \underline{\omega}>}e^{i<\underline{\omega}_j, \underline{x}>}$, for all $\underline{\omega} \in \Lambda$. Hence $e^{i<\underline{\omega}_j, \underline{\omega}>} = 1$, for all $\underline{\omega} \in \Lambda$, and the vectors $\underline{\omega}_j$ must lie in the dual lattice Λ'. The dual lattice is constructed by choosing a basis dual to that of Λ.

If Λ is generated by $\underline{\omega}_1$ and $\underline{\omega}_2$ choose $\underline{\omega}_1'$, $\underline{\omega}_2'$ such that $<\underline{\omega}_i',\underline{\omega}_j> = 2\pi\delta_{ij}$, $i = 1, 2$. Then Λ' is the lattice generated by $\underline{\omega}_1'$ and $\underline{\omega}_2'$.

Now suppose $e^{i<\underline{\omega}',\underline{x}>}$ belongs to $\eta(\Lambda)$. If r is a rotation-reflection that leaves $\eta(\Lambda)$ invariant, we must have $e^{i<\underline{\omega}',r^{-1}(\underline{x}+\underline{\omega})>} = e^{i<r\underline{\omega}',\underline{\omega}>}e^{i<r\underline{\omega}',\underline{x}>} = e^{i<\underline{\omega}',r^{-1}\underline{x}>}$, for all $\underline{x}$. Consequently we must have $r\underline{\omega}'$ in the dual lattice as well. We denote the group of rotations and reflections which leave Λ-periodic functions invariant by $\mathfrak{D}(\Lambda)$. There are three types of lattices in the plane generated by basic vectors $\underline{\omega}_1$ and $\underline{\omega}_2$ of equal length. From now on we drop the primes and use $\underline{\omega}_1$ and $\underline{\omega}_2$ to denote the basic vectors of the dual lattice. These are shown in FIGURE 1.

The kernel $\eta(\Lambda)$ consists of the wave functions $\psi_j(\underline{x}) = e^{i<\underline{\omega}_j,\underline{x}>}$, where the vectors $\underline{\omega}_j$ lie on the circle of critical wave vectors. The radius $\sqrt{<\underline{\omega}_1,\underline{\omega}_1>}$ is determined in the process of investigating the onset of instability.[12] Let $\psi_1(\underline{x}) = e^{i<\underline{\omega}_1,\underline{x}>}$. The other basis vectors of $\eta(\Lambda)$ are obtained by operating on ψ_1 by

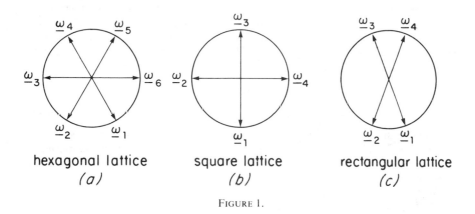

hexagonal lattice
(a)

square lattice
(b)

rectangular lattice
(c)

FIGURE 1.

the rotations and reflections in $\mathfrak{D}(\Lambda)$. The vectors $\underline{\omega}_j$ are the vertices of a regular polygon, which we shall call *the fundamental polygon for the lattice Λ*.

THEOREM 2. Let T_g be the representation of $\mathcal{E}(\Lambda)$ on $\eta(\Lambda)$. The action of T_g is as follows:

(i) If T_a is a translation, $T_a\psi_j = e^{i<\underline{\omega}_j,\underline{a}>}\psi_j$;

(ii) If g is a rotation-reflection in $\mathfrak{D}(\Lambda)$, T_g acts as a permutation on the ψ_j, permuting them in the same way as g permutes the vertices of the fundamental polygon;

(iii) $T_g\psi_j = \overline{\psi}_j$, if $g\omega_j = -\underline{\omega}_j$.

Proof: (i) $(T_a\psi_j)(\underline{x}) = e^{i<\underline{\omega}_j,\underline{x}+\underline{a}>} = e^{i<\underline{\omega}_j,\underline{a}>}\psi_j(\underline{x})$;

(ii) $T_r\psi(\underline{x}) = e^{i<\underline{\omega}_j,r^{-1}\underline{x}>} = e^{i<r\underline{\omega}_j,\underline{x}>} = \psi_{r(j)}$;

(iii) If $r\underline{\omega}_j = -\underline{\omega}_j$ then $T_r e^{i<\underline{\omega}_j,\underline{x}>} = \overline{\psi}_j$. $\square$

For example, suppose Λ is the hexagonal lattice of FIGURE 1(a). The symmetry group of the hexagon is generated by the permutations $\alpha = (123456)$ and $\beta = (26)(35)$. The corresponding elements of $\mathfrak{D}(\Lambda)$ are a rotation through 60° and a reflection across the $\underline{\omega}_1 - \underline{\omega}_4$ axis. Therefore $T(\alpha)\psi_1 = \psi_2$, $T(\alpha)\psi_2 = \psi_3, \ldots$ $T(\alpha)\psi_6 = \psi_1$, while $T(\beta)\psi_1 = \psi_1$, $T(\beta)\psi_2 = \psi_6, \ldots$.

If w is a vector in $\eta(\Lambda)$, we write $w = \sum_j z_j \psi_j$, where w is real iff $z_j = \bar{z}_k$, whenever $\psi_j = \overline{\psi}_k$.

THEOREM 3. For the hexagonal lattice, the action of $\mathcal{E}(\Lambda)$ on the components $(z_1, \ldots, z_6)$ is as follows:

   (i) $\alpha(z_1, \ldots, z_6) = (z_6, z_1, z_2, \ldots, z_5)$;
   (ii) $\beta(z_1, \ldots, z_6) = (z_1, z_6, z_5, z_4, z_3, z_2)$;
   (iii) $T_a(z_1, \ldots, z_6) = (e^{i<\underline{\omega}_1 \cdot \underline{a}>}z_1, \ldots, e^{i<\underline{\omega}_6 \cdot \underline{a}>}z_6)$.

Furthermore, for real vectors we require $z_4 = \bar{z}_1$, $z_5 = \bar{z}_2$, $z_6 = \bar{z}_3$.

THEOREM 3 follows from THEOREM 2, and the simple observation that the components $z_1 \ldots$ and the basis vectors $\psi_1 \ldots$ must transform contragradiently.

Using THEOREM 3, we can compute the general mapping $F(z)$ which is covariant relative to $\mathcal{E}(\Lambda)$. From $T(\alpha)F(z) = F(T(\alpha)z)$ we get $F_6(z_1, \ldots, z_6) = F_1(z_6, z_1, \ldots, z_5)$, and so forth. So once $F_1(z_1, \ldots, z_6)$ is determined, the other components of the mapping can be found by cyclic permutation of the arguments. From (ii) of THEOREM 3 we get,

$$e^{i<\underline{\omega}_1 \cdot \underline{a}>} F_1(z_1, \ldots z_6) =$$

$$F_1(e^{i<\underline{\omega}_1 \cdot \underline{a}>} z_1, \ldots, e^{i<\underline{\omega}_6 \cdot \underline{a}>} z_6). \tag{7}$$

We break F_1 into linear, quadratic, cubic terms. The only linear term satisfying Equation 7 is easily seen to be $F_1(z_1, \ldots, z_6) = az_1$. Then from cyclic permutation we get $F_j(z_1, \ldots, z_6) = az_j$, so the only linear covariant mapping is a scalar multiple of the identity.

When F_1 is homogeneous of degree 2, we get from (7),

$$e^{i<\underline{\omega}_1 \cdot \underline{a}>} z_j z_k = e^{i<\underline{\omega}_j \cdot \underline{a}>} z_j e^{i<\underline{\omega}_k \cdot \underline{a}>} z_k.$$

For this to hold for all a, we must have $\underline{\omega}_1 = \underline{\omega}_j + \underline{\omega}_k$. The only possible solution is $\underline{\omega}_2 + \underline{\omega}_6 = \underline{\omega}_1$, so the general quadratic term is $F_1 = bz_2z_6$, and hence $F_2 = bz_3z_1$, $F_3 = bz_4z_2, \ldots$.

In the case of cubic mappings we are led by similar arguments to the condition $\underline{\omega}_1 = \omega_j + \underline{\omega}_k + \underline{\omega}_\ell$, whose only possible solutions are $\underline{\omega}_j = \underline{\omega}_1$ (say) and $\omega_k + \omega = 0$. Thus the cubic covariant terms are $z_1^2 z_4$, $z_2 z_5 z_1$, $z_3 z_6 z_1$.

The covariance with respect to β implies that F_1 must be symmetric in $\{z_2, z_6\}$ and $\{z_3, z_5\}$. Therefore, the general cubic term is $F_1(z_1, \ldots, z_6) = az_1^2 z_4 + bz_1(z_2z_5 + z_3z_6)$. By similar arguments we could construct covariant terms of arbitrary degree, but it is sufficient to know only the lowest-order nonvanishing nonlinear term in order to resolve the bifurcation problem.

The structure of the bifurcation equations can be determined for the other lattices by similar arguments. For the square and rectangular lattices the dimension of the kernel of $G_u(\lambda_c, 0)$ is four, while for the hexagonal lattice it is six. The reduced bifurcation equations may be obtained in each case from the generating functions,

$$\Lambda_6 : F_1(\Lambda, z_1, \ldots, z_6) = \lambda z_1 + a z_1^2 z_4 + b z_1 (z_2 z_5 + z_3 z_6) \qquad \text{(8a)}$$

$$\Lambda_2, \Lambda_4 : F_1(\lambda, z_1, \ldots, z_4) = \lambda z_1 + a z_1^2 z_3 + b z_1 z_2 z_4 \qquad \text{(8b)}$$

This is the case when the quadratic term in Λ_6 vanishes. If it should not vanish, the generating function for the reduced bifurcation equations in Λ_6 is

$$\Lambda_6 : F_1(\lambda_1 z_1, \ldots, z_6) = \lambda z_1 + c z_2 z_6 \qquad \text{(8c)}$$

The complete set of bifurcation equations is obtained by cyclically permuting the variables $z_1, z_2, \ldots$. The coefficients a, b, c depend on the lattice (in fact on the angle θ between $\underline{\omega}_1$ and $\underline{\omega}_2$) and on the original physical parameters of the problem.

Information about the stability of the bifurcating solutions can be determined in terms of the parameters a and b. By stability here we mean the formal linearized stability of a solution relative to disturbances within the same lattice class. Certainly a necessary condition for stability is that the solution is stable with respect to all disturbances in the same lattice class.

THEOREM 4. Let the bifurcation equations for a square or rectangular lattice be generated by Equation 8. Then the following are necessary conditions for the stability of rolls or squares/rectangles,

> stable rolls: $b < a < 0$

> stable squares/rectangles: $a + b < 0, a - b < 0$

For the hexagonal lattice, where the bifurcation equations are generated by (8a), the necessary conditions for stability are,

> stable rolls: $b < a < 0$

> stable hexagons: $a < b, a + 2b < 0$

In the case of (8c), a transcritical branch of hexagonal solutions occurs at the branch point. The bifurcating solutions are unstable on both sides of criticality, but the subcritical branch has only one unstable mode.[26]

The theory of subcritical bifurcation in convection problems is discussed in detail by Joseph.[12] The essential result is that the transcritical branch of purely hexagonal solutions may bend back and become stable, yielding subcritical stable hexagonal solutions.

In order to obtain any results on pattern selection in the supercritical cases, we must determine the dependence of a and b on the lattice. This will enable us to compare simultaneously the stability of the solutions in various lattices.

THEOREM 5. There exists a function $q(\theta)$, called the lattice function, such that

> 1) $q(\theta) = A_0 + A_2 \cos 2\theta + A_4 \cos 4\theta \ldots$

> 2) $a = 3q(0), b = 6q(\theta)$

where θ is the angle between the basic vectors of the lattice, and the coefficients $A_0, A_2, \ldots$ depend only on the basic physical parameters of the problem. (For the full proof of THEOREM 5, including the construction of the lattice function, see Ref. 6.)

Recall that all of the above analysis applies whenever the original problem is covariant with respect to the groups of rigid motions in the plane. In some cases it is possible to show that $q(\theta)$ is constant, while in others, that q has weight two; that is, $q(\theta) = A \cos 2\theta + B$. In this case, it is a simple matter to translate the conditions of THEOREM 4 into conditions on A and B, and the pattern selection diagram in FIGURE 2 holds.

A necessary condition for stability of a given pattern is that the parameter values A and B lie in the corresponding sector. Rolls are the only pattern uniquely selected on this basis, while the regions for squares and hexagons overlap. Note that rolls are the only possible stable pattern in the supercritical case if $A = 0$. In this case, the equations transform as a scalar. A simple example would be $u_t = \Delta u + \lambda u + f(u)$ in the plane. This is also the situation in the case of the integral equation approach to phase transitions.[20,21]

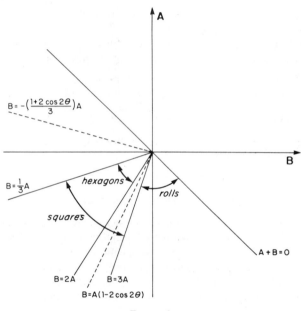

FIGURE 2.

BIFURCATION IN THE PRESENCE OF $O(3)$

In this section we shall show how the program can be carried out when the group in question is $O(3)$, the group of rotations of the sphere.[27] There are a number of classical bifurcation problems in which this situation arises, for example, the buckling of a perfectly uniform spherical shell,[22] or the onset of convection in a spherical mass. The latter problem and its possible connection with convection in the earth's mantle has been discussed recently by Busse.[7]

Our first task is to construct the 2nd- or 3rd-order covariant mappings $B(v, w)$ or $B(u, v, w)$. It is important to keep in mind that these are completely symmetric mappings: $B(v, w) = B(w, v)$.

The irreducible representations of $O(3)$ are of dimension $2\ell + 1$, $\ell = 0, 1, \ldots$, and are denoted by D^ℓ. To construct quadratic mappings, we consider the tensor product $D^\ell \otimes D^\ell$ acting on $N \otimes N$. Let us recall the Clebsch-Gordon series[31]:

$$D^\ell \otimes D^\ell = D^{2\ell} \oplus D^{2\ell+1} \oplus \ldots \oplus D^0 \qquad (9)$$

This means that $N \otimes N$ decomposes into a direct sum of invariant irreducible subspaces:

$$N \otimes N = V^{2\ell} \oplus V^{2-1} \oplus \ldots \oplus V^\ell \oplus \ldots \oplus V^0$$

One of these subspaces; namely, V^ℓ, transforms like D^ℓ. Since we require $B(D^\ell v, D^\ell w) = D^\ell B(v, w)$, this is precisely the subspace we want. The components of our covariant mapping B then form a basis for this subspace.

Now in the above decomposition $V^{2\ell}$ contains symmetric tensors $V^{2\ell-1}$ anti-symmetric tensors, and so forth. Accordingly, V^ℓ will be symmetric iff ℓ is even. Therefore, *for odd ℓ there is no quadratic term* so we must go to cubic terms.

Let us consider the case of even ℓ. We construct the mapping B using Lie algebra methods, which are well known in the theory of angular momentum coupling in quantum mechanics. Since B can be assumed to be completely symmetric, we can work with polynomials and write,

$$F_i(z_1, \ldots, z_n) = \sum_{j,k} a_{ijk} z_j z_k$$

We begin with the infinitesimal generators of the rotation group J_1, J_2, J_3. These satisfy the commutation relations,

$$[J_i, J_j] = \epsilon_{ijk} J_k$$

where ϵ_{ijk} is the completely antisymmetric tensor. Now put,

$$J^\pm = \pm J_2 + iJ_1, \quad J^3 = -iJ_3$$

These operators satisfy the commutation relations,

$$[J^+, J^-] = 2J^3, \quad [J^3, J^\pm] = \pm J^\pm \qquad (10)$$

The operators $J^\pm$ are the familiar ladder operators of quantum mechanics. Let N be an irreducible (real) invariant vector space which transforms like D^ℓ under $SO(3)$. Then the complexified space $N + iN$ has a basis f_m^ℓ such that,

$$J_3 f_m = m f_m$$

$$J_\pm f_m = \beta_{\pm m} f_{m\pm1}$$

$$f_m = (-1)^{-m} f_{-m} \qquad (11)$$

where $-\ell \le m \le \ell$ and $\beta_m = \sqrt{(\ell - m)(\ell + m + 1)}$ (see Ref. 17, Chap. 7).

We now represent N as the vector space of linear polynomials in $z_{-\ell}, \ldots, z_\ell$, where the variables z_m act as the f_m in (11). Denote by $K[z_{-\ell}, \ldots, z_\ell]$ the ring of polynomials in the independent variables $z_{-\ell}, \ldots, z_\ell$; K is isomorphic to the algebra of symmetric tensors over N. We now extend the operators J_3 and $J_\pm$ to be *derivations* over K:

$$J(\alpha f + \beta g) = \alpha Jf + \beta J_g$$

$$J(fg) = fJ_g + (Jf)_g \tag{12}$$

where $f, g \in K$ and α and β are scalars. It is natural to extend the J's in this way since they are Lie derivatives. We now require the polynomials F_m to transform in the same way as the z_m:

$$J_3 F_m = mF_m \quad J_{\pm} F_m = \beta_{\pm m} F_{m \pm 1}$$

Consider first the quadratic terms in F_m. The linear term of F_m is a scalar times z_m; since the representation is irreducible, the linear term must be a scalar multiple of the Identity. For quadratic terms,

$$J_3(z_j z_k) = (J_3 z_j) z_k + z_j (J_3 z_k)$$

$$= j z_j z_k + k z_j z_k$$

$$= (j + k) z_j z_k$$

Since F_m is to be a sum of quadratic terms and $J_3 F_m = mF_m$, we require $(j + k) = m$, so,

$$F_m(z_{-\ell}, \ldots, z_\ell) = \sum_{m_1 + m_2 = m} \alpha_{m_1 m_2 m} z_{m_1} z_{m_2}$$

In particular (when ℓ is even),

$$F_\ell = \alpha_0 z_\ell z_0 + \alpha_1 z_\ell z_{\ell-1} + \cdots + \alpha_{\ell/2} (z_{\ell/2})^2$$

Furthermore $J_+ F_\ell = \beta_\ell F_\ell = 0$, and this condition gives us a set of linear equations for the coefficients $\alpha_0, \ldots, \alpha_{\ell/2}$. For example, in the case $\ell = 2$ we have,

$$F_2 = az_2 z_0 + bz_1^2$$

$$J_+ F_2 = a\beta_0 z_2 z_1 + 2bz_1 z_2 \beta_1$$

$$= (a\beta_0 + 2b\beta_1) z_1 z_2 = 0$$

whence $a\beta_0 + 2b\beta_1 = 0$. This determines a and b, hence F_2, up to a scalar multiple. Once F_ℓ is known, we get $F_{\ell-1}$ from $J_- F_\ell = \beta_{\ell-1} F_{\ell-1}$, and so forth. In this way we can construct all the F_m's.

This procedure extends immediately to higher-order terms. For example, to get 3rd-order terms we write,

$$F_\ell = \sum_{i+j+k=\ell} A_{ijk} z_i z_j z_k$$

and apply $J_+ F_\ell = 0$ to get a linear system of equations for the A_{ijk}.

For $\ell = 1$, there is only one solution, but for $\ell = 3$ there are two independent solutions. In fact, the condition $J_+ F_\ell = 0$ leads to five equations in seven unknowns.

For the quadratic case, the general mapping is given in terms of the Clebsch-Gordon coefficients,

$$F_m(z_{-\ell}, \ldots z_\ell) = \sum_{m_1 + m_2 = m} C(\ell, m; \ell_1 m_2 \mid \ell, m) z_{m_1} z_{m_2}$$

or the Wigner $3 - j$ coefficients[17,31]:

$$= (-1)^m \sum_{m_1 + m_2 = m} \begin{pmatrix} \ell & \ell & \ell \\ m_1 & m_2 & -m \end{pmatrix} z_{m_1} z_{m_2}$$

Special Results

Let me discuss the solutions of the bifurcation equations for low values of ℓ. For $\ell = 1, 2$ the only solutions of the bifurcation equations are axisymmetric. There is a two-parameter family of such solutions, and these are orbitally stable, when they appear supercritically, and unstable, when they appear subcritically.

For $\ell = 3$ there are two independent covariant mappings,

1) $F_3 = z_3(z_0^2 - 2z_1 z_{-1} + 2z_2 z_{-2} - 2z_3 z_{-3})$

2) $F_3 = 9 \sqrt{\dfrac{60}{7}} z_3^2 z_{-3} - 9 \sqrt{\dfrac{60}{7}} z_3 z_2 z_{-2}$

$$+ 3 \sqrt{\dfrac{60}{7}} z_3 z_1 z_{-1} - 3 \sqrt{10} z_2 z_1 z_0$$

$$+ \dfrac{30}{\sqrt{7}} z_2^2 z_{-1} + \sqrt{7} z_1^3$$

This means the bifurcation equations take the form $\lambda z = AF(z) + BG(z)$, where the parameters A and B depend on the external physical constants of the problem, and we again have the possibility of mechanisms for pattern selection.

For $\ell = 4$, there are many solutions to the bifurcation equations in this case.[7] Two special ones are:

1) Axisymmetric solutions: $z_{\pm 1} = \cdots = z_{\pm 4} = 0$, $z_0 \neq 0$. The eigenvalues of the Jacobian are,

$$-1, 0, 0, \dfrac{20}{9}, \dfrac{20}{9}, \dfrac{10}{3}, \dfrac{10}{3}, \dfrac{-5}{9}, \dfrac{-5}{9}$$

The axisymmetric solutions are thus unstable on both sides of criticality, with 3 unstable subcritical modes.

2) Octahedral symmetry:

$$z_4 = \dfrac{5}{\sqrt{14}} = z_{-4} \qquad z_0 = \sqrt{5} \qquad z_{\pm 1} = z_{\pm 2} = z_{\pm 3} = 0$$

This solution was found by Busse. The eigenvalues are,

$$0, 0, 0, -1, \frac{20}{7}, \frac{20}{7}, \frac{20}{7}, \frac{5}{7}, \frac{5}{7}$$

This solution thus has one unstable subcritical mode; the subcritical branch can therefore bend back and regain stability.

For $\overset{\backprime}{\ell} = 6, 8$, there are also special solutions for these cases.[7] Busse used an extremum principle to determine the physically relevant solutions. We shall discuss that principle below. It indicates that the axisymmetric solutions are not the physically relevant ones. Busse conjectures that for $\ell = 6$ the relevant solution has the symmetry of a dodecahedron.

Gradient Structure of the Bifurcation Equations

The extremum principle is the following. For even ℓ, the reduced bifurcation equations can be written in the form,

$$\lambda z_m = a F_m(z_{-\ell} \ldots \overset{\backprime}{z\ell}) \tag{13}$$

where
$$F_m(z_{-\ell} \ldots z_\ell) = \sum_{m_1 + m_2 + m = 0} (-1)^m \begin{pmatrix} \ell & \ell & \ell \\ m_1 & m_2 & -m \end{pmatrix} z_{m_1} z_{m_2}$$

Consider the function,

$$p(z) = \frac{1}{3} \sum_{-\ell}^{\ell} F_m \bar{z}_m = \frac{1}{3} \sum_{-\ell}^{\ell} (-1)^m F_m z_{-m}$$

$$= \sum_{m_1 + m_2 + m_3 = 0} \begin{pmatrix} \ell & \ell & \ell \\ m_1 & m_2 & m_3 \end{pmatrix} z_{m_1} z_{m_2} z_{m_3}$$

For even ℓ, the Wigner coefficients are completely symmetric and therefore,

$$F_m(z_{-\ell}, \ldots z_\ell) = \frac{\partial p}{\partial z_m}$$

Consequently our reduced bifurcation equations have a gradient structure, and this fact is independent of the structure of the original equations $G(\lambda, u)$; it is a purely group theoretic result.

Therefore the bifurcation equations (13) are the Euler-Lagrange equations for the minimax problem,

$$\min_{|z| = 1} p(z) \tag{14}$$

where
$$|z|^2 = \sum_{-\ell}^{\ell} z_m \bar{z}_m = \sum_{-\ell}^{\ell} (-1)^m z_m z_{-m}$$

The function p is a third-order invariant for $O(3)$. That is, $p(T_g z) = p(z)$ for all $g \in O(3)$. In terms of the infinitesimal generators, this is equivalent to $J^{\pm} p = J_3 p = 0$. The norm $|z|^2$ is the second-order invariant.

Leon Green (School of Mathematics, University of Minnesota) and I have succeeded in casting the bifurcation problem in a slightly different way. Consider the Clebsch-Gordon series,

$$D^{\ell/2} \otimes D^{\ell/2} = D^{\ell} \oplus D^{\ell-1} \oplus \cdots \oplus D^{0} \tag{15}$$

and the associated representation,

$$U_g A = D^{\ell/2}(g) A D^{\ell/2}(g^{-1})$$

on $(\ell + 1) \times (\ell + 1)$ symmetric matrices A. This representation is unitary relative to the inner product $(A, B) = \operatorname{tr} AB^*$. Furthermore, the third-order invariant (there is only one) is,

$$P(A) = \frac{1}{3} \operatorname{tr} A^2 A^*$$

Now the highest-weight space, the one that transforms like D^{ℓ} in (15), consists of symmetric tensors $(A = A^+)$, so we may rephrase our bifurcation problem as,

$$\min \frac{1}{3} \operatorname{tr} A^3$$

subject to $\operatorname{tr} A^2 = 1$ and $\operatorname{tr} AB_j = 0$, where the B_j are symmetric matrices that lie in the lower-weight invariant subspaces. In particular, $\operatorname{tr} AI = \operatorname{tr} A = 0$. For $\ell = 2$, we get the bifurcation equations $A^2 = \lambda A + \gamma I$.

So far we have only been able to apply this approach in the case $\ell = .2$ (which we have solved completely). However, it is interesting because of its similarity to Michel's approach to symmetry breaking problems in physics.[16]

FURTHER PROBLEMS FOR INVESTIGATION

In the preceeding sections we have presented some of the specific details involved in applying group theory to bifurcation problems. There are several directions that are worthwhile to pursue.

The Bénard Problem

Extend the analysis here to three-space dimensions. Such an analysis would be of interest, for example, in the qualitative theory of phase transitions. Many substances are known to possess more than one crystalline form (sulphur possesses both a monoclinic and a triclinic form). What mechanisms govern the selection of crystallographic patterns in such cases as these? And what are the possible stable bifurcating patterns? Attempts to explain phase transitions (liquid to crystalline form) on the basis of bifurcation theory go back to Kirkwood and Monroe[14] and are continuing in the present.[20,21] In this approach, the BBGKY hierarchy is truncated by a closure hypothesis (e.g., the Kirkwood superposition) to arrive at a finite set (usually one or two) of nonlinear integral-differential equations. The bifurcation points of these equations are then taken to represent phase transitions.

Landau's Theory of Second-Order Phase Transitions

Landau's theory is a phenomenological approach that accurately predicts the symmetry changes that a crystalline substance may undergo continuously.[3,16] It fails, however, to give any accurate predictions of the critical exponents at the transition. Nevertheless it is a useful theory in practice. It is precisely a theory of spontaneous symmetry-breaking transitions. The approach can be described very simply as follows. Consider a "thermodynamic potential" $\Phi(\lambda, u)$, where u lies in a finite vector space η; the components of u are called the *order parameters*. Assume that we have a representation T_g acting on η and that $\Phi(\lambda, T_g u) = \Phi(\lambda, u)$, $g \in \mathcal{G}$. Suppose for $\lambda < \lambda_{cr}$, Φ takes on its minimum at u_0, where $T_g u_0 = u_0$ for all $g \in \mathcal{G}$, but that for $\lambda > \lambda_{cr}$ several critical points $u_1, \ldots, u_p$ branch from u_0 (which is no longer stable). The questions are: which of the u_i are stable and what are the isotropy subgroups of the stable branch?

Bifurcation in the Presence of $O(3)$

As you can see from the analysis in the preceding section, there is a difficult problem involved in finding the stable bifurcating solutions for each fixed ℓ. For even ℓ, the problem reduces to the variational problem (Equation 14). Since the unit sphere in finite dimensions is compact, there is no question about the existence of the minimum. The question is, what is its isotrophy subgroup? For $\ell = 2$, the minimum point is axisymmetric, while for $\ell = 4$, its symmetry group is O (the symmetry group of the cube). This was conjectured by Busse[7] on the basis of numerical results. For $\ell = 6$, Busse conjectures that the symmetry group is that of the dodecahedron, but a complete analysis is still lacking. It would be interesting to solve the problem for general even ℓ.

Symmetry Breaking in Chemical and Biological Systems

Whether bifurcation theory will ultimately give any account for the evolution of chemical and biological structures remains to be seen, but there is certainly an intense current interest in this approach. Here one tries to model the chemical processes in a biological network by a system of reaction-diffusion equations, and to describe the emergence of spatial or temporal patterns from a homogeneous state as a symmetry-breaking transition associated with the instability of the old state.[1,2,9,18,19,30] Undoubtedly one could construct simple models that exhibit the bifurcation of crystallographic states, which certainly exist in nature. The real question is whether such models aid in our understanding of such phenomena.

REFERENCES

1. AUCHMUTY, J. F. G. & G. NICOLIS. 1975. Bifurcation analysis of nonlinear reaction-diffusion equations–I. Bull. of Math. Biol. **37**: 323–363.
2. AUCHMUTY, J. F. G. & G. NICOLIS. 1974. Dissipative structures, catastrophes, and pattern formation: A bifurcation analysis. Proc. Nat. Acad. Sci. USA **71**: 2748–2751.

3. BIRMAN, J. L. 1967. Symmetry changes, phase transitions, and ferroelectricity. Ferroelectricity. Elsevier.
4. JARIĆ, M. V. & J. L. BIRMAN. 1977. New algorithms for the Molien function. J. Math. Phys. **18:** 1456–1458.
5. JARIĆ, M. V. & J. L. BIRMAN. 1977. Calculation of the Molien generating function for invariants of space groups. J. Math. Phys. **18:** 1459–1465.
6. BUSSE, F. 1967. The stability of finite amplitude cellular convection and its relation to an extremum principle. J. Fluid Mech. **30:** 625–650.
7. BUSSE, F. 1975. Patterns of convection in spherical shells. J. Fluid Mech. **72:** 67–85.
8. FIFE, P. 1973. "The Bénard problem for general fluid dynamical equations and remarks on the Boussinesq approximations. Indiana Univ. Math. J. **20.**
9. FIFE, P. 1976. Pattern formation in reacting and diffusing systems. J. Chem. Phys. **64:** 554–564.
10. GOLUBITSKY, M. An introduction to catastrophe theory. SIAM Review.
11. HAKEN, H. 1977. Synergetics: An Introduction. Springer-Verlag, New York.
12. JOSEPH, D. D. 1976. Stability of Fluid Motions. Vol. I & II. Springer-Verlag, Berlin.
13. KIRCHGÄSSNER, K. & H. KIELHÖFER. 1973. Stability and bifurcation in fluid mechanics. Rocky Mountain J. Math. **3:** 275–318.
14. KIRKWOOD, J. G. & E. MONROE. 1941. Statistical mechanics of fushion. J. Chem. Phys. **9.**
15. KOSCHMIEDER, E. L. 1974. Bénard convection. Ad. Chem. Phys. Vol. 26. 1. Prigogine & S. A. Rice, Eds. Wiley, New York.
16. MICHEL, L. 1975. Les brisures spontanées de symétrie en physique. J. Phys. **36:** C7 41, C7–51.
17. MILLER, W. 1972. Symmetry Groups and Their Applications. Academic Press, New York.
18. OTHMER, H. G. 1977. Models yield insights and problems in biological pattern formation. SIAM News **10,** No. 3.
19. OTHMER, H. G. & L. E. SCRIVEN. 1974. Nonlinear aspects of dynamic pattern in cellular networks. J. Theor. Biol. **43:** 83–112.
20. RAVECHÉ, H. J. & C. A. STUART. 1975. Towards a molecular theory of freezing. J. Chem. Phys. **63.**
21. RAVECHÉ, H. J. & C. A. STUART. 1976. Bifurcation of solutions with crystalline symmetry. J. Math. Phys. **17:** 1949–1953.
22. SATTIER, D. 1975. Branching and stability for nonlinear shells. Proc. IUTAM/IMU Symp. (Marseilles) Springer Lecture Notes.
23. SATTINGER, D. H. 1976. Pattern formation in convective phenomena. Proc. Turbulence and Navier Stokes Equations, Springer Lect. Notes, **565.**
24. SATTINGER, D. H. 1977. Group representation theory and branch points of nonlinear functional equations. SIAM J. Math. Anal. **8:** 179–201.
25. SATTINGER, D. H. 1978. Group representation theory, bifurcation theory, and pattern formation. J. Funct. Anal. **28:** 58–101.
26. SATTINGER, D. H. 1977. Selection mechanisms for pattern formation. Arch. Rat. Mech. Anal. **66:** 31 42.
27. SATTINGER, D. H. 1977. Group theoretic methods in bifurcation theory. Proc. 23rd Conf. Army Math.
28. SEGEL, L. A. 1965. Nonlinear problems in hydrodynamic stability. In Nonequilibrium Thermodynamics, Variational Techniques, and Stability. Univ. Chicago Press.
29. THOM, R. 1975. Structural Stability and Morphogenesis. D. H. Fowler, Trans. W. A. Benjamin, Reading, Mass.
30. TURING, A. M. 1952. The chemical basis of morphogenesis. Phil. Trans. Roy. Soc. London. **B237:** 37–72.
31. WIGNER, E. 1959. Group Theory and Atomic Spectra. Academic Press, New York.

APPLICATIONS OF BIFURCATION THEORY
IN THE ANALYSIS OF SPATIAL AND
TEMPORAL PATTERN FORMATION

H. G. Othmer

Department of Mathematics
Rutgers University
New Brunswick, N. J. 08903

INTRODUCTION

Problems that arise in a physical or biological context often give rise to the following type of deterministic mathematical model. The temporal evolution of the state variables u is governed by an autonomous evolution equation of the form,

$$\frac{du}{dt} = G(u, p) \tag{1}$$

and the steady states, if any exist, are solutions of,

$$G(u, p) = 0 \tag{2}$$

where G is a map $D \times I \to B$, for B a real Banach or Hilbert space, $D \subseteq B$, and $I \subseteq R$. For instance, G could represent an operator that arises from the Navier-Stokes equations or from a system of nonlinear reaction-diffusion equations, and p is either a fixed or slowly varying parameter.

In many cases there is a distinct or basic solution u_0, such as the conduction solution in a fluid layer heated from below or a uniform steady state in a reacting system, that exists for all $p \in I$. The evolution of small disturbances of u_0 is usually governed by the spectrum $\sigma(G_u)$ of the linearization G_u of G, and the critical values p_0 of p are those for which σ intersects the imaginary axis. Under appropriate hypotheses on G, general bifurcation theorems[1,2,3] can be used to show that, whenever zero is an eigenvalue of odd algebraic multiplicity of G_u, other steady-state solutions bifurcate from (u_0, p_0). If G_u has a simple pair of complex conjugate eigenvalues at p_0, then generalizations of the Hopf theorem can often be used to prove the existence of a bifurcating time-periodic solution.

The problem of spatial pattern formation in the context of developing biological systems is that of *assigning specific states to an ensemble of cells, whose initial states are relatively similar, such that the resulting ensemble of states forms a well-defined spatial pattern.*[4] In the reaction-diffusion theory of pattern formation originated by Turing,[5] the first nonuniform spatial distribution of morphogens is generated from an instability of the uniform steady state, and further refinement of the pattern can be accomplished by secondary and tertiary bifurcations. The successive bifurcations are symmetry breaking in the sense that bifurcating solutions have lower symmetry than the solutions from which they bifurcate.

For reasons that will later be elaborated, the presence of nontrivial symmetry in the problem often leads to degenerate eigenvalues, and when the multiplicity is

0077-8923/78/0316-0064 $1.75/1 © 1979, NYAS

even, it can happen that no solutions bifurcate at the critical parameter value. When bifurcation does occur, it will be shown that there may be several bifurcating solutions, of which some exist on both sides of the critical p, and others, only for $p > p_0$ or $p < p_0$. By systematically applying group theoretic techniques, the bifurcation equations are greatly simplified, and it will be seen that the existence of some branches of solutions can be established by reducing the problem to one in which the multiplicity is odd. In essence, this reduction is possible for the problems treated here because whenever (1) and (2) are invariant under some group of transformations, there is a system of invariant subspaces for (1), each of which can be labeled by the subgroup that leaves all of its elements invariant.

CONSEQUENCES OF GROUP INVARIANCE

Let Ω be a bounded region of R_2 with a smooth boundary $\partial\Omega$, and consider the problem,

$$\frac{\partial c}{\partial t} = \mathfrak{D}\Delta c + R(c, p), \text{ in } \Omega$$

$$c(\mathbf{r}, 0) = c_0(\mathbf{r}), \text{ in } \Omega$$

(3)

with Neumann or Dirichlet boundary conditions. Here $c = (c_1, \ldots, c_n)^T$, $\mathfrak{D}$ is an $n \times n$ diagonal matrix with $\mathfrak{D}_{ii} > 0$, Δ is the Laplacian for Ω, and $R(c, p)$ is a C^2-smooth nonlinear function of c and p; c is to be regarded as an element of $L_2(\Omega)$, with the inner product $<u, v>_2 = \int_\Omega <u, v> d\Omega$.* For a fixed choice of boundary conditions (3) can be cast in the form of (1), and we can show that the initial value problem has a unique solution.[6]

Let $M \equiv \{c \mid 0 \leq c_i \leq C_i, i = 1, \ldots, n\}$ and suppose that the kinetics $R(c, p)$ are such that M is positively invariant under the flow of the ordinary differential equation $\dot{c} = R(c, p)$. Then solutions of (3) have the property that if $c_0(\mathbf{r}) \in M$ $\forall$ $\mathbf{r} \in \overline{\Omega}$, then $c(\mathbf{r}, t) \in M$ $\forall$ $(\mathbf{r}, t) \in \overline{\Omega} \times [0, \infty)$.[7] Further, suppose that there exist one or more solutions c^s of $R(c, p) = 0$ for all p of interest, and, when (3) has Dirichlet data, assume that $c = c^s$ on $\partial\Omega$ for one of the c^s. As a result, (3) has a uniform steady state $c = c^s$ for either type of data. Finally, set $u = c - c^s$ and rewrite (3) as,

$$\frac{\partial u}{\partial t} = \mathfrak{D}\Delta u + K(p)u + F(u, p)$$

$$u(\mathbf{r}, 0) = u_0(\mathbf{r})$$

(4)

where

$$(K(p))_{ij} = \frac{\partial R_i}{\partial c_j} \mid c_j = c_j^s$$

---

*Throughout, $<\cdot, \cdot>$ and $\| \cdot \|$ denote the Euclidean inner product and norm, respectively.

and $\| F(u, p) \| \sim o(\| u \|)$, as $\| u \| \to 0$. The linear stability properties of c^s are governed by the spectrum of the linear operator $\mathfrak{D}\Delta + K(p)$.

For any domain Ω there is a maximal group $G \subseteq O(2)$ of orthogonal transformations that map Ω into itself (G may consist of only the identity).† We call G the symmetry group of the domain and define a unitary representation of G on $L_2(\Omega)$ as follows: if $g \in G$ and,

$$\mathbf{r} \to \mathbf{r}' = g\mathbf{r}$$

then   (5)

$$(T_g f)(\mathbf{r}) \to f(g^{-1}\mathbf{r}).$$

This defines the *unitary operator* $T_g : L_2(\Omega) \to L_2(\Omega)$. A function $f \in L_2(\Omega)$ is said to be invariant under a group $H \subseteq G$ if $T_g f = f$ for all $g \in H$. H is called the symmetry group of f or the isotrophy subgroup of G at f. Equation 4 is said to be invarient under G if T_g commutes with the operations in (4) for all $g \in G$. Thus, it is necessary that,

$$T_g \left[\frac{\partial u}{\partial t} - \mathfrak{D}\Delta u - K(p)u - F(u, p) \right]$$

$$= \frac{\partial}{\partial t} T_g u - \mathfrak{D}\Delta T_g u - K(p) T_g u - F(T_g u, p).$$

Since $\mathbf{r}$ enters explicitly only in Δ, and since Δ is invariant under all orthogonal transformations, the differential equation is, in fact, invariant under $O(2)$. The boundary conditions on u are always homogeneous and are therefore also invariant under $O(2)$. If the initial function $u_0(\mathbf{r})$ is invariant under a group $H \subseteq G$, the whole problem is invariant under H, and we call H the symmetry group of (4). This leads to the following conclusion concerning *conservation of symmetry*.

PROPOSITION 1. If $u_0(\mathbf{r})$ is invariant under a group $H \subseteq G$, the solution $u(\mathbf{r}, t)$ of (4) is invariant under H for all $t \geq 0$.

Proof: We introduce Green's function $G(\mathbf{r}, \xi, t - \tau)$ for the problem.

$$\frac{\partial u}{\partial t} = \mathfrak{D}\Delta u$$

with homogeneous boundary conditions, and write the solution of (4) as,

$$u(\mathbf{r}, t) = \int_\Omega G(\mathbf{r}, \xi, t) u_0(\xi) \, d\xi$$

$$+ \int_0^t \int_\Omega G(\mathbf{r}, \xi, t - \tau)[K(p)u(\xi, \tau) + F(u(\xi, \tau), p)] \, d\xi d\tau$$

The conclusion follows by applying T_g, $g \in H$, to both sides and invoking the uniqueness for the initial value problem. □

†Group-theoretical terms used here can be found in Reference 8 or 9.

In the steady-state version of (**4**),

$$\mathcal{D}\Delta U + K(p)u + F(u, p) = 0 \tag{6}$$

the initial condition is absent, and the symmetry group of (**6**) coincides with the symmetry group G of the domain. Consequently, if a nontrivial solution of (**6**) is to be unique, it must be invariant under G, i.e., $T_g u = u$ for all $g \in G$. Stated otherwise, a unique solution necessarily transforms according to the identity representation of G. If there exists a solution u of (**6**) that is not invariant under G, then every transform $T_g \bar{u}$ of $\bar{u}$ is also a solution, and so the entire orbit $\{T_g \bar{u}\}$ of $\bar{u}$ under G is contained in the set of all solutions of (**6**). Since T_g is isometric, all solutions on a given orbit have the same norm. In general, there may be several disjoint orbits in the set of nontrivial solutions that bifurcate at a degenerate eigenvalue.

The symmetry group G of Ω is always compact (and often finite), and therefore the representation $g \rightarrow T_g$ is always a direct sum of finite-dimensional *irreducible representations* (irreps).[10] Since $G \subseteq O(2)$, every irrep is at most 2-dimensional, but there can be two or more inequivalent 2-dimensional irreps. If we denote the irreps by $\Gamma_1, \Gamma_2, \ldots$, then any representation has the decomposition $\Gamma = \Gamma_1 \oplus \Gamma_2 \oplus \Gamma_3 \ldots$, where one or more Γ_p's may be repeated infinitely often. The set $\{y_j \bar{\phi}_k\}$, where $\{y_j\}$ is a basis for an n-dimensional Euclidean space and the $\bar{\phi}_k$'s are the eigenfunctions of Δ for the given boundary conditions, is a basis for $L_2(\Omega)$. For the present we can set $n = 1$ and dispense with the y_j's. Now $\Delta \bar{\phi}_k = \alpha_k \bar{\phi}_k$, and Δ commutes with T_g, so whenever $\bar{\phi}_k$ is an eigenfunction belonging to α_k, $T_g \bar{\phi}_k$ is also an eigenfunction belonging to α_k. If $\bar{\phi}_k$ transforms according to an irrep of dimension 2 or greater, then α_k is necessarily a multiple (or degenerate) eigenvalue. Even if all irreps are 1-dimensional, α_k may still be degenerate.

Suppose that α_k is an m-fold degenerate eigenvalue, and let L^k be the m-dimensional subspace of $L_2(\Omega)$ spanned by the degenerate $\bar{\phi}_k$'s. The ϕ_k's carry an m-dimensional representation of G that may or may not be reducible. If it is, a symmetry-adapted basis $\{\phi_k\}$ for L^k can be constructed from the $\{\bar{\phi}_k\}$ such that the representation is fully reduced.[11,12] Each ϕ_k in the symmetry-adapted basis transforms according to one and only one irrep of G. This change of basis can be made at every degenerate eigenvalue α_k, and, in the symmetry-adapted basis, $L_2(\Omega)$ has the following decomposition into orthogonal subspaces:

$$L_2(\Omega) = L_1 \oplus L_2 \oplus \ldots \tag{7}$$

Here L_p is the linear span of all those ϕ_k, $k = 1, 2, \ldots$, that transform according to Γ_p. For any fixed k, there may be several p's for which $L^k \cap L_p \neq \phi$, and dim $(L^k \cap L_p)$ can be an integer multiple of the dimension of Γ_p. If Γ_p is s-dimensional and dim $(L^k \cap L_p) = s\ell$, $\ell \geq 1$, the $s\ell$ corresponding ϕ_k's can be grouped into ℓ sets, each of which is closed under the transformation T_g.

With each L_j in (**7**) one can associate the maximal subgroup G_j of G under which every element of L_j is invariant; G_j is the symmetry group of functions in L_j. Clearly if $G_j \subset G_k$, then all functions in $L_j \cup L_k$ are invariant under G_j. Corresponding to the chain of subgroups $G_j \subset G_k \subset \cdots \subset G$ is the subspace $H \equiv L_j \cup L_k \cup \cdots \cup L_1$, where L_1 contains all functions that are invariant under G; H contains all functions whose symmetry group is at least G_j, and, by PROPOSITION 1, if the initial data lie in H, then the solution lies in H for all $t \geq 0$. If (4) is linear, then all modes evolve independently, and for $t \geq 0$ the solution lies in the subspace spanned by those ϕ_k that are present in the initial data. This is no longer true for nonlinear problems, but the foregoing shows that symmetry considerations can be used to predict which Fourier modes may be excited by nonlinear coupling within a subset of modes, irrespective of the precise form of the nonlinearity.

BIFURCATION AT A MULTIPLE ZERO EIGENVALUE

Hereafter write (4) in the form,

$$\frac{du}{dt} = L(p)u + F(u,p) \tag{8}$$

Let $|\epsilon|$ be small, and suppose that for $|p - p_0| \leq O(\epsilon)$, $\sigma(L)$ can be written $\sigma(L) = \sigma_1(L) \cup \sigma_2(L)$, where,

$$\sigma_1(L) \equiv \{\lambda\epsilon\sigma(L) \,|\, |\operatorname{Re}\lambda| \leq \gamma\}$$
$$\sigma_2(L) \equiv \{\lambda\epsilon\sigma(L) \,|\, \operatorname{Re}\lambda < -\gamma\} \tag{9}$$

and γ is a positive constant. If $\lambda\epsilon\sigma(L)$, then λ satisfies the determinantal equation,

$$|K(p) + \alpha_k \mathfrak{D} - \lambda I| = 0, \qquad k = 1,\ldots \tag{10}$$

where α_k runs through the eigenvalues of Δ. Suppose that there is a k_0 for which α_{k_0} is an eigenvalue of multiplicity m, and that there is a simple eigenvalue $\lambda^0(p)$ of $K(p) + \alpha_{k_0}\mathfrak{D}$ such that $\lambda^0(p_0) = 0$ and $\lambda^0\epsilon\sigma_1(L)$ for $|p - p_0| \leq O(\epsilon)$. Assume that all other eigenvalues of $K(p) + \alpha_{k_0}\mathfrak{D}$ and all eigenvalues for $k \neq k_0$ are contained in $\sigma_2(L)$ when $|p - p_0| < O(\epsilon)$. Let $y(p)$ be the eigenvector of $K + \alpha_{k_0}\mathfrak{D}$ that corresponds to λ^0. Then the eigenfunctions of L for $\lambda\epsilon\sigma_1(L)$ have the form,

$$\psi_j = y(p)\phi_j, \; j = 1,\ldots,m \tag{11}$$

Under the preceding hypotheses, the multiplicity arises solely from the multiplicity of α_{k_0}. In general, it can also happen that $K(p_0) + \alpha_k\mathfrak{D}$ has a zero eigenvalue for several distinct α_k's, but these cases are not treated here.

Define projections P_1 and P_2, corresponding to the above decomposition of σ, as follows:

$$P_1 u \equiv \sum_{j=1}^{m} <\psi_j^*, u>_2 \psi_j$$
$$P_2 \equiv I - P_1 \tag{12}$$

Here ψ_j^* is the eigenfunction adjoint to ψ_j, normalized so that $<\psi_j^*, \psi_k>_2 = \delta_{ij}$.

To construct solutions of (8) for which $||u||_2 \sim O(\epsilon)$, scale u as follows:

$$u = \epsilon(v + \overline{w}) = P_1 u + P_2 u$$

Now (8) can be written,

$$\frac{dv}{dt} = L_1 v + \epsilon F_1(v, \overline{w}, p)$$

$$\frac{d\overline{w}}{dt} = L_2 \overline{w} + \epsilon F_2(v, \overline{w}, p)$$

(13)

where L_1 and L_2 are restrictions of L to $P_1 L_2(\Omega)$ and $P_2 L_2(\Omega)$, respectively, and

$$F_1(v, \overline{w}, p, \epsilon) = P_1 \frac{F(\epsilon(v + \overline{w}), p)}{\epsilon^2}$$

$$F_2(v, \overline{w}, p, \epsilon) = P_2 \frac{F(\epsilon(v + \overline{w}), p)}{\epsilon^2}$$

It can be shown from the second equation of (13) that if $||\overline{w}(0)||_2$ is sufficiently small and γ is sufficiently large, then $||\overline{w}(t)||_2 \sim O(\epsilon)$ uniformly in t. Assume that this is the case, and scale $\overline{w}$ again by setting $\overline{w} = \epsilon w$; then (13) can be written,

$$\frac{dv}{dt} = L_1 v + \epsilon F_1(v, \epsilon w, p, \epsilon)$$

$$\frac{dw}{dt} = L_2 w + F_2(v, \epsilon w, p, \epsilon)$$

(14)

If $F(u, p)$ is analytic in a neighborhood of $(0, p_0)$, then,

$$F(u, p) = Q(u, u, p) + C(u, u, u, p) + \cdots$$

and so,

$$\epsilon F_1(v, \epsilon w, p, \epsilon) = \epsilon\{Q^1(v, v, p)\} + \epsilon^2\{2Q^1(v, w, p) + C^1(v, v, v, p)\} + O(\epsilon^3)$$

$$F_2(v, \epsilon w, p, \epsilon) = Q^2(v, v, p) + \epsilon\{2Q^2(v, w, p) + C^2(v, v, v, p)\} + O(\epsilon^2)$$

(15)

where Q, Q^i and C, C^i are homogeneous of second and third degree, respectively. It is clear from (15) that, to $O(\epsilon^2)$, v evolves independently of w, while, to $O(\epsilon)$, w is "forced" by v through the term $Q^2(v, v, p)$. This separation of scales between the v and w equations makes it possible to solve sequentially for the terms in an expansion of v and w. We shall treat the steady-state case in the remainder of this section.

The steady-state equations are,

$$L_1 v + \epsilon F_1(v, \epsilon w, p, \epsilon) = 0$$

$$L_2 w + F_2(v, \epsilon w, p, \epsilon) = 0$$

(16)

and since L_2 is invertible, it can be shown that the second equation is uniquely solvable for $w = W(v, p, \epsilon)$. The bifurcation equation is obtained by substituting this function into the first equation. Define $p_1(\epsilon)$ by the relation $p - p_0 = \epsilon^k p_1(\epsilon)$ and use (15) in (16). The result is that the bifurcation equation can be written,

$$(L_{10} + \epsilon^k p_1 L_{11} + \cdots)v + \epsilon\{Q_0{}^1(v, v, p_0) + \epsilon^k p_1 Q_1{}^1(v, v, p_0) + \cdots\} +$$
$$\epsilon^2\{2Q_0{}^1(v, w, p_0) + c_0{}^1(v, v, p_0) + \cdots\} + O(\epsilon^3) = 0 \tag{17}$$

Here the added subscripts on L_1, Q^1, etc. denote partial derivatives with respect to p. By hypothesis, $L_{10}v(0) = 0$, and so (17) reduces to

$$\epsilon^k p_1 L_{11} v_0 + \epsilon Q_0{}^1(v_0, v_0, p_0) + O(\epsilon^2) = 0 \tag{18}$$

where $p_1 \equiv p_1(0)$ and $v_0 \equiv v(0)$. Write $v_0 = \Sigma a_j \psi_j(p_0)$ and $v_0^* = \Sigma a_j \psi_j^*(p_0)$, where the a_j's are unknown (real) amplitudes, and suppose that the system,

$$p_1 L_{11} v_0 + Q_0{}^1(v_0, v_0, p_0) = 0$$
$$\langle v_0^*, v_0 \rangle = 1 \tag{19}$$

has a nontrivial solution (p_1, v_0). The following shows when this solution can be extended to a solution of (18) for $\epsilon \neq 0$.

PROPOSITION 2. Suppose that (19) has a nontrivial solution (p_1, v_0) at which the linear system,

$$\left[p_1 L_{11} + 2 \frac{\partial Q}{\partial v} \right] x + (L_{11} v_0) y = 0 \tag{20}$$
$$\langle v_0^*, x \rangle_2 = 0$$

has no nontrivial solution (y, x). Then there exists an analytic solution $[p_1(\epsilon), v(\epsilon)]$ of (17) for ϵ sufficiently small, such that $p - p_0 = \epsilon p_1(\epsilon)$ and $[p_1(0), v(0)] = (p_1, v_0)$.

The proof involves a simple application of an implicit function theorem.

Of course there may be several solutions for (19) that satisfy PROPOSITION 2, and each gives a bifurcating branch of solutions for (16). Moreover, if there exists a $\tilde{v}$ for which $L_{11}\tilde{v} \neq 0$, $Q_0{}^1(\tilde{v}, \tilde{v}, p_0) = 0$, and which satisfies,

$$p_1 L_{11}\tilde{v} + 2Q_0{}^1(\tilde{v}, W, p_0) + C_0{}^1(\tilde{v}, \tilde{v}, p_0) = 0 \tag{21}$$

then if the corresponding linear problem has no nontrivial solution, (16) has an analytic solution $[p_1(\epsilon), v(\epsilon)]$ such that $p - p_0 = \epsilon^2 p_1(\epsilon)$ and $(p_1(0), v(0)) = (p_1, \tilde{v})$. Consequently, it can happen that there are branches that exist on both sides of p_0 and branches that exist only for $p > p_0$ or $p < p_0$. This is in contrast to the case of bifurcation at a simple eigenvalue.

The solutions can be written,

$$v = v_0 + \epsilon v_1 + \epsilon^2 v_2 + \cdots$$
$$w = w_0 + \epsilon w_1 + \epsilon^2 w_2 + \cdots \tag{22}$$
$$p = p_0 + \epsilon p_1 + \epsilon^2 p_2 + \cdots$$

and the various coefficients can be found by solving the following sequence of linear problems:

For ϵ^0
$$\begin{cases} L_{10}v_0 = 0 \\ L_{20}w_0 + Q_0^2(v_0, v_0, p_0) = 0 \end{cases}$$

For ϵ^1
$$\begin{cases} L_{10}v_1 + p_1 L_{11}v_0 + Q_0^1(v_0, v_0, p_0) = 0 \\ L_{20}w_1 + p_1 L_{21}w_0 + 2Q_0^2(v_0, v_1, p_0) + p_1 Q_1^2(v_0, v_0, p_0) \\ \qquad + 2Q_0^2(v_0, w_0, p_0) + C_0^2(v_0, v_0, p_0) = 0 \end{cases}$$

(23)

For ϵ^2
$$\begin{cases} L_{10}v_2 + p_2 L_{11}v_1 + p_1^2 L_{12}v_0 + 2Q_0^1(v_0, v_1, p_0) \\ \qquad + p_1 Q_1^1(v_0, v_0, p_0) + 2Q_0^1(v_0, w_0, p_0) \\ \qquad + C_0^1(v_0, v_0, v_0, p_0) = 0 \end{cases}$$

$\vdots$

The condition that $\langle v_0^*, v_0 \rangle_2 = 1$ derives from the definition $\epsilon = \langle v_0^*, u \rangle_2$, and this implies that,

$$\sum_{j=1}^{m} a_j^2 = 1$$

(24)

$$\sum_{j=1}^{m} a_j \langle \psi_j^*(p_0), v_k \rangle_2 = 0, \qquad k \geq 1$$

Now recall that $\psi_j = y\phi_j$, and define the following quantities:

$$\Omega_1 = \left\langle y^*, \frac{\partial K}{\partial p} y \right\rangle \qquad\qquad \Omega_2 = \langle y^*, Q(y, y, p_0) \rangle$$

$$\Lambda_1 = \left\langle y^*, \frac{\partial Q}{\partial p}(y, y, p_0) \right\rangle \qquad \Lambda_2 = \langle y^*, C(y, y, y, p_0) \rangle \qquad (25)$$

$$\Gamma_{ijk} = \int_\Omega \phi_i \phi_j \phi_k \, d\Omega \qquad\qquad \Gamma_{ijk\ell} = \int_\Omega \phi_i \phi_j \phi_k \phi_\ell \, d\Omega$$

An easy calculation shows that the solvability condition for the $O(\epsilon)$ equation reduces to,

$$p_1 \Omega_1 a_i + \Omega_2 \sum_{j,k=1}^{m} \Gamma_{ijk} a_j a_k = 0$$

(26)

$$\sum_{j=1}^{m} a_j^2 = 1, \, i = 1, \ldots m$$

It is noteworthy that there is a virtually complete separation in these equations into factors which depend on the geometry of the domain and those which depend on the chemical kinetics. The integral Γ_{ijk} depends only on Ω, and thus can be evaluated once and for all for any given domain.[11,13] The quantities Ω_1 and Ω_2 appear to depend only on the chemical kinetics, but, in fact, depend on Ω through the dependence of y on α_{k_0}. Nonetheless, once the kinetics are fixed, Ω_1 and Ω_2 become known functions of α_{k_0}.

The bifurcation equations (26) apply equally well when α_{k_0} is a simple eigenvalue and in this case they reduce to $p_1\Omega_1 + \Omega_2\Gamma_{111} = 0$. If the solution of this is nonzero then bifurcation occurs, and the solution u is asymptotic to $\epsilon y\phi_1$. Even if the multiplicity m is greater than 1, there may exist solutions to (26) of the form $(p_1, 0, \ldots u_k, \ldots, 0)$ in which only one mode has a nontrivial amplitude. According to PROPOSITION 2, any isolated solution of (26) gives rise to the first term in an expansion of v and $p - p_0$ along a nontrivial solution of (17). Given any such solution, it can be shown that the successive Equations 23, in conjunction with the orthogonality conditions of (24), uniquely determine $w_0, v_1, \ldots$. Clearly if $(p_1, a_1, \ldots, a_m)$ is a solution of (26), so is $(-p_1, -a_1, \ldots, -a_m)$.

If $\Omega_1 \neq 0$, and there exists a vector $\tilde{a} = (\tilde{a}_1, \ldots, \tilde{a}_m)$, not identically zero, for which $\Omega_2 \Sigma \Gamma_{ijk}\tilde{a}_j\tilde{a}_k = 0$, then $p_1 \equiv 0$ and the $O(\epsilon^2)$ solvability condition must be checked to see whether any nontrivial solutions bifurcate at $O(\epsilon^2)$. This condition leads to the following system of equations:

$$p_2\Omega_1 a_i + \Lambda_1 \sum_{j,k=1}^{m} \Gamma_{ijk}a_ja_k + \Lambda_2 \sum_{j,k,\ell=1}^{m} \Gamma_{ijk\ell}a_ja_ka_\ell +$$

$$\sideset{}{^*}\sum_{p,q} \Lambda_{pq} \sum_{j,k,\ell=1}^{m} \Gamma_{jkq}\Gamma_{i\ell q}a_ja_ka_\ell = 0 \qquad (27)$$

$$\sum_{j=1}^{m} a_j^2 = 1, \; i = 1, 2, \ldots m$$

Here

$$\Lambda_{pq} \equiv -2 \frac{\langle y_{pq}^*, Q(y, y, p_0)\rangle \langle y^*, Q(y_{pq}, y, p_0)\rangle}{\lambda_{pq}},$$

and λ_{pq} and y_{pq} are the pth eigenvalue and eigenvector, respectively, of $K + \alpha_q \mathfrak{D}$. The sum Σ^* runs over all indices except those that belong to the degenerate eigenfunctions at $q = k$. The kinetics enter in (27) only via the factors Ω_1, Λ_1, Λ_2, and Λ_{pq}, and, as before, they can be evaluated treating α_q as a parameter once the kinetics are specified.

At an m-fold degenerate eigenvalue there are m^3 constants Γ_{ijk} and m^4 constants $\Gamma_{ijk\ell}$, but many of these vanish for symmetry reasons when the eigenfunctions ϕ_j are symmetry adapted. For instance, suppose that $\phi_i(\phi_j)$ transforms according to the irrep Γ_i (Γ_j) of G. Then the product $\phi_i\phi_j$ transforms according to the direct-product representation $\Gamma_i \otimes \Gamma_j$. Since this representation is generally reducible,

$$\Gamma_i \otimes \Gamma_j = \sum_p \gamma_{ijp} \Gamma_p$$

and so $\Gamma_{ijk} \equiv 0$ for all modes ϕ_k that transform according to a Γ_p for which $\gamma_{ijp} = 0$. Stated more symmetrically, $\Gamma_{ijk} \equiv 0$ unless $\Gamma_i \otimes \Gamma_j \otimes \Gamma_k$ contains the identity representation. Further simplification is possible because $\phi_i\phi_j = \phi_j\phi_i$ and so only the symmetric portion $(\Gamma_i \otimes \Gamma_j)_S$ of $\Gamma_i \otimes \Gamma_j$ is required. Finally, the number of integrals that have to be evaluated is further reduced by the fact that $\Gamma_{ijk} = \Gamma_{\sigma(ijk)}$ for all permutations $\sigma(ijk)$ of (ijk). All the foregoing applies equally well to $\Gamma_{ijk\ell}$. An example will help to illustrate the procedure.

Consider a square domain Ω with C_{4v} symmetry and periodic boundary conditions, and for illustrative purposes suppose that Ω is approximated by a 16-cell regular square lattice.[11]‡ The eigenvalues of the discrete Laplacian are $0, -2, -4, -6$, and -8, and are of multiplicity 1, 4, 6, 4, and 1, respectively. C_{4v} has five inequivalent irreps, four of which are 1-dimensional ($\Gamma_1, \ldots \Gamma_4$) and one that is two-dimensional (Γ_5). The subgroups of C_{4v} and their relationship under inclusion are as follows:

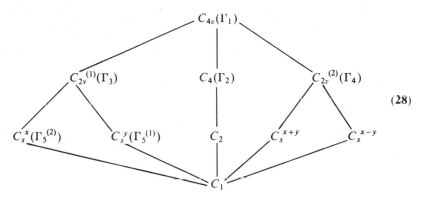

(28)

In $|C_{2v}^{(1)}$ the x and y axes are lines of symmetry while in $C_{2v}^{(2)}$ the lines $x = \pm y$ are lines of symmetry; C_s^x is a subgroup of order 2 with elements $\{e, \sigma_x\}$, where σ_x is the reflection across the x-axis. The other subgroups of order 2 are labeled accordingly and all other designations are standard. A parenthetical Γ_k appears beside the maximal subgroup under which the ϕ_j's that transform according to Γ_k are invariant, and $\Gamma_5^{(1)}(\Gamma_5^{(2)})$ denotes the first, second row of Γ_5. The pairing between Γ_k and the subgroups is done using a representation table.[11]

According to PROPOSITION 1 and the foregoing decomposition of the group, the following sets of modes span subspaces that are invariant under the dynamics:

$$\{\Gamma_1, \Gamma_2\}, \{\Gamma_1, \Gamma_3\}, \{\Gamma_1, \Gamma_4\}, \{\Gamma_1, \Gamma_3, \Gamma_5^{(1)}\}, \{\Gamma_1, \Gamma_3, \Gamma_5^{(2)}\}, \{\Gamma_1, \ldots, \Gamma_5\}$$

‡Ω can be regarded as one cell in an infinite planar lattice and the functions ϕ_j, symmetry adapted to the point group C_{4v}, transform according to the identity representation of the translation subgroup of the full space group.

Clearly the steady-state equations must reflect this invariance. Elsewhere these subsets of modes are called *consistent subsets*.[11] For the 16-cell system there are sixteen eigenfunctions ϕ_k, and these are grouped according to their eigenvalue and the representation according to which they transform, as follows:

		Γ_1	Γ_2	Γ_3	Γ_4	Γ_5
$\alpha_k = 0$		ϕ_1	—	—	—	—
-2		ϕ_2	—	ϕ_5	—	$\{\phi_9, \phi_{10}\}$
-4		ϕ_3	—	—	ϕ_8	$\{\phi_{13}, \phi_{14}\}, \{\phi_{15}, \phi_{16}\}$
-6		—	ϕ_4	—	ϕ_7	$\{\phi_{11}, \phi_{12}\}$
-8		—		ϕ_6		—

The first (second) function of each pair under Γ_5 transforms according to $\Gamma_5^{(1)}(\Gamma_5^{(2)})$.

We say that a subset of eigenfunctions belonging to a degenerate eigenvalue α_k is consistent if that subset is closed, under formation of products, relative to the set of eigenfunctions belonging to α_k. In general, products of functions from a consistent subset at a fixed α_k will involve functions that belong to other α_k's, but these other functions are irrelevant to the structure of the bifurcation equations at α_k. Suppose that the degenerate eigenvalue is $\alpha_k = -4$. The consistent subsets of modes belonging to this eigenvalue are,

$$\{\phi_3\}, \{\phi_3, \phi_8\}, \{\phi_3, \phi_{13}, \phi_{15}\}, \{\phi_3, \phi_{14}, \phi_{16}\}, \{\phi_3, \phi_8, \phi_{13}, \ldots, \phi_{16}\}$$

Since the first, third, and fourth sets contain an odd number of functions, the bifurcation equations at $\alpha_k = -4$ must at least have solutions of the following form:

(i) $a_3 \neq 0$, all others a_i's $= 0$,

(ii) At least one of $a_3, a_{13}, a_{15} \neq 0$,

(iii) At least one of $a_3, a_{14}, a_{16} \neq 0$,

It may happen that all solutions under (ii) and (iii) already appear in (i). Further information on solutions can be gotten by setting all a_i, except a_3 and a_8, equal to zero and solving the resulting 2-dimensional problem.

The full set of bifurcation equations at $O(\epsilon)$ is

$$p_1 \Omega a_3 + a_{13}a_{15} + a_{14}a_{16} = 0$$
$$p_1 \Omega a_8 + a_{13}a_{16} + a_{14}a_{15} = 0$$
$$p_1 \Omega a_{13} + a_3 a_{15} + a_8 a_{16} = 0$$
$$p_1 \Omega a_{14} + a_{13}a_{16} + a_8 a_{15} = 0 \tag{29}$$
$$p_1 \Omega a_{15} + a_3 a_{13} + a_8 a_{14} = 0$$
$$p_1 \Omega a_{16} + a_3 a_{14} + a_8 a_{13} = 0$$

and solutions are subject to the normalization $a_3^2 + a_8^2 + a_{13}^2 + a_{14}^2 + a_{15}^2 + a_{16}^2 = 1$. The constant $\Omega \equiv 2\Omega_1/\Omega_2$ contains all the kinetic information at $O(\epsilon)$. It will be assumed that $\Omega \neq 0, \infty$. One finds that solutions of these equations lie on two disjoint orbits, each of length four, having the following structure:

$$O_1:(a_3, a_8, a_{13}, a_{14}, a_{15}, a_{16}) = \begin{cases} \gamma(1, 0, 1, 0, 1, 0) \\ \gamma(1, 0, 0, 1, 0, 1) \\ \gamma(1, 0, -1, 0, -1, 0) \\ \gamma(1, 0, 0, -1, 0, -1) \end{cases}$$

(30)

$$O_2:(a_3, a_8, a_{13}, a_{14}, a_{15}, a_{16}) = \begin{cases} \gamma(1, 1, 1, 1, 1, 1) \\ \gamma(1, -1, -1, 1, -1, 1) \\ \gamma(1, -1, 1, -1, 1, -1) \\ \gamma(1, 1, -1, -1, -1, -1) \end{cases}$$

Here $\gamma = -p_1\Omega/2$ and $p_1 = \pm 2/\sqrt{3}\,|\,\Omega\,|$ $(p_1 = \pm 2/\sqrt{6}\,|\,\Omega\,|)$ on $O_1(O_2)$. The solutions on these orbits give the complete set of solutions that bifurcate at $O(\epsilon)$. Notice that solutions with C_s^x symmetry $(a_3 = a_{14} = a_{16} \neq 0$, others zero) lie on the same orbit as those with C_s^y symmetry $(a_3 = a_{13} = a_{16} \neq 0$, others zero). This happens because ϕ_3 transforms according to the identity representation, and the other modes are partners that transform according to Γ_5.

There is no solution at $O(\epsilon)$ with only a_3 nonzero, and therefore this branch of solutions bifurcates at $O(\epsilon^2)$ or higher. Since ϕ_3 transforms according to the identity representation, the orbit through this solution is of length one. There may also be other branches of solutions that bifurcate at $O(\epsilon^2)$, but this possibility is not explored here.

STABILITY OF THE BIFURCATING SOLUTIONS

Unlike the case of a simple eigenvalue, there are no general rules concerning the stability or instability of solutions that bifurcate at a multiple eigenvalue. For instance, it is easy to construct examples in which branches that exist on both sides of p_0 are either stable on both sides or unstable on both sides. To decide stability of solutions on a given branch $(\hat{v}, \hat{w})$, write $v = \hat{v} + \xi_1$, $w = \hat{w} + \xi_2$, and linearize (14). The result is the system:

$$\frac{d\xi_1}{dt} = \left[L_{10} + \epsilon\left(L_{11} + 2\frac{\partial Q_0}{\partial v}\right)\right]\xi_1 + O(\epsilon^3, \xi_1^2, \xi_2^2)$$

(31)

$$\frac{d\xi_2}{dt} = \left[2\frac{\partial Q_0^2}{\partial v} + 3\frac{\partial C_0^2}{\partial v}\right]\xi_1 + L_{20}\xi_2 + O(\epsilon, \xi_1^2, \xi_2^2)$$

It follows that to order ϵ, stability is determined by the spectral problem,

$$\left[L_{10} + \epsilon \left(L_{11} + 2 \frac{\partial Q_0^1}{\partial v} \right) \right] \xi_1 = \lambda \xi_1 \tag{32}$$

Both ξ_1 and λ can be expanded as follows:

$$\xi_1 = \xi_1^0 + \epsilon \xi_1^1 + \cdots$$

$$\lambda = \epsilon \lambda_1 + \epsilon^2 \lambda_2 + \cdots \tag{33}$$

where $\xi_1^0 = \Sigma b_i \psi_i$. This leads to the m-dimensional system:

$$\lambda_1 b_i - \sum_j b_j \left\langle \psi_i^* \left(I_{11} + 2 \frac{\partial Q_0^1}{\partial v} \right) \psi_j \right\rangle, \tag{34}$$

After a reduction like that which led to (26), we find that λ_1 is an eigenvalue of the matrix whose (i, j)th element is,

$$p_1 \Omega_1 \delta_{ij} + 2\Omega_2 \sum_{k=1}^{m} \Gamma_{ijk} a_k \tag{35}$$

which is, of course, what would be obtained by linearizing (26) directly. It turns out that at this order the eigenvalues can be readily computed for the branches of solutions on O_1 and O_2 and the results are as follows:

For O_1: the largest eigenvalue is zero and therefore all solutions are marginally stable.

For O_2: all solutions are unstable.

Consequently the next term in the expansion of λ must be computed to determine the stability of solutions on O_1, but this will not be pursued here.

DISCUSSION

Both the scaled version of the transient equations and the bifurcation equations for steady-state solutions are applicable in general, whether or not the domain has some symmetry or multiple eigenvalues exist. These equations can be used to simplify computations for specific kinetic mechanisms, even when the critical eigenvalue is simple.[14] When the multiplicity is greater than one and symmetries are present, systematic application of group theoretic techniques will reduce bifurcation equations to their simplest form, enabling us to identify certain classes of solutions for which the multiplicity is odd and bifurcation is guaranteed by general theorems. The analysis can be extended to the transient equations, and the result is a set of nonlinear evolution equations for the amplitudes of the symmetry-adapted eigenfunctions.[11,12,15]. This is essential for the case in which the critical eigenvalue is complex (and degenerate), rather than real. In that case, we obtain a set of $2m$ amplitude equations (m complex conjugate pairs) that are to be solved for the amplitudes and phases of the m degenerate modes. Just as in the steady-state case, classes of solutions can be identified for which bifurcation is guaranteed.

Details of the analysis will be reported elsewhere (for other applications of group theory to bifurcation problems, see Refs. 13, 16, 17, and for related work on multiple eigenvalue problems, see Refs. 18, 19, 20).

Generally, systems do not show perfect symmetry, and it is important to determine how the solution set changes when Equation 8 is perturbed by a term that reduces the symmetry of the equation. The easiest case to study arises when (8) is perturbed by a one-parameter family of linear differential operators $\mu \tilde{L}_1$ densely defined in $L_2(\Omega)$, of order less than or equal to two. This leads to a two-parameter bifurcation problem in the parameters $p - p_0$ and μ, and we can study changes in the set of nontrivial solutions as the parameters vary in a full neighborhood of $(0, 0)$. Depending on the choice of $\tilde{L}_1$, the degeneracy may or may not disappear completely when $\mu \neq 0$, and group theoretic techniques can be used to determine how the splitting occurs. Analysis of the bifurcation equations enables us to find all loci in the $(p - p_0, \mu)$ space along which secondary bifurcations occur near $(0, 0)$. One example of such analysis, in which symmetry plays no role, is given in Reference 6.

Acknowledgment

This research was supported by a Rutgers Research Council Faculty Fellowship and by an NIH grant (GM-21558).

References

1. Krasnosel'skii, M. A. 1964. Topological Methods in the Theory of Nonlinear Integral Equations. Macmillan.
2. Rabinowitz, P. H. 1971. J. Funct. Anal. 7: 487–513.
3. Ize, J. 1975. Bifurcation theory for Fredholm operators. Mem. Am. Math. Soc. 174. The American Mathematical Society.
4. Wolpert, L. 1969. J. Theor. Biol. 25: 1–47.
5. Turing, A. M. 1952. Philos. Trans. R. Soc. London. B237: 37–72.
6. Ashkenazi, M. & H. G. Othmer. 1977. J. Math. Biol. To be published.
7. Chueh, K. N., C. C. Conley & J. A. Smoller. 1977. Indiana Univ. Math. 26(2): 373–392.
8. Lomont, J. S. 1959. Applications of Finite Groups. Academic Press, New York.
9. Hamermesh, M. 1962. Group Theory. Addison-Wesley.
10. Sugiura, M. 1975. Unitary Representations and Harmonic Analysis. Wiley.
11. Othmer, H. G. & L. E. Scriven. 1974. J. Theor. Biol. 43: 87–112.
12. Gmitro, J. I. 1969. University of Minnesota. Ph.D. Thesis.
13. Schiffman, Y. 1977. Preprint.
14. Auchmuty, J. F. G. & G. Nicolis. 1975. Bull. Math. Biol. 37(4): 323–365.
15. Schwab, T. H. 1975. Univ. of Minnesota. Ph.D. Thesis.
16. Ruelle, D. 1973. Arch. Ration. Mech. Anal. 51(2): 136–152.
17. Sattinger, D. H. 1977. SIAM J. Appl. Math. To appear.
18. Sather, D. 1970. Arch. Ration. Mech. Anal. 36(1): 47–64.
19. Krasnosel'skii, M. A. 1970. Sov. Math. Dokl. 11 (6): 1609–1613.
20. Dancer, E. N. 1971. Proc. London Math. Soc. 23: 699–734.

THE BIFURCATION OF QUADRATIC FUNCTIONS*

John Guckenheimer

Department of Mathematics
University of California
Santa Cruz, California 95060

The subject of this paper is the qualitative behavior under iteration of a real quadratic function: $f(x) = ax^2 + bx + c$. This means that we study the *orbits* of f, defined to be the sequence $\{f^n(x)\}_{n \geq 0}$ with $f^n = f \circ \cdots \circ f$. Of particular interest are periodic points and their stability. A point x is *periodic of period n* if $f^n(x) = x$. A periodic point of period n is stable or unstable according to whether

$$\lambda = \frac{d f^n}{dx}(x)$$

satisfies $|\lambda| < 1$ or $|\lambda| > 1$. We call λ the multiplier of the periodic orbit. Our primary goal will be the study of the dependence of periodic points on the coefficients of f. If we regard as equivalent those quadratic maps that differ by a linear change of coordinates, then there is only one parameter which must be specified to determine a quadratic map f. Moreover, the changes in the numbers of points of different periods occur as this parameter varies over a finite, closed interval. Two choices of 1-parameter families of f that capture all of the bifurcation behavior and that have been used commonly in the literature[7] are

$$f(x; \mu) = \mu - x^2 \qquad \mu \in [-\tfrac{1}{4}, 2], \tag{1}$$

$$f(x; \mu) = \mu x(1 - x) \qquad \mu \in [0, 4]. \tag{2}$$

Equation 2 has received particular attention recently as a model for the population dynamics of a density-dependent population with discrete generations. For convenience, we shall study (1) rather than (2), and refer to it as *the* family of quadratic functions. Our purpose here is to survey what is known about the bifurcations of the family of quadratic functions and state some unresolved questions and conjectures. Much of the theory is applicable to more general families of functions, but we feel that the attention received by quadratic functions warrants a survey of their particular properties.

The first result concerning the quadratic family is an old result that comes from an analysis of quadratic functions as maps of the (extended) complex plane.

THEOREM. (Fatou,[2] Julia[6]) Let f be a quadratic function. Then f has at most one stable periodic orbit.

A quadratic function has a single *critical point*, a point where its derivative vanishes. In this theorem, the critical point plays a crucial role. Moreover, it continues to play an essential part in the analysis of other phenomena as we proceed. The proof of the theorem gives a constructive way of finding the stable periodic orbit of a quadratic function, if it exists. The orbit of the critical point

*This work was partially supported by the National Science Foundation.

0077-8923/78/0316-0078 $1.75/1 © 1979, NYAS

must tend asymptotically to the stable periodic orbit, so one need merely follow the orbit of the critical point.[1,2,5] Thus, the bifurcations of the family (1) can be studied by observing how the orbit of the critical point 0 changes as a function of the parameter μ.

The next results we mention deal with the changes that occur in individual periodic orbits as they undergo bifurcations—either appearance, disappearance, or changes in stability as μ is varied. First, one has an easy, but basic proposition.

PROPOSITION. Suppose $x(\mu)$ is a periodic point of period n for the 1-parameter family $f(x; \mu)$. If $(d/dx)\, f^n(x; \mu_0) \neq 1$, then $x(\mu)$ is a smooth function of μ in a neighborhood of μ_0.

The proof is just an application of the implicit function theorem to the equation $f^n(x; \mu) = x$.

COROLLARY. Changes in the number of periodic orbits occur only at periodic orbits with multiplier 1.

For real maps of the interval, the condition $|\lambda| = 1$ for neutral periodic orbits can be met in precisely two ways: λ is 1 or -1. For each of these cases, there is a *generic* pattern to the bifurcations of periodic orbits. These are described by the next two theorems.

THEOREM.[4] Suppose that $f: I \times I \to I$ is a family of smooth maps and that there is a point p, parameter ν, and integer n such that

(1) $f^n(p, \nu) = p$,

(2) $\dfrac{d}{dx}\, f^n(p; \nu) = 1$,

(3) $\dfrac{d^2}{dx^2}\, f^n(p, \nu) > 0$,

(4) $\dfrac{d}{d\mu}\, f^n(p, \nu) > 0$.

Then there are intervals (ν_1, ν) and (ν, ν_2) and $\epsilon > 0$ such that:

(1) If $\mu \in (\nu_1, \nu)$, then $f^n(x; \mu)$ has no fixed points in $(p - \epsilon, p + \epsilon)$.

(2) If $\mu \in (\nu, \nu_2)$, then $f^n(x; \mu)$ has two fixed points in $(p - \epsilon, p + \epsilon)$. One is stable and one is unstable.

This kind of bifurcation is called a *saddle-node*.

THEOREM.[4] Suppose $f: I \times I$ is a family of smooth maps and that there is a point p, parameter ν, and integer n such that

(1) $f^n(p; \nu) = p$,

(2) $\dfrac{d}{dx}\, f^n(p, \nu) = -1$,

(3) $\dfrac{d^3}{dx^3}\, f^{2n}(p; \nu) < 0$,

(4) $\dfrac{d^2}{dx\, d\mu}\, f^n(p; \nu) > 0$.

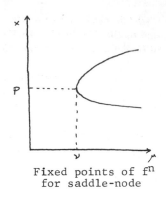

Fixed points of f^n
for saddle-node

Fixed points of f^{2n}
for flip

FIGURE 1.

Then there are intervals (ν_1, ν) and (ν, ν_2) and $\epsilon > 0$ such that:

(1) If $\mu \in (\nu_1, \nu)$, f^{2n} has exactly one fixed point in $(p - \epsilon, p + \epsilon)$. This fixed point is stable.

(2) If $\mu \in (\nu, \nu_2)$, then f^{2n} has three fixed points in $(p - \epsilon, p + \epsilon)$. The largest and smallest lie in a stable periodic orbit of period $2n$, and the middle point lies in an unstable periodic orbit of period n.

This kind of bifurcation is called a *flip*.

FIGURE 1 shows the locus of fixed points of f^n and f^{2n} for the saddle-node and flip bifurcations. The proofs of the theorems are straightforward applications of the implicit function theorem. If any of the inequalities in the hypothesis are reversed, the conclusion is changed by reversing the stability of the orbits and/or orientation on the μ-axis. Note that the bifurcation of the family (2) at $\mu = 1$ is *not* generic. However, there are no counterexamples known to the following conjecture:

CONJECTURE: The bifurcations of periodic orbits of the family $f(x;\mu) = \mu - x^2$ are all generic saddle-nodes or flips.

The preceding results summarize most of the information known about the bifurcations of the quadratic family that are local in the parameter μ. We now pass to questions that deal with the nature of the whole bifurcation set. There are some answers, some conjectures, and some questions related to the nature of the bifurcation set.

One result concerns the order in which bifurcations of periodic orbits of various periods occur. In the quadratic family (1), f has no periodic orbits at all for $\mu < -\frac{1}{4}$. For every x, $f(x) < x$ and $f^n(x) \to -\infty$. For $\mu > 2$, one can prove that all of the fixed points of f^n are real. The fundamental theorem of algebra gives the number of these as 2^n. Therefore, the number of fixed points of f^n changes from 0 to 2^n as μ increases from $-\frac{1}{4}$ to 2. One can count easily the number of periodic orbits represented by these 2^n points. If m_i is the number of periodic orbits of least period i, then we have

$$\sum_{i \mid n} m_i = 2^n$$

with the sum being over all divisors of n (including 1 and n). TABLE 1 gives the number of periodic orbits of period i for $n \leq 16$. Presuming that the minimal number of bifurcations to create these periodic orbits occurs, the number of bifurcations of periodic orbits of period n can also be calculated. This is discussed below, and the numbers for $n \leq 16$ are given in TABLE 1. The present question of interest is the order in which the bifurcations of periodic orbits for different periods occur. This question is partially answered by the analysis below.

PROPOSITION: For fixed μ, suppose γ_1 and γ_2 are two periodic orbits of $f^n(x; \mu)$ and that γ_1 contains a point that is larger than all of the points in γ_2. If γ_1 and γ_2 were created by different bifurcations, then the bifurcation that gave rise to γ_1 occurred for a larger value of μ than the bifurcation that gave rise to γ_2.

This proposition relates the order of periodic points for fixed parameter value to the order in which bifurcations occur. The argument is based upon an analysis of how the critical point of f behaves. This analysis extends beyond the proposition. If we consider the map $f(x; 2) = 2 - x^2$, there are 2^n fixed points of f^n. These points can be partitioned into orbits so that the proposition applies to determine the order in which bifurcations created these periodic orbits.

There is a combinatorial procedure for doing this partitioning. Number the fixed points of f^n in increasing order, using binary numbers from 0 to 2^{n-1}. Next perform a transformation T to binary sequences of length n to get new binary sequences of length n: $T(a_1 a_2 \ldots a_n) = b_1 b_2 \ldots b_n$ with $a_1 = b_1$ and $a_{i-1} + a_i = b_i \pmod 2$ for $2 \leq i \leq n$. The map T has the property that it transforms the binary sequences of points in the same orbit into binary sequences that are cyclic permutations of one another. This analysis yields the following theorem:

THEOREM.[4] There is a set $B \subset I$ and a map $\pi: B \to \mathbb{Z}$ such that, if $\mu \in B$, then the quadratic map $f(x; \mu)$ has a bifurcation of a periodic orbit of period $\pi(\mu)$. The map π has the properties that (1) $\Sigma s_\mu \pi(\mu) = 2^n$, the sum being over all $\mu \in B$ such that $\pi(\mu)$ divides n. Here s_μ is 1 or 2 according to whether the bifurcation is a flip or saddle-node. (2) There is an algorithm (described above) for determining the map π on the inverse image of a finite interval.

The formula in this theorem can be used to calculate the last line of TABLE 1. The number of bifurcations of period n orbits equals the number of period n orbits with negative multiplier.

CONJECTURE. The set B of the preceding theorem contains all bifurcations of periodic orbits.

An approach to this conjecture is inspired by the work of Myrberg.[9] He also studies the behavior of the critical point 0 of the quadratic family $f(x; \mu)$. Let

TABLE 1

THE NUMBER OF PERIODIC ORBITS AND BIFURCATIONS FOR VARIOUS VALUES OF n

$n =$	1	2	3	4	5	6	7	8	9	10	11	12	13	14	15	16
No. of periodic orbits	2	1	2	3	6	9	18	30	56	99	186	335	630	1161	2182	4080
No. of bifurcations		1	1	1	2	3	5	9	16	28	51	93	170	315	585	1091	2048

us denote $G_n(\mu) = f^n(0; \mu)$. Then $G_n(\mu)$ gives the location of the nth iterate of the critical point of f. We say that f has a *critical periodic orbit* at a zero of G_n. Each bifurcation in the set B gives rise to one periodic orbit which becomes critical for some larger value of μ. If the bifurcation is a saddle-node, only one of the two orbits can become critical.

CONJECTURE. G_n has exactly the same number of zeros as the number of bifurcations of fixed points of f^n.

The solutions of the equation $G_n(\mu) = 0$ also satisfy an equation of the form

$$\mu = \sqrt{\mu \pm \sqrt{\mu \pm \cdots \pm \sqrt{\mu}}}$$

where the choice of signs is determined by the relative order of points in the orbit. Real solutions exist only for those choices of sign which correspond to the fact that μ must be the largest point in a critical periodic orbit. The conjecture would be proved if one established that there can be only one real solution of the radical equation for each particular choice of signs.

These last two conjectures are closely related. The first of the two counts periodic orbits with multiplier 1, and the second counts periodic orbits with multiplier 0. It is reasonable to guess that, at each bifurcation, a periodic orbit arises for which the multiplier is a decreasing function of the parameter μ. If this could be proved, it would establish the validity of both conjectures.

When the quadratic function f has an attracting periodic orbit, it is "structurally stable" in the sense that it has the following properties. If there are periodic orbits whose periods are not a power of 2, then f has an invariant Cantor set in which periodic points are dense. This Cantor set will be topologically conjugate to a "subshift of finite type." It and the attracting periodic orbit constitute the nonwandering set of f. The nonwandering set does not change qualitatively with small changes in the parameter.

Thus the changes in the qualitative dynamics of f occur when there is no attracting periodic orbit. When do changes happen? There are two ways of asking this question:

CONJECTURE. There is an open, dense set of $\mu \in [-\frac{1}{4}, 2]$ for which the quadratic family (1) has an attracting periodic orbit.

QUESTION: Is there a set of μ of positive measure in $[-\frac{1}{4}, 2]$ for which the quadratic family has no attracting periodic orbit?

Opinion about the answer to this last question is divided. There are piecewise linear families of interval maps which yield different answers to this question. Define $g_i(x; \mu)$ on $[-1, 1]$, $i = 1, 2$, by the equations

$$g_1(x; \mu) = \mu - (1 + \mu)|x| \qquad\qquad \mu \in [0, 1]$$

$$g_2(x; \mu) = \begin{cases} 2\mu - 1 & \text{if} & |x| \le 1 - \mu \\ 1 - 2|x| & \text{if} & 1 \ge |x| \ge 1 - \mu \end{cases}, \mu \in [0, 1].$$

The bifurcation set of the family g_1 has positive measure while the bifurcation set of the family g_2 has measure zero. This provides an indication that the answer to the question must involve estimates that rely upon the smoothness (and nondegeneracy?) of the critical point.

There are two other fascinating observations about the detailed structure of the bifurcation set of quadratic maps that are based largely upon numerical data and still require proof. One of these is due to Mira[8] and the other to Feigenbaum.[3]

Mira's observation he calls the "embedded box." For some values of the parameter, say for μ_0, he notes that $f(x; \mu_0)$ eventually maps the critical point into a nonattractive periodic orbit γ. If γ has period n, then f^n maps a subinterval onto itself so that the image completely covers the subinterval twice in the same way $f(x; 2) = 2 - x^2$ covers the interval $[2, 2]$. This is indicated in FIGURE 2.

In order for the embedded box to appear, between the parameter value for which γ has a critical periodic orbit and μ_0, the map f^n must undergo the same sequence of bifurcations the whole family f exhibits. This can be expressed by

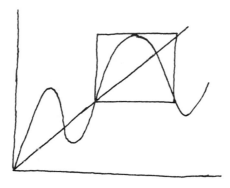

FIGURE 2.

saying that there is a subinterval J of $[-\frac{1}{4}, 2]$ and an order-preserving map ρ of $J \cap B$ onto B such that $\pi \circ \rho(\mu) = (1/n)\pi(\mu)$. This demonstrates some of the internal structure of the set B and map π.

PROBLEM: Find conditions (like the existence of the map ρ) that characterize the period map π for the bifurcations of the quadratic family.

Unlike the previous question, this problem is a combinatorial one.

Feigenbaum's observation is the only one that deals with metric properties of the bifurcation set for quadratic maps. Each bifurcation in the set B is contained in a sequence of flips which occur as the period of the attractive periodic orbits repeatedly doubles. Feigenbaum observes numerically that the lengths of the intervals between successive flips decrease at a geometric rate. The rate is $\delta = 4.669201609103 \ldots$ and depends only upon the fact that the quadratic family has a nondegenerate critical point. Feigenbaum arrived at this conclusion by applying "renormalization" ideas of physics to the study of interval maps.

Here we shall recast Feigenbaum's observations in a more geometric language and arrive at a restatement of these observations as a conjecture. Our approach differs a great deal from that of Feigenbaum. The intuitive idea is that, as the period n of an attracting periodic orbit doubles with a change in the parameter, the behavior of f^n repeats itself for the map f^{2n} on a smaller scale. The rescaling is observed to approach multiplication by α where $\alpha^{-1} = -2.5029078750957$ .... As one continues looking at $f^{2^m n}$ on smaller and smaller scales, the behavior should be dominated by the values of f near the critical point. If one looks at the limit μ_0 of a sequence of flips, then the map $f(x; \mu_0)$ should embody the rescaling properties previously seen *without* further changes in the parameter value.

Consider the quadratic family f with the parameter value chosen so that there is exactly one periodic orbit of period 2^n for each $n > 0$. The *kneading sequence*[7] of f is the binary sequence defined by $b_i = 0$ or 1 as $f^i(0)$ is less than or greater than 0. For the particular parameter value chosen for f, the kneading sequence of f can be calculated. It is defined recursively by $b_1 = 0$ and

$$
b_i = \begin{cases} b_{i-2^n} \text{ if } 2^n + 1 \le i < 2^{n+1} \\ 1 - b_{2^n} \text{ if } i = 2^{n+1}. \end{cases}
$$

The section of the sequence from $i = 2^n + 1$ to 2^{n+1} is generated by repeating the first 2^n terms and then changing the last. The first 32 terms are 1,0,11,1010, 10111011, 1011101010111010. The kneading sequence of f^2 is obtained by taking the subsequence of $\{b_i\}$ of terms with even indices. This results in a sequence which is obtained from $\{b_i\}$ by changing every term. From this, we conclude that $f(x)$ and $-f^2(-x)$ have the same kneading sequences. This leads to the further conclusion that f and f^2 are topologically conjugate: there is a homeomorphism h such that $f \circ f = h \circ f \circ h^{-1}$. The map h is the rescaling transformation of Feigenbaum. His results lead to the following conjecture.

CONJECTURE: Let f be the quadratic function which has exactly one periodic orbit of period 2^n for each $n > 0$. Then there is a unique homeomorphism h defined on a suitable interval which has the properties (1) $f \circ f(x) = h \circ f \circ h^{-1}(x)$, and (2) h is differentiable at 0 with derivative $\alpha = -2.5029 \ldots$.

The final question that we raise concerns the bifurcations of a family of diffeomorphisms of the plane that are related to the quadratic functions. Henon[5] considered the map $G: \mathbb{R}^2 \to \mathbb{R}^2$ defined by $G(x,y) = (y, \epsilon x + v - y^2)$ (in slightly different coordinates.) From numerical observations, he asserted that there were values of the parameters for which G has a "strange attractor." At the same time, Newhouse[10] proves that there is a large set of parameter values for which G has an infinite number of attracting periodic orbits, most of which must have very long periods. His work casts doubt about the conclusions which Henon draws from his numerical work despite the precision used in the computations. An analog of one of the question stated above for quadratic functions might be phrased as follows.

QUESTION: Consider the map $G_{\mu,\epsilon}(x,y) = (y, \epsilon x + \mu - y^2)$. Is there a set of positive measure in (μ, ϵ) space for which $G_{\mu,\epsilon}$ possesses an infinite, attracting set with a dense orbit.

The map G is closely related to the quadratic functions. Indeed, when $\epsilon = 0$, the map G has constant rank 1 and behaves exactly like the quadratic function with respect to the y coordinate. In this sense, the quadratic function is a singular limit of diffeomorphisms of the plane. One obtains the following proposition from this perspective.

PROPOSITION: Assume that the family of quadratic functions has generic bifurcations. If $N > 0$, then there is an $\bar{\epsilon} > 0$ such that the order of bifurcations of periodic orbits of period less than N for the family $G_{\mu,\epsilon}$ $\mu \in \mathbb{R}$ variable and $0 < \epsilon < \bar{\epsilon}$ fixed, agrees with the order of bifurcations for the family of quadratic functions.

Newhouse's results indicate that one cannot omit the bound on N in this proposition. Unlike a quadratic function that can have only one attracting periodic orbit, one expects there to be values of (μ, ϵ) for which $G_{\mu,\epsilon}$ has an infinite number of attracting periodic orbits. This can only happen if the curves in (μ, ϵ) space describing the bifurcations of various periodic orbits cross one another. Indeed, one may be able to calculate the slopes of these curves and thereby prove directly the existence of (μ, ϵ) for which $G_{\mu,\epsilon}$ has an infinite number of attracting periodic orbits.

PROBLEM: For fixed $\epsilon > 0$, calculate the order of bifurcations of periodic orbits of $G_{\mu,\epsilon}(x, y) = (y, \epsilon x + \mu - y^2)$ as μ varies. (This is expected to depend upon ϵ.)

REFERENCES

1. BROLIN, H. 1965. Invariant sets under iteration of rational functions. Arkiv. Mat. Astron. Fys. **6:** 103–144.
2. FATOU, M. P. 1919. Sur les équations fonctionelles, Bull. Societé Math. de France **47:** 161–271; and 1920. *Ibid.* **48:** 33–94, 208–314.
3. FEIGENBAUM, M. J. Quantitative universality for a class of nonlinear transformations. Unpublished.
4. GUCKENHEIMER, J. 1977. On the bifurcations of maps of the interval. Inventiones Math. **39:** 165–178.
5. HENON, M. 1976. A two-dimensional mapping with a strange attractor, Comm. Math. Phys. **50:** 69–78.
6. JULIA, C. 1918. Memoire sur l'iteration des fonctions rationelles, J. Math. Pure Appl. **4:** 47–245.
7. MILNOR, J. & W. THURSTON. On iterated maps of the interval, I and II. Unpublished.
8. MIRA, C. 1975. Accumulations de bifurcations et "structures boîtes emboîtees" dan les récurrences et transformations pouctuelles. VIIe International Conference on Nonlinear Oscillations, Berlin.
9. MYRBERG, P. J. 1963. Iteration der reellen polynome zweiten grades III. Ann. Acad. Sci. Fenn. **336/3:** 1–18.
10. NEWHOUSE, S. The abundance of wild hyperbolic sets and non-smooth stable sets for diffeomorphisms. Unpublished.

BIFURCATION FROM COMPLETELY UNSTABLE
FLOWS ON THE CYLINDER

Zbigniew Nitecki

Department of Mathematics
Tufts University
Medford, Massachusetts 02155

INTRODUCTION

Bifurcation theory, in the topological study of smooth dynamical systems, concerns changes in the orbit structure of a cascade or flow that varies continuously with a parameter. Global bifurcation phenomena for parametrized dynamical systems on closed manifolds (i.e., compact manifolds without boundary) have been studied intensively in the past few years, but relatively little consideration has been given to the corresponding phenomena on open manifolds (i.e., noncompact, but finite-dimensional and paracompact, without boundary).

In this paper, we shall consider a class of flows on the open manifold $M = S^1 \times R$ (the open annulus, cylinder, or punctured plane) from the point of view of structural stability and bifurcation. These flows have a simple orbit structure, whose behavior under perturbation can be easily understood. They illustrate some differences between closed and open systems—especially the important role played by separatrices or prolongational limit sets in global stability and bifurcation questions on open manifolds.

If a flow is not structurally stable, it may nevertheless be stable in the weaker sense that it necessarily occurs in the transition between different structurally stable classes of flows. Thom[1] has formulated this idea precisely, for singularities of smooth maps, in the notion of an "unfolding" (i.e., a parametrized family of maps which is structurally stable among such families). This view was adopted by Sotomayor[2] in his study of flows on closed surfaces. He formulated definitions of these concepts for flows in higher dimensions in a subsequent paper.[3] More recently, this view has been adopted by Newhouse and Peixoto,[4] and others.

After considering some definitions, we shall formulate and prove three results. The main result characterizes the one-parameter unfolding of a flow near a structurally unstable separatrix (THEOREM A). Using this local result, a global one-parameter unfolding of a flow (in our class) with a single structurally unstable separatrix (THEOREM B) can be constructed. However, when several such separatrices are present (as, for example, in certain "explosive" flows considered in two earlier papers[5]), an unfolding requires several parameters. In this case, a parametrized family through the given system will be constructed, which will be shown to be stable in a weaker sense (THEOREM C).

ORBIT STRUCTURE OF THE FLOW

Let us consider flows on the cylinder $M = S^1 \times R$, whose velocity is everywhere length 1 and depends only on the first coordinate. In polar coordinates,

0077-8923/78/0316-0086 $1.75/1 © 1979, NYAS

such a flow is given by the differential equations,

$$\theta' = \sin f(\theta)$$
$$r' = \cos f(\theta) \tag{1}$$

where $f(\theta)$ is a smooth function and θ, $-\infty < \theta < +\infty$, is given in radians and for some integer a (the degree of f as a map of the circle), $f(\theta) - a\theta$ is continuously differentiable with period 2π.

The vector field described by (1) is vertical precisely where $f(\theta)$ is an integer multiple of π. We shall assume that the angles at which (1) is vertical form a finite set, given by $0 \le \theta_1 < \cdots < \theta_n < 2\pi$.

The orbit structure given by (1) can easily be determined by looking at the behavior of $f(\theta)$ at vertical values. If the vector field is never vertical, let Δ be the "left" of an orbit during one circuit around the circle,

$$\Delta = \int_0^{2\pi} \cot\,[f(\theta)]\,d\theta.$$

One of two possibilities occurs: 1) If $\Delta = 0$, all orbits are periodic; 2) if $\Delta \ne 0$, the orbits spiral along the cylinder, with $r \to +\infty$ in one time-direction and $r \to -\infty$ in the other.

If $f(\theta) = k\pi$ for $\theta = \theta_i$, then the line $\theta = \theta_i$ is an orbit of (1), with r increasing if and only if k is even. These θ_i lines separate the cylinder M into strips $\theta_i \le \theta \le \theta_{i+1}$, called *canonical regions,* where the orbit structure is one of two types:

(a) If $f(\theta_i) = f(\theta_{i+1})$, the flow on the strip $\theta_i \le \theta \le \theta_{i+1}$ is *parallelizable*— that is, the foliation by orbits is homeomorphic to the foliation of the strip by parallel lines (see FIGURE 1).

(b) If $f(\theta_i) \ne f(\theta_{i+1})$, the orbits form a Reeb-like foliation of the strip (see FIGURE 2).

There are four possible phase portraits obtained from FIGURE 2 by reflection about either a horizontal or a vertical axis, depending on whether f is an *even* multiple of π and which edge represents the higher value of f.

The condition in FIGURE 2 has been studied in various contexts by several authors: Nemytskii[6] refers to (b) as an "improper saddle," while Markus[7] refers to either edge of (b) as a "separatrix." A characterization which is easily adapted to a more general theory was formulated by Ura[8] and Auslander[9].

We say that $y \in M$ belongs to the (first positive) *prolongational limit* set of $x \in M$ under a flow φ, if there exist points $x_i \to x$ and times $t_i \to +\infty$ such that $\varphi(t_i, x_i) \to y$. In FIGURE 2, it is easy to see that every point on the right edge is in the prolongational limit of the left-hand edge. We denote the prolongational limit set of x by $J^+(x)$.

On a closed manifold, the orbit structure of a flow is usually described by reference to its *nonwandering set,* which can be characterized as the set $\Omega(\varphi)$ of points $x \in M$ for which $x \in J^+(x)$. On a closed manifold, it is always possible to take $x_i = x$ in the definition of $J^+(x)$ in order to show that Ω is never empty, in particular, that $\Omega \ne \phi$ on closed manifolds.

By contrast, one can show that the only case of (1) for which $\Omega \ne \phi$ is when

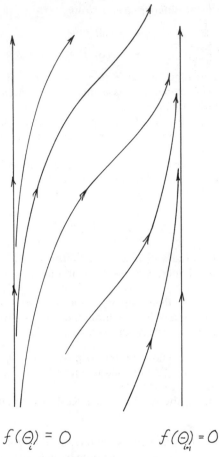

$$f(\underset{i}{\theta}) = 0 \qquad\qquad f(\underset{i+1}{\theta}) = 0$$

FIGURE 1.

there are no vertical values of $f(\theta)$ and $\Delta = 0$. In general, when there are vertical values, one can see from the above description that $y \in J^+(x)$ if and only if x belongs to the "earlier" edge and y belongs to the "later" edge of a Reeb-like canonical region.

According to Nemytskii a flow with $\Omega = \phi$ is called *completely unstable.* This is derived from Lagrange stability of orbits, rather than structural stability. In fact, we shall see that most choices of $f(\theta)$ make (**1**) structurally stable.

There is, however, one case in which we clearly have structural instability (see FIGURE 3). Here, we have a "chain" of prolongations, $y \in J^+(x)$, $z \in J^+(y)$, where x belongs to the left edge, y to the middle vertical, and z to the right edge. It is clear that, by a small push to the right, near y, we obtain a flow for which some orbits travel from a position near the left edge to a position near the right edge. Such a case arises when two equal vertical values of $f(\theta)$ are separated by a different vertical value; i.e., when $f(\theta_{i-1}) = f(\theta_{i+1}) \neq f(\theta_i)$. The "unfolding" in this case will be considered later.

STRONG AND WEAK TOPOLOGIES

When the cylinder M is compact, a space of continuous functions (real valued or with values in a normed fiber bundle) on M is usually given the topology of uniform convergence in which a neighborhood of the function $f(x)$ consists of those $g(x)$ that satisfy the uniform estimate,

$$\| f(x) - g(x) \| < \epsilon, \quad x \in M$$

The same idea is used to topologize the continuously differentiable functions on the closed manifold M by imposing this requirement on the first-order partial derivatives as well as the values of the functions f and g. Thus, the space $\mathfrak{X}^1(M)$ of continuously differentiable vector fields on the closed manifold M is given the

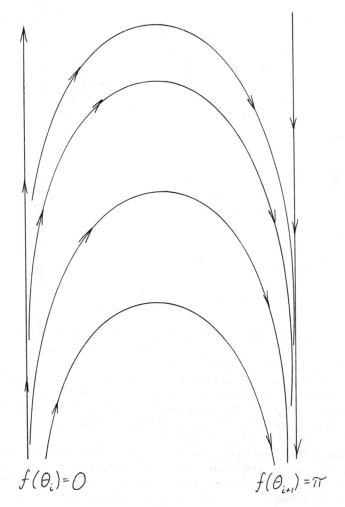

$$f(\theta_i) = 0 \qquad\qquad f(\theta_{i+1}) = \pi$$

FIGURE 2.

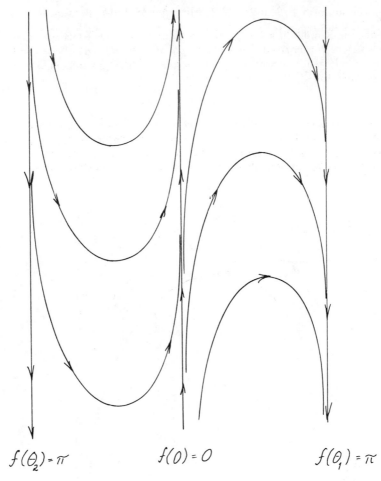

$$f(\theta_2) = \pi \qquad\qquad f(0) = 0 \qquad\qquad f(\theta_1) = \pi$$

FIGURE 3.

C^1 *topology* by regarding each vector field as a map into the tangent bundle TM. This topology on velocity vector fields, $\dot\varphi = \partial\varphi/\partial t$, induces a topology on flows, $\varphi(t, x)$ that is independent of the choice of norm $\| \ \|$.

When M is not compact, the topology has several natural extensions. The topology of *uniform convergence on compacta*—also called the *compact-open* or *weak* topology—is generated by weak neighborhoods specified by a constant $\epsilon >$ 0, and a compact subset $K \subset M$: a weak neighborhood of $f(x)$ consists of $g(x)$ where $\| f(x) - g(x) \| < \epsilon, x \in K$ holds on K.

The graphs of $g(x)$ in a weak neighborhood of $f(x)$ lie near the graph of f in the region over K, but have arbitrary behavior outside of K (see FIGURE 4). One can, of course, control the behavior of $g(x)$ on M by simply imposing the uniform estimate on M; however, the resulting *uniform topology* varies with the choice of norm. The uniform neighborhoods corresponding to all possible choices of a (continuous) norm generate the *Whitney, or strong topology,* in which a neighborhood of $f(x)$ is determined by a continuous function $\epsilon(x) > 0$, and consists of those $g(x)$ that satisfy the pointwise estimate, $\| f(x) - g(x) \| < \epsilon(x)$, $x \in M$. Since $\epsilon(x)$ need not be globally bounded away from zero, the graphs of functions in a strong neighborhood of $f(x)$ are generally asymptotic to that of $f(x)$ (see FIGURE 5).

The weak and strong topologies, unlike the uniform topology, are independent of the choice of norm.[10] The same device as before leads to two topologies—the weak C^1 and strong C^1—for the space $\mathfrak{X}^1(M)$ of continuously differentiable vector fields (hence flows) on the open manifold M.

We are interested in comparing the orbit structure of flows in a neighborhood of a given flow. Two flows are regarded as equivalent if they are *topologically conjugate*—that is, if there exists a homeomorphism h taking orbits of one flow onto orbits of the other. We do not require h to respect the time parameter.

A flow is *structurally stable* if some neighborhood of the flow consists of flows topologically conjugate to it. This notion, of course, depends on the topology on $\mathfrak{X}^r(M)$. It is well known that a nontrivial notion of structural stability requires

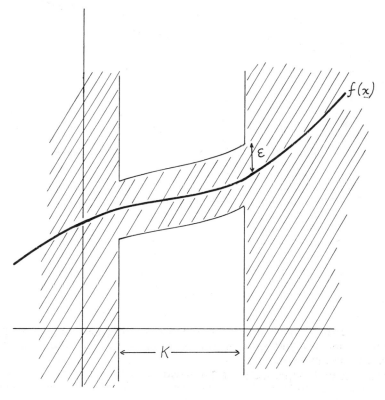

FIGURE 4.

control over at least the first derivatives, and it is easy to see that the weak topology leads to a vacuous notion of structural stability because of the lack of control
outside of K. Thus, structural stability questions on open manifolds are most
naturally studied in the strong C^1 topology (an exception is the theorem of
Osipov,[16] which is formulated in the uniform C^1 topology for a fixed metric).

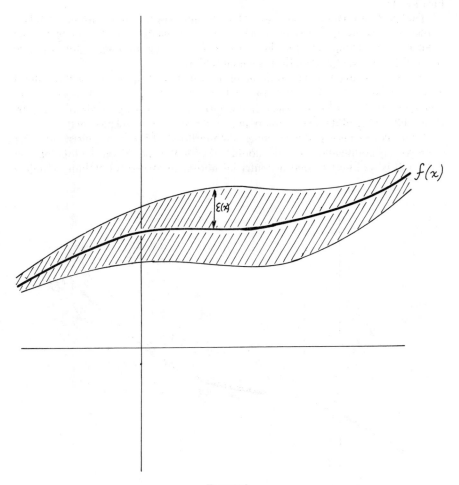

FIGURE 5.

The structurally stable flows given by Equation 1 can be characterized by an
easy adaptation of arguments given for the plane by Collins[11] and Krych.[12] We
state the result without proof.

PROPOSITION 1. Suppose φ is the flow described by

$$\theta' = \sin f(\theta)$$
$$r' = \cos f(\theta)$$

where $f(\theta)$ is C^1 and takes vertical values on the finite (possibly empty) set, $0 \leq \theta_1 < \cdots \leq \theta_n < 2\pi$. Then φ is structurally stable in the C^1 topology if *and only if* (i.e., "precisely if") *either* (a) *or* (b) holds:

(a) $\Delta \neq 0$, in case the set is empty;
(b) if $n > 1$, then whenever $f(\theta_{i-1}) = f(\theta_{i+1})$, we must also have $f(\theta_i) = f(\theta_{i-1})$, where $\theta_{-1} = \theta_n - 2\pi, \theta_{n+1} = \theta_1 + 2\pi$.

When (b) breaks down, we have the case considered at the end of the previous section. In this case, we try to discover whether the behavior of a flow is necessary in a transition between different structurally stable flows. In the context of flows on closed manifolds, Sotomayor[3] has formulated several versions of this notion. We shall consider two of these notions, both of which involve the concept of a k-parameter family of flows.

Note that a continuously differentiable map $X:M \times I^k \to TM$, which assigns to a point $x \in M$ and a k-tuple $u = (u_1, \ldots, u_k)$, $0 \leq u_i \leq 1$, of real numbers, a vector $X(x, u) = X_u(x)$ in the tangent space at x defines for each parameter u a vector field $X_u( ) \in \mathfrak{X}^1(M)$. If the map $u \to X_u$ is continuous (in the strong C^1 topology on $\mathfrak{X}^1(M)$), we call X a *strong* (k-parameter) *family of flows*. A strong one-parameter family of flows will be called a *strong arc* in $\mathfrak{X}^1(M)$. The strong C^1 topology defined earlier applies to the collection $\mathfrak{X}_k^1(M)$ of all k-parameter families, regarded as C^1 maps from $M \times I^k$ to TM.

Following Sotomayor, let us call a flow φ with a velocity vector field $X(x)$ *stable of codimension k* if there exists a strong k-parameter family $X(x, u)$, with $X(x, u_0) = X(x)$ for some $u_0 \in I^k$ such that every strong k-parameter family $Y(x, u)$ near $X(x, u)$ exhibits a flow conjugate to φ. A stronger notion (k-parameter stability) requires that the family $X(x, u)$ be *structurally stable* in the sense that for each $u \in I^k$, the flow of $Y_u(x)$ is conjugate to the flow $X_{\alpha(u)}(x)$, where $\alpha: I^k \to I^k$ is a continuous bijection. We shall construct a structurally stable arc through Equation 1, when there is only one unstable separatrix, and show stability of codimension n, when there are n such separatrices.

To characterize which maps $X: M \times I^k \to TM$ are continuous in the strong topology, we define the *support* of X (supp X) as the closure of the set of points $x \in M$ where $X_{\bar{u}}(x) \neq X_u(x)$ for some pair of parameters $u, \bar{u} \in I^k$.

PROPOSITION 2. If $X: M \times I^k \to TM$ is a C^1 map such that $X(x, u)$ belongs to the tangent space at x, then the map $u \to X_u( )$ is

(i) always continuous in the weak C^1 topology;
(ii) continuous in the strong C^1 topology if and only if X has a compact support.

Proof: The continuity in the weak topology is an easy consequence of the uniform continuity of the continuous functions X, $\partial X/\partial x$, and $\partial X/\partial u$ on any compact set $K \times I$. On the other hand, if $X: I^k \to \mathfrak{X}^1(M)$ is continuous in the strong topology, its image in $\mathfrak{X}^1(M)$ is compact, and so can be covered by a finite number of strong neighborhoods—in particular, for any continuous function $\epsilon(x) > 0$, there is a finite set of $\epsilon(x)$-dense parameters u such that every $v \in I^k$ satisfies,

$$\| X(x, u) - X(x, v) \| < \epsilon(x), \quad \text{for all} \quad x \in M$$

for one of these u. By the triangle inequality, it follows that for each $\epsilon(x)$ there exists an integer $N = N(\epsilon)$ such that for any two parameter values, $v, \bar{v} \in I^k$, we have

$$\| X(x, v) - X(x, \bar{v}) \| < N[\epsilon(x)], \quad x \in M$$

Fix $u_0 \in I^k$. If X has a noncompact support, there exist $x_i \in M$ with no accumulation points, and $u_i \in I^k$ such that

$$\epsilon_i = \| X(x_i, u_0) - X(x_i, u_i) \| > 0$$

But then, for a function $\epsilon(x) > 0$ satisfying $\epsilon(x_i) < \epsilon_i/i$, we must have $N(\epsilon) > i$, for all i—a contradiction. The converse of (ii) follows from the fact that on supp X the strong and weak topologies agree (if supp X is compact). □

Parametrized vector fields have a dual character: first, an arc $X(x, u)$ of vector fields defines a single flow on the product space $M \times I$, stratified by the levels of u, and second, this arc can be viewed as a path in the space $\mathfrak{X}^1(M)$. We have stressed the second view, which is the more natural one for studying the topology of the space $\mathfrak{X}^1(M)$; however, we point out that in certain applications, the former view is the more natural—for example, when u represents energy for a Hamiltonian flow, as in Devaney's paper.[13] When M is compact, the distinction between these two views is rather academic, since both naturally lead to the same mathematical results; however, the result above shows that for open M, the former approach leads to the weak topology, while the latter leads to the strong topology.

Unfolding a Separatrix

In this section we shall construct a structurally stable strong arc of vector fields defined near a separatrix where condition (b) of PROPOSITION 1 breaks down. The construction is basically a parametrized version of a procedure noted by Collins[11] and Krych.[14] Assume that $f(0) = 0$ and that there exist two values of θ, $\theta_1 < 0$ and $\theta_2 > 0$ such that $f(\theta_1) = f(\theta_2) = \pi$ and $0 < f(\theta) < \pi$ for $\theta_1 < \theta < 0$ and $0 < \theta < \theta_2$. The phase portrait of Equation 1 on the strip $\theta_1 < \theta < \theta_2$ is shown in FIGURE 3. The analysis in this section will clearly suit, with minor variations, any of the various cases in which structural stability breaks down at a separatrix of (1).

Under our assumptions $f(\theta)$ has a local minimum at $\theta = 0$. This can be "unfolded" by considering the family of functions defined by,

$$f_u(\theta) = f(\theta, u) = f(\theta) + u - \frac{1}{2}, \quad 0 \le u \le 1.$$

For $u < \frac{1}{2}$, $f_u(\theta)$ has two zeroes, one on either side of $\theta = 0$, while for $u > \frac{1}{2}$, $f_u(\theta)$ has no zeroes in the strip $\theta_1 \le \theta \le \theta_2$. If we substitute $f_u(\theta)$ for $f(\theta)$ in (1), we obtain the one-parameter family of vector fields defined by

$$\theta' = \sin f(\theta, u)$$
$$r' = \cos f(\theta, u)$$

(2)

which, for $u < \frac{1}{2}$, replaces the vertical orbit $\theta = 0$ with a parallelizable canonical region (FIGURE 6), while for $u > \frac{1}{2}$, the vertical orbit at $\theta = 0$ disappears, and the strip $\theta_1 \le \theta \le \theta_2$ becomes a parallelizable canonical region (FIGURE 7).

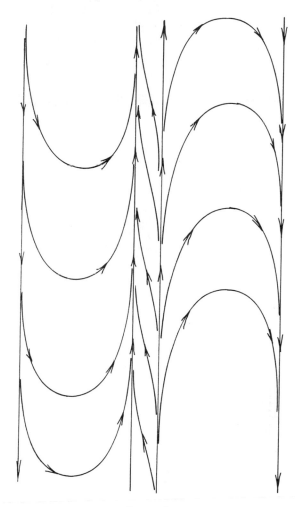

FIGURE 6.

The arc defined by Equation 2 is not strong, since its support includes arbitrary values of r. But we might ask what features of the succession of conjugacy types represented by (2) can be achieved by a stable strong arc through (1).

To analyze the local behavior of a strong arc through (1), we isolate the separatrix $\theta = 0$ from the two nearest separatrices ($\theta = \theta_1$, $\theta = \theta_2$) by picking $0 < \delta < \min\{|\theta_1|, |\theta_2|\}$, restricting our attention to the strip $N = [-\delta, \delta] \times R$. The boundary of the strip consists of two vertical lines,

$$\partial^- N = \{-\delta\} \times R$$

$$\partial^+ N = \{+\delta\} \times R$$

We note that by our hypotheses,

(i) $\theta' > 0$ on $\partial^+ N \cup \partial^- N$;

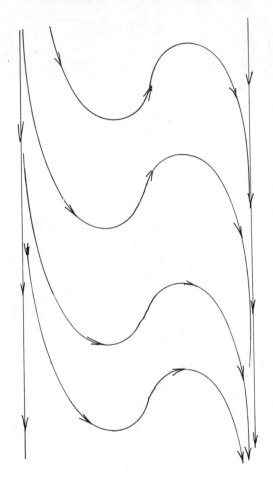

FIGURE 7.

by picking δ sufficiently small, we can also assume that,

(ii) $r' > 0$ on N.

Note that by (i), the vector field is transverse to the boundary of N, pointing into N on $\partial^- N$ and out of N on $\partial^+ N$. Now, pick $k > 0$ and consider two horizontal segments in N defined by,

$$T^+ = [-\delta, \delta] \times \{k\}$$
$$T^- = [-\delta, \delta] \times \{-k\}$$

By (ii), these segments are transverse to the flow, and they divide N into three parts:

$$N^+ = [-\delta, \delta] \times [k, +\infty)$$

$$N^0 = [-\delta, \delta] \times [-k, k]$$
$$N^- = [-\delta, \delta] \times (-\infty, -k]$$

Assumption (ii) implies that the vector field points into N^+ along T^+ and out of N^- along T^- (see FIGURE 8).

We shall analyze the orbit structure of flows which satisfy (i), (ii), and,

(iii) There exists a forward semiorbit from T^+ that remains in N^+ and a backward semiorbit from T^- that remains in N^-.

For the flow (1), such a semiorbit is given by the line $\theta = 0$ in N^+ (resp. N^-).

To analyze the orbit structure of flows satisfying (i), (ii), and (iii), we distinguish two points, $A \in T^+$, and $B \in T^-$, as follows: $A = (\alpha, k) \in T^+$, where α is the greatest lower bound of those θ for which the forward orbit of (θ, k)

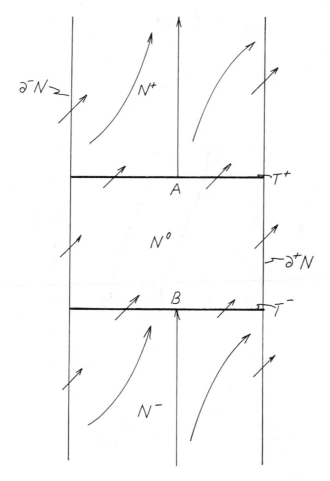

FIGURE 8.

crosses $\partial^+ N$. Since orbits crossing $\partial^+ N$ do so transversally, this set of θ is open, so that A is the rightmost point of T^+ whose forward semiorbit stays in N^+. Similarly, $B = (\beta, k) \in T^-$ is the leftmost point of T^- whose backward semiorbit stays in N^-, or equivalently, β is the least upper bound of those θ for which the backward orbit of $(\theta, -k)$ crosses $\partial^- N$.

This analysis is based on the Poincaré map ϕ which assigns to each point of T^- the point at which its orbit leaves N^0. Since there are no rest points in N^0, every orbit leaves N^0 by either T^+ or $\partial^+ N^0 = \partial^+ N \cap N^0$.

The Poincaré map is a homeomorphism to its image. Its domain, $T^+ \cup \partial^+ N^0$, is a "broken" interval that can be ordered from left to right and then top to bottom. We claim that the topological conjugacy type of a vector field on N satisfying conditions (i), (ii), and (iii) is determined simply by the relation of $\phi(B)$ and A.

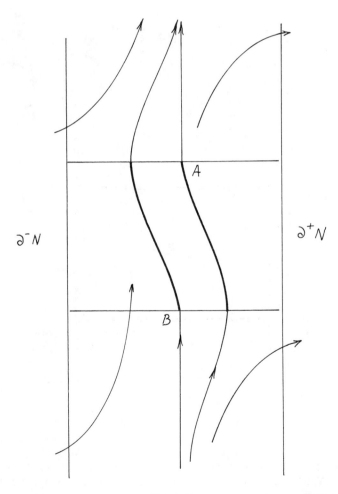

FIGURE 9.

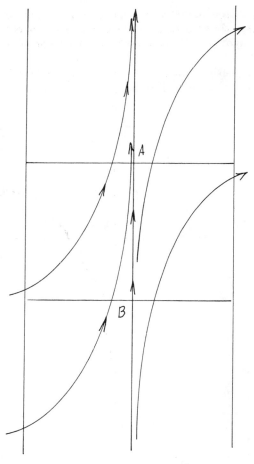

FIGURE 10.

CASE 1. If $\phi(B) < A$, then the orbit of B enters N^+ to the left of the orbit of A, and hence remains in N^+ for all time. Similarly, the backward orbit of A crosses T^- at $\phi^-(A) > B$, so it is trapped in N^- for all time. The orbits of A and B bound a parallelizable canonical region. Every orbit to the right of this region stays in N for $t \to -\infty$, and leaves N via ∂^+N in forward time, while every orbit to the left of this region enters N via ∂^-N and then remains in N with forward time (see FIGURE 9).

CASE 2. If $\phi(B) = A$, then the full orbit of A (also for B) is the only full orbit contained in N. Orbits to the right leave N via ∂^+N in positive time while the orbits to the left cross ∂^-N in negative time (see FIGURE 10).

CASE 3. If $\phi(B) > A$, then the forward orbit of B leaves N at some point $(\delta, b) \in \partial^+N$, while the backward orbit of A leaves N at $(-\delta, a) \in \partial^-N$. These two orbits separate N into three regions: all orbits above and/or to the left of the orbit of A cross $\partial^- N$ at a point above a and flow up without escaping N in forward time; similarly orbits below and/or to the right of the orbit of B leave N in forward time at a point of $\partial^+ N$ below b, but stay in N for all backward time (see Figure 11).

These cases are clearly exhaustive. That each case determines a unique con-

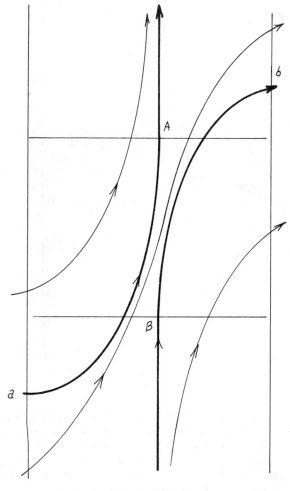

Figure 11.

jugate class on N is demonstrated by methods of Neumann[15] and Collins.[11] (Neumann and Collins do not establish the results referred to, but their methods can be used to do so.) However, we can simply note that, in each case, the piece-wise linear curve

$$\Gamma = \partial^+ N^+ \cup T^+ \cup \partial^- N^0 \cup \partial^- N^-$$

where $\partial^\alpha N^\beta = \partial^\alpha N \cap N^\beta, \alpha = +, -, \beta = +, -$, or 0, intersects each orbit in N exactly once, so that every point in N is determined by a point of Γ together with a time. A homeomorphism of Γ which takes T^+ to itself and A and $\phi(B)$ for one flow to the corresponding points of another flow easily extends to a conjugacy in N.

This analysis suggests how to construct a structurally stable strong arc through Equation 1.

THEOREM A. A strong arc $X(x, u)$ in $\mathfrak{X}^1(N)$ is structurally stable if, for every $u \in I$, it satisfies: assumptions (i), (ii), and (iii), above, and

(iv) supp $X \cup N^0$;
(v) $\phi_u(B) < A$ for $u < \frac{1}{2}$;
(vi) $\partial\phi_u(B)\partial u > 0$ for $u = \frac{1}{2}$;
(vii) $\phi_u(B) > A$ for $u > \frac{1}{2}$.

Proof: Since the above analysis applies to every vector field $X_u( )$ along the arc, our proof has two steps. First, we show that for any strong arc $Y(x, u)$ sufficiently near $X(x, u)$, assumptions (i)–(iii) apply to $Y_u( )$ for every $u \in I$; then, we show that each nearby Y has a unique parameter value for which $\phi_u(B) = A$.

Assumptions (i) and (ii) are trivially stable, but we must find a neighborhood of X for which (iii) holds. By hypothesis (iv), $X(x, u)$ is independent of u for $x \in N^+$. We emphasize this by substituting $X(x)$ for $X_u(x)$. We shall find a function $\epsilon(x) > 0$ such that any vector field Y satisfying,

$$\| X(x) - Y(x) \| < \epsilon(x), \quad x \in N^+$$

has a positive semiorbit from T^+ contained in N^+. The other half of (iii) is similar.

Pick any point $C_0 \in T^+$ to the left of A. We shall show that for Y near X, the positive Y-semiorbit of C_0 remains in N^+. Draw horizontal segments at unit intervals across N^+:

$$T_i^+ = [-\delta, \delta] \times \{k + i\}, \quad i = 1, \ldots$$

Let N_i^+ denote the part of N^+ between T_i^+ and T_{i+1}^+:

$$N_i^+ = [-\delta, \delta] \times [k + i, k + i + 1].$$

The X-orbit of A crosses T_i^+ at a point A_i. We wish to insure that the Y-orbit of C_0 remains to the left of A_i. Define a sequence of constants $\epsilon_i > 0$ and points $C_i \in T_i^+$: by induction, let C_{i+1} be a point on T_{i+1}^+ halfway between the X-orbits of C_i and A_{i+1}, and pick $\epsilon_i > 0$ so that, for every vector field Y with $\| X(x) - Y(x) \| < \epsilon_i$ on N_i^+, the Y-orbit of C_i crosses T_{i+1}^+ to the left of C_{i+1}. If a vector field Y on N satisfies $\| X - Y \| < \epsilon_i$ on $N_i^+, i = 1, \ldots,$ then the Y-orbit of C_0 crosses T_i^+ to the left of C_i and A_i, and so it never leaves N^+. Since the N_i^+ are locally finite, there exists a continuous function $\epsilon(x) > 0$ such that $\epsilon(x) < \epsilon_i$ on N_i^+ for all i. This establishes (iii).

Since $Y_u(x)$ satisfies (i), (ii), and (iii), the analysis applied earlier to X can now be applied to Y_u. Defining points $A(Y_u) \in T^+$ and $B(Y_u) \in T^-$, we see that the conjugacy type of Y_u is determined entirely by the relation of $\phi_{Y,u}[B(Y_u)]$ to $A(Y_u)$.

Naively, we might argue that these are equal for a unique value of u by noting that, since $\partial\phi_{X,u}(B)/\partial u > 0$ for $u = \frac{1}{2}$, we can find neighborhoods $U \subset I$ of $u = \frac{1}{2}$ and $J \subset T^-$ of B such that $\partial\phi_{X,u}(x)/\partial u$ is bounded away from zero, say by $m_0 > 0$, for all $x \in J$ and all $u \in U$. Now, for Y_u near X_u, $B(Y_u)$ must be near $B(X_u)$ and $A(Y_u)$ must be near $A(X_u)$, so that $\phi_{Y,u}[B(Y_u)]$ crosses $A(Y_u)$ with positive u-velocity for some u near $\frac{1}{2}$.

The difficulty with this argument is that the support of an arc Y_u near X_u, while compact, may be arbitrarily large. In particular, we cannot assume it is contained in N^0. Thus, $A(Y_u)$ and $B(Y_u)$, unlike $A(X_u)$ and $B(X_u)$, can vary with u.

We prove the uniqueness of the bifurcation value as follows: first, we show that $A(Y)$ and $B(Y)$ are continuous as Y varies in the strong topology. This allows us, by keeping Y_u close to X_u, to keep $A(Y_u)$ very near $A(X)$ and $B(Y_u)$ very near $B(X)$ for all $u \in I$. This insures, in turn, that all crossings $\phi_u(B) = A$ which occur, do so for u very near $\frac{1}{2}$, say inside U as before. Then we need to show that for $\partial Y_u/\partial u$ near $\partial X_u/\partial u$ on all of N, we can make $\partial A(Y_u)/\partial u$ and $\partial B(Y_u)/\partial u$ so small that $\partial\phi_{Y,u}[B(Y_u)]/\partial u - \partial A(Y_u)/\partial u$ is smaller than m_0, so that the crossing does indeed occur with positive u-velocity, and so is unique.

To show that $A(Y)$ varies continuously with Y, we note that C_0 in the last part of the argument for (iii) is an arbitrary lower bound for $A(Y)$, so that $A(Y)$ is lower semicontinuous. To establish upper semicontinuity, we pick a curve γ arbitrarily near the X-orbit of A, but to the right of it and transverse to the flow. It is easy to see that any Y-orbit crossing γ also crosses ∂^+N, for Y near X, so that γ forms an arbitrary upper bound for $A(Y)$, and upper semicontinuity is also established.

To show that $\partial A(Y_u)/\partial u$ can be made arbitrarily small for Y_u near X_u, we control $\partial Y_u/\partial u$ on the pieces N_i^+ of N^+. Let ϕ_i denote the Poincaré map back from T_i^+ to T_{i-1}^+, and let m_i denote its Lipschitz constant; let $M_i = (m_1 + 1)(m_2 + 1)$ $\cdots (m_i + 1)$. For Y_u sufficiently near X on N_i^+ assume that Lip $\phi_{i,Y,u} < m_i + 1$. Since $\partial X_u/\partial u = 0$ on N_i^+, we can control Y_u and $|\partial Y_u/\partial u|$ on N_i^+, so that, for every $x_i \in T_i^+$,

$$\left|\frac{\partial\phi_{i,Y,y}}{\partial u}\right| < \frac{\epsilon_1}{2^i M_i}$$

Now, if Y_u is an arc subject to all these controls, then, picking i so that T_i^+ is above the support of Y_u, we locate $A_i(Y) \,\epsilon\, T_i^+$ (independent of u) such that,

$$A(Y_u) = \phi_{1,Y,u}\phi_{2,Y,u} \cdots \phi_{i,Y,u}[A_i(Y)]$$

It is then easy to check that,

$$\left|\frac{\partial A(Y_u)}{\partial u}\right| \leq M_i \sum_{k=1}^{i} \left|\frac{\partial\phi_{i,Y,u}}{\partial u}\right|$$

$$< \epsilon_1/2 + \cdots + \epsilon_1/2^i < \epsilon_1$$

Similar arguments for B and N^- give $|\partial B(Y_u)/\partial u| < \epsilon_2$. Thus, we see that we can insure that $B(Y_u) \in J$ for all i, and,

$$\frac{\partial \phi_{Y,u}[B(Y_u)]}{\partial u} - \frac{\partial A(Y_u)}{\partial u} > m_0 - \epsilon_2 \operatorname{Lip}[\phi_{u,Y}(x)] - \epsilon_1$$

which is positive for ϵ_1 and ϵ_2 small.

This establishes that, for any arc Y_u sufficiently near X_u, $\phi_u(B)$ crosses A at a unique value u_0 of $u \in I$, so that letting $\alpha: I \to I$ preserve order, with $\alpha(u_0) = \frac{1}{2}$, we see that Y_u is conjugate to $X_{\alpha(u)}$. □

GLOBAL STABILITY

Theorem A makes it quite easy to construct globally a stable arc of flows on $M = S^1 \times R$ through Equation 1 in the case where structural stability breaks down at a single separatrix. We simply isolate the unstable separatrix, as done above, by a strip N, and construct an arc $X(x, u)$ in $\mathfrak{X}^1(M)$ that is stable on N and whose support is contained in N. The strip N is contained in a union of two canonical regions of (1), which we call P. For any arc $Y(x, u)$ near $X(x, u)$, the conjugacies on N clearly extend to P, while the orbit structure on $M - P$ is the same, by the arguments of Collins[11]. Combining these two observations, by means of Neumann's results[15], we obtain a proof of THEOREM B.

THEOREM B. If in Equation 1 there exists a unique separatrix, $\theta = \theta_1$, such that $f(\theta_{i-1}) = f(\theta_{i+1}) \neq f(\theta_i)$, then there exists a structurally stable arc X_u of flows on M such that $X_{1/2}$ is precisely (1).

The succession of global conjugacy types along this arc is slightly different from that described in the previous section along the weak arc (Equation 2). For $u < \frac{1}{2}$ and $u = \frac{1}{2}$, the type in P is illustrated in FIGURES 6 and 3, respectively. As u approaches $\frac{1}{2}$ from below, a successively narrower parallelizable band, bounded by the orbits of A and B, separates two Reeb-like regions, and the prolongational relations in P are,

$$B \in J^+[\theta = \theta_{i-1}], \quad \{\theta = \theta_{i+1}\} = J^+[A]$$

However, after u crosses $u = \frac{1}{2}$, not every orbit in P crosses the strip N, so that P is *not* parallelizable, but divides into three regions. The orbit of A has $r \to +\infty$ in both time directions, and the region above it is parallelizable. Similarly, the orbit of B has $r \to -\infty$ in both time directions, and the region below it is parallelizable. The region between, bounded by $\theta = \theta_{i-1}$, $\theta = \theta_{i+1}$, and the orbits of A and B, is a canonical region for which we still have the relations $B \in J^+[\theta_{i-1}]$ and $\{\theta = \theta_{i+1}\} = J^+[A]$. This is shown in FIGURE 12. Of course, the orbit structure on other canonical regions is unaffected by this arc.

If the unstable situation occurs at several separatrices, then we cannot hope to construct a stable arc through (1), because for most arcs, the "crossings" in different strips will occur at different parameter values. However, by introducing more parameters, this obstruction to stability can be eliminated. While it seems likely that this is the only obstacle to stability, a proof of stability would require an involved analysis of global orbit structure, which we shall not attempt here. Instead, we content ourselves with a weaker result.

THEOREM C. Suppose that the situation of the previous theorem occurs at precisely n separatices. Then (1) is stable of codimension n.

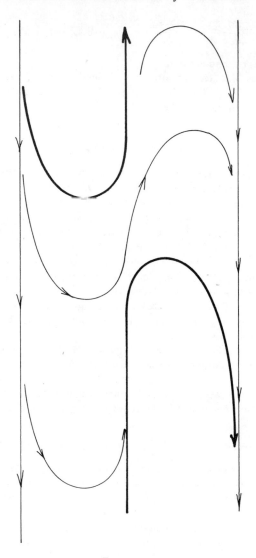

FIGURE 12.

Proof: We set up strips $N_1, \ldots, N_n$ that isolate the n unstable separatrices, as in the previous section. These results allow us to construct a one-parameter family $X^i(x, s)$, $0 \leq s \leq 1$, of vector fields on each N_i, with support interior to N_i, which is stable as an arc on N_i. We define an n-parameter family $X(x, u_1, \ldots, u_n)$ by,

$$X(x, u_1, \ldots, u_n) = X^i(x, u_i), \, x \in N^i$$

$$X(x, u_1, \ldots, u_n) = X(x), \qquad x \in M - \cup N_i$$

Now, let A_i, B_i denote the points in N_i corresponding to A and B in the previous section and ϕ^i, the Poincaré map for X^i. Then, given X, we define a map $g: I^n \to R^n$ by assigning to the n-tuple $u_1, \ldots, u_n$ an n-tuple whose ith entry is $\phi_{Y}{}^i[B_i(Y_{u_i})] - A_i(Y_{u_i})$. For our original flow (Equation 1), we saw that g maps I^n onto a neighborhood of the origin. By the continuity arguments of the previous section, g, for Y near X, still maps I^n onto some neighborhood of the origin. In particular, the origin is hit by some parameter value, and at this value, we have a conjugate of $X_{1/2}$. $\square$

STABILITY PROPERTIES OF WEAK ARCS

If we adopt the view described earlier, that weak arcs in $\mathfrak{X}^1(M)$ represent stratified dynamical systems on $M \times I$, then it is natural to control perturbations of weak arcs by strong neighborhoods in $\mathfrak{X}^1(M \times I)$. Thus, a perturbation Y_u of the weak arc X_u will be a weak arc for which,

$$\| X(x,u) - Y(x,u) \| < \epsilon(x) \quad \text{for every } x \in M, u \in I$$

One might expect in this case that the arc described by Equation 2 should be structurally stable. However, we note that by adding a small push to the right, near $\theta = 0$, for $|r|$ large, it is possible to make orbits far above and far below $r = 0$ cross from $\partial^- N$ to $\partial^+ N$ before the strip in the middle shrinks to a single orbit. Thus, we pass from FIGURE 6 to FIGURE 7 without going through FIGURE 3. In fact, we can construct a stable weak arc X_u such that X_u is itself a C^1 structurally stable vector field for every value of u, as follows. We leave the vector field alone in N^0. Pick $B \in T^-$, $\phi(B) < A$ on T^+, and assume that for $u = 0$, the full orbits of A and B bound a parallelizable strip separating $\partial^- N$ from $\partial^+ N$ (see FIGURE 6). However, instead of moving A and $\phi(B)$, we add a right-hand component to X_0 far up N^+ (and far down N^-) so that, for sufficiently large i (independent of u) and $u = \frac{1}{2} - 1/n$, the orbit of B crosses $T_i{}^+$ at $B_i = 1 - 1/(n + i)$, and the orbit of A crosses $T_i{}^+$ at $A_i = 1 - 1/(n + i + 1)$. Similarly, we insure that, at $u = \frac{1}{2} - 1/n$, $A_{-i} = 1/(n + i)$, $B_{-i} = 1/(n + i + 1)$.

Thus, as u approaches $\frac{1}{2}$ from below, the band takes on an "S" shape, and for $u > \frac{1}{2}$, every orbit crosses from $\partial^- N$ to $\partial^+ N$ with a large upward shift in r for u near $\frac{1}{2}$. Note that, at the bifurcation value $u = \frac{1}{2}$, there are still no orbits crossing, because if we consider the stratified flow on $N \times I$, the crossings from $\partial^- N \times I$ to $\partial^+ N \times I$ are transverse, hence they form an open set—so if an orbit were to cross at $u = \frac{1}{2}$, it would also cross for u slightly below $u = \frac{1}{2}$, which is impossible. To see the stability of this situation, note that a perturbation of X_u for $u > \frac{1}{2}$ will still have orbits crossing from $\partial^- N$ to $\partial^+ N$, for $|r|$ large. Thus, we can determine the canonical regions, and hence the conjugacy type, quite easily for Y_u near X_u.

This sketch can be made into a theorem, but space does not permit the necessary elaboration. We wish, however, to point out the fact that weak arcs can pass stably between different stable conjugacy types without necessarily hitting any structurally unstable flows.

Accessibility

We are primarily interested in the possibility of passing from a completely un-stable flow to one with nonempty nonwandering set. Since we are dealing with a class of flows without rest points, it is natural to demand that our arcs satisfy $X(x, u) \neq 0$ globally, for all u. This is much in the spirit of an open extension of the work of Asimov[17] on nonsingular deformation of nonsingular flows on closed manifolds.

A weak arc is nothing more than a smooth homotopy. Thus obstructions to joining two flows with a weak arc are simply the invariants of a smooth homotopy of nonvanishing vector fields. For example, we note that the index of the vector field about any nonbounding circle in $S^1 \times M$ is an invariant; since any periodic orbit of a nonvanishing vector field on $S^1 \times M$ must be such a circle, we see that Equation 1 cannot be deformed out of the class of completely unstable flows if $f(\theta)$ has a degree different from zero (e.g., $f(\theta) = n\theta$, $n \neq 0$). On the other hand, it is clear that if a nonvanishing vector field on $S^1 \times M$ has periodic orbits, it can be weakly deformed into one depending only on θ, and hence to, say, (1) with $f(\theta) = \pi/4$, which is completely unstable.

For a strong arc, the case is quite different, since the two ends of such an arc can differ only on a bounded set. A flow on $S^1 \times M$ whose nonwandering set consists of finitely many periodic orbits can be strongly deformed to a completely unstable flow provided the uppermost and lowermost of these orbits do not both attract or repel and that the vector field on the closed annulus bounded by them is homotopic to the vector field $r' = 0$, $\theta' = 1$. On the other hand, a flow with an unbounded set of hyperbolic periodic orbits—for example, $r' = \cos r$, $\theta' = \sin r$—cannot be strongly deformed to a completely unstable flow.

On a 3-dimensional (or higher) manifold, it is always possible, by deforming the flow inside a solid torus $D^2 \times S^1$ embedded in the manifold, to introduce a periodic orbit of the form $(0, 0) \times S^1$, so that complete instability can always be deformed away. It might be interesting, however, to ask: When can the nonwander-ing set be deformed away? That is, what are the conditions on a Morse-Smale flow φ (on an open manifold) that insure the existence of nonwandering points along any strong arc through φ?

REFERENCES

1. THOM, R. 1975. Structural Stability and Morphogenesis, An Outline of a General Theory of Models. D. H. Fowler, Trans. W. A. Benjamin, Inc., Reading, Mass.
2. SOTOMAYOR, J. 1974. Generic one parameter families of vectorfields. Publ. IHES, No. 43. Paris.
3. SOTOMAYOR, J. 1973. Structural stability and bifurcation theory. *In* Dynamical Systems. M. M. Peixoto, Ed. Academic Press, New York. 549–560.
4. NEWHOUSE, S. & M. M. PEIXOTO. 1976. There is a simple arc joining any two Morse-Smale flows. *In* Trois Etudes en Dynamique Qualitative, Astérisque Soc. Math. Fr. **31:** 15–41. See also Newhouse, S. & J. Palis. *Ibid.* **31:** 43–140.
5. NITECKI, Z. 1977. Explosions in completely unstable flows; I. Preventing Explosions, II. Some Examples. Trans. Am. Math. Soc. In press.
6. NEMYTSKII, V. V. & V. V. STEPANOV. 1960. Qualitative Theory of Differential Equations. S. Lefschetz *et al.*, Trans. Princeton Univ.

7. MARKUS, L. 1969. Parallel dynamical systems. Topology **8**: 47–57.
8. URA, T. 1953. Sur les courbes définies par les équations différentielles dans l'espace à m dimensions. Ann. Sci. École Norm. Sup. **70**: 287–360.
9. AUSLANDER, J. 1964. Generalized recurrence in dynamical systems. *In* Contributions to Differential Equations. John Wiley **3**: 55–74.
10. HIRSCH, M. W. 1976. Differential Topology, Chap. 2. Springer-Verlag, New York.
11. COLLINS, J. 1977. Structural stability of completely unstable flows in the plane. Ph.D. Thesis. Tufts University.
12. KRYCH, M. 1975. O strukturalnej stabilności dyfeomorfizmów i pól wektorowych rozmaitości niezwartych. Ph.D. Thesis. Univ. of Warsaw.
13. DEVANEY, R. L. 1977. Homoclinic orbits to hyperbolic equilibria. This volume.
14. KRYCH, M. 1977. A generic property of non-vanishing vectorfields on R^2. Bull. Acad. Pol. Sci. **25**: 361–368.
15. NEUMANN, D. A. 1977. Completely unstable flows on 2-manifolds. Trans. Am. Math. Soc. **225**: 211–226.
16. OSIPOV, J. S. 1976. Structural stability of noncompact Anosov flows and hyperbolic motions in Kepler's problem. Soviet Math. Dokl. **17**: 1389–1393.
17. ASIMOV, D. 1975. Homotopy of non-singular vector fields to structurally stable ones. Ann. Math. **102**: 55–65.

HOMOCLINIC ORBITS TO HYPERBOLIC EQUILIBRIA*

Robert L. Devaney

Department of Mathematics
Tufts University
Medford, Massachusetts 02155

INTRODUCTION

Since the time of Poincaré and Birkhoff, it has been known that the orbit structure of a dynamical system near a homoclinic point or orbit is extremely complicated. Briefly, a homoclinic orbit is one that is asymptotic to an invariant set in both the forward and backward time directions. However, it is only recently that this complicated structure has begun to be understood. The first, and most important, result along this line is that due to Smale.[1] He has shown that, near a transverse homoclinic orbit asymptotic to a hyperbolic closed orbit of a vector field, one can distinguish a compact invariant set for the flow which admits a Poincaré map topologically conjugate to the shift automorphism on finitely many *symbols*. Roughly speaking, the invariant set that Smale has discovered consists of infinitely many long periodic orbits, together with nonperiodic orbits, some of which wind densely throughout the invariant set. More importantly, the Smale result assigns a *code word* to each orbit in the set, and this code word completely describes the qualitative behavior of the orbit. Moser[2] also considers this phenomenon.

Smale's result has had far-reaching applications. Sitnikov[19] and Alekseev[3] have used the shift automorphism to find capture, escape, and oscillatory orbits in the *restricted three body problem*. Transverse homoclinic orbits occur in the *Störmer problem*[4] and the *anisotropic Kepler problem*,[5] both physically important classical mechanical systems. The Smale result gives the existence of a rich orbit structure in each case. In the four body problem, it appears that homoclinic behavior may help to explain the behavior of orbits discovered by Mather and McGehee[6] which become unbounded in finite time.

In this paper, we attempt to survey the rather complete results which have been obtained for orbits that are homoclinic to hyperbolic equilibrium points. Surprisingly, this case is more complicated than the case Smale has considered—a hyperbolic closed orbit. We also restrict our attention to Hamiltonian systems, as this case turns out to be the easiest to describe. Silnikov[7] has dealt with non-Hamiltonian systems, and has a fairly complete picture of the generic structure there.

HOMOCLINIC ORBITS IN HAMILTONIAN SYSTEMS

We shall deal exclusively with Hamiltonian systems with two degrees of freedom, although many of our results extend to higher-dimensional systems. On R^4,

*Partially supported by NSF Grant MCS 77-00430.

0077-8923/78/0316-0108 $1.75/1 © 1979, NYAS

we take coordinates $(x, y) = (x_1, x_2, y_1, y_2)$ together with a sufficiently smooth real-valued function H, called the *Hamiltonian*. The system is then given by the differential equations,

$$\dot{x} = \frac{\partial H}{\partial y}$$

$$\dot{y} = -\frac{\partial H}{\partial x} \tag{1.1}$$

As is well known, H is constant along the solutions of (1.1); i.e., H is a first integral for the system.

We are primarily interested in equilibrium points, or zeroes, of (1.1). Note that the system has equilibria at critical points of H; i.e., at points where grad $H = 0$. We henceforth assume that grad H vanishes at the origin, and that, for simplicity, $H(\underline{0}) = 0$. The level set, or energy surface $H^{-1}(0)$, is not quite a submanifold of R^4, as $\underline{0}$ is a critical point for H. However, if we assume that 0 is the only critical point in $H^{-1}(0)$, then $H^{-1}(0) - \underline{0}$ is a smooth submanifold.

The system (1.1) may be written in the form,

$$\dot{z} = Az + \text{higher-order terms} \tag{1.2}$$

where A is a 4×4 matrix. Because of the Hamiltonian character of the system, if λ is an eigenvalue of A, then so are $-\lambda$, $\bar{\lambda}$, and $-\bar{\lambda}$. These eigenvalues are called the *characteristic exponents* of the system. The equilibrium point is called *hyperbolic* if all of the characteristic exponents have nonzero real parts.

In the hyperbolic case, it follows that two of the characteristic exponents have negative real parts and two have positive real parts. So R^4 splits as a direct sum $E^s \oplus E^u$, where E^s is the generalized eigenspace for A corresponding to eigenvalues with negative real parts, and E^u is the generalized eigenspace corresponding to the eigenvalues with positive real parts. There is also a nonlinear analog of this splitting for a hyperbolic equilibrium. There exists a 2-dimensional, immersed, invariant manifold W^s, the *stable manifold*, having the following properties:

1) W^s is tangent at $\underline{0}$ to E^s;
2) All orbits in W^s tend asymptotically toward $\underline{0}$;
3) $W^s \subset H^{-1}(\underline{0})$.

A similar invariant manifold W^u, the *unstable manifold*, passes through 0 tangent to E^u and consists of orbits tending asymptotically toward $\underline{0}$ as $t \to -\infty$. The local phase portrait for a system with one degree of freedom is shown in FIGURE 1.

Now suppose $x \in W^s \cap W^u$. As both W^s and W^u are invariant under the flow, the entire orbit through x must lie in $W^s \cap W^u$. Such an orbit is called a *homoclinic orbit*. One usually assumes that W^s meets W^u transversely along a homoclinic orbit, but herein lies one of the difficulties associated with orbits homoclinic to equilibria. The dimensions of W^s and W^u add up to exactly the dimension of the ambient manifold, and so W^s can never meet W^u transversely along a 1-dimensional orbit. This implies that, for non-Hamiltonian systems, W^s generically misses W^u. In the Hamiltonian case, however, the stable and unstable manifolds may meet transversely along an orbit within the codimension 1 energy surface. Such homoclinic orbits are then called *nondegenerate*.

Shift Automorphism

Our aim is to study the complicated orbit structure that arises near a non-degenerate homoclinic orbit to a hyperbolic equilibrium point. In the next section, we reduce this problem to the study of a certain diffeomorphism that arises as a Poincaré map on a local transversal section. This mapping is topologically conjugate to the shift automorphism, which we now describe.

Let A denote a finite or countable set; A is called the *alphabet*. If A has n elements, we define the *symbol space* S_n:

$$S_n = \{(s) = (\ldots s_{-2}, s_{-1}, s_0 . s_2, \ldots) \mid s_j \in A\} \qquad (2.1)$$

whose elements consist of doubly infinite sequences of elements of A. When A is finite, S_n may be topologized in a natural way so that S_n is homeomorphic to a

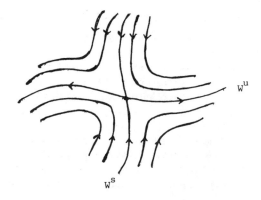

FIGURE 1. The flow near a hyperbolic equilibrium point.

Cantor set, and hence is compact. When A is infinite, S_∞ is noncompact. One may compactify S_∞ by allowing one- and two-sided terminating sequences of the form,

$$(\infty, s_{k+1}, s_{k+2}, \ldots, s_{j-1}, \infty) \qquad (2.2)$$

where k and j are integers with $-\infty \le k \le 0, 0 \le j \le \infty$ (Moser[2] has more details on this construction).

The shift automorphism on S_n is defined by,

$$[\sigma(s)]_k = (s)_{k-1} \qquad (2.3)$$

That is,

$$\sigma(\ldots s_{-2}, s_{-1}, s_0 . s_1, s_2 \ldots) = (\ldots s_{-2}, s_{-1} . s_0, s_1, s_2 \ldots) \qquad (2.4)$$

so that σ shifts the decimal point one place to the left. When A is finite, σ can be shown to be a homeomorphism of S_n which satisfies:

1) Periodic points (i.e., repeating sequences) are dense in S_n;
2) σ is topologically transitive (i.e., there is a dense orbit).

When A is infinite, we define σ on the terminating sequences exactly as above, provided $s_0 \neq \infty$; if $s_0 = \infty$, $\sigma(s)$ is undefined. Hence σ is no longer a homeomorphism. Thus, for the shift on infinitely many symbols, we must sacrifice either the compactness of S_∞, or the σ-invariance of S_∞.

SHIFT AUTOMORPHISM NEAR A HOMOCLINIC ORBIT

In this section, we show how the shift automorphism arises, as a Poincaré map, near a nondegenerate homoclinic orbit asymptotic to a hyperbolic equilibrium. We need, however, one further assumption on the system. An equilibrium point is called a *saddle-focus* if the characteristic exponents are $\pm(\alpha \pm i\beta)$, with $\alpha,\beta > 0$. The assumption that α is nonzero guarantees that the equilibrium is hyperbolic. The assumption that β is nonzero implies that orbits spiral into and away from the origin. As we shall show, in a later section, this spiraling is necessary for the existence of the shift. This is one more complication in the case of orbits homoclinic to an equilibrium.

We choose coordinates $(x, y) = (x_1, x_2, y_1, y_2)$ near the equilibrium so that the system (1.1) may be written,

$$\dot{x}_1 = \alpha x_1 + \beta x_2 + A_1(x, y)$$
$$\dot{x}_2 = -\beta x_1 + \alpha x_2 + A_2(x, y)$$
$$\dot{y}_1 = -\alpha y_1 + \beta y_2 + A_3(x, y)$$
$$\dot{y}_2 = -\beta y_1 - \alpha y_2 + A_4(x, y) \tag{3.1}$$

where the A_i are smooth and vanish together with their first partial derivatives at the origin. In addition, assume that $A_1(0, y) = 0 = A_2(0, y)$ and $A_3(x, 0) = 0 = A_4(x, 0)$. This means that, locally, W^s is given by the plane $x = 0$ while W^u is given by $y = 0$.

Suppose γ is a nondegenerate homoclinic orbit. To study nearby orbits, we construct a pair of local transversal sections together with the associated Poincaré maps. First define,

$$\sum{}^s = \sum{}^s(\delta_1, r_1) = \{(x, y) \mid |y| = \delta_1, |x| \le r_1\}$$

$$\sum{}^u = \sum{}^u(\delta_2, r_2) = \{(x, y) \mid |x| = \delta_2, |y| \le r_2\} \tag{3.2}$$

Both $\sum^s$ and $\sum^u$ are solid tori which are transverse to the flow if δ_i and r_i are small enough; $\sum^s$ and $\sum^u$ meet W^s and W^u in circles which we denote by σ^s and σ^u, respectively. Also, there is defined a Poincaré map $\Phi_1 : \sum^s - \sigma^s \to \sum^u - \sigma^u$ obtained by following orbits leaving $\sum^s$ until their first intersection with $\sum^u$. For later purposes, we also define,

$$\sum_{\epsilon}^{s} = \sum^{s} \cap \, H^{-1}(\epsilon)$$

$$\sum_{\epsilon}^{u} = \sum^{u} \cap \, H^{-1}(\epsilon). \tag{3.3}$$

Since H is constant along orbits, Φ_1 maps $\sum_{\epsilon}^{s}$ to $\sum_{\epsilon}^{u}$.

Now γ meets σ^s and σ^u at points which we denote by x^s and x^u. Let B be a small open disk about x^u in $\sum^{u}$, and let $B_\epsilon = B \cap H^{-1}(\epsilon)$. By following orbits leaving B until they meet $\sum^{s}$, there is defined a second Poincaré map $\Phi_2 : B \rightarrow \sum^{s}$. Again we have $\Phi_2(B_\epsilon) \subset \sum_{\epsilon}^{s}$. The composition $\Phi = \Phi_1 \cdot \Phi_2$ thus takes $B - \sigma^u$ into $\sum^{u}$ (FIGURE 2).

Now let $\Lambda_\epsilon = \bigcap_{n=-\infty}^{\infty} \Phi^n(B_\epsilon - \sigma^u)$. Clearly, Λ_ϵ is Φ-invariant and contained in $H^{-1}(\epsilon)$. Concerning Λ_0, our main result is

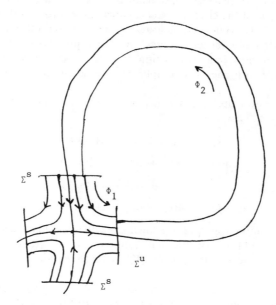

FIGURE 2. The construction of the Poincaré maps Φ_1 and Φ_2 near a hyperbolic equilibrium point.

THEOREM 1. Λ_0 is a hyperbolic set for Φ on which Φ is topologically conjugate to the shift on infinitely many symbols. That is, there is a homeomorphism $h: \Lambda_0 \rightarrow S_\infty$ such that the following diagram commutes.

$$
\begin{array}{ccc}
\Lambda_0 & \xrightarrow{\ \Phi\ } & \Lambda_0 \\
\downarrow{\scriptstyle h} & & \downarrow{\scriptstyle h} \\
S_\infty & \xrightarrow[\ \sigma\ ]{} & S_\infty
\end{array}
$$

THEOREM 1 was first proven by the author.[8]

COROLLARY 1. Near a nondegenerate homoclinic orbit to a saddle-focus, there exist infinitely many long periodic solutions in $H^{-1}(0)$.

In fact, each sequence in S_∞ of the form $(\ldots k, k, k, \ldots)$ corresponds to a fixed point of Φ, that is, to an orbit in $H^{-1}(0)$ which closes up after passing through $\sum^s$ and $\sum^u$ exactly once. The time periods of these orbits tend to infinity, however.

The fact that Λ_0 is a hyperbolic set does not guarantee that the shift on infinitely many symbols persists on nearby energy levels. Because of the noncompactness of Λ_0, we can only assert the following corollary.

COROLLARY 2. Given $\epsilon > 0$, there is an $N = N(\epsilon)$, a compact, invariant subset Γ_ϵ of $H^{-1}([-\epsilon, \epsilon])$, and a homeomorphism $\tau : \Gamma_\epsilon \to S_N \times [-\epsilon, \epsilon]$ such that the following diagram commutes:

$$
\begin{array}{ccc}
\Gamma_\epsilon & \xrightarrow{\ \ \Phi\ \ } & \Gamma_\epsilon \\
\Big\downarrow{\scriptstyle\tau} & & \Big\downarrow{\scriptstyle\tau} \\
S_N \times [-\epsilon, \epsilon] & \xrightarrow[(\sigma, \mathrm{id})]{} & S_N \times [-\epsilon, \epsilon].
\end{array}
$$

Also, τ is level-preserving in the sense that τ maps $H^{-1}(e) \cap \Gamma_\epsilon$ onto $S_N \times \{e\}$.

The proof of this corollary is an immediate consequence of the fact that any invariant subset of Λ_0 conjugate to $S_N \subset S_\infty$ is a compact, hyperbolic set which therefore persists. □

BLUE SKY CATASTROPHES

The results of the previous section raise some intriguing questions. On $H^{-1}(0)$ there exist infinitely many fixed points for the Poincaré map; i.e., all points in $\sum_0^s$ which correspond to sequences of the form $(\ldots k, k, k, \ldots)$ in S_∞. Since each of the corresponding closed orbits are hyperbolic within $H^{-1}(0)$, it is well known that they must lie on smooth one-parameter families of closed orbits that cross nearby energy levels. However, if we look at any fixed energy level $H^{-1}(\epsilon)$ with $\epsilon \neq 0$, we detect only finitely many fixed points for Φ via COROLLARY 2. So the question is, how do these families of closed orbits disappear? Or, more generally, how are the various fixed points related to each other?

These questions have a surprisingly simple answer—all of the fixed points for Φ given by THEOREM 1 lie on the same family of closed orbits. We shall discuss how this happens more fully later on. First, however, we discuss some weaker results. The following theorem was apparently known to Birkhoff.[9]

THEOREM 2. Let γ be a nondegenerate homoclinic orbit to a hyperbolic equilibrium point (not necessarily a saddle-focus). Then there exists a one-parameter family of closed orbits γ_τ that converges to γ as the parameter τ approaches infinity. Moreover, τ may be chosen to be the period of the closed orbit.

The phenomenon of a family of closed orbits whose periods tend to infinity has been given the descriptive title of *blue sky catastrophe* by Abraham.[10] Here we have the only known *stable* occurrence of a blue sky catastrophe in the theory of Hamiltonian systems.[11]

In the case of a saddle-focus, one can say much more about the family of closed orbits. Any closed orbit in a system with two degrees of freedom must have at least two characteristic multipliers equal to 1. The remaining two multipliers are of the form λ, λ^{-1}, where λ may be either positive or negative real (the hyperbolic case), or else complex of absolute value 1 (the elliptic case). Denote the characteristic multipliers of the family in THEOREM 2 by λ_τ, λ_τ^{-1}, where λ_τ satisfies $|\lambda_\tau| \geq 1$, Im $\lambda_\tau \geq 0$. Then we have the following result.[11]

THEOREM 3. If γ is a nondegenerate homoclinic orbit to a saddle-focus, then λ_τ satisfies,

$$\limsup_{\tau \to \infty} \lambda_\tau = +\infty$$

$$\liminf_{\tau \to \infty} \lambda_\tau = -\infty$$

That is, since λ_τ depends continuously on τ, the multipliers oscillate between large positive reals and large negative reals. However, the only way λ_τ can change from the positive to the negative real axis is for λ_τ to traverse the top half of the unit circle.

COROLLARY 3. Let μ be an nth root of unity. Under the hypotheses of THEOREM 3, there exists a sequence of positive reals $\{\tau_i\} \to \infty$ such that, for each i, either $\lambda_{\tau_i} = \mu$ or $\lambda_{\tau_i}^{-1} = \mu$.

Orbits whose multipliers are nth roots of unity are bifurcation orbits that have been studied by Meyer.[12] Generically, one or two additional families of closed orbits branch off from γ_τ at such orbits. Hence our original family spawns many more families as $\tau \to \infty$.

We return to the fixed points of Φ. The proof of THEOREM 2 shows that they all lie along the same family of closed orbits which we have denoted by γ_τ. They are all hyperbolic. In between successive crossings of $H^{-1}(0)$ by γ_τ, one may show that H has a unique maximum or minimum along γ_τ, or, equivalently, that there exists a unique orbit with $\lambda_\tau = 1$. Hence the intersection of γ^τ with the cross section $\sum^s$ is an infinite "spiral" that converges to x^s and is everywhere transverse to $\sum_0^s$ (FIGURE 3).

Some of the families branching off γ_τ at nth roots of unity apparently cross $H^{-1}(0)$ also. It would be interesting to know which periodic sequences in S_∞ arise as such branching families; i.e., which periodic sequences correspond to closed orbits that lie on families which at some time have bifurcated away from γ_τ. One might also turn this question around: which nth roots of unity give rise to branching families of closed orbits which eventually meet $H^{-1}(0)$, and thus are part of the shift? Thus the entire question is one of the genealogy of the families of closed orbits. We remark that this entire question is reminiscent of work of Guckenheimer[13] on quadratic maps. He has traced the occurrence and disappearance of periodic points for a one-parameter family of maps of the unit interval of the form $f_b(x) = bx(1 - x)$, where b is a real parameter, $0 \leq b \leq 4$.

AN EXAMPLE

THEOREM 1 is false without the assumption of spiraling. This surprising fact leads to a basic question. Let Λ be an invariant set for a (not necessarily

Hamiltonian) flow and let γ be a homoclinic orbit to Λ. Under what conditions on γ and/or Λ do we have an invariant set nearby whose Poincaré map is conjugate to the shift automorphism? Recalling Smale's result for closed orbits, we might conjecture that the transversality of W^s and W^u along γ, together with some sort of recurrence on Λ itself, might give the result. However, the example below shows that this is not the case, as the invariant set is an equilibrium point, and thus has all the recurrence we need. This example was first shown to me by Uhlenbeck,[14] who discovered it in her work on minimal surfaces. Later Moser informed me that this system is a classical system first studied by Neumann[15] in 1858.

We first construct a system on the tangent bundle of the 2-dimensional sphere TS^2, and then project this system to TP^2. Let $(x, \dot{x}) \in TS^2$, where $x, \dot{x} \in R^3$, $|x| = 1$, $x \cdot \dot{x} = 0$ and let A be the 3×3 diagonal matrix with entries 0, λ_1,

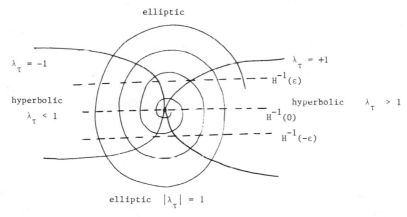

FIGURE 3. The spiral represents the set of fixed points of the Poincare map Φ. The horizontal lines represent the energy levels $H^{-1}(\epsilon)$.

λ_2 along the diagonal. Assume that $0 < \lambda_1 < \lambda_2$. The potential energy is given by $V(x) = - |Ax|^2$. Newton's law then gives the differential equation,

$$\nabla_{\dot{c}} \dot{c} = -\text{grad} \, \nabla \qquad (4.1)$$

where ∇ is the Levi-Civita connection on S^2. In $(x, \dot{x})$ coordinates, the vector field on TS^2 is given by,

$$X = (\dot{x}, - |\dot{x}|^2 x + 2A^2 x - 2 |Ax|^2 x) \qquad (4.2)$$

This system has two equilibrium solutions at $(\pm 1, 0, 0, 0, 0, 0)$. The characteristic exponents at these equilibria are easily computed and found to be $\pm \sqrt{2\lambda_1}$ and $\pm \sqrt{2\lambda_2}$. Hence both points are hyperbolic. Note that (4.2) is invariant under reflection through the origin in R^3, and hence (4.1) projects to a mechanical system on TP^2. The two equilibria above project to the same point in TP^2, which we denote by p.

We claim that there are transversal homoclinic orbits to p, but that X is a completely integrable system. It is known that there can be no subsystems conjugate to a shift in a real analytic, completely integrable system (Ref. 2, p. 107).

The integrability of X follows from the fact that, for $j = 1, 2$:

$$\Phi_j(x, \dot{x}) = x_j^2 + \frac{1}{2} \sum_{\substack{k=0 \\ k \neq j}}^{2} \frac{(\dot{x}_j x_k - \dot{x}_k x_j)^2}{\lambda_k^2 - \lambda_j^2} \qquad (4.3)$$

are integrals in involution for the system. This is a straightforward computation using (**4.2**).

The Poincaré maps Φ_1 and Φ_2 are not linearly independent on $x_1 = 0$ and $x_2 = 0$. However, the level set $\Sigma = \Phi_1^{-1}(0) \cap \Phi_2^{-1}(0)$ is smooth away from these two sets, and its topology may be analyzed. This is important, since $W^s(p) = W^u(p) = \Sigma$. That is, every orbit in $W^s(p)$ is actually homoclinic, and these homoclinic orbits fill out the entire level set Σ. The proof of this fact is due to the author.[16]

Now the orbits of X which project entirely into $x_1 = 0$ or $x_2 = 0$ play a special role. On S^2, these orbits project to curves which connect $(1, 0, 0)$ to $(-1, 0, 0)$. All other orbits in Σ project to curves which begin and end at the same point. Careful analysis of this fact leads to the proof that the orbits which project to $x_1 = 0$ and $x_2 = 0$ are nondegenerate homoclinic orbits. The remaining homoclinic orbits are degenerate in the sense that W^s and W^u do not meet transversely in the energy surface along them.

Recent Results

Since the Smale result, there have been several other papers dealing with the structure of a Hamiltonian system near a transverse homoclinic orbit. Churchill and Rod[16] have shown that one may weaken the transversality assumptions of Smale and still find the shift automorphism nearby. Easton[17] has studied examples of orbits homoclinic to an invariant torus in Hamiltonian systems with several degrees of freedom. In certain cases he has also found shifts present. We remark that the flow on the torus in his case is the translation flow. This assumption is similar in spirit to our requirement of spiraling in the equilibrium point case.

Blue sky catastrophes were first noticed by Henrard.[18] His work was motivated by the numerical work of Strömgren, who conjectured the existence of transverse homoclinic orbits in the restricted three body problem as well as the accompanying family of closed orbits. Also, Churchill and Rod have recently found blue sky catastrophes near orbits homoclinic to partially hyperbolic, partially elliptic equilibria. Apparently, in this case, one need not get the oscillatory behavior of the characteristic multipliers as we find in Theorem 3.

There are very few results known about neighborhoods of nontransverse homoclinic orbits. Such orbits do occur in mechanical systems—recent results of Moser indicate that the Hamiltonian character of the flow sometimes forces such orbits to persist under perturbation. One expects, however, more complicated behavior in this case.

REFERENCES

1. SMALE, S. 1965. Diffeomorphisms with many periodic points. *In* Differential and Combinatorial Topology. S. S. Cairns, Ed.: 63–80. Princeton Univ. Press. Princeton, N.J.

2. MOSER, J. 1973. Stable and Random Motions in Dynamical Systems. Princeton Univ. Press.

3. ALEKSEEV, V. M. 1968–1969. Quasi-random dynamical systems I, II, III. Math USSR Sbornik. **5:** 73–128; **6:** 505–560; **7:** 1–43.

4. DRAGT, A. J. & J. M. FINN. 1976. Insolubility of trapped particle motion in a magnetic dipole field. J. Geophys. Res. **81:** 2327.

5. DEVANEY, R. 1977. Collision orbits in the anisotropic Kepler problem. To appear.

6. MATHER, J. N. & R. McGEHEE. 1975. Solutions of the collinear four body problem which become unbounded in finite time. *In* Dynamical Systems, Theory and Applications. J. Moser, Ed. Springer-Verlag Lecture Notes in Physics, No. 37.

7. SILNIKOV, L. P. 1970. A contribution to the problem of the structure of an extended neighborhood of a rough equilibrium state of saddle-focus type. Math USSR Sbornik. **10:** 91–102.

8. DEVANEY, R. 1976. Homoclinic orbits in Hamiltonian systems. J. Diff. Eq. **21:** 431–438.

9. BIRKHOFF, G. D. 1917. Dynamical systems with two degrees of freedom. Trans. Am. Math. Soc. **18:** 199–300.

10. PALIS, J. & C. PUGH. 1974. Fifty problems in dynamical systems. *In* Dynamical Systems—Warwick. A. Manning, Ed. Springer-Verlag Lecture Notes in Mathematics, No. 468. 345–353.

11. DEVANEY, R. 1977. Blue sky catastrophes in reversible and Hamiltonian systems. Indiana Univ. Math. J. **26:** 247–263.

12. MEYER, K. 1970. Generic bifurcation of periodic points. Trans. Am. Math. Soc. **149:** 95–107.

13. GUCKENHEIMER, J. 1977. On bifurcation of maps of the interval. Inv. Math. **39:** 165–178.

14. UHLENBECK, K. 1976. Minimal two-spheres and tori in S^k. To appear.

15. NEUMANN, C. 1858. De problemate quondam mechanico. J. Reine Angewandte Math. **55:** 46–63.

16. CHURCHILL, R. C. & D. ROD. 1976. Pathology in dynamical systems I, II. J. Diff. Eq. **21:** 39–65, 66–112.

17. EASTON, R. W. 1976. Homoclinic phenomena in Hamiltonian systems with several degrees of freedom. To appear.

18. HENRARD, J. 1973. Proof of a conjecture of E. Strömgren. Celestial Mech. **7:** 449–457.

19. SITNIKOV, K. 1960. Existence of oscillating motions for the three body problem. Dokl. Akad. Nauk SSSR **133:** 303–306.

UPPER SEMICONTINUITY OF
TOPOLOGICAL ENTROPY

Anthony Manning

Mathematics Institute
University of Warwick
Coventry, CV4 7 AL, England

The natural setup in which to consider bifurcation theory is, so far as this paper is concerned, continuous maps of the unit interval I into $C(X, X)$, the space of continuous maps of a compact topological space X to itself. The space X is the state space and $f \subset C(X, X)$ represents a system with discrete time. (We use discrete time for simplicity; all that we say has its analog for continuous time.) If the system depends on parameters, say on p belonging to the parameter space P, we write g_p and consider the continuous map $g: P \to C(X, X)$, $p \mapsto g_p$. Work[6] on stability and genericity of bifurcations has required X to be a differentiable manifold, each g_p to be C^2, C^3, or even better and g itself to be at least C^1.

In this article we shall discuss topological entropy so no differentiability assumptions are necessary. Following the definition of Adler, Konheim, and McAndrew[1] we fix $f \in C(X, X)$ and choose an open cover $\mathfrak{U}$ of X. Then $f^{-i}\mathfrak{U}$ is again an open cover and we can consider the refinement $\mathfrak{U} \vee f^{-1}\mathfrak{U} \vee \ldots \vee f^{-n+1}\mathfrak{U}$. Define $N_n(f, \mathfrak{U})$ to be the minimum number of open sets of

$$\bigvee_{i=0}^{n-1} f^{-i}\mathfrak{U}$$

required to cover X. Then the topological entropy of f with respect to $\mathfrak{U}$, $h(f, \mathfrak{U})$, is the growth rate of $N_n(f, \mathfrak{U})$, i.e.,

$$h(f, \mathfrak{U}) = \lim_{n \to \infty} n^{-1} \log N_n(f, \mathfrak{U}).$$

In proving that this is a limit (and not just a lim sup) Adler, Konheim, and McAndrew show that $n^{-1} \log N_n(f, \mathfrak{U})$ has a weak monotonicity property, namely, for any r, $(rn)^{-1} \log N_{rn}(f, \mathfrak{U})$ decreases monotonically to the limit $h(f, \mathfrak{U})$. The topological entropy of f, $h(f)$, is defined as the limit of $h(f, \mathfrak{U})$ as the cover $\mathfrak{U}$ is made finer and finer. For this it is sufficient to consider only a confinal family of covers, e.g., if the space X is metric, covers $\mathfrak{U}_\delta$ of X by all open sets of diameter δ; then $h(f) = \lim_{\delta \to 0} h(f, \mathfrak{U}_\delta)$.

To see the significance of this definition we interpret $N_n(f, \mathfrak{U}_\delta)$ as follows: X can be covered by $N_n(f, \mathfrak{U}_\delta)$ open sets taken from

$$\bigvee_{i=0}^{n-1} f^{-i}\mathfrak{U}_\delta.$$

If x and y both belong to a typical such set, $U_{j_0} \cap f^{-1} U_{j_1} \cap \ldots \cap f^{-n+1} U_{j_{n-1}}$, then $f^i x$ and $f^i y$ belong to U_{j_i} and so the orbits of x and y display the same behavior as far as the cover $\mathfrak{U}_\delta$ is concerned. The term $N_n(f, \mathfrak{U}_\delta)$ is the number of different types of behavior the system can display from time 0 to $n - 1$; $h(f, \mathfrak{U}_\delta)$ is the growth rate of this, i.e., the richness of the system when δ is the accuracy with which we can measure difference of behavior.

118

We should remark that $h(f)$ is cruder than Kolmogorov's entropy $h_\mu(f)$ defined for an f-invariant measure μ in that we have simply counted the sets $U_{j_0} \cap \ldots \cap f^{-n+1} U_{j_{n-1}}$ rather than assign them values based on their measures. Moreover, $\mathfrak{U}$ was a cover rather than a partition. However, by the theorem of Dinaburg (extended by Goodman[2] and Goodwyn), the topological entropy $h(f)$ = $\sup_\mu h_\mu(f)$ taken over all invariant Borel measures μ.

Several people have investigated the semicontinuity of the function $h: C(X, X) \to R, f \to h(f)$. (This is quite separate from semicontinuity results for measure-theoretic entropy where it is the measure μ and not the transformation f that is allowed to vary.) Misiurewicz and Szlenk[5] studied piecewise monotone maps of the interval and found that, on the space of such maps, h is an upper semicontinuous function. Moreover, on the space of C^2 maps $f: I \to I$ if f satisfies the nondegeneracy condition that $f'(x) = 0$ implies $f''(x) \neq 0$ then h is lower semicontinuous at f. However, Misiurewicz has given examples to show that h is in general not upper semicontinuous[4] nor lower semicontinuous.[3]

THEOREM. For a fixed cover $\mathfrak{U}$ the map $h(\cdot, \mathfrak{U}): C(X, X) \to R, f \to h(f, \mathfrak{U})$ is upper semicontinuous.

Proof. Given $\epsilon > 0$ and $f \in C(X, X)$ choose n so that $n^{-1} \log N_n(f, \mathfrak{U}) < h(f, \mathfrak{U}) + \epsilon$. Then there are $N_n(f, \mathfrak{U})$ sets out of

$$V_{i=0}^{n-1} \, f^{-i} \mathfrak{U}$$

that cover X. These sets determine a Lebesgue number that describes how much they overlap. Thus f has a neighborhood V say in $C(X, X)$ so that if $g \in V$ then, for $i = 0, \ldots, n - 1$, f^i and g^i are close enough to ensure that the corresponding sets in

$$V_{i=0}^{n-1} \, g^{-i} \mathfrak{U}$$

still cover X. These might not be a minimal cover so for $g \in V$ we have $N_n(g, \mathfrak{U}) \leq N_n(f, \mathfrak{U})$. Hence

$$h(g, \mathfrak{U}) \leq n^{-1} \log N_n(g, \mathfrak{U}) \leq n^{-1} \log N_n(f, \mathfrak{U}) < h(f, \mathfrak{U}) + \epsilon,$$

which is the statement of upper semicontinuity of $h(\cdot, \mathfrak{U})$ at f as required.

In Misiurewicz's example[4] where h is not upper semicontinuous, he constructs a sequence $f_n \to f$ such that $h(f) = 0$ but $h(f_n) > \epsilon$ for all n. The entropy of f_n is found in something like a horseshoe that is transported round a solid torus by f_n. However, the size of the arms of the horseshoe decreases as n increases so that $h(\cdot, \mathfrak{U})$ is able to be upper semicontinuous for fixed $\mathfrak{U}$.

If $g: I \to C(X, X)$ is continuous then the function $t \to h(g_t, \mathfrak{U})$ is upper semicontinuous. If g is a path in the space of diffeomorphisms of a compact manifold M starting at a Morse-Smale system and leaving that region of structural stability at some time $t = b$ then Newhouse and Palis[6] have investigated the systems g_t for $t > b$ for generic arcs g and we might ask how entropy changes at such a bifurcation. In the most complicated examples they found $\Omega(g_b)$ was like $\Omega(g_t)$ for $t < b$ together with a single orbit of nontransversal intersection of stable and unstable manifolds. $\Omega(g_b | \Omega(g_b))$ is finite so $h(g_b) = 0$. Thus, for any fixed $\mathfrak{U}$, $h(g_t, \mathfrak{U}) \to 0$ as $t \to b+$. On the other hand we do not know whether

$h(g_t) \rightarrow 0$ as $t \rightarrow b+$. If g is a path in the space of diffeomorphisms that crosses at $t = b$ from one region of structural stability to another then we see from the theorem that $h(g_b, \mathcal{U})$ is at least the larger of $h(g_t, \mathcal{U})$ for $t < b$ and $t > b$. If by changing a parameter a system bifurcates from low complexity to high then the complexity of the system at the time of bifurcation (as observed by some $\mathcal{U}$) is high.

REFERENCES

1. ADLER, R., A. KONHEIM & M. MCANDREW. 1965. Topological entropy. Trans. Amer. Math. Soc. **114:** 309–319.
2. GOODMAN, T. N. T. 1971. Relating topological entropy and measure entropy. Bull. Lond. Math. Soc. **3:** 176–180.
3. MISIUREWICZ, M. 1971. On non-continuity of topological entropy. Bull. Acad. Polon. Sci. Sér. Sci Math, Astronom. Phys. **19:** 319–320.
4. MISIUREWICZ, M. 1973. Diffeomorphism without any measure with maximal entropy. Bull. Acad. Polon. Sci. Sér. Sci. Math. Astronom. Phys. **21:** 903–910.
5. MISIUREWICZ, M. & W. SZLENK. 1977. Entropy of piecewise monotone mappings. In press.
6. NEWHOUSE, S. & J. PALIS. 1973. Bifurcations of Morse-Smale dynamical systems. In Symposium on Dynamical Systems. 1971. M. M. Peixoto, Ed.: 303–366. Academic Press. New York, N.Y.

STRUCTURAL STABILITY AND
BIFURCATION THEORY

S. E. Newhouse

Department of Mathematics
University of North Carolina
Chapel Hill, North Carolina 27514

We consider a family of vector fields $\{X_\mu\}$ on a compact manifold M where μ is in some parameter space P. If he prefers, the reader may replace M by a closed disk D^n in the n-dimensional Euclidean space $\mathbb{R}^n$ and assume each X_μ is nowhere tangent to the boundary of D^n. For simplicity, we assume each X_μ is C^∞ and the mapping $(\mu, x) \to X_\mu(x)$ is C^∞.

Bifurcation theory is concerned with points μ where the orbit structure of X_μ changes. There are generally two types of analyses:

(1) Local—dealing with systems near a critical point or periodic orbit. In this case one uses implicit function theorem methods, degree methods, and so forth.

(2) Global—dealing with nontrivial recurrence, interaction between different kinds of orbits, or the relation between orbits in one region and those in a distant region.

These two categories are not meant to be exhaustive or mutually exclusive, but are merely a crude way of separating general methods in bifurcation theory. It frequently happens that a local analysis when dim $P > 1$ forces one to study nontrivial recurrence in a small region. For example, an attracting fixed point of a diffeomorphism f_{μ_0} may give rise to an attracting figure eight for f_{μ_1} with μ_1 near μ_0 and then yield homoclinic points and much nontrivial recurrence for f_{μ_2} with μ_2 near μ_1. This can happen in a generic 2-parameter family of diffeomorphisms of $\mathbb{R}^2$, and hence, in a generic two parameter family of vector fields in $\mathbb{R}^3$ as was pointed out by Takens.[20] For other examples the reader should see the paper of Holmes and Marsden in these proceedings. The complete analysis of this phenomenon has not yet been done.

For category (1) above there are many results with dim $P > 1$. We refer the reader, for example, to the papers of Hale and Scanlon in these proceedings. On the other hand, there are few results for category (2) with dim $P > 1$. Indeed, at the present time the only complete treatment of systems with nontrivial recurrence and dim $P > 1$ is in the work of Guckenheimer and Williams on Lorenz attractors.

We thank J. Scanlon for emphasizing to us at this meeting that degree methods are often used to continue periodic solutions over a large area as in the work of Alexander and Yorke[1] and the lecture of P. Rabinowitz and as such are not confined to local questions.

Let us now describe some global bifurcation theory results for families X_μ with μ in the closed unit interval $[0, 1]$.

First recall that vector fields X and Y on M are called topologically equivalent if there is a homeomorphism $h: M \to M$ carrying orbits of X onto those of Y. The

121

0077-8923/78/0316-0121 $1.75/1 © 1979, NYAS

field X is *structurally stable* if whenever Y is C^1 near X it follows that X and Y are topologically equivalent. Thus a structurally stable vector field has its whole orbit structure persistent under small C^1 changes.

There is a well-known conjecture due to Palis and Smale that the structurally stable vector fields coincide with those satisfying Axiom A and strong transversality.[14,15] Here it is not necessary to define these terms. We shall pretend the conjecture is true and thereby avoid many technicalities in the statements of the theorems below. The reader seeking precision is referred to the references at the end of this article.

Let $\Sigma = \Sigma(M)$ denote the set of structurally stable vector fields on M. The main attributes of the elements of Σ are that (1) their orbit structures are well understood[2] and (2) there are many different examples.[17]

This gives rise to the hope that elements of Σ can be used to model physical phenomena. However, relatively little has been done on this at the global level.

One basic question is the following. Suppose X_0 and X_1 are in Σ. If X_μ is a "generic" curve between X_0 and X_1, what is the orbit structure of each X_μ? This is a difficult question and at present there are only partial results.

The simplest structurally stable vector fields are those with only finitely many critical points and closed orbits. These are the Morse-Smale vector fields. Denote the set of Morse-Smale vector fields by MS.

Morse-Smale systems are obviously quite special, but there is a class of vector fields in which they are abundant. This is the class of gradient vector fields. Recall that a gradient vector field X on M is one whose solution curves form the orthogonal trajectories of the level sets of a real function f on M. More precisely, if M is in some Euclidean space $\mathbb{R}^n$, g is an inner product on $\mathbb{R}^n$, and $f: M \rightarrow \mathbb{R}$ is a C^2 real function, we may define $X = \text{grad}_g f$ by the formula

$$g(X(x), Y) = - \, df(x)(Y)$$

for every $x \in M$ and vector Y tangent to M at x. Here $df(x)$ is the derivative of f at x.

One could picture f as a "height" function. In FIGURE 1, we suppose M is a 2-sphere or torus in $\mathbb{R}^3$ with the usual dot product, and f is the height of M along the vertical direction.

Fixing g, and an integer $r \geq 2$, consider the topology of uniform C^r convergence on the space $C^r(M, \mathbb{R})$ of C^r functions from M to $\mathbb{R}$. This gives a topology on the set of gradient systems $\text{grad}_g f \in C^r(M, \mathbb{R})$.

THEOREM 1. (Palis-Smale.[12,18]) The set of structurally stable (Morse-Smale) gradient systems on M is open and dense in the space of all gradient systems on M.

While gradient systems are special, they played an important part in the development of Axiom A systems[19] which led to general structural stability theorems.[14,15] Conceptually they are still very important. For in some sense, one may think of a general structurally stable vector field as a Morse-Smale gradient system with some critical points replaced by periodic orbits or by sets with infinitely many dense orbits.

Together with Palis, we considered a "generic" curve $\{X_\mu\}$, $0 \leq \mu \leq 1$, of vector fields with X_0 in MS. Let $b = b(X_\mu) = \inf \{\mu : X_\mu \in \Sigma\}$ be the first bifurcation point of $\{X_\mu\}$. Assume $0 < b < 1$. Let $L(X_\mu)$ be the set of α and ω limit points of X_μ.

THEOREM 2.[6,7] For most $\{X_\mu\}$ with X_0 in MS and $L(X_b)$ having finitely many orbits, one can describe X_b explicitly. In many cases, one can describe X_μ for $b < \mu < b + \epsilon$ and ϵ small.

The condition that $L(X_b)$ has only finitely many orbits means that one does not create much recurrence at X_b. It may be a generic condition for arcs starting in MS. That is, there is no known open set of arcs $\{X_\mu\}$ with X_0 in MS and $L(X_b)$ having infinitely many orbits.

References 6 and 7 actually deal with diffeomorphisms, but one may translate the results to vector fields.

We define a curve $\{X_\mu\}$, $0 \le \mu \le 1$, to be *stable* if for $\{Y_\mu\}$ near $\{X_\mu\}$, there is a homeomorphism $\eta:[0, 1] \rightarrow [0, 1]$ such that for each μ there is a topological equivalence h_μ from X_μ to $Y_{\eta(\mu)}$ varying continuously with μ. If the h_μ terms exist, but do not vary continuously with μ, we say $\{X_\mu\}$ is *mildly stable*.

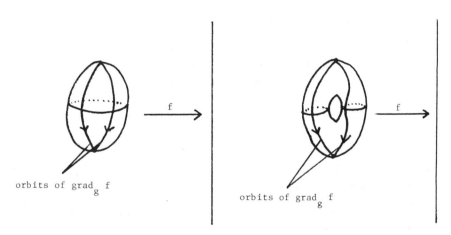

orbits of grad$_g$ f   orbits of grad$_g$ f

FIGURE 1.

Stable and mildly stable arcs have the property that the variation of orbit structure along the arc persists under small perturbation of the arc.

Let A be the set of one parameter families of vector fields $\{X_\mu\}$, $0 \le \mu \le 1$, such that $X_0 \in$ MS and $L(X_\mu)$ consists of finitely many orbits for each μ.

THEOREM 3.[8] (1) There are necessary and sufficient conditions for mild stability and stability for $\{X_\mu\}$ in A. (2) An open dense set of arcs of gradient vector fields consists of stable elements.

The second statement of Theorem 3 is the one-parameter analog of Theorem 1. It should extend to some higher dimensional families of gradient systems. However, John Guckenheimer pointed out that, in view of his paper,[3] it is unlikely that it extends to all finite dimensional familities of gradient systems.

THEOREM 4.[4,9]   Any two structurally stable vector fields on a manifold of dimension less than four can be joined by a mildly stable arc.

All of the vector fields in the arcs of Theorems 3 and 4 can be described explicitly. In References 4 & 9 it was not asserted that the arcs described were mildly stable. However, this may be proved with the methods in Reference 8.

From Theorem 4 one can give mildly stable arcs where a strange attractor is created. That is, there are open sets U in the space $\chi^\infty(M)$ of C^∞-vector fields on any manifold M, $\dim M > 2$, with the following properties. There is a C^∞ codimension one submanifold H in U so that $U - H$ consists of two open connected sets C_1 and C_2. The elements of C_1 are structurally stable with only periodic attractors, and the elements of C_2 are structurally stable with strange attractors. The elements of H are all topologically equivalent and have only periodic attractors. Thus a curve of vector fields in U that crosses H from C_1 to C_2 will have a strange attractor suddenly appear. If strange attractors are related to turbulence as Ruelle and Takens have suggested,[16] this would provide models for an abrupt transition to turbulence. This behavior has not yet been related to explicit physical systems, and it would be good to do this.

Let us describe a specific example of how the preceding behavior occurs. We do this with a curve of diffeomorphisms on $\mathbb{R}^3$. This can be carried over to vector fields on $\mathbb{R}^4$ by the usual suspension construction.[11] We shall use one dimensional attractors in $\mathbb{R}^3$. If one uses one dimensional attractors in $\mathbb{R}^2$ as described in References 10 and 13, then the vector field examples may be given in $\mathbb{R}^3$.

Let S^1 be the unit circle of all complex numbers z with modulus 1, and let D^2 be the closed unit disk of complex numbers w with $|w| \leq 1$. Define the mapping $f: S^1 \times D^2 \to S^1 \times D^2$ by

$$f(z, w) = \left(z^2, \frac{z}{2} + \frac{w}{4}\right).$$

Then $f(S^1 \times \{0\})$ is a circle in $S^1 \times D^2$ which wraps around $S^1 \times \{0\}$ twice, and $f(S^1 \times D^2)$ meets each disk $\{z\} \times D^2$ in two circles each having radius $1/4$ (FIGURE 2).

Also, $\Lambda = \bigcap_{n \geq 0} f^n(S^1 \times D^2)$ is a set which is locally the product of a Cantor set and an interval. This well-known example was first discovered by Smale.[19] We modify it through a curve of diffeomorphisms of $S^1 \times D^2$. Write $z = e^{i\theta}$ for $0 \leq \theta \leq 2\pi$. Let $\psi(\theta)$ be a C^∞ function so that $\psi(\theta) = 1$ for $|\theta| \leq \frac{1}{2}$, $\psi(\theta) = 0$ for $|\theta| \geq \frac{3}{4}$, and $|\psi(\theta)| \leq 1$ for all θ. Consider the mapping

$$\zeta(\theta) = \psi(\theta)\, 2\theta + [1 - \psi(\theta)][\theta + \tfrac{1}{2}(\theta - 1)^2]$$

and the curve of mappings $\phi_\mu(\theta) = (1 - \mu)\, 2\theta + \mu\zeta(\theta)$. Finally, let

$$f_\mu(e^{i\theta}, w) = \left(e^{i\phi_\mu(\theta)}, \frac{e^{i\theta}}{2} + \frac{w}{4}\right).$$

Then $f_0 = f$, and when $\mu = 1$ a new fixed point is introduced at $z_0 = (e^i, \frac{2}{3} e^i)$. This fixed point is a saddle node in the sense of Reference 6. For $\mu > 1$ and near 1, f_μ will have two new fixed points near z_0. One will be a sink (that is, asymptotically stable), and the other will be a saddle point. If we let $z_{1\mu}$ be the sink, then almost all points are f_μ asymptotic to $z_{1\mu}$ for $\mu > 1$, and almost all points are f_1 asymptotic to z_0. But for $\mu < 1$, almost all points are spread out uniformly along

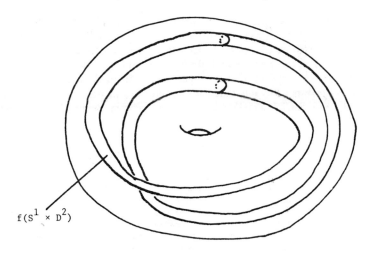

$f(S^1 \times D^2)$

$f(S^1 \times D^2) \cap \{z\} \times D^2$

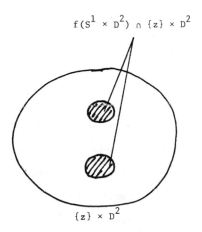

$\{z\} \times D^2$

FIGURE 2.

the strange attractor $\Lambda_\mu = \bigcap_{n \geq 0} f_\mu(S^1 \times D^2)$. The diffeomorphisms f_μ may be extended to $\mathbb{R}^3$ by standard methods.

We shall now state a result that naturally occurs with the creation of homoclinic orbits. First recall that if γ is a periodic orbit of a vector field X, an orbit γ_1 of X is homoclinic to γ if it is both forward and backward asymptotic to γ. It is well known that homoclinic points give rise to complicated and interesting motions.[19]

Denote by div X, the divergence of the vector field X.

THEOREM 6.[5] If a generic curve $\{X_\mu\}$ on M^3 creates a homoclinic orbit at $\mu = \mu_0$ and $\mathrm{div} X_{\mu_0} < 0$, then there are $\bar{\mu}$ terms near μ_0 so that X_μ has infinitely many attracting periodic orbits.

In closing, we mention a general problem. Given a vector field X, it is of interest to know the asymptotic behavior of most orbits (say with respect to Lebesgue measure). In the case of Axiom A, this is described by the Bowen-Ruelle-Sinai Theory.[2] Beyond the Axiom A case, there are few results at present. Recent numerical results on certain polynomial mappings of the plane obtained by S. Feit indicate that, except for a set of small measure, orbits are either asymptotic to attracting period orbits or to sets that behave like Axiom A attractors. The lecture by D. Ruelle at this meeting gives some idea of how to make this precise.

REFERENCES

1. ALEXANDER, J. & J. YORKE. In press. Global bifurcation of periodic orbits.
2. BOWEN, R. 1975. Equilibrium states and the Ergodic Theory of Anosov Diffeomorphisms. Springer Lecture Notes Vol 470
3. GUCKENHEIMER, J. 1973. Bifurcation and catastrophe. In Dynamical Systems, M. Peixoto, Ed.: 95–111. Academic Press. New York, N.Y.
4. NEWHOUSE, S. 1974. On simple arcs between structurally stable flows. Dyn. Sys., Warwick; 1975. Springer Lecture Notes Vol. 468: 209–234.
5. NEWHOUSE, S. 1977. The abundance of wild hyperbolic sets and non-smooth stable sets for diffeomorphisms. I.H.E.S. Preprint.
6. NEWHOUSE, S. & J. PALIS. 1973. Bifurcations of Morse-Smale dynamical systems. Dynamical Systems. M. Peixoto, Ed.: 303–366. Academic Press. New York, N.Y.
7. NEWHOUSE, S. & J. PALIS. 1976. Cycles and bifurcation theory. Astérisque 31: 41–141.
8. NEWHOUSE, S., J. PALIS & F. TAKENS. In press. Stable arcs of dynamical systems I, II.
9. NEWHOUSE, S. & M. PEIXOTO. 1976. There is a simple arc joining any two Morse-Smale flows. Astérisque 31: 15–43.
10. NEWHOUSE, S., D. RUELLE & F. TAKENS. In press. Strange attractors near constant vector fields on $T^m, m \geq 3$.
11. NITECKI, Z. 1971. Differentiable Dynamics—An Introduction to the Orbit Structure of Diffeomorphisms. MIT Press, Cambridge, Mass.
12. PALIS, J. & S. SMALE. 1970. Structural stability theorems. In Global Analysis, Proc. AMS Symp. in Pure Math. 14: 223–233.
13. PLYKIN, R. 1974. Sources and sinks of A-diffeomorphisms of surfaces. Mat Sbornik No. 24: 243–264.
14. ROBINSON, R. C. 1976. Structural stability of C^1 diffeomorphisms. J. Diff. Eq. 22: 28–73.
15. ROBINSON, R. C. 1974. Structural stability of C^1 flows. Dynamical Systems, Warwick. Springer Lecture Notes in Math. Vol. 468: 262–277.
16. RUELLE, D. & F. TAKENS. 1971. On the nature of turbulence. Commun. Math. Phys. 20: 167–192; 23: 343–344.
17. SHUB, M. & D. SULLIVAN. 1975. Homology theory and dynamical systems. Topology 14: 109–132.
18. SMALE, S. 1961. On gradient dynamical systems. Ann. Math. 74:199–205.
19. SMALE, S. 1967. Differentiable dynamical systems. Bull. Amer. Math. Soc. 73: 747–817.
20. TAKENS, F. 1974. Forced oscillations and bifurcations. Applications of Global Analysis, Comm. 3 Math. Inst. Rikjsuniversität Utrecht.

AN ANALYSIS OF IMPERFECT BIFURCATION*

M. Golubitsky†

Department of Mathematics
Courant Institute
New York, New York 10012

D. Schaeffer‡

Mathematics Research Center
University of Wisconsin-Madison
Madison, Wisconsin 53706

In this article we outline an application of the Thom-Mather singularity theory of C^∞ mappings to a problem in bifurcation theory. A detailed description of the results stated here may be found in Reference 1. Unless otherwise stated all mappings are assumed to be germs of C^∞ functions near the origin.

We view a bifurcation problem as a germ of a C^∞ map $G: \mathbb{R}^n \times \mathbb{R} \to \mathbb{R}^m$ with $G(0, 0) = 0$ where $x \in \mathbb{R}^n$ denotes the state of a system and $\lambda \in \mathbb{R}$ the load or bifurcation parameter. The associated bifurcation diagram is the set $\{G(x, \lambda) = 0\}$.

PROBLEM. Find a classification scheme for all small perturbations of G up to some suitable equivalence.

Note that this is in reality two problems. One must first find a definition for equivalent bifurcation problems and second, perform the classification. Before presenting the technical notion of equivalence, we discuss a simple example.

EXAMPLE. Let $n = m = 1$ and $G(x, \lambda) = x^3 - \lambda x$. The associated bifurcation diagram is the familiar pitchfork. See FIGURE 1. The physical model for G that we consider is a finite element analogue of the Euler beam problem illustrated in FIGURE 2. This system, consisting of two rigid rods of unit length connected by frictionless pins, is subjected to a compressive force λ, which is resisted by a torsional spring of unit strength. The state of the system is the angular derivation x from the horizontal. (We assume that a linear change in the λ coordinate has been made so that $\lambda = 0$ is the bifurcation point.) If we let $V(x, \lambda)$ denote the potential energy then, as is well known,

$$\frac{\partial V}{\partial x}(x, \lambda) = G(x, \lambda) + 0(|x, \lambda|^4).$$

Next we subject this system to two types of imperfections:

(i) An initial curvature b. Here we assume that the spring is at rest at some angular deviation from 0 (FIGURE 3i).

(ii) A central load a (FIGURE 3ii).

*This work was partially supported by the National Science Foundation.
†Present address: Department of Mathematics, Queens College, Flushing, N.Y. 11367.
‡Present address: Department of Mathematics, Duke University, Durham, N.C. 27706.

127

0077–8923/78/0316–0127 $1.75/1 © 1979, NYAS

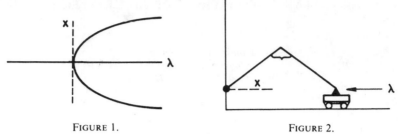

FIGURE 1.

FIGURE 2.

FIGURE 3.

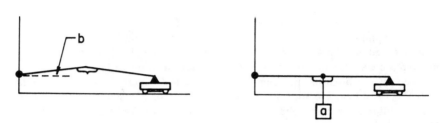

FIGURE 4.

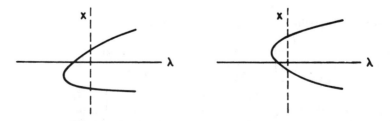

FIGURE 5.

Again it is well known that the presence of either of these imperfections separately will yield bifurcation diagrams similar to those in FIGURE 4. The question that we ask is what happens when both of the above imperfections are present? For example, if the initial curvature b is exactly offset by a central load a, then the bifurcation diagram is as in FIGURE 5. Of course these diagrams assume the exact relation "$a = -b$". Varying these imperfections yields either the diagrams in FIGURE 4 or those in FIGURE 6.

One should observe that the bifurcations diagrams in FIGURE 4 differ from those in FIGURE 6 is a physically discernible way. Moving λ quasistatically in the diagrams of FIGURE 6 can create a hysteresis cycle while in the diagrams of FIGURE 4 no such behavior is possible.

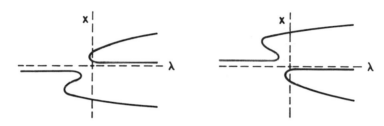

FIGURE 6.

The theorems that we shall describe state that no matter what new imperfections are added to the system no new qualitative behavior will appear.

Our technical notion of equivalence is:

DEFINITION. Two bifurcation problems G, $\overline{G}$: $\mathbb{R}^n \times \mathbb{R} \rightarrow \mathbb{R}^n$ are *contact equivalent* if

$$G(x, \lambda) = T_{x,\lambda} \cdot \overline{G}(X(x, \lambda), \Lambda(\lambda)) \tag{1}$$

where for each (x, λ) $T_{x,\lambda}$ is an invertible $m \times m$ matrix and $X(\cdot, 0)$, $\Lambda(\cdot)$ are orientation preserving changes of coordinates with $X(0, 0) = 0$ and $\Lambda(0) = 0$.

COMMENTS. (1) $\partial V / \partial x$ in the example above and $G(x, \lambda) = x^3 - \lambda x$ are contact equivalent. (2) Clearly λ is a distinguished variable in the above change of coordinates. This is demanded on physical grounds; one cannot have the applied force λ depend on the state x. (3) The bifurcation diagrams associated to G and $\overline{G}$ can be obtained from one another by a λ-preserving change of coordinates. (4) One could allow $T_{x,\lambda}$ to be a nonlinear coordinate change for each (x, λ) with $T_{x,\lambda}(0) = 0$ but no generality would be gained as proved by Mather.[2]

A standard approach to classifying small perturbations in singularity theory is given by the so called universal unfolding. We demonstrate its use by the following.

PROPOSITION A. $F(x, \lambda, p, q) = x^3 - \lambda x + px^2 + q$ is a universal unfolding of $x^3 - \lambda x$ (relative to contact equivalence); that is, let $H(x, \lambda, \epsilon) = x^3 - \lambda x + \epsilon_1 h_1(x, \lambda, \epsilon) + \cdots + \epsilon_l h_l(x, \lambda, \epsilon)$. Then for each fixed ϵ near 0, $H(\cdot, \cdot, \epsilon)$ is contact equivalent to $F(\cdot, \cdot, p, q)$ for some p and q. Moreover, the dependence of p and q on ϵ may be chosen to be smooth.

NOTE 1. It is not hard to show that the universal perturbation parameters p and q depend nonsingularly on a and b in the above example. This fact along with Proposition A proves that all qualitative behavior can be obtained by appropriate choices of a and b (up to contact equivalence).

PROPOSITION B. The bifurcation diagrams associated with $F(\cdot, \cdot, p, q)$ are

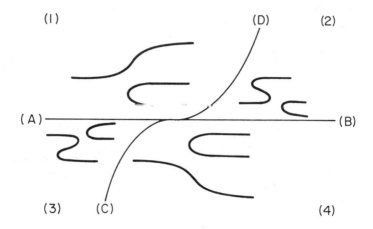

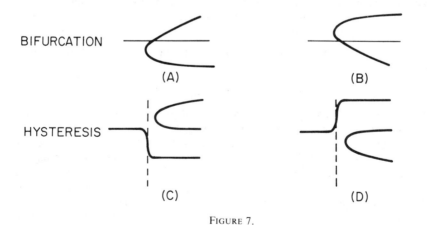

FIGURE 7.

codified (up to contact equivalence) by FIGURE 7. The separating curves are $q = 0$ and $q = p^3/27$.

NOTE 2. FIGURE 7 shows that for (p, q) near $(0, 0)$ only the diagrams associated to one of the four connected open regions or one at the four separating branches occur thus giving the desired classification of bifurcation diagrams near $x^3 - \lambda x = 0$.

NOTE 3. The diagrams associated with branches (C) and (D) are not usually considered to have bifurcation. They are not stable under small perturbation as a small perturbation can create a diagram exhibiting hysteresis. We propose the name *hysteresis point* for those points (p, q) on branches (C) and (D). The name *bifurcation point* for the points $(p, 0)$ is clear.

NOTE 4. Proposition A states that a one-parameter family of perturbations of $x^3 - \lambda x$ (ϵ is a single variable) may be represented (up to contact equivalence) by a curve through the origin in pq-space. Since the separating curves of Proposition B are tangent to second order at the origin a typical curve will enter only the regions 1 and 3. In order to observe regions 2 and 4 one must consider two-parameter perturbations. This observation seems consistent with the engineering literature and is substantiated in that both perturbations when considered separately in the model problem yield diagrams from regions 1 and 3.

We now state our main theorems, which draw heavily on standard techniques from singularity theory. Given a bifurcation problem $G: \mathbb{R}^n \times \mathbb{R} \to \mathbb{R}^m$, let $O_G = \{\bar{G} | \bar{G}$ is contact equivalent to $G\}$. Think of O_G as an infinite dimensional "submanifold" of $C^\infty(\mathbb{R}^n \times \mathbb{R}, \mathbb{R}^n)$, the space of all smooth bifurcation problems. Let TG be the tangent space to O_G at G; TG is computed as follows: choose a curve $G_t(x, \lambda)$ in O_G with $G_0 = G$. Compute $(dG_t/dt)|_{t=0}$ to obtain a function in TG; in fact, TG is the totality of such functions. For example, if $m = n = 1$, then $G_t(x, \lambda)$ in O_G implies

$$G_t(x, \lambda) = T(x, \lambda, t)\, G(X(x, \lambda, t), \Lambda(\lambda, t)), \qquad (2)$$

where $X(x, \lambda, 0) = x$, $\Lambda(\lambda, 0) = \lambda$, and $T(x, \lambda, 0) = 1$ as $G_0 = G$. Differentiating (2) with respect to t (indicated by $\dot{}$) and evaluating at $t = 0$ yields

$$\dot{G}_0(x, \lambda) = \dot{T}(x, \lambda, 0) \cdot G(x, \lambda) + \frac{\partial G}{\partial x}(x, \lambda)\, \dot{X}(x, \lambda, 0) + \frac{\partial G}{\partial \lambda}(x, \lambda)\, \dot{\Lambda}(\lambda, 0).$$

Hence

$$TG = \left\{ a(x, \lambda)\, G + b(x, \lambda)\, \frac{\partial G}{\partial x} + c(\lambda)\, \frac{\partial G}{\partial \lambda} \right\},$$

where a, b, and c are arbitrary smooth functions.

DEFINITION. A bifurcation problem G has *codimension* k if there exists a k-dimensional vector subspace K of $C^\infty(\mathbb{R}^n \times \mathbb{R}, \mathbb{R}^m)$ such that $TG \oplus K = C^\infty(\mathbb{R}^n \times \mathbb{R}, \mathbb{R}^m)$.

THEOREM A. Suppose G is a bifurcation problem with codimension k. Let $g_1(x, \lambda), \ldots, g_k(x, \lambda)$ be a basis for K. Then

$$F(x, \lambda, \alpha) = G(x, \lambda) + \alpha_1 g_1(x, \lambda) + \cdots + \alpha_k g_k(x, \lambda)$$

is a universal unfolding for $G(x, \lambda)$; that is, let

$$H(x, \lambda, \epsilon) = G(x, \lambda) + \epsilon_1 h_1(x, \lambda, \epsilon) + \cdots + \epsilon_l h_l(x, \lambda, \epsilon).$$

Then for each ϵ near 0, $H(\cdot, \cdot, \epsilon)$ is contact equivalent to $F(\cdot, \cdot, \alpha)$ for some α. Moreover the dependence of α on ϵ may be chosen to be smooth.

One needs only Taylor's Theorem to show that Theorem A implies Proposition A. For if $G = x^3 - \lambda x$, then

$$TG = \{a \cdot (x^3 - \lambda x) + b \cdot (3x^2 - \lambda) - c \cdot x\}$$
$$= \{\overline{a}(x, \lambda)x^3 + \overline{b}(x, \lambda)(3x^2 - \lambda) + \overline{c}(\lambda)x\}.$$

We claim that $K = \text{span } \{1, x^2\}$ is a complementary subspace to TG. If so, then we are done. Given an arbitrary C^∞ function $f(x, \lambda)$ we have by Taylor's Theorem

$$f(x, \lambda) = f_0(\lambda) + f_1(\lambda)x + f_2(\lambda)x^2 + f_3(x, \lambda)x^3$$
$$\equiv f_0(\lambda) + f_2(\lambda)x^2 \pmod{TG}.$$

Since $\lambda \equiv 3x^2 \pmod{TG}$ we have that $f(x, \lambda) \equiv P_0 + P_2x^2 \pmod{TG}$ where P_0 and P_2 are constants. Thus K is as claimed.

We now give a first step towards the analysis of perturbed bifurcation diagrams. Let $F(x, \lambda, \alpha)$ be a universal unfolding of $G(x, \lambda)$. As described in Note 3 the diagrams associated to points α, which are bifurcation or hysteresis points, are not stable under small perturbation. In general there is a third category of unstable points which we call double limit points (FIGURE 8).

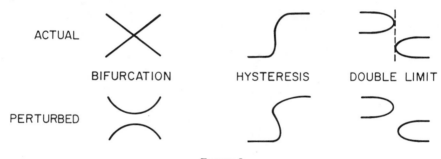

FIGURE 8.

THEOREM B. Let $\Sigma = \{\alpha \mid \alpha$ is one of the three unstable points listed above$\}$. Then any two α terms in the same connected component of the complement of Σ correspond to contact equivalent bifurcation problems.

We analyse the special case $F(x, \lambda, p, q) = x^3 - \lambda x + px^2 + q$. For $F(x_0, \lambda_0, p, q)$ to be a bifurcation point the equations $F = F_x = F_\lambda = 0$ must be satisfied at (x_0, λ_0). The condition $F_\lambda = 0$ implies that $x_0 = 0$ and $F = 0$ implies that $q = 0$. Next observe that for $F(x_0, \lambda_0, p, q)$ to be a hysteresis point the equation $F = F_x = F_{xx} = 0$ must be satisfied at (x_0, λ_0). $F_{xx} = 0$ implies that $x_0 = -p/3$ while $F_x = 0$ implies that $\lambda = -p^2/3$. Finally $F = 0$ implies that $q = p^3/27$. It is not hard to show that double limit points cannot occur in this example. To complete FIGURE 7 we must choose only one (p, q) from each open region and graph the associated bifurcation diagram.

We mention one last result.

THEOREM C. If $G: \mathbb{R}^n \times \mathbb{R} \to \mathbb{R}^m$ is a bifurcation problem of finite codimension, then G is contact equivalent to a polynomial in x and λ. The universal unfolding $F(x, \lambda, \alpha)$ may be chosen to be a polynomial in x, λ, and α. The set Σ is an algebraic variety of codimension one in α space.

REFERENCES

1.   GOLUBITSKY, M. & D. SCHAEFFER. In press. A general analysis of imperfect bifurcation via singularity theory. Commun. Pure Appl. Math.
2.   MATHER, J. 1968. Stability of C^∞ mappings, III: Finitely determined map germs. Inst. Hautes Etudes Sci. Publ. Math. **36:** 127–156.

EXTENSION AND INTERPOLATION
OF CATASTROPHES*

L. Markus

Department of Mathematics
University of Minnesota
Minneapolis, Minnesota 55455

MOTIVATIONS FOR INTERPOLATION

Let $f(x^1, x^2, \ldots, x^n)$ be a real C^∞-function for x in the real n-space $\mathbb{R}^n$, or in some other differentiable n-manifold M. That is, M is a connected, separable, metrizable n-manifold, without boundary, but with a maximal C^∞-atlas of local charts or coordinates that are typically denoted by $(x) = (x^1, \ldots, x^n)$. We consider C^∞-functions

$$f: M \to \mathbb{R},$$

often indicated by the notation $f(x)$.

A critical point $x_0 \in M$ of $f(x)$, where $df(x_0) = 0$, is nondegenerate in case the Hessian matrix $H = (H_{ij}) = (\partial^2 f / \partial x^i \partial x^j)$ is nonsingular at x_0. Near such a nondegenerate critical point x_0, the qualitative structure of the function f is precisely that of a homogeneous quadratic polynomial, namely,

$$\overline{f}(\overline{x}) = H_{ij} \overline{x}^i \overline{x}^j.$$

Here the transformed function $\overline{f}$, where

$$\overline{f}(\overline{x}) = f(x) - f(x_0),$$

is obtained by interpreting f in terms of new local coordinates $\overline{x}^i = \overline{x}^i(x^1, \ldots, x^n)$ for $i = 1, \ldots, n$, centered about x_0 in M, and upon introducing a shifted coordinate scale $\overline{y} = y - f(x_0)$ in the range space $\mathbb{R}$. In other words $f(x)$ is equivalent to $\overline{f}(\overline{x})$ near x_0, upon a right composition of $\overline{f}(\overline{x})$ with $\overline{x} = \overline{x}(x)$ and a left composition with $\overline{y}^{-1}(\overline{f}) = \overline{f} + f(x_0)$.

We call $f(x)$ a *Morse function* in M in case each of its critical points in nondegenerate. In this case $f(x)$ can be shown to have the property of local structural stability, as defined later. Essentially this means that each small perturbation of $f(x)$ to some $g(x) = f(x) + \epsilon(x)$, yields a Morse function $g(x)$ in M; and moreover the corresponding critical points of f and g are of the same qualitative structure.

Suppose that we have prescribed two Morse functions $f_0(x)$ and $f_1(x)$ on M, and we seek to interpolate a one-parameter family of real functions between these

*The first few parts of this research article were taken from papers of E. C. Zeeman,[2] and the remaining parts were taken from papers of some others, I forget who. This research was partially supported by Grant No. MCS 76-06003 A02 from the National Science Foundation, and The Lobachevsky Institute.

0077-8923/78/0316-0134 $1.75/1 © 1979, NYAS

two. That is, we seek a function-family $f(x, \alpha)$ in C^∞ for $(x, \alpha) \in M \times \mathbb{R}$ with

$$f(x, 0) \equiv f_0(x) \quad \text{and} \quad f(x, 1) \equiv f_1(x).$$

Certainly such an interpolation $f(x, \alpha)$ always exists, for instance, $f(x, \alpha) = (1 - \alpha) f_0(x) + \alpha f_1(x)$. However, if $f_0(x)$ and $f_1(x)$ possess different types of critical points, then in any interpolating function-family there must necessarily occur functions that are not Morse functions. Hence we seek some guiding principle in the construction of interpolating function-families. In order to make this interpolation problem meaningful and useful, we shall demand that the function-family $f(x, \alpha)$, treated as a whole, shall be structurally stable in the sense explained below.

STRUCTURAL STABILITY OF FUNCTION-FAMILIES

Let M be a differentiable n-manifold, called the state space, and C be a differentiable r-manifold, called the control space. Denote by $\mathcal{F}(M \times C)$ the set of all real C^∞-functions

$$f: M \times C \to \mathbb{R}.$$

As a notational device we shall often indicate a function $f \in \mathcal{F}(M \times C)$ by its typical local representative $f(x, \alpha)$ expressed in terms of a state local chart $(x^1, \ldots, x^n)$ in M and a control local chart $(\alpha^1, \ldots, \alpha^r)$ in C. Also $f(x, \alpha)$ can be considered as an r-parameter family of real functions on M.

We impose the Whitney C^∞-topology on the space $\mathcal{F}(M \times C)$. That is, a neighborhood $\mathcal{U}$ of f in $\mathcal{F}(M \times C)$ is defined by any positive integer k (an order of differentiation) and any positive function $\epsilon(x, \alpha) \in \mathcal{F}(M \times C)$ (a modulus of uniformity) so that $g \in \mathcal{U}$ means:

$$| D^k(f - g) | < \epsilon \quad \text{for all} \quad 0 \leq k \leq l$$

everywhere on $M \times C$. Here D^k indicates any partial covariant derivative of total order k in (x, α), and the pointwise norm inequalities are taken with reference to an arbitrary but fixed Riemann metric on $M \times C$. Then $\mathcal{F}(M \times C)$ is topologized as a Baire space, in fact when $M \times C$ is compact the topology arises from a complete metric on $\mathcal{F}(M \times C)$. In the Baire space $\mathcal{F}(M \times C)$ a subset $\mathcal{S}$ is generic in case $\mathcal{S}$ contains a countable intersection of open-dense subsets of $\mathcal{F}(M \times C)$. A generic set $\mathcal{S}$ is itself dense in $\mathcal{F}(M \times C)$ since its complement is contained within a subset of first category in the sense of Baire.

Definition 1. Consider function-families

$$f_1: M_1 \times C_1 \to \mathbb{R} \quad \text{and} \quad f_2: M_2 \times C_2 \to \mathbb{R},$$

or $f_1 \in \mathcal{F}(M_1 \times C_1)$, $f_2 \in \mathcal{F}(M_2 \times C_2)$ as above. We say that f_1 and f_2 are structurally equivalent, written $f_1 \sim f_2$, in case there exists a commutative diagram:

$$
\begin{array}{ccccc}
C_1 & \xleftarrow{\ \pi\ } & M_1 \times C_1 & \xrightarrow{\ f_1 \times \pi\ } & \mathbb{R} \times C_1 \\
\alpha \downarrow & & \beta \downarrow & & \gamma \downarrow . \\
C_2 & \xleftarrow{\ \pi\ } & M_2 \times C_2 & \xrightarrow{\ f_2 \times \pi\ } & \mathbb{R} \times C_2
\end{array}
$$

Here α, β, and γ are C^∞-diffeomorphisms and π is the appropriate projection map.

Remark 1. In the special case where $M_1 = M_2 = \mathbb{R}^n$ and $C_1 = C_2 = \mathbb{R}^r$ we can write the diffeomorphisms α, β, γ in terms of the global cartesian coordinates $(x^1, \ldots, x^n)$ and $(\alpha^1, \ldots, \alpha^r)$. In the usual notation the required diffeomorphisms and maps for the equivalence $f_1 \sim f_2$ can be indicated as follows:

$$\alpha : C_1 \to C_2 \qquad\qquad \text{or} \quad \overline{\alpha} = \overline{\alpha}(\alpha)$$

$$\beta : M_1 \times C_1 \to M_2 \times C_2 \quad \text{or} \quad \overline{x} = \overline{x}(x, \alpha)$$

$$\overline{\alpha} = \overline{\alpha}(\alpha)$$

$$\gamma : \mathbb{R} \times C_1 \to \mathbb{R} \times C_2 \qquad \text{or} \quad \overline{y} = \overline{y}(y, \alpha)$$

$$\overline{\alpha} = \overline{\alpha}(\alpha).$$

Then the commutativity of the diagram asserts that

$$\overline{y} f_1(x, \alpha) = f_2(\overline{x}(x, \alpha), \overline{\alpha}(\alpha))$$

or

$$f_1 = \overline{y}^{-1} \circ f_2(\overline{x}(x, \alpha), \overline{\alpha}(\alpha)).$$

Remark 2. In case distinguished points $(x_1, \alpha_1) \in M_1 \times C_1$ and $(x_2, \alpha_2) \in M_2 \times C_2$ are also to be respected, then we require that

$$\alpha : \alpha_1 \to \alpha_2 \quad \text{and} \quad \beta : (x_1, \alpha_1) \to (x_2, \alpha_2).$$

In these terms we introduce the concept of local structural equivalence of f_1 near (x_1, α_1) with f_2 near (x_2, α_2), written $f_1 \sim f_2$ near $(x_1, \alpha_1) \to (x_2, \alpha_2)$. This means that a commutative diagram, as above, exists for restrictions of f_1 and f_2 to some suitable neighborhoods of (x_1, α_1) in $M_1 \times C_1$ and (x_2, α_2) in $M_2 \times C_2$.

Remark 3. More refined types of equivalence relations can be defined by restricting the nature of the diffeomorphism γ as it acts on $\mathbb{R}$. For instance $f_1 \tilde{\oplus} f_2$ requires that γ preserve the orientation of $\mathbb{R}$, that is $\partial \overline{y} / \partial y > 0$ everywhere on $\mathbb{R} \times C$. Even further we could take γ to be a translation on $\mathbb{R}$, that is, $\overline{y} = y + \overline{y}(\alpha)$. However, in this paper, we shall use primarily the concept in Remark 2, referred to as local structural equivalence of f_1 near $(x_1, \alpha_1) \in M_1 \times C_1$ with f_2 near $(x_2, \alpha_2) \in M_2 \times C_2$.

Definition 2. A function-family $f(x, \alpha) \in \mathcal{F}(M \times C)$ is locally structurally stable, about a given point $(x_1, \alpha_1) \in M \times C$, in case for every neighborhood $N = N_x \times N_\alpha$ of (x_1, α_1) in $M \times C$ there exists a neighborhood $\mathcal{U}$ of f in $\mathcal{F}(M \times C)$, such that: for each $g(x, \alpha) \in \mathcal{U}$ there exists a point $(x_2, \alpha_2 \in N$ so that $f|N \sim g|N$ near $(x_1, \alpha_1) \to (x_2, \alpha_2)$. Further, $f(x, \alpha)$ is locally structurally stable on $M \times C$ in case f is locally structurally stable about each point of $M \times C$. We denote the collection of all such locally structurally stable function-families by $\mathfrak{F}^{TZ}(M \times C)$, and call these the Thom-Zeeman function-families on $M \times C$.

Remark 4. Let $f(x, \alpha) \in \mathfrak{F}(M \times C)$ be noncritical at the point (x_1, α_1), that is, the real function $f(x, \alpha_1)$ on M has a nonvanishing gradient at $x = x_1$. Then f is locally structurally equivalent to the linear function $g(x, \alpha) = x^1$ near $(x_1, \alpha_1) \to (0, 0)$.

Further, let $f(x, \alpha) \in \mathfrak{F}(M \times C)$ have a critical point at (x_1, α_1) which is nondegenerate, that is, $\partial f / \partial x^i = 0$ and $\det (H_{ij}) = \det (\partial^2 f / \partial x^i \partial x^j) \neq 0$ at (x_1, α_1). Then f is locally structurally equivalent to the homogeneous quadratic polynomial $g(x, \alpha) = H_{ij} x^i x^j$ near $(x_1, \alpha_1) \to (0, 0)$. In both these cases $g(x, \alpha)$ does not depend on the parameter α within the appropriate locality.

In each of these two cases f is necessarily locally structurally stable about (x_1, α_1). Thus the force of the definition, concerning local structural stability in $M \times C$, refers to the behavior of $f(x, \alpha)$ near the critical points where the Hessian matrix is singular.

Remark 5. Consider the special case where C collapses to a single point, that is $M \times C = M$. In this case $\mathfrak{F}^{TZ}(M \times C)$ consists precisely of the Morse functions on M, and so we write $\mathfrak{F}^{TZ}(M \times C) = \mathfrak{F}^M(M)$. We recall that the set $\mathfrak{F}^M(M)$ of Morse functions on M is open and dense in $\mathfrak{F}(M)$.

We shall next define a subset $\mathfrak{F}^{\pitchfork}(M \times C) \subset \mathfrak{F}(M \times C)$ by means of certain transversality conditions. These transversal function-families in $\mathfrak{F}^{\pitchfork}(M \times C)$ will coincide with the locally structurally stable function-families of $\mathfrak{F}^{TZ}(M \times C)$, at least under certain hypotheses explained later.

First we study the local analysis where $M = \mathbb{R}^n$ and $C = \mathbb{R}^r$. Consider a function-family $f \in \mathfrak{F}(\mathbb{R}^n \times \mathbb{R}^r)$ and use this to define a map

$$f^{\dagger} : \mathbb{R}^n \times \mathbb{R}^r \to m : (x, \alpha) \to f^{\dagger}_{(x,\alpha)}(\cdot),$$

where

$$f^{\dagger}_{(x,\alpha)}(\xi) = f(x + \xi, \alpha) - f(x, \alpha)$$

for ξ near the origin in $\mathbb{R}^n$. Here $f^{\dagger}_{(x,\alpha)}(\cdot)$ denotes the germ of a C^{∞}-function, an element in the ring m of all C^{∞}-germs of maps $(\mathbb{R}^n, 0) \to (\mathbb{R}, 0)$.

The group $\mathcal{G}$, consisting of all germs of C^{∞}-diffeomorphisms of $(\mathbb{R}^n, 0) \to (\mathbb{R}^n, 0)$, acts on m according to the usual transformations of the coordinate ξ, as in the right equivalence. Also $\mathcal{G}$ acts on the Lie group J^k of all k-jets of germs of m, and we shall be interested in the $\mathcal{G}$-orbit of the k-jet $f_{(x,\alpha)}{}^k \subset J^k$ of the germ $f^{\dagger}_{(x,\alpha)}(\cdot) \in m$.

Definition 3. The function-family $f(x, \alpha) \in \mathfrak{F}(\mathbb{R}^n \times \mathbb{R}^r)$ is transversal at the point (x_1, α_1) in case:

(i) the germ $f^{\dagger}_{(x_1,\alpha_1)}(\cdot)$ has a finite determinacy $\leq (r + 2)$, and

(ii) the map $\mathbb{R}^n \times \mathbb{R}^r \to J^{r+2} : (x, \alpha) \to f_{(x,\alpha)}{}^{r+2}$ is transversal at (x_1, α_1) to the $\mathcal{G}$-orbit of $f_{(x_1,\alpha_1)}{}^{r+2}$. We shall abbreviate the stipulation of these two conditions by:

$$f \pitchfork \text{ at } (x_1, \alpha_1).$$

Remark 6. It should be noted that the required transversality is always guaranteed (when $r \geq 0$) whenever $f(x, \alpha_1)$ is noncritical at $x = x_1$. Thus we need consider only the case where x_1 is a critical point of $f(x, \alpha_1)$, and here $f_{(x_1,\alpha_1)}{}^k$ lies in the jet space $I^k \subset J^k$ of all k-jets with vanishing gradients at $\xi = 0$.

For each such jet η, say determining $f_{(x_1,\alpha_1)}^{\dagger}(\cdot)$, it is known[2] that for all integers k satisfying

$$\det \eta \leq \operatorname{codim} \eta + 2 \leq k$$

the $\mathcal{G}$-orbit of η in I^k has the same geometric codimension. Since I^k has codimension n in J^k, the demanded transversality can be possible only when

$$r \geq \operatorname{codim} \eta \geq (\det \eta - 2).$$

This explains why we require that the $(r + 2)$-jet of $f_{(x_1,\alpha_1)}^{\dagger}(\cdot)$ be determining, that is, $\det f_{(x_1,\alpha_1)}^{\dagger}(\cdot) \leq r + 2$, and that the transversality condition be met in the jet space J^{r+2}.

Lemma. Consider $f(x, \alpha)$ and $g(x, \alpha)$ in $\mathcal{F}(\mathbb{R}^n \times \mathbb{R}^r)$ and assume the local structural equivalence.

$$f \sim g \text{ near } (x_1, \alpha_1) \rightarrow (x_2, \alpha_2).$$

Then $f \pitchfork$ at (x_1, α_1) if and only if $g \pitchfork$ at (x_2, α_2).

Proof. Recall that the local structural equivalence of $f(x, \alpha)$ and $g(x, \alpha)$ is based on the diffeomorphisms $\overline{\alpha} = \overline{\alpha}(\alpha)$ in $\mathbb{R}^r$, $\overline{x} = \overline{x}(x, \alpha)$ in $\mathbb{R}^n$, and $\overline{y} = \overline{y}(y, \alpha)$ in $\mathbb{R}$, with

$$f(x, \alpha) = \overline{y}^{-1} \circ g(\overline{x}(x, \alpha), \overline{\alpha}(\alpha)).$$

Since the parameter α is not a differentiation variable in computing the jets of $f_{(x,\alpha)}^{\dagger}(\cdot)$ or $g_{(\overline{x},\overline{\alpha})}^{\dagger}(\cdot)$, the dependence on α is irrelevant and we can assume that $\overline{\alpha} = \alpha$ is the identity map. In particular $\alpha_1 = \overline{\alpha}(\alpha_1) = \alpha_2$ and $x_2 = \overline{x}(x_1, \alpha_1)$.

Assume $f \pitchfork$ at (x_1, α_1) so that: (i) the germ $f_{(x_1,\alpha_1)}^{\dagger}(\cdot)$ has finite determinancy $\leq (r + 2)$, and (ii) the map $\mathbb{R}^n \times \mathbb{R}^r \rightarrow J^{r+2} : (x, \alpha) \rightarrow f_{(x,\alpha)}{}^{r+2}$ is transversal at (x_1, α_1) to the $\mathcal{G}$-orbit of $f_{(x_1,\alpha_1)}{}^{r+2}$. Now suppose that $\overline{y} = \overline{y}(y, \alpha) = y$ is the identity map on $\mathbb{R}$ for all α, and compute

$$f_{(x,\alpha)}^{\dagger}(\xi) = g(\overline{x}(x + \xi, \alpha), \alpha) - g(\overline{x}(x, \alpha), \alpha)$$

so

$$f_{(x,\alpha)}^{\dagger}(\xi) = g(\overline{x}(x, \alpha) + \beta_{x\alpha}(\xi), \alpha) - g(\overline{x}(x, \alpha), \alpha)$$

where we define

$$\beta_{x\alpha}(\xi) \equiv \overline{x}(x + \xi, \alpha) - \overline{x}(x, \alpha).$$

Then

$$f_{(x,\alpha)}^{\dagger}(\xi) = g_{(\overline{x},\overline{\alpha})}^{\dagger}(\beta_{x\alpha}(\xi)).$$

This shows that $f_{(x,\alpha)}^{\dagger}(\cdot)$ and $g_{(\overline{x},\overline{\alpha})}^{\dagger}(\cdot)$ are right equivalent under the diffeomorphism

$$\xi \rightarrow \beta_{x\alpha}(\xi) \text{ of } (\mathbb{R}^n, 0),$$

which depends smoothly on (x, α). In this case $f^\dagger_{(x_1, \alpha_1)}(\cdot)$ and $g^\dagger_{(x_2, \alpha_2)}(\cdot)$ have the same order of jet determinancy, say $\leq (r + 2)$, and also the same $\mathcal{G}$-orbits in J^{r+2}. Hence $g \pitchfork$ at (x_2, α_2) and the lemma is proved whenever $\overline{y} = y$ is the identity.

In any case $\overline{y} = \overline{y}(y, \alpha)$ and we compute

$$f^\dagger_{(x,\alpha)}(\xi) = \overline{y}_\alpha^{-1} g(\overline{x}(x, \alpha) + \beta_{x\alpha}(\xi), \alpha) - \overline{y}_\alpha^{-1} g(\overline{x}(x, \alpha), \alpha),$$

for a diffeomorphism $\overline{y}_\alpha^{-1}$ depending smoothly on the parameter α. At $\alpha = \alpha_2$ the diffeomorphism $y_{\alpha_2}^{-1}$ of $\mathbb{R}$, defined through composition on the left, merely induces a fixed linear diffeomorphism of J^{r+2}. Thus we see that $f^\dagger_{(x_1, \alpha_1)}(\cdot)$ and $g^\dagger_{(x_1, \alpha_1)}(\cdot)$ still have the same jet determinancy.

Next suppose that $\overline{y}_\alpha^{-1} = \overline{y}_{\alpha_2}^{-1}$ is a fixed diffeomorphism. Then the corresponding linear automorphism of J^{r+2} makes conjugates of the jet extensions defined by the two maps

$$(x, \alpha) \rightarrow \overline{y}_\alpha^{-1} \circ g(\overline{x}(x, \alpha) + \beta_{x\alpha}(\xi), \alpha) - \overline{y}_\alpha^{-1} \circ g(\overline{x}(x, \alpha), \alpha)$$

and

$$(x, \alpha) \rightarrow g^\dagger_{(\overline{x}, \overline{\alpha})}(\beta_{x\alpha}(\xi)).$$

But the first map is assumed to be transversal at (x_2, α_2) to the $\mathcal{G}$-orbit of the $(r+2)$-jet of

$$\overline{y}_{\alpha_2}^{-1} \circ g(x_2 + \beta_{x_2\alpha_2}(\xi), \alpha_2) - \overline{y}_{\alpha_2}^{-1} \circ g(x_2, \alpha_2).$$

Thus the second map must also be transversal at (x_2, α_2) to the $\mathcal{G}$-orbit obtained from $g^\dagger_{(x_2, \alpha_2)}(\beta_{x_2\alpha_2}(\xi))$ which is the same as the $\mathcal{G}$-orbit of $g^\dagger_{(x_2, \alpha_2)}(\xi)$. In this case $g \pitchfork$ at (x_2, α_2), as required.

But if we finally replace $\overline{y}_{\alpha_2}^{-1}$ by the variable diffeomorphism $\overline{y}_\alpha^{-1}$, with $\alpha \rightarrow \alpha_2$, then the $(r+2)$-jets for the corresponding maps will be close approximates. So even in this general case we find the required transversality for $g^\dagger_{(\overline{x}, \overline{\alpha})}(\xi)$ in J^{r+2}, and hence we conclude that $g \pitchfork$ at (x_2, α_2). □

Definition 4. A function-family $f \in \mathcal{F}(M \times C)$ is transversal, or $f \in \mathcal{F}^\pitchfork (M \times C)$, in case: at each point $(x_0, \alpha_0) \in M \times C$, and in each pair of local charts $(x^1, \ldots, x^n)$ and $(\alpha^1, \ldots, \alpha^r)$ containing (x_0, α_0), the corresponding local function-family $f(x, \alpha)$ is transversal at (x_0, α_0).

Remark 7. Since any two local representatives $f_1(x, \alpha)$ and $f_2(\overline{x}, \overline{\alpha})$ of the same function-family are locally structurally equivalent, the preceding lemma asserts that $f_1 \pitchfork$ at (x_0, α_0) if and only if $f_2 \pitchfork$ at (x_0, α_0). Hence the above defini-

---

†The author confronted E. C. Zeeman with a partial proof of this theorem and demanded that Zeeman demonstrate confidence in its validity by taking a pledge on his "scout's honor." The resulting outcome was Zeeman's proof of the theorem and the preceding lemma.

tion that $f \in \mathcal{F}^\pitchfork (M \times C)$ is meaningful. In fact, the same concept could be defined for real functions on a fiber bundle having base C and fiber M.

THEOREM (Scout's Honor†). Let M be a differentiable n-manifold and let C be a differentiable r-manifold with $0 \leq r \leq 5$. Then

$$\mathcal{F}^{TZ}(M \times C) = \mathcal{F}^\pitchfork (M \times C).$$

Proof. Assume $f(x, \alpha) \in \mathcal{F}(M \times C)$ belongs to $\mathcal{F}^\pitchfork (M \times C)$. Then, near any noncritical point or any nondegenerate critical point, the qualitative structure of $f(x, \alpha)$ is that of a linear or a quadratic function, and there $f(x, \alpha)$ must be locally structurally stable. Near a degenerate critical point (x_0, α_0) the function-family $f(x, \alpha)$ must be locally structurally equivalent to one of the finite number of elementary catastrophe types, since $r \leq 5$. This is one of the main results[2] of catastrophe theory, and the long hard proof proceeds via the classification of universal unfoldings. At any rate, the technical results of catastrophe theory guarantee that $f(x, \alpha)$ is also locally structurally stable about the degenerate critical point (x_0, α_0). Thus $f(x, \alpha) \in \mathcal{F}^{TZ}(M \times C)$.

Conversely, assume $f(x, \alpha) \in \mathcal{F}^{TZ}(M \times C)$. Suppose $f(x, \alpha)$ were not transversal about some point $(x_1, \alpha_1) \in M \times C$. For each neighborhood N of (x_1, α_1) in $M \times C$, there exists a neighborhood $\mathcal{U}$ of f in $\mathcal{F}(M \times C)$ such that: each $g \in \mathcal{U}$ admits a point $(x_2, \alpha_2) \in N$ so that $f|N \sim g|N$ near $(x_1, \alpha_1) \to (x_2, \alpha_2)$. But $\mathcal{F}^\pitchfork (M \times C)$ is known to be dense in $\mathcal{F}(M \times C)$, when $r \leq 5$.[2] Thus we can require that $g \in \mathcal{F}^\pitchfork (M \times C)$.

Since f is locally structurally equivalent to g near $(x_1, \alpha_1) \to (x_2, \alpha_2)$, the preceding lemma shows that $f \pitchfork$ at (x_1, α_1), which contradicts the above supposition. Therefore $f(x, \alpha) \in \mathcal{F}^\pitchfork (M \times C)$, as required. □

The above theorem is closely related to Thom's classification of the elementary catastrophes for function-families $f(x, \alpha)$ in $\mathcal{F}(M \times C)$. We recall part of this famous result in a version formulated by E. C. Zeeman.[1,2]

THEOREM (Thom-Zeeman). Let M be a differentiable n-manifold and C a differentiable r-manifold, for $n \geq 1$ and $0 \leq r \leq 5$. Then $\mathcal{F}^\pitchfork (M \times C)$ is open and dense in $\mathcal{F}(M \times C)$. Moreover, near any degenerate critical point of a function-family $f \in \mathcal{F}^\pitchfork (M \times C)$, f is locally structurally equivalent to one of a finite number of types called the elementary catastrophes.

For $n = 1$ and $r = 1$ there is only the fold catastrophe described by $f(x, \alpha) = x^3 + \alpha x$ near $x = 0$, $\alpha = 0$; and for $r = 2$ there is the fold and also the cusp catastrophe described by the "Thom polynomial" $f(x, \alpha) = x^4 + (\alpha^2)x^2 + (\alpha^1)x$. For $n > 1$ these Thom polynomials in $x = x^1$ must be supplemented by nondegenerate quadratic forms in the remaining $(n - 1)$ state coordinates.

INTERPOLATION AND EXTENSION IN $\mathcal{F}^{TZ}(M \times C)$

The restriction $\hat{f}$ of a Morse function f may not remain a Morse function, for instance, $f(x^1, x^2) = (x^1)^2 - (x^2)^2$ on $\mathbb{R}^2$ restricts to $\hat{f}(\hat{x}) \equiv 0$ on the line $\mathbb{R}:(x^1) = (x^2) = \hat{x}$. In this section we consider the inverse problem of extending a function-family $\hat{f}(\hat{x}, \hat{\alpha})$, given on submanifolds $\hat{M}$ and $\hat{C}$ of the state space M

and the control space C. The extension of $\hat{f} \in \mathcal{F}^{TZ}(\hat{M} \times \hat{C})$ to some $f \in \mathcal{F}^{TZ}(M \times C)$ is to maintain the property of local structural stability.

To be more specific, let M be a differentiable n-manifold, and C a differentiable r-manifold, for $n \geq 1$ and $0 \leq r \leq 5$. Let $\hat{M}$ be a submanifold of M and $\hat{C}$ a submanifold of C. We shall always assume that every submanifold is a C^∞-submanifold that is topologically embedded as a closed subset. This implies that $\hat{M}$ in M, or $\hat{C}$ in C, has a tubular neighborhood τ of positive, but varying radial thickness. To construct τ it is convenient to place a complete Riemann metric on M and then use the geodesics normal to $\hat{M}$ to define normal distances and corresponding local coordinates in τ. It is not necessary that $\hat{M}$ have a product normal bundle in M, or that τ be a topological product of $\hat{M}$ and a linear space, since all our constructions in τ will be radially symmetric and depend only on the normal distance ρ to $\hat{M}$. In any local chart in τ we shall use coordinates $(\hat{x}, \hat{y})$, where $\hat{x}$ symbolizes local coordinates on $\hat{M}$ and $\hat{y}$ symbolizes the appropriate set of normal coordinates out from $\hat{M}$, wherein $\rho^2 = \Sigma(\hat{y})^2$.

Lemma 1. Let $\hat{M}_1, \hat{M}_2, \ldots, \hat{M}_p$ be disjoint submanifolds of the differentiable n-manifold M, and let C be a differentiable r-manifold for $0 \leq r \leq 5$. Let there be given function-families $\hat{f}_i \in \mathcal{F}^{TZ}(\hat{M}_i \times C)$ for each $i = 1, 2, \ldots, p$. Then there exists an extension $f \in \mathcal{F}^{TZ}(M \times C)$ such that $f \equiv \hat{f}_i$ on each $\hat{M}_i \times C$.

Proof. We first take $p = 1$ and discuss only $\hat{M}_1 = \hat{M}$, since all constructions will be localized within disjoint tubular neighborhoods of the submanifolds $\hat{M}_1, \hat{M}_2, \ldots, \hat{M}_p$. Let τ be a tubular neighborhood of $\hat{M}$, with radial thickness greater than $t(\hat{x})$ at each point $\hat{x} \in \hat{M}$. Here we can take $t(\hat{x})$ to be a C^∞-function on $\hat{M}$, satisfying the bounds $0 < t(\hat{x}) < 1$.

Given $\hat{f}(\hat{x}, \alpha) \in \mathcal{F}^{TZ}(\hat{M} \times C)$ we first seek an extension in $\mathcal{F}^{TZ}(\tau \times C)$. For this purpose consider the function-family $\hat{f}(\hat{x}, \alpha) + \rho^2$ for $\hat{x} \in \hat{M}$ and $0 \leq \rho < t(\hat{x})$. On $\hat{M}$ where $\rho = 0$ the only critical points are just those of $\hat{f}(\hat{x}, \alpha)$, as supplemented by the nondegenerate quadratic form ρ^2 in the normal coordinates $\hat{y}$. But the only degenerate critical points of $\hat{f}$ on $\hat{M}$ are elementary catastrophes and hence locally structurally stable. Thus $\hat{f}$ and also $[\hat{f} + \rho^2]$ are transversal at each point of $\hat{M} \times C$. But for $\rho > 0$ we compute the radial derivative

$$\frac{\partial}{\partial \rho} [\hat{f}(\hat{x}, \alpha) + \rho^2] = 2\rho,$$

and so there are no other critical points within the tube τ.

Next extend $[\hat{f}(\hat{x}, \alpha) + \rho^2]$ to any differentiable function $\tilde{f}(x, \alpha) \in \mathcal{F}(M \times C)$ (perhaps after shrinking the radial thickness of the tube τ slightly). Since $r \leq 5$, $\mathcal{F}^{TZ}(M \times C)$ is open-dense in $\mathcal{F}(M \times C)$ and so we can approximate $\tilde{f}(x, \alpha)$ by some $\bar{f}(x, \alpha) \in \mathcal{F}^{TZ}(M \times C)$. Moreover, such an extension $\tilde{f}$ and subsequent approximation $\bar{f}$ can be achieved globally on $M \times C$, while simultaneously matching local data in several disjoint tubes $\tau_1, \tau_2, \ldots, \tau_p$. Hence our initial restriction to $p = 1$ causes no loss in generality in the proof. The nature of the perturbation $(\tilde{f} - \bar{f})$ in $\mathcal{F}(M \times C)$ will be specified in a moment.

Take a real-valued "blending function" $\varphi(\sigma)$ in C^∞ on $0 < \sigma \leq 1$, monotonically nonincreasing from $\varphi(\sigma) \equiv 1$ for $0 \leq \sigma \leq \frac{1}{3}$ to $\varphi(\sigma) \equiv 0$ for $\frac{2}{3} \leq \sigma \leq 1$.

Now define the required function-family

$$f(x, \alpha) = \varphi\left(\frac{\rho}{t(x)}\right)\tilde{f}(x, \alpha) + \left[1 - \varphi\left(\frac{\rho}{t(x)}\right)\right]\bar{f}(x, \alpha)$$

for $0 \leq \rho < t(x)$ in the tube τ, and

$$f(x, \alpha) = \bar{f}(x, \alpha) \text{ elsewhere on } M \times C.$$

Clearly $f(x, \alpha) \in \mathfrak{F}(M \times C)$ and $f(\hat{x}, \alpha) \equiv \hat{f}(\hat{x}, \alpha)$ on $\hat{M} \times C$. In fact, $f \equiv \tilde{f}$ for $0 \leq \rho \leq t/3$ and $f \equiv \bar{f}$ for $\rho \geq 2t/3$ and outside τ. Thus $f(x, \alpha)$ is locally structurally stable everywhere on $M \times C$ with the possible exception of points within the shell $t/3 \leq \rho \leq 2t/3$ in τ.

In order to examine $f(x, \alpha)$ in the shell $t/3 \leq \rho \leq 2t/3$ we compute the radial derivative

$$\frac{\partial f}{\partial \rho} = \varphi\left(\frac{\rho}{t}\right)\frac{\partial \tilde{f}}{\partial \rho} + \left[1 - \varphi\left(\frac{\rho}{t}\right)\right]\frac{\partial \bar{f}}{\partial \rho}$$

$$+ \varphi'\left(\frac{\rho}{t}\right) \cdot \frac{1}{t} \cdot \tilde{f} - \varphi'\left(\frac{\rho}{t}\right) \cdot \frac{1}{t} \cdot \bar{f}.$$

Thus

$$\frac{\partial \bar{f}}{\partial \rho} > \min\left(\frac{\partial \tilde{f}}{\partial \rho}, \frac{\partial \bar{f}}{\partial \rho}\right) - \left|\varphi'\left(\frac{\rho}{t}\right)\right| \cdot \frac{1}{t} \cdot |\tilde{f} - \bar{f}|.$$

Now we specify the perturbation $(\tilde{f} - \bar{f})$ in $M \times C$ by demanding that

$$|\tilde{f} - \bar{f}| < \frac{1}{2} \rho \frac{t(x)}{\max |\varphi'|}$$

and

$$\left|\frac{\partial \tilde{f}}{\partial \rho} - \frac{\partial \bar{f}}{\partial \rho}\right| < \rho \quad \text{in the shell } \frac{t}{3} \leq \rho \leq 2t/3.$$

In this case

$$\frac{\partial f}{\partial \rho} > \rho - \frac{1}{2} \rho = \frac{1}{2} \rho > 0,$$

in the shell, for all $\alpha \in C$, and so $f(x, \alpha)$ is structurally stable about each point of $M \times C$. Thus $f(x, \alpha) \in \mathfrak{F}^{TZ}(M \times C)$, as required. □

Lemma 2. Let M be a differentiable n-manifold and let $\hat{C}_1, \hat{C}_2, \ldots, \hat{C}_q$ be disjoint submanifolds of a differentiable r-manifold C, for $0 \leq r \leq 5$. Let there be given function-families $\hat{f}_j(x, \hat{\alpha}_j) \in \mathfrak{F}^{TZ}(M \times \hat{C}_j)$ for each $j = 1, 2, \ldots, q$. Then there exists an extension $f(x, \alpha) \in \mathfrak{F}^{TZ}(M \times C)$ such that $f \equiv \hat{f}_j$ on each $M \times \hat{C}_j$.

Proof. As in Lemma 1 we shall first take $q = 1$ and discuss only the case $\hat{C}_1 = \hat{C}$, since all constructions will be localized within disjoint tubular neighborhoods of the submanifolds $\hat{C}_1, \hat{C}_2, \ldots, \hat{C}_q$ of C. Let τ be a tubular neighborhood of $\hat{C}$, having radial thickness greater than $t(\hat{\alpha})$ at each point $\hat{\alpha}$ of $\hat{C}$. Here we take $t(\hat{\alpha})$ to be a C^∞-function on $\hat{C}$, satisfying the bounds $0 < t(\hat{\alpha}) < 1$. We designate a typical chart of local coordinates in τ by $(\hat{\alpha}, \hat{\beta}) = \alpha$, with $\hat{\alpha}$ on $\hat{C}$ and $\hat{\beta}$ indicating the coordinates normal to $\hat{C}$ in τ, so the radial distance from $\hat{C}$ is $\rho = |\Sigma(\hat{\beta})^2|^{1/2}$.

Given the function-family $\hat{f}_1(x, \hat{\alpha}) = \hat{f}(x, \hat{\alpha})$ in $\mathcal{F}^{TZ}(M \times \hat{C})$ we trivially extent $\hat{f}$ into $M \times \tau$ by $\hat{f}(x, \hat{\alpha}, \hat{\beta}) \equiv \hat{f}(x, \hat{\alpha})$. By general differential topology we further extend $\hat{f}(x, \alpha)$ to some global function-family $\tilde{f}(x, \alpha) \in \mathcal{F}(M \times C)$ so that $\tilde{f} = \hat{f}$ in $M \times \tau$ (possibly after shrinking the radial thickness of τ). Since $\mathcal{F}^{TZ}(M \times C)$ is open dense in $\mathcal{F}(M \times C)$, because $r \leq 5$, we can approximate $\tilde{f}(x, \alpha)$ by some $\overline{f}(x, \alpha) \in \mathcal{F}^{TZ}(M \times C)$. The precise nature of the approximation of $\tilde{f}$ to $\hat{f}$ in $\mathcal{F}(M \times C)$ will be explained later.

Note that such a global extension $\tilde{f}$, and subsequent approximation $\overline{f}$, can be made to match appropriately the local data in each of the disjoint tubes $\tau_1, \ldots, \tau_q$. Hence our initial assumption that $q = 1$ does not restrict the generality of the proof.

Next we proceed to blend $\tilde{f}$ into $\overline{f}$ within the tube τ, concentrating the changes within the shell $t(\hat{\alpha})/3 \leq \rho \leq 2t(\hat{\alpha})/3$. The validity of this process will depend on the smallness of the perturbation $(\tilde{f} - \overline{f})$ as explained later. Take any real-valued "blending function" $\varphi(\sigma)$ in C^∞ on $0 \leq \sigma \leq 1$, monotonically nonincreasing from $\varphi(\sigma) \equiv 1$ and $0 \leq \sigma \leq \frac{1}{3}$ to $\varphi(\sigma) \equiv 0$ on $\frac{2}{3} \leq \sigma \leq 1$. Now define the desired function-family

$$f(x, \alpha) = \hat{f}(x, \alpha) \text{ on } 0 \leq \rho \leq t(\hat{\alpha})/3$$

$$f(x, \alpha) = \varphi\left(\frac{\rho}{t}\right) \hat{f}(x, \alpha) + \left[1 - \varphi\left(\frac{\rho}{t}\right)\right] \overline{f}(x, \alpha)$$

in the shell $t(\hat{\alpha})/3 \leq \rho \leq 2t(\hat{\alpha})/3$

$$f(x, \alpha) = \overline{f}(x, \alpha) \text{ on } 2t/3 \leq \rho \text{ and outside the tube } \tau.$$

Clearly $f(x, \alpha) \in \mathcal{F}(M \times C)$ and also $f = \hat{f}$ on $M \times \hat{C}$.

As the last and crucial step we must prove that $f(x, \alpha) \in \mathcal{F}^{TZ}(M \times C)$. In this argument we use the characterization that $\mathcal{F}^{TZ}(M \times C) = \mathcal{F}^\pitchfork(M \times C)$, and we verify that $f(x, \alpha)$ is locally transversal at each point $(x_0, \alpha_0) \in M \times C$. That is, at each (x_0, α_0) we must verify that (i) the germ $f^\dagger_{(x_0, \alpha_0)}(\xi) = f(x_0 + \xi, \alpha_0) - f(x_0, \alpha_0)$ is $(r + 2)$-determinate and (ii) the map $(x, \alpha) \to f_{(x,\alpha)}^{r+2}$ is transversal at (x_0, α_0) to the orbit of $f_{(x_0,\alpha_0)}^{r+2}$ in the jet-space J^{r+2}. For $0 \leq \rho \leq t(\hat{\alpha})/3$ within τ, we note that $f = \hat{f}$ and so the local transversality conditions are satisfied. The same holds in $2t/3 \leq \rho$ and outside τ where $f = \overline{f}$. Thus the only verification required is at points $(x_0, \alpha_0) = (x_0, \hat{\alpha}_0, \hat{\beta}_0)$ that lie within the shell $t(\hat{\alpha})/3 \leq \rho \leq 2t(\hat{\alpha})/3$ with $x \in M$.

To assume the required transversality condition at points of the shell, we restrict the size of the perturbation $(\tilde{f} - \overline{f})$ in $\mathcal{F}(M \times C)$. To make this procedure precise take an open neighborhood $\mathcal{U}$ of $\hat{f}(x, \hat{\alpha})$ in $\mathcal{F}^{TZ}(M \times \hat{C})$, say as prescribed by some inequalities: $g \in \mathcal{U}$ in case,

$$| D^k(\hat{f} - g)| < \epsilon(x, \hat{\alpha})$$

for $k = 0, 1, \ldots, l$ and all $(x, \hat{\alpha}) \in M \times \hat{C}$. Thus, whenever $(\tilde{f} - \bar{f})$ is sufficiently small in $\mathfrak{F}(M \times C)$, we conclude that $\bar{f}(x, \hat{\alpha}, \hat{\beta}_0)$ lies in $\mathfrak{U}$, for each fixed $\hat{\beta}_0$, while $(\hat{\alpha}, \hat{\beta}_0) \in \tau$. But within the specified shell of the tube τ we note that the x-derivatives of f are convex combinations of the corresponding derivatives of $\hat{f}$ and $\bar{f}$. Therefore

$$f(x, \hat{\alpha}, \hat{\beta}_0) \in \mathfrak{U} \subset \mathfrak{F}^{TZ}(M \times \hat{C})$$

for each fixed $\hat{\beta}_0$. Hence $f(x, \hat{\alpha}, \hat{\beta}_0)$ is locally transversal at $(x_0, \hat{\alpha}_0) \in M \times \hat{C}$, for each fixed $\hat{\beta}_0$. Then, a fortiori, $f(x, \hat{\alpha}, \hat{\beta})$ is locally transversal at $(x_0, \hat{\alpha}_0, \hat{\beta}_0)$ for $x_0 \in M$ and $(\hat{\alpha}_0, \hat{\beta}_0)$ within the shell in τ. Thus $f(x, \alpha) \pitchfork (x_0, \alpha_0)$ for all points $(x_0, \alpha_0) \in M \times C$, and so $f(x, \alpha) \in \mathfrak{F}^{TZ}(M \times C)$ as required. □

An important instance of the construction of Lemma 2 arises when C is the line $\mathbb{R}$ (or the circle S^1) and $\hat{C}_1$ and $\hat{C}_2$ are two distinct points of C. Then $\hat{f}_1(x)$ and $\hat{f}_2(x)$ are Morse functions on the n-manifold M. The extension $f(x, \alpha) \in \mathfrak{F}^{TZ}(M \times C)$ then provides an interpolation between these two Morse functions. Since the dimension of C is $r = 1$, only fold catastrophes occur in the function-family $f(x, \alpha)$, regardless of the dimension n of M. This result then answers the interpolation problem posed in the first section of this paper.

The next theorem asserts the existence of the solution to a general extension problem. The proof follows immediately from the above two lemmas.

THEOREM 1. Let M be a differentiable n-manifold with disjoint submanifolds $\hat{M}_1, \ldots, \hat{M}_p$; and let C be a differentiable r-manifold, for $0 \leq r \leq 5$, with disjoint submanifolds $\hat{C}_1, \ldots, \hat{C}_q$. Let there be given function-families $\hat{f}_{ij}(\hat{x}_i, \hat{\alpha}_j) \in \mathfrak{F}^{TZ}(\hat{M}_i \times \hat{C}_j)$ for each $i = 1, \ldots, p$ and $j = 1, \ldots, q$. Then there exists an extension $f(x, \alpha) \in \mathfrak{F}^{TZ}(M \times C)$ such that $f \equiv \hat{f}_{ij}$ on each $\hat{M}_i \times \hat{C}_j$.

SPECIAL EXTENSION PROBLEMS

We consider various special extension problems in $\mathfrak{F}^{TZ}(M \times C)$, particularly for zero and one dimensional manifolds $\hat{C}$ of C.

THEOREM 2. Let M be a differentiable n-manifold and C a differentiable r-manifold, for $0 \leq r \leq 5$. Let $\hat{C}_1, \ldots, \hat{C}_q$ be distinct submanifolds of C, each with dimension zero (each $\hat{C}_j$ a point), and assign corresponding function-families $\hat{f}_j(x, \hat{\alpha}_j) \in \mathfrak{F}^{TZ}(M \times \hat{C}_j)$ for $j = 1, \ldots, q$ (each $\hat{f}_j$ a Morse function on M). Then there exists an extension $f(x, \alpha) \in \mathfrak{F}^{TZ}(M \times C)$ of $f \equiv \hat{f}_j$ on $\hat{C}_j$, with the property that f possesses only fold catastrophes. If some $\hat{C}_j$ are 1-manifolds (while the others are points), then we can require that the extension f possess only fold and cusp catastrophes.

Proof. In the first case, when $\hat{C}_j$ are all points and each $\hat{f}_j$ is a Morse function on M, we shall interpolate between each $\hat{f}_j$ and an arbitrarily selected global Morse function $F(x)$ on M. Take disjoint ball neighborhoods $\tau_j \subset C$ around the points $\hat{C}_j$, and we indicate the construction within τ_1 near $\hat{C}_1$. For simplicity write $\hat{f}_1 = \hat{f}$ and $\tau_1 = \tau, \hat{C}_1 = \hat{C}$.

Just as in the proof of Theorem 1, make a differentiable extension $\tilde{f}(x, \alpha)$ by interpolating, along each radial ray in τ, between $\hat{f}$ and F. Then approximate $\tilde{f}$ by $\bar{f}$ along each individual ray leading outwards from $\hat{C}$ in τ, and moreover construct $\bar{f}$ so that it is transversal at each point of the ray, as developed along that ray. Since the function-family $\bar{f}$ along each radial coordinate in τ depends on only one control parameter, we note that along each ray $\bar{f}$ can possess only fold catastrophes. Note that the construction of $\bar{f}$ should be made radially symmetric in τ, with the same behavior outwards along each individual ray, and so $\bar{f}$ is a differentiable function-family in $\mathcal{F}(M \times C)$. Finally, as in Theorem 1, blend $\bar{f}$ into $\hat{f}$ near the center of τ, and into F near the boundary of τ, to obtain $f \in \mathcal{F}^{TZ}(M \times C)$. Then f is radially symmetric in τ and so has only fold catastrophes, although such catastrophe points can form $(r-1)$-spheres about the points $\hat{C}_j$ in C.

In the second case let $\hat{C}_1 = \hat{C}$ be a 1-manifold, either a diffeomorph of a line or a circle. We consider the case where $\hat{C}$ is a circle S^1; the case of a line $\mathbb{R}$ is similar but simpler. Then take a tubular neighborhood τ centered on $\hat{C}$ so τ is either a topological product $S^1 \times \mathbb{R}^{r-1}$ or the nonorientable fiber bundle over S^1. We seek to extend the given $\hat{f} \in \mathcal{F}^{TZ}(M \times \hat{C})$ across the tube τ so as to blend into a prescribed Morse function $F(x)$ on M.

Assume temporarily that $r = 2$ and construct $\tilde{f}$, then $\bar{f}$, and finally $f \in \mathcal{F}^{TZ}(M \times C)$ by interpolating outwards along each ray normal to $\hat{C}$ in the cylindrical or Möbius band τ. While $\hat{f}$ has only fold catastrophes on $\hat{C}$, the extention f depends on two control parameters in τ and so f possesses only fold and cusp catastrophes in τ.

For the general case where $r \geq 2$, we make the same construction for f within each cylindrical or Möbius 2-band through $\hat{C}$ in τ. Then observe that the resulting function-family f is radially symmetric around $\hat{C}$ in τ and possesses only fold and cusp catastrophes in τ, but such catastrophe points may fill $(r-1)$-manifolds near $\hat{C}$, in fact, topological products $S^1 \times S^{r-2}$ (or a nonorientable fiber bundle over S^1 when $r = 3$ or 4). $\square$

The first nontrivial problem on global extensions arises when $\hat{C}$ is a diffeomorph of a circle S^1 in C. Then $\hat{f} \in \mathcal{F}^{TZ}(M \times \hat{C})$ has only fold catastrophes, and there exists an extension $f \in \mathcal{F}^{TZ}(M \times C)$ possessing only fold and cusp catastrophes. We pose the problem: *Under what circumstances does there exist an extension $f \in \mathcal{F}^{TZ}(M \times C)$ with no cusp catastrophes?*

If $C = \hat{C} \times S^1 = S^1 \times S^1$ is a torus T^2, then it is clear that there always exists such an extension $f \in \mathcal{F}^{TZ}(M \times T^2)$ that has no cusps. This is easy to see if we take $\hat{C}$ to be a meridan circle on T^2 and then use an extension that is invariant along each longitude circle. [The same type of argument shows that if $\hat{C}$ is a 2-manifold and $C = \hat{C} \times S^1$, then only folds and cusps need occur for some extension $f \in \mathcal{F}^{TZ}(M \times C)$].

Remark 8. We shall assume that $\hat{C}$ is a diffeomorph of a circle S^1, which bounds a smooth disk D on the 2-surface C. Thus, for all practical purposes, provided we are mainly interested in the compact closure $\bar{D}$, we can take C to be compact.

But it is also sensible to assume that M is compact, or at least require that the set of critical points of f be compact in $M \times C$. For otherwise the possible cusp points of f could be "pulled out to infinity on M," and thus excised from M. For instance, consider the standard cusp catastrophe given by $f(x, \alpha^1, \alpha^2) = x^4 + (\alpha^2)^2 x^2 + (\alpha^1)x$ for $x \in M = \mathbb{R}$ and $\alpha = (\alpha^1, \alpha^2) \in C = \mathbb{R}^2$. Then the cusp at $x = 0$, $\alpha = 0$ can be excised by deleting the closed subset of $M \times C$

defined by some spike such as $x \geq [(\alpha^1)^2 + (\alpha^2)^2] + [(\alpha^1)^2 + (\alpha^2)^2]^{1/9}$, and then introducing new coordinates $\bar{x} = -\log[\,|\,\alpha\,|^2 + (\,|\,\alpha\,|^2)^{1/9} - x]$ on each remaining vertical fiber, each of which is still diffeomorphic to $\mathbb{R}$.

Hence we shall usually assume that $M \times C$ is compact, although preliminary local analysis may sometimes ignore this assumption.

Remark 9. As a topological analytical tool we shall examine the critical manifolds M_f of a function-family $f \in \mathscr{F}^{TZ}(M \times C)$. The set of all critical points, where $\partial f/\partial x^i = 0$, is a closed subset of the space $M \times C$, and each component is known to be a submanifold M_f of dimension r. The catastrophe points on M_f, or their images on C, are precisely those points where the projection map $\pi\colon M_f \to C$ has rank less than r.

If M is not compact, then the critical set can have countably many components in $M \times C$. For instance, consider the Morse function $f(x) = \sin x$ on $M = \mathbb{R}$. Also in such a case the catastrophe points on C may be dense. For instance, take $M = \mathbb{R}$ and $C = S^1$ and use the function-family with fold points on each critical manifold M_f (see FIGURE 1), with the countably many components of the critical set each similar manifolds but each raised one unit along the line $M = \mathbb{R}$ and each turned through an irrational rotation above the circle $C = S^1$.

In order to avoid such complications we shall usually assume that $M \times C$ is compact, so that there are only a finite number of compact critical manifolds for each $f \in \mathscr{F}^{TZ}(M \times C)$. In particular, if C is a circle S^1, then each critical manifold M_f is also a compact 1-manifold and must be a circle in $M \times S^1$. Then the only catastrophe points on $M_f = S^1$ are the turning points where $M_f = S^1$ has a singular projection into $C = S^1$.

We analyse a few geometric configurations that illustrate various types of extensions when $M = S^1$ (or the covering space $\mathbb{R}$), and $\hat{C} = S^1$ which bounds a smooth disk D in a 2-surface C.

Case 1A. Take $M = \mathbb{R}$, $\hat{C} = S^1$ and $\hat{f} \in \mathscr{F}^{TZ}(\mathbb{R} \times S^1)$ as indicated in FIGURE 1A. Along each vertical fiber, for $x \in \mathbb{R}$, the function $\hat{f}(x, \hat{\alpha}_0)$ has a positive derivative—except within the "lips" bounded by the critical manifold $M_{\hat{f}}$. There are only two fold points for $\hat{f}$ on $M_{\hat{f}}$, and $M_{\hat{f}}$ is a topological circle that projects down into a proper subset of the circle $\hat{C}$ (so the projection map $M_{\hat{f}} \to \hat{C}$ has degree zero).

Here $\hat{f}$ admits an extension $f \in \mathscr{F}^{TZ}(\mathbb{R} \times \tilde{D})$ which has only fold catastrophes (where $\tilde{D}$ is a neighborhood of the compact closure $\bar{D}$). To define f we first construct the critical manifold M_f, which consists of a "tongue stuck between the lips," as sketched in the second picture in FIGURE 1A. The set of catastrophe points on the tongue M_f projects to an arc in $\bar{D}$, consisting of fold points and having both its ends on the boundary $\hat{C}$. Then the construction of f is easily completed so that f has no other critical points in $\mathbb{R} \times \bar{D}$ except for the tongue M_f.

If $\hat{f}$ is periodic in x along $\mathbb{R}$, so we can take $M = S^1$ and $\hat{f} \in \mathscr{F}^{TZ}(S^1 \times S^1)$, then there must be other critical points in $M \times \hat{C}$, but we shall take all of these others to be nondegenerate critical points. Then the construction of $f \in \mathscr{F}^{TZ}(S^1 \times \tilde{D})$, having only folds, is still valid. In fact, if we double the disk $\bar{D}$ to form a sphere S^2 with identifications along the equator $\hat{C} = S^1$, then we can find the required $f \in \mathscr{F}^{TZ}(S^1 \times S^2)$.

Case 1B. The configurations for $M_{\hat{f}}$ in FIGURE 1B illustrate situations analogous to that of case 1A. In both of these configurations there is an extension $f \in \mathcal{F}^{TZ}(M \times \tilde{D})$ with only fold catastrophes. It is interesting to note that the projection map $M_f \rightarrow \hat{C}$ has degrees zero and one, in the two situations pictured here.

Case 2A. In FIGURE 2A we encounter a new phenomenon. Here $M = \mathbb{R}$ and $\hat{C} = S^1$, as before. Also $M_{\hat{f}}$ is a topological circle that projects onto $\hat{C}$ with a map of degree one, and there exist just two fold points for the given $\hat{f} \in \mathcal{F}^{TZ}(\mathbb{R} \times S^1)$. But we shall show, under certain hypotheses, that every extension

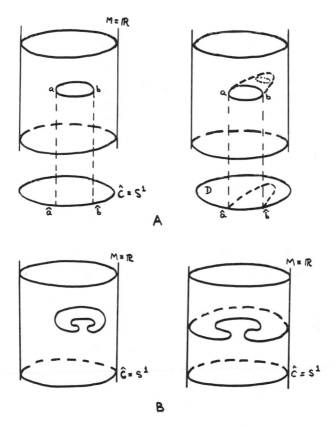

FIGURE 1. Extensions with only folds.

$f \in \mathcal{F}^{TZ}(\mathbb{R} \times D)$ must possess at least one cusp catastrophe. As an instance of the restrictive hypotheses, we assume that f has a compact critical set in $\mathbb{R} \times \overline{D}$ (or else we take $M = S^1$).

Take any appropriate extension $f \in \mathcal{F}^{TZ}(\mathbb{R} \times \tilde{D})$ and we shall usually examine its restriction to $\mathbb{R} \times \overline{D}$ wherein the critical set is compact. Consider the critical manifold M_f passing through the curve $M_{\hat{f}}$, so M_f is a compact surface with the boundary $M_{\hat{f}}$.

Examine the curve M_f over the circle $\hat{C}$. The two fold points a and b on M_f project to points $\hat{a}$ and $\hat{b}$ on $\hat{C}$. In detail, the curve M_f provides a 3-fold covering of the arc $\hat{a}\hat{b}$, and we refer to the upper, middle, and lower arcs of M_f over $\hat{a}\hat{b}$.

The set of catastrophe points of f in $\mathbb{R} \times \bar{D}$ is compact, and we consider the nonempty component Σ through the fold point $a \in M_f$. We suppose that f has only fold catastrophes in $\mathbb{R} \times \bar{D}$. Then the geometry of such folds shows that Σ is

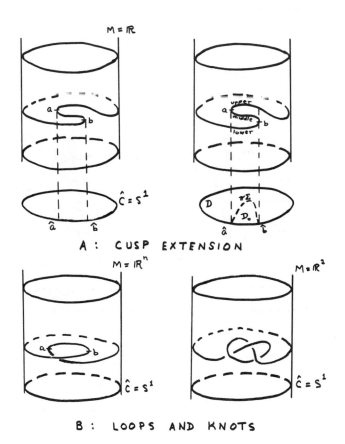

A : CUSP EXTENSION

B : LOOPS AND KNOTS

FIGURE 2. Extensions with cusps.

topologically a simple arc, a homeomorph of a compact interval. One end of Σ starts at the fold point a, and the other ends at the fold point $b \in M_f$. Both ends of Σ must lie on the boundary curve M_f, as is seen from standard arguments about limit-sets.

Now we make the assumption Σ contains all the catastrophes of f in $\mathbb{R} \times \bar{D}$ (that is, the catastrophe set is connected), and furthermore that the projection $\pi\Sigma$ is a simple arc in $\bar{D}$ joining $\hat{a}$ to $\hat{b}$. We believe that these assumptions can be greatly relaxed, or even eliminated, by introducing more elaborate topological machinery.

Under the above assumptions, the arc $\pi\Sigma$ cuts off a simply-connected region D_0 in the disk D. We define the sheets of M_f above D_0 to be the components of the noncatastrophe points of f in $M_f \cap (\mathbb{R} \times D_0)$. By monodromy arguments, each sheet is a simple covering of D_0 and has its boundary contained in $\Sigma \cap M_f$. In more detail, there are three sheets of M_f above D_0, called the upper, middle, and lower, in accord with the boundary covering arcs of M_f over $\hat{a}\hat{b}$.

Now observe that the curve Σ on M_f has two banks, at least locally near the endpoint a where the upper and middle sheets meet. But the banks of Σ near the point b yield a meeting of the middle and lower sheets of M_f. Because each sheet of M_f above D_0 is simply-connected, the same two sheets must constitute the banks of Σ everywhere along Σ. This contradiction shows that Σ cannot be a compact arc consisting only of fold points. Therefore f must have a cusp point in $\mathbb{R} \times \bar{D}$, when we admit the above hypotheses.

Remark 10. If $M = \mathbb{R}^n$, for $n \geq 2$, and $\hat{C} = S^1$, then the above argument based on upper and lower sheets is inapplicable. If fact, the curve M_f in $\mathbb{R}^n \times S^1$ can be deformed from a wave-shape (as in FIGURE 2A) to a loop-shape (as in FIGURE 2B), without modifying the problem significantly.

Moreover, for $n = 2$ the additional complication of knots in M_f makes this case especially intractible.

REFERENCES

1. THOM, R. 1972. Stabilité structurelle et morphogénese. Benjamin. New York.
2. TROTMAN, D. & E. C. ZEEMAN. 1974. The classification of elementary catastrophes of codimension ≤ 5. Univ. Warwick Notes; 1976. Lecture Notes in Math. **525**. Springer-Verlag. New York, N.Y.
3. ZEEMAN, E. C. Private communications.

FACTORIZATION THEOREMS AND REPEATED BRANCHING OF SOLUTIONS AT A SIMPLE EIGENVALUE*

D. D. Joseph

Department of Aerospace Engineering and Mechanics
University of Minnesota
Minneapolis, Minnesota 55455

STABILITY AND BIFURCATION IN $\mathbb{R}_1$†

In this paper I prove factorization theorems which show that, under certain typical hypotheses, the stability of steady and time-periodic solutions can change only at a turning point or at a point of bifurcation.

It is best to start in $\mathbb{R}_1$ where the hypotheses require C_2 smoothness plus a transversality (strict crossing) condition of Hopf's type. I show that these hypotheses imply that points of bifurcation are just double points of plane curves, that other types of bifurcation (double points with higher-order contact, triple points, and higher multiple points) are arbitrarily excluded. The same conclusions hold for general problems in Banach space when the eigenvalue of the Fréchet derivative on the solution branch is algebraically simple. The conclusion applies to symmetry-breaking steady bifurcation of steady solutions of partial differential equations of evolution type and to symmetry breaking τ-periodic bifurcations of τ-periodic solutions of partial differential equations of evolution type.

We consider an evolution equation in $\mathbb{R}_1$ of the form

$$u_t + F(\mu, u) = 0 \tag{1}$$

where $F(\cdot, \cdot)$ has two continuous derivatives in $\mathbb{R}_1 \times \mathbb{R}_1$. It is conventional in the study of stability of bifurcation to arrange things so that

$$F(\mu, 0) = 0 \quad \text{for all} \quad \mu \in \mathbb{R}_1. \tag{2}$$

But we shall not require (2). Instead we require that equilibrium solutions of (1) have $u = \epsilon, \epsilon_t = 0$, and

$$F(\mu, \epsilon) = 0. \tag{3}$$

The study of bifurcation of equilibrium solutions of the autonomous problem (1) is equivalent to the study of singular points of the plane curves (3).

In our study of equilibrium solutions (3) it is desirable to introduce the following classification of points:

(i) A *regular point* of $F(\mu, \epsilon) = 0$ is one for which the implicit function theorem works,

$$F_\mu \neq 0 \quad \text{or} \quad F_\epsilon \neq 0. \tag{4}$$

*This work was supported by the National Science Foundation under Grant 19047.
†This section elaborates on joint work initiated by S. Rosenblat[8] and D. D. Joseph.[5]

If (4) holds, then, we can find a unique curve $\mu = \mu(\epsilon)$ or $\epsilon = \epsilon(\mu)$ through the point.

(ii) A *regular turning point* is a point at which $\mu_\epsilon(\epsilon)$ changes sign and $F_\mu(\mu, \epsilon) \neq 0$.

(iii) A *singular point* of the curve $F(\mu, \epsilon) = 0$ is a point at which

$$F_\mu = F_\epsilon = 0 \qquad (5)$$

(iv) A *double point* of the curve $F(\mu, \epsilon) = 0$ is a singular point through which pass two and only two branches of $F(\mu, \epsilon) = 0$ possessing distinct tangents.

(v) A *singular turning (double) point* of the curve $F(\mu, \epsilon)$ is a double point at which μ_ϵ changes sign.

(vi) A *cusp point* of the curve $F(\mu, \epsilon) = 0$ is a point of higher order contact between two branches of the curve. The two branches of the curve have the same tangent at a cusp point.

(vii) A *higher order singular point* of the curve $F(\mu, \epsilon) = 0$ is a singular point at which all three second derivatives of $F(\mu, \epsilon)$ are null.

I am going to connect the study of stability to the study of bifurcation under the "strict crossing" hypothesis introduced by Hopf and used in almost all studies of bifurcation and stability. I will explain what is meant by "strict crossing" in due course; for now it will suffice to remark that this hypothesis restricts the study of bifurcation to *double points;* cusp points and higher order singular points are excluded.

It is necessary to be precise about double points. Suppose (μ_0, ϵ_0) is a singular point. Then equilibrium curves passing through the singular points satisfy

$$2F(\mu, \epsilon) = F_{\mu\mu}\delta\mu^2 + 2F_{\epsilon\mu}\delta\epsilon\delta\mu + F_{\epsilon\epsilon}\delta\epsilon^2 + o(\delta\mu^2 + \delta\epsilon\delta\mu + \delta\epsilon^2) = 0 \qquad (6)$$

where $\delta\mu = \mu - \mu_0$, $\delta\epsilon = \epsilon - \epsilon_0$, $F_{\mu\mu} = F_{\mu\mu}(\mu_0, \epsilon_0)$, and so forth. In the limit, as $(\mu, \epsilon) \rightarrow (\mu_0, \epsilon_0)$ the Equation 6 for the curves $F(\mu, \epsilon) = 0$ reduce to the quadratic equation.

$$F_{\mu\mu}d\mu^2 + 2F_{\epsilon\mu}d\epsilon d\mu + F_{\epsilon\epsilon}d\epsilon^2 = 0 \qquad (7)$$

for the tangents to the curve. We find that

$$\begin{bmatrix} \mu_\epsilon^{(1)} & (\epsilon_0) \\ \mu_\epsilon^{(2)} & (\epsilon_0) \end{bmatrix} = -\frac{F_{\epsilon\mu}}{F_{\mu\mu}}\begin{bmatrix} 1 \\ 1 \end{bmatrix} + \sqrt{\frac{D}{F_{\mu\mu}^2}}\begin{bmatrix} 1 \\ -1 \end{bmatrix} \qquad (8)$$

or

$$\begin{bmatrix} \epsilon_\mu^{(1)} & (\mu_0) \\ \epsilon_\mu^{(2)} & (\mu_0) \end{bmatrix} = -\frac{F_{\epsilon\mu}}{F_{\epsilon\epsilon}}\begin{bmatrix} 1 \\ 1 \end{bmatrix} - \sqrt{\frac{D}{F_{\epsilon\epsilon}^2}}\begin{bmatrix} 1 \\ -1 \end{bmatrix} \qquad (9)$$

where

$$D = F_{\epsilon\mu}^2 - F_{\mu\mu}F_{\epsilon\epsilon}. \qquad (10)$$

If $D < 0$ there are no real tangents through (μ_0, ϵ_0) and the point (μ_0, ϵ_0) is an isolated point solution of $F(\mu, \epsilon) = 0$.

We shall consider the case when (μ_0, ϵ_0) is *not* a higher order singular point. Then (μ_0, ϵ_0) is a double point if and only if $D > 0$. If $D = 0$ then the slope at the singular point of higher order contact is given by (3) or (4). If $D > 0$ and $F_{\mu\mu} \neq 0$, then there are two tangents with slopes $\mu_\epsilon^{(1)}(\epsilon_0)$ and $\mu_\epsilon^{(2)}(\epsilon_0)$ given by (8). If $D > 0$ and $F_{\mu\mu} = 0$, then $F_{\epsilon\mu} \neq 0$ and

$$d\epsilon[2d\mu F_{\epsilon\mu} + d\epsilon F_{\epsilon\epsilon}] = 0 \tag{11}$$

and there are two tangents $\epsilon_\mu(\mu_0) = 0$ and $\mu_\epsilon(\epsilon_0) = -F_{\epsilon\epsilon}/2F_{\epsilon\mu}$. If $\epsilon_\mu(\mu_0) = 0$ then $F_{\mu\mu}(\mu_0, \epsilon_0) = 0$. So all possibilities are covered in the following two cases:

(i) $D > 0, F_{\mu\mu} \neq 0$ with tangents $\mu_\epsilon^{(1)}(\epsilon_0)$ and $\mu_\epsilon^{(2)}(\epsilon_0)$.
(ii) $D > 0, F_{\mu\mu} = 0$ with tangents $\epsilon_\mu(\mu_0) = 0$ and $\mu_\epsilon(\epsilon_0) = -F_{\epsilon\epsilon}/2F_{\epsilon\mu}$.

Now I am going to connect stability and bifurcation. To study the stability of the solution $u = \epsilon$, we study the linearized equation

$$Z_t + F_\epsilon(\mu, \epsilon) Z = 0$$

by the spectral method

$$Z = e^{-\gamma t} Z'$$

where

$$\gamma = F_\epsilon(\mu, \epsilon). \tag{12}$$

The solution $u = \epsilon$ is stable when $\gamma > 0$ and is unstable when $\gamma < 0$.

THEOREM 1 (Factorization Theorem). For every equilibrium solution $F(\mu, \epsilon) = 0$ for which $\mu = \mu(\epsilon)$ we have

$$\gamma(\epsilon) = F_\epsilon(\mu(\epsilon), \epsilon) = -\mu_\epsilon(\epsilon) F_\mu(\mu(\epsilon), \epsilon) \equiv -\mu_\epsilon \hat{\gamma}(\epsilon). \tag{13}$$

The proof of Theorem 1 follows from (12) and the equation

$$\frac{dF(\mu(\epsilon), \epsilon)}{d\epsilon} = F_\epsilon(\mu(\epsilon), \epsilon) + \mu_\epsilon(\epsilon) F_\mu(v(\epsilon), \epsilon) = 0. \tag{14}$$

This type of factorization may be proved for the stability of bifurcation solutions in Banach spaces more complicated than $\mathbb{R}_1$.[3,5,6,7] But the theorem is most easily understood in $\mathbb{R}_1$. One of the main implications of the factorization theorem is that $\gamma(\epsilon)$ *must change sign as ϵ is varied across a regular turning point.* This implies that the solution $u = \epsilon, \mu = \mu(\epsilon)$ is stable on one side of a regular turning point and is unstable on the other side.

THEOREM 2. (A) Any point (μ_0, ϵ_0) of the curve $\mu = \mu(\epsilon)$ for which $\hat{\gamma}(\epsilon_0) = 0$ is a singular point. (B) Any point (μ_0, ϵ_0) of the curve $\epsilon(\mu)$ for which $\gamma(\mu_0) = 0$ is a singular point.

The proof of (A) follows from (13) and the proof of (B) from

$$\gamma(\mu) = F_\epsilon(\mu, \epsilon(\mu)), \quad \frac{dF}{d\mu} = F_\mu + \epsilon_\mu F_\epsilon = 0. \tag{15}$$

The next theorem connects the hypothesis of strict loss of stability to bifurcation into double points.

THEOREM 3. Suppose that (μ_0, ϵ_0) is a singular point and (A) $\gamma_\epsilon(\epsilon_0) \neq 0$ or (B) $\gamma_\mu(\mu_0) \neq 0$. Then (μ_0, ϵ_0) is a double point.

In case (A) we find from (13) that at the singular point $(\mu(\epsilon_0), \epsilon_0)$

$$\gamma_\epsilon(\epsilon_0) = F_{\epsilon\epsilon} + \mu_\epsilon F_{\epsilon\mu} = -\mu_\epsilon^2 F_{\mu\mu} - \mu_\epsilon F_{\epsilon\mu} \neq 0 \tag{16}$$

Equation 16 shows that the characteristic quadratic (7) holds at $(\mu(\epsilon_0), \epsilon_0)$. Since there is a curve through this point, $D \geq 0$ and we need to show that $D \neq 0$. We shall assume that $D = F_{\epsilon\mu}^2 - F_{\mu\mu} F_{\epsilon\epsilon} = 0$ and show that this assumption contradicts (16). We first note that (16) implies that not all three of the second derivatives of F are null at $(\mu(\epsilon_0), \epsilon_0)$. If $F_{\mu\mu} F_{\epsilon\epsilon} \neq 0$ and $D = 0$ then (8) becomes $\mu_\epsilon(\epsilon_0) = -F_{\epsilon\mu}/F_{\mu\mu}$ and (16), may be written as $F_{\epsilon\epsilon} - F_{\epsilon\mu}^2/F_{\mu\mu} = -D/F_{\mu\mu} \neq 0$. So $D \neq 0$ after all. If $F_{\mu\mu} F_{\epsilon\epsilon} = 0$ and $D = 0$ then $F_{\epsilon\mu} = 0$ and (16) may be written as $\gamma_\epsilon = F_{\epsilon\epsilon} = -\mu_\epsilon^2 F_{\mu\mu} \neq 0$. So $D \neq 0$ after all.

In case (B) we solve $F(\mu, \epsilon) = 0$ for $\epsilon(\mu)$. At the singular point (μ_0, ϵ_0), we have strict loss of stability.

$$\gamma_\mu = F_{\epsilon\mu}(\mu_0, \epsilon(\mu_0)) + \epsilon_\mu(\mu_0) F_{\epsilon\epsilon}(\mu_0, \epsilon(\mu_0)) \neq 0 \tag{17}$$

and $D \geq 0$. Assuming $D = 0$ we find, using (9), that

$$\epsilon_\mu = -F_{\epsilon\mu}/F_{\epsilon\epsilon} \tag{18}$$

if $F_{\epsilon\epsilon} \neq 0$. Then (17) and (18) imply that $\gamma_\mu = 0$, and if $D = 0$ and $F_{\epsilon\epsilon} = 0$, then $F_{\epsilon\mu} = 0$ so that $\gamma_\mu = 0$. So $D = 0$ is inconsistent with $\gamma_\mu \neq 0$ and $D > 0$ after all.

The analysis of bifurcation in $\mathbb{R}_1$ just given shows that double point bifurcation is implied by a strict crossing hypothesis of the Hopf type. The situation is more complicated when these hypotheses are relaxed. If $\gamma_\epsilon = 0$ when $\gamma = 0$ we may get cusp bifurcation; or if all three second derivatives vanish, then the cubic equation can give a triple point (three real roots for the slopes) or no bifurcation (two complex conjugate roots). If third derivatives also vanish we face the problem of classifying the roots of a quartic. For example, we may get four bifurcating branches.

It is possible to make precise statements about the stability of solutions near double points of bifurcation. All of the possibilities for the stability of double point bifurcation can be described by the cases (A) and (B), which were fully specified under (11). In case (A) two curves $\mu^{(1)}(\epsilon)$ and $\mu^{(2)}(\epsilon)$ pass through the double point (μ_0, ϵ_0). In case (B) two curves, $\epsilon^{(1)}(\mu)$ (with $\epsilon_\mu^{(1)}(\mu_0) = 0$) and $\mu^{(2)}$, pass through the double point. The eigenvalue $\gamma^{(1)}$ belongs to the curve with superscript (1) and $\gamma^{(2)}$ to the curve with superscript (2).

THEOREM 4. Suppose (μ_0, ϵ_0) is a double point. Then, in case (A),

$$\gamma^{(1)}(\epsilon) = -\mu_\epsilon^{(1)}(\epsilon) \{\hat{s}\sqrt{D}(\epsilon - \epsilon_0) + o(\epsilon - \epsilon_0)\}, \tag{19}$$

and

$$\gamma^{(2)}(\epsilon) = \mu_\epsilon^{(2)}(\epsilon) \{\hat{s}\sqrt{D}(\epsilon - \epsilon_0) + o(\epsilon - \epsilon_0)\}, \tag{20}$$

where $\hat{s} = F_{\mu\mu}/|F_{\mu\mu}|$ and D and $F_{\mu\mu}$ are evaluated at $\epsilon = \epsilon_0$. And in case (B),

$$\gamma^{(1)}(\mu) = s\sqrt{D}(\mu - \mu_0) + o(\mu - \mu_0) \tag{21}$$

and

$$\gamma^{(2)}(\epsilon) = -s\mu_\epsilon^{(2)}(\epsilon)\{\sqrt{D}(\epsilon - \epsilon_0) + o(\epsilon - \epsilon_0)\} \tag{22}$$

where $s = F_{\epsilon\mu}/|F_{\epsilon\mu}|$

Proof. If $\mu = \mu/(\epsilon)$ we have (13) in the form,

$$\gamma(\epsilon) = -\mu_\epsilon(\epsilon) F_\mu(\mu(\epsilon), \epsilon)$$

$$= -\mu_\epsilon(\epsilon)\{F_{\mu\mu}(\mu_0, \epsilon_0)\mu_\epsilon(\epsilon_0) + F_{\epsilon\mu}(\mu_0, \epsilon_0)(\epsilon - \epsilon_0) + o(\epsilon - \epsilon_0)\} \tag{23}$$

The formulas (19) and (20) arise from (23) when $\mu_\epsilon(\epsilon_0)$ is replaced with the values given by (8). If $\epsilon = \epsilon(\mu)$ with $\epsilon_\mu(\mu_0) = 0$ then $F_{\mu\mu}(\mu_0, \epsilon_0) = 0$, $F_{\epsilon\mu}{}^2(\mu_0, \epsilon_0) = D$, and

$$\gamma(\mu) = F_\epsilon(\mu, \epsilon(\mu)) = F_{\epsilon\mu}(\mu_0, \epsilon_0)(\mu - \mu_0) + o(\mu - \mu_0)$$

$$= s\sqrt{D}(\mu - \mu_0) + o(\mu - \mu_0).$$

Theorem 4 gives an exhaustive classification relating the stability of solutions near a double point to the slope of the bifurcation curves near that point. The result may be summarized as follows. Suppose $|\epsilon - \epsilon_0| > 0$ is small. Then (19) and (20) show that $\gamma^{(1)}(\epsilon)$ and $\gamma^{(2)}(\epsilon)$ have the same (different) sign if $\mu_\epsilon^{(1)}(\epsilon)$ and $\mu_\epsilon^{(2)}(\epsilon)$ have different (the same) sign. A similar conclusion can be drawn from (21) and (22). The possible distributions of stability of solutions is sketched in FIGURE 1.

Almost all the work in the theory of bifurcation is restricted to problems satisfying hypothesis (2). Then $F(\mu, 0) = F_\mu(0, 0) = F_{\mu\mu}(0, 0) = 0$. The strict crossing hypothesis, introduced by Hopf, states that $\gamma_\mu^{(1)}(0) = F_{\mu\epsilon}(0, 0) < 0$. Then we get $D > 0$ and $\gamma^{(2)}(\epsilon) = -\mu_\epsilon^{(2)}(\epsilon)\gamma_\mu^{(1)}(0)\{(\epsilon - \epsilon_0) + o(\epsilon - \epsilon_0)\}$.

We now suppose that $F(\cdot, \cdot)$ has four continuous partial derivatives and show what happens to the stability of solutions bifurcating at a cusp point of second-order contact and at a triple point. When $\mu = \mu(\epsilon)$ all derivatives $\mathfrak{F}(\epsilon) \equiv F(\mu(\epsilon), \epsilon) = 0$ vanish. Then we have (14).

$$\frac{d^2\mathfrak{F}}{d\epsilon^3} = F_{\epsilon\epsilon} + 2\mu_\epsilon F_{\epsilon\mu} + \mu_\epsilon{}^2 F_{\mu\mu} + \mu_{\epsilon\epsilon} F_\mu = 0, \tag{24}$$

$$\begin{aligned}\frac{d^3\mathfrak{F}}{d\epsilon^3} = {} & F_{\epsilon\epsilon\epsilon} + 3\mu_\epsilon F_{\epsilon\epsilon\mu} + 3\mu_\epsilon{}^2 F_{\epsilon\mu\mu} + \mu_\epsilon{}^3 F_{\mu\mu\mu} + 3\mu_{\epsilon\epsilon} F_{\epsilon\mu} \\ & + 3\mu_{\epsilon\epsilon}\mu_\epsilon F_{\mu\mu} + \mu_{\epsilon\epsilon\epsilon} F_\mu = 0,\end{aligned} \tag{25}$$

$$\begin{aligned}\frac{d^4\mathfrak{F}}{d\epsilon^4} = {} & F_{\epsilon\epsilon\epsilon\epsilon} + 4\mu_\epsilon F_{\epsilon\epsilon\epsilon\mu} + 6\mu_\epsilon{}^2 F_{\epsilon\epsilon\mu\mu} + 4\mu_\epsilon{}^3 F_{\epsilon\mu\mu\mu} + \mu_\epsilon{}^4 F_{\mu\mu\mu\mu} + 4\mu_{\epsilon\epsilon\epsilon} F_{\epsilon\mu} \\ & + 4\mu_{\epsilon\epsilon\epsilon}\mu_\epsilon F_{\mu\mu} + 3\mu_{\epsilon\epsilon}{}^2 F_{\mu\mu} + 6\mu_{\epsilon\epsilon} F_{\epsilon\epsilon\mu} + 12\mu_\epsilon\mu_{\epsilon\epsilon} F_{\epsilon\mu\mu} \\ & + 6\mu_{\epsilon\epsilon}\mu_\epsilon{}^2 F_{\mu\mu\mu} + \mu_{\epsilon\epsilon\epsilon\epsilon} F_\mu = 0\end{aligned} \tag{26}$$

When $\epsilon = \epsilon(\mu)$, $\mathit{f}(\mu) = F(\mu, \epsilon(\mu)) = 0$, and

$$\frac{d^2\mathit{f}}{d\mu^2} = F_{\mu\mu} + 2\epsilon_\mu F_{\epsilon\mu} + \epsilon_\mu{}^2 F_{\epsilon\epsilon} + \epsilon_{\mu\mu} F_\epsilon = 0,$$

then

$$\frac{d^3 \ell}{d\mu^3} = F_{\mu\mu\mu} + 3\epsilon_\mu F_{\epsilon\mu\mu} + 3\epsilon_\mu{}^2 F_{\epsilon\epsilon\mu} + \epsilon_\mu{}^3 F_{\epsilon\epsilon\epsilon} + 3\epsilon_{\mu\mu} F_{\epsilon\mu} + 3\epsilon_{\mu\mu} \epsilon_\mu F_{\epsilon\epsilon}$$
$$+ \epsilon_{\mu\mu\mu} F_\epsilon = 0, \tag{27}$$

$$\frac{d^4 \ell}{d\mu^4} = F_{\mu\mu\mu\mu} - 4\epsilon_\mu F_{\epsilon\mu\mu\mu} + 6\epsilon_\mu{}^2 F_{\epsilon\epsilon\mu\mu} + 4\epsilon_\mu{}^3 F_{\epsilon\epsilon\epsilon\mu} + \epsilon_\mu{}^4 F_{\epsilon\epsilon\epsilon\epsilon} + 4\epsilon_{\mu\mu\mu} F_{\epsilon\mu}$$
$$+ 4\epsilon_{\mu\mu\mu} \epsilon_\mu F_{\epsilon\epsilon} + 3\epsilon_{\mu\mu}{}^2 F_{\epsilon\epsilon} + 6\epsilon_{\mu\mu} F_{\epsilon\mu\mu} + 12\epsilon_\mu \epsilon_{\mu\mu} F_{\epsilon\epsilon\mu} + 6\epsilon_{\mu\mu} \epsilon_\mu{}^2 F_{\epsilon\epsilon\epsilon}$$
$$+ \epsilon_{\mu\mu\mu\mu} F_\epsilon = 0.$$

At a cusp point $F - F_\epsilon = F_\mu = D = 0$. In case (Λ), $F_{\mu\mu} \neq 0$ $\mu_\epsilon(\epsilon_0) = -F_{\epsilon\mu}/F_{\mu\mu}$, (24) is satisfied identically, (25) becomes

$$F_{\epsilon\epsilon\epsilon} + 3\mu_\epsilon(\epsilon_0) F_{\epsilon\epsilon\mu} + 3\mu_\epsilon{}^2(\epsilon_0) F_{\epsilon\mu\mu} + \mu_\epsilon{}^3 F_{\mu\mu\mu} = 0,$$

and the coefficient of $\mu_{\epsilon\epsilon}$ in (26) vanishes, leaving a quadratic equation for the

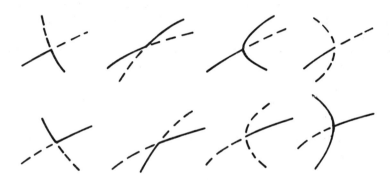

FIGURE 1. Stability of solutions in the neighborhood of double point bifurcation.

curvature $\mu_{\epsilon\epsilon}$

$$\mu_{\epsilon\epsilon}{}^2 + 2\mu_{\epsilon\epsilon}\xi/F_{\mu\mu} + \zeta/F_{\mu\mu} = 0 \tag{29}$$

where

$$\xi = F_{\epsilon\epsilon\mu} + 2\mu_\epsilon F_{\epsilon\mu\mu} + \mu_\epsilon{}^2 F_{\mu\mu\mu}$$

and

$$3\zeta = F_{\epsilon\epsilon\epsilon\epsilon} + 4\mu_\epsilon F_{\epsilon\epsilon\epsilon\mu} + 6\mu_\epsilon{}^2 F_{\epsilon\epsilon\mu\mu} + 4\mu_\epsilon{}^3 F_{\epsilon\mu\mu\mu} + \mu_\epsilon{}^4 F_{\mu\mu\mu\mu}.$$

Equation (29) has two roots

$$\begin{pmatrix} \mu_{\epsilon\epsilon}{}^{(1)} & (\epsilon_0) \\ \mu_{\epsilon\epsilon}{}^{(2)} & (\epsilon_0) \end{pmatrix} = -\frac{\xi}{F_{\mu\mu}} \begin{pmatrix} 1 \\ 1 \end{pmatrix} + \sqrt{\frac{\mathfrak{D}_1}{F_{\mu\mu}{}^2}} \begin{pmatrix} 1 \\ -1 \end{pmatrix}, \tag{30}$$

where

$$\mathfrak{D}_1 = \xi^2 - F_{\mu\mu}\zeta.$$

In case (B), $F_{\mu\mu} = 0$ and $F_{\epsilon\epsilon} \neq 0$; and since $D = 0$, $F_{\epsilon\mu} = 0$ and $\epsilon_\mu(\mu_0) = 0$. Equation 27 then shows that $F_{\mu\mu\mu} = 0$ and (28) reduces to a quadratic equation for the curvature $\epsilon_{\mu\mu}(\mu_0)$:

$$\epsilon_{\mu\mu}{}^2 + 2\epsilon_{\mu\mu}F_{\epsilon\mu\mu}/F_{\epsilon\epsilon} + F_{\mu\mu\mu\mu}/3F_{\epsilon\epsilon} = 0 \tag{31}$$

Equation 31 has two roots

$$\begin{pmatrix} \epsilon_{\mu\mu}{}^{(1)} & (\mu_0) \\ \epsilon_{\mu\mu}{}^{(2)} & (\mu_0) \end{pmatrix} = -\frac{F_{\epsilon\mu\mu}}{F_{\epsilon\epsilon}}\begin{pmatrix} 1 \\ 1 \end{pmatrix} + \sqrt{\frac{\mathfrak{D}_2}{F_{\epsilon\epsilon}{}^2}}\begin{pmatrix} 1 \\ -1 \end{pmatrix}, \tag{32}$$

where

$$\mathfrak{D}_2 = F_{\epsilon\mu\mu}{}^2 - F_{\epsilon\epsilon}F_{\mu\mu\mu\mu}/3F_{\epsilon\epsilon}.$$

At a point of second-order contact the two curves have common tangents and different real-valued curvatures. It follows that $\mathfrak{D}_1 > 0$ or $\mathfrak{D}_2 > 0$ at a point of second-order contact.

Restricting our attention to a point of second-order contact, we expand the factor $F_\mu(\mu(\epsilon), \epsilon)$ into a series of powers of $(\epsilon - \epsilon_0)$ and find that in case (A)

$$\begin{pmatrix} \gamma^{(1)} & (\epsilon) \\ \gamma^{(2)} & (\epsilon) \end{pmatrix} = -\frac{1}{2}\hat{s}\sqrt{\mathfrak{D}_1}\begin{pmatrix} \mu_\epsilon{}^{(1)} & (\epsilon) \\ -\mu_\epsilon{}^{(2)} & (\epsilon) \end{pmatrix}(\epsilon - \epsilon_0)^2 + O(\epsilon - \epsilon_0)^3. \tag{33}$$

In case (B), we expand $\gamma(\mu) = F_\epsilon(\mu, \epsilon(\mu))$ into a series of powers in $(\mu - \mu_0)$ and find that

$$\begin{pmatrix} \gamma^{(1)} & (\mu) \\ \gamma^{(2)} & (\mu) \end{pmatrix} = -\frac{1}{2}s\sqrt{\mathfrak{D}_2}\begin{pmatrix} 1 \\ -1 \end{pmatrix}(\mu - \mu_0)^2 + O(\mu - \mu_0)^3. \tag{34}$$

It follows from (32) and (34) that the stability of any branch passing through a cusp point of second order changes sign if and only if $\mu_\epsilon(\epsilon)$ does. The possible distributions of stability at a cusp point are exhibited in FIGURE 2.

We turn next to the case in which all second order derivatives of $F(\mu, \epsilon) = 0$ are null at a singular point. Confining our attention to the case in which $F_{\mu\mu\mu} \neq 0$ we may write (25) as

$$(\mu_\epsilon - \mu_\epsilon{}^{(1)})(\mu_\epsilon - \mu_\epsilon{}^{(2)})(\mu_\epsilon - \mu_\epsilon{}^{(3)})$$

$$= \mu_\epsilon{}^3 + 3\mu_\epsilon{}^2\frac{F_{\epsilon\mu\mu}}{F_{\mu\mu\mu}} + 3\mu_\epsilon\frac{F_{\epsilon\epsilon\mu}}{F_{\mu\mu\mu}} + \frac{F_{\epsilon\epsilon\epsilon}}{F_{\mu\mu\mu}}$$

$$= 0 \tag{35}$$

where $\mu_\epsilon{}^{(1)}$, $\mu_\epsilon{}^{(2)}$ and $\mu_\epsilon{}^{(3)}$ are values of $\mu_\epsilon(\epsilon)$ at $\epsilon = \epsilon_0$. It follows from (35) that

$$\frac{F_{\epsilon\epsilon\mu}}{F_{\mu\mu\mu}} = \frac{1}{3}(\mu_\epsilon{}^{(1)}\mu_\epsilon{}^{(2)} + \mu_\epsilon{}^{(1)}\mu_\epsilon{}^{(3)} + \mu_\epsilon{}^{(2)}\mu_\epsilon{}^{(3)}) \tag{36}$$

and

$$\frac{F_{\epsilon\mu\mu}}{F_{\mu\mu\mu}} = -\frac{1}{3}(\mu_\epsilon^{(1)} + \mu_\epsilon^{(2)} + \mu_\epsilon^{(3)}). \tag{37}$$

If the three roots of (35) are real and distinct three bifurcating solutions pass through the singular point (μ_0, ϵ_0). The stability of these branches may be determined from the sign of

$$\begin{aligned}
\gamma(\epsilon) &= -\mu_\epsilon(\epsilon) F_\mu(\mu(\epsilon), \epsilon) \\
&= -\tfrac{1}{2}\mu_\epsilon(\epsilon)\{\xi(\epsilon - \epsilon_0)^2 + O(\epsilon - \epsilon_0)^3\} \\
&= -\tfrac{1}{2}\mu_\epsilon(\epsilon)F_{\mu\mu\mu}\{\tfrac{1}{3}(\mu_\epsilon^{(1)}\mu_\epsilon^{(2)} + \mu_\epsilon^{(1)}\mu_\epsilon^{(3)} + \mu_\epsilon^{(2)}\mu_\epsilon^{(3)}) \\
&\quad - \tfrac{2}{3}\mu_\epsilon(\epsilon_0)(\mu_\epsilon^{(1)} + \mu_\epsilon^{(2)} + \mu_\epsilon^{(3)}) + \mu_\epsilon^2(\epsilon_0)\}(\epsilon - \epsilon_0)^2 \\
&\quad + O(\epsilon - \epsilon_0)^3.
\end{aligned}$$

We find that

$$\begin{bmatrix} \gamma(\epsilon)^{(1)} \\ \gamma(\epsilon)^{(2)} \\ \gamma(\epsilon)^{(3)} \end{bmatrix} = -\tfrac{1}{6}F_{\mu\mu\mu}\begin{bmatrix} \mu_\epsilon^{(1)}(\epsilon)(\mu_\epsilon^{(1)} - \mu_\epsilon^{(2)})(\mu_\epsilon^{(1)} - \mu_\epsilon^{(3)}) \\ \mu_\epsilon^{(2)}(\epsilon)(\mu_\epsilon^{(1)} - \mu_\epsilon^{(2)})(\mu_\epsilon^{(3)} - \mu_\epsilon^{(2)}) \\ \mu_\epsilon^{(3)}(\epsilon)(\mu_\epsilon^{(1)} - \mu_\epsilon^{(3)})(\mu_\epsilon^{(2)} - \mu_\epsilon^{(3)}) \end{bmatrix}(\epsilon - \epsilon_0)^2$$

$$+ O(\epsilon - \epsilon_0)^3 \tag{38}$$

where it may be assumed, without loss of generality, that $\mu_\epsilon^{(1)} > \mu_\epsilon^{(2)} > \mu_\epsilon^{(3)}$. The

FIGURE 2. Stability of solutions bifurcating at a cusp point of second order.

distribution of stability of the three distinct branches is easily determined from (38). We leave further deductions about bifurcation and stability at a singular point where the second derivatives are all null as an exercise for the interested reader. It will suffice here to remark that the stability of a branch passing through such a point can change if and only if $\mu_\epsilon(\epsilon)$ changes sign there.

In the next two theorems, I give a global characterization of the stability of equilibrium solutions. I first note that $\gamma \neq 0$ at regular points of the curve $F(\mu, \epsilon) = 0$ at which $\mu_\epsilon \neq 0$. The factorization theorem $\gamma = -\mu_\epsilon(\epsilon)F_\mu(\mu(\epsilon), \epsilon)$ shows that γ can change sign at a stationary regular point $(\mu_\epsilon = 0, F_\mu \neq 0)$ only if it be a turning point.

THEOREM 5. Assume that all singular points of solutions of $F(\mu, \epsilon) = 0$ are double points. The stability of such solutions must change at each regular turning point and at each singular point (which is not a turning point) and only at such points.

Theorem 5 gives a fairly complete catalogue of the stability of solution on connected branches of $F(\mu, \epsilon) = 0$. But solutions of $F(\mu, \epsilon) = 0$ need not be connected (see FIGURE 3 for a typical example). It is, however, possible to relate the stability of equilibrium solutions of isolated branches that pierce the line $\mu = $ const with solutions of $F(\mu, \epsilon) = 0$. Label these points in an increasing sequence $\epsilon_1 < \epsilon_2 < \cdots < \epsilon_n$. Then $F(\mu, \epsilon_1) = F(\mu, \epsilon_2) = \cdots = F(\mu, \epsilon_n) = 0$. Of course,

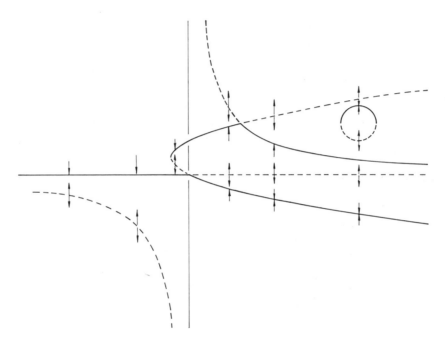

FIGURE 3. Bifurcation, stability, and domains of attraction of equilibrium solutions of

$$\frac{du}{dt} = u(9 - \mu u)(\mu + 2u - u^2)([\mu - 10]^2 + [u - 3]^2 - 1).$$

The equilibrium solution $\mu = 9/u$ in the third quadrant and the circle are isolated solutions that cannot be obtained by bifurcation analysis.

$F(\mu, \epsilon)$ is of one sign between any two successive zeros. It follows that if $F_\epsilon(\mu, \epsilon_l) < 0$, then $F_\epsilon(\mu, \epsilon_{l+1}) \geq 0$ and $F_\epsilon(\mu, \epsilon_{l-1}) \geq 0$. Suppose, for example, that $F_\epsilon(\mu, \epsilon_{l-1}) = 0$, then $\gamma(\epsilon_{l-1}) = 0$ so that (μ, ϵ_{l-1}) is a singular point or a regular stationary point, $\mu_\epsilon(\epsilon_{l-1}) = 0$. The stability of solutions at regular stationary points is completely described by Theorem 1 and at singular double points by Theorem 5. Isolated (conjugate) singular points may be ignored because $D < 0$ implies that $F(\mu_0, \epsilon_0) = 0$ is an extreme value of $F(\mu, u)$, which does not change sign as ϵ is varied across ϵ_0 at a fixed $\mu = \mu_0$. In any event, it will usually be possible to shift μ slightly so that $F_\epsilon(\mu, \epsilon_l) \neq 0$ at each and every piercing point.

THEOREM 6. If $F_\epsilon(\mu, \epsilon_l) \neq 0$ on each and every piercing point of solutions of $F(\mu, \epsilon) = 0$ and the line $\mu = \text{const}$, then the sign of $\gamma(\epsilon)$ at such points is a sequence of alternating sign. If the solution (μ, ϵ_l) of $F(\mu, \epsilon) = 0$ is stable (unstable) then the solutions (μ, ϵ_{l-1}) and (μ, ϵ_{l+1}) are unstable (stable).

Theorem 6 is an obvious global extension in $\mathbb{R}_1$ of a local theorem of H. Weinberger,[9] which holds in a general Banach space. Similar conclusions have been derived by Benjamin[1,2] in interesting applications of the theory of Leray-Schauder Degree to steady solutions of the Navier Stokes Equations.

FACTORIZATION THEOREMS[3,5,6,7]

Consider the evolution equation

$$\frac{dV}{dt} + F(t, \mu, V) = 0, \tag{39}$$

where V is an element of a Banach space X and $F(t, \mu, V)$ is analytic map from $\mathbb{R}_1 \times \mathbb{C} \times X$ to $Y \supset X$. Two types of problems will be considered. In the first the external data is steady and $F(t, \mu, V) = F(\mu, V)$ is independent t. In this case (39) is an autonomous problem. In the second case the external data is T-periodic and $F(t, \cdot, \cdot) = F(t + T, \cdot, \cdot)$.

In the applications I have in mind $V \in X$ corresponds to a solutions $V(x, t)$ of a partial differential equation defined over a region $\mho$ of space. It is convenient to introduce a scalar product

$$(u, v) - (\overline{v, u}) = \int_\mho u(x) \overline{v}(x) \, d\mho, \tag{40}$$

where the overbar designates complex conjugate. I shall discuss three different types of equilibrium solutions of (39):

(a) Steady solutions of autonomous problems

$$V = u(\epsilon), \quad \mu = \mu(\epsilon),$$

(b) Limit cycle solutions of autonomous problems (Hopf's type)

$$V = u(s, \epsilon) = u(s + 2\pi, \epsilon), \quad s = \omega(\epsilon)t, \quad \mu = \mu(\epsilon),$$

(c) nT-periodic solutions of T-periodic problems[4]

$$V = u(t, \epsilon) = u(t + nT, \epsilon), \quad \mu = \mu(\epsilon), \quad n = 1, 2, 3, 4.$$

I have assumed that the equilibrium solutions can be defined parametrically through a parameter $\epsilon = f(u)$ where f is a linear functional that will distinguish different solutions. For example, f can be taken as a projection of the solution into a certain null-space, ϵ is then an amplitude and the quantities u, μ, ω are known to be analytic in ϵ when F is.

The equilibrium solutions satisfy

$$F(\mu(\epsilon), \quad u(\epsilon)) = 0, \tag{41a}$$

$$\omega(\epsilon)\mathring{u}(s, \epsilon) + F[\mu(\epsilon), \quad u(s, \epsilon)] - 0, \tag{41b}$$

$$\mathring{u}(t, \epsilon) + F[t, \mu(\epsilon), \quad u(t, \epsilon)] = 0. \tag{41c}$$

Differentiating (**41**) with respect to ϵ we get

$$\mu_\epsilon \underline{F}_\mu[\mu(\epsilon), \underline{u}(\epsilon)] + \underline{F}_u[\mu(\epsilon), \underline{u}(\epsilon)|u_\epsilon] = 0, \tag{42a}$$

$$\omega_\epsilon \mathring{\underline{u}} + \mu_\epsilon \underline{F}_\mu[\mu(\epsilon), \underline{u}(s, \epsilon)] + \mathfrak{g} \, u_\epsilon = 0, \tag{42b}$$

$$\mu_\epsilon \underline{F}_\mu[\mu(\epsilon), \underline{u}(t, \epsilon)] + \mathbb{J} \, u_\epsilon = 0. \tag{42c}$$

where

$$\mathfrak{g}(\cdot) = \omega(\epsilon)(\mathring{\cdot})(s, \epsilon) + \underline{F}_u[\mu(\epsilon), \underline{u}(s, \epsilon)|(\cdot)],$$

$$\mathbb{J}(\cdot) = (\mathring{\cdot})(t, \epsilon) + F_u[t, \mu(\epsilon), \underline{u}(t, \epsilon)|(\cdot)],$$

and

$$\mathrm{dom}\, \underline{F}_u[\mu(\epsilon), \underline{u}(\epsilon)| \cdot] = \underline{X}$$

$$\mathrm{dom}\, \mathfrak{g} = \underline{X}_{2\pi} = \underline{X} \cap 2\pi\text{-periodic functions of } s,$$

$$\mathrm{dom}\, \mathbb{J} = \underline{X}_{nT} = \underline{X} \cap nT\text{-periodic functions of } t.$$

We define scalar products on $X_{2\pi}$, and X_{nT}

$$<\underline{u}, \underline{v}>_{2\pi} \equiv \frac{1}{2\pi} \int_0^{2\pi} (\underline{u}(s, \epsilon), \underline{v}(\underline{s}, \epsilon)) \, d\underline{s},$$

$$<\underline{u}, \underline{v}>_{nT} \frac{1}{nT} \int_0^{nT} [\underline{u}(t, \epsilon), \underline{v}(t, \epsilon)] \, ds.$$

The linearized equations for small disturbances Z of equilibrium solution u are

$$\mathring{\underline{Z}} + \underline{F}_u[\mu(\epsilon), \underline{u}(\epsilon)|Z] = 0, \tag{43a}$$

$$\mathring{\underline{Z}} + \underline{F}_u[\mu(\epsilon), \underline{u}(s, \epsilon)|Z] = 0, \tag{43b}$$

$$\mathring{\underline{Z}} + \underline{F}_u(t, \mu(\epsilon), \underline{u}(t, \epsilon)|Z] = 0. \tag{43c}$$

Spectral problems for these equations may be formed using the method of Floquet. Setting $\underline{Z} = e^{-\gamma t} \underline{\zeta}$ we get

$$-\gamma\underline{\zeta} + \underline{F}_u[\mu(\epsilon), \underline{u}(\epsilon)|\underline{\zeta}] = 0, \tag{44a}$$

$$-\gamma\underline{\zeta} + \mathfrak{g} \, \underline{\zeta} = 0, \underline{\zeta}(s) = \underline{\zeta}(s + 2\pi), \tag{44b}$$

$$-\gamma\underline{\zeta} + \mathbb{J} \, \underline{\zeta} = 0, \underline{\zeta}(t) = \underline{\zeta}(t + nT). \tag{44c}$$

The adjoint spectral problems are

$$-\overline{\gamma}\underline{\zeta}^* + \underline{F}_u^*(\mu(\epsilon), \underline{u}(\epsilon)|\underline{\zeta}^*) = 0, \tag{45a}$$

$$-\overline{\gamma}\underline{\zeta}^* + \mathfrak{g} \, {}^*\underline{\zeta}^* = 0, \qquad \underline{\zeta}^*(s) = \underline{\zeta}^*(s + 2\pi), \tag{45b}$$

$$-\overline{\gamma}\underline{\zeta}^* + \mathbb{J}^* \underline{\zeta}^* = 0, \qquad \underline{\zeta}^*(t) = \underline{\zeta}^*(t + nT). \tag{45c}$$

Now we can start to work out the factorization. We form scalar products of (**42**) with adjoint eigenfunctions and find that Equations 42 are solvable only if

$$\gamma(\underline{u}_\epsilon, \underline{\zeta}^*) + \mu_\epsilon(\underline{F}_\mu, \underline{\zeta}^*) = 0, \tag{46a}$$

$$\gamma <\underline{u}_\epsilon, \underline{\zeta}^*>_{2\pi} + \mu_\epsilon <\underline{F}_\mu, \underline{\zeta}^*>_{2\pi} = 0, \tag{46b}$$

$$\gamma <\underline{u}_\epsilon, \underline{\zeta}^*>_{nT} + \mu_\epsilon <\underline{F}_\mu, \underline{\zeta}^*>_{nT} = 0. \tag{46c}$$

In deriving (46b) we made use of the equation

$$\mathcal{J} \, \mathring{u}(s, \epsilon) = 0, \tag{47}$$

which follows, by differentiation with respect to s, from (41b). Comparing (47) with (45b), we find that

$$\gamma <\mathring{u}, \underline{\zeta}^*>_{2\pi} = 0 \tag{48}$$

The first hypothesis (H.1) for the factorization

$$\gamma(\epsilon) = -\mu_\epsilon(\epsilon) \, \hat{\gamma}(\epsilon) \tag{49}$$

is that the inner products in the first term of each of Equations (46) does not vanish. Then using analyticity with respect to ϵ we conclude that coefficient of μ_ϵ in the equations that arise when (46) and (49) are combined must vanish.‡ Hence

$$\hat{\gamma}(\epsilon) = (\underline{F}_\mu, \underline{\zeta}^*)/(\underline{u}_\epsilon, \underline{\zeta}^*) \tag{50a}$$

$$\hat{\gamma}(\epsilon) = <\underline{F}_\mu, \underline{\zeta}^*>_{2\pi}/<\underline{u}_\epsilon, \underline{\zeta}^*>_{2\pi}, \tag{50b}$$

$$\hat{\gamma}(\epsilon) = <\underline{F}_\mu, \underline{\zeta}^*>_{nT}/<\underline{u}_\epsilon, \underline{\zeta}^*>_{nT} \tag{50c}$$

The second hypothesis (H.2) is that γ is an algebraically simple eigenvalue of the operators $F_u(\mu, \underline{u}| \cdot)$, $\mathcal{J}$ and $\mathbb{J}$ and these operators are Fredholm with compact resolvents from $\underline{Y}, \underline{Y}_{2\pi}, \underline{Y}_{nT}$ to $\underline{X}, \underline{X}_{2\pi}, \underline{X}_{nT}$.

THEOREM 7. (Factorization Theorem). If (H.1) and (H.2) hold then (49) and (50) are valid along with the decompositions

$$\underline{\zeta} = b(\epsilon)\{\underline{u}_\epsilon(\epsilon) + \mu_\epsilon \underline{q}(\epsilon)\} \tag{51a}$$

$$\underline{\zeta} = b(\epsilon)\left\{ -\frac{\omega_\epsilon(\epsilon)}{\gamma(\epsilon)} \, \mathring{\underline{u}}(s, \epsilon) + \underline{u}_\epsilon + \mu_\epsilon \underline{q}(s, \epsilon) \right\} \tag{51b}$$

$$\underline{\zeta} = b(\epsilon)\{\underline{u}_\epsilon(t, \epsilon) + \mu_\epsilon \underline{q}(t, \epsilon)\} \tag{51c}$$

where $\underline{q}(\epsilon), \underline{q}(s, \epsilon)$ and $\underline{q}(t, \epsilon)$ satisfy

$$-\hat{\gamma}\underline{u}_\epsilon + \underline{F}_\mu[\mu(\epsilon), \underline{u}(\epsilon)] + (\gamma\underline{q} - \underline{F}_u[\mu(\epsilon), \underline{u}(\epsilon)|\underline{q}] = 0 \text{ and } (\underline{q}, \underline{\zeta}^*) = 0, \tag{52a}$$

$$-\hat{\gamma}\underline{u}_\epsilon(s, \epsilon) + \underline{F}_\mu[\mu(\epsilon), \underline{u}(s, \epsilon)] + (\gamma - \mathcal{J})\underline{q} = 0,$$

$$<\underline{q}, \underline{\zeta}^*>_{2\pi} = 0, \text{ and } \underline{q}(s, \epsilon) = \underline{q}(s + 2\pi, \epsilon), \tag{52b}$$

$$-\hat{\gamma}\underline{u}_\epsilon + \underline{F}_\mu[\mu(\epsilon), \underline{u}(t, \epsilon)] + (\gamma - \mathbb{J})\underline{q} = 0,$$

$$<\underline{q}, \underline{\zeta}^*>_{nT} = 0, \text{ and } \underline{q}(t, \epsilon) = \underline{q}(t + nT, \epsilon). \tag{52c}$$

‡Assuming $\mu(\epsilon)$ is not constant on a branch.

Proof. (H.2) asserts that $\underline{F}_u$, $\mathcal{g}$, and $\mathbb{J}$ are Fredholm operators with compact resolvents. Since γ is to be an algebraically simple eigenvalue, each of the three Equations 52 is uniquely solvable for q on the complement of the null space associated with γ when the Equations 50 hold. The orthogonality of q and $\underline{\zeta}^*$ and u and $\underline{\zeta}^*$ Equation 48 imply that the normalizing factors $b(\epsilon)$ in (51) may be expressed as

$$\underline{\zeta} = \frac{(\underline{\zeta}, \underline{\zeta}^*)}{(\underline{u}_\epsilon, \underline{\zeta}^*)} [\underline{u}_\epsilon(\epsilon) + \mu_\epsilon \underline{q}(\epsilon)], \tag{53a}$$

$$\underline{\zeta} = \frac{<\underline{\zeta}, \underline{\zeta}^*>_{2\pi}}{<\underline{u}_\epsilon, \underline{\zeta}^*>_{2\pi}} \left[-\frac{\omega_\epsilon(\epsilon)}{\gamma(\epsilon)} \underline{\mathring{u}}(s, \epsilon) + \underline{u}_\epsilon(s, \epsilon) + \mu_\epsilon \underline{q}(s, \epsilon) \right], \tag{53b}$$

$$\underline{\zeta} = \frac{<\underline{\zeta}, \underline{\zeta}^*>_{nT}}{<\underline{u}_\epsilon, \underline{\zeta}^*>_{nT}} [\underline{u}_\epsilon(t, \epsilon) + \mu_\epsilon \underline{q}(t, \epsilon)]. \tag{53c}$$

This completes the proof of the factorization under the hypotheses (H.1) and (H.2).

The hypothesis (H.2) is insufficient for the discussion of turning points on the bifurcation curve for periodic solutions (Case II). At such points the factorization, (49) shows that $\gamma(\epsilon) = 0$ and a difficulty is already apparent in (52b). This difficulty may be traced to the fact that $\gamma(\epsilon) = 0$ cannot be an algebraically simple eigenvalue of $\gamma(\epsilon)$. The following theorem was proved by Joseph[6]:

The algebraic multiplicity of the eigenvalue $\gamma(\epsilon) = 0$ of $\mathcal{g}$ (ϵ) is at least two. Relative to such an eigenvalue, we have at least a two-link Jordan chain

$$\mathcal{g} \, \underline{\mathring{u}}(s, \epsilon) = 0,$$

$$\mathcal{g} \{-(\underline{u}_\epsilon(s, \epsilon) + \mu_\epsilon \underline{q}(s, \epsilon)/\omega_\epsilon(\epsilon)\} = \underline{\mathring{u}}(s, \epsilon)$$

whenever $\omega_\epsilon \neq 0$ when $\gamma(\epsilon) = 0$. If $\omega_\epsilon = 0$ when $\gamma = 0$, then the geometric multiplicity of $\gamma(\epsilon) = 0$ is at least two, and $\mathring{u}$ and $\underline{u}_\epsilon + \mu_\epsilon \underline{q}$ are both eigenfunction on the null space of $\mathcal{g}$ (ϵ).

If we suppose that $\omega_\epsilon \neq 0$ when $\gamma(\epsilon) = 0$ then the algebraic multiplicity of $\gamma(\epsilon) = 0$ is two and the geometric multiplicity is one. In this case the Riesz-Schauder Theory[6] implies that $<\underline{\zeta}, \underline{\zeta}^*>_{2\pi} = 0$. At such points the factorization theorem still holds but the normalizing factor *cannot* be taken as $<\underline{\zeta}, \underline{\zeta}^*>_{2\pi}/ <\underline{u}_\epsilon, \underline{\zeta}^*>_{2\pi}$.

When $\mu_\epsilon(\epsilon) = 0$ we find that $\gamma(\epsilon) = \mu_\epsilon(\epsilon)\hat{\gamma}(\epsilon) = 0$ and

$$\underline{\zeta} = b\underline{u}_\epsilon(\epsilon), \tag{54a}$$

$$\underline{\zeta} = b\left\{ \frac{-\omega_\epsilon(\epsilon)}{\mu_\epsilon(\epsilon)\hat{\gamma}(\epsilon)} \underline{\mathring{u}}(s, \epsilon) + \underline{u}_\epsilon(s, \epsilon) \right\}, \tag{54b}$$

$$\underline{\zeta} = b\underline{u}_\epsilon(t, \epsilon). \tag{54c}$$

If $\omega_\epsilon \neq 0$, then in case II, $\underline{\zeta} \propto \underline{\mathring{u}}(s, \epsilon)$

We may now define, as in $\mathbb{R}_1$, a regular turning point as a point at which $\mu_\epsilon(\epsilon)$ changes sign and $\hat{\gamma}(\epsilon) \neq 0$ where $\hat{\gamma}(\epsilon)$ is given by (50).

STABILITY AND SYMMETRY-BREAKING BIFURCATION

I shall now specify the assumptions which insure that problems (a) and (c) under (40) are essentially problems in $\mathbb{R}_1$. What we need to insure is that γ is real-valued and algebraically simple at points at which $\gamma = 0$. If we assume that $\gamma = 0$ at singular points, then we cannot break the temporal symmetry pattern of a solution through bifurcation; instead we get new steady solutions as a bifurcation of old steady solutions and new T-periodic solutions from old T-periodic solutions. When $\gamma = 0$ at criticality, bifurcation from steady solutions to T-periodic solutions or from T-periodic solutions to nT-periodic solutions ($n = 3, 4$) or tori is impossible. In fluid mechanics or, more generally, in problems governed by partial differential equations, such bifurcations will lead to new patterns of spatial symmetry.

I shall consider the T-periodic problem (III) and require conditions that allow only double-point, T-periodic bifurcation. For this we need a strict crossing condition under the hypothesis:

(H.3) Let d be an open rectangle in the (μ, ϵ) plane and suppose that $\underline{u}(t) = \underline{u}(t + T)$ for some point in d and dom $\mathbb{J} = \underline{X}_T$ in d. Let γ satisfying (H.1) and (H.2) be real-valued and suppose that all of the other eigenvalues of $\mathbb{J}$ have positive real parts in d.

I have already proved the factorization theorem (Theorem 7) under the assumptions (H.1) and (H.2). This theorem is the equivalent of Theorem 1, which gives the form of the factorization $\mathbb{R}_1$. My aim now is to show that under the hypothesis (H.3) we get the equivalent of Theorems 2 through 5. These theorems concern the behavior of solutions near a singular point (μ_0, ϵ_0). Various definitions of the amplitude ϵ are possible and all the good ones are equivalent. In the present context, we use (H.1) to justify the following definitions: Let $\underline{u}[t, \mu(\epsilon), \epsilon] = \underline{u}[t + T, \mu(\epsilon), \epsilon] \equiv \underline{u}(\epsilon)$ and, in the case where $\mu_\epsilon(\epsilon) = \infty$, $\underline{u}[t, \mu, \epsilon(\mu)] = \underline{u}[t + T, \mu, \epsilon(\mu)] \equiv \underline{u}(\mu)$. By $\underline{u}_0$ I mean $u(\epsilon_0) = \underline{u}(\mu_0)$ where, in a loose notation which confuses the values of a function with the function, I write $\underline{u}(\epsilon) = \underline{u}(\mu)$, $\underline{\zeta}(\epsilon)$, and $\underline{\zeta}^*(\epsilon) = \underline{\zeta}^*(\mu)$. The amplitude ϵ may be defined in any convenient way.

It is useful in the analysis to introduce the T-periodic operators from $\underline{X}_T$ to $\underline{Y}_T$:

$$\underline{G}(\mu, \underline{u}) = \mathring{\underline{u}} + \underline{F}(t, \mu, \underline{u}) = 0, \tag{55}$$

$$\underline{G}_u(\mu, \underline{u} \mid \underline{v}) = \mathbb{J}\underline{v}, \tag{56}$$

$$\underline{G}_\mu(\mu, \underline{u}) - \underline{F}_\mu(t, \mu, \underline{u}), \tag{57}$$

$$\underline{G}_{uu}(\mu, \underline{u} \mid \underline{v}_1 \mid \underline{v}_2) = \underline{F}_{uu}(t, \mu, \underline{u} \mid \underline{v}_1 \mid \underline{v}_2) \tag{58}$$

$$= G_{uu}(\mu, \underline{u} \mid \underline{v}_2 \mid \underline{v}_1),$$

$$\underline{G}_{\mu\mu}(\mu, \underline{u}) = \underline{F}_{\mu\mu}(t, \mu, \underline{u}), \tag{59}$$

$$\underline{G}_{u\mu}(\mu, \underline{u} \mid \underline{v}) = \underline{F}_{u\mu}(t, \mu, \underline{u} \mid \underline{v}). \tag{60}$$

A singular point is a point where

$$\mathcal{G}_\mu \equiv <\underline{G}_\mu(\mu_0, \underline{u}_0), \underline{\zeta}_0^* >_T = 0 \tag{61}$$

and

$$\mathcal{G}_u \equiv \, <\underline{G}_{\dot{u}}[\mu_0, \underline{u}_0 \,|(\cdot)], \underline{\zeta}_0^* >_T = 0. \tag{62}$$

Consider T-periodic solutions $\underline{u}(\epsilon)$ of

$$\underline{G}[\mu(\epsilon), \underline{u}(\epsilon)] = 0. \tag{63}$$

Differentiating with respect to ϵ, we get

$$\mu_\epsilon \underline{G}_\mu[\mu(\epsilon), \underline{u}(\epsilon)] + \underline{G}_u[\mu(\epsilon), \underline{u}(\epsilon) \,| \underline{u}_\epsilon] = 0 \tag{64}$$

and

$$\underline{G}_u[\mu(\epsilon), \underline{u}(\epsilon) \,|\underline{u}_{\epsilon\epsilon}] + \mu_{\epsilon\epsilon} \underline{G}_\mu[\mu(\epsilon), \underline{u}(\epsilon)] +$$
$$\mu_\epsilon^2 \underline{G}_{\mu\mu}[\mu(\epsilon), \underline{u}(\epsilon)] + 2\mu_\epsilon \underline{G}_{u\mu}[\mu(\epsilon), \underline{u}(\epsilon) \,| \underline{u}_\epsilon] +$$
$$\underline{G}_{uu}[\mu(\epsilon), \underline{u}(\epsilon) \,| \underline{u}_\epsilon \,| \underline{u}_\epsilon] = 0. \tag{65}$$

Consider T-periodic solutions $\underline{u}(\mu)$ of

$$\underline{G}[\mu, \underline{u}(\mu)] = 0. \tag{66}$$

Differentiating with respect to μ, we get

$$\underline{G}_\mu[\mu, \underline{u}(\mu)] + \underline{G}_u[\mu, \underline{u}(\mu) \,| \underline{u}_\mu] = 0 \tag{67}$$

and

$$\underline{G}_u[\mu, \underline{u}(\mu) \,| \underline{u}_{\mu\mu}] + \underline{G}_{\mu\mu}[\mu, \underline{u}(\mu)]$$
$$+ 2\underline{G}_{u\mu}[\mu, \underline{u}(\mu) \,| \underline{u}_\mu] + \underline{G}_{uu}[\mu, \underline{u}(\mu) \,| \underline{u}_\mu \,| \underline{u}_\mu] = 0. \tag{68}$$

At the singular point (μ_0, ϵ_0) we find from (67) and (68) that

$$\mu_\epsilon^2(\epsilon_0) < \underline{G}_{\mu\mu}(\mu_0, \underline{u}_0), \underline{\zeta}_0^* >_T + 2\mu_\epsilon(\epsilon_0) < \underline{G}_{\mu\mu}[\mu_0, \underline{u}_0 \,| \underline{u}_\epsilon(\epsilon_0)], \underline{\zeta}_0^* >_T$$
$$+ \, <\underline{G}_{uu}(\mu_0, \underline{u}_0 \,| \underline{u}_\epsilon(\epsilon_0) \,| \underline{u}_\epsilon(\epsilon_0)), \underline{\zeta}_0^* >_T = 0 \tag{69}$$

and

$$<\underline{G}_{\mu\mu}(\mu_0, \underline{u}_0), \underline{\zeta}_0^* >_T + 2 < G_{u\mu}[v_0, \underline{u}_0 \,| \underline{u}_\mu(\mu_0)], \underline{\zeta}_0^* >_T$$
$$+ \, <\underline{G}_{uu}[\mu_0, \underline{u}_0 \,| \underline{u}_\mu(\mu_0) \,| \underline{u}_\mu(\mu_0)], \underline{\zeta}_0^* >_T = 0. \tag{70}$$

Equations 69 and 70 are the starting place for the derivation of a characteristic quadratic equation, like (7), for determining the tangents to the curves intersecting at a double point. They are not suitable for this purpose as yet because the quantities $\underline{u}_\epsilon(\epsilon_0)$ and $\underline{u}_\mu(\mu_0)$ depend on the branch on which they are evaluated. But we may use the factorization theorem to derive characteristic quadratic equations of the appropriate form. With this aim in mind we turn to (53c), which may be written as

$$\underline{\zeta}(\epsilon) = \frac{<\underline{\zeta}, \underline{\zeta}^* >_T}{<\underline{u}_\epsilon, \underline{\zeta}^* >_T} \{\underline{u}_\epsilon(\epsilon) + \mu_\epsilon(\epsilon) \underline{q}(\epsilon)\} \tag{71}$$

$$= \frac{<\underline{\zeta}, \underline{\zeta}^* >_T}{\epsilon_\mu <\underline{u}_\epsilon, \underline{\zeta}^* >_T} \epsilon_\mu \{\underline{u}_\epsilon + \mu_\epsilon(\epsilon) \underline{q}(\epsilon)\}$$

$$= \frac{<\underline{\zeta}, \underline{\zeta}^* >_T}{<\underline{u}_\mu, \underline{\zeta}^* >_T} \{\underline{u}_\mu(\mu) + \underline{q}(\mu)\} = \underline{\zeta}(\mu),$$

where, in keeping with our notational convention, we have suppressed the T-periodic dependence of $\underline{\zeta}$, $\underline{\zeta}^*$, $\underline{u}$, q on t. At the singular point (μ_0, ϵ_0), $\underline{\zeta}_0$ satisfying $\underline{G}_u(\mu_0, \underline{u}_0 \,|\, \underline{\zeta}_0) = 0$ and $\underline{q}_0$ satisfying

$$F_\mu(\mu_0, \underline{u}_0) - \underline{G}_u(\mu_0, \underline{u}_0 \,|\, \underline{q}_0) = 0, \quad <\underline{q}_0, \underline{\zeta}_0^*>_T = 0 \tag{72}$$

are determined by the point (μ_0, ϵ_0) and are independent of the branch passing through the point. We next introduce slope parameters

$$\tilde{\mu}_\epsilon(\epsilon) = \mu_\epsilon(\epsilon)\, \frac{<\underline{\zeta}(\epsilon), \underline{\zeta}^*(\epsilon)>_T}{<\underline{u}_\mu(\mu), \underline{\zeta}^*(\mu)>_T} \tag{73a}$$

and

$$\tilde{\epsilon}_\mu(\mu) = \frac{<\underline{\zeta}(\mu), \underline{\zeta}^*(\mu)>_T}{<\underline{u}_\mu(\mu), \underline{\zeta}^*(\mu)>_T}. \tag{73b}$$

Combining (71) evaluated at $\epsilon = \epsilon_0$ with (69) we find that

$$\tilde{\mu}_\epsilon^2(\epsilon_0)\, \mathcal{G}_{\mu\mu} + 2\tilde{\mu}_\epsilon(\epsilon_0)\, \mathcal{G}_{u\mu} + \mathcal{G}_{uu} = 0, \tag{74}$$

where

$$\mathcal{G}_{\mu\mu} = <\underline{G}_{\mu\mu}(\mu_0, \underline{u}_0), \underline{\zeta}^*_0>_T + <\underline{G}_{uu}(\mu_0, \underline{u}_0 \,|\, \underline{q}_0 \,|\, \underline{q}_0), \underline{\zeta}^*_0>_T$$
$$- 2<\underline{G}_{u\mu}(\mu_0, \underline{u}_0 \,|\, \underline{q}_0), \underline{\zeta}^*_0>_T,$$

$$\mathcal{G}_{u\mu} = <\underline{G}_{u\mu}(\mu_0, \underline{u}_0 \,|\, \underline{\zeta}_0), \underline{\zeta}^*_0>_T - <\underline{G}_{uu}(\mu_0, \underline{u}_0 \,|\, \underline{q}_0 \,|\, \underline{\zeta}_0), \underline{\zeta}^*_0>_T$$

and

$$\mathcal{G}_{uu} = <\underline{G}_{uu}(\mu_0, \underline{u}_0 \,|\, \underline{\zeta}_0 \,|\, \underline{\zeta}_0), \underline{\zeta}^*_0>_T$$

depend only on (μ_0, ϵ_0) and not on the branch passing through (μ_0, ϵ_0). Combining (71) with (70) we find that

$$\mathcal{G}_{\mu\mu} + 2\tilde{\epsilon}_\mu(\mu_0)\, \mathcal{G}_{u\mu} + \tilde{\epsilon}_\mu^2(\mu_0)\, \mathcal{G}_{uu} = 0. \tag{75}$$

To complete the transformation of the problem of T-periodic bifurcation into the framework of the analysis in $\mathbb{R}_1$ we need to obtain formulas expressing strict crossing. The following perturbation formulas hold at (μ_0, ϵ_0):

$$\gamma_\mu(\mu_0)\underline{\zeta}_0 = \underline{G}_u(\mu_0, \underline{u}_0 \,|\, \underline{\zeta}_v) + \underline{G}_{\mu u}(\mu_0, \underline{u}_0 \,|\, \underline{\zeta}_0)$$
$$+ \underline{G}_{uu}(\mu_0, \underline{u}_0 \,|\, \underline{u}_\mu(\mu_0) \,|\, \underline{\zeta}_0)$$

and

$$-\hat{\gamma}_\epsilon \underline{u}_\epsilon + \mu_\epsilon \underline{G}_{\mu\mu}(\mu_0, \underline{u}_0) + \underline{G}_{u\mu}(\mu_0, \underline{u}_0 \,|\, \underline{u}_\epsilon)$$
$$+ (\gamma_\epsilon \underline{q}_0 - \mu_\epsilon \underline{G}_{u\mu}(\mu_0, \underline{u}_0 \,|\, \underline{q}_0) - \underline{G}_{uu}(\mu_0, \underline{u}_0 \,|\, \underline{u}_\epsilon \,|\, \underline{q}_0))$$
$$- \underline{G}_u(\mu_0, \underline{u}_0 \,|\, \underline{q}_\epsilon) = 0.$$

Hence

$$\hat{\gamma}_\epsilon <\underline{u}_\epsilon, \underline{\zeta}^*_0>_T + \mu_\epsilon <\underline{G}_{\mu\mu}, \underline{\zeta}^*_0>_T + <\underline{G}_{u\mu}(\mu_0, \underline{u}_0 \,|\, \underline{u}_\epsilon), \underline{\zeta}_0^*>_T$$
$$- \mu_\epsilon <\underline{G}_{u\mu}(\mu_0, \underline{u}_0 \,|\, \underline{q}_0), \underline{\zeta}^*_0>_T - <\underline{G}_{uu}(\mu_0, \underline{u}_0 \,|\, \underline{u}_\epsilon \,|\, \underline{q}_0), \underline{\zeta}^*_0>_T = 0 \tag{76}$$

and

$$\gamma_\mu <\underline{\zeta}_0, \underline{\zeta}^*_0> \; = \; <\underline{G}_{uu}(\mu_0, \underline{u}_0 | \underline{\zeta}_0), \underline{\zeta}^*_0 >_T$$

$$+ \; <\underline{G}_{uu}(\mu_0, \underline{u}_0 | \underline{u}_\mu | \underline{\zeta}_0), \underline{\zeta}^*_0 >_T. \quad (77)$$

Introducing (71) and (73a) into (76), we find that

$$\hat{\gamma}_\epsilon(\epsilon_0) \; = \; \tilde{\mu}_\epsilon(\epsilon_0) \, g_{\mu\mu} \; - \; g_{\mu\mu}. \quad (78)$$

Introducing (71) and (73b) into (77) we find that

$$\gamma_\mu(\mu_0) \; = \; g_{\mu\mu} \; + \; \tilde{\epsilon}_\mu(\mu_0) \, g_{uu}. \quad (79)$$

We have now completed the reduction to $\mathbb{R}_1$ and can prove Theorems 1 through 5 for T-periodic solutions. The proof of these theorems, summarized as Theorem 8 follows, word for word, the proof given for $\mathbb{R}_1$ in the first section when we replace F_ϵ, F_μ, $F_{\mu\mu}$, $F_{\epsilon\mu}$, $F_{\epsilon\epsilon}$, and $D = F_{\epsilon\mu}^2 - F_{\epsilon\epsilon} F_{\mu\mu}$ with $\mathcal{G}_u$, $\mathcal{G}_\mu$, $\mathcal{G}_{\mu\mu}$, $\mathcal{G}_{u\mu}$, $\mathcal{G}_{uu}$, and $D = \mathcal{G}^2_{u\mu} - \mathcal{G}_{uu} \mathcal{G}_{\mu\mu}$.

THEOREM 8. Under the assumptions (H.1) and (H.2) the following factorization holds:

$$\gamma(\epsilon) \; = \; -\mu_\epsilon(\epsilon) \, \hat{\gamma}(\epsilon)$$

where $\hat{\gamma}(\epsilon)$ is given by (50c). The term $\gamma(\epsilon)$ must change sign as ϵ is varied across a regular turning point. Any point (μ_0, ϵ_0) of the curve $\mu = \mu(\epsilon)$ for which $\gamma(\epsilon_0) = 0$ is a singular point. If (H.3) holds and $\gamma_\epsilon(\epsilon_0) \neq 0)$ or $\gamma_\mu(\mu_0) \neq 0$, then (μ_0, ϵ_0) is a double point. The stability of the two solutions bifurcating at the double point is determined in the linearized approximation by

$$\gamma^{(1)}(\epsilon) \; = \; -\mu_\epsilon^{(1)}(\epsilon) \{ \hat{s} \sqrt{\tilde{D}} (\epsilon - \epsilon_0) + o(\epsilon - \epsilon_0) \}$$

and

$$\gamma^{(2)}(\epsilon) \; = \; \mu_\epsilon^{(2)}(\epsilon) \{ \hat{s} \sqrt{\tilde{D}} (\epsilon - \epsilon_0) + o(\epsilon - \epsilon_0) \}$$

where $\hat{s} = \mathcal{G}_{\mu\mu} / |\mathcal{G}_{\mu\mu}|$, or by

$$\gamma^{(1)}(\mu) \; = \; s \sqrt{\tilde{D}} (\mu - \mu_0) + O(\mu - \mu_0)$$

and

$$\gamma^{(2)}(\epsilon) \; = \; -s\mu_\epsilon^{(2)}(\epsilon) \{ \sqrt{\tilde{D}} (\epsilon - \epsilon_0) + o(\epsilon - \epsilon_0),$$

where $s = \mathcal{G}_{uu} / |\mathcal{G}_{uu}|$. Assume that all singular points of $\underline{G}(\mu, \underline{u}) = 0$ are double points. The stability of such T-periodic solutions must change at each regular turning point and at each singular point (which is not a turning point) and only at such points.

Theorem 8 holds for steady solutions. I think that theorem 8 implies that the results on stability of steady solutions which have been given by Benjamin,[1] using topological degree, hold also for steady and T-periodic solutions under (H.1), (H.2), and (H.3). For the index of a solution I get

$$i \; = \; \text{sign} \; \frac{<\underline{G}_u(\mu, \underline{u} | \underline{\zeta}), \underline{\zeta}^* >_T}{<\underline{\zeta}, \underline{\zeta}^* >_T} \; = \; \text{sign} \left\{ -\mu_\epsilon(\epsilon) \; \frac{<F_\mu(\mu, \underline{u}), \underline{\zeta}^* >_T}{<\underline{\zeta}, \underline{\zeta}^* >_T} \right\},$$

so that i = sign γ when $\gamma \neq 0$ and, more generally, $i = [i(+) + i(-)]/2$ where i is the index of a solution at a point of the bifurcation curve and $i(+)$ are limiting values on one and the other side of the point. Then $i = 0$ at double points that are not turning points, and $i = \pm 1$ at turning points that are double points.

REFERENCES

1. BENJAMIN, T. B. 1976. Application of Leray-Schauder degree theory to problems of hydrodynamic stability. Math. Proc. Cambridge Phil. Soc. **79**: 373.
2. BENJAMIN, T. B. 1978. Bifurcation phenomena in steady flows of a viscous fluid. Part 1, Theory and Part 2, Experiments. Proc. R. Soc. **359**: 1.
3. CRANDALL, M. & P. H. RABINOWITZ. In press. The Hopf bifurcation theorem in infinite dimensions. Arch. Rational Mech. Anal.
4. IOOSS, G. & D. D. JOSEPH. 1977. Bifurcation and Stability of nT-periodic solutions branching from T-periodic solutions at points of resonance. Arch. Rational Mech. Anal. **66**: 135.
5. JOSEPH, D. D. 1976. Stability of fluid motions, I and II. Springer Tracts in Nat. Phil. **27** and **28**.
6. JOSEPH, D. D. 1977. Factorization theorems, stability, and repeated bifurcation. Arch. Rational Mech. Anal. **66**: 99.
7. JOSEPH, D. D. & D. A. NIELD. 1975. Stability of bifurcating time-periodic and steady solutions of arbitrary amplitude. Arch. Rational Mech. Anal. **58**: 369.
8. ROSENBLAT, S. 1977. Global aspects of Hopf bifurcation and stability. Arch. Rational Mech. Anal. **66**: 119.
9. WEINBERGER, H. In press. The stability of solutions bifurcating from steady or periodic solutions. Intern. Symp. on Dynamical Systems, Gainsville, 1976. Academic Press, New York, N.Y.

THE NONOCCURRENCE OF SECONDARY
BIFURCATION WITH HAMMERSTEIN OPERATORS

George H. Pimbley

Theoretical Division
Los Alamos Scientific Laboratory
Los Alamos, New Mexico 87545

We write the standard eigenvalue problem for Hammerstein's operator as follows:

$$\lambda x(s) = \int_0^1 K(s,t) f(t, x(t)) d\sigma(t), \quad f(t,0) \equiv 0. \qquad (1)$$

Such problems often result, for example, when we convert semilinear elliptic differential equations with boundary conditions to integral equation form, in which case the kernel $K(s,t)$ is the appropriate Green's function.

We single out two kinds of nonlinearity for Equation 1:

Superlinear: $0 < \dfrac{f(t,x)}{x} < f_x(t,x), \quad x \neq 0$

$$(2)$$

Sublinear: $\quad 0 < f_x(t,x) < \dfrac{f(t,x)}{x}, \quad 0 \le t \le 1, -\infty < x < \infty.$

We shall dwell on the superlinear case, with occasional remarks about the sublinear case.

Otherwise we assume that $K(s,t)$ is continuous symmetric and positive definite, that $f(s,-x) = -f(s,x)$, and that $f(s,x)$ is continuous and differentiable in x as many times as needed. The term $\sigma(t)$ is monotonic increasing.

Under these conditions it is known that problem (1) has a branch $x_\lambda^{(j)}(s)$ of solutions, parametrized by $\lambda > 0$, which bifurcates from the trivial solution at the jth eigenvalue μ_j^0 of the linearized problem:

$$\mu h(s) = \int_0^1 K(s,t) f_x(t, x(t)) h(t) d\sigma(t), \qquad (3)$$

centered at $x(t) \equiv 0.$[1,2] We *assume here that* μ_j^0 *has unit multiplicity.* In the superlinear case this branch of solutions tends locally to the right (i.e. to values $\lambda > \mu_j^0$); the opposite is true for the sublinear case.

Problem (3) is assumed to have a complete system of eigenvalues $\mu_k^{(x)}$, with $0 < \mu_{k+1}^{(x)} < \mu_k^{(x)}, k = 1, 2, 3, \ldots$ which depend on x. If $x = x_\lambda^{(j)}$ then the eigenvalues $\mu_k^{(x)}$ are parametrized by λ.

Another linearized problem will be of interest in what follows, namely,

$$\mu \hat{h}(s) = \int_0^1 K(s,t) \frac{f(t, x(t))}{x(t)} \hat{h}(t) d\sigma(t). \qquad (4)$$

Problem (4) has a complete system of eigenvalues $\xi_k^{(x)}$ with $0 < \xi_{k+1}^{(x)} \le \xi_k^{(x)},$

0077-8923/78/0316-0168 $1.75/1 © 1979, NYAS

$k = 1, 2, 3, \ldots$. It is readily proven that on the jth branch, i.e., $x \equiv x_\lambda^{(j)}$, we have $\xi_j^{(x_\lambda)} \equiv \lambda$.[3]

THEOREM 1. In the superlinear (sublinear) case, there can be no secondary bifurcation (e.g., formation of twigs) on the jth branch $x_\lambda^{(j)}$ provided that

$$\mu_{j+1}^{(x_\lambda)} < \lambda < \mu_j^{(x_\lambda)} \quad (\mu_j^{(x_\lambda)} < \lambda < \mu_{j-1}^{(x_\lambda)}). \tag{5}$$

Note 1. In (5), $x_\lambda \equiv x_\lambda^{(j)}$ in the superscript.

Note 2. It can be seen by comparing problems (3) and (4) that we always have $\lambda < \mu_j^{(x_\lambda)}$ ($\mu_j^{(x_\lambda)} < \lambda$) in the superlinear (sublinear) case. Thus the right (left) inequality in (5) is already satisfied. The question of secondary bifurcation really hinges therefore on whether we have $\mu_{j+1}^{(x_\lambda)} < \lambda$ ($\lambda < \mu_{j-1}^{(x_\lambda)}$) in the superlinear (sublinear) case.[1]

THEOREM 2. Sufficient for the nonoccurence of secondary bifurcation on the jth branch $x_\lambda^{(j)}$ is the restriction of that branch (if possible) to a region of the space $C(0, 1)$ defined by the inequality

$$F_j(x) = \max_{0 \le s \le 1} \frac{x(s) f_x(s, x(s))}{f(s, x(s))} - \frac{\mu_j^{(x)}}{\mu_{j+1}^{(x)}} < 0 \quad \text{(superlinear)},$$

$$\hat{F}_j(x) = \min_{0 \le s \le 1} \frac{x(s) f_x(s, x(s))}{f(s, x(s))} - \frac{\mu_j^{(x)}}{\mu_{j-1}^{(x)}} > 0 \quad \text{(sublinear)}. \tag{6}$$

Proof for the Superlinear Case. We let K stand for the linear integral operator generated by the kernel $K(s, t)$ and we let H be its square root, i.e., $H^2 = K$. Then using symmetrized versions of operators in (3) and (4),

$$\mu_{j+1}^{(x)} = \frac{\mu_{j+1}^{(x)}}{\mu_j^{(x)}} \cdot \min_{h_1,\ldots,h_{j-1}} \max_{h \perp h_1,\ldots,h_{j-1}} \frac{(H f_x Hh, h)}{(h, h)}$$

$$\le \max_{h \perp Z_1,\ldots,Z_{j-1}} \frac{\mu_{j+1}^{(x)}}{\mu_j^{(x)}} \frac{(H f_x Hh, h)}{(h, h)}$$

$$= \max_{h \perp Z_1,\ldots,Z_{j-1}} \frac{\mu_{j+1}^{(x)}}{\mu_j^{(x)}} \frac{\displaystyle\int_0^1 f_x(t, x(t))\{Hh\}^2 d\sigma(t)}{\displaystyle\int_0^1 |h|^2 d\sigma(t)}$$

$$< \max_{h \perp Z_1,\ldots,Z_{j-1}} \frac{\displaystyle\int_0^1 \frac{f(t, x(t))}{x(t)} \{Hh\}^2 d\sigma(t)}{\displaystyle\int_0^1 |h|^2 d\sigma(t)} = \xi_j^{(x)} = \lambda$$

using inequality (6) in the last step; $Z_1, \ldots, Z_{j-1}$ are eigenelements of the operator $H \frac{f}{x} H$, which is the symmetrized operator in problem (4). In this proof, $x \equiv x_\lambda^{(j)}$ everywhere. Use has been made of the fact that setwise, $C(0, 1) \subset L_2(0, 1)$. This ends the proof.

Corollary. If $\mu_j{}^0$ is simple, the region of no secondary bifurcation in $C(0, 1)$ contains a spherical neighborhood of the origin.

The integral equation theory is carried forward from this point in a recently published paper by the author, with certain expository difficulties.[4] For better exposition here, the author now resorts to discretization. Subdividing the interval as follows:

$$0 \le t_1 \le t_2 < \cdots < t_N \le 1,$$

let us define the Stieltjes integrator function $\sigma(t)$ as follows: $\sigma(t) \equiv b_k$, $t_i < t < t_{i+1}$, where $0 < b_i < b_{i+1}$. Then an N-dimensional analogue of problem (1) is

$$\lambda x_i - \sum_{k=1}^{N} a_{ik} f_k(x_k) \quad i = 1, \ldots, N \tag{7}$$

Our results with Equation 7 are generally indicative of the results for Equation 1. Moreover, discretized results with Equation 7 are more immediately extendable to a discretization of a Hammerstein equation such as (1) but with a multiple integral over a domain of higher dimension.

Of course in (7) we define $a_{ik} = K(s_i, t_k)$ and $f_k(x_k) = \Delta_k f(t_k, x_k)$, with $\Delta_k = b_k - b_{k-1}$ and $x_k = x(t_k)$. We can write conditions on the nonlinear functions $f_k(x_k)$ that are completely analogous to those above for $f(s, x)$:

oddness: $f_k(-x_k) = -f_k(x_k)$

monotonicity: $f'_k(x_k) > 0$

superlinearity: $0 < \dfrac{f_k(x_k)}{x_k} < f'_k(x_k)$

(sublinearity): $0 < f'_k(x_k) < \dfrac{f_k(x_k)}{x_k}, \quad -\infty < x_k < \infty. \tag{8}$

Additionally it will be nice to assume power-law asymptotic nonlinear behavior for $f_k(x_k)$:

$$\lim_{x_k \to \infty} \frac{f_k(x_k)}{x_k{}^n} = c_k, \quad \begin{array}{l} n = \text{the asymptotic power} \\ n > 1, \, k = 1, \ldots, N. \end{array} \tag{9}$$

In the superlinear case, which is our stated interest here, it is convenient to assume that

$$f'''_k(x_k) > 0, \quad k = 1, \ldots, N \quad 0 < x_k < \infty$$

$$\left[\frac{x_k f'_k(x_k)}{f_k(x_k)}\right]''' < 0 \text{ whenever } \left[\frac{x_k f'_k(x_k)}{f_k(x_k)}\right]'' \ge 0 \tag{10}$$

An example of such a function is $f_k(x_k) = x_k + a_k x_k{}^3$ with $a_k > 0, k = 1, \ldots, N$.

One can write a discretized linear problem analogous to (3), as follows:

$$\mu h_i = \sum_{k=1}^{N} a_{ik} f'_k(x_k) h_k, \quad i = 1, \ldots, N \tag{11}$$

with positive eigenvalues $\mu_l^{(x)}$, $l = 1, \ldots, N$, and with $0 < \mu_{l+1}^{(x)} < \mu_l^{(x)}$, where x represents the vector $x = \{x_k\}$.

The functional $F_j(x)$ defining the no-bifurcation region for the jth branch $x_\lambda^{(j)}$ also has a discrete form, analogous to (6):

$$F_j(x) = \max_{1 \leq i \leq N} \frac{x_i f_i'(x_i)}{f_i(x_i)} - \frac{\mu_j^{(x)}}{\mu_{j+1}^{(x)}} < 0 \quad \text{(superlinear)}$$

$$\hat{F}_j(x) = \min_{1 < i \leq N} \frac{x_i f_i'(x_i)}{f_i(x_i)} - \frac{\mu_j^{(x)}}{\mu_{j-1}^{(x)}} > 0 \quad \text{(sublinear)} \qquad (12)$$

where of course $x = \{x_i\}$ is a vector in the space R_N. Functionals (12) define no-bifurcation regions in R_N as indicated by the inequalities in (12).

Note 3. Consider the locus of all points x in R_N such that

$$f_1'(x_1) = f_2'(x_2) = \cdots = f_N'(x_N), \; x_i > 0, \; i = 1, \ldots, N. \qquad (13)$$

Along this locus the ratio

$$\frac{\mu_j^{(x)}}{\mu_{j+1}^{(x)}} \left(\frac{\mu_j^{(x)}}{\mu_{j-1}^{(x)}} \right)$$

is constant. For functions $f_i(x_i)$ sufficiently nonlinear that

$$\min_{1 \leq i \leq N} \frac{x_i f_i'(x_i)}{f_i(x_i)}$$

exceeds this constant ratio, locus (13) cannot remain in the no-bifurcation region, as can be seen from (12). This would be true if

$$n > \frac{\mu_j^{(x)}}{\mu_{j+1}^{(x)}}$$

in the superlinear case with power law asymptotic behavior, (see Equation 9). A corresponding property would hold for the same locus in the sublinear case. Since the functions $f_i'(x_i)$, $i = 1, \ldots, N$, are even in the variable x_i, we see that there are similar locii, apart from (13) in the positive orthant, along which the ratio $\mu_j^{(x)}/\mu_{j+1}^{(x)}$ is constant. This shows the importance of a delineation of the no-bifurcation region.

Hence the positive orthant (cone) K_N of the space R_N is likely to contain in its interior a region where the functional $F_j(x)$ is positive; if the branch $x_\lambda^{(j)}$ of solutions of (7) remains in the interior of K_N, secondary bifurcation may happen. This causes us to be interested in the boundary of K_N as a place to identify points of the no-bifurcation region.

It turns out that the suborthants

$$K_j^{(i)}, \; i = 1, \ldots, \frac{N!}{j!(N-j)!}$$

each spanned by sets of j positive axes $x_{n1}, x_{n2}, \ldots, x_{nj}$ of R_N, are crucial in defining the no-bifurcation region for the jth branch $x_\lambda^{(j)}$.

For simplicity we restrict attention to that suborthant $K_k^{(1)}$ spanned by the axes $x_1, \ldots, x_j$: we should proceed similarly with any other suborthant $K_j^{(i)}$. Indeed by reordering the equations in problem (7), and relabeling the variables, considerations relative to any positive suborthant $K_j^{(i)}$ can be transformed to the same considerations relative to $K_j^{(1)}$.

A representative ray properly interior to $K_j^{(1)}$ would be characterized by direction components $\nu_i > 0$, $i = 1, \ldots, j$; $\nu_i = 0$, $i = j + 1, \ldots, N$; $\nu_1^2 + \nu_2^2 + \cdots + \nu_j^2 = 1$, as the locus

$$x_1 = \rho\nu_1, x_2 = \rho\nu_2, \ldots, x_j = \rho\nu_j, 0 < \rho < \infty. \tag{14}$$

So as to study the ratio $\mu_j^{(x)}/\mu_{j+1}^{(x)}$ using variational principles for eigenvalues, we symmetrize problem (11) and center it along the representative ray (14). Its matrix becomes

$$A(\nu_1, \ldots, \nu_j)$$

$$= \begin{pmatrix} a_{11}f_1'(\rho\nu_1) & \cdots & a_{1j}\sqrt{f_1'(\rho\nu_1)f_j'(\rho\nu_j)} & a_{1N}\sqrt{f_1'(\rho\nu_1)f_N'(0)} \\ \vdots & \ddots & \vdots & \vdots \\ a_{j1}\sqrt{f_1'(\rho\nu_1)f_j'(\rho\nu_j)} & \cdots & a_{jj}f_j'(\rho\nu_j) & \cdots \\ a_{N1}\sqrt{f_1'(\rho\nu_1)f_N'(0)} & \cdots & \cdots & \ddots \\ & & & a_{NN}f_N'(0) \end{pmatrix} \tag{15}$$

Using matrix (15) in the symetrized linearization of problem (11) we can prov the following result:

THEOREM 3. Let $A_u^{(j)}$ and $A_L^{(j)}$ be respectively the matrices

$$A_u^{(j)} = \begin{pmatrix} a_{11} & \cdots & a_{1N} \\ \vdots & \ddots & \vdots \\ a_{j1} & \cdots & a_{jj} \end{pmatrix} \quad A_L^{(j)} = \begin{pmatrix} a_{j+1,j+1} & \cdots & a_{j+1,N} \\ \vdots & \ddots & \vdots \\ a_{N,j+1} & \cdots & a_{NN} \end{pmatrix}$$

with $A_u^{(j)}$ having *least* (positive) eigenvalue u_j^{Au}, and $A_L^{(j)}$ having greatest (positive) eigenvalue μ_j^{AL}. Further, let $\underline{f}'(x_1) = \min_{1 \leq i \leq j} f_i'(x_1)$ and $\overline{f}'(x_1) = \max_{j+1 \leq i \leq N} f_i'(x_1)$. Then the main diagonal of the suborthant $K_j^{(1)}$, i.e., the ray with direction components $\nu_1 = \nu_2 = \cdots \nu_j$, and $0 \leq \rho < \infty$, is properly interior to the no-bifurcation region provided that

$$\frac{\mu_j^{Au}}{\mu_{j+1}^{AL}} > \frac{\overline{f}'(0)}{\underline{f}'(0)}, \quad \text{and} \quad \frac{f_i'''(0)}{f_i'''(0)} \geq \frac{2}{3} \tag{16}$$

for all $i, l = 1, \ldots, j$.

Proof. Using variational principles and matrix **(15)**,

$$
\frac{\mu_j^{(x)}}{\mu_{j+1}^{(x)}} \geq \frac{\displaystyle\min_{u_1^2+\cdots+u_j^2=1} \sum_{i=1}^{j}\sum_{k=1}^{j} a_{ik}\sqrt{f_i'(\rho\nu_1)f_k'(\rho\nu_1)}\,u_i u_k}{\displaystyle\max_{\nu_{j+1}^2+\cdots+\nu_N^2=1} \sum_{i=j+1}^{N}\sum_{k=j+1}^{N} a_{ik}\sqrt{f_i'(0)f_k'(0)}\,v_i v_k}
$$

$$
\geq \frac{\displaystyle\min_{u_1^2+\cdots+u_j^2=1} \sum_{i=1}^{j}\sum_{k=1}^{j} a_{ik} u_i u_k}{\displaystyle\max_{\nu_{j+1}^2+\cdots+\nu_N^2=1} \sum_{i=j+1}^{N}\sum_{k=j+1}^{N} a_{ik} v_i v_k} \cdot \frac{f'(\rho\nu_1)}{\overline{f}'(0)}
$$

$$
\geq \frac{\mu_j^{Au}}{\mu_{j+1}^{Au}} \cdot \frac{f'(\rho\nu_1)}{\overline{f}'(0)}. \tag{17}
$$

By an elementary calculation, it turns out that

$$
\left[\frac{x_i f_i'(x_i)}{f_i(x_i)}\right]''\bigg|_{x_i=0} = \frac{2}{3}\frac{f_i'''(0)}{f_i'(0)} > 0 \quad i = 1,\ldots,N \tag{18}
$$

(see **(8)** and **(10)**), and that for $1 \leq i \leq J$, using **(10)**,

$$
\frac{2}{3}\frac{f_i'''(0)}{f_i'(0)} < \frac{\mu_j^{Au}}{\mu_{j+1}^{AL}}\frac{f_l'''(0)}{f_k'(0)} \quad \begin{array}{l} \text{for any } l = 1,\ldots,j \\ \text{and any } k = j+1,\ldots,N. \end{array} \tag{19}
$$

Thus

$$
\frac{d^2}{d\rho^2}\left[\max_{1\leq i\leq j} \frac{\rho\nu_i f_i'(\rho\nu_i)}{f_i(\rho\nu_i)}\right]\bigg|_{\rho=0} < \frac{d^2}{d\rho^2}\left[\frac{\mu_j^{Au}}{\mu_{j+1}^{AL}}\frac{f'(\rho\nu_1)}{\overline{f}'(0)}\right]\bigg|_{\rho=0}
$$

where we have used **(18)** and **(19)**. Then from **(10)** and the consequent divergence of the second derivatives as $\rho > 0$ increases, we see that the two curves (FIGURE 1) start separately at $\rho = 0$, and remain separate for $\rho > 0$. In FIGURE 1,

$$
\Lambda_j = \frac{\mu_j^{Au}}{\mu_{j+1}^{AL}} \cdot \frac{f'(0)}{\overline{f}'(0)} > 1
$$

by **(16)**. Thus

$$
\max_{1\leq i\leq j} \frac{\rho\nu_i f_i'(\rho\nu_i)}{f_i(\rho\nu_i)} < \frac{\mu_i^{(x)}}{\mu_{j+1}^{(x)}}, \quad \begin{array}{l} 0 \leq \rho < \infty \\ \nu_1 = \nu_2 = \cdots = \nu_j. \end{array}
$$

and the functional $F_j(x)$ in **(12)** is negative on the main diagonal of $K_j^{(1)}$.

Off the main diagonal of $K_j^{(1)}$ we have the estimate:

$$\frac{\mu_j^{(x)}}{\mu_{j+1}^{(x)}} \geq \frac{\displaystyle\max_{u_1^2+\cdots+u_j^2=1}\sum_{i=1}^{j}\sum_{k=1}^{j} a_{ik}\sqrt{f_i'(\rho v_i)f_k'(\rho v_k)}\,u_i u_k}{\displaystyle\max_{v_{j+1}^2+\cdots+v_N^2=1}\sum_{i=j+1}^{N}\sum_{k=j+1}^{N} a_{ik}\sqrt{f_i'(\rho v_i)f_k'(\rho v_k)}\,v_i v_k}$$

$$\rightarrow L_n(v_1,\ldots,v_N) < \infty \quad \text{as} \quad \rho \rightarrow \infty$$

where $L_n(v_1,\ldots,v_N)$ is a finite asymptotic level, n being the asymptotic power of the nonlinearity defined in (9).

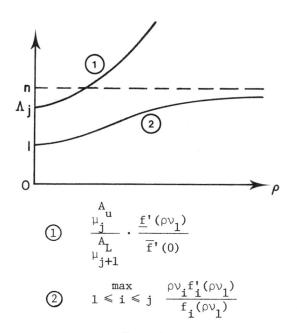

$$\text{①} \quad \frac{\mu_j^{A_u}}{\mu_{j+1}^{A_L}} \cdot \frac{f'(\rho v_1)}{f'(0)}$$

$$\text{②} \quad \max_{1 \leq i \leq j} \frac{\rho v_i f_i'(\rho v_1)}{f_i(\rho v_1)}$$

FIGURE 1.

By perturbing the direction components $v_1,\ldots,v_N$, in a neighborhood of the diagonal of the suborthant $K_j^{(1)}$, on the unit sphere $v_1^2 + \cdots + v_N^2 = 1$, we can produce a bundle of rays such that $F_j(\rho v_1,\ldots,\rho v_N) < 0$, by continuity of the curves (1) and (2) of FIGURE 1, during such perturbation.

This proves the Theorem as stated. Likewise are the diagonals of the (positive) suborthants $K_j^{(i)}$, $i = 2, 3, \ldots, N$ properly contained in the no-bifurcation region under conditions similar to (16) involving other submatrices for $A_u^{(j)}$, $A_L^{(j)}$. Also, since the functions $f_i'(\rho v_i)$ and $f_i(\rho v_i)/(\rho v_i)$ are even in ρv_i, suborthants bounded by negative half axes, or by mixed negative and positive half axes in R_N, have diagonals which are also properly contained in the no-bifurcation region each time the diagonal of $K_j^{(1)}$ is properly contained therein (see Equation 12).

Note 4. When $j = 1$, the orthants $K_1^{(1)}$, $K_1^{(2)}, \ldots, K_1^{(N)}$ are the positive coordinate axes of R_N. When $j = 2$, the orthants $K_2^{(1)}, K_2^{(2)}, \ldots, K_2^{(N(N-1)/2)}$ are just the positive coordinate planes of R_N, etc.

Note 5. The successive conditions (**16**) written for $j = 1, 2, \ldots, N$, for the proper containment of the main diagonals of the orthants $K_j^{(1)}$ is the no-bifurcation region, are compatible, i.e., they can be satisfied simultaneously. These conditions can be regarded as being on the eigenvalues $\mu_j^{A_u}$, $\mu_{j+1}^{A_L}$ of, respectively, the jth order, $(N - j)$th order, principal minors, *or* as being on the numbers $f_1'(0), f_2'(0), \ldots, f_N'(0)$, *or* on both the eigenvalues and the numbers.

Theorem 3 merely gives conditions under which an opening occurs through the bifurcation region for the evolution of the branch $x_\lambda^{(j)}$ without secondary bifurcation (in the discretized case). This does not yet mean that $x_\lambda^{(j)}$ enters the opening. How can $x_\lambda^{(j)}$ possibly lie in the no-bifurcation passageway? It is necessary to find out how the matrix of the problem (**7**), or the functions $f_k(x_k)$, or both, can be altered so as to effectuate conditions (**16**), and then to see if this alteration also brings $x_\lambda^{(j)}$ into the resulting passageway.

The first condition in (**16**) can be effectuated by multiplying $A_u^{(j)}$ by a sufficiently high positive number σ. Thus the nonlinear system (**7**) to be studied would be written

$$
\begin{pmatrix}
\sigma a_{11} & \cdots & \sigma a_{1j} & \cdots & a_{1N} \\
\vdots & & \vdots & & \vdots \\
\sigma a_{j1} & \cdots & \sigma a_{jj} & & \vdots \\
\vdots & & & & \vdots \\
a_{N1} & & \cdots & & a_{NN}
\end{pmatrix}
\begin{pmatrix}
f_1(x_1) \\
\vdots \\
f_j(x_j) \\
\vdots \\
f_N(x_N)
\end{pmatrix}
= \lambda
\begin{pmatrix}
x_1 \\
\vdots \\
x_j \\
\vdots \\
x_N
\end{pmatrix}
\tag{20}
$$

Or, since $\lambda = 0(\sigma)$, it is better to write

$$
\begin{pmatrix}
a_{11} & \cdots & a_{1j} & \cdots & \sigma^{-1}a_{1N} \\
\vdots & & \vdots & & \vdots \\
a_{j1} & \cdots & a_{jj} & & \sigma^{-1}a_{jN} \\
\vdots & & \vdots & & \vdots \\
\sigma^{-1}a_{N1} & \cdots & \sigma^{-1}a_{Nj} & \cdots & \sigma^{-1}a_{NN}
\end{pmatrix}
\begin{pmatrix}
f_1(x_1) \\
\vdots \\
f_j(x_j) \\
\vdots \\
f_N(x_N)
\end{pmatrix}
= \lambda'
\begin{pmatrix}
x_1 \\
\vdots \\
x_j \\
\vdots \\
x_N
\end{pmatrix}
\tag{21}
$$

In the limit as $\sigma \to \infty$, the jth order problem

$$
\begin{pmatrix}
a_{11} & \cdots & a_{1j} \\
\vdots & & \vdots \\
a_{j1} & \cdots & a_{jj}
\end{pmatrix}
\begin{pmatrix}
f_1(x_1) \\
\vdots \\
f_j(x_j)
\end{pmatrix}
= \lambda'
\begin{pmatrix}
x_1 \\
\vdots \\
x_j
\end{pmatrix}
\tag{22}
$$

becomes isolated. Problem (**22**) has j branches of nontrivial solutions (assuming the eigenvalues of the linearization of (22) are simple when evaluated at the origin). All of these branches are contained in the subspace $R_j \subset R_N$; R_j has $K_j^{(1)}$ for a positive orthant.

The jth branch $y_{\lambda'}^{(j)}$ of isolated problem (22) has no secondary bifurcations since in (5), $\mu_{j+1} \to 0$ as $\sigma \to \infty$, and in (12) $\mu_j^{(y)}/\mu_{j+1}^{(y)}$ becomes infinite, so that $F_j(y_{\lambda'}^{(j)}) < 0$. However, $|y_{\lambda'}^{(j)}|$, the vector consisting of components, each of which is the absolute value of the corresponding component of $y_{\lambda'}^{(j)}$, need not lie in the no-bifurcation region for problem (21); i.e., $|y_{\lambda'}^{(j)}|$ need not go through the passageway unless σ is sufficiently large. As $\sigma \to \infty$, the bifurcation region for problem (21) is enlarged (cf. (12)), and in the limit includes the entire sub-orthant $K_j^{(1)}$.

With $|y_{\lambda'}^{(j)}|$ and thus $y_{\lambda'}^{(j)}$ properly contained in the no-bifurcation region of (21) for sufficiently large σ (see paragraph above Note 4), since the no-bifurcation region is an open set, and since $x_{\lambda'}^{(j)} \to y_{\lambda'}^{(j)}$ as $\sigma \to \infty$ uniformly on any finite interval, σ can be taken sufficiently large that $x_{\lambda'}^{(j)}$ is also in the no-bifurcation region of problem (21).

We can state the following result:

THEOREM 4. Suppose that the coefficients a_{ij} of isolated problem (22) be adjusted that the jth branch $y_{\lambda'}^{(j)}$ is defined as emanating from a bifurcation point of unit multiplicity. Then for sufficiently high σ there is no secondary bifurcation on the jth branch $x_{\lambda'}^{(j)}$ of solutions of problem (20).

Theorem 4 shows how the jth branch of solutions of problem (20) might be made free of secondary bifurcation, where $j = 1, 2, \ldots, N$. How can we assure that *all* N branches (assuming that all eigenvalues μ_j^0 or primary bifurcation points are of unit multiplicity) are free of secondary bifurcation? What would the kernel look like if this were to be the case?

In succession for $j = 1, 2, \ldots, N$ we can perturb the entries a_{ik} of the matrix of problem (22) with the object of making the branches $y_{\lambda'}^{(j)}$ lie as close as possible to the diagonals of the successive suborthants $K_j^{(1)}$. Then working backwards, for $j = N - 1$, we can find σ_{N-1} sufficiently high that the no-bifurcation region defined in (12) includes $y_{\lambda'}^{(N-1)}$ in its interior, and also high enough that the branch $x_{\lambda'}^{(N-1)}$ of problem (20) is close enough to $y_{\lambda'}^{(N-1)}$ to be in the no-bifurcation region also. (We do not concern ourselves with $x_{\lambda'}^{(N)}$, since in the superlinear case this is free of secondary bifurcation; this is because $\mu_{N+1}^{(x)} = 0, x = x_{\lambda}^{(N)}$.) Having redefined the matrix entries of problem (20) to take into account this constant multiplier σ_{N-1} for A'', we now find σ_{N-2} sufficiently high that the no-bifurcation region defined in (12), with $j = N - 2$, includes $y_{\lambda'}^{(N-2)}$ in its interior; also we make σ_{N-2} high enough that the branch $x_{\lambda}^{(N-2)}$ of problem (20) is close enough to $y_{\lambda'}^{(N-2)}$ to be in the no-bifurcation region as well. And so on, back to $j = 1$.

The matrix (a_{ik}) that results from this process has been affected by the need to have the successive branches $y_{\lambda'}^{(j)}$ of problem (22) close to the diagonals of the successive orthants $K_j^{(1)}, j = 1, 2, \ldots, N$. Also because of the compounding of the multipliers $\sigma_j, j = N - 1, N - 2, \ldots, 3, 2, 1$, the resulting matrix (a_{ik}) is peaked at a_{11}, and the diagonal elements, a_{kk} decrease monotonically with k. (If orthants $K_j^{(i)}$ had been used, with $1 \le i \le N$, rather than the orthants $K_j^{(1)}$, then the peak would have been elsewhere on the diagonal.)

Such matrices are similar to those that are obtained by discretizing one-dimensional Green's functions.

REFERENCES

1. BAZLEY, N. W. & G. H. PIMBLEY. 1974. A region of no secondary bifurcation for nonlinear Hammerstein operators. Z. Angew. Math. Phys. **25:** 743–751.
2. PIMBLEY, G. H. 1963. The eigenvalue problem for sublinear Hammerstein operators with oscillation kernels. J. Math Mech. **12:** 577–803.
3. PIMBLEY, G. H. 1969. Eigenfunction Branches of Nonlinear Operators, and their Bifurcations. Lecture Notes in Mathematics. Vol. **104:** 60–62. Springer-Verlag. New York.
4. PIMBLEY, G. H. 1977. Superlinear Hammerstein operators and the nonocurrence of secondary bifurcation. SIAM J. Applied Math. **33:** 298–322.

BIFURCATIONS AND BIOLOGICAL OBSERVABLES

Robert Rosen

Department of Physiology and Biophysics
Faculty of Medicine
Dalhousie University
Halifax, Nova Scotia
Canada B3H 4H7

INTRODUCTION

It is a truism that biological systems are complex. It has further come to be regarded as axiomatic that complex systems in general are "counter-intuitive." To set the stage for the subsequent discussion, it will be helpful to study these two propositions a bit further, and establish some relationships between them.

The term *complex* is almost as hard to define as is life itself. Many approaches to complexity attempt to treat it as if it were an intrinsic property of a system, or class of systems, related somehow to entropy or "information." These approaches seek to obtain a single quantitative measure of complexity in terms of the number of elements, interactions, or operations required to characterize some aspect of system behavior. I would rather suggest that complexity is not an intrinsic property of a system—it must also reflect something about the manner in which we, as observers, can interact with the system.

Roughly, then, I would suggest that complexity is a property of system descriptions rather than a property of the systems themselves. Indeed, we may say that a system *appears* complex when it is possible to generate many apparently independent descriptions of its behaviors. Each such independent description must arise out of a different process for observing the system, and hence out of a distinct available mechanism that enables us to interact with the system. For example, a stone usually appears simpler than an organism, because we have only a few ways to interact with the stone, and many ways to interact with the organism. As we increase our capabilities for interaction with the stone, its complexity grows; as we narrow our capabilities to interact with an organism, its complexity diminishes. Thus, complexity appears as a *contingent,* rather than an *intrinsic,* property and ultimately reflects interactive capabilities reflected in observation or measurement. It is these capabilities that provide the elements for corresponding system descriptions.

With this as background, let us consider what is meant by the proposition that *complex systems are counter-intuitive.* Roughly, such a proposition connotes the absence of an expected implication between two or more aspects of system behavior. If each such system behavior arises from a particular mode of interaction or observation and generates a corresponding system description, then the assertion of counter-intuitive behavior implies a logical independence between these modes of descriptions. Let us be more precise about this: imagine a class K of systems, each of which can be described in two ways. Suppose that a body of experience exists which indicates that in some available subclass $K' \subset K$, one of

178

0077–8923/78/0316–0178 $1.75/1 © 1979, NYAS

these descriptions implies something about the other. We then come to expect that such an implication will hold for every system in K; such an expectation provides the intuitive basis for relating the two descriptions. However, as soon as we encounter a system not in K', the implication relation breaks down, and we can say that such a system behaves in a counter-intuitive manner.

Such a breakdown of an expected implication between modes of system descriptions is what we shall call a *bifurcation*. In general, a bifurcation indicates a situation in which distinct modes of system descriptions are logically independent; i.e., in which properties of one description do not imply corresponding properties of another description. As will be seen, this usage subsumes the traditional mathematical definition, but also substantially extends its scope. In general terms, a bifurcation manifests a situation in which the incompleteness of a given mode of system description becomes manifest, and hence must be supplemented or replaced by another. In the study of natural systems, such as biological organisms, the fundamental problems all ultimately concern the interrelationships of different modes of description. For this purpose, the notion of bifurcation, or of the absence of bifurcation, becomes a crucial tool. In the sections that follow, we wish to explore some of the ramifications of such ideas, treating bifurcation phenomena in the context of natural systems, but arising out of the comparison of differing modes of description of such systems.

SYSTEM OBSERVATION AND DESCRIPTION

In the preceding section, we suggested that each mode of description of a system arises from a corresponding behavior or interaction of the system which we can observe; conversely, each mode of observing a system generates a system description. In this section, we shall develop some general properties of the relation between system interactions (observations) and the descriptions to which they give rise.

The basic unit of system description and of system measurement is a single observable. Intuitively, an observable of a system is a quantity that can induce dynamics in some appropriate meter; i.e., in some other system with which the given one can interact. A system can be regarded, at least in part, as simply a collection of observables; i.e., as a family of capabilities to induce dynamics in other systems.

A closely related concept is that of *state*. For present purposes, it is sufficient to regard a state s of a system S as connoting the specific dynamics which S can induce in any particular meter at a specific instant of time. That dynamics, its corresponding attractor, or some parameter value associated with that attractor, represents the *value* of the observable in question, evaluated on s.

Modern physics is dominated by the proposition that all "physically real" events involve observables evaluated on states. Furthermore, it is usually supposed that it suffices to consider observables that take their values in real numbers; i.e., the attractor states of all meters can be effectively parameterized by real numbers. Thus for our purposes, an observable is simply a mapping $f:S \rightarrow \mathbb{R}$ from states to real numbers. The image $f(S)$ is what is usually denoted as the *spectrum* of the observable f. More generally, however, an observable represents a mapping from a set of states of a system S to a set of attractor states of some other system with which S can interact.

Clearly, if our only access to the system S were through the single observable f, we could not distinguish two states s and s' for which $f(s) = f(s')$. Thus, we would, in fact, not be observing S itself, but rather a quotient set S/R_f, where R_f is the equivalence relation on S defined by writing sR_fs' is and only if $f(s) = f(s')$. By definition, there is a 1-1 correspondence between S/R_f and the spectrum $f(S)$; in these circumstances, $f(S)$ would be *the state space* of S.

Now $f(S)$ is a set of real numbers, and hence comes equipped with a variety of natural structures. In particular, $f(S)$ is a metric space. We can employ such structures in $f(S)$ to *impute* corresponding structures to S/R_f, and thence to S itself. In particular, via the metric structure on $f(S)$, we can say that two states s, s' of S are "close" if the corresponding values $f(s)$, $f(s')$ are close in $f(S)$. It cannot be too strongly emphasized, however, that such a topological structure is not intrinsic to S, but is imputed to S through a process of system description derived ultimately from observation in the fashion we have described.

Suppose now that we are given another observable g. We can repeat the above argument—g can be regarded as a mapping $g:S \to \mathbb{R}$ with spectrum $g(S)$; this spectrum is in 1-1 correspondence with the quotient set S/R_g, and we can impute another topology to S/R_g, and thence to S, through the metric properties of $g(S)$. Using the observable g alone then, we have another "state space" representation of S.

We can now ask: how does the description of S obtained from the observable f compare with the corresponding description obtained with the observable g? We shall consider here primarily the metric properties, in the following form: if $f(s')$ is close to $f(s)$, under what circumstances will it be true that $g(s')$ is also close to $g(s)$? Stated another way: If s' "approximates" s under f, when will it also approximate s under g?

This kind of question is closely related to the compatibility of the mapping g with the equivalence relation R_f; i.e., the capability of g to distinguish states indistinguishable (or approximately so) under f. Basically, we proceed as follows: Given a state s in S, consider the set of all states s' for which $|f(s) - f(s')|$ is small. Look at the image under g of this set of states. If this image lies in a sufficiently small neighborhood of $g(s)$, we shall say that s (or more accurately, its equivalence class under f) is a *stable point* of g with respect to f. It is clear that the set of all stable points of g with respect to f comprise an open set in S/R_f, under the topology coming from $f(S)$. The complement of the set of stable points will be called the *bifurcation set* of g with respect to f. Obviously, at a bifurcation point, the proximity of two states s, s' of S as viewed by the observable f does not imply their proximity as viewed by g; at a stable point this implication does obtain. Thus, on the stable points the f-description may be replaced by the g-description; on the bifurcation points it may not be. Stated another way: On the stable points, the g-description conveys essentially the same "information" as does the f-description, and is hence redundant to it; on the bifurcation points, the g-description conveys "new information," distinct from that conveyed by the f-description.

In the discussion of the previous paragraph, we can interchange the roles of f

and g, and obtain the dual concept of the stable points and bifurcation points of f with respect to g. These represent complementary subsets of S/R_g, or $g(S)$, and are thus generally quite different from the corresponding subsets of S/R_f considered in the preceding paragraph. Thus, given a pair of descriptions, we obtain two distinct notions of stability and bifurcation, depending on which of the descriptions is chosen as the reference.

To illustrate these concepts, let us look at a well-known mathematical example. We may describe a dynamical system in two distinct ways: 1) in terms of a vector field on a manifold, or 2) in terms of the attractors of the system. Invariably the vector field description is taken as the reference; thus we say that two dynamical systems are close if their vector fields are close in some appropriate norm. The problem of structural stability revolves around determining when it is the case that two dynamical systems whose vector fields are close are also close in terms of their attractors. The implications of structural stability (e.g., in terms of the "robustness" of dynamical descriptions of real systems) are well known.

On the other hand, we may interchange the roles of the two descriptions and refer the vector-field description to that involving attractors. Intuitively, we would then ask: Under what circumstances is it true that closeness in terms of attractors implies closeness of the corresponding vector fields in some norm? Such a question has profound implications; e.g., for modeling and simulation; it arises naturally out of the preceding considerations, but, as far as we know, it does not appear to have received any systematic study.

Before proceeding to some simple applications of these ideas, let us draw one elementary consequence from them. Suppose f and g are two observables such that the bifurcation sets of f with respect to g, and of g with respect to f, are empty. Under these circumstances, it is appropriate to say that the two observables are *equivalent;* i.e., with respect to metric properties, the two observables are everywhere interchangeable. Moreover, in these circumstances, the relation between f and g is one of *conjugacy;* i.e., we can establish a commutative diagram of the form,

$$
\begin{array}{ccc}
S & \xrightarrow{\ f\ } & S/R_f \\
\phi \downarrow & & \downarrow \psi \\
S & \xrightarrow[\ g\ ]{} & S/R_g
\end{array}
\qquad (1)
$$

which allows us to "translate" the f-description into the g-description. This is again what we would expect from a study of purely mathematical examples of the concepts of stability and bifurcation. This equivalence (between *observables,* rather than between states) will become important to us subsequently.

To conclude this section, we may note that the results we have obtained for descriptions arising from single observables may be generalized to descriptions involving any number of observables. To illustrate this, let us indicate how we may construct a more comprehensive description from a pair of observables f, g than that arising from either observable alone.

The utilization of a pair of observables essentially allows us to define a new equivalence relation R_{fg} on S, where $R_{fg} = R_f \cap R_g$. We now observe that we can always define a mapping,

$$\theta : S/R_{fg} \to f(S) \times g(S)$$

which is in general 1-1 and *into*. This map arises from the fact that every equivalence class in S/R_{fg} is the intersection of a unique class in S/R_f and a unique class in S/R_g; we associate each of these classes with the corresponding elements in $f(S)$ and $g(S)$, respectively. The image of this map corresponds to a 2-dimensional "state space," in which the observables f and g play the role of "state variables." It may be noted that the map θ is onto if and only if every R_f-class intersects every R_g-class, and conversely; i.e., if and only if the respective bifurcation sets are maximal. If these bifurcation sets are both empty, then as we would expect, the image of θ collapses to a 1-dimensional subset of $f(S) \times g(S)$. This kind of representation can be extended in an obvious way to any number of observables. In each case, we obtain a topological space in which the arguments given above can be repeated word for word. Fuller details may be found in a forthcoming monograph.[1]

APPLICATION: ARE BIOLOGICAL DESCRIPTIONS REDUCIBLE TO PHYSICAL ONES?

The prevailing idea among many biologists is that all biological descriptions are ultimately effectively reducible to physical ones, or, even more strongly, that all biological descriptions are effectively derivable from physical ones. We now wish to explore how this idea can be tested, in the context of the discussion presented above.

To begin, let us cast the reductionistic hypothesis in terms of observables. A biological system, such as an organism, is surely also a physical system, and hence may be described in terms of the traditional observables with which physics is concerned. A dominant role here is played by the energy (Hamiltonian) of the system. On the other hand, the biological behaviors of the organism can be described phenomenologically; the ingredients for such descriptions are new observables of a fundamentally biological character. The reductionistic assertion is that each phenomenological observable arising in a biological description can be expressed in terms of the underlying physical observables. From this we can conclude that *such biological observables cannot bifurcate with respect to the underlying physical ones.* If such a bifurcation can be demonstrated, we could conclude that the corresponding biological observables could not be expressed as functions of the underlying physical ones, but rather comprise a logically independent mode of description of our system on the bifurcation points.

As we have defined it, observables are simply quantities which are capable of inducing dynamics on other systems, such as meters. If we are given a system S, then the characterization of S as a physical system is obtained by causing the states of S to interact with appropriate meters, which, in effect, evaluate physical observables on these states. In this way, as indicated above, we obtain a description of S as a physical system.

Now let us suppose that we use the states of S to induce dynamics on some other system S'. By definition, this dynamics must be expressible in terms of one or more observables of S. The question is whether the observables of S responsible for inducing the dynamics on S' are the same ones as we measure when we characterize the states of S with our meters (or more generally, are definite functions of these observables).

To answer this kind of question, we must generate two descriptions of S which are to be compared. One is already given   it arises from the set of meters through which we physically characterize the states of S. Further, we can obtain a second description of S through the fact that its states induce dynamics in S'. For we can also characterize the states of S' through the employment of the same meters that characterize the states of S. Schematically, we have a diagram of the following form:

Here the dotted arrow represents the association of a state of S with an asymptotic state (attractor) of S' generated by the induced dynamics. If we interpret the description of these asymptotic states of S' as also being descriptions of the states of S, which we require. One of these is obtained directly from our meters; the other involves the dynamics induced by S in S', and hence essentially involves those observables of S which generate this dynamics.

Clearly, if the second description bifurcates with respect to the first, it follows that the dynamics induced by S on S' involves observables distinct from those measured by our meters. Furthermore, these new observables are not reducible to those we measure directly, at least on the bifurcation points. These "new" observables must then enter as independent elements of system description on exactly the same footing as those defined by our meters. Indeed, we may use the dynamics induced by S on S' to construct a *new meter* in terms of which the observables of S that generate the dynamics may be defined. Such techniques of "bio-assay" are, in fact, widely used to measure the activities of organic substances, such as hormones.

It then becomes an empirical question to determine whether the observables manifested in biological interactions are distinct from those appearing in our physical descriptions of the system. A good place to look for such new "biological" observables is in situations involving specificity or discrimination mechanisms. We have suggested that primary genetic processes would provide good candidates for the isolation of such observables (although the character of that argument was quite different).[2] More recently, Comorosan has applied the same circle of ideas to an empirical study of the observables involved in enzyme-substrate interactions.[3] From his work, he concludes that simple substrates for enzymes may exist in classes of states which appear indistinguishable to our physical meters, but which may be split (discriminated) by enzymes. The enzymic discrimination appears as a small, but significant, modification of reaction rate. Independent experimental confirmation of this work has been reported,[4] and further study would be desirable.

The implications of such considerations for reductionism in biology are obvious. It should be stressed, however, that there is nothing "unphysical" about such new biological observables, just as there was nothing "unphysical" about, say, spin. It is perhaps not surprising to find that interactions between complex systems reveal capabilities not manifested on interaction with simple systems. The concept of bifurcation of descriptions provides an explicit probe of these capabilities, and of how they are logically interrelated.

ABSENCE OF BIFURCATION: MODELING AND SIMILARITY

Here we wish to consider some of the ramifications of diagrams like (1), which represent the context for the development of concepts of similarity between systems.

In general, suppose that $f, g : A \rightarrow B$ are conjugate maps related by a commutative diagram of the form,

$$
\begin{array}{ccc}
A & \xrightarrow{\ f\ } & B \\
\phi \downarrow & & \downarrow \psi \\
A & \xrightarrow[\ g\ ]{} & B
\end{array}
\qquad (2)
$$

We shall interpret such a diagram as follows: the expression $b = f(a)$ is *invariant* to:

    (i)   the replacement of a by $\phi(a)$;
    (ii) the replacement of f by g;
    (iii) the replacement of b by $\psi(b)$.

We introduce the following terminology: for a in A, the element $\phi(a)$ will be called its *corresponding element;* likewise for b in B, $\psi(b)$ will be its corresponding element. Then the assertion of conjugacy between f and g means precisely that corresponding elements are mapped by g onto corresponding elements. The invariance of the relation $b = f(a)$ to these replacements is the abstract analog to the *law of corresponding states* in thermodynamics.

Now let us suppose that S is a system and that we are given two descriptions $F = (f_1, \ldots, f_m)$ and $G = (g_1, \ldots, g_n)$ of S, where the f_i and g_j are observables. Suppose further that the set of bifurcation points of the G-description with respect to the F-description is empty. Then, as we have seen, there is a sense in which the F-description implies the G-description on every state of S. We saw also that (under mild assumptions) each such description can be regarded as giving rise to a manifold in which the f_i and g_j, respectively, can be regarded as local coordinates. Under these circumstances, we can write a functional relation of the form,

$$
(g_1, \ldots, g_n) = \Phi(f_1, \ldots, f_m)
\qquad (3)
$$

valid for every state s in S. Such a relation can be regarded as an *equation of state* for S. It should be noted that the mapping Φ is not itself a system observable, but rather, expresses a relation between observables (i.e., between descriptions of S).

In general, a single function of m variables can be regarded as a one-parameter family of functions of $m-1$ variables indexed by the range of one of the arguments.

Thus in particular, $\phi(f_1, \ldots, f_m)$ can be regarded as a one-parameter family of functions $\Phi_r(f_2, \ldots, f_m)$, where the index r runs through the spectrum $f_1(S)$ of f_1. Suppose all the functions in this one-parameter family are conjugates (the condition under which this is true can be expressed in terms of the bifurcation set of f_1 with respect to each of the other f_i), then given any two elements r, r' in spectrum of f_1, there is a diagram of the form,

$$
\begin{array}{ccc}
X & \xrightarrow{\Phi_r} & Y \\
\phi_{r,r'} \downarrow & & \downarrow \psi_{r,r'} \\
X & \xrightarrow{\Phi_{r'}} & Y
\end{array}
$$

with the properties we have noted above (here X is the manifold determined by the observables $f_2, \ldots, f_m$, Y is the manifold determined by the observables $g_1, \ldots, g_n$).

In terms of S, this process amounts to regarding S as being composed of a one-parameter family of (sub)systems S_r, each of which is described by the corresponding equation of state determined by Φ_r. The assertion, that the Φ_r are all conjugate, implies intuitively that all the S_r are *similar*. This in turn, means that the replacement of a given S_r by an $S_{r'}$ can be "annihilated" by replacing corresponding elements by corresponding elements in the equation of state; i.e., by coordinate transformations in the domain and range of the Φ_r.

We can imagine this process continued in such a way that the original function Φ can be expressed as a p-parameter family of conjugate functions of $m - p$ variables of the form,

$$\Phi_{r_1 \cdots r_p}(f_{m-p+1}, \ldots, f_m)$$

and that p is maximal for this property (i.e., any set of $p + 1$ of the observables f_i gives rise to nonconjugate functions). Then S has accordingly been decomposed into a p-parameter family of systems $S_{r_1 \cdots r_p}$, and all of these systems are similar. Once again, this means that an arbitrary transition $(r_1, \ldots, r_p) \to (r'_1, \ldots, r'_p)$ can be annihilated by replacing corresponding elements by corresponding elements. It may be noted that the above considerations can be regarded as an abstract form of the Buckingham Π-Theorem,[5] which was originally obtained through very special dimensional arguments.

Another way to express the above construction is: The bifurcation set of the description of S obtained from the observables $(f_{m-p+1}, \ldots, f_m)$ with respect to the description obtained from the observables $(f_1, \ldots, f_p)$ is empty and the value p is maximal for this property.

Now let us introduce some picturesque, but not entirely unjustified, terminology. We shall call the set of parameter values $r_1, \ldots, r_p$ the *genome* of the corresponding system $S_{r_1 \cdots r_p}$, the domain of the map $\Phi_{r_1 \cdots r_p}$ will be called the set of *environments* of the system, and the range of this map will be called the set of *phenotypes*. Then the equation of state asserts simply that a specification of a particular genome in a particular environment uniquely determines the corresponding phenotype.

The Thompson Theory of Transformations[6] asserts essentially that closely related species are similar. In our terminology, this translates into the assertion that if the genomes of two systems are close in some appropriate norm, then the phenotypes are corresponding. This formulation permits a number of interesting and potentially testable conclusions to be drawn. Space does not permit a fuller discussion here, but details may be found in Rosen.[7] Moreover, the above discussion allows us to extend the functional concepts of genome and phenotype to non-biological systems. This permits, for example, a better understanding of the significance of nonbiological morphogenetic metaphors, such as critical phenomena and diffusion-reaction mechanisms, which seem at first sight to be devoid of any plausible genetic component.

Another facet of the above formulation is that: If we keep the genome fixed and modify the environment, we obtain a corresponding change in phenotype governed by the equation of state; likewise, if we keep the environment fixed and modify the genome, we again obtain a change in phenotype. Such phenotypic changes can be regarded as *adaptations* of the phenotype to the imposed modifications. They are imposed by the requirements of invariance of the equations of state. In general, if we study what varies when a perturbation is imposed on a biological system, we obtain a theory of adaptation; if we study what remains invariant under such a perturbation, we obtain a theory of homeostasis. What we wish to stress is that adaptation and homeostasis represent different aspects (descriptions) of the same basic phenomena.

Let us now turn briefly to the concept of modeling and model systems. Intuitively, we may say that some structure S' is a model for another system S if, in a given set of circumstances, S' may be substituted for S with no observable change; i.e., if this substitution is *invisible* under the given circumstances. Thus, given an observable $f: S \to \mathbb{R}$, we may replace a state s by any state s' for which $f(s) = f(s')$; under these circumstances we may say that s' models s. Likewise, we may replace an observable by a conjugate observable, if we simultaneously replace each element of the domain and range by the corresponding elements. Similarly, we may replace a system by a *similar* system. It is clear that the formalism we have developed above, based on the comparison of system descriptions, is a general theory of modeling. As noted earlier, the appearance of bifurcations indicates where a modeling relation breaks down. However, it is equally significant to note where such a relation is maintained—this is the province of the concept of similarity.

Let us conclude with one final variation of the diagram (2). Let us suppose we have a pair of conjugate relations of the form,

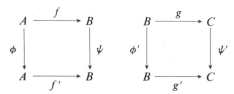

We may say that f and f' are similar, and g and g' are similar. This means that we can replace f by f' and g by g', as long as we also replace elements with corresponding elements, and observe no change. However, let us compose the maps f and g, and ask whether the composite gf is conjugate to the composite $g'f'$. It is evident that the conjugacy of the composites does not follow in general from the conjugacy of the factors. Moreover, even if the composites are conjugate, the elements of A and C, which correspond in the diagrams, considered separately will no longer in general correspond when the maps are composed. This simple observation throws another kind of light on the problem of reductionism; namely, that, in general, the availability of models (descriptions) of the components of a composite system does not allow us to construct a model for the composite system itself. Conversely, a model for a composite system does not allow us to construct models for the components. Hence we see, in a particularly stark way, the difficulties faced in the analysis and the synthesis (design) of complex systems. Furthermore, the failure of conjugacy to be preserved under composition of mappings raises some deep questions of a category-theoretic nature. In category theory, composition of mappings is the basic operation, which is preserved by *functors*. Since conjugacy is not preserved under composition, it cannot be a *functorial relation*, and hence is not "natural" in the category-theoretic sense. Therefore it appears that category theory cannot by itself provide us with an appropriate tool for the analysis and synthesis of complex systems.

REFERENCES

1. ROSEN, R. 1978. Fundamentals of Measurement and the Representation of Natural Systems. American Elsevier Publ. Co. In press.
2. ROSEN, R. 1960. A quantum-theoretic approach to genetic problems. Bull. Math. Biophys. **22:** 227–225.
3. COMOROSAN, S. 1976. Biological observables. Progr. Theor. Biol. Academic Press. **4:** 1–31.
4. BASS, G. E. & J. E. CHEVENEY. 1976. Irradiation induced rate enhancements for the LDH-pyruvate reaction. Int. J. Quantum Chem. (Quantum Biol. Symp. **3:** 247–250.
5. BUCKINGHAM, E. 1915. On Physically Similar Systems. Physiol. Rev. **4:** 345–370.
6. THOMSPON, D. W. 1917. On Growth and Form. Cambridge University Press.
7. ROSEN, R. 1978. Dynamical similarity and the theory of biological transformations. Bull. Math. Biol. In press.

THE RANDOM CHARACTER OF BIFURCATIONS
AND THE REPRODUCIBLE PROCESSES OF
EMBRYONIC DEVELOPMENT

Hans Meinhardt

Max-Planck-Institut für Virusforschung
D-7400 Tübingen, Federal Republic of Germany

INTRODUCTION

A fascinating aspect of the development of higher organisms is the formation of a huge diversity of structures in a precise spatial arrangement. The final pattern cannot be contained in a latent form in the egg, but must emerge in a series of consecutive decisions by which further pathway of some cells is altered with respect to the others. There are indications that the spatially determined patterns of cell differentiation are controled by a "prepattern" (primary pattern).[1] Such a prepattern may consist, for instance, of an inhomogenious distribution of a particular substance that influences cell differentiation.

A first attempt to explain the formation of such a prepattern by mutual interactions of substances and diffusion was undertaken by Turing[2] in 1952 and this concept has been further elaborated by Prigogine and Nicolis,[3] Martinez[4] and Edelstein.[5] By considering known principles of development—fields of inhibition and possibility of local induction—Gierer and Meinhardt[6-8] have shown that among the many possible pattern-forming reactions, a selection of only a few biologically reasonable reactions is possible.

The basic idea is that a local inhomogeneity in the distribution of a substance is amplified by a strong autocatalytic feedback and that the emerging pattern is stabilized by the antagonistic effect of a substance with a higher diffusion range. Such a reaction has bifurcating properties; that is, if the field (the area in which a reaction can take place) is small compared with the range of an *autocatalytic* substance, only a homogeneous equilibrium of two substances is possible. With increasing size (most biological systems grow during development) homogeneous distribution will become unstable and a high concentration of the autocatalytic and inhibitory substances will appear at one boundary. However, the side that will carry the high concentration will depend on small external or internal asymmetries, or even on random fluctuations.

If the process of development consists of a series of such bifurcations, the question arises: How can we be assured that a subpattern, say a limb, has the correct orientation with respect to the whole organism? In this article, we shall illustrate our proposed answer[8]: Once a pattern is formed during development, it can provide an asymmetry which will orient the following subpattern in such a way that a chain of such reactions can lead to a deterministic result.

0077–8923/78/0316–0188 $1.75/1 © 1979, NYAS

PATTERN FORMATION BY AUTOCATALYSIS AND LATERAL INHIBITION

The development of structures from more or less homogeneous initial conditions is a very common biological phenomenon. This is also true in the nonliving world. For example, a sharply contoured bolt of lightning may arise from a diffuse cloud, or rain water, homogeneously distributed over land, can wash out well-defined rivers. Common to all these pattern-forming processes is the fact that a small deviation from uniformity has a strong feedback for further growth. However, to obtain a stable pattern, self-amplification must be complemented by an antagonistic effect that spreads out more rapidly into the surroundings, preventing an "infection" with the autocatalytic reaction there.

To translate this idea into more biochemical terms, consider two substances: an autocatalytic activator and an inhibitor. The autocatalytic activator $a(x, t)$ should enhance, directly or indirectly, its own production. Simultaneously, it should also enhance the production of an antagonist, the inhibitor $h(x, t)$. If the total area is small such that by diffusion, a rapid equilibration of both substances is assured, an interaction should occur that leads to stable equilibrium of both substances. To avoid oscillations in time, a rapid adaptation of the inhibitor to a fluctuation in the activator concentration is necessary. That means the inhibitor must have a higher turnover rate. However, if the field is at least as large as the activator range (i.e., the mean distance an activator molecule can diffuse between production and decay) and the diffusion of the inhibitor, which provides lateral inhibition, is more rapid, a stable pattern can be formed. A small local activator increase can no longer be completely compensated for by a corresponding increase of inhibitor, since additional inhibitor diffuses rapidly away, thus suppressing activator production in the surrounding area. By contrast, at the incipient activation center, activation will further increase.

We have derived a criterion for determining which interaction will lead to a pattern and which will not.[6,7] A simple possible interaction would be,

$$\frac{\partial a}{\partial t} = c\,\frac{a^2}{h} - \mu a + D_a\,\frac{\partial^2 a}{\partial x^2} + \rho_0 \tag{1a}$$

$$\frac{\partial h}{\partial t} = ca^2 - \nu h + D_h\,\frac{\partial^2 h}{\partial x^2} + \rho_1 \tag{1b}$$

It is intuitively reasonable that activator production must be nonlinear since the normal decay $-\mu a$ has to be overcome by autocatalysis. The terms ρ_0 and ρ_1 refer to small-production terms, which are independent of the existing activator and inhibitor concentrations. These terms become important at low activator or inhibitor concentrations. The formation of a strongly patterned distribution of a and h according to this equation is shown in FIGURE 1a, b. The inhibitor can also have a quadratic influence on activator production, but according to the general criterion,[6,7] inhibitor production could have then a linear dependence on activator concentration,

$$\frac{\partial a}{\partial t} = c\,\frac{a^2}{h^2} - \mu a + D_a\,\frac{\partial^2 a}{\partial x^2} + \rho_0 \tag{2a}$$

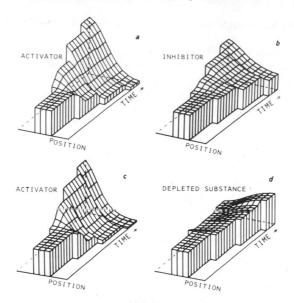

FIGURE 1. *Pattern formation by autocatalysis and lateral inhibition:* a) and b) Generation of a graded concentration profile of an autocatalytic substance—the activator (a) and its more diffusible antagonist, the inhibitor, according to Equation 1. Concentrations of both substances are plotted as a function of space and time. A growing linear array of cells with margins impermeable to both substances is assumed. In a small area, both substances are in stable equilibrium, since any increase in activator concentration will lead to a corresponding increase in the inhibitor concentration which, in turn, regulates activator concentration. Pattern formation starts quite abruptly after the size of the order of the activator range has been exceeded. Homogeneous distribution is then no longer stable, since a small local activator increase (caused even by random fluctuation) can increase still further by auto-catalysis. This is because the additionally produced inhibitor diffuses rapidly into the sur-roundings, thus regulating activator production. High concentrations of activator and in-hibitor appear at one end, since there is only space for one activator slope in the field—a central maximum would require space for two slopes. The maximum remains at the bound-ary also during further growth. The result is a stable graded distribution of both substances, which can be used for the supply of "positional information,"[17] enabling the appropriate determination of cells according to their positions. c) and d) A similar pattern can emerge if an inhibitory effect is obtained from a substrate which is consumed in the autocatalysis (Equation 4).

$$\frac{\partial h}{\partial t} = ca - \nu h + D_h \frac{\partial^2 h}{\partial x^2} + \rho_1 \tag{2b}$$

This allows the molecular interpretation,[8] that the activator decays into an inac-tive molecule, which acts as an inhibitor, because it is in competition with the activator molecules. A very symmetric interaction, which can lead to a pattern, is,

$$\frac{\partial a}{\partial t} = c \frac{a^2}{h^4} - \mu a + D_a \frac{\partial^2 a}{\partial x^2} \tag{3a}$$

$$\frac{\partial h}{\partial t} = c \frac{a^2}{h^4} - \nu h + D_h \frac{\partial^2 h}{\partial x^2} \tag{3b}$$

Lateral inhibition need not be provided by a real inhibitory substance, but can be mediated by the depletion of a substrate $s(x, t)$, which is consumed in auto-catalysis. A simple possible interaction would be,

$$\frac{\partial a}{\partial t} = a^2 s - \mu a + D_a \frac{\partial^2 a}{\partial x^2} \tag{4a}$$

$$\frac{\partial s}{\partial t} = c_0 - a^2 s - \nu s + D_s \frac{\partial^2 s}{\partial x^2} \tag{4b}$$

A pattern-forming process, according to these equations, is shown in FIGURE 1c, d. An equation which allows the interpretation, that the substrate is converted from an inactive into an active form in autocatalysis, is,

$$\frac{\partial a}{\partial t} = ca^2 s^2 - \mu a + D_a \frac{\partial^2 a}{\partial x^2} \tag{5a}$$

$$\frac{\partial s}{\partial t} = c_0 - ca^2 s^2 - \nu s + D_s \frac{\partial^2 s}{\partial x^2} \tag{5b}$$

The activator-depleted substrate mechanism has some properties different from the activator-inhibitor mechanism.[9] On the basis of experimental observations, this allows, in some cases, a decision as to which mechanism is realized.

As shown in FIGURE 1, a patterned distribution of both substances can emerge only if the size of the field is at least as large as the activator range. Due to the lateral inhibition mechanism, high and low activator concentrations will appear at a maximum distance from each other; that is, at one end or the other. Therefore, a monotonic profile of both substances is guaranteed; this fits well into the observation that many developing systems obtain a "polarity" (i.e., a systematic change of some tissue property from one end to the other).

In other systems, pattern formation starts, not because a certain size is exceeded, but rather, because the system seems to be dormant until a certain biochemical event has taken place, for example, the fertilization of an egg or with

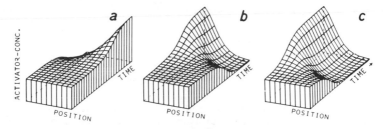

FIGURE 2. *Orientation of the pattern:* a) and b) If only some random fluctuations are present ($< 1\%$ difference in the const. c of Equation 1) and the size of the field is of the order of the activator range, chance will determine which of the two sites becomes activated. The final pattern is, except for orientation, independent of the initiating perturbations. c) If, in addition to the fluctuation, a systematic asymmetry is present (1% difference in c of Equation 1 across the field is assumed), the orientation of the pattern is predictable, and the time necessary to reach a final steady state is much reduced.

the inductive contact of one tissue with another. In the model, a pattern formation can be suppressed by a small constitutive (activator-independent) inhibitor production (ρ_1 in Equation 1b; a high constitutive activator production, ρ_0 in Equation 1a, suppresses the pattern formation, as well).

Up to now, only random fluctuation has been assumed as a means of initiating transition from a semistable homogeneous equilibrium into a strongly patterned steady state. Consequently, the side at which the activation will appear is random (FIGURE 2a, b). However, any asymmetry in a parameter, which influences activator or inhibitor production, can orient the evolving pattern. This may be a difference in temperature, in pH, in the supply of nutrients, in the removal of waste products, or in the distribution of important granules that sink due to gravity.

Indeed, most biological pattern-forming processes take place in an asymmetric environment. For example, the egg cells of many insects are enlarged by a discharge of cytoplasm of nursery cells into the anterior side of an egg cell [10] A hen's egg rotates in the uterus and the direction of this rotation will decide the orientation of the embryo,[11]—the ultimate reason for this effect is the position of the blastoderm with respect to gravity.[12] The location of the dorsoventral axis and the oral field of a sea urchin embryo can be shifted by a gradient of KCN.[13] The latter two examples demonstrate that weak or unspecific stimuli can orient a pattern, supporting the view that semistable equilibrium is pushed in one or another direction.

A classical example for an initially very symmetric egg is that of the brown algae *Fucus*. Here, the outgrowth of the first structure, a rhizoid, can be oriented by almost any stimulus, such as differences in temperature, light, pH, or voltage.[14,15] In the absence of any stimuli, this outgrowth appears to be random, but much delayed. This is in agreement with the proposed model, since a systematic difference between the two sides would accelerate the pattern-formating process considerably (FIGURE 2c).

Polarizing influences can be incorporated into equations in several ways. If the "sources" of activator or inhibitor production—like ribosomes or messenger RNA's—are not homogeneously distributed, the constant c becomes space dependent, and one can show that in a reaction according to Equation 1 or 4, the orientation (not the final concentration of activator) depends on source density.[6] An external supply of activator or inhibitor can be taken into account by using appropriate values for ρ_0 and ρ_1.

THE ORIENTING EFFECT OF ONE PATTERN FORMATION UPON THE NEXT
IN A SEQUENCE OF PATTERN-FORMING EVENTS

The first pattern is, of course, not sufficient to determine every detail of an organism along one axis, since cells can respond to graded distribution with only limited accuracy. If a structure to be formed is periodic, such as the initiation sites of leaves in a growing plant, the same signal can be used repeatedly. If, in the model, growth proceeds to a size in which the distance of some cells from an activated center is large, inhibitor concentration will drop below a threshold concentration so that a small constitutive activator production will become autocatalytic, and a second activation center will be formed. Periodic patterns can be formed due to the inhibiting influence of the established center upon the developing center (FIGURE 3). For example, the spacing of hydra buds[8] or plant leaves[9] can be understood in these terms. The generation of a very regular periodic structure is possible despite the fact that the first maximum may have emerged from random fluctuations and that also the following ones obey the same rules.

Another possibility for a finer subdivision is to form a secondary pattern within a group of cells that have been specified by the first pattern. The orientation of the subpattern will again be a bifurcating process but, of course, a subpattern, such as a limb bud[16] or a retina, must have a correct orientation in an organism. Since the proposed reaction can amplify very small differences, asymmetry remaining from the first pattern can be used to orient the subpattern. Let

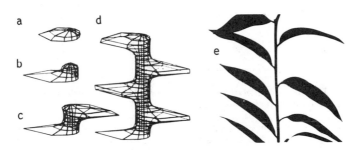

FIGURE 3. *Formation of an ordered structure out of random fluctuations:* Simulation of an alternate leaf pattern [8,9] A cylindrical arrangement of cells that grow by doubling the uppermost ring of cells is assumed. The first activator maximum is initiated by random fluctuations at a random position (a). After further growth (b), the next maximum appears at a predictable position on the opposite side, although the cells there are subject to the same fluctuation. The repulsive influence of the inhibitor, which emanates from the first maximum, is much stronger then the randomizing influence of the fluctuations. A regular alternating pattern of activator peaks is formed (d). A biological example is given (e). Under the assumption of different geometries, an opposite (decussate) or parallel arrangement of leaves can be simulated.[9]

us assume that the first pattern has been generated (FIGURE 4). Absolute inhibitor concentration, with its graded distribution over a large area, is a convenient possibility for the supply of "positional information."[17] This means that cells, which detect a certain concentration range of inhibitor, will be determined in a particular direction—many experiments with early insect development are quantitatively explainable under this assumption.[18] Further, there is evidence that, after such determination, the tissue at the borders between two differently determined cell populations becomes impermeable to certain substances.[19] Thus, a prerequisite for the formation of "subfields" in which a subpattern can be formed is given. If now a substance of the first pattern-forming system diffuses, at least to some extent, into a subpattern, and has some cross reaction, the subpattern will be oriented. For example, an (graded) inhibitor of the first pattern-forming system may compete with that of the subpattern. It has therefore an activating influence on the subpattern. This results in a parallel orientation of the subpattern with respect to the first system (FIGURE 4c).

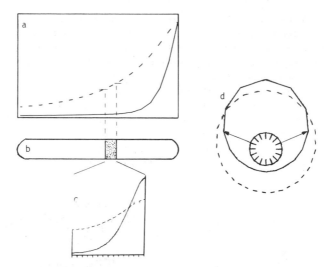

FIGURE 4. *Formation of a subpattern:* A first activator (——) and inhibitor (- - -) pattern should have been formed, perhaps out of fluctuations as shown in FIGURE 1. A group of cells within this field (dots in b), exposed to a particular inhibitor concentration, under-goes a determination to form a particular structure (e.g., a segment of insect larvae). For the determination of finer subdivisions, additional prepatterns are required. A subpattern within this group of cells can be positioned parallel to the first (c), if the inhibitor of the first pattern competes with that of the subpattern. On the other hand, gradients orthogonal to the first can be formed (see FIGURE 6). If the subpattern is something like a slide of a more-or-less tube-shaped blastoderm (as it is in insects), the cells to be organized are arranged in a ring (indicated by the central circle (d); activator and inhibitor concentrations are given by the distance of the lines to that circle). The resulting pattern *must be* bilaterally sym-metric; a particular concentration is present twice, on the left and on the right side (note the two arrows). Therefore, all structures are formed in pairs, like the eyes, antennas, wings, or legs.

An experimental system in which the orientation of a subpattern is analyzed in great detail is the determination of the axis of the amphibian retina.[20-26] Hunt and Jacobson[23,24] have excised primordial eyecups at different stages and reimplanted them in different orientations. Electrophysiological mapping of the developing retinotectal connection was used to determine which internal orientation the retina has developed. They found a completely normal retinotectal mapping after an early rotation of the eyecup, whereas a reversed orientation develops if the rotation is made only a few hours later. In terms of the model[8] (FIGURE 5), at the moment of an early rotation, the imposed asymmetry in the retina is weak such that it can be reversed by a polarizing influence of the surrounding tissue. However, when more time has elapsed, the autocatalytically amplified asymmetry will be stronger, so it can no longer be reversed by an external influence, but will increase despite a reversed external influence. In agreement with the model, this is an either-or effect, but no intermediate forms have been observed, such as a reversed anterior part and a normal posterior part of the retina. This behavior is a property of the prepattern-forming system and does not result from an irreversible programming of individual cells. This is supported by the finding that a reprogramming is possible after the critical time if the geometry or the boundary is changed, for example, by an incision in the retina.[25]

An important support for the assumption of the autocatalytic nature of sub-pattern formation comes from the observation that the transition from the reversible to the irreversible form can also happen in tissue culture in the absence of any polarizing effect from the surrounding tissue.[24] According to the model, once initiated, the gradient formation is autonomous and no longer requires an external stimulus. Through the autocatalytic amplification, the gradient becomes so strong that it is no longer reversible by a reversed external influence (FIGURE 5g).

From the model, one would expect the critical parameter that decides whether

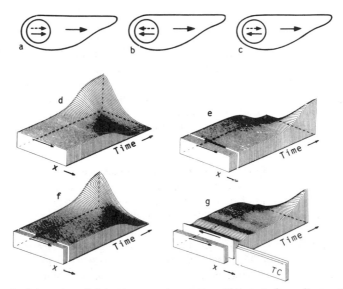

FIGURE 5. *Orientation of the retina—a subpattern—within an embryo:* Schematic drawing (a–c) of experimental observations after rotation of an amphibian eyecup[20-26] together with a simulation (d–f).[8] a) Undisturbed development: the physical anterior-posterior orientation of the tadpole is indicated by a solid arrow; the retina develops a corresponding specificity (broken arrow). d) Model: formation of an activator gradient under the influence of the surrounding tissue; only one axis is considered here. The influence is simulated by a small graded (1% difference from end to end) external supply of inhibitor. b) If the eye is rotated after a critical time, the retina will develop a reversed specificity. The animal would later see everything upside down and front-back reversed. e) Model: after rotation (the moment is marked by a cleft in the space-time plot), the already developed activator gradient can no longer be reversed by the final external influence; the final activator pattern has a slope reversed in relation to the environment. c) If the rotation of the eye is made earlier, the specification of the retina will be normal despite its reversed physical orientation. Correspondingly, in the simulations (f), the incipient activator gradient is reoriented by the external influence, and the resulting gradient is normal. g) However, if the retina is removed at a time when the specification is normally reversible, kept for some time in tissue culture (TC), and is later reimplanted in a reversed orientation, the gradient is no longer reversible.[24] By the autocatalysis, which is inherent to this pattern-forming mechanism, the activator gradient becomes steeper and, therefore, irreversible during this critical period, even without an external influence.

the retina is still in the reversible or already in the irreversible stage is its diameter, since the maximum asymmetry that can be imposed depends critically on the size (FIGURE 1) of the retina.

If the orientation of a subpattern is determined by an overall property of the first pattern, one would expect that the polarizing influence is not restricted to the normal location of this structure. Indeed, a retina grafted into the belly of a tadpole would be oriented in exactly the same way as it is in its normal position.[23] However, there is an interesting exception: if the retina is reimplanted at the ventral midline, no orientation takes place indicating that the anterior-posterior coordinate system is not completely independent of the dorsoventral axis.[23]

How to Generate an Orthogonal Coordinate System

Most cell determination takes place in a cell sheet, in a 2-dimensional array of cells, such as the spherical blastoderm or the more-or-less cuplike retina in the above example. The linear approximation used so far has to be complemented by taking into account the second dimension. A second gradient system, which, at best, is orthogonal to the first gradient system, would be necessary. But how can the cells know what a right angle is? In a 2-dimensional field of the size of the activator range, a reaction of the proposed type will lead to a graded distribution along one axis, but with an almost constant concentration perpendicular to it (FIGURE 6a–c).[8] The gradient will preferably be oriented in the direction of the largest extension of the field.

To see how an orthogonal gradient can be formed, we have to take into account the action of this gradient. It is used as a signal to cause a determination of cells in different directions. Let us say the cells which have been exposed to the high concentration of activator are then in the determination stage B, the others remain in stage A (FIGURE 6d). Again, we have to assume that the diffusion rate is low between cells of different determination stages. This leads to a degeneration of the pattern in this first direction—the pattern is no longer necessary, since its purpose was to determine the cells. With this "compartmentalization,"[27,28] the longest extension of these subfields is now perpendicular to the first, and the pattern will reorient (or a new one can be formed) parallel to the borderlines of the first subdivisions, that is, orthogonal to the first gradient (FIGURE 6e,f). Gierer[29] has proposed a similar mechanism by assuming that the cells become polarized by the first gradient, and an anisotropic diffusion enables the formation of a second orthogonal pattern.

Such an organization of a 2-dimensional field would predict a sequential determination first of 1 dimension (possibly the largest), and later of the second dimension. The existence of a sequential determination can be concluded from ligation experiments made by Sander[30,31] with eggs of the leafhopper (*Euscelis*): The results are schematically shown in FIGURE 6g–i. After a transverse ligation at the preblastoderm stage, almost all segments are formed—the cells have "learned" which segment they have to form. However, a longitudinal ligation leads to a reorganization of the dorsoventral axis: In each portion a complete embryo will be formed. This indicates that, in this dimension, the prepattern can still be modified and no irreversible process has yet taken place.

Another example of the sequential determination of the two coordinate-systems can be found in the determination of the amphibian retina which has been mentioned above. Székely[26] has shown that the dorsoventral axis is fixed *after* the anterior-posterior axis: a 180°-rotation at a critical time can lead to a reversed anterior-posterior, but normally oriented, dorsoventral organization. This indicates that the dorsoventral axis is fixed, but that the anterior-posterior axis is not.

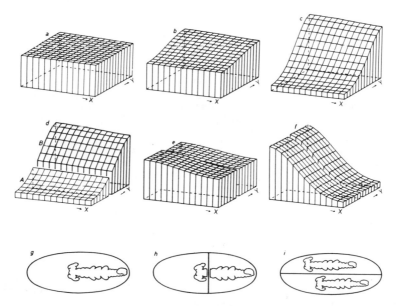

FIGURE 6. *Sequential formation of orthogonal gradients:* a)–c) In a 2-dimensional field, a graded distribution in 1 dimension with a constant concentration profile in the other dimension can be formed.[8] This gradient can be used for the determination of cells. If, after such determination, the diffusion rate is suppressed between cells of a different determination state (d), the pattern degenerates (e) and is reformed in the direction of the largest extension of the new fields, that is, orthogonal to the first gradient. g)–i) Experimental evidence for the sequential determination of two orthogonal axes: A schematic drawing of the results of Sander[30,31] with ligations of the egg of the leafhopper (*Euscelis*). g) The germ band (the proper embryo) after an undisturbed development of the egg. h) A ligation at the preblasto-derm stage reveals that the anterior-posterior determination is almost completed: Nearly all segments are formed even after such a severe operation. i) However, if a longitudinal ligation is made at the same stage, a reorganization in the dorsoventral dimension occurs, and two complete embryos are formed.

It is also interesting that, after experimental manipulations, like the grafting together of eye fragments, the first axis can have an irregular shape.[25] Then however, although the second axis is curved, it is still orthogonal to the first. This supports the view that orthogonality is not solely derived by the polarizing influence of the surrounding tissue, but that strong internal mechanisms for the generation of orthogonal coordinate systems exist.

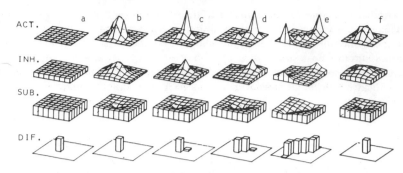

FIGURE 7. *Formation of a filament of differentiated cells*[32]: a) One cell is assumed to be initially differentiated (high y-concentration, bottom row). According to Equation 6, this triggers a high activator concentration (top row), and simultaneously depletes the substance s (b). Since activator production depends on s, the activator maximum is shifted; a neighboring cell becomes activated, and finally differentiated (c). The direction of the first shift depends, in the absence of a systematic difference, on local fluctuations. However, the further shift of the activator is no longer arbitrary, but will go to the end of this incipient line and not at the side. Long filaments of differentiated cells can be formed (d, e). However, without any fluctuation, and, in a symmetrical environment, the direction of growth is undecidable and development will stop (f).

BRANCHING STRUCTURES

Another example of the influence of one decision upon the next in determining an ordered development can be found in the formation of netlike patterns, such as blood vessels or veins in leaves. With a discussion of such systems, we also illustrate the importance of random fluctuations.

We have shown that differentiation of cells along a line, as well as branching, can be achieved by the following interactions[32]:

$$\frac{\partial a}{\partial t} = ca^2 \frac{s}{h} - \mu a + D_a \nabla^2 a + \rho_0 y \qquad \text{(6a)}$$

$$\frac{\partial h}{\partial t} = ca^2 s - \nu h + D_h \nabla^2 h + \rho_1 y \qquad \text{(6b)}$$

$$\frac{\partial s}{\partial t} = c_0 - \gamma s - \epsilon y s + D_s \nabla^2 s \qquad \text{(6c)}$$

$$\frac{\partial y}{\partial t} = da - ey + y^2/(1 + fy^2) \qquad \text{(6d)}$$

The principle of line formation is illustrated in FIGURE 7: By the interaction of the activator a and the inhibitor h (Equation 6a, b), a local high concentration is formed which causes the irreversible switch from a low to a high y-concentration—this represents a differentiation (Equation 6d). The purpose of the "net" could be the removal of a substance s (Equation 6c) upon which the activator depends (Equation 6a). The result is a shift of the activator peak into a neighboring cell, which, in turn, differentiates, and so on.

Without the assumption of random fluctuation (e.g., the constant c in Equation 6a, b), the system could run into a situation in which a decision cannot be made. To give an example, if the systems starts from one cell (as in the development of the blood vessels of a chick embryo from "blood islands"[33]), the direction into which the line will grow could not be chosen since every direction is equally favorable. The differentiated cell would be surrounded by a ring of cells with high activator concentration, but no further development would be possible since the activated cells compete with each other via the inhibitor, and none reach the concentration necessary for the differentiation of the next cell (FIGURE 7f). Such a system is very sensitive to local differences, and if the concentration of s is slightly higher at one side, the corresponding cell will win the competition, and the line will grow in the direction of the highest concentration of the substance to be removed—an optimal decision. But in the absence of any systematic difference, random fluctuations will lead to an arbitrary selection of one of the surrounding cells (FIGURE 7b). If two differentiated cells are already present, the selection of

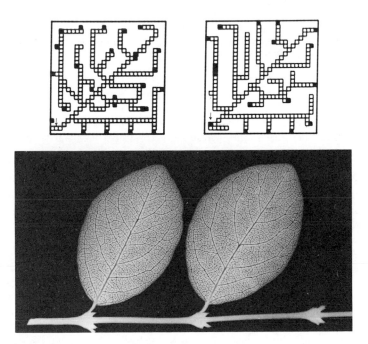

FIGURE 8. *Netlike pattern and the influence of random fluctuations:* In the simulations (top) one differentiated cell (↓) was assumed to be initially present. The differentiations (□) occur along a diagonal line, since a location at a distance from the margins is favored because no substance s can be obtained from beyond. A high activator concentration is indicated by ■. Branches try to grow out at 90°, but are deflected by the margins or by earlier branches, and appear, therefore, at about 45°. Since local fluctuations can be decisive for the precise location and orientation of a branch, two patterns will be similar, but not identical (the simulation is made with max. 3% random fluctuation in the const. c of Equation 6a, b). Correspondingly, the vein pattern of two leaves (bottom), of the same tree will never be identical, although they contain presumably the same genetic information.

the next cell is no longer random: Due to the depletion of the substance s around the differentiated cells, the activator maximum will be shifted to that the neighboring cell which is most remote from the already differentiated cell—that is, the cell at the end, not the one at the site of the incipient line. In this way the filament is systematically elongated in a straight line.

However, similar decision problems arise during the initiation of a branch. According to the model, a high activator concentration can be formed in an established line (by $\rho_0 y$ in Equation 6a). Since a depression of s is centered around each line, the activator maximum is pushed away from the line and a branch is initiated, usually at 90°. But to which side should the branch extend? Usually such a decision is unimportant; it can be chosen at random. For example, it is unimportant whether the first lateral branch of a midvein in a leaf extends to the left or to the right. Once a first decision has been made, say to the left, then the concentration of s will drop on the left side due to the new branch, and the probability is greatly increased that the next branch will extend to the right. The random element here implies that two patterns, generated by the same mechanism, will be similar, but not identical. Indeed, the leaves of the *same* tree, which are assumed to be governed by the same genetic information, are not identical (FIGURE 8). Similarly, the details of neural connections within genetically identical species of the water flea *(Daphnia)* are different.[34]

CONCLUSION

The proposed interaction provides a basis for the transition from a homogeneous to a patterned distribution of substances after a field has reached a critical size. At this point, the system is extremely sensitive to external influences. Since the homogeneous distribution is no longer stable, a pattern must be formed. Only the orientation of the pattern depends on external influences, but the pattern itself is determined by the diffusion rates, turnover, and interactions of the substances, which must be coded for by the genes. Therefore, reproducible pattern formation is guaranteed, although the pattern may have been initiated by random fluctuations.

The extreme sensitivity of the system at the moment that homogeneous distribution becomes unstable allows optimal decisions. For instance, the growth of a vein can be oriented into the largest available free space, and the oral field of a sea urchin embryo can be located at an area of maximum oxygen supply. Local optimization requires little genetic information, since details do not have to be programmed. It is, as the proposed mechanism shows, much easier to code for the overall pattern and to leave the details to be chosen by local optimization, or at random if no optimum can be found.

After a pattern is formed, it is very insensitive to perturbations. The activator maximum is shielded by a surrounding cloud of inhibition. Self-regulation of a pattern is due to the antagonist interaction of both substances—an inherent property of the system. Possibilities for a growing organism to develop into finer and finer subdivisions have been provided: for orthogonal gradients to organize the other dimension, for subpatterns and their orientations, for the formation of periodic structures or of filaments and their ramifications. In all these cases, a decision, which may have been randomly made, determines the next step so that an ordered development occurs despite the fact that each step is based on a process with bifurcating properties.

Summary

Systems of coupled biochemical reactions with bifurcating properties have been proposed for the explanation of pattern formation during embryonic development. After a critical size of a field is exceeded, a high concentration of an autocatalytic substance appears at one boundary. It is, however, random at which boundary this occurs. Nonetheless, small asymmetries can force development in a predictable direction. Therefore, a pattern, once generated, can serve as the initial asymmetry for the development of a "subpattern." Coordinated development of the parts in relation to the total organism is thus assured with this type of "chain reaction." Examples which illustrate the bifurcating character of biological pattern formation and its control by small asymmetries have been given and compared with simulations of proposed pattern-forming reactions. Such examples are the orientation of the axes of the retina or limbs in amphibia, or branching pattern of veinlike structures.

References

1. WADDINGTON, C. H. 1962. Principles of Embryology. Macmillan.
2. TURING, A. 1952. The chemical basis of morphogenesis. Philos. Trans. R. Soc. London Ser. B. **237**: 37–72.
3. PRIGOGINE, I. & G. NICOLIS. 1971. Biological order, structures and instabilities. Q. Rev. Biophys. **4**: 107–148.
4. MARTINEZ, H. M. 1972. Morphogenesis and chemical dissipative structure, a computer simulated case study. J. Theor. Biol. **36**: 479–501.
5. EDELSTEIN, B. B. 1972. The Dynamics of cellular differentiation and associated pattern formation. J. Theor. Biol. **37**: 221–243.
6. GIERER, A. & H. MEINHARDT. 1972. A theory of biological pattern formation. Kybernetik **12**: 30–39.
7. GIERER, A. & H. MEINHARDT. 1974. Biological pattern formation involving lateral inhibition. Lect. Math. Life Sci. The American Mathematical Society. **7**: 163–183.
8. MEINHARDT, H. & A. GIERER. 1974. Applications of a theory of biological pattern formation based on lateral inhibition. J. Cell Sci. **15**: 321–346.
9. MEINHARDT, H. 1977. Models for the ontogenetic development of higher organism. Rev. Physiol. Biochem. Pharmacol. **80**: 47–104.
10. MAHOWALD, A. P. 1972. Oogenesis. In Developmental Systems: Insects. S. J. Counce & C. H. Waddington, Eds. Academic Press. **1**: 1–49.
11. VINTENBERGER, P. & J. CLAVERT. 1970. Sur le determinisme de la symetrie bilaterale chez les oiseaux. R.C. Soc. Biol. Paris. **154**: 1072–1076.
12. KOCHAV, Sh. & H. EYAL-GILADI. 1971. Bilateral symmetry in chick embryo determination by gravity. Science **171**: 1027–1029.
13. PEASE, D. C. 1941. Echinoderm bilateral determination in chemical concentration gradients. J. exp. Zool. **86**: 381–404; **89**: 329–345; **89**: 347–356.
14. CHILD, C. M. 1946. Organizers in the development and the organizer concept. Physiol. Zool. **19**: 89–148.
15. JAFFE, F. 1968. Localisation in the developing Fucus egg and the general role of localizing currents. Adv. Morphog. **7**: 295–328.

16. HARRISON, R. G. 1921. On relations of symmetry in transplanted limbs. J. Exp. Zool. **32**: 1–136.
17. WOLPERT, L. 1969. Positional information and the spatial pattern of cellular differentiation. J. Theor. Biol. **25**: 1–47.
18. MEINHARDT, H. 1977. A model of pattern formation in insect embryogenesis. J. Cell Sci. **23**: 117–139.
19. LAWRENCE, P. A. 1970. Polarity and patterns in the postembryonic development of insects. Adv. Insect Physiol. **7**: 197–266.
20. STONE, L. S. 1960. Polarization of the retina and development of vision. J. Exp. Zool. **145**: 85–93.
21. GAZE, R. M., M. JACOBSON & G. SZÉKELY. 1965. On the formation of connexion by compound eyes in Xenopus. J. Physiol. (London) **176**: 409–417.
22. JACOBSON, M. 1968. Development of neuronal specificity in retinal ganglion cell of Xenopus. Dev. Biol. **17**: 202–218.
23. HUNT, R. K. & M. JACOBSON. 1972. Development and stability of positional information in Xenopus retinal ganglion cells. Proc. Nat. Acad. Sci. U.S.A. **69**: 780–783.
24. HUNT, R. K. & M. JACOBSON. 1973. Specification of positional information in retinal ganglion cells of Xenopus: assays for analysis of the unspecified state. Proc. Nat. Acad. Sci. U.S.A. **70**: 507–511.
25. HUNT, R. K. & M. JACOBSON. 1973. Neuronal locus specificity: altered pattern of spatial deployment in fused fragments of embryonic Xenopus eyes. Science. **180**: 509–511.
26. SZÉKELY, G. 1957. Regulationstendenzen in der Ausbildung der "funktionalen Spezifität" der Retinaanlage bei *Triturus* vulgaris. Wilhelm Roux Arch. Entwicklungsmech. Org. **150**: 48–60.
27. GARCIA-BELLIDO, A., P. RIPOLL & G. MORATA. 1973. Developmental compartmentalization of the wing disc of Drosophila. Nature, New Biol. **245**: 251–253.
28. CRICK, F. H. C. & P. A. LAWRENCE. 1975. Compartments and polyclones in insect development. Science. **189**: 340–347.
29. GIERER, A. 1977. Biological features and physical concepts of pattern formation, exemplified by hydra. Curr. Top. Dev. Biol. **11**: 17–59.
30. SANDER, K. 1959. Analyse des ooplasmatischen Reaktionssystems von Euscelis plebejus Fall. (Cicadina) durch Isolieren und Kombinieren von Keimteilen. I. Mitt.: Die Differenzierungsleistungen vorderer und hinterer Eiteile. Wilhelm Roux Arch. Entwicklungsmech. Org. **151**: 430–497.
31. SANDER, K. 1971. Pattern formation in longitudinal halves of leaf hopper eggs (Homoptera) and some remarks on the definition of 'Embryonic regulation.' Wilhelm Roux Arch. Entwicklungsmech. Org. **167**: 336–352.
32. MEINHARDT, H. 1976. Morphogenesis of lines and nets. Differentiation **6**: 117–123.
33. ROMANOFF, A. L. 1960. The avian embryo. Macmillan.
34. MACAGNO, E. R., V. LOPRESTI & C. LEVINTHAL. 1973. Structure and development of neuronal connections in isogenetic organisms: Variations and similarities in the optic system of *Daphnia magna*. Proc. Nat. Acad. Sci. **70**: 57–61.

TEMPORAL AND SPATIAL ORDER IN BIOCHEMICAL SYSTEMS

Benno Hess and Theodor Plesser

Max-Planck-Institut für Ernährungsphysiologie
4600 Dortmund,
Federal Republic of Germany

NONEQUILIBRIUM STATES IN BIOCHEMICAL SYSTEMS

In 1955, the observation of oscillating enzyme reactions brought the first direct experimental evidence that biological systems can function beyond some threshold of instability, the source of which is highly nonlinear kinetics in biochemical processes.[1,2] Nonlinearity arises from the kinetics of multiple types of positive and negative feedback interactions of biochemical processes in living cells, which are organized in the form of enzyme cycles, such as energy metabolism, protein biosynthesis, or fatty acid synthesis and break down, among others. The rates of these processes are commonly controled by allosteric and/or interconvertible enzymes, which respond to small changes in substrate products and controling ligands in a cooperative manner by changing their conformation states, as well as their chemical structure. In addition, it should be noted that the organization of transport of chemical particles in living systems implies, not only free diffusion, but, to a large extent, transport through membrane, which is facilitated by membrane-bound enzymes. Recently, electrical field effects were observed as an additional large-scale force affecting the transport of charged particles and controling the microenvironment of catalytic centers.[3]

Experimental observations, as well as model studies, have shown that, under proper dynamic conditions, oscillatory states are maintained in various systems composed of single or multiple enzyme reactions. Depending on the turnover of the system, multiple steady states with monotone or overshooting transitions from one state to another, rotation on a limit cycle around an unstable singular point,[4] as well as chaotic behavior,[5] have been described. For the nonhomogeneous case, transport processes by diffusion and active transport coupled to enzyme reactions have been found to settle on dynamic states which might yield the occurrence of standing or moving chemical waves, depending on the geometry and size of the system.[6,7,8,9]

Recently, major experimental and model studies focused on the mechanism of oscillation in the peroxidase reaction, the membrane-bound papain reaction, the glycolytic oscillation, as well as the oscillation of the synthesis of cyclic AMP in dictyostelium discoideum.[2] In the context of this presentation, we would like to discuss more recent studies on the mechanism of glycolytic oscillation.

OSCILLATORY STATES OF GLYCOLYSIS

Glycolytic oscillations have first been demonstrated in intact yeast cells and yeast cell extracts and, later, also in the extract of beef heart.[1,2] The detailed

203

0077–8923/78/0316–0203 $1.75/1 © 1979, NYAS

analysis of this phenomenon revealed that in its oscillatory state, the concentration and fluxes of all intermediates of a glycolytic reaction pathway oscillate with periods ranging from seconds to minutes, depending on the turnover and the enzyme composition of the system. In studying this system, it was soon recognized that the enzyme, phosphofructokinase (PFK), is the primary source of the oscillatory state. By a periodic change in the activity of the enzyme, the glycolytic process is periodically pulsed, and a time pattern of concentration changes is generated. The change in the activity of PFK is, to a large extent, controled by the generation and consumption rate of adenosinetriphosphate (ATP), which is one of the two substrates, and, at the same time, an inhibitor of PFK. If the ATP concentration in the system rises due to the activity of other enzymes in the chain, the activity of PFK is self-inhibited. In addition, the enzyme activity is strongly activated by the allosteric ligand, adenylic acid (AMP). The kinetics of PFK can be described on the basis of the allosteric theory of Monod et al.[10] In addition to the function of PFK in maintaining an oscillatory state of glycolysis, the function of the allosteric pyruvate kinase (PK) must be considered. Both enzymes share with phosphoglycerate kinase the same adenosine substrates. This leads to a synchronization and modification of their pulsed activity change.

Soon after the identification of the role of PFK in generating glycolytic oscillations, biochemical experiments were complemented by a series of analyses of dynamic models. Based on a phenomenological description of a Michaelis Menten model of PFK, Higgins,[11] and later Sel'kov,[12] showed that a feedback activation of PFK is a necessary condition for explaining the destabilizing interaction leading to sustained temporal oscillations. Both models stress the importance of the autocatalytic property of model structure as a source of nonlinearity in its kinetic equations. The role of bifurcations in these and other biochemical systems has been discussed by Glansdorff and Prigogine[6], and also by Gurel.[13,14]

With the recognition of the allosteric properties of PFK of yeast, it became obvious that the source of the autocatalytic property of PFK was not only feedback activation by a metabolic product of the reaction system, but, in addition, a consequence of the multiple-state function of the PFK. This observation led to the one-substrate model of Goldbeter and Lefever[15] based on the allosteric theory of Monod et al.,[10] which implies that PFK is biochemically composed of at least two subunits (=protomers) which interact and maintain two conformation states defined by their different affinities to the substrates. An extension of the allosteric model was given for enzymes with n protomers.[16] With a constant influx rate v_{in} and a linear sink reaction with the rate constant k_{out}, the differential equations are,

$$\frac{d\alpha}{d\tau} = v - v(\alpha, \gamma)$$

$$\frac{d\gamma}{d\tau} = \epsilon[v(\alpha, \gamma) - \kappa\gamma]$$

(1)

where
$$\alpha = \frac{S}{K_R}, \gamma = \frac{M}{K_{R,M}}, \tau = \frac{V_{R,max}}{K_R} t,$$

$$v = \frac{V_{in}}{V_{R,max}}, \kappa = \frac{k_{out}}{V_{R,max}} \cdot K_{R,M},$$

S is the substrate concentration in the catalytic process of the enzyme, and M is the allosteric modifier, or effector, of the enzyme. In this model, the product P of the enzyme reaction is the modifier. The rate law $v(\alpha, \gamma)$ for concerted allosteric interactions[10,18] is given by,

$$v(\alpha, \gamma) = \frac{\alpha(1 + \alpha)^{n-1} + qLc\alpha(1 + c\alpha)^{n-1}}{(1 + \alpha)^n + L(1 + c\alpha)^n} \tag{2}$$

where $q = V_{T,\max}/V_{R,\max}$, $c = K_R/K_T$, $0 < c < 1$, $V_{T,\max}$ and $V_{R,\max}$ are the maximal catalytic rates of the T- and R-state, respectively, and c is the nonexclusive binding coefficient, expressed as the ratio of the Michaelis constants for S in the T- and R-state. For the allosteric modifier, the following holds,

$$L = L_0 \left(\frac{1 + c_M\gamma}{1 + \gamma}\right)^n \quad \text{and} \quad c_M = \frac{K_{R,M}}{K_{T,M}}$$

For $0 < c_M < 1$, M functions as activator and for $1 < c_M < \infty$, M functions as inhibitor. The allosteric constant L_0 is the equilibrium constant of the R- and T-state in a substrate and modifier-free solution.

An analysis of Equations 1 shows that there is at most one singularity for $S > 0$ and $P > 0$. A necessary condition for instability at this singularity is given by $c_M < 1$. If this and the other parameters are chosen so that the system becomes unstable at this singularity, a limit cycle exists. Any solution of the differential equations approaches the limit cycle if the initial values of the solution are positive. This was proved by computer simulation, therefore, sustained oscillations are only possible in this model, if the product is an allosteric activator of the enzyme.

FIGURE 1 shows the domains for different dynamic states of the system as a function of the normalized input rate ν and the normalized output rate constant κ. It is characteristic for this model that the oscillatory domain is, in part, surrounded by stable states. Therefore it is possible for a range of κ to cross the oscillatory domain by increasing the input rate. This is also found in experiments with cell-free yeast extracts.[4] Recently, biochemical experiments for various input conditions were compared with corresponding computer simulations of the Goldbeter-Lefever model.[19] In the comparison of amplitude, period, response to stochastic input, and periodic entrainment, as well as phase-shift analysis, no serious discrepancy was detected and the conclusion is that: *The dynamic behavior of a complex system can be reduced to the molecular properties of a single protein species operating as a master enzyme in a biochemical pathway.*

The analyses of Equations 1, with a rate law derived from the sequential model[21] by assuming two protomers, shows that limit cycle oscillations are also observed comparable to those in the concerted model of Monod, *et al.*, if the product is an allosteric activator of the enzyme.[2,17] However, a full understanding of the overall dynamics of the glycolytic pathway can only be obtained by a quantitative comparison of the model and experiments under conditions in which a two-substrate model is the basis of the rate laws, and other regulatory enzymes in the glycolytic reaction sequence are analyzed. We have developed a three-enzyme model with five metabolite intermediates, as shown in FIGURE 2.[17] The system is open for the substrates fructose-6-phosphate (F6P) and phosphoenol pyruvate (PEP); the products fructosediphosphate (FDP) and pyruvate (PYR)

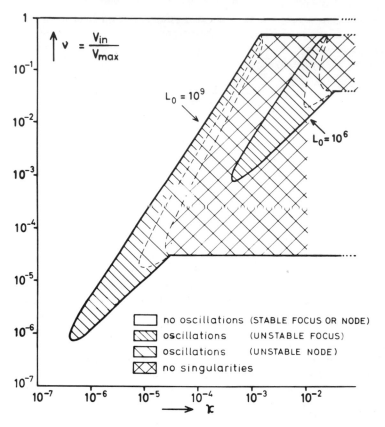

FIGURE 1. Dynamic states of the Goldbeter-Lefever model $K_{R,S}/K_{R,P} = 1, 0; n = 2.$[20]

accumulate. The total quantity of the adenosine phosphates, AMP + ADP + ATP, remains constant. Therefore, the system has four independent state variables: F6P, PEP, and two of the adenosine phosphates. The PFK catalyzes a quasi-irreversible phosphate transfer reaction in the first part of glycolysis as follows:

$$\text{ATP} + \text{F6P} \xrightarrow{\text{PFK}} \text{ADP} + \text{FDP} \qquad \text{(i)}$$

The quasi-irreversible reaction catalyzed by PK restores ATP from ADP in the second part of the glycolytic pathway, thus,

$$\text{ADP} + \text{PEP} \xrightarrow{\text{PK}} \text{ATP} + \text{PYR} \qquad \text{(ii)}$$

Adenylate kinase (AK) catalyzes the reversible reaction,

$$2\,\text{ADP} \underset{k_{-1}}{\overset{k_1}{\rightleftharpoons}} \text{AMP} + \text{ATP} \qquad \text{(iii)}$$

This enzyme allows an equilibration of ATP, ADP, and AMP, the latter one being the allosteric activator of PFK of yeast.

A two-substrate rate law for Reactions i and ii was derived for the concerted model as an analog to the one-substrate case (in collaboration with H. J. Wieker of this laboratory).[22] The derivation of Equation 3 holds, in the limiting situation, when the enzyme species are in equilibrium at any time;

$$V = V_{R,\max} \cdot v(\alpha_1, \alpha_2, \gamma_1, \gamma_2) \tag{3}$$

where

$$v = \frac{g_R \cdot \alpha_1 \cdot \alpha_2 R^{n-1} + q \cdot L \cdot g_T \cdot c_1 \alpha_1 \cdot c_2 \alpha_2 T^{n-1}}{R^n + L \cdot T^n}$$

$$\alpha_j = \frac{S_j}{K_{R,j}}, \; \gamma_j = \frac{M_j}{K_{R,M_j}} \; (M_j = P_j),$$

$$q = \frac{V_{T,\max}}{V_{R,\max}}, \; c_j = \frac{K_{R,j}}{K_{T,j}} \; (0 \le c_j \le 1),$$

$$R = 1 + \alpha_1 + \alpha_2 + g_R \cdot \alpha_1 \cdot \alpha_2,$$

$$T = 1 + c_1 \alpha_1 + c_2 \alpha_2 + g_T \cdot c_1 \alpha \cdot c_2 \alpha_2.$$

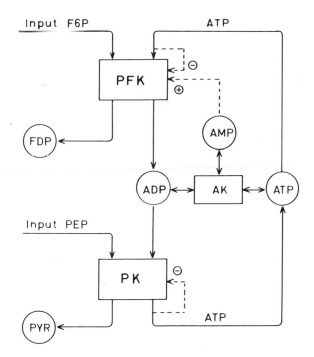

FIGURE 2. Three-enzyme model for glycolytic oscillations.[17]

For m independent binding sites per protomer, we have,

$$L = L_0 \prod_{j=1}^{m} \left(\frac{1 + c_{M,j} \cdot \gamma_i}{1 + \gamma_j} \right)^n$$

where

$$c_{M,j} = \frac{K_{R,M_j}}{K_{T,M_j}},$$

$$0 \leq c_{M,j} < 1 \text{ (activator)},$$

$$1 < c_{M,j} \leq \infty \text{ (inhibitor)},$$

and g_R and g_T are the affinity coefficient for the two substrates at the R and T state.

The parameters used for the investigation of this system by computer simulation for PFK and PK are given in TABLE 1. The allosteric activation of PK by FDP is minimized by high FDP concentrations. The equilibrium constant for the AK reaction,

$$K_{eq} = \frac{k_1}{k_{-1}}$$

is set to 0.45, and the concentration of the adenosine phosphate pool is fixed to 5.0 mM. The maximal rate $V_{R,max}$ (PFK) is set equal to $5.0 \text{ mM} \cdot \text{s}^{-1}$. The remaining three parameters: the input rate V_{in}, the maximal rate activity $V_{R,max}$ (PK), and the rate of AK reaction (expressed by the rate constant k_1) are varied in the different computer runs.

As is characteristic for the concerted model with product activation, the domain where oscillations are possible is surrounded by nonoscillatory states (FIGURE 3). The period decreases by a factor of about 40, if V_{in} and $V_{R,max}$ (PFK) are increased by a factor 500. The parameter $V_{R,max}$ (PK) corresponds to the sink constant κ in Equations 1.

TABLE 1
CONCERTED TWO-SUBSTRATE MODELS

PFK Parameters		PK Parameters	
$K_{R,F6P}$	1.0 mM	$K_{R,PEP}$	0.2 mM
c_{F6P}	5×10^{-4}	c_{PEP}	4.10^{-3}
$K_{R,ATP}$	6×10^{-2} mM	$K_{R,ADP}$	50.0 mM
c_{ATP}	1.0	c_{ADP}	1.0
g_R	10.0	g_R	1.0
g_T	1.0	g_T	1.0
$K_{R,AMP}$	2.5×10^{-2}	$K_{R,FDP}$	0.2 mM
$c_{M,AMP}$	1.9×10^{-2}	$c_{M,FDP}$	10^{-2}
L_0	4×10^6	L_0	3.9×10^3
$V_{R,max}$	5.0 mM/s	$V_{R,max}$	—
q	0.0	q	1.0
n	2	n	2

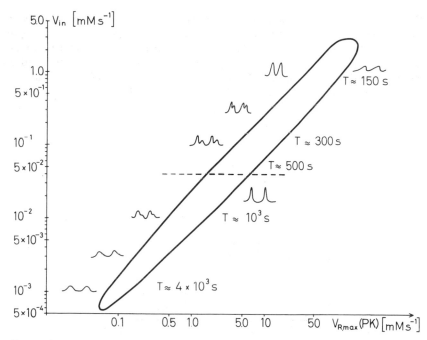

FIGURE 3. Oscillatory domain for the system in FIGURE 2. The wavelike lines illustrate the time course of ADP.[17]

The wavelike lines in FIGURE 3 show the concentration changes of ADP. All these patterns are well known from experiments with cell-free extracts of yeast. The oscillatory domain was calculated with a rate constant $k_1 = 10^{-2}$ $(mM \cdot s^{-1})$. A decrease of this constant by a factor of 10 leads to suppression of oscillations, because of the time lag between the production of ADP and the corresponding allosteric activation of PFK by AMP. The calculations show that for sustained oscillations this time lag must be less than 10% of the period. If this time lag is shortened by increasing k_1 to values larger than 1.0, the oscillatory domain is reduced to input rates below 4×10^{-2}; this value is indicated in FIGURE 3 by a broken horizontal line.

In the nonoscillatory region the system can assume one stable and two unstable states. The instabilities are due to the fact that the adenosine phosphates are totally stored in AMP or in ATP, and therefore ADP and ATP, or ADP and AMP tend to zero. In the first case, the rates of PFK and PK go to zero, and F6P and PEP accumulate. In the second case, the reaction rate of PK tends to zero and PEP accumulates. These three states occur also in the oscillatory domain of FIGURE 3, if the initial concentrations of the variables are chosen in the stable or unstable regions of FIGURE 4. This figure shows that the system has three regions of attraction for the time course of the variables in their positive range.

The results obtained by stability analysis of homogeneous biochemical reaction systems soon led to a study of the conditions of instability with respect to diffusion. Here a figure of merit is given by the critical length λ_c separating the regimes dominated by chemical reactions and transport processes.[6] This figure depends on the square root of the ratio of the transport rate/chemical reaction rate, indicating a delicate cooperation and balance between both rates, and the experiment and model conditions for a dissipative state structure. In an analysis of a phenomenological model of glycolysis, a range of $10^{-4} < \lambda_c < 10^{-2}$ cm was

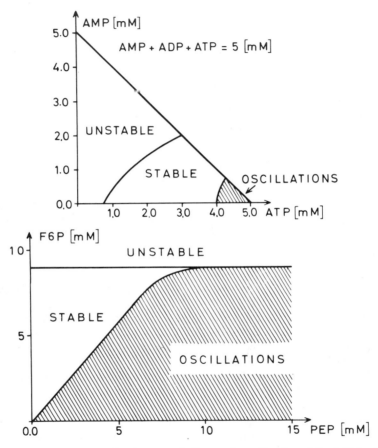

FIGURE 4. Dynamic states for the system in FIGURE 2 as a function of the initial values of variables, if the system parameters are in the oscillatory range.[17]

found, illustrating the magnitude of the expected space structure relative to molecular dimensions.[6] The analysis of a more realistic 1-dimensional model for an allosteric enzyme oscillator, such as PFK, yielded a critical length in the order of 10^{-1} cm for fixed boundaries.[7] This value was recently confirmed and extended for a 2-dimensional transport coupled to the oscillating activity states of the same enzyme model by Chance and Hess. The computed critical length in these analyses is large compared with the size of a single yeast cell in which an oscillatory state of glycolysis has been recorded.[9] However, using a highly concentrated cell-free extract of yeast, which permits observation of larger sizes of a glycolytic reaction territory, the occurrence of time-dependent spatial states of oscillating glycolysis has been demonstrated.[9] This was observed by a 2-dimensional recording of the absorption changes of reduced pyridinnucleotides as indicator of the dynamic state of glycolysis.

The Existence of Stable Periodic Solutions in Dynamic Systems

It is known that most results from the theoretical understanding of biochemical oscillations are based on linear stability analysis and computer simulation. If the existence and stability of periodic solutions are in question, however, these methods may lead to uncertain results. Therefore, proved mathematical theorems are necessary. In fact, only the *stable* periodic solutions guarantee the possibility of experimental observation.

A summary of a qualitative theory of ordinary differential equations in n dimensions is given in a paper by Cronin.[23] This paper emphasizes the general aspect of stability in biological and biochemical systems. Hopf's theorem[24] and its extension by Marsden and McCraken[25] are recent developments, which should be instrumental in the model analysis of biochemical systems. For the n-dimensional systems, based on peeling (bifurcation), global stability theorem due to Gurel[26,27] gives qualitative results in detecting the existence of stable oscillatory solutions. Recently Erle et al.[28] proved a theorem for the existence of *stable* periodic solutions for a plane autonomous system. This theorem is an extension of the Poincaré-Bendixon theorem on the existence of periodic solutions. Furthermore the theorem is used for the formulation of five conditions for the rate law and sink reaction in a biochemical system, which guarantee the existence of observable periodic solutions in a 2-dimensional system. These five conditions in biochemical terms are as follows:

1. The catalytic rate increases if substrate and/or product concentration increase; this rate is bounded to a finite value. The sink reaction is a monotone increasing function of product concentration, but independent of substrate concentration.
2. The rate for a substrate concentration equal to zero is less than the input rate, as fulfilled by definition.
3. For very high substrate concentration, there exists a range in product concentration, where the enzyme produces more product than the sink reaction removes.
4. There exists a product concentration defining a break-even point above which the rate of the enzyme reaction is always less than the sink rate.
5. There exists an unstable equilibrium point.

These conditions hold for Equations 1 with the rate law (Equation 2). Having the conditions at hand it is easy to construct systems with peculiar properties.

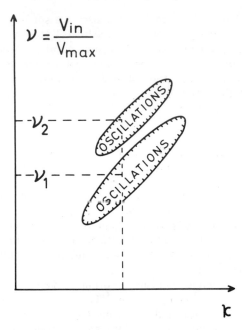

FIGURE 5. Oscillatory domains for a system that fulfills the five conditions for the rate law and sink reaction in a biochemical system, and has a sink reaction with two activity plateaus.

For example, in FIGURE 5 it is demonstrated that two distinct oscillatory domains are possible in the input-output plane. This phenomenon is observed if the sink reaction has two plateaus. Also from the above list of conditions, it should be emphasized that both the sink reaction and the rate law may play an important role for the occurrence of oscillations in biochemical systems.

REFERENCES

1. HESS, B. & A. BOITEUX. 1971. Oscillatory phenomena in biochemistry. Annu. Rev. Biochem. **40**: 237–258.
2. HESS, B., A. GOLDBETER & R. LEFEVER. Temporal, spatial and functional order in regulated biochemical and cellular systems. Adv. Chem. Phys. In press.
3. NAZAREA, A. D. Electric fields and self-coherent patterns and structures in chemical systems: Large scale effects in biological applications. Adv. Chem. Phys. In press.
4. HESS, B., A. BOITEUX & J. KRÜGER. 1969. Cooperation of glycolytic enzymes. Adv. Enzyme Regul. **7**: 149–167.
5. RÖSSLER, O. E. In this volume.
6. GLANSDORFF, P. & I. PRIGOGINE. 1971. Thermodynamic Theory of Structure, Stability and Fluctuations. Wiley Interscience.
7. GOLDBETER, A. 1973. Patterns of spatio temporal organization in an allosteric enzyme model. Proc. Nat. Acad. Sci. U.S.A. **70**: 3255–3258.
8. GOLDBETER, A. & G. NICOLIS. 1976. An allosteric enzyme model with positive feedback applied to glycolytic oscillations. Prog. Theor. Biol. **4**: 65–160.

9. HESS, B., A. BOITEUX, H.-G. BUSSE & G. GERISCH. 1975. Spatio-temporal organization in chemical and cellular systems. Adv. Chem. Phys. **29:** 137–168.
10. MONOD, J., J. WYMAN & J.-P. CHANGEUX. 1965. On the nature of allosteric transitions: A plausible model. J. Mol. Biol. **12:** 88–118.
11. HIGGINS, J. 1967. The theory of oscillating reactions. Ind. Eng. Chem. **59:** 18–62.
12. SEL'KOV, E. E. 1968. Self-oscillations in glycolysis. Eur. J. Biochem. **4:** 79–86.
13. GUREL, O. 1972. Bifurcation theory in biochemical dynamics. *In* Analysis and Simulation of Biochemical Systems. H. C. Hemker & B. Hess, Eds. Vol. **25:** 81–89. FEBS. North-Holland, Amsterdam.
14. GUREL, O. 1975. Limit cycles and bifurcations in biochemical dynamics. Bio-Systems **7:** 83–91.
15. GOLDBETER, A. & R. LEFEVER. 1972. Dissipative structures for an allosteric model. Biophys. J. **12:** 1302–1315.
16. PLESSER, T. 1975. Faraday symposium 9: Physical chemistry of oscillatory phenomena. Chem. Soc. (London). 225.
17. PLESSER, T. 1977. Dynamic states of allosteric enzymes. *In* VII Internationale Konferenz über nichtlineare Schwingungen 1975. (Abhandlungen der Akademie der Wissenschaften der DDR.) G. Schmidt, Ed. **2:** 273–280.
18. DALZIEL, K. 1968. A kinetic interpretation of the allosteric model of Monod, Wyman and Changeux. FEBS Lett. **1:** 346–348.
19. BOITEUX, A., A. GOLDBETER & B. HESS. 1975. Control of oscillating glycolysis of yeast by stochastic, periodic, and steady source of substrate: A model and experimental study. Proc. Nat. Acad. Sci. U.S.A. **72:** 3829–3833.
20. SCHWARZMANN, V. 1973. Numerische Analyse Biochemischer Modelle der Oszillierenden Glykolyse. Max-Planck-Institute für Ernährungsphysiologie. Dortmund, Federal Republic of Germany. Thesis.
21. KOSHLAND, D. E., G. NEMETHY, & D. FILMER. 1966. Comparison of experimental binding data and theoretical models in proteins containing subunits. Biochem. **5:** 365–385.
22. WIEKER, H.-J. 1975. Wasserstoffionen als isosterische und allosterische Effektoren kooperativer Protein-Ligand-Wechselwirkungen Habilitationsschrift. Ruhr-Universität. Bochum, Federal Republic of Germany.
23. CRONIN, J. 1977. Some mathematics of biological oscillations. SIAM Rev. **19:** 100–138.
24. HOPF, E. 1942. Abzweigung einer periodischen Lösung von einer stationären Lösung eines Differentialsystems. Ber. Verh. Sächs. Akad. Wiss. Leipzig, Math.-Naturwiss. Kl. **95:** 3–22.
25. MARSDEN, J. & M. MCCRAKEN. 1976. The Hopf Bifurcation and Its Applications. Springer-Verlag, New York.
26. GUREL, O. 1975. Peeling and nestling of a striated singular point. Collect. Phenom. **2:** 89–97.
27. GUREL, O. 1976. Peeling studies of complex dynamic systems. *In* Simulation of Systems. L. Decker, Ed.: 53–58.
28. ERLE, D., K. H. MAYER & TH. PLESSER. 1977. The existence of stable limit cycle for enzymes with positive feedback. Abstr. B2-2 450 11th FEBS Meet. Copenhagen.

PATHOLOGICAL CONDITIONS RESULTING FROM INSTABILITIES IN PHYSIOLOGICAL CONTROL SYSTEMS*

Leon Glass and Michael C. Mackey

Department of Physiology
McGill University
Montreal, Quebec, Canada
H3G 1Y6

1. Introduction

A large number of human diseases are characterized by changes in the qualitative dynamics of physiological control systems: Systems that normally oscillate, stop oscillating, or begin to oscillate in a new and unexpected fashion, and systems that normally do not oscillate, begin oscillating. These changes in qualitative dynamics often have a sudden onset, and in many instances it has not been possible to identify the factors that lead to the disease. By *dynamical disease* we mean a disease that occurs in an intact physiological control system operating in a range of control parameters that leads to abnormal dynamics and human pathology.[41] In this paper, the changes in qualitative dynamics associated with the onset of the disease are identified with bifurcations in the dynamics of mathematical models of the physiological control systems. We shall consider in some detail dynamical diseases in the respiratory and haematopoietic systems.

Our starting point is the ordinary differential equation

$$\frac{dx}{dt} = \lambda - \gamma x \qquad (1.1)$$

where x is a variable of interest, λ is a production rate for x, γ is the destruction rate of x, and t is the time. For λ and γ constant, $x \rightarrow \lambda/\gamma$ in the limit $t \rightarrow \infty$. However, in many physiological systems λ and γ at t may depend on x and/or x_τ (the value of x at a time $t - \tau$, where τ is the time lag). We show that instabilities analogous to those found in pathological conditions in humans can be reproduced by assuming that λ and γ in Equation 1.1 are appropriate nonlinear functions of x and/or x_τ.

2. Respiration

Respiratory oscillations in mammals are generated in the brainstem. Several groups have shown that this region is essential for respiration, and that cells located in the brainstem fire phasically during the respiratory cycle. Several dif-

*This research has been supported by a grant from the National Research Council of Canada and the Cancer Research Society of Montreal.

ferent classes of cells have been identified (e.g., inspiratory cells and expiratory cells which fire during inspiration and expiration, respectively), but the number of different classes of cells, their anatomical location, and interconnections are not agreed upon by workers in the field.[74] A number of mathematical models of the respiratory oscillator have been suggested.[11,14]

Experimental studies have shown that both the frequency and amplitude of the respiratory oscillations can be modulated by a variety of factors including activity in the cerebral cortex, pH and concentrations of CO_2 and O_2 in arterial blood and cerebrospinal fluid, and the amount of stretching in the intercostal muscle in the chest.[24,74] In healthy humans, these inputs act to maintain arterial concentrations of O_2 and CO_2 at constant levels.

Respiratory Disorders[32]

Rapid shallow breathing (panting, tachypnea, or polypnea) occurs in a variety of pathological conditions, for example, as a result of pain in structures moved by breathing, during fever, and under severe hypoxia (low oxygen tension) of long duration. In dogs, panting is a normal response to heat stress, and brief periods of panting (frequency 300–400 min^{-1}) alternate with periods of normal breathing (frequency 20–40 min^{-1}).[63] Superimposed on the panting rhythm may be an occasional deep breath to give a *sighing* pattern.

There are a variety of patterns in which periods of apnea alternate with periods of breathing. We call these *apneic patterns.* Apneic patterns are referred to by clinicians generically as "periodic breathing." The variety of apneic patterns that have been described include Cheyne-Stokes respiration, Biot breathing, and infant apnea.

Cheyne-Stokes breathing is characterized by a regular waxing and waning of breathing amplitude separated by periods of distinct apnea (FIGURE 1). This is the most common apneic pattern encountered clinically, and is often found in obese patients, patients with congestive heart failure, and patients with certain neurological deficits. It is also seen in normal humans after arrival at high altitude. It is interesting that a regular waxing and waning of breathing amplitude *without* apnea (a "wavy" pattern), is more commonly observed than Cheyne-Stokes respiration[65] and is not necessarily associated with pathological conditions (FIGURE 1).

Biot breathing refers to alternating periods of breathing with apnea. The regular alternations of Cheyne-Stokes respiration are absent, and marked irregularity is observed. Biot breathing is often observed just prior to death.

Infant apnea refers to the pronounced periods of apnea found in most premature and many full term infants (FIGURE 2).[66] The apnea generally occurs during rapid eye movement sleep. It has been speculated that there is a causal relation between the sudden infant death syndrome and infant apnea.

Although we have classified apneic patterns into a small number of discrete classes, intermediate patterns also exist. Unfortunately, extended records of pathological breathing patterns are not generally available.

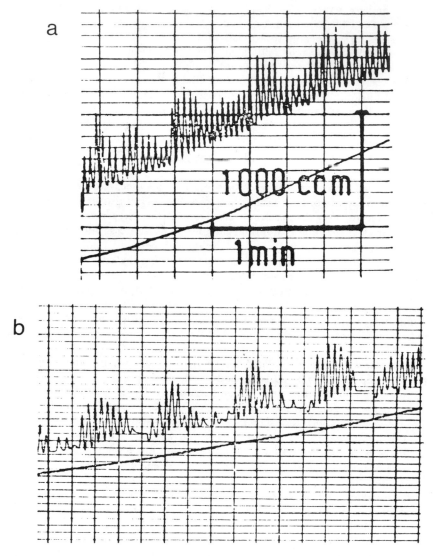

FIGURE 1. (a) A wavelike respiration pattern. (b) Cheyne-Stokes respiration in a 29-year-old man (5 horizontal divisions = 1 min, 10 vertical divisions = 1 liter). (From Sprecht and Fruhman.[65] By permission of *Bulletin Européen de Physiopathologie Respiratoire*.)

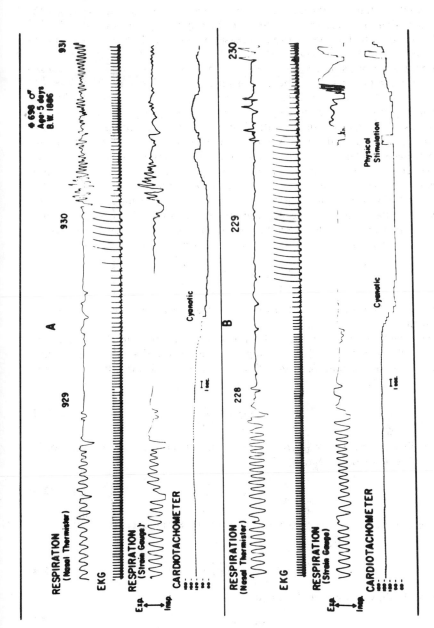

FIGURE 2. Apneic episodes leading to severe bradycardia and cardiac arrhythmia in a 5-day-old infant: (A) breathing spontaneously returned, (B) physical stimulation was app ied. (From Steinschneider.[66] By permission of *The Canadian Foundation for the Study of Infant Deaths*.)

Mathematical Models of Respiratory Disorders

Theoretical studies of the mechanism of Cheyne-Stokes respiration ascribe the slow oscillations observed to instabilities in the respiratory control system.[4,37,47,48] It is known that the total ventilation increases monotonically as the CO_2 concentration in arterial blood increases. However, since the blood is oxygenated in the lungs but the receptors, which are sensitive to the CO_2 concentration, are believed to be present in the brainstem, there is an inherent time lag τ in the respiratory control system. Several investigators have developed complex systems of differential-delay equations to describe the production, transport, and elimination of CO_2 in humans.[37,47,48] Since the mathematical properties of these complex systems of equations are not easily deduced, we have proposed a simplified schematic model which displays similar qualitative features to the more complex models.[41]

The ventilation V at time t is assumed to depend on $x(t - \tau)$, the CO_2 concentration at time $t - \tau$. We also assume that CO_2 elimination is proportional to the product of CO_2 concentration (x) and ventilation. Experimental studies indicate that ventilation is an increasing monotonic function of CO_2 concentration.[24] Assuming that the dependence of the ventilation on CO_2 concentration is described by the Hill function $V = V_{max}x_\tau^n/(\theta^n + x_\tau^n)$, we obtain,

$$\frac{dx}{dt} = \lambda - \frac{\alpha V_{max} x_\tau{}^n x}{\theta^n + x_\tau{}^n} \tag{2.1}$$

where V_{max} is the maximum ventilation and n, θ and α are parameters chosen to agree with experimental data.[41]

The stability of (2.1) in the neighborhood of the steady state can be analyzed (at the steady state $dx/dt = 0$). Denoting the values of x and V at the steady state by x_0 and V_0, and setting $S_0 = (dV/dx)_{x_0}$ and $\alpha = \lambda/x_0 V_0$, the first-order equation in x and x_τ is,

$$\frac{dx}{dt} = -\frac{\lambda}{x_0}(x - x_0) - \frac{\lambda S_0}{V_0}(x_\tau - x_0) \tag{2.2}$$

The stability criteria for (2.2) are well known.[21] In general, for the first-order linear differential-delay equation,

$$\frac{dz}{dt} = Az + Bz_\tau \tag{2.3}$$

the eigenvalues of the steady state $z = 0$ have negative real parts if and only if,

$$A\tau < 1$$
$$A\tau < -B\tau < \sqrt{(A\tau)^2 + a_1{}^2} \tag{2.4}$$

where $a_1 \in (0, \pi)$ is the root of the equation,

$$a \cot a = A\tau \tag{2.5}$$

and $a_1 = \pi/2$ if $A\tau = 0$. Applying (2.4) to determine the stability of the steady state of (2.1), we find that the steady state will be stable provided,

$$\frac{\lambda S_0 \tau}{V_0} < \sqrt{\left(\frac{\lambda\tau}{x_0}\right)^2 + a_1^2} \tag{2.6}$$

where a_1 is found by solving,

$$a \cot a = -\frac{\lambda\tau}{x_0} \tag{2.7}$$

Further analysis requires numerical values for the parameters in (2.6) and (2.7). Approximate values of the parameters for normal humans are readily obtained from the experimental literature.[41] We take,

$$x_0 = 40 \text{ mm Hg}$$

$$\lambda = 6 \text{ mm Hg/min}$$

$$V_0 = 7 \text{ liter/min}$$

$$S_0 = 4 \text{ liter/min mm Hg}$$

$$\tau = 0.25 \text{ min} \tag{2.8}$$

From (2.8) we find that $\lambda\tau/x_0 = 0.0375$. A numerical solution of (2.7) gives $a_1 = 1.5943$. Since $a_1 \sim \pi/2$, and for the parameters in (2.8), $a_1 \gg \lambda\tau/x_0$, the condition for a stable steady state can be given approximately as,

$$S_0 > \frac{\pi V_0}{2\lambda\tau} \tag{2.9}$$

The period of the oscillation is about 4τ.

FIGURE 3 shows numerical integration of (2.1) for two values of S_0 in the unstable region with the other parameters given in (2.8). Choosing a value of S_0, the parameters in (2.1) can be found from the relations,

$$n = \frac{x_0 V_{max} S_0}{V_0(V_{max} - V_0)}$$

$$\theta = x_0 \frac{(V_{max} - V_0)^{1/n}}{V_0} \tag{2.10}$$

Note that both a "wavy" pattern and apneic pattern are found in the figure.

These results are of interest for several reasons. Investigators have noted an increased CO_2 sensitivity (S_0) and delay time in patients displaying Cheyne-Stokes respiration.[3,33] Further, the numerical values computed using (2.8) for the unstable regime fall in a range outside accepted values for normals. For example, given V_0, λ, τ in (2.8) instability is predicted for $S_0 > 7.44$ 1/min/mm Hg, which is above normal.[24] However, the crudeness of the model makes detailed quantitative comparisons questionable. Other workers, who have investigated stability for mathematical models of respiration, have noted an approximate hyperbolic relation between S_0 and τ, leaving the other parameters of the system fixed.[37,47,48] The boundary separating stable and unstable regions in the S_0, τ-plane, given by (2.9), is a hyperbola.

The Piecewise Linear Case

Numerical integration of (2.1) indicates that, in the unstable regime, the oscillations approach a stable limit cycle oscillation. The problem of determining the asymptotic behavior of time-delay differential equations is complex, and only

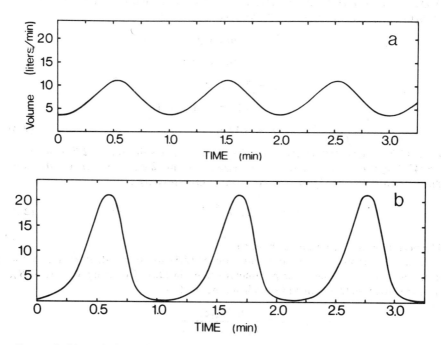

FIGURE 3. Numerical solutions to Equation 2.1. The time course of $x(t)$ as computed from (2.1), using parameters listed in (2.8) with the exception of S_0. (a) S_0 = 7.7 1/min mm Hg. (b) S_0 = 10.0 1/min mm Hg. With these values of S_0, n and θ in (2.1) were computed using (2.10). (From Mackey and Glass.[41] By permission of *Science*.)

limited results have been achieved to date.[6,23,44] In the limit $n \to \infty$, Equation 2.1 becomes piecewise linear, and explicit solutions can be constructed.

In the limit $n \to \infty$, (2.1) can be written,

$$\frac{dx}{dt} = \begin{cases} \lambda - \gamma x, & x_\tau > \theta \\ 0, & x_\tau = \theta \\ \lambda, & x_\tau < \theta \end{cases} \qquad (2.11)$$

where $\lambda/\gamma < \theta$. Assume that for $t < 0$, $x = x_0$, where $x_0 < \theta$. For $t > 0$, we assume that (2.11) holds. By direct integration of (2.11), the time course of x can be explicitly computed. From (2.11), x initially increases and at some time t_0, $x = \theta$. For $t > t_0$, the temporal evolution of x is given by

$$x(t) = \theta + \lambda (t - t_0), \qquad\qquad t_0 \leq t \leq t_1$$

$$x(t) = (\theta + \lambda\tau)e^{-\gamma(t-t_1)} + \frac{\lambda}{\gamma}\left[1 - e^{-\gamma(t-t_1)}\right], \quad t_1 \leq t \leq t_2$$

$$x(t) = \theta e^{-\gamma(t-t_2)} + \frac{\lambda}{\gamma}\left[1 - e^{-\gamma(t-t_2)}\right], \qquad t_2 \leq t \leq t_3 \qquad \textbf{(2.12)}$$

$$x(t) = (\theta - \lambda/\gamma)e^{-\gamma\tau} + \frac{\lambda}{\gamma} + \lambda(t - t_3), \qquad t_3 \leq t \leq t_4$$

where:
$$t_1 - t_0 = \tau$$

$$t_2 - t_1 = \frac{1}{\gamma} \log\left[\frac{\theta + \lambda\tau - \lambda/\gamma}{\theta - \lambda/\gamma}\right]$$

$$\text{(2.13)}$$

$$t_3 - t_2 = \tau$$

$$t_4 - t_3 = \frac{(\theta - \lambda/\gamma)(1 - e^{-\gamma\tau})}{\lambda}$$

During the ascending and descending phases, Equation 2.12 is continuous and differentiable. The derivatives are discontinuous at $t = t_1$ and $t = t_3$. Note that $x(t_4) = \theta$, so that for all times $t > t_4$ the solution repeats with a cycle period of,

$$T = 2\tau + \frac{1}{\gamma} \log\left[\frac{\theta + \lambda\tau - \lambda/\gamma}{\theta - \lambda/\gamma}\right] + \frac{(\theta - \lambda/\gamma)(1 - e^{-\gamma\tau})}{\lambda} \qquad \textbf{(2.14)}$$

The solution defined by Equations 2.12 and 2.13 is also a solution of Equation 2.11 with $\tau_m = \tau + mT$. As a consequence, for τ fixed, there is a family of solutions, so the solution is not unique.

3. Haematopoiesis

In the normal mammal, circulating levels of the formed elements of the blood (the white and red blood cells, platelets, and lymphocytes) are maintained at fairly constant levels. In response to various assaults by the environment, however, one or more of these blood cell types may change their relative concentrations in a transient fashion.[73]

It is generally believed that there exists a self-maintaining *pluripotential stem cell* population (PPSC) in the marrow capable of producing committed cells for the erythroid, myeloid, or thromboid lines. These populations are not self-maintaining but depend on a cellular flux from the PPSC for their continued integrity. Cells at the committed level undergo four to five effective divisions (nuclear divisions for the myeloid and erythroid series; cytoplasmic divisions for the thromboid series) before losing their nuclei to enter a maturation phase. Cells are then released from this marrow maturational compartment to enter the blood as a mature white blood cell, red blood cell, or platelet.[73]

There is control operating in the erythroid series between the circulating erythrocyte and the PPSC. Decreases in oxygen levels lead to the release of a substance called *erythropoietin* (EP), which acts to increase the cellular flux from the PPSC.[1,67] A number of investigators have looked for similar regulators in the myeloid and thromboid series, and to date the existence of putative *granulopoietins* (GP)[59] and *thrombopoietins* (TP)[7] has been claimed. In addition to this stimulatory effect on cell production, there have been numerous reports concerning the partial isolation of mitotic inhibitory substances in the myeloid and erythroid lines, which are termed *chalones*.[28,62] These chalones appear to be produced by mature granulocytes and erythrocytes, and inhibit proliferation at the myeloblast or erythroblast stage.

In addition to the feedback from circulating blood cells to the PPSC, control mechanisms are believed to exist within the PPSC itself, acting to control cell population numbers.[1] Although the details are not clear, it seems that the PPSC regulates its size by adjusting the rate at which cells enter active proliferation on the basis of the number of resting (G_0) phase cells.[31,46,53,69]

Dynamic Hematologic Disorders

In the hematology literature, there are a number of well-documented pathologies characterized by periodic oscillations in the formed elements of the blood in an apparently constant environment, and in the absence of any clinical intervention. Pathologies in which blood elements remain approximately constant at abnormal levels are also well known.

Cyclical neutropenia is characterized by an oscillation in circulating neutrophil numbers from normal to low values (FIGURE 4).[17] In humans, the majority of cases display a period in the range of 17 to 28 days. All grey collies have a similar disorder, with an oscillatory period of 11 to 12 days.[7,10] In both humans and grey collies, a concomitant oscillation of all the formed blood elements, with the exception of the lymphocytes, is observed.[9,10,22] These elements oscillate with the same period as the neutrophils, but with phase differences consistent with the known differences in maturation times for each of the cell types. Thus, this disorder is more appropriately termed *periodic haematopoiesis* (PH). In *aplastic anemia* (AA), there are severely depressed levels of all circulating elements of the blood, as well as a hypoplastic marrow. There is experimental evidence that the defect giving rise to PH and AA is contained in the PPSC,[2] and a common mechanism has been hypothesized to underly both diseases.[38,39]

Chronic myelogenous leukemia (CML) is a neoplastic disorder of the haematopoietic system generally characterized by a massive increase in circulating cells of the myeloid and thromboid series, and approximately normal erythroid elements.[73] In the past decade, reports in the clinical literature establish the existence of an interesting and provocative periodic variant, *periodic CML* (PCML). In the handful of patients in which PCML has been found, the peripheral leukocyte and thrombocyte counts oscillate around elevated levels with a period of 30 to 70 days depending on the patient (FIGURE 5).[5,12,58,68] Oscillations are commonly noted in all of the myeloid and thromboid series elements of these patients, and in two patients there is clear evidence of oscillations in the erythrocyte levels.[5,58] As in PH and AA, the PPSC is implicated as the source of the defect giving rise to CML and PCML.[73]

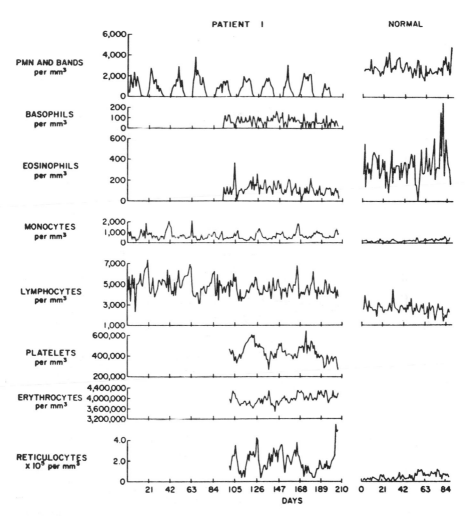

FIGURE 4. An example of human cyclic neutropenia (left-hand side) compared to a normal volunteer. The marked periodicity in the neutrophil count, with a period of 20 days, is also reflected in the monocytes, lymphocytes, platelets, and reticulocytes to a significant degree (period-gram analysis). (From Guerry et al.[17] By permission of Journal of Clinical Investigation.)

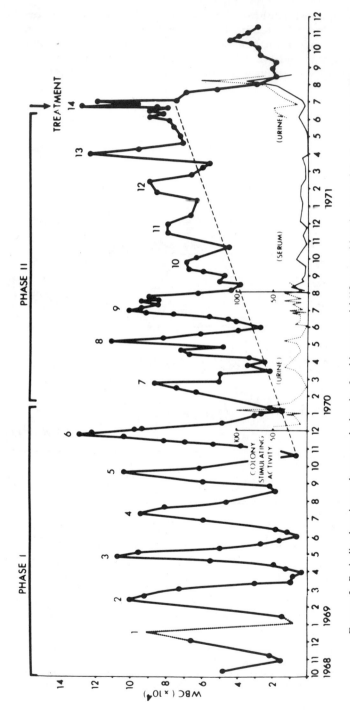

FIGURE 5. Periodic chronic myelogenous leukemia. In this young girl (12 years old at the start of the figure) with CML, the white blood cell count spontaneously oscillates with a period of approximately 70 days in the absence of any therapy. (From Gatti *et al.*[12] By permission of *Blood.*)

Cyclical thrombocytopenia is a rare disease in which rhythmic changes in platelets and megakaryocytes from normal to low values are observed to occur with a period of about 28 days.[34] There are documented cases in the literature in which the presence of these platelet cycles has persisted for up to 12 years.

Mathematical Models of Dynamic Hematological Diseases

In this section we consider two problems: the first is related to peripheral control over circulating cell numbers (e.g., as exercised via erythropoietin), and the second is related to control within the PPSC compartment. In both cases, our models are similar to models proposed by others for coupled stem cell-peripheral control systems in granulopoiesis,[25,60,71] erythropoiesis,[27] and thrombopoiesis.[15]

Peripheral Control in Haematopoiesis

We first consider a simple model for the control of peripheral blood cell numbers via a humoral feedback mechanism. Let $x(t)$ be the concentration of circulating cells (cells/kg) and assume that cells are randomly lost from the circulation at a rate $\gamma(\text{day}^{-1})$ proportional to their concentration. To reproduce the effects of poietin feedback control from the circulating population of cells, we assume that the flux (λ in cells/kg/day) into the circulation from the stem cell compartment depends on x at time $t - \tau$, and thus the dynamics of $x(t)$ is governed by,

$$\frac{dx}{dt} = \lambda(x_\tau) - \gamma x \tag{3.1}$$

We have examined two forms for $\lambda(x)$:

$$\lambda(x) = \begin{cases} \dfrac{\lambda_0 \theta^n}{\theta^n + x^n} & (3.2) \\[3ex] \dfrac{\lambda_1 \theta^n x}{\theta^n + x^n} & (3.3) \end{cases}$$

where n, θ (cells/kg), λ_0 (day^{-1}), and λ_1 (kg/day-cell) are parameters.[41] The control function (3.2) is a simple monotone decreasing function of x for $x \geq 0$, while that of (3.3) is a single-humped function.

Combining Equation 3.1 with the equations for λ thus gives two possible equations governing the evolution of $x(t)$,

$$\frac{dx}{dt} = \frac{\lambda_0 \theta^n}{\theta^n + x_\tau{}^n} - \gamma x \tag{3.4}$$

and

$$\frac{dx}{dt} = \frac{\lambda_1 \theta^n x_\tau}{\theta^n + x_\tau{}^n} - \gamma x \tag{3.5}$$

An equation similar to (3.4) has been proposed for the control of erythropoiesis.[6,70]

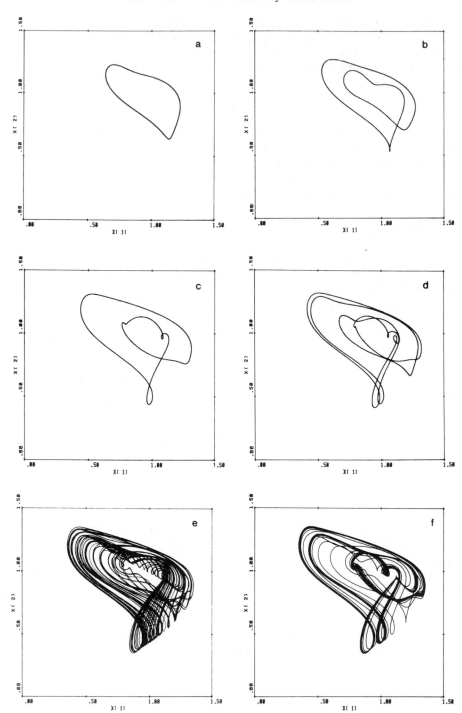

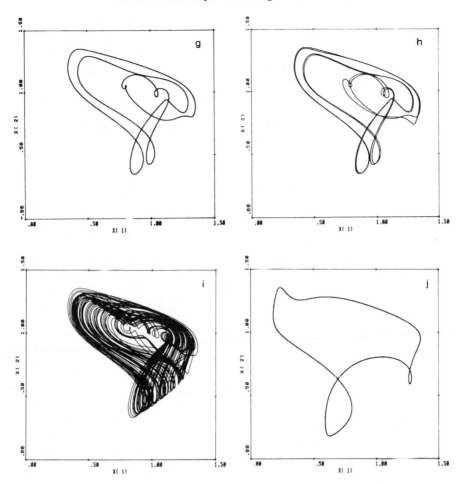

FIGURE 6. Bifurcating solutions to Equation 3.5. Here we show the numerically deter-mined solutions to this equation in the form of phase plots of $x_\tau = x(2)$ versus $x = x(1)$. The integrations were carried out with a step size of 0.05 using a predictor-corrector method, assuming $\gamma = 1$, $\lambda_1 = 2$, $\theta = 1$, $\tau = 2$, and an initial condition on x and x_τ of 0.50.

(a) $n = 7$, $100 \le t \le 150$ (b) $n = 7.75$, $150 \le t \le 200$
(c) $n = 8.50$, $200 \le t \le 250$ (d) $n = 8.79$, $300 \le t \le 400$
(e) $n = 9.65$, $300 \le t \le 600$ (f) $n = 9.69715$, $300 \le t \le 800$
(g) $n = 9.6975$, $300 \le t \le 400$ (h) $n = 9.76$, $300 \le t \le 400$
(i) $n = 10.0$, $300 \le t \le 600$ (j) $n = 20.0$, $300 \le t \le 400$

The nonzero steady states x_0 of these equations may be calculated and the local behavior of the solutions examined near x_0 as in the section on respiratory models. The results of these computations indicate that increases in n, τ, or λ_0, or decreases in γ or λ_1 may lead to a loss of stability at x_0 and the appearance of oscillatory solutions about x_0.

To investigate the behavior of (3.4) and (3.5) away from the region of applicability of this linear analysis, we have numerically integrated the equations using either a predictor-corrector or a Runge-Kutta integration scheme. For Equation 3.4, we have found only two qualitatively different behaviors: 1) either a stable steady state or 2) a stable limit cycle oscillation—for any set of parameters only one or the other behavior is found.

However, the qualitative behavior of (3.5) in response to parameter changes are quite different. To illustrate this behavior we assume that $\gamma = 1$, $\lambda_1 = 2$, $\theta = 1$ (so $r_0 = 1$), and $\tau = 2$. Equation 3.5 was integrated starting from an initial condition $x(t) = 0.50$, $-\tau < t < 0$, using a predictor-corrector integration routine with a step size of $\Delta = 0.05$ for various values of n.

The linear analysis of (3.5) indicates that with these parameters, $x_0 = 1$ should be stable for $n < 5.0404$, and, as stability is lost for $n \sim 5.0404$, periodic solutions of period $T \sim 5.49$ should appear. Numerical solutions of (3.5) in the neighborhood of $n = 5.04$ bear out the accuracy of this analysis, and indicate that the periodic solutions are stable.

In FIGURE 6 we show the dynamics of (3.5) in the $x_\tau - x$ phase plane for several values of n. Notice that as n is increased, the oscillation undergoes a sequence of bifurcations. Further, this sequence is analogous to the sequence of bifurcations observed in a class of finite-difference equations in 1-dimension.[35,42,43,45] Notice that the oscillatory patterns in FIGURE 6a and b are analogous to the "period 2" oscillation; FIGURE 6b and 6c are analogous to the "period 4" oscillation; FIGURE 6d is analogous to the "period 8" oscillation; FIGURE 6g is analogous to the "period 3" oscillation; FIGURE 6h is analogous to the "period 6" oscillation; FIGURE 6e and 6i are analogous to the "chaotic regimes." In FIGURE 6f we observe that the "period 3" oscillation has almost "condensed" out of the chaotic regime. FIGURE 6j shows a stable oscillation, which appears for large n.

At the moment we are not aware of any clinical data concerning circulating levels of blood elements that displays similar sequences of bifurcations in their qualitative dynamics. However, (3.4) can be used[40] to describe the periodic fluctuations in circulating red blood cell numbers observed in an auto-immune hemolytic anemia[52] (increased cell destruction rate γ).

Control Within The PPSC

To examine the possible origins of PH within the PPSC, we assume that[38,39]: 1) cells in the PPSC are either proliferating or in the resting phase (G_0) cells, 2) cells travel through proliferation to undergo mitosis at a fixed time τ (days) from their time of entry into the proliferative phase, 3) all cells enter G_0 upon the completion of mitosis, and 4) cells in G_0 exit randomly to differentiate either irreversibly into one of the haematopoietic lines (myeloid, erythroid, or thromboid) at a rate α (day^{-1}) or reenter proliferation at a rate β (day^{-1}) propor-

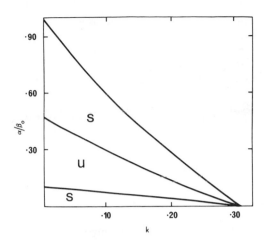

FIGURE 7. Linear stability analysis for Equation 3.7. Here we show the predicted regions of local stability of the steady state x_0 of Equation 3.7, assuming $n = 3$, $\tau = 2.22$ days, $\beta_0 = 1.77$ days^{-1}; u denotes the region where x_0 is not stable according to the linear analysis, while s denotes stability. For all parameter values $(k, \alpha/\beta_0)$ above the top curve, a positive steady state does not exist.

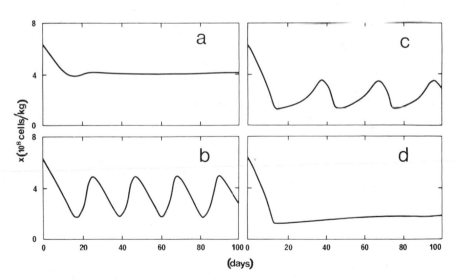

FIGURE 8. Numerical solutions to Equation 3.6: Here we illustrate the time course of $x(t)$ as computed from Equation 3.6 using a predictor-corrector integration scheme with an integration step size of 0.05, $n = 3$, $\alpha = 0.05$ day^{-1}, $\beta_0 = 1.77$ day^{-1}, $\tau = 2.22$ days, $\theta = 1.98 \times 10^8$ cells/kg, and an initial condition $x(0) = 6.43 \times 10^8$ cells/kg. (a) $k = 0.20$ day^{-1}, (b) $k = 0.25$, (c) $k = 0.28$, and (d) $k = 0.29$. The behavior in (a) and (d) has been interpreted as reflecting the pattern of aplastic anemia, while that of (b) and (c) is similar to periodic haematopoiesis.[39]

tional to their concentration (in certain pathological states, proliferating phase cells die at a rate k (day^{-1}) proportional to their concentration).

We further assume that control in the PPSC is exercised over the rate β of cell reentry into proliferation and that β is a monotonic decreasing function of G_0 cells. Calling the population density of the G_0 cells x (cells/kg), we obtain,

$$\frac{dx}{dt} = \frac{2\beta_0 \theta^n x_\tau}{\theta^n + x_\tau{}^n} \exp(-k\tau) - \alpha x - \frac{\beta_0 \theta^n x}{\theta^n + x^n} \qquad (3.6)$$

where θ (cells/kg), β_0 (day^{-1}), and n are parameters characterizing the PPSC.

PH and AA. Based on several lines of evidence, there is good reason to believe that the strange dynamics of PH is intimately connected with the death of cells from the proliferating phase of the PPSC. Further, at least some cases of AA may involve the death of proliferating PPSC cells.[2]

It is possible to estimate the values of the parameters characterizing a PPSC population in a normal ($k = 0$) state. Depending on the values of these parameters, the linear analysis of the stability of the nonzero steady state x_0 of (**3.6**) predicts two possible responses in the PPSC to increases in k (FIGURE 7). For humans, taking $n = 3$, $\tau = 2.22$ days, $\alpha = 0.05$ day^{-1}, and $\beta_0 = 1.77$ days, an increase in k leads to a depression of the steady state x_0. At about $k = 0.235$ day^{-1}, the steady state is no longer stable and periodic solutions of period 19.00 days are predicted. At $k = 0.287$ day^{-1} a stable steady state reappears. For $0.235 < k < 0.287$, numerical studies indicate stable limit cycle oscillation. These behaviors are illustrated in FIGURE 8. Since both depression of cell densities and periodic dynamics can be accounted for in this single model and the numerical values obtained are reasonable, it has been suggested that both AA and PH can be accounted for by varying rates of cell destruction during proliferation of stem cells.[39]

4. MISCELLANEOUS DYNAMICAL DISEASES

During the course of our research we have become aware of a large number of other pathological conditions with striking dynamical behavior. Although some of the conditions are well known to the layman, others are obscure and often not even accurately diagnosed. We mention some representative examples, and recent references, which can be consulted for extensive bibliographies.

Cardiac Arrhythmias[36,61]

The heart is capable of beating in a variety of different regular and irregular oscillatory patterns. These patterns can be visualized by the electrocardiogram (ECG) shown in FIGURE 9. Normally the cardiac cycle is initiated by the sinoatrial (S-A) node that acts as a pacemaker, initiating activity that travels through the cardiac tissue to the atrioventricular (A-V) node. Excitation at the S-A node leads to contraction of the atria and excitation at the A-V node leads to ventricular contraction.

However, most heart tissue is capable of generating spontaneous rhythms, and in many pathological conditions the S-A node partially or completely loses its role as the pacemaker. A few pathologies will be illustrated. FIGURE 9a shows an ECG taken from a patient in which every fourth pulse generated by the S-A node is ineffective in driving the ventricular rhythm (the Wenkebach phenomenon). In the same patient, at a slightly higher pulse rate (FIGURE 9b) every other pulse generated by the S-A node is ineffective. In another pathology, the ventricles display high-frequency contraction, apparently out of control of the S-A node. This condition (ventricular tachycardia) may have sudden onset and cessation (FIGURE 9c). Ventricular fibrillation is characterized by "chaotic, irregular, and disorganized ventricular activity,"[36] and usually leads to death within a few minutes (FIGURE 9d).

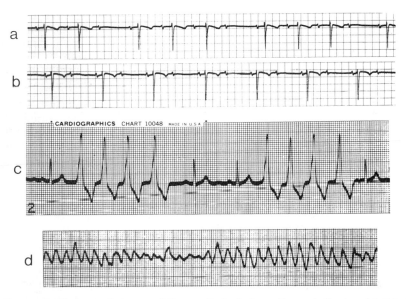

FIGURE 9. Electrocardiograms of four pathological cardiac arrhythmias: (a) 2:1 A-V block, (b) The Wenkebach phenomenon, (c) Ventricular tachycardia, (d) Ventricular fibrillation. (From Lindsay and Budkin.[36] By permission of *Year Book Medical Publishers*.)

There is an intriguing and widely scattered theoretical literature dealing with mechanisms of cardiac arrhythmias. In an early paper, van der Pol and Mark[54] generated oscillatory patterns similar to those shown in FIGURE 9 by coupling nonlinear oscillators with different natural frequencies. There has been additional theoretical work that attributes other arrhythmias to abnormal patterns of wave propagation in the excitable tissue of the heart.[29,49,72]

Psychological and Neurological Disorders

There are a variety of psychological and neurological disorders including insomnia, epilepsy, manic-depressive episodes, and schizophrenia that have been reported to display a variety of regular and irregular oscillatory patterns with periodicities ranging from several days to several years.[57] Theoretical work has been done on the analysis of a proposed mechanism for periodic catatonic schizophrenia.[8]

Cancer

There is a large body of theoretical work that ascribes cancerous growth to instabilities in the dynamics of the cell cycle,[18,19] or in feedback control systems regulating mitosis.[16,64] It has been observed that cancer, if untreated, does not necessarily grow "without limit."[30] Further, some cancerous growth, (e.g., the CML referred to above) may have a periodic time course (see the section on haematopoiesis).

Miscellaneous

There are a large number of other diseases that can display regular and irregular oscillatory dynamics; e.g., periodic synoviosis (swelling of the articulated joints), periodic peritonitis (swelling of the gut), periodic fever, and periodic pancreatitis.[55-57] Although it is plausible that some of these diseases arise from instabilities in feedback control mechanisms similar to those we have discussed here,[51] we are not aware of detailed theoretical analyses. Finally, the importance of periodic factors in both normal and pathological conditions in humans have been emphasized in the work of Halberg et al.[20]

5. DISCUSSION

Physiological control systems are extremely complex. Moreover, there are large differences in the details of the structure of these systems among different species, and even different members of the same species. As a consequence, those interested in studying dynamical diseases face numerous experimental difficulties. Experimentation on normal and diseased humans must be of an extremely limited kind. Since human experimentation is difficult, most of the systematic information that has been gathered about physiological control systems is from other mammals. Although principles of organization may be the same in these animals as in humans, extrapolation of quantitative results to humans is not easily done. Even more subtle problems arise. Many experimental paradigms that have been adopted by physiologists tend to obscure variability in a single animal during the course of an experiment, as well as the differences among different animals. Typically, data from a number of animals are lumped together, and average data are reported. Since diseases tend to be rare, abnormal dynamical behavior (if it occurs at all), which might be most relevant from the standpoint of dynamical disease, is often disregarded or lumped with enough other data to give "good statistics." One experimental approach to the study of dynamical disease is to develop animal models of the disease in which animals, after a variety of manipulations, display similar qualitative dynamics to those displayed in the human counterpart of the disease. Clearly, studies such as these can not be used to identify the cause of the disease in humans, but simply to generate plausible hypotheses.

The existence of classifiable dynamical diseases in humans suggests a correspondingly rich theory of bifurcations in nonlinear ordinary, partial, and functional differential equations which model physiological control systems. At this point a sufficient body of data is not yet available for actual testing of theories of dynamical diseases. In our view, close collaboration between theorists and clinicians is needed to clarify the bases of these dynamical diseases.

REFERENCES

1. BOGGS, D. R. 1966. Homeostatic regulatory mechanisms of hematopoiesis. Ann. Rev. Physiol. **28:** 39–53.
2. BOGGS, D. R. & S. S. BOGGS. 1976. The pathogenesis of aplastic anemia: A defective pluripotential hematopoietic stem cell with inappropriate balance of differentiation and self-replication. Blood. **48:** 71–76.
3. BROWN, H. W. & F. PLUM. 1961. The neurologic basis of Cheyne-Stokes respiration. Am. J. Med. **30:** 849–860.
4. CHERNIACK, N. S. & G. S. LONGOBARDO. 1973. Cheyne-Stokes breathing, an instability in physiologic control. N. Engl. J. Med. **288:** 952–957.
5. CHIKKAPPA, G., G. BORNER, H. BURLINGTON, A. D. CHANANA, E. P. CRONKITE, S. ÖHL, M. PAVELEC & J. S. ROBERTSON. 1976. Periodic oscillation of blood leukocytes, platelets, and reticulocytes in a patient with chronic myelocytic leukemia. Blood. **47:** 1023–1030.
6. CHOW, S. N. 1974. Existence of periodic solutions of autonomous functional differential equations. J. Diff. Eqn. **15:** 350–378.
7. COOPER, G. W. 1970. The regulation of thrombopoiesis. In Regulation of Hematopoiesis. A. S. Gordon, Ed. Appleton-Century-Crofts. **2:** 1611–1629.
8. CRONIN-SCANLON, J. 1974. A mathematical model for catatonic schizophrenia. In Ann. N.Y. Acad. Sci. O. Gurel, Ed. **231:** 112–122.
9. DALE, D. C., D. W. ALLING & S. M. WOLFF. 1972. Cyclic hematopoiesis: The mechanism of cyclic neutropenia in grey collie dogs. J. Clin. Invest. **51:** 2197–2204.
10. DALE, D. C., S. B. WARD, H. R. KIMBALL & S. M. WOLFF. 1972. Studies of neutrophil production and turnover in grey collie dogs with cyclic neutropenia. J. Clin. Invest. **51:** 2190–2196.
11. FELDMAN, J. L. & J. D. COWAN. 1975. Large-scale activity in neural nets. II. A model for the brainstem respiratory oscillator. Biol. Cybern. **17:** 39–51.
12. GATTI, R. A., W. W. ROBINSON, A. S. DEINARE, M. NESBIT, J. J. McCULLOUGH, M. BALLOW & R. A. GOOD. 1973. Cyclic leukocytosis in chronic myelogenous leukemia. Blood. **41:** 771–782.
13. GAVOSTO, F. 1974. Granulopoiesis and cell kinetics in chronic myeloid leukemia. Cell Tissue Kinet. **7:** 151–163.
14. GEMAN, S. & M. MILLER. 1976. Computer simulation of brainstem respiratory activity. J. Appl. Physiol. **41:** 931–938.
15. GRAY, W. M. & J. KIRK. 1971. Analysis by analogue and digital computers of the bone marrow stem cell and platelet control systems. In Computers for Analysis and Control in Medical and Biological Research. I.E.E. Publications, London: 120–124.
16. GREENSPAN, H. P. 1974. On the self-inhibited growth of cell cultures. Growth. **81:** 81–95.
17. GUERRY, D., D. C. DALE, M. OMINE, S. PERRY & S. M. WOLFF. 1973. Periodic hematopoiesis in human cyclic neutropenia. J. Clin. Invest. **52:** 3220–3230.
18. GUREL, O. 1972. Bifurcation models of mitosis. Physiol. Chem. Phys. **4:** 139–152.
19. GUREL, O. 1975. Limit cycles and bifurcations in biochemical dynamics. BioSystems. **7:** 83–91.
20. HALBERG, F., R. LAURO & F. CARADENTE. 1976. Autorhythmometry. Ric. Clin. Lab. **6:** 207–250.
21. HAYES, N. D. 1950. Roots of the transcendental equation associated with a certain difference-differential equation. J. London Math. Soc. **25:** 226–232.

22. HOFFMAN, J. H., D. GUERRY & D. C. DALE. 1974. Analysis of cyclic neutropenia using digital band-pass filtering techniques. J. Interdiscip. Cycle Res. **5:** 1–18.

23. KAPLAN, J. L. & J. A. YORKE. 1975. On the stability of a periodic solution of a differential delay equation. SIAM J. Math. Anal. **6:** 268–282.

24. KELLOGG, R. H. 1964. Central chemical regulation of respiration. *In* Handbook of Physiology. Section 3, Vol. 1. W. O. Fenn & H. Rahn, Eds. American Physiological Society, Washington, D.C. 507–534.

25. KING-SMITH, E. A. & A. MORLEY. 1970. Computer simulation of granulopoiesis: Normal and impaired granulopoiesis. Blood. **36:** 254–262.

26. KIRK, J., J. S. ORR & C. S. HOPE. 1968. A mathematical model of red blood cell and bone marrow stem cell control mechanisms. Br. J. Haematol. **15:** 35–46.

27. KIRK, J., J. S. ORR & J. FORREST. 1970. The role of chalone in the control of the bone marrow stem cell population. Math. Biosci. **6:** 129–143.

28. KIVILAAKSO, E. & T. RYTOMAA. 1971. Erythrocytic chalone, a tissue specific inhibitor of cell proliferation in the erythron. Cell Tissue Kinet. **4:** 1–9.

29. KRINSKII, V. 1968. Fibrillation in excitable media. Probl. Kyburn. **20:** 59.

30. LAIRD, A. K. 1964. Dynamics of tumor growth. Br. J. Cancer. **18:** 490–502.

31. LAJTHA, L. G., C. W. GILBERT & E. GUZMAN. 1971. Kinetics of haemopoietic colony growth. Br. J. Haematol. **20:** 343–354.

32. LAMBERTSEN, C. J. 1974. Abnormal types of respiration. *In* Handbook of Medical Physiology. 13th Edit. V. Mountcastle, Ed.: 1522–1537. The C. V. Mosby Co., St. Louis.

33. LANGE, R. L. & H. H. HECHT. 1962. The mechanism of Cheyne-Stokes respiration. J. Clin. Invest. **41:** 42–52.

34. LEWIS, M. L. 1974. Cyclic thrombocytopenia: A thrombopoietin deficiency? J. Clin. Pathol. **27:** 242–246.

35. LI, T. Y. & J. A. YORKE. 1975. Period three implies chaos. Am. Math. Mon. **82:** 985–992.

36. LINDSAY, A. E. & A. BUDKIN. 1975. The Cardiac Arrhythmias, 2d Edit., Yearb. Med. Publ., Chicago.

37. LONGOBARDO, G. S., N. S. CHERNIACK & A. P. FISHMAN. 1966. Cheyne-Stokes breathing produced by a model of the human respiratory system. J. Appl. Physiol. **21:** 1839–1846.

38. MACKEY, M. C. 1978. Dynamic haematological disorders of stem cell origin. *In* Cellular Mechanisms of Reproduction and Aging. J. Vassileva-Popova, Ed. Plenum Press, New York. In press.

39. MACKEY, M. C. 1978. A unified hypothesis for the origin of aplastic anemia and periodic haematopoiesis. Blood **51:** 941–956.

40. MACKEY, M. C. 1978. A comparison between two models for the control of erythrocyte production. In preparation.

41. MACKEY, M. C. & L. GLASS. 1977. Oscillation and chaos in physiological control systems. Science. **197:** 287–289.

42. MAY, R. M. 1974. Biological populations with nonoverlapping generations: Stable points, stable cycles, and chaos. Science. **186:** 645–647.

43. MAY, R. M. 1976. Simple mathematical models with very complicated dynamics. Nature. **261:** 459–467.

44. MAY, R. M., G. R. CONWAY, M. P. HASSEL & T. R. E. SOUTHWOOD. 1974. Time delays, density-dependence and single species oscillations. J. Anim. Ecol. **43:** 747–770.

45. MAY, R. M. & G. F. OSTER. 1976. Bifurcations and dynamic complexity in simple ecological models. Am. Nat. **110:** 573–599.

46. MCCULLOCH, E. A., T. W. MAK, G. B. PRICE & J. E. TILL. 1974. Organization and communication in populations of normal and leukemic hemopoietic cells. Biochem. Biophys. Acta. **355:** 260–299.

47. MILHORN, H. T., R. BENTON, R. ROSS & A. C. GUYTON. 1965. A mathematical model for the human respiratory control system. Biophys. J. **5:** 27–46.

48. MILHORN, H. T. & A. C. GUYTON. 1965. An analog computer analysis of Cheyne-Stokes breathing. J. Appl. Physiol. **20:** 328–333.

49. MOE, G. K. 1975. Evidence for re-entry as a mechanism of cardiac arrhythmias. Rev. Physiol. Biochem. Pharmacol. **72:** 56–82.
50. MORLEY, A. A. 1969. Blood cell cycles in polycythaemia vera. Aust. Ann. Med. **18:** 124–126.
51. MORLEY, A. A. 1970. Periodic diseases, physiological rhythms and feedback control. Aust. Ann. Med. **19:** 244–249.
52. ORR, J. S., J. KIRK, K. G. GRAY & J. R. ANDERSON. 1968. A study of the interdependence of red cell and bone marrow stem cell populations. Br. J. Haematol. **15:** 23–24.
53. PATT, H. M. & M. A. MALONEY. 1972. Relationships of bone marrow cellularity and proliferative activity: A local regulatory mechanism. Cell Tissue Kinet. **5:** 303–309.
54. VAN DER POL, B. & J. VAN DER MARK. 1928. The heart beat considered as a relaxation oscillation, and an electrical model of the heart. Philos. Mag. **6:** 763–775.
55. REIMANN, H. A. 1963. Periodic Diseases. F. A. Davis, Philadelphia.
56. REIMANN, H. A. 1975. Clinical insight on the nature of periodic diseases. Mod. Med. April. 40–48.
57. RICHTER, C. P. 1965. Biological Clocks in Medicine and Psychiatry. C. C. Thomas. Springfield, Ill.
58. RODRIGUEZ, A. R. & C. L. LUTCHER. 1976. Marked cyclic leukocytosis-leukopenia in chronic myelogenous leukemia. Am. J. Med. **60:** 1041–1047.
59. ROTHSTEIN, G., E. H. HUGL, P. A. CHERVENICK, J. W. ATHENS & J. MACFARLANE. 1973. Humoral stimulators of granulocyte production. Blood. **41:** 73–78.
60. RUBINOW, S. I. & J. L. LEBOWITZ, 1975. A mathematical model of neutrophil production and control in normal man. J. Math. Biol. **1:** 187–225.
61. RUSHMER, R. F. 1961. Cardiovascular Dynamics, 2d Edit., W. B. Saunders, Philadelphia.
62. RYTÖMAA, T. & K. KIVINIEMI. 1967. Regulation system of blood cell production. *In* Control of Cellular Growth in Adult Organisms. H. Tier & T. Rytömaa, Eds. Academic Press, London. 106–139.
63. SCHMIDT-NIELSEN, K. 1972. How Animals Work. Cambridge University Press.
64. SHYMKO, R. M. & L. GLASS 1976. Cellular and geometric control of tissue growth and mitotic instability. J. Theor. Biol. **63:** 355–374.
65. SPECHT, H. & G. FRUHMAN. 1972. Incidence of periodic breathing in 2000 subjects without pulmonary or neurological disease. Bull Physiol. Pathol. Respir. **8:** 1075–1082.
66. STEINSCHNEIDER, A. 1974. The concept of sleep apnea as related to SIDS. *In* SIDS 1974, Proceedings of the Francis E. Camps International Symposium on Sudden and Unexpected Deaths in Infancy. Canadian Foundation for the Study of Infant Deaths, Toronto. 177–190.
67. STOHLMAN, F. 1971. Control mechanisms in erythropoiesis. *In* Regulation of Erythropoiesis. A. S. Gordon, M. Condorelli & C. Peschle, Eds. Il Ponte, Milano. 71–88.
68. VODOPICK, H., E. M. RUPP, C. L. EDWARDS, F. A. GOSWITZ & J. J. BEAUCHAMP. 1972. Spontaneous cyclic leukocytosis and thrombocytosis in chronic granulocytic leukemia. N. Engl. J. Med. **286:** 284–290.
69. VOS, O. 1972. Multiplication of haemopoietic colony forming units (CFU) in mice after x-irradiation and bone marrow transplantation. Cell Tissue Kinet. **5:** 341–350.
70. WAZEWSKA-CZYZEWKKA, M. & A. LASOTA. 1976. Matematyczme problemy dynamiki ukladu krwinek czerwonych. Roczniki Polskiego Towarzystwa Matematycznego, Seria III. Matematyka Stosowana **VI:** 23–40.
71. WHELDON, T. E. 1975. Mathematical models of oscillatory blood cell production. Math. Biosci. **24:** 289–305.
72. WIENER, N. & A. ROSENBLUETH. 1946. A mathematical formulation of the problem of conduction of impulses in a network of excitable elements, specifically in cardiac muscle. Arch. Inst. Cardiol. Mexico. **16:** 205–265.
73. WINTROBE, M. M. 1976. Clinical Hematology. 7th Edit., Lea & Febiger, Philadelphia.
74. WYMAN, R. J. 1977. Neural generation of the breathing rhythm. Annu. Rev. Physiol. **39:** 417–448.

SPATIAL, CHIRAL, AND TEMPORAL SELF-ORGANIZATION THROUGH BIFURCATION IN "BIOIDS," OPEN SYSTEMS CAPABLE OF A GENERALIZED DARWINIAN EVOLUTION

Peter Decker

Chemical Institute
Veterinary School
D 3000 Hannover
Federal Republic of Germany

INTRODUCTION

In this paper I report experimental work on the kinetics of the autocatalytic formation of sugars from formaldehyde (formol reaction). This reaction has been studied as an example of self-organizing chemical open systems capable of a generalized Darwinian evolution ("*Bioids*"),[9,13,15] i.e., of cumulative acquisition and conservation of information[8,9,14] through stepwise bifurcations.

I found that the feedback structure of this reaction is formally equivalent to an *autocatalytic, second-order in products* (ASOP) reaction[12,21]:

$$A + 2B = 3B. \tag{1}$$

Cooperative feedback (autocatalysis more than first order in products),

$$A + (1 + \gamma)B = (2 + \gamma)B, \tag{2}$$

can induce nontrivial behavior in open chemical systems: bifurcations or symmetry-breaking instabilities[18] involving periodic and nonperiodic fluctuations in time (oscillations and chaos[29]) or in space[27] (chemical waves and turbulence), and the spontaneous production of molecular asymmetry through kinetic bistability in open systems.[10,11]

It appears significant that the formol reaction because of its ASOP kinetics, at least in principle, also might be capable of such features. Indeed, under prebiological conditions formaldehyde could arise by photo-oxidation from a methane atmosphere. The switching-on of the autocatalytic reaction in this open system has been interpreted as the acquisition and conservation of the first bit of evolutionary information by a formol bioid—the first ancestor of more complex systems including terrestrial life.[8,9,13-15] Therefore, in contrast to purely mathematical models and model reactions like the Zhabotinsky reaction[41] or the Bray-Liebafsky reaction,[36] here we have for the first time a chemical system that may be directly related to the origins of life and lifelike systems.

BIFURCATION AND THE DARWINIAN PRINCIPLE

Bifurcation means a qualitative change in the topological structure of a dynamical system as a consequence of a shift in some critical parameter.[24] Since we can regard changing parameters as new variables, we may consider bifurcations

0077-8923/78/0316-0236 $1.75/1 © 1979, NYAS

alternatively as a creation of new features in the phase portrait of a system through the introduction of a new variable.

The term "bioid" has been defined as an open system that can exist in several steady states.[6,15] It has been proposed as the most general concept of systems capable of a stepwise generalized Darwinian evolution[9,13] by transitions (mutations) through thermodynamical instabilities from one steady state into another.[13,15] An autocatalytic reaction in a flow reactor is the simplest model of a system capable of such evolution[5]: it can mutate between two different steady states, the reacting state and the "extinct" state, depending on the balance between overflow rate and the growth rate of the autocatalyst. Similarly, two autocatalysts competing for a common substrate represent a model for selection of the fittest.[15,23]

For the start of an autocatalytic reaction, however, at least one molecule of the autocatalyst must be present. Therefore, in chemical bifurcations we have an additional aspect: The autocatalytic species represents *molecular information* which is amplified, i.e., made redundant, and indefinitely conserved by the system. It represents a self-organizing material feedback channel.[9,13 15] The addition of new feedback structures entailing bifurcation is equivalent to an increase of information dynamically stored in the system.

In terrestrial life information is stored in a standard format (code) in a linear polynucleotide matrix that is capable of self-replication and can perform an instructed synthesis of standardized catalysts—the protein enzymes.[16] The latter govern a complex metabolic network that provides energy and exactly the necessary monomers. I believe that this rather involved system arose from simpler bioids, which stored information "informally," i.e., in the structure of a self-stabilizing network of chemical feedback reactions. It is the aim of our Hannover program[9,13,15] to elucidate which reactions are capable of acquiring and accumulating the first bits of evolutionary information.

NONLINEAR FEEDBACK: A SOURCE OF NONTRIVIAL PHENOMENA

Autocatalysis more than first order in products (2) in open chemical systems is a potent source of nontrivial phenomena other than simple steady states. Now in the formol reaction we found kinetics equivalent to autocatalysis second order in products (ASOP) (1). Therefore we are interested in what behavior an ASOP can induce if combined in an open system with conventional reactions, simple first-order parallel reactions between substrate and product, and inflow and overflow involving external sources and sinks:

$$\xrightarrow{a_1} A \qquad A \xrightarrow{d} B \qquad\qquad\qquad A \xrightarrow{a_2}$$
$$\text{shunt}$$

$$A + 2B \xrightarrow{c} 3B$$

$$\xrightarrow{b_1} B \quad \Big|\quad B \xrightarrow{p} A \qquad\qquad B \xrightarrow{b_2}$$
$$\text{inflow} \qquad \text{pumping} \qquad\qquad \text{ASOP} \qquad\qquad \text{overflow.}$$

FIGURE 1 shows the scheme of such a general ASOP system and some subsets thereof. If we assume a concentration of 1.0 in the pool feeding the inflow of A and B, and in the auxiliary compound C in Figure 1b, then we can describe the general system by two simultaneous differential equations (where A and B indicate the concentrations of A and B):

$$\frac{dA}{dt} = a_1 - Aa_2 - Ad + Bp - AB^2c \tag{3}$$

$$\frac{dB}{dt} = b_1 - Bb_2 + Ad - Bp + AB^2c \tag{4}$$

The nullclines, Equations 3 and 4, obtained from zero motion in (3) and (4) show that maximally three steady states can exist in the positive quadrant, where all

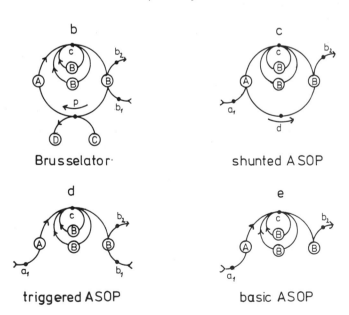

FIGURE 1. Different types of open systems involving autocatalysis second order in products (ASOP). Small dots (·) represent reactions; A, B, C, D are reactants; a_i, b_i, c, d, p are reaction constants.

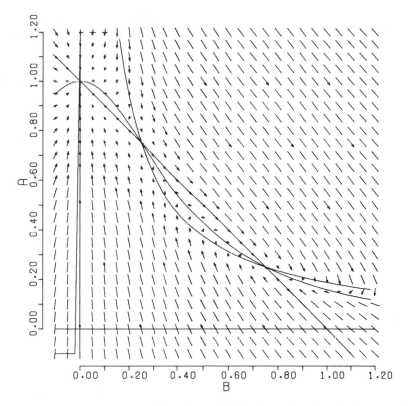

FIGURE 2. Computer drawing of isoclines and directions of trajectories in a complete ASOP system (FIGURE 1a) with three steady states.

three intersections are on a linear isocline obtained for $d(A + B)/dt = 0$ (FIGURE 2).

$$A = \frac{a_1 + Bp}{B^2c + a_2 + d}, \tag{5}$$

$$A = \frac{-b_1 + Bb_2 + Bp}{B^2c + d}. \tag{6}$$

The nature of the steady states follows from the sign of the roots of the characteristic equation of the coefficient matrix of the linearized system after shifting the steady state to the origin.[2] Especially, a steady state is unstable if one of the roots has a positive real part. This is the case if one of the two following relations is true:

$$a_2 + b_2 + d + p + cB_s^2 - 2cA_sB_s < 0, \tag{7}$$

$$a_2b_2 + a_2p + b_2d + b_2cB_s^2 - 2a_2cA_sB_s < 0, \tag{8}$$

where A_s and B_s are the coordinates of the steady state. The general expressions for A_s and B_s can be obtained as solutions of cubic equations; however, in the special cases with no overflow of substrate we obtain second-order equations affording uncomplicated relations among parameters, which represent threshold values for instability as given in TABLE 1. In this case there is only one steady state at finite values of A and B. Because of the overflow of the product B, the system is bounded in the open interval $B > 0$ at finite values of A. Instability of the steady state means the existence of at least one limit cycle, since trajectories originating from infinity and from the instable steady state must have a sink. Depending on the parameters p, d and b_1 there are three kinds of oscillatory systems:

With $p > 0$ and $d = b_1 = 0$ we have Prigogines[18] "Brusselator," FIGURE 1b, which we may describe as a "pumped ASOP" with inflow and overflow of B, and with an auxiliary pumping reaction

$$B + C = A + D. \tag{9}$$

By this reaction, as in laser action, B is pumped to an energetic state A. At a critical value, A periodically discharges through the cooperative ASOP reaction back to B.

With $p = 0$ and positive d or b_1 (FIGURES 1c & d) we have two systems[39] that may reflect a chemically more natural situation: a steady supply of a substrate A, which affords the product B through an ASOP. At low concentrations of autocatalyst B the substrate A tends to accumulate. A sudden periodical discharge (FIGURE 3) occurs at a critical value, initiated by B from the first-order shunt d or from the inflow b_1. Either d or b_1 secure a finite concentration of B for a restart of the discharge. Since the limit cycles in systems FIGURE 1b, c, and d are asymptotically stable, I anticipate they may not disappear if we introduce a small

TABLE 1

STEADY STATES AND THRESHOLD VALUE FOR INSTABILITY ($R < 0$)
NECESSARY FOR OSCILLATIONS IN THE SYSTEMS IN FIGURE 1

System	Steady States		Threshold Value
	A_s	B_s	R
(b) Brusselator	$\dfrac{p}{cB_s}$	$\dfrac{b_1}{b_2}$	$b_2 + c\dfrac{b_1^2}{b_2} - p$
(c) Shunted ASOP	$\dfrac{a_1}{B_s^2 + d}$	$\dfrac{a_1}{b_2}$	$\dfrac{\left(c\dfrac{a_1^2}{b_2^2} + d\right)^2}{c\dfrac{a_1^2}{b_2^2} - d}$
(d) Triggered ASOP	$\dfrac{a_1}{cB_s^2}$	$\dfrac{a_1 + b_1}{b_2}$	$c\dfrac{(a_1 + b_1)^3}{a_1 - b_1} - b_2^3$
(e) Basic ASOP	$\dfrac{a_1}{cB_s^2}$	$\dfrac{a_1}{b_2}$	$ca_1^2 - b_2^3$

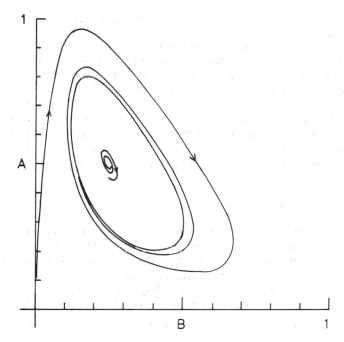

FIGURE 3. Oscillations in a shunted ASOP (FIGURE 1c); $a_1 = 0.25$; $b_2 = 1.0$; $c = 6.5$; $d = 0.0937$.

overflow a; therefore, the general system FIGURE 1a should also allow oscillatory solutions.

If $p = d = b_1 = 0$, we have a limiting case ($\alpha = 1$)[35] of Selkov's oscillator.[35] However, here the system is bounded only in a closed set in the 1st quadrant, and A can escape into infinity if $B \to 0$.

SPATIAL STRUCTURES

Such systems also can exhibit spatial structures,[18,39] and the Zhabotinsky reaction[41] has shown that such structures also can occur in chemical systems. I must, however, remark that the kinds of structures studied so far fall short with regard to the main problem of prebiological self-organization: the formation of *solitons*,[22] i.e., of *spatial structures that are independent of the boundaries* of the system. Such solitons should grow to a limiting size (equifinality) beyond which they become unstable, resulting in a formal equivalent to cell division or spore formation.[9] I suggest that mathematical models of such systems should be sought after. We can hope that such models may turn out to be not too complex, since we have seen how many unexpected qualitative features can appear through the addition of new variables. In 1971 I proposed that such models should consider differences in diffusibility, and viscous or insoluble intermediates and/or end products.[14]

THE ACQUISITION OF ENVIRONMENTAL INFORMATION

A bifurcation in a chemical system produced by new molecular information, i.e., a new feedback, becomes selected according to environmental conditions. As an example, consider the *three possible states* in a *"competitive system"* of two autocatalytic reactions I and II competing for one substrate in an open system: I reacting, II reacting, and both extinct. Accordingly, there are *three classes of possible environmental conditions* selecting and supporting the respective state. Such a selection process has been interpreted as an acquisition of environmental information, considered as a *mapping* of a class of possible environments onto the respective state which is selected by environments belonging to that class.[14]

STOCHASTIC INFORMATION: THE ORIGIN OF MOLECULAR ASYMMETRY

Instead of environmental information, the acquired information may also represent a purely random choice between exactly equivalent cases.[14] This is the case in a special class of bioid models: bistable systems that can produce chiral compounds.[10] Such chemical flip-flops bifurcate into one of two equivalent asymmetrical states. As an example let us consider a *"hypercompetitive system"* (FIGURE 4), an open system where two ASOP reactions compete for one substrate[11]:

$$\rightarrow A; \quad A + 2B_1 \rightarrow 3B_1; \quad A + 2B_2 \rightarrow 3B_2; \quad B_1 \rightarrow; \quad B_2 \rightarrow; \quad A \rightarrow. \quad (10)$$

inflow   ASOP reactions   overflow

Here the symmetrical steady state is unstable even if the competing reaction constants are exactly equal, as would be the case if the competing autocatalysts were chiral antipodes produced from a common symmetrical substrate. In such a system, using a natural racemic mixture of autocatalysts, we move exactly along a separatrix toward an unstable saddle point. Since here the symmetrical state is unstable, local microscopic fluctuations in the respective concentrations of the antipodes become amplified into macroscopic dimensions and eventually one of the antipodes dominates its competitor. Therefore the system must spontaneously switch into one of two equivalent asymmetrical states, where it autonomously affords an optically active product, i.e., a stoichiometric surplus of one antipode, without induction or interaction by an external asymmetrical agent like an enzyme, or circularly polarized light, etc. It would work even when the reactions are not absolutely stereoselective. Several similar systems have been devised in the last years,[6,7,10,34,40] but so far we have no working chemical model. Such a model would resolve Pasteur's 140-year-old problem on how biological molecular asymmetry arose. Indeed, such a solution appears more germane than previous theories based on the action of optically active quartz, circularly polarized light,[20] or a cosmic disparity of right and left.[37] Since we found ASOP-like kinetics in the formol bioid where the resulting sugars are indeed antipodes, this appears to be the first reaction which, at least in principle, might be capable of such a spontaneous creation of asymmetry by bifurcation. For closed reacting systems the principle appears to have been understood as early as 1932 by W. H. Mills;[26] F. C. Frank[17] stated it clearly in 1953 but there was almost no resonance among stereochemists.[10]

Many models of chemical automata have been considered by Rössler.[32] Systems exhibiting almost periodic and nonperiodic motions, including catastrophic[38] and chaotic behavior[25] have recently evoked much interest. Rössler[9] proposed a chemically realistic model that should produce true chaos, i.e., nonperiodic oscillations, and indeed, shortly later Olsen and Degn[29] reported an actual enzymatic reaction producing chaos. Such systems are chemical versions of the classical 3-body-problem, the escape phenomenon, and secular motions in celestial mechanics.[4]

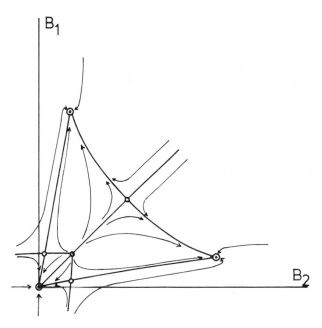

FIGURE 4. Phase portrait of a hypercompetitive system (Equation 10) projected onto the B_1,B_2-plane.[11] The origin ("extinct state") is stable. Introduction of a racemic mixture of $B_1 + B_2$ moving the system along the diagonal beyond the instable source induces further evolution along the separatrix toward the instable saddle point. From there, by bifurcation, the system switches into one of two equivalent steady states with preferential production of B_1 or B_2 respectively.

COOPERATIVE KINETICS IN THE FORMOL REACTION

We were interested in the formol reaction, since among all compounds that can arise in a prebiological scenario from a primeval methane atmosphere, formaldehyde is the sole compound that is known to react autocatalytically and has been shown in the reacting state to be capable of subsisting as a "bioid" even on a minute supply of substrate.[8,9,13-15] Continuous monitoring of the consumption of formaldehyde in a batch reaction, inoculated with a minute amount of glycolic aldehyde as autocatalyst, produces a decay curve with a very characteristic shape;

a slow inclination is followed by a nearly linear slope almost down to the abscissa [21] (FIGURE 7). This is significantly different from the symmetrical S-shaped decay of the substrate in a simple autocatalytic reaction of the type A + B = 2B. Earlier authors considered mechanisms equivalent to such simple auto-catalysis[31,1] involving combinations of aldol condensations, Lobry de Bruyn rearrangements and retro aldol splitting (FIGURE 5). The decay curve was analyzed by means of the curve-fitting program "Checkmat."[3] We first assumed parallel reactions, a simple first-order shunt, and autocatalysis first, second, and third order in products. As a best fit to our experimental data we obtained the following reaction constants:

$$A = B \quad : \quad 0.0021; \quad A + B = 2B \quad : \quad 0.045 \; ;$$

$$A + 2B = 3B \quad : \quad 1.09 \quad ; \quad A + 3B = 4B \quad . \quad 0.0026;$$

We see that the program allocated by far the largest constant, 1.09, to the ASOP reaction. This shows that the dominant mechanism in the formol system is equivalent to an ASOP reaction. No significant conclusions can be drawn from the other constants, which only improve the fit; in fact, the real mechanism is much more complex. Later we introduced more realistic kinetics, involving for-maldehyde (F), glycol aldehyde (A2), glyceric aldehyde (A3), and calcium ions as the co-catalyst (FIGURE 6), and obtained an equally good fit with our data (FIGURE 7). Here the ternary ASOP reaction is mediated by a calcium complex with two molecules of A3; F reacts with one of those ligands affording two mole-cules of A2; A2 is recycled into A3 by reaction with F. This scheme also reflects

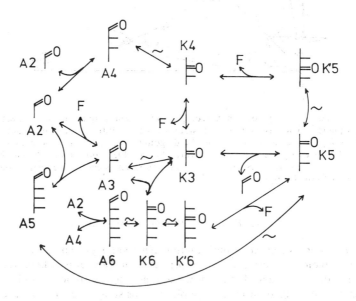

FIGURE 5. Formation of sugars from formaldehyde through conventional autocatalytic cycles involving aldol reactions: F, formaldehyde; A2, glycol aldehyde; K3, A3, Ketotriose, Aldotriose, etc.

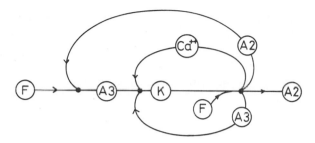

FIGURE 6. Schematical ASOP kinetics in the formol reaction involving a calcium complex (K) with two triose ligands (A3). A reaction of the complex with formaldehyde (F) affords two molecules of glycolic aldehyde (A2) from one of the ligands. A2 is recycled by reaction with a second molecule F. Small dots represent reactions.

only the kinetic aspect of a reaction mechanism, which certainly is much more complex and so far remains obscure. However, this demonstrates the role of metal complexes as mediators in quasi-third-order kinetics and as potential precursors of stereoselective enzyme catalysis, as I first proposed in 1975.[11] Indeed, a reaction sequence involving a complexing metal ion Me as "enzyme" in a Michaelis-Menten-like system

$$Me + A + 2B \rightleftharpoons MeAB_2 \rightarrow MeB_3 \rightleftharpoons Me + 3B \qquad (11)$$

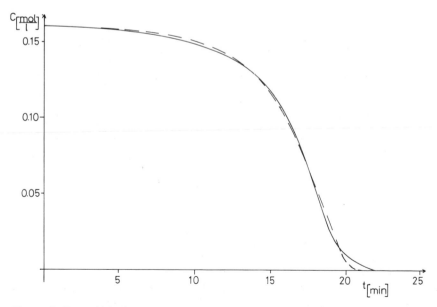

FIGURE 7. Formaldehyde consumption in the course of the formol reaction: (- - - -) experimental; (—) computed using kinetics of FIGURE 6 and constants computed by the program "Checkmat."[3,21]

approaches third-order kinetics in the case of weak complexes and nonlimiting Me concentration.[21]

A Phenomenon of Hysteresis

Additional evidence of more than first-order feedback came from the observation of hysteresis[12]: Let us consider an ASOP in a flow reactor with inflow and overflow of product and substrate (where the inflow of the product, i.e., autocatalyst, secures that the reactor never becomes entirely extinct). Then the steady concentration of substrate depends on reaction conditions in such a manner (Figure 8) that there is a region where the system exhibits bistability, i.e., there are two branches corresponding to *two alternative stable states,* "reacting" and (nearly) "extinct," connected by an instable branch with negative slope. If we shift the parameter up and down, we should obtain a hysteresis loop. (This would be not the case with a simple autocatalysis of type A + B = 2B). We implemented this experiment with the formol reaction: Having monitored the concentration of unreacted formaldehyde in the overflow, we steadily lowered the concentration of the co-reactant, OH^- or Ca^{2+} (by applying a concentration gradient to the reactor inflow) and then increased it again. Doing so we indeed observed a characteristic hysteresis loop (Figure 9).

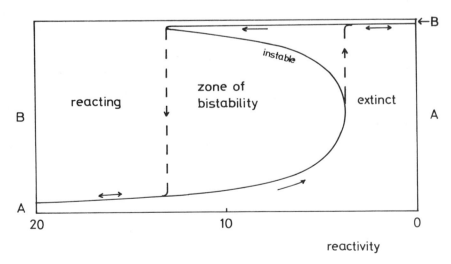

FIGURE 8. Multiple steady states in a triggered ASOP as a function of reactivity c:

$$0 = dA/dt = a_1 - a_2A - cAB^2; \quad 0 = dB/dt = b_1 - b_2B + cAB^2$$

where $a_2 = b_2 = 1$ (common overflow) and $b_1 = 0.02\ a_1$. A describes a hysteresis loop when c is shifted up and down.

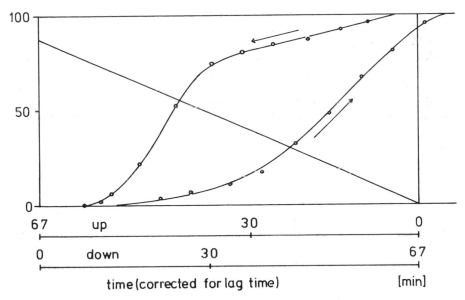

FIGURE 9. Observed hysteresis loop in the formol reaction in a flow reactor. The reactivity was shifted up and down using a NaOH-concentration gradient in the reactor inflow.[12] 157 mmol formaldehyde, 1.6 mmol glycol aldehyde, 72 meq Ca-acetate, 50°C, reactor half-time 10 min.

THE PROBLEM OF THE NEXT BIT

We may interpret the double feedback in the reaction A + 2B = 3B as an acquisition of *two* bits of information allowing a more complex behavior. In the hysteresis experiment the first bit—"reacting" or "not reacting"—is externally supplied by triggering with product; the second bit provides a memory of whether the gradient is moving up or down.

It appears not unreasonable to suspect that in prebiological evolution processes the formol bioid may play a role even beyond the trivial switching on of the first bit. It is the first known reaction appearing realizable in a prebiological scenario, which, at least in principle, could produce nontrivial phenomena representing important evolutionary acquisitions: oscillations, spatial structure and production, and maintenance of molecular asymmetry.

The next bit should involve additional feedback loops among the many possible reactions between the resulting sugars, ammonia, and other ingredients of the "primeval soup," known in nutritional chemistry as "Maillard reaction" or "non-enzymatic browning." Colored compounds as photosensitizers could couple the system to solar energy; macromolecular condensation products might mediate individuation into solitons, allowing parallel evolution along different pathways.[8,9,13–15] Products of the Maillard reaction, melanoidins, can form complexes with metal ions which possibly were the "prenucleoprotic" precursors of metalloenzymes.[28]

The elucidation of self-organizing feedback structures in such a reaction network poses many analytical problems. The first steps of bioid evolution, however, should involve fast reactions capable to compete with dilution and parasitic reactions. Therefore we can hope that those first steps are not "rare events" and might well be observable under appropriate conditions.

In principle, every new feedback is represented by an additional nonlinear equation and may double the maximal number of possible steady states; this number depends on the nonlinearity of the system (FIGURE 10). However, in multicomponent systems true steady states and stable oscillations are rare[30] and motions of higher order[19] should prevail[9]: slow drifts superposed by regular or

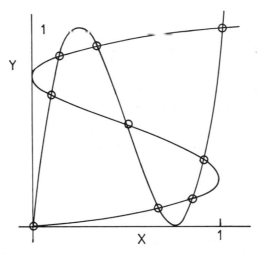

FIGURE 10. Isoclines and multiplicity of steady states in nonlinear systems.[9] Two third-order differential equations give maximally 9 steady states; e.g., $dx/dt = 16y^3 - 24y^2 + 9y - x$; and $dy/dt = 16x^3 - 24x^2 + 9x - y$.

chaotic oscillations and casual "catastrophic" instabilities—a behavior similar to biological objects.

In conclusion, most of the work done on origins of life in the last decades has been centered on abiogenic syntheses. The problem of how all this information could become conserved and accumulated has been simply reduced to a hypothetical spontaneous creation of a working nucleoprotic system including its metabolic network. Our results support the rather heretic opinion that the historical course of evolution—accumulation and conservation over time of information—began with simpler bioids, and involved consecutive bifurcations in open chemical systems as an essential mechanism. The evolution of the nucleoprotic system may represent a standardization process at a rather advanced stage, in which the exclusive use of proteins as catalysts was introduced. Admittedly, in the terrestrial system this was an essential step on the way to the tremendous capacity of information storage in present life.

REFERENCES

1. BRESLOW, R. 1959. On the mechanism of the formose reaction. Tetrahedron Lett. Nr. **21**: 22.
2. CESARI, L. 1971. Asymptotic Behavior and Stability Problems in Ordinary Differential Equations. Springer. New York.
3. CURTIS, A. R. & E. M. CHANCE. 1972. Numerical methods for simulation and optimization. *In* Analysis and Simulations of Biochemical Systems. H. C. Hemker & B. Hess, Eds. FEBS. North-Holland Publishing Co. Amsterdam.
4. DAVIS, H. T. 1962. Introduction to Nonlinear Differential and Integral Equations. Dover Publishing, Inc. New York.
5. DENBIGH, K. G., M. HICKS & F. M. PAGE. 1947. The kinetics of open reaction systems. Trans. Faraday Soc. **44**: 479.
6. DECKER, P. 1973. Evolution in open systems: Bistability and the origin of molecular asymmetry. Nature New Biol. **241**: 72.
7. DECKER, P. 1973. Possible resolution of racemic mixtures by bistability in "Bioids," open systems which can exist in several steady states. J. Mol. Evol. **2**: 137.
8. DECKER, P. 1973. Begann Evolution mit Formaldehyd? Umschau **73**: 733.
9. DECKER, P. 1974. Evolution in offenen Systemen: prize essay submitted to the Bavarian Academy of Sciences; reprints (170 pp.) available from Documentation Center on the Origin of Life, Universitätsbibliothek, Ulm, FRG.; cf. DECKER, P. 1975. Nachr. Chem. Techn. **23**: 167.
10. DECKER, P. 1974. The origin of molecular asymmetry through the amplification of "stochastic information" (noise) in bioids, open systems which can exist in several steady states. J. Mol. Evol. **4**: 49.
11. DECKER, P. 1975. Evolution in bioids: Hypercompetitivity as a source of bistability and a possible role of metal complexes as prenucleoprotic mediators of molecular asymmetry. Origins of Life **6**: 211.
12. DECKER, P. 1976. The formaldehyde bioid—an autocatalysis second order in products (ASOP), producing hysteresis. 10th Intern. Congr. Biochemistry, Hamburg 1976, Abstr. No. 07-5-106.
13. DECKER, P. & W. HEIDMANN. 1977. Evolution in open systems: Acquisition and conservation of information in bioids. *In* Origin of Life. Proc. 2nd ISSOL & 5th ICOL Meeting, Kyoto, 1977. H. Noda, Ed.: 617. Center Acad. Publ. Japan, Tokyo, 1978 (JSSP No. 01312-1104).
14. DECKER, P., A. SPEIDEL & W. NICOLAI. 1971. On the origin of information in biological systems and in bioids. *In* Proc. 1st European Biophysics Congress, Baden/Vienna. E. Broda *et al.*, Eds. Vol. 4: 515–521. Verl. Wiener Med. Akad. Vienna.
15. DECKER, P. & A. SPEIDEL. 1972. Open systems which can mutate between several steady states ("Bioids") and a possible role of the autocatalytic condensation of formaldehyde. Z. Naturforsch. **27b**: 257.
16. EIGEN, M. 1971. Molecular self-organization and the early stages of evolution. Quart. Rev. Biophysics **4**: 149.
17. FRANK, F. C. 1953. On spontaneous asymmetric synthesis. Biochim. Biophys. Acta **11**: 459.
18. GLANSDORFF, P. & I. PRIGOGINE. 1971. Thermodynamics of Structure, Stability and Fluctuations. Wiley-Interscience. New York.
19. GÜREL, O. 1973. A classification of the singularities of (X, f). Mathematical Systems Theory **7**: 154.
20. HARADA, K. 1970. Origin and development of optical activity of organic compounds on the primeval earth. Naturwiss. **57**: 114.
21. HEIDMANN, W., P. DECKER & R. POHLMANN. 1977. Cooperative kinetics in the formaldehyde bioid. *In* Origin of Life.: 625. See Ref. 13.
22. LIBCHABER, A. & G. TOULOUSE. 1976. Le retour des solitons. La Recherche **7**: 1027.
23. LOTKA, A. J. 1956. Elements of Mathematical Biology. Dover Publ. Inc. New York.
24. MARSDEN, J. E. & M. MCCRACKEN. 1976. The Hopf Bifurcation and its Applications. Springer. New York.

25. MAY, R. M. 1976. Simple mathematical models with very complicated dynamics. Nature **261**: 459.
26. MILLS, W. H. 1932. Some aspects of stereochemistry. Chemistry and Industry **1932**: 750.
27. NICOLIS, G. & J. F. G. AUCHMUTY. 1974. Dissipative structures, catastrophes, and pattern formation: A bifurcation analysis. Proc. Nat. Acad. Sci. U.S.A. **71**: 2748.
28. NISSENBAUM, A., D. H. KENYON & J. ORO. 1975. On the possible role of organic melanoidin polymers as matrices for prebiotic activity. J. Mol. Evol. **6**: 253.
29. OLSEN, L. F. & H. DEGN. 1977. Chaos in an enzyme reaction. Nature **267**: 177.
30. PEIXOTO, M. M. 1959. On structural stability. Ann. Math. **69**: 199.
31. PFEIL, E. & G. SCHROTH. 1952. Kinetik und Reaktionsmechanismus der Formaldehyd-Kondensation. Chem. Berichte **85**: 293.
32. ROESSLER, O. E. 1972. Grundschaltungen von flüssigen Automaten und Relaxations-systemen. Z. Naturforsch. **216**: 455.
33. ROESSLER, O. E. 1976. Chaotic behaviour in simple reaction systems. Z. Naturforsch. **31a**: 259.
34. SEELIG, F. F. 1970. Systems-theoretic model for the spontaneous formation of optical antipodes in strongly asymmetric yield. J. Theor. Biol. **31**: 355.
35. SELKOV, E. E. 1967. Self-oscillations in glycolysis 1. A simple kinetic model. European J. Biochem. **4**: 79.
36. SHARMA, K. R. & R. M. NOYES. 1976. Oscillations in chemical systems. 13. A detailed molecular mechanism for the Bray-Liebafsky reaction of iodate and hydrogen peroxide. J. Amer. Chem. Soc. **98**: 4345.
37. THIEMANN, W. 1975. Life and chirality beyond the earth. Origins of Life **6**: 475.
38. THOM, R. 1972. Stabilité structurelle et morphogenèse. Essay d'une theorie générale des modéles. W. A. Benjamin Inc. Reading, Mass.
39. TYSON, J. J. 1973. Properties of two-component bimolecular and trimolecular chemical reaction systems. J. Chem. Phys. **59**: 4164.
40. VITAGLIANO, V. & A. VITAGLIANO. 1976. A simple kinetic model of prebiotic asymmetric synthesis. Gazz. Chimica Ital. **106**: 509.
41. ZAIKIN, A. N. & A. M. ZHABOTINSKY. 1970. Concentration wave propagation in two-dimensional liquid-phase self-oscillating system. Nature **225**: 535.

EFFECT OF FLUCTUATIONS ON
BIFURCATION PHENOMENA

G. Nicolis and J. W. Turner

Faculté des Sciences de l'Université Libre de Bruxelles
Campus Plaine
1050 Brussels, Belgium

INTRODUCTION

Bifurcation theoretic problems involving applications in physics, chemistry, biology deal typically with equations of evolution with a limited number of macroscopic observables. Such equations (*deterministic* rate laws) are usually taken for granted on the basis of vague references to the laws of large numbers in probability theory, according to which statistical averages provide a satisfactory description of the time course of macrovariables. Therefore, everything happens as though, in any small volume element ΔV, the rate of change of a macrovariable were the result of a collective effect determined by the values of all the variables, which, in turn, sense their effect. We call this notion the *mean-field* view of bifurcation phenomena. This lumping together of all but the macroscopic degrees of freedom ignores *fluctuations* from average behavior. Our purpose in this paper is to assess the influence of these fluctuations in a variety of problems involving the onset of cooperative behavior associated with bifurcation.

Consider a typical bifurcation phenomenon shown in FIGURE 1, (see e.g., the paper by M. Herschkowitz-Kaufman and T. Erneux in this volume). At $\lambda = \lambda_c$, a stable state, branch (*a*) loses its stability and is succeeded for $\lambda > \lambda_c$ by two simultaneously stable branches (*b*) and (*c*). At the critical point $\lambda = \lambda_c$, the system has to choose between coalescing solutions. But nothing in the equations of evolution justifies preference for any particular choice. It is not unreasonable to expect that fluctuations will play an important role in this critical choice.

Suppose next that we have a range of $\lambda > \lambda_c$ values in which two bifurcating states (*b*) and (*c*) coexist in macroscopic amounts. A particular system will exist for the most part in the neighborhood of either (*b*) or (*c*). Yet the statistical average will give a value on a branch (*a''*), which is close to the unstable branch (*a'*). This value is completely irrelevant to the observed behavior. Again, one would be justified in expecting that fluctuations around this average would play a prominent role.

We shall illustrate the progressive breakdown of the deterministic description associated with bifurcation and the concomitant existence of simultaneously stable states. We shall present some results obtained from a master equation for analyzing model systems in which spatial coordinates can be ignored. We shall deal, successively, with the exact form of a steady-state solution in the large volume limit, the coexistence region of simultaneously stable states, and the transitions between these states. Finally we shall present some general comments on the nature of the solutions of master equations around bifurcation points leading to limit

251

0077/8923/78/0316–0251 $1.75/1 © 1979, NYAS

cycles and to spatial patterns, and on the comparison between bifurcation phenomena and equilibrium phase transitions.

STOCHASTIC ANALYSIS OF SIMPLEST BIFURCATION—THE SCHLÖGL MODEL

The simplest bifurcation predicted by equations of evolution of the reaction-diffusion type is in systems involving one internal variable and a cubic rate law. This phenomenon, which is also related to the well-known cusp catastrophe,[14] is best illustrated by the following chemical model due to Schlögl[12]:

$$A + 2X \underset{k_2}{\overset{k_1}{\rightleftharpoons}} 3X$$

$$X \underset{k_4}{\overset{k_3}{\rightleftharpoons}} B \tag{2.1}$$

where X denotes a variable intermediate, the concentration of A, and B is supposed to be controled externally. Let n_X be an extensive variable denoting the number of particles of X. The macroscopic rate equation associated with (2.1) is,

$$\frac{dn_X}{dt} = -k_2 n_X^3 + k_1 A n_X^2 - k_3 n_X + k_4 B \tag{2.2}$$

We introduce the following scaled variables and parameters:

$$n_X = a(1 + x)$$

$$\frac{k_1 A}{k_2} = 3a$$

$$\frac{k_3}{k_2} = (3 + \delta)a^2 \tag{2.3}$$

$$\frac{k_4 B}{k_2} = (1 + \delta')a^3$$

Equation 2.2 takes the form:

$$\frac{dx}{dt} = -x^3 - \delta x + \delta' - \delta \tag{2.4}$$

Equation 2.4 admits a steady-state solution, which, when written as,

$$\delta' = x^3 + \delta(x + 1) \tag{2.5}$$

presents some striking analogies with the van der Waals equation in the neighborhood of liquid-vapor transition, provided δ', δ, and x are taken as analogs, respectively, of pressure, temperature, and specific volume.

FIGURE 2 describes the dependence of the roots of Equation 2.5 on the parameters δ and δ'. At the cusp singularity $\delta = \delta' = 0$ there is a triple root $x = 0$ (i.e., $n_X = a$). Suppose now the cusp is approached along the path $\delta = \delta'$. Then,

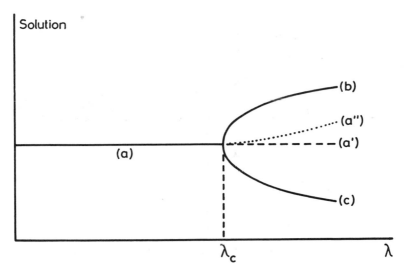

FIGURE 1. A typical bifurcation diagram.

for $\delta > 0$, $x_0 = 0$ is the only real solution of Equation 2.5. If, on the other hand, $\delta < 0$, then, in addition to $x_0 = 0$, we have the two real roots,

$$x_\pm = \pm \sqrt{\delta}, \quad \delta < 0 \qquad (2.6)$$

This is shown in FIGURE 3. We see that $\delta = \delta' = 0$ is a bifurcation point and for $\delta < 0$ we have multiple steady states. A linearized stability analysis shows that

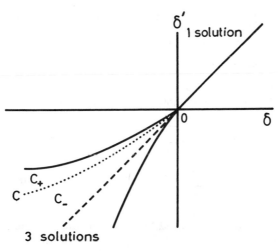

FIGURE 2. Dependence of the deterministic steady-state solution on parameters δ, δ'; C = coexistence line, calculated from stochastic theory.

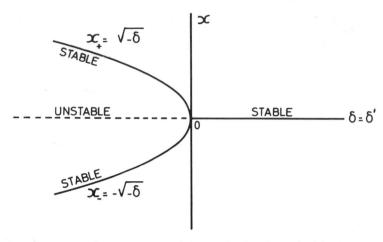

FIGURE 3. Bifurcation in the Schlögl model.

$x_0 = 0$ is unstable, whereas $x_\pm$ are both asymptotically stable. Following the usual terminology of phase transitions we call $|x_\pm|$ the *order parameter*.

We now want to investigate the behavior of fluctuations in the neighborhood of a bifurcation point. To this end, we shall describe a probability function $P(X, t)$, where X denotes possible values of a number of particle of X, i.e., any integer from 0 to ∞. We assume that space coordinates can be lumped together. Then, the evolution of $P(X, t)$ is determined by a simple birth-and-death process described by the *master equation,*

$$\frac{dP(X, t)}{dt} = k_1 A(X - 1)(X - 2)P(X - 1) - k_1 A X(X - 1)P(X)$$

$$+ k_2(X + 1)X(X - 1)P(X + 1) - k_2 X(X - 1)(X - 2)P(X)$$

$$+ k_3(X + 1)P(X + 1) - k_3 XP(X)$$

$$+ k_4 B P(X - 1) - k_4 B P(X) \qquad (2.7)$$

Assuming that this equation has been solved, we can establish connection with macroscopic properties by taking statistical averages, for instance,

$$<X> = \sum_{X=0}^{\infty} XP(X, t) \quad \text{(mean value)} \qquad (2.8a)$$

$$<\delta X^2> = \sum_{X=0}^{\infty} (X - <X>)^2 P(X, t) \quad \text{(variance)} \qquad (2.8b)$$

etc. . . .

Finite-difference equations like (2.7) are best handled in the generating function representation:

$$F(s, t) = \sum_{X=0}^{\infty} s^X P(X, t) \qquad (2.9)$$

From this definition we can see that the behavior of statistical averages of the sort shown in Equation 3 2.8 is related to the s-derivatives of F evaluated at $s = 1$.

We can easily convert Equation 2.7 into an equation for $F(s, t)$ at the steady state. Using the notation of (**2.3**), we have

$$s^2\left(\frac{d^3F}{ds^3} - 3a\frac{d^2F}{ds^2}\right) + (3 + \delta)a^2\frac{dF}{ds} - (1 + \delta')a^3F = 0 \qquad (2.10)$$

We are interested in the solutions of Equation 2.10 for δ, δ' around the bifurcation point in a macroscopic system. This will be expressed formally by the *thermodynamic limit:*

$$A, a \rightarrow \infty \qquad (2.11)$$

Equation 2.10 can be solved exactly as it leads, after a Laplace transform, to an equation of the type satisfied by hypergeometric functions of the kind $_3F_2$. Upon obtaining a solution which is regular at $s = 1$, and inverting the Laplace transform, we obtain an integral representation of the form[9]:

$$F(s) = C \int_C \varphi(t)e^{af(t,s)} dt, \qquad (2.12)$$

where
$$f(t) = t - t\log t + t\log 3 + (t - ik)\log(t - ik)$$
$$+ (t + ik)\log(t + ik) - (ti\lambda)\log(t - i\lambda)$$
$$- (t + i\lambda)\log(t + i\lambda) + t\log s$$

$$k^2 = \frac{1 + \delta'}{3} - \frac{1}{4a^2}$$

$$\lambda^2 = (3 + \delta) - \frac{1}{4a^2}$$

$$\varphi(t) = t^{-1/2}\frac{k^2}{t^2 + k^2}\frac{t^2 + \lambda^2}{\lambda^2} + \cdots$$

The leaders refer to terms which are negligible for our purpose when $a \rightarrow \infty$ and C is a constant. This form is suitable for asymptotic evaluation when a is very large.

The following results are obtained, assuming that the cusp is approached along $\delta = \delta'$.

(1) For $\delta > 0$: the function $f(t)$ has a single maximum. A straightforward application of the method of steepest descent gives,

$$\lim_{a\to\infty}\frac{<\delta X^2>}{a} = \frac{4}{\delta} + 1, \quad \delta > 0 \qquad (2.13)$$

Thus, as $\delta \rightarrow 0$ the variances diverge. In the language of equilibrium phase transitions, the divergence exponent is "classical" as it is equal to unity.[11]

(2) For $\delta < 0$: the function $f(t)$ has two maxima, but one of them dominates for large a. The dominant contribution turns out to be associated with the macroscopic solution $x_- = -\sqrt{-\delta}$ and leads to,

$$\lim_{a \to \infty} \frac{<dX^2>}{a} = \frac{2}{-\delta} - \frac{3}{\sqrt{-\delta}} + 1, \quad -1 \le \delta < 0 \qquad (2.14)$$

Again, the divergence exponent is classical.

(3) For $\delta = 0$: at the bifurcation point the function $f(t)$ has vanishing first-, second-, and third-order derivatives. A modified steepest-descent evaluation of Equation 2.12 yields,

$$\lim_{a \to \infty} \frac{<\delta X^2>}{a^{3/2}} = 4 \frac{\Gamma\left(\frac{3}{4}\right)}{\Gamma\left(\frac{1}{4}\right)}, \quad \delta = 0 \qquad (2.15)$$

We see that the central limit theorem applies only to cases (1) and (2), but not *at* the bifurcation point. Still, the deterministic equations of evolution make sense even at that point. The reason is that the equation for the average value, generated by master Equation 2.7, differs from the deterministic law, Equation 2.2, by terms proportional to $<\delta \dot{X}^2>/<X>^2$ and $<\delta X^3>/<X>^3$. By Equation 2.15, these terms are still negligible in the thermodynamic limit.

THE COEXISTENCE REGION

So far we have seen that the model reaction, Equation 2.1, gives rise to a *sharp transition* when the bifurcation point is crossed. The new solutions $x_{\pm}$ emerge smoothly beyond the transition point (see Equation 2.6) and provide system properties formally analogous to those characterizing second-order phase transitions.

In this section we shall analyze the structure of the steady-state solution, Equation 2.12, in the case where two deterministic solutions $x_{\pm}$, coexist in macroscopic amounts. The first work in this area was reported in a pioneering paper by Landauer.[6] (See also paper by this author in this volume.) In the probabilistic language, this would mean that if P is written in the form,

$$P \propto \exp[aU(X)] \qquad (3.1)$$

then the *stochastic potential* $U(X)$ has two equal-height maxima:

$$U(\hat{X}_+) = U(X_-) = U_{max} \qquad (3.2a)$$

where $\hat{X}_-, \hat{X}_+$ are, respectively, the smallest and largest root of the equation,

$$\frac{dU(\hat{X})}{d\hat{X}} = 0 \qquad (3.2b)$$

Actually P can be evaluated asymptotically for large a in the form:

$$P(x) = N^{-1}\varphi(x) \exp[aU(x)] \qquad (3.3)$$

Here x is the excess variable, where,

$$x = \frac{X - a}{a} \qquad (3.4a)$$

N^{-1} is the normalization factor; where,

$$N = \int_{-\infty}^{+\infty} \varphi(x) e^{aU(x)} dx \qquad (3.4b)$$

and the functions φ and U are given, respectively, by the expressions $\varphi(t)$ and $f(t, s)$ of Equation 2.12 evaluated at $s = 1$. In this form, it can be verified that the extrema of $U(x)$ coincide with the roots of the deterministic equation (see Equations 2.2 and 2.5).

It is instructive to express $U(x)$ in the neighborhood of the bifurcation point. Expanding Equation 3.2 in x for δ' and δ small we get,

$$U(x) = \frac{\delta' - \delta}{4} x + \frac{\delta - 3\delta'}{16} x^2 + \frac{\delta'}{8} x^3 - \frac{1}{16} x^4 + \cdots \qquad (3.5)$$

Equation 3.2b then reads,

$$\hat{x}^3 - \frac{3}{2} \delta \hat{x}^2 + \frac{3\delta' - \delta}{2} \hat{x} + (\delta - \delta') = 0 \qquad (3.6)$$

For small δ', δ the roots of this equation reduce to phenomenological values (Equation 2.6) on the line $\delta = \delta'$.

TABLE 1.

THE PHASE COEXISTENCE LINE AS GIVEN BY THE MASTER EQUATION

δ'	-0.9	-0.7	-0.5	-0.25
δ	-1.277	-0.864	-0.569	-0.264

Returning to the general case, Equations 3.2 represent a transcendental relationship between δ' and δ which can be solved numerically. TABLE 1 gives some illustrative results. From this table and (3.6) it is clearly seen that the coexistence line is not the symmetry axis of the cusp curve of FIGURE 2.

It is worth pointing out that there is a qualitative difference between the stochastic potential $U(x)$ (Equation 3.5) and the deterministic potential $V(x)$, which generates the macroscopic rate law. From Equation 2.4, the latter is found to be,

$$V = \frac{x^4}{4} + \delta \frac{x^2}{2} + (\delta - \delta')x \qquad (3.7)$$

This is very similar to a Landau-Ginzburg Hamiltonian.[8] In particular, it is symmetric about the axis of symmetry $\delta = \delta'$ of the cusp curve.

In applications of catastrophe theory, a "Maxwellian convention" requiring equal values of V is frequently used to determine the transition line between multiple steady states.[7] In the light of our stochastic analysis, we see that this convention does not reflect the correct structure of probability distribution. We are witnessing a breakdown in the deterministic description of transition phenomena in the coexistence region.[10]

The evaluation of the first few moments of the probability distribution (Equation 3.3) sheds further light on this problem. We find,

$$\frac{<\delta X^2>}{a^2} = c(\hat{x}_+ - \hat{x}_-)^2 + O\left(\frac{1}{a}\right)$$

$$\frac{<\delta X^3>}{a^3} = \tilde{c}(\hat{x}_+ - \hat{x}_-)^3 + O\left(\frac{1}{a}\right)$$

(3.8)

where c, $\tilde{c}$ are nonvanishing constants.[3] These relations clearly demonstrate that stationary probability distribution is not symmetric around the mean value. Moreover, the variances are now of the same order of magnitude as the fluctuations. As a result, the equation for the average $<X>$ deduced from the master equation will differ from the macroscopic rate law.

Although the type of probability distribution used in Equation 3.3 is the result of the asymptotic evaluation of Equation 2.12, it is instructive to carry the consequences of the limit $a \to \infty$ somewhat further. Consider first values of δ, δ' to the left of the coexistence line region C_+ in FIGURE 2. The dominant solution is then $\hat{x} = \hat{x}_+$, and P reduces to a Gaussian around $\hat{x}_+$. Using the well-known result,

$$\lim_{a \to \infty} (2\pi\alpha^{-1})^{-1/2} \exp[-(x - \hat{x}_+)^2/(2\alpha - 1)] = \delta(x - \hat{x}_+) \quad (3.9)$$

We conclude that in the above-mentioned range of values of δ and δ', the stationary probability $P(x)$ behaves as,

$$\lim_{a \to \infty} P(x) = \delta(x - \hat{x}_+) \equiv P_+, \; \delta, \delta' \in C_+ \quad (3.10a)$$

Similarly, in the region enclosed by the cusp curve, denoted by C_- in FIGURE 2, we have,

$$\lim_{a \to \infty} P(x) = \delta(x - \hat{x}_-) \equiv P_-, \; \delta, \delta' \in C_- \quad (3.10b)$$

Starting from the region C_+ or C_-, imagine now that we get closer to coexistence line C (FIGURE 2). The corresponding probability distributions (Equations 3.10) remain stationary measures during this limiting procedure. We therefore expect the simultaneous existence of *two* stationary measures in this case. Thus, in Equation 3.3, we can explicitly carry out the limit $a \to \infty$ on the coexistence curve C. Because of the existence of two equal-height maxima, a steepest-descent evaluation of the norm N (Equation 3.4b) will give two contributions:

$$N = \int_{-\infty}^{\hat{x}_0} \varphi(x)e^{aU(x)}dx + \int_{\hat{x}_0}^{\infty} \varphi(\hat{x})e^{aU(\hat{x})}d\hat{x} \quad (3.11)$$

where $\hat{x}_0$ is the root of Equation 3.2b between x_+ and $\hat{x}_-$. In each of the terms of Equation 3.11, a simple steepest-descent calculation can be performed; this yields,

$$N = \left[\varphi(\hat{x}_+)\sqrt{-\frac{2}{U''(\hat{x}_+)a}} + \varphi(\hat{x}_-)\sqrt{-\frac{2}{U''(\hat{x}_-)a}}\right] e^{aU_{max}} \quad (3.12)$$

The probability distribution (Equation 3.3) takes the form:

$$P(x) = \left[\varphi(\hat{x}_+) \sqrt{-\frac{2}{U''(\hat{x}_+)a}} + \varphi(\hat{x}_-) \sqrt{-\frac{2}{U''(\hat{x}_-)a}} \right]^{-1} \varphi(x) e^{a[U(x) - U_{max}]}$$

(3.13)

From this expression it is obvious that in the limit $a \to \infty$, $P(x)$ vanishes everywhere except at $\hat{x}_+$ or $\hat{x}_-$. At these points its value goes to infinity, because of the factor $a^{-1/2}$ in the normalization constant. The properly normalized limiting distribution function is therefore,

$$P(x) = C_+ \delta(x - \hat{x}_+) + C_- \delta(x - \hat{x}_-)$$

(3.14)

with,

$$C_+ = \frac{\varphi(\hat{x}_\pm) \sqrt{\frac{-2}{U''(\hat{x}_\pm)}}}{\varphi(\hat{x}_+) \sqrt{\frac{-2}{U''(\hat{x}_+)}} + \varphi(\hat{x}_-) \sqrt{\frac{-2}{U''(\hat{x}_-)}}}$$

(3.15)

This result is reminiscent of the decomposition of Gibbs states in extremal KMS states familiar in statistical mechanics.[2] Note however that, in contrast to the decomposition found in the Ising model, where the weights of the extremal states are equal, we have unequal weights in this case. This reflects, once again, the skewness of the probability distribution, which is already apparent from Equation 3.5.

TRANSITIONS BETWEEN STATES

As stated earlier, the stationary solution $P(X)$ of Equation 2.7 can exhibit one or two peaks according to the values of δ and δ'. In particular, if both δ and δ' are negative and lie within the cusp curve of FIGURE 2, $P(X)$ is a bimodal distribution, and the relative importance of each peak depends on whether (δ, δ') lies in C_+ (right peak higher) or in C_- (left peak higher).

Let us assume that the system has initially a number of particles X greater than X_{min}, for which $P(X)$ is a minimum. It turns out that for any a, however large, the transition probability to a state with a number of particles less than X_{min} is equal to 1. Indeed it is known[4] that the transition probability from state m to state 0 in a birth-and-death process whose infinitesimal rates are, respectively, λ_j and M_j, is given by,

$$\frac{\sum_{i=1}^{\infty} \prod_{j=1}^{i} \frac{\mu_j}{\lambda_j}}{1 + \sum_{i=1}^{\infty} \prod_{j=1}^{i} \frac{\mu_j}{\lambda_j}}$$

(4.1)

whence by a slight modification, we find that the transition probability from a state S to a state F is given by,

$$\omega_{S \to F} = \frac{\mu_F P(F) \left[\dfrac{1}{\lambda_S P(S)} + \dfrac{1}{\lambda_{S+1} P(S+1)} + \cdots \right]}{\lambda + \mu_F P(F) \left[\dfrac{1}{\lambda_F P(F)} + \dfrac{1}{\lambda_{F+1} P(F+1)} + \cdots \right]} \tag{4.2}$$

as,

$$\frac{P(X+1)}{P(X)} = \frac{\lambda_x}{\mu_{X+1}} \tag{4.3}$$

As the tail of $P(X)$ decreases monotonically, both series diverge, consequently, $\omega_{S \to F} = 1$. A more complete picture requires the computation of the mean first passage time from state S to state F.[12] Here too, an exact solution can be given, namely,[4]

$$\tau = \sum_{k=1}^{S-F} \frac{1}{Ms+k} \left[1 + \frac{1}{P(F+k)} \left(P(F+k+1) + P(F+k+2) + \cdots \right) \right] \tag{4.4}$$

It can be shown that for increasing a, we have,

$$\frac{1}{a} \log \tau \sim \frac{1}{a} \log \frac{P_{\max,+}}{P_{\min}} + O\left(\frac{\log a}{a}\right) \tag{4.5}$$

Consequently,

$$\lim_{a \to \infty} \frac{1}{a} \log \tau = C(\delta, \delta') \tag{4.6}$$

where

$$C(\delta, \delta') = \beta \log \beta - \beta - \alpha \log \alpha + \alpha - (\beta - \alpha) \log 3$$

$$+ 2 \sum_{i=1}^{2} \sum_{j=1}^{2} \epsilon_{ij} [p_{ij} \log p_{ij} \cos \Theta_{ij} - p_{ij} \Theta_{ij} \sin \Theta_{ij}], \tag{4.7}$$

$$p_{ij} = [x_i^2 + \zeta_j]^{1/2}, \quad \Theta_{ij} = \tan^{-1} \frac{1}{x_i} \zeta_j^{1/2},$$

$$x_1 = \alpha_1, \quad x_2 = \beta, \quad \zeta_1 = \frac{1+\delta'}{3}, \quad \zeta_2 = 3 + \delta,$$

$$\epsilon_{11} = \epsilon_{22} = 1, \quad \epsilon_{12} = \epsilon_{22} = -1,$$

α and β = second and third roots of $x^3 + \delta x + \delta - \delta' = 0$, and $C(\delta, \delta')$ shows no singular behavior as we cross the coexistence line (see FIGURE 2), although a significant change may occur in the order of magnitude of τ for large values of a. For instance, by numerical computation, we have,

$$a = 2000, \quad \delta' = -0.5, \quad \delta = -0.5, \quad \tau = 2.3058 \times 10^3$$

$$a = 2000, \quad \delta' = -0.5, \quad \delta = -0.57, \quad \tau = 1.1633 \times 10^{15}$$

The second value of δ lies very close to the coexistence line. However (4.7) is too weak an asymptotic result for this change to appear. Should the starting state S lie to the left of X_{min} and, furthermore, $S - X_{min} \sim O(a)$ and $X_{max} = F \sim O(a)$, then one can show that τ is $O(1)$. Should however $X_{max, -} - F = O(a)$, then τ is again infinite when $a \rightarrow \infty$. This behavior of τ is to be compared with that of τ as obtained from the deterministic equation, namely,

$$\dot{x} = - x^3 - \delta x + \delta' - \delta \tag{4.8}$$

Clearly if S lies to the right of the middle zero of the right-hand side of (4.8), any state to the left cannot be reached. However if S lies to the left, and the final state is not the left zero of the right-hand side of (4.8), then τ is finite. If the final state is the left zero, then τ again diverges.

CONCLUSION

We have seen in a simple model that, when a bifurcation point is crossed, some new and unexpected features appear in the behavior of fluctuations as described by the stochastic master equation. Of particular interest is the decomposition of the steady-state probability distribution in extremal states (Equation 3.14) related to stable solutions of deterministic equations. This however does not imply that stochastic effects become irrelevant in the thermodynamic limit. On the contrary, the all-important question of the *relative weight* of the probability peaks rests entirely on the behavior of fluctuations. True, if the evolution of the system is followed in time, we may find that stochastic behavior remains close to the deterministic trajectory for an extended period of time.[5] Eventually however, "diffusion" effects take over and drive the system to the solutions previously discussed, as long as the long time limit $t \rightarrow \infty$ is taken before the thermodynamic limit $a \rightarrow \infty$.

A somewhat related question concerns the nature of individual states of the random process. We have shown that, as long as a is finite (see especially Equation 4.2), we are dealing with a Markov chain with a single irreducible ergodic set of states. In the limit $a \rightarrow \infty$, on the other hand, a deep change takes place in the sense that a "state" is now to be defined in terms of the continuous (intensive) variable x, rather than the initial, discrete (extensive) variable X. A finite transition in this state space really implies an infinite jump in the X space. Hence, the nature of the various states—including the values $x_\pm$ on which $P(x)$ is centered—must be examined with great caution.

It is tempting to extrapolate some of these results for problems involving bifurcation of limit cycles. Numerical simulations on the trimolecular model reaction (see the paper by Herschkowitz-Kaufman and Erneux in this volume) show the existence of a steady-state probability distribution for systems of finite size. Since the deterministic stable solutions in this problem are time dependent and differ by a continuous parameter, the phase, the analog of Equation 3.14 would be,

$$P(x) = \int_0^{2n} C(\varphi)\delta[x - \hat{x}(t; \varphi)]d\varphi \tag{5.1}$$

Work being done to establish this relation analytically is now in progress.

The behavior of space-dependent fluctuations in connection with the bifurcation of spatial patterns is a most interesting area. Most of the results in this area are based on approximations equivalent to the truncation of moment equations. So far there are no exact, or at least systematic, solutions of the master equation. A different approach based on the notions underlying renormalization group theory of critical phenomena has recently been developed.[1] When applied to the Schlögl reaction scheme, it leads to a divergence law of variances according to nonclassical exponents, when the bifurcation point is approached from the region $\delta > 0$. The connection between these results and the master-equation approach is one of the major open problems of fluctuation theory.

REFERENCES

1. DEWEL, G., D. WALGRAEF & P. BORCKMANS. 1977. Z. Phys. **B28**: 235.
2. EMCH, G. 1972. Algebraic Methods in Statistical Mechanics and Quantum Field Theory. Wiley.
3. HORSTHEMKE, W., M. MALEK-MANSOUR & L. BRENIG. 1977. Z. Phys. **B28**: 135.
4. KARLIN, S. 1969. A first Course in Stochastic Processes. Academic Press, New York.
5. KURTZ, T. G. 1971. J. Appl. Prob. **8**: 344.
6. LANDAUER, R. 1962. J. Appl. Phys. **33**: 2209.
7. LU, Y.-C. 1976. Singularity Theory/Catastrophe Theory. Springer-Verlag, Berlin.
8. MA, S. 1976. Modern Theory of Critical Phenomena. Benjamin, Reading, Mass.
9. NICOLIS, G. & J. W. TURNER. 1977. Physica **89A**: 326.
10. NICOLIS, G. & R. LEFEVER. 1977. Phys. Lett. **62A**: 469.
11. NITZAN, A., P. ORTOLENA, J. DEUTCH & J. ROSS. 1974. J. Chem. Phys. **61**: 1056.
12. SCHLÖGL, F. 1971. Z. Phys. **248**: 446.
13. OPPENHEIM, I., K. SHULER & G. WEISS. 1977. Physica **88A**: 191.
14. THOM, R. 1972. Stabilité Structurelle et Morphogénèse. Benjamin, New York.

BIFURCATING WAVES*

J. F. G. Auchmuty

*Department of Mathematics
Indiana University
Bloomington, Indiana 47401*

INTRODUCTION

Bifurcation theory may be considered as the theory of finding new classes of solutions of equations near known classes of solutions. These new solutions may be of a similar type as in ordinary bifurcation or they may be different as in the Hopf bifurcation of periodic solutions from stationary solutions of ordinary or partial differential equations. Here another possibility will be described, namely, the bifurcation of wave-like solutions of an equation from stationary solutions.

These bifurcating waves will be obtained as solutions of weakly coupled parabolic systems of equations defined on a bounded axisymmetric domain. The requirement of axisymmetry for the domain puts these waves into a similar context to the other papers in this volume on bifurcation under symmetry by Othmer and Sattinger. The restriction to weakly coupled parabolic systems is not absolutely necessary but they are particularly amenable to analysis. Other classes of evolutionary partial differential equations or integro-differential equations where there are both spatial and time dependencies may allow similar phenomena.

Weakly coupled parabolic systems of equations arise in many parts of biology, chemistry, and chemical engineering and are often called "reaction-diffusion" or "interaction-diffusion" equations. For background on these equations see Aris,[2] Auchmuty,[3] Fife,[8] Nicolis and Prigogine,[13] or Tyson.[16] Turing[15] in his remarkable paper on morphogenesis suggested that there might be travelling wave solutions of his particular system of reaction-diffusion equations on a circle. Such solutions were constructed for the "Brusselator" by Auchmuty and Nicolis,[4] and some similar solutions have been numerically computed by Erneux and Herschkowitz-Kaufman.[6] Here, a more extensive theory of such waves will be developed.

The waves to be described here are different from the plane waves described by Kopell and Howard,[12] Ortoleva and Ross,[14] and others. A recent survey of such solutions has been given by Fife.[8] As will be seen, both the types of solution desired and the methods of construction are quite different, although in both theories one is constructing wavelike solutions of reaction-diffusion equations by bifurcation-theoretic methods.

In this paper, the bifurcating waves will also be time-periodic solutions of the equations. Consequently they could also be constructed by using the methods of Hopf bifurcation theory. Such methods, however, do not usually indicate when the time-periodic solution is in fact a traveling wave solution of the equations. The theory to be described here guarantees the wavelike character of the solutions and may be generalized to domains with more than one periodic symmetry (such as a torus) where the bifurcating waves need not be time periodic.

*This research was partially supported by the National Science Foundation.

0077-8923/78/0316-0263 $1.75/1 © 1979, NYAS

The problems are formulated below, after which some spectral theory of the linearized problem will be described. The construction of bifurcating waves is done next, and problems regarding their stability are described in the concluding section. Thanks are due to T. Erneux, M. Herschkowitz-Kaufman, G. Nicolis, and H. F. Weinberger for helpful comments on various aspects of this paper.

WEAKLY COUPLED PARABOLIC SYSTEMS ON AXISYMMETRIC DOMAINS

Our interest is in studying certain special solutions of weakly coupled parabolic systems of equations defined on a bounded, axisymmetric domain in $\mathbb{R}^n$. A domain is a connected, open set and the closure of a set Ω will be denoted $\overline{\Omega}$.

When $n = 2$, let $T(\theta)$ be the rotation of $\mathbb{R}^2$ about the origin through an angle θ. A domain Ω in $\mathbb{R}^2$ is said to be axisymmetric about the origin if $T(\theta)\Omega = \Omega$ for all $0 \le \theta \le 2\pi$. In $\mathbb{R}^2$, we shall only consider domains of the form $\Omega = \Omega_0 \times [0, 2\pi]$ where Ω_0 is a bounded open interval in $\mathbb{R}^1$, $0 \in \overline{\Omega}_0$ and where $(y, 0)$ and $(y, 2\pi)$ are identified for all y in Ω.

For $n \ge 3$, the x_n-axis will be chosen to be the axis of symmetry. A domain Ω in $\mathbb{R}^n$ will be said to be axisymmetric about the x_n-axis if each rotation R of $\mathbb{R}^n$ that maintains the x_n-axis as invariant is a homeomorphism of Ω onto itself. When Ω is axisymmetric, it may be written in the form $\Omega_0 \times [0, 2\pi]$ where Ω_0 is a bounded domain in $\mathbb{R}^{n-1}$ and $(y, 0)$ and $(y, 2\pi)$ are identified for all y in Ω_0. We shall only consider axisymmetric domains whose closure does not intersect the x_n-axis.

Given such a domain Ω, let Ω_1 be the projection of Ω onto the hyperplane $x_1 = 0$. Let $\Omega_0(\overline{\Omega}_0)$ be the intersection of $\Omega_1(\overline{\Omega}_1)$ with the half-space $x_2 \ge 0$. One can then write

$$\Omega = \{(x_2 \sin\theta, x_2 \cos\theta, x_3, \ldots, x_n) : (x_2, x_3, \ldots, x_n) \in \Omega_0\}.$$

Henceforth, we shall assume that Ω_0 (and hence Ω) has a smooth boundary $\partial\Omega_0$ $(\partial\Omega)$ and when $x \in \Omega$ we shall write $x = (y, \theta)$ with $y = [(x_1^2 + x_2^2)^{1/2}, x_3, \ldots, x_n]$ and θ defined by the solution of $x_1 = r\sin\theta$, $x_2 = r\cos\theta$, $r = (x_1^2 + x_2^2)^{1/2}$. The components of y will be denoted y_i, $1 \le i \le n - 1$.

Simple examples of such axisymmetric domains include annuli in $\mathbb{R}^2$ and toroidal figures or the region between two cylinders in $\mathbb{R}^3$. When $\Omega_0 = (R_1, R_2)$, $R_1 > 0$, the above construction yields an annulus of inner and outer radii, R_1 and R_2. When Ω_0 is an open, bounded, connected set in $\mathbb{R}^2$ that does not intersect the x_2-axis, the construction yields a torus of uniform cross-section Ω_0. If Ω_0 is a ball, Ω is the usual torus, but when Ω_0 is a rectangle, Ω represents the annular region between two cylinders. A degenerate case is a circle in $\mathbb{R}^2$. This may be considered to be the limiting case of an annulus whose inner and outer radii are equal.

The equations to be studied are

$$\frac{\partial u_i}{\partial t} = L_i u_i + f_i(y, u_1, u_2, \ldots, u_m, \mu), \tag{1}$$

defined for $x = (y, \theta) \in \Omega$, $t \ge 0$ and $1 \le i \le m$. Each L_i is defined by

$$L_i w(x) = \sum_{j,k=1}^{n-1} \frac{\partial}{\partial y_j} \left(a_{ijk}(y) \frac{\partial w}{\partial y_k} \right) + a_{inn}(y) \frac{\partial^2 w}{\partial \theta^2}, \tag{2}$$

and one would like to find real-valued functions $u_1, u_2, \ldots, u_m$, defined on $\overline{\Omega} \times (0, \infty)$, which satisfy these equations.

Let $u : \overline{\Omega} \times (0, \infty) \to \mathbb{R}^m$ be the function whose components are u_i. Assume that for each i, j, and k, $a_{ijk}(y)$ is a real-valued, Hölden continuous function on $\overline{\Omega}$ which is independent of θ, and is symmetric in j and k; $a_{ijk}(y) = a_{ikj}(y)$. For each i, assume that there exists $\alpha_i > 0$ so that

$$\sum_{j,k=1}^{n-1} a_{ijk}(y)\xi_j\xi_k + a_{inn}(y)\xi_n^2 \geq \alpha_i |\xi|^2$$

for all ξ in $\mathbb{R}^n$ and y in $\overline{\Omega}_1$. Then each L_i is said to be uniformly elliptic.

In (1) μ is a parameter lying in an open subset Λ of $\mathbb{R}^1$. Assume $f : \overline{\Omega}_1 \times \mathbb{R}^m \times \Lambda \to \mathbb{R}^m$ is the function whose components f_i obey

(F1) $f_i(y, 0, \mu) = 0$ for all i, y, μ,

(F2) f_i is continuous,

(F3) f_i is twice continuously differentiable on $\mathbb{R}^m$ for each (y, μ) in $\overline{\Omega}_1 \times \Lambda$ and all i;

    (i) $(\partial f_i / \partial u_j)(\cdot, u, \mu)$ is bounded on $\overline{\Omega}_1$, for each (u, μ) in $\mathbb{R}^m \times \Lambda$ and each i and j and

    (ii) $(\partial^2 f_i / \partial u_j \partial u_k)(\cdot, u, \mu)$ is uniformly bounded on $\overline{\Omega}_1$ for u in any bounded subset of $\mathbb{R}^m$, for all μ, i, j, k.

The problem is to solve (1) subject to the following periodicity and boundary conditions

$$u_i(y, 0, t) = u_i(y, 2\pi, t),$$

$$\frac{\partial u_i}{\partial \theta}(y, 0, t) = \frac{\partial u_i}{\partial \theta}(y, 2\pi, t), \tag{3}$$

where $(y, t) \in \Omega_1 \times [0, \infty)$ and $1 \leq i \leq m$ and

$$u_i(y, \theta, t) = 0 \quad \text{for } 1 \leq i \leq m_1$$

$$\frac{\partial u_i}{\partial \nu}(y, \theta, t) = 0 \quad \text{for } m_1 + 1 \leq i \leq m \tag{4}$$

when

$$(y, \theta, t) \in \partial\Omega_1 \times [0, 2\pi] \times (0, \infty).$$

Here $\partial/\partial\nu$ is the directional derivative in the direction of the outward normal on $\partial\Omega$ and $0 \leq m_1 \leq m$. When $m_1 = 0$, Neumann conditions hold for each component u_i and when $m_1 = m$ then Dirichlet conditions hold for each u_i. Conditions (4) will be written $B_i u_i(x, t) = 0$ on $\partial\Omega_1 \times [0, 2\pi] \times (0, \infty)$.

In this paper our interest is in constructing and analyzing solutions of (1)–(4) of the form

$$u_i(x, t) = v_i(y, \theta - ct) \quad 1 \leq i \leq m. \tag{5}$$

Such solutions represent waves rotating about the axis of symmetry. They obey

$$\sum_{j,k=1}^{n-1} \frac{\partial}{\partial y_j} \left(a_{ijk}(y) \frac{\partial v_i}{\partial y_k} \right) + a_{inn}(y) \frac{\partial^2 v_i}{\partial y_n^2} + c \frac{\partial v_i}{\partial y_n} + f_i(y, v, \mu) = 0 \qquad (6)$$

Here $1 \le i \le m$, $y_n = \theta - ct \in \mathbb{R}^1$ and $y \in \Omega_1$. The boundary conditions become

$$B_i v_i(y, y_n) = 0 \quad \text{for } (y, y_n) \text{ in } \partial \Omega_1 \times \mathbb{R}^1 \qquad (7)$$

and

$$v(y, y_n) \qquad (8)$$

must be periodic of period 2π in y_n for each y in $\overline{\Omega}_1$.

This is a semilinear, weakly coupled elliptic system of equations. (Weakly coupled means that the ith equation only depends on v_j for $j \ne i$ through the lowest order terms.) Using standard methods this will be converted to a compact operator equation.

For each i, consider the problem of solving

$$- \sum_{j,k=1}^{n-1} \frac{\partial}{\partial y_j} \left(a_{ijk}(y) \frac{\partial w}{\partial y_k} \right) - a_{inn}(y) \frac{\partial^2 w}{\partial y_n^2} + w = g(y, y_n) \text{ on } \Omega_1 \times \mathbb{R}^1 \quad (9)$$

subject to

$$B_i w(y, y_n) = 0 \qquad (10)$$

for (y, y_n) in $\partial \Omega_1 \times \mathbb{R}^1$ and $w(y, y_n)$ being periodic of period 2π in y_n for each y in Ω_1.

Assume g is a continuous function on $\overline{\Omega}_1 \times \mathbb{R}^1$, which is periodic of period 2π in y_n. Then from standard results on elliptic equations,[1] this equation has a solution w which has continuous first partial derivatives on $\Omega_1 \times \mathbb{R}^1$. More precisely one has the following.

Let $C^0(\overline{\Omega}_1 \times \mathbb{R}^1)$ be the set of all continuous and bounded functions on $\overline{\Omega}_1 \times \mathbb{R}^1$, and let $P^0(\overline{\Omega})$ be the subset of $C^0(\overline{\Omega}_1 \times \mathbb{R}^1)$ consisting of all functions which are periodic of period 2π in y_n. This is a Banach space under the norm

$$\| u \|_0 = \sup_{y \in \overline{\Omega}_1 \times \mathbb{R}^1} | u(y) |.$$

Similarly, let $C^1(\overline{\Omega}_1 \times \mathbb{R}^1)$ be the subset of $C^0(\overline{\Omega}_1 \times \mathbb{R}^1)$ of all functions whose first partial derivatives are all continuous and bounded on $\overline{\Omega}_1 \times \mathbb{R}^1$. This is a Banach space under the norm

$$\| u \|_1 = \sum_{|\alpha|=1} \sup_{y \in \overline{\Omega}_1 \times \mathbb{R}^1} | D^\alpha u(y) | + \| u \|_0.$$

Again let $P^1(\overline{\Omega})$ be the subset of $C^1(\overline{\Omega}_1 \times \mathbb{R}^1)$ of all functions that are periodic of period 2π in y_n. The result is the following:

PROPOSITION 1. For each g in $P^0(\overline{\Omega})$, there exists a unique w in $P^1(\overline{\Omega})$ that solves (9)–(10) in a distributional sense. Define $G_i: P^0(\overline{\Omega}) \to P^1(\overline{\Omega})$ by $w = G_i g$, then G_i is a linear, compact map.

The proof of this follows from the results in Morrey.[17] Let $X_0(X_1)$ be the Cartesian product of m copies of $P^0(\overline{\Omega})(P^1(\overline{\Omega}))$. The elements of X_0 or X_1 will be written $u = (u_1, u_2, \ldots, u_m)$. They are Banach spaces under the norms

$$\| u \|_0 = \sup_{1 \leq i \leq m} \| u_i \|_0,$$

$$\| u \|_1 = \sup_{1 \leq i \leq m} \| u_i \|_1.$$

Define the maps $F : \Lambda \times X_0 \to X_0$, $D : X_1 \to X_0$ and $G : X_0 \to X_1$ by

$$F(\mu, u) = (f_1(\cdot, u, \mu) + u_1, \ f_2(\cdot, u, \mu) + u_2, \ldots, \ f_m(\cdot, u, \mu) + u_m)$$

$$Du = \left(\frac{\partial u_1}{\partial y_n}, \frac{\partial u_2}{\partial y_n}, \ldots, \frac{\partial u_m}{\partial y_n} \right)$$

and

$$Gu = (G_1 u_1, G_2 u_2, \ldots, G_m u_m)$$

where the f_i and G_i terms are defined as earlier. The term D is a continuous, linear operator; G is a compact linear operator while $F(\mu, \cdot)$ is a continuous and bounded mapping for each μ in Λ.

Equations 6–8 may be rewritten as

$$v = cGDv + GF(\mu, v),$$

or

$$v = c\Lambda v + \mathcal{g}(\mu, v). \tag{*}$$

where A and $\mathcal{g}(\mu, \cdot)$ are compact maps of X_1 into itself.

This will be the basic equation for our bifurcation analysis. From our construction one sees that a solution of this nonlinear operator equation will be a wavelike solution of Equations 1–4 rotating with wave-velocity c about the x_n-axis. Moreover, if $v(y, \theta - ct)$ is a solution of (6) or (*) then so also is $v(y, \theta + \varphi - ct)$ for any φ. Thus, the solutions of (6) or (*) will only be determined to within a phase. Finally, note that "steady-state" or time-invariant solutions of (1)–(4) will be solutions of (6) or (*) with $c = 0$.

BIFURCATION POINTS AND SPECTRAL THEORY

Assumption (F1) implies that $v \equiv 0$ is a solution of our equations for all values of μ. Our interest will be in nonzero solutions that bifurcate from the zero solution.

A triple (μ, c, v) will be said to be a solution of (6)–(8) or, equivalently, of (*), if v is a solution of the equations when μ, c are specified. Let

$$\mathcal{S}_0 = \{(\mu, c, 0) : \mu \in \Lambda, c \in \mathbb{R}^1\}$$

and

$$\mathcal{S} = \{(\mu, c, v) \in \Lambda \times \mathbb{R}^1 \times X_1 : (\mu, c, v) \text{ is a solution of } (*)\}.$$

Then $\mathcal{S}$ is the set of all solutions and $\mathcal{S}_0$ is the set of trivial solutions. An element of $\mathcal{S} - \mathcal{S}_0$ will be called a nontrivial solution. A point $(\mu, c_0, 0)$ in $\mathcal{S}_0$ is a bifurcation point for (*) if there is a nontrivial solution of (*) in every neighborhood of $(\mu, c_0, 0)$.

When $\mathcal{G}(\mu, \cdot)$ is Fréchet differentiable at 0, a necessary condition for $(\mu, c_0, 0)$ to be a bifurcation point is that 0 lie in the spectrum of the operator $L_0 = I - c_0 A - K(\mu)$ with $K(\mu) = D_v \mathcal{G}(\mu, 0)$ being the Fréchet derivative of $\mathcal{G}(\mu, \cdot)$ at 0. In this section we shall describe the differentiability of $\mathcal{G}$ and the spectrum of L_0.

A mapping $\mathcal{R}: X_1 \rightarrow X_1$ is said to be Fréchet differentiable at a point u in X_1 provided there exists a continuous linear operator $B: X_1 \rightarrow X_1$ such that

$$\lim_{\|h\| \to 0} \frac{\|\mathcal{R}(u + h) - \mathcal{R}(u) - Bh\|}{\|h\|} = 0.$$

We shall write $B = D_u \eta(u)$ when this derivative exists.

PROPOSITION 2. When f obeys (F1)–(F3), $\mathcal{G}(\mu, \cdot)$ is Fréchet differentiable on X_1 and its derivative at each point is a compact map of X_1 into itself.

PROOF. It suffices to prove this with $m = 1$ and at 0 in X_1. We shall show that $D_v \mathcal{G}(\mu, 0) = K(\mu) = GF_1(\mu)$ where $F_1(\mu): X_1 \rightarrow X_0$ is the multiplication operator defined by

$$F_1(\mu)v(y) = \left(1 + \frac{\partial f}{\partial u}(y, 0, \mu)\right) v(y) = \tilde{f}_{11}(y, \mu)v(y).$$

From (F3), the function $\tilde{f}_{11}(\cdot, \mu)$ is defined and bounded on $\bar{\Omega}$ so $F_1(\mu)$ is a continuous map. Now

$$\|\mathcal{G}(\mu, h) - \mathcal{G}(\mu, 0) - GF_1(\mu)h\|_1 = \|G[F(\mu, h) - F(\mu, 0) - F_1(\mu)h]\|_1$$

$$\leq M \|F(\mu, h) - F(\mu, 0) - F_1(\mu,)h\|_0,$$

as G is a bounded map of X_0 into X_1. However,

$$\sup_{y \in \bar{\Omega}} |f(y, h(y), \mu) - f(y, 0, \mu) - \frac{\partial f}{\partial u}(y, 0, \mu)h(y)| \leq K \sup_{y \in \bar{\Omega}} |h(y)|^2,$$

using Taylor's expansion and part (ii) of (F3).

Therefore

$$\frac{\|F(\mu, h) - F(\mu, 0) - F_1(\mu)h\|_0}{\|h\|_0} \leq K \|h\|_0,$$

so

$$\lim_{\|h\|_1 \to 0} \frac{\|\mathcal{G}(\mu, h) - \mathcal{G}(\mu, 0) - GF_1(\mu)h\|_1}{\|h\|_1}$$

$$\leq M \lim_{\|h\|_0 \to 0} \frac{\|F(\mu, h) - F(\mu, 0) - F_1(\mu)h\|_0}{\|h\|_0} \leq KM \lim_{\|h\|_0 \to 0} \|h\|_0 = 0.$$

Thus, $\mathcal{G}$ is Fréchet differentiable at 0 and its derivative at 0 is $K(\mu) = GF_1(\mu)$. The function $K(\mu)$ is compact as $F_1(\mu)$ is continuous and G is compact.

When $m > 1$, the multiplication operator $F_1(\mu)$ is an $m \times m$ matrix of functions whose components are

$$\tilde{f}_{jk}(y, \mu) = \frac{\partial f_j}{\partial y_k}(y, 0, \mu) + \delta_{jk}.$$

Since $K(\mu)$ is compact, 0 lies in the spectrum $\sigma(L_0)$ of L_0 if and only if it is an eigenvalue of finite multiplicity. Thus $0 \in \sigma(L_0)$ implies that there is a w in X_1 such that

$$w = c_0 A w + K(\mu) w. \tag{11}$$

Reversing the construction of the last section, we find that w is a distributional solution of

$$\mathcal{C}_i(c_0, \mu) w = \sum_{j,k=1}^{n-1} \frac{\partial}{\partial y_j} \left(a_{ijk}(y) \frac{\partial w_i}{\partial y_k} \right) + a_{inn}(y) \frac{\partial^2 w_i}{\partial y_n^2}$$

$$+ c_0 \frac{\partial w_i}{\partial y_n} + \sum_{l=1}^{m} f_{il}(y, \mu) w_l = 0 \tag{12}$$

on $\overline{\Omega}_1 \times \mathbb{R}^1$, for $1 \le i \le m$ and subject to (7)–(8). Here, and henceforth, we shall identify functions on Ω with their periodic extensions in y_n and $f_{il}(y, \mu) = \tilde{f}_{il}(y, \mu) - \delta_{il}$.

Equations 11 or 12 are nonstandard eigenvalue problems. One would like to find those values of μ and c_0 for which these linear equations have nontrivial solutions.

Since w is periodic of period 2π in y_n, write

$$w_j(y) = \sum_{k=-\infty}^{\infty} d_{jk}(y_1, \ldots, y_{n-1}) e^{iky_n}, \quad 1 \le j \le m \tag{13}$$

Substituting into (12) one finds that each d_{ik} satisfies

$$M_i(k, \mu) d_{ik}(y) = \sum_{j,l=1}^{n-1} \frac{\partial}{\partial y_j} \left(a_{ijl}(y) \frac{\partial d_{ik}}{\partial y_l} \right) - k^2 a_{inn}(y) d_{ik}$$

$$+ \sum_{l=1}^{m} f_{il}(y, \mu) d_{lk}(y) = -ik c_0 d_{ik}(y) \tag{14}$$

on Ω_1 for $1 \le i \le m$ and each k.

The boundary conditions become

$$B_i d_{ik}(y) = 0 \quad y \in \partial\Omega_1, \ 1 \le i \le m.$$

Thus (12) has a nontrivial solution if and only if there is a nonzero integer k such that the elliptic system of operators $M_i(k, \mu)$ has a purely imaginary eigenvalue, or if the system $M_i(0, \mu)$ has a zero eigenvalue.

To describe the spectral theory of these operators we shall use the following definitions. Let $\mathbb{Z}$ be the set of all integers and $L^2(\Omega_1)$ is the Hilbert-space of all square-integrable, complex-valued, Lebesgue-measurable functions on Ω_1 with the inner product

$$<u, v> = \int_{\Omega_1} u(y)\overline{v(y)}dy.$$

The norm on $L^2(\Omega_1)$ is given by

$$|u| = <u, u>^{1/2}$$

$H_j^1(\Omega_1)$ is the completion of the C^∞ complex-valued functions on Ω_1 which obey

$$B_j u(y) = 0 \quad \text{on } \partial\Omega_1$$

with respect to the norm

$$||u||_1 = \left[\int_{\Omega_1} \left(\sum_{l=1}^{n-1} \left| \frac{\partial u}{\partial y_l}(y) \right|^2 + |u(y)|^2 \right) dy \right]^{1/2}.$$

It is a Hilbert space with respect to the inner product

$$<u, v>_1 = \int_{\Omega_1} \left[\sum_{l=1}^{n-1} \frac{\partial u}{\partial y_l} \frac{\overline{\partial v}}{\partial y_l} + u\bar{v} \right] dy.$$

Define

$$H_1 = \prod_{i=1}^{m} H_i^1(\Omega_1)$$

to be the Cartesian product of the spaces $H_i^1(\Omega_1)$ and define an inner product on H_1 by

$$<u, v>_1 = \sum_{i=1}^{m} <u_i, v_i>_1,$$

where u_i is the ith component of u. Similarly let H_0 be the Cartesian product of m copies of $L^2(\Omega_1)$. It is a Hilbert space under the inner product

$$<u, v> = \sum_{i=1}^{m} <u_i, v_i>.$$

Define the usual norms on H_0 and H_1, namely,

$$|u| = <u, u>^{1/2} \quad ||u|| = <u, u>_1^{1/2}$$

Consider the sesquilinear forms $\mathcal{L}_i : H_i^1(\Omega_1) \times H_i^1(\Omega_1) \to \mathbb{C}$ defined by

$$\mathcal{L}_i(u, v) = \sum_{j=1}^{n-1} \sum_{l=1}^{n-1} \int_{\Omega_1} a_{ijl}(y) \frac{\partial u}{\partial y_l} \frac{\overline{\partial v}}{\partial y_j} dy$$

and $\mathcal{L}: H_1 \times H_1 \to \mathbb{C}$ defined by

$$\mathcal{L}(u, v) = \sum_{i=1}^{m} \mathcal{L}_i(u_i, v_i).$$

The assumptions in the previous section on the coefficients a_{ijl} imply that each $\mathcal{L}_i$ is a continuous sesquilinear form on its domain and so, consequently, is $\mathcal{L}$. Moreover, each $\mathcal{L}_i$ is also symmetric and semibounded on a dense subset of $L^2(\Omega_1)$, so the same holds for $\mathcal{L}$. Then, as described in Kato,[11] one may associate with $\mathcal{L}$ a closed, densely defined self-adjoint linear operator $\mathcal{L}_0 : D_0(\subset H_0) \to H_0$ defined by

$$\mathcal{L}(u, v) = \langle \mathcal{L}_0 u, v \rangle \quad \text{for all } u, v \text{ in } H_1.$$

Consider the closed, densely defined linear operator $M(k, \mu) : D_0(\subset H_0) \to H_0$ given by

$$M(k, \mu)u = -\mathcal{L}_0 u - k^2 A_n u + F(\mu)u.$$

Here A_n and $F(\mu)$ are the continuous linear operators on H_0 whose jth components are the functions

$$a_{jnn}u_j \quad \text{and} \quad \sum_{l=1}^{m} f_{jl}(\cdot, \mu)u_l,$$

respectively.

From (14), one sees that the pure imaginary eigenvalues of $M(k, \mu)$ are connected with the possible points at which waves bifurcate. The following theorem describes the spectrum of $M(k, \mu)$.

THEOREM 3. The spectrum $\sigma[M(k, \mu)]$ of $M(k, \mu)$ is a discrete set of isolated eigenvalues of finite multiplicity contained in a rectangular region

$$\Sigma = \{\lambda = \sigma + i\xi : \sigma \leq c_1 - c_2 k^2, |\xi| \leq c_1\}$$

where c_1 and c_2, are positive constants independent of k. If λ lies in $\sigma[M(k, \mu)]$ so does $\bar{\lambda}$ and when λ lies in the resolvent set, $[M(k, \mu) - \lambda]^{-1}$ is compact.

PROOF. The operator A_n is bounded on H_0 since each a_{jnn} is bounded on $\bar{\Omega}_1$. The uniform ellipticity of each L_j implies that

$$\langle A_n u, u \rangle = \sum_{j=1}^{m} \int_{\Omega_1} a_{jnn}(y)u_j(y)\overline{u_j(y)}\,dy \geq \alpha |u|^2$$

where $\alpha > 0$.

Condition (F2) implies that $F(\mu)$ is a bounded linear operator on H_0. Then there exists $C(\mu)$ so that

$$|\langle F(\mu)u, u \rangle| \leq C(\mu)|u|^2 \quad \text{for all} \quad u \text{ in } H_0.$$

The uniform ellipticity of each L_j also yields

$$\mathcal{L}(u, u) = \sum_{j=1}^{n-1} \sum_{l=1}^{n-1} \sum_{i=1}^{m} \int_{\Omega_1} a_{ijl}(y) \frac{\partial u_i}{\partial y_j} \frac{\overline{\partial u_i}}{\partial y_l} \, dy$$

$$\geq \alpha_0(\| u \|^2 - | u |^2),$$

where $\alpha_0 = \min_{1 \leq i \leq m} \alpha_i > 0$.

Hence, if ν is a constant, one has

$$\text{Re} < -[M(k, \mu) - \nu]u, u > = \text{Re} < -M(k, \mu)u, u > + \nu | u |^2$$

$$\geq \alpha_0(\| u \|^2 - | u |^2)$$

$$+ [\nu + k^2\alpha - C(\mu)] | u |^2$$

upon substituting for $M(k, \mu)$ and using the above inequalities.

Choosing ν large enough, the right-hand side will be positive for any nonzero u. Given f in H_0, consider the problem of finding those u such that

$$- <[M(k, \mu) - \nu]u, v > = <f, v > \quad \text{for all} \quad v \text{ in } H_1.$$

When ν is large enough, the Lax-Milgram theorem implies that this has a unique solution u for each f in H_0 and there exists a constant C so that

$$\| u \|_1 \leq C | f |.$$

Hence $[M(k, \mu) - \nu]^{-1}$ is a continuous map of H_0 into H_1. Rellich's theorem implies that this is a compact map of H_0 into itself, so $[M(k, \mu) - \lambda]^{-1}$ is compact for all λ in its resolvent set (see Kato's theorem III 6.29).[11] This compactness implies that $\sigma[M(k, \mu)]$ consists only of isolated eigenvalues of finite multiplicity.

Suppose $\lambda = \sigma + i\xi$ lies in the resolvent set of $M(k, \mu)$. Then for each f in H_0, there exists u in H_1 such that

$$M(k, \mu)u - \lambda u = f.$$

Taking the inner product of this with u and equating real parts one obtains

$$\text{Re} < M(k, \mu)u, u > - \sigma < u, u >$$

$$= \text{Re} < f, u > - \sigma | u |^2$$

$$\geq \alpha_0(\| u \|^2 - | u |^2) + [k^2\alpha - C(\mu)] | u |^2 - | f | | u |$$

or

$$\{\sigma - [C(\mu) + \tfrac{1}{2} - k^2\alpha]\} | u |^2 \leq \tfrac{1}{2} | f |^2 \tag{15}$$

Similarly equating imaginary parts, one obtains

$$\text{Im} < M(k, \mu)u, u > - \xi < u, u > = \text{Im} < f, u >$$

or

$$\xi | u |^2 \leq C(\mu) | u |^2 + | f | \, | u |.$$

Rearranging, this becomes

$$[\xi - C(\mu) - \tfrac{1}{2}] \, |\, u\, |^2 \le \tfrac{1}{2} |\, f\, |^2. \tag{16}$$

Also one has

$$\xi \, |\, u\, |^2 \ge -C(\mu) \, |\, u\, |^2 - |\, f\, | \, |\, u\, |$$

or

$$[\xi + C(\mu) + \tfrac{1}{2}] \, |\, u\, |^2 \ge -\tfrac{1}{2} |\, f\, |^2 \tag{17}$$

From (15) one sees that $\sigma > C(\mu) + \tfrac{1}{2} - k^2\alpha$ implies an *a priori* bound on $|\, u\, |$, so any such λ must lie in the resolvent set of $M(k, \mu)$. Similarly (16) or (17) shows that λ must lie in the resolvent set of $M(k, \mu)$ if $\text{Im}\,\lambda > C(\mu) + \tfrac{1}{2}$ or $\text{Im}\,\lambda < -C(\mu) - \tfrac{1}{2}$, respectively. Hence, the spectrum lies in Σ with $c_1 = C(\mu) + \tfrac{1}{2}$ and $c_2 = \alpha$.

Finally, if λ lies in $\sigma[M(k, \mu)]$ so does $\overline{\lambda}$ as $M(k, \mu)$ is a real operator or $\overline{M(k, \mu)u} = M(k, \mu)\overline{u}$.

COROLLARY. For each μ, there exists K such that $|\, k\, | > K$ implies $\sigma[M(k, \mu)]$ does not intersect the imaginary axis $I = \{\lambda = i\xi : -\infty < \xi < \infty\}$.

PROOF. This follows from the fact that $\lambda \in \sigma[M(k, \mu)]$ implies $\text{Re}\,\lambda \le C(\mu) + \tfrac{1}{2} - \alpha k^2$ with $\alpha > 0$. If k is large enough, $\text{Re}\,\lambda$ is always negative.

For later usage, it is worth remarking the $F(\mu)$ considered as a map from Λ into the set of all bounded linear operators on H_0 is continuous from condition (F3). Hence, $M(k, \mu)$ for μ in Λ is a family of closed operators with a common domain, and from results in Kato,† the eigenvalues of $M(k, \mu)$ depend continuously on μ.

The relationship between the spectrum of $M(k, \mu)$ and that of $I - cA - K(\mu)$ is given by the following result.

THEOREM 4. Suppose $k \ne 0$ and v is an eigenvector corresponding to a purely imaginary eigenvalue $i\nu$ of $M(k, \mu)$. Define $v_+ = \overline{v}e^{iky_n}$ and $v_- = ve^{-iky_n}$, then v_+ and v_- are eigenvectors corresponding to the eigenvalue 0 of $L_0 = I - \nu k^{-1}A - K(\mu)$.

PROOF. From (12) one has

$$\mathcal{Q}_i(c, \mu)v_- = M_i(k, \mu)v_- - ickv_- = e^{-iky_n}[M_i(k, \mu)v - ickv] = 0$$

for all i, provided $\nu = ck$ or $c = \nu k^{-1}$.

Similarly $\mathcal{Q}_i(c, \mu)v_+ = 0$ for all i when $c = \nu k^{-1}$. Since (12) is equivalent to (11), one has the result.

COROLLARY 1. Suppose $k \ne 0$ and $v = v_1 + iv_2$ is an eigenvector corresponding to a purely imaginary eigenvalue $i\nu$ of $M(k, \mu)$. Then $w_1(y, y_n) = v_1(y)\cos(ky_n) + v_2(y)\sin(ky_n)$ and $w_2(y, y_n) = v_1(y)\sin(ky_n) - v_2(y)\cos(ky_n)$ are two linearly independent real eigenfunctions of $L_0 = I - \nu k^{-1}A - K(\mu)$ corresponding to the eigenvalue 0.

PROOF. With these definitions $w_1 = \text{Re}(\overline{v}e^{iky_n})$ and $w_2 = \text{Im}(\overline{v}e^{iky_n})$. But from the theorem $w = w_1 + iw_2$ is an eigenfunction of L_0, and L_0 is a real operator, so $L_0w_i = 0$ for $i = 1, 2$.

Also

†See Reference 11, section IV. 3.

$$<w_1, w_2> = \int_{\Omega_1} \int_0^{2\pi} w_1(y)\overline{w_2(y)}\,dy = 0$$

so w_1, w_2 must be linearly independent.

Define

$$\Sigma_1(\mu) = \bigcup_{k=1}^{\infty} \sigma[M(k,\mu)].$$

COROLLARY 2. $\Sigma_1(\mu) \cap I$ consists of a finite number of points with no point of accumulation and the set

$$\mathcal{C} = \{c_j : c_j = \nu_j k^{-1} \text{ with } i\nu_j \in \sigma[M(k,\mu)], k \neq 0\}$$

is a finite, bounded set.

PROOF. From Theorem 3, $\sigma[M(k,\mu)]$ is a discrete set with no finite point of accumulation. Hence $\sigma[M(k,\mu)] \cap I$ is a finite set for each k and μ. The corollary to Theorem 3 implies that $\sigma[M(k,\mu)] \cap I$ is empty if $|k| > K$, so the first statement holds. The second statement then follows immediately. In particular $c \in \mathcal{C}$ implies $-c \in \mathcal{C}$ and $|c| \leq c_1$, where c_1 is the constant in the statement of Theorem 3.

The set $\mathcal{C}$ is the set of possible bifurcating wavespeeds at μ. One sees that for given μ there are only finitely many such speeds and that they are bounded. Note that in this analysis we have not had to exclude $\nu = 0$, so that 0 may be a bifurcating wavespeed. It is also worth noting that if for particular k and μ,

$$\sigma[M(k,\mu)] \cap I = \{\pm i\nu_j : 1 \leq j \leq m\},$$

where $\nu_j \neq \nu_l$ for $j \neq l$, and $\nu_j \neq 0$ for all j, then these $2m$ eigenvalues correspond to $2m$ different bifurcating wavespeeds; namely, $c_j = \pm \nu_j k^{-1}$, $1 \leq j \leq m$. (If $\nu_j = 0$ for some j, then there only are $(2m - 1)$ eigenvalues of $\sigma[M(k,\mu)]$ on I and there will only be $(2m - 1)$ bifurcating wavespeeds.)

Finally, one may define the adjoint operators $M^*(k,\mu)$, A^*, and $K^*(\mu)$ in the usual way. They will have similar properties to $M(k,\mu)$, A, and $K(\mu)$. In particular, $\sigma[M^*(k,\mu)]$ obeys Theorem 3 and if $v^* = v_1^* + iv_2^*$ is an eigenvector corresponding to a purely imaginary eigenvalue $i\nu$ of $M^*(k,\mu)$ with $k \neq 0$, then one may construct two linearly independent real eigenfunctions of L_0^* corresponding to the eigenvalue 0.

BIFURCATION OF WAVES

We have seen above that wavelike solutions of our parabolic system correspond to nontrivial solutions (μ, c, v) of the equation

$$v = cAv + \mathcal{G}(\mu, v), \tag{*}$$

where A and $\mathcal{G}(\mu, \cdot)$ are compact maps of X_1 into itself.

This is a two-parameter bifurcation problem in that one must find the values of both μ and c for which there are nontrivial solutions. For simplicity, assume henceforth that μ enters linearly so that

$$f(y, v, \mu) = g_1(y, v) + \mu g_2(y, v)$$

and write

$$\mathcal{G}(\mu, v) = \mathcal{G}_1(v) + \mu \mathcal{G}_2(v)$$

$$L(\mu, c) = I - cA - K(\mu) \tag{18}$$

Here $K(\mu)$ is the Fréchet derivative of $\mathcal{G}(\mu, \cdot)$ at 0 in X_1 and we shall write $\mathcal{N}(\mu, v) = \mathcal{G}(\mu, v) - K(\mu)v$. Then (*) may be rewritten as

$$L_0 v = [I - c_0 A - K(\mu_0)]v = (c - c_0)Av + (\mu - \mu_0)K_2 v + \mathcal{N}(\mu, v) \tag{19}$$

with $K_2 = D_v \mathcal{G}_2(0)$. The only possible bifurcation points for (*) occur when L_0 is singular.

When L is any linear map, $N(L)$ and $R(L)$ will denote its kernel and range, respectively. If Y is a subspace of a vector space X, then dim Y is the algebraic dimension of Y and codim Y is the dimension of the complement of Y in X (when it exists).

The basic result on the bifurcation of nontrivial solutions is the following.

THEOREM 5. Suppose $\mathcal{G}: \Lambda \times X_1 \rightarrow X_1$ defined by (18) is continuously Fréchet differentiable on an open neighborhood of $(\mu_0, 0)$ and $L_0: X_1 \rightarrow X_1$ is defined by (19) and dim $N(L_0)$ = codim $R(L_0)$ = 2. Let φ_1, φ_2 span $N(L_0)$ and φ_1^*, φ_2^* span the complement Y of $R(L_0)$ in X_1. Assume further that

$$\det \begin{pmatrix} <A\varphi, \varphi_1^*> & <K_2\varphi, \varphi_1^*> \\ <A\varphi, \varphi_2^*> & <K_2\varphi, \varphi_2^*> \end{pmatrix} \neq 0 \tag{20}$$

for some φ in $N(L_0)$. Then there exists an open neighborhood I of 0 in $\mathbb{R}^1$ and continuous functions $v: I \rightarrow X_1$, $c: I \rightarrow \mathbb{R}$ and $\mu: I \rightarrow \Lambda$ obeying $v(0) = 0$, $c(0) = c_0$ and $\mu(0) = \mu_0$ such that $[\mu(\epsilon), c(\epsilon), v(\epsilon)]$ is a nontrivial solution of (*) for $\epsilon(\neq 0)$ in I.

PROOF. Write $X_1 = X \oplus N(L_0)$. Choose $\varphi(\neq 0)$ in $N(L_0)$ and look for solutions of (*) of the form

$$v(\epsilon) = \epsilon[\varphi + w(\epsilon)]$$

where $w(\epsilon) \in X$. Define

$$F(w, c, \mu, \epsilon) = L_0 w - (c - c_0)A(\varphi + w)$$

$$- (\mu - \mu_0)K_2(\varphi + w) - \epsilon^{-1}\mathcal{N}[\mu, \epsilon(\varphi + w)].$$

If $v(\epsilon)$ is a solution of (19) then $w(\epsilon)$ must be a zero of F for given c, μ. Here $F: X \times \mathbb{R} \times \Lambda \times \mathbb{R} \rightarrow X_1$ is a continuous map that is continuously Fréchet differentiable near $(0, c_0, \mu_0, 0)$. Moreover

$$F(0, c_0, \mu_0, 0) = 0$$

and the derivatives of F are

$$D_w F(0, c_0, \mu_0, 0) = L_0, \quad D_c F(0, c_0, \mu_0, 0) = -A\varphi,$$

$$D_\mu F(0, c_0, \mu_0, 0) = -K_2\varphi.$$

The theorem now follows by invoking an implicit function theorem. We shall use the formulation given in Fife.‡ The only requirement left to verify is his requirement that if $P: X_1 \to Y$ is the canonical projection, then $PD_\lambda F(0, c_0, \mu_0, 0)$ is $1 - 1$ and onto, where $\lambda = (c, \mu)$.

Using the above expressions for $D_c F$ and $D_\mu F$ one sees that this holds if and only if

$$\det \begin{pmatrix} <A\varphi, \varphi_1{}^*> & <K_2\varphi, \varphi_1{}^*> \\ <A\varphi, \varphi_2{}^*> & <K_2\varphi, \varphi_2{}^*> \end{pmatrix} \neq 0,$$

where the pairing $<,>$ is defined as in the last section and $\varphi_1{}^*$ and $\varphi_2{}^*$ span Y. This is condition (**19**) and the result follows.

COROLLARY 1. Suppose $\mathcal{G}: \Lambda \times X_1 \to X_1$ is k-times continuously Fréchet differentiable (resp. real analytic) in an open neighborhood of $(\mu_0, 0)$ and the other assumptions of Theorem 5 hold. Then the nontrivial solutions $[\mu(\epsilon), c(\epsilon), v(\epsilon)]$ are $(k - 1)$-times continuously Fréchet differentiable (resp. real analytic) on I.

This follows from the natural extension of Fife's result; see Crandall and Rabinowitz[5] for analogous theorems in the case of the Hopf bifurcation.

REMARKS

(1) The term φ may be any nonzero element of $N(L_0)$. From Corollary 1 to Theorem 4, however, one has that

$$w_1(y, y_n) = w_2(y, y_n + \pi/2).$$

In other words, the two linearly independent real eigenfunctions of L_0 differ only by a phase. Moreover, any real linear combination $\alpha w_1 + \beta w_2$ of w_1 and w_2 with $\alpha^2 + \beta^2 = 1$, differs only by a phase from w_1 (or w_2). Thus the choice of φ in the construction determines both a phase factor and a normalization of ϵ.

(2) We have not excluded $c_0 = 0$. Moreover, as mentioned in the beginning, time-invariant solutions correspond to solutions of (*) with $c = 0$. Thus, if the function c defined in the theorem is identically zero, the resulting solutions are, in fact, bifurcating time-invariant (or steady-state) solutions of the equations.

(3) The condition (**20**) guaranteeing bifurcation is a transversality condition and when it holds one may solve recursively for the power series expansions of the eigenvalues $\lambda(\mu, c)$ and eigenvectors $v(\mu, c)$ of $L(\mu, c)$ near (μ_0, c_0).

(4) The corollary justifies the use of power series expansions to approximate the bifurcating waves as was done by Auchmuty and Nicolis.[4]

(5) This result is a local result, giving the existence of such waves near the bifurcation point. It would be very interesting to prove some results on the global behavior of these families of waves.

STABILITY OF BIFURCATING WAVES

The bifurcating waves constructed in the last section are special solutions of our parabolic system. In practice, they will only be observed if they are stable solutions of the initial value problem for the system.

‡See Reference 7, theorem 1.

Here we shall make a few comments on the linear stability analysis of these waves. Full details will be published elsewhere. For a discussion of the validity of linear stability analyses for parabolic equations see Henry.[9]

The solution $u \equiv 0$ of (1)–(4) is said to be linearly stable provided that the eigenvalues λ of

$$S_i(\mu)w = L_i w_i + \sum_{l=1}^{m} f_{il}(y,\mu)w_l = \lambda w_i \quad \text{in } \Omega, \quad 1 \le i \le m \tag{21}$$

subject to (3)–(4); all obey $\text{Re } \lambda < 0$. Let the set of all such eigenvalues be denoted $\Sigma(\mu)$. Then by using Fourier expansions of w in terms of y_n as in Equation 13, one sees that

$$\Sigma(\mu) = \bigcup_{k=0}^{\infty} \sigma[M(k,\mu)].$$

The zero solution is said to be linearly unstable if there is a λ in $\Sigma(\mu)$ with $\text{Re } \lambda > 0$.

As μ varies along an interval in Λ, the eigenvalues of $M(k,\mu)$ vary continuously with μ (see remark after Theorem 3). Suppose the zero solution is stable when $\mu < \hat{\mu}$ and is unstable when $\mu > \hat{\mu}$, and μ is close enough to $\hat{\mu}$. Then for some value (or values) of k, at least one eigenvalue of $M(k,\mu)$ must have crossed the imaginary axis as μ increased through $\hat{\mu}$. If this happened for $k = 0$, one does not necessarily obtain a bifurcating wave from the theorem of the last section, so we shall henceforth only consider the case $k \neq 0$.

Suppose that $i\nu$ is an eigenvalue of (geometric) multiplicity m of $M(k,\hat{\mu})$; then the next lemma shows that it is an eigenvalue of (geometric) multiplicity at least $2m$ for $S(\hat{\mu})$.

LEMMA 1. Suppose that $i\nu \in \sigma[M(k,\hat{\mu})]$ and that $w^{(1)}, w^{(2)}, \ldots, w^{(m)}$ are m linearly independent eigenfunctions of $M(k,\hat{\mu})$ corresponding to the eigenvalue $i\nu$. Then $\{w^{(j)}e^{\pm iky_n}: 1 \le j \le m\}$ are $2m$ linearly independent eigenfunctions of $S(\hat{\mu})$ corresponding to the eigenvalue $i\nu$.

PROOF. One has $S_i(\hat{\mu})(we^{\pm iky_n}) = e^{\pm iky_n}M_i(k,\hat{\mu})w = i\nu we^{\pm iky_n}$ if w is an eigenfunction corresponding to the eigenvalue $i\nu$. Hence each $w^{(j)}e^{\pm iky_n}$ is an eigenfunction of $S(\hat{\mu})$ corresponding to $i\nu$. Their linear independence is easily verified.

This shows that these bifurcating waves can only arise at degenerate pairs of complex conjugate eigenvalues of the linear stability problem and causes the complications in treating the stability of these bifurcating waves. Nevertheless, Floquet theory as described by Joseph and Sattinger[10] or Crandall and Rabinowitz[5] may be used to analyze the stability of the waves. Moreover, the numerical simulations of Erneux and Herschkowitz-Kaufman[6] indicate that such solutions do appear in particular systems and, consequently, waves of this type appear to be stable solutions for some equations in some parameter ranges.

REFERENCES

1. AGMON, S. 1966. Lectures on Elliptic Boundary Value Problems. Van Nostrand. New York.

2. ARIS, R. 1975. The Mathematical Theory of Reaction and Diffusion in Porous Catalysts. Oxford University Press. Oxford.
3. AUCHMUTY, J. F. G. In press. Qualitative effects of diffusion in chemical systems. *In* Lectures on Mathematics in the Life Sciences. Amer. Math. Soc. Providence.
4. AUCHMUTY, J. F. G. & G. NICOLIS. 1976. Bifurcation analysis of reaction-diffusion equations III. Chemical oscillations. Bull. Math. Bio. **38:** 325–350.
5. CRANDALL, M. G. & P. H. RABINOWITZ. 1976. The Hopf bifurcation theorem. MRC Report 1604. Univ. of Wisconsin.
6. ERNEUX, T. & M. HERSCHKOWITZ-KAUFMAN. 1977. Rotating waves as asymptotic solutions of a model chemical reaction. J. Chem. Phys. **66:** 248–253.
7. FIFE, P. C. 1974. Branching phenomena in fluid dynamics and chemical reaction diffusion theory. *In* Eigenvalues of Non-linear Problems.: 25–83. CIME. Cremonese. Rome.
8. FIFE, P. C. 1978. Asymptotic states for equations of reaction and diffusion. Bull. Amer. Math. Soc. **84:** 693–726.
9. HENRY, D. 1976. Geometric Theory of Semilinear Parabolic Equations. Chap. 9. Lecture Notes. Univ. of Kentucky.
10. JOSEPH, D. D. & D. H. SATTINGER. 1972. Bifurcating time-periodic solutions and their stability. Arch. Rat. Mech. Anal. **45:** 79–109.
11. KATO, T. 1966. Perturbation Theory for Linear Operators. Sect. IV.3. Springer-Verlag. New York.
12. KOPELL, N. & L. N. HOWARD. 1973. Plane wave solutions to reaction-diffusion equations. Stud Appl. Math. **52:** 291–328.
13. NICOLIS, G. & I. PRIGOGINE. 1977. Self-Organization in Nonequilibrium Systems. John Wiley. New York.
14. ORTOLEVA, P. & J. ROSS. 1974. On a variety of wave phenomena in chemical reactions. J. Chem. Phys. **60:** 5090–5107.
15. TURING, A. 1952. The chemical basis of morphogenesis. Phil. Trans. R. Soc. **B237:** 37–72.
16. TYSON, J. J. 1976. The Belousov-Zhabotinskii reaction. *In* Lecture Notes in Biomathematics. Vol. 10. Springer-Verlag. New York.
17. MORREY, C. B., JR. 1966. Multiple Integrals in the Calculus of Variations. Springer-Verlag. Berlin.

OSCILLATIONS, BISTABILITY, AND ECHO WAVES IN MODELS OF THE BELOUSOV-ZHABOTINSKII REACTION*

John J. Tyson†

Institut für Biochemie und Experimentelle Krebsforschung
Universität Innsbruck
A-6020 Innsbruck, Austria

INTRODUCTION

Under appropriate conditions the cerium catalyzed oxidation of malonic acid by bromate in acid medium exhibits a numer of interesting phenomena, including sustained periodic oscillations,[1,2] the coexistence of alternative stable steady states,[3,4] nonperiodic ("chaotic") oscillations,[5,6] and spatiotemporal waves of chemical activity.[7-9] Some insight into the mechanism that generates such diverse behavior can be gained by examining simple models of this reaction. Because of the wide separation of time scales, which occurs naturally in this problem, it is possible to derive analytic expressions in terms of system parameters for characteristic properties of models, such as the waveform and period of oscillations, the domain of existence of multiple steady states, and the conditions for periodic "echo" waves in an excitable medium.

SOME MECHANISTIC DETAILS

The chemical mechanism of the Belousov-Zhabotinskii (BZ) reaction has received much attention by a number of investigators.[2,10-18] The principle overall reaction is,[15]

$$2BrO_3^- + 3CH_2(COOH)_2 + 2H^+ \rightarrow 2BrCH(COOH)_2 + 3CO_2 + 4H_2O \quad (1)$$

This reaction can be divided into three major processes[2,14,19]:

(A) The reduction of BrO_3^- to Br_2 by a series of oxygen atom transfers, and the subsequent bromination of malonic acid by Br_2.

(B) The reduction of BrO_3^- to Br_2 by an autocatalytic process involving one-electron transfers among free radical oxybromine intermediates. The electron is supplied by Ce^{+3}. As in (A), Br_2 adds to malonic acid.

(C) The oxidation of organic and bromo-organic compounds by Ce^{+4}. This process regenerates Ce^{+3} and Br^-.

The mechanism of (A) is well known[14,20]; namely,

(R3) $\qquad\qquad BrO_3^- + Br^- + 2H^+ \rightarrow HBrO_2 + HOBr$

*Supported by a National Institutes of Health Followship (5 F32 CA05152-02) from the National Cancer Institute.

†Present address: Department of Biology, Virginia Polytechnic Institute and State University, Blacksburg, Virginia 24061.

0077-8923/78/0316-0279 $1.75/1 © 1979, NYAS

(R2) $HBrO_2 + Br^- + H^+ \rightarrow 2HOBr$

(R1) $HOBr + Br^- + H^+ \rightarrow Br_2 + H_2O$

The bromination of malonic acid,

(R8) $Br_2 + CH_2(COOH)_2 \rightarrow BrCH(COOH)_2 + Br^- + H^+$

is rate limited by enolization of the acid.[21] Bromine adds rapidly to the C=C double bond. The mechanism of **(B)** is currently under debate.[17, 18, 22] It seems clear, however, that $HBrO_2$ is produced autocatalytically; that is,

(G) $2Ce^{+3} + BrO_3^- + HBrO_2 + 3H^+ \rightarrow 2Ce^{+4} + 2HBrO_2 + H_2O$

and its accumulation is eventually limited by the disproportionation reaction,

(R4) $2HBrO_2 \rightarrow BrO_3^- + HOBr + H^+$

The oxidation of organic compounds by Ce^{+4} is a complicated reaction. Kasperek and Bruice[12] have studied the kinetics of the two overall reactions,

(R9) $CH_2(COOH)_2 + 6Ce^{+4} + 2H_2O \rightarrow$

$$6Ce^{+3} + HCOOH + 2CO_2 + 6H^+$$

(R10) $BrCH(COOH)_2 + 4Ce^{+4} + 2H_2O \rightarrow$

$$4Ce^{+3} + Br^- + HCOOH + 2CO_2 + 5H^+$$

Noyes and Jwo[23] have estimated that the number of bromide ions produced per ceric ion consumed on oxidation of mixtures of organic and bromo-organic compounds is,

$$h = -\frac{d[Br^-]}{d[Ce^{+4}]} = 0.5 - 1.0 \qquad (2)$$

Rate laws for the reactions discussed here are collected in TABLE 1. More complete reviews of the mechanism of the BZ reaction are available.[24,25]

OSCILLATIONS

Field and Noyes have suggested that the mechanism just outlined can be reduced to five essential steps.[19] Let $A = BrO_3^-$, $B = BrMA$, $P = HOBr$, $X = HBrO_2$, $Y = Br^-$, $Z = Ce^{+4}$. Their model is:

(M1) $A + Y \rightarrow X + P$ $k_1 = k_{R3}[H^+]^2$

(M2) $X + Y \rightarrow 2P$ $k_2 = k_{R2}[H^+]$

(M3) $A + X \rightarrow 2X + 2Z$ $k_3 = k_{R5}[H^+]$

(M4) $2X \rightarrow A + P$ $k_4 = k_{R4}$

(M5) $B + Z \rightarrow hY$ $k_5 = k_{R10}/(K_{R10} + [B])$

Step **(M1)** is the rate-limiting step of process **(A)**. If $[Ce^{+3}] \gg [Ce^{+4}]$, then

TABLE 1.

RATE LAWS FOR VARIOUS STEPS IN THE BZ REACTION.[14]

Reaction	Initial Rate	Rate Constants
R1	$-\dfrac{d[Br^-]}{dt} = k_{R1}[HOBr][Br^-][H^+]$	$k_{R1} = 8 \times 10^9\ M^{-2} s^{-1}$
R2	$-\dfrac{d[Br^-]}{dt} = k_{R2}[HBrO_2][Br^-][H^+]$	$k_{R2} = 2 \times 10^9\ M^{-2} s^{-1}$
R3	$-\dfrac{d[Br^-]}{dt} = k_{R3}[BrO_3^-][Br^-][H^+]^2$	$k_{R3} = 2\ M^{-3} s^{-1}$
R4	$-\dfrac{1}{2}\dfrac{d[HBrO_2]}{dt} = k_{R4}[HBrO_2]^2$	$k_{R4} = 4 \times 10^4\ M^{-2} s^{-1}$
G	$+\dfrac{d[HBrO_2]}{dt} = k_{R5}[HBrO_2][BrO_3^-][H^+]\,F([Ce^{+3}],[Ce^{+4}])$	$k_{R5} = 10^4\ M^{-2} s^{-1}$ $F(\cdot,\cdot)$ = some function
R8	$+\dfrac{d[Br^-]}{dt} = k_{R8}[H^+][CH_2(COOH)_2]$	$k_{R8} = 10^{-2}\ M^{-1} s^{-1}$
R9	$-\dfrac{d[Ce^{+4}]}{dt} = \dfrac{k_{R9}[Ce^{+4}][CH_2(COOH)_2]}{K_{R9} + [CH_2(COOH)_2]}$	$k_{R9} = 0.53\ s^{-1}$ $K_{R9} = 0.53\ M^{-1}$
R10	$-\dfrac{d[Ce^{+4}]}{dt} = \dfrac{k_{R10}[Ce^{+4}][BrCH(COOH)_2]}{K_{R10} + [BrCH(COOH)_2]}$	$K_{R10} = 7 \times 10^{-2}\ s^{-1}$ $K_{R10} = 0.2\ M$

$F = 1$ (see TABLE 1) and (M3) represents the rate law and stoichiometry of the autocatalytic step (G). Step (M2) ensures that the one-electron and two-electron pathways are mutually exclusive: if [Br⁻] is large, then [HBrO₂] is small and vice versa. Step (M4) limits the autocatalytic production of HBrO₂ and (M5) represents the regeneration of Br⁻ by Ce⁺⁴ oxidation of organic compounds. Notice that, since the reactions are considered irreversible, HOBr has no effect on the dynamics.[26]

Let us suppose that $A = $ [BrO₃⁻] and $B = $ [BrMA] are time independent, then (M1)–(M5) imply that,

$$\frac{dX}{dT} = k_1 AY - k_2 XY + k_3 AX - 2k_4 X^2$$

$$\frac{dY}{dT} = -k_1 AY - k_2 XY + hk_5 BZ$$

$$\frac{dZ}{dT} = 2k_3 AX - k_5 BZ \tag{3}$$

where X, Y, Z are concentrations in moles liter⁻¹ and T is time in seconds. These equations can be simplified considerably by introducing dimensionless variables x, y, z, t, defined by,

$$x = \alpha X, \quad y = \beta Y, \quad z = \gamma Z, \quad t = \delta T \tag{4}$$

The kinetic equations in dimensionless form are,

$$\epsilon_1 \frac{dx}{dt} = \epsilon_1 \frac{\alpha}{\delta} \left(k_1 A \frac{y}{\beta} - k_2 \frac{xy}{\alpha\beta} + k_3 A \frac{x}{\alpha} - 2k_4 \frac{x^2}{\alpha^2} \right)$$

$$\epsilon_2 \frac{dy}{dt} = \epsilon_2 \frac{\beta}{\delta} \left(-k_1 A \frac{y}{\beta} - k_2 \frac{xy}{\alpha\beta} + hk_5 B \frac{z}{\gamma} \right)$$

$$\epsilon_3 \frac{dz}{dt} = \epsilon_3 \frac{\gamma}{\delta} \left(2k_3 A \frac{x}{\alpha} - k_5 B \frac{z}{\gamma} \right) \tag{5}$$

where the time-scale factor ϵ_i is introduced to ensure that the right-hand side of the ith equation is $O(1)$ whenever the dependent variables x, y, z, are $O(1)$. Notice that the ith variable changes appreciably only on a time scale of order ϵ_i.

It is not commonly appreciated that, depending on our choice of $\alpha, \beta, \gamma, \delta$, the time-scale factors may take on drastically different relations with respect to each other. For instance, Field and Noyes[19] introduced a scaling convention for which $\epsilon_1 \ll \epsilon_3 \ll \epsilon_2$. I introduced a different convention for which $\epsilon_1 \ll \epsilon_2 \ll \epsilon_3$, and argued that this convention led to much simpler portraits of trajectories in phase space.[26] Now I would like to introduce a third convention for which $\epsilon_2 \ll \epsilon_1 \ll \epsilon_3$, and show that the phase portraits are simpler still. Let,

$$\alpha = \frac{2k_4}{k_3 A} \simeq 10^6 \text{M}^{-1}, \; \beta = \frac{k_2}{k_3 A} \simeq 2 \times 10^7 \text{ M}^{-1},$$

$$\gamma = \frac{k_4 k_5 B}{(k_3 A)^2} \simeq 20 \text{ M}^{-1}, \; \delta = k_5 B \simeq 4 \times 10^{-3} \text{ sec}^{-1} \tag{6}$$

where we have used $A = B = 10^{-2}$ M in the order of magnitude estimates. The kinetic equations become,

$$\epsilon \, \frac{dx}{dt} = \mu y - xy + x - x^2$$

$$\epsilon' \, \frac{dy}{dt} = -\mu y - xy + fz$$

$$\frac{dz}{dt} = x - z \qquad (7)$$

where, $\epsilon = \dfrac{k_5 B}{k_3 A} \cong 4 \times 10^{-5}, \; \epsilon' = \epsilon \dfrac{\alpha}{\beta} \cong 2 \times 10^{-6}$

$$\mu = \frac{2k_1 k_4}{k_2 k_3} \cong 10^{-5}, \; f = 2h \cong 1 \qquad (8)$$

In this case, $\epsilon' < \epsilon \ll 1$, and it is reasonable to take $[\mathrm{Br}^-]$ as the fast variable rather than $[\mathrm{HBrO_2}]$. This leads to considerable simplification of analysis because now the slow manifold is defined by,

$$\epsilon' \to 0, \; y = \frac{fz}{x + \mu} \qquad (9)$$

rather than by the clumsy quadratic equation for $x = x(y)$ obtained by letting $\epsilon \to 0$. As $\epsilon' \to 0$, system (7) reduces to the planar system,

$$\epsilon \, \frac{dx}{dt} = x(1 - x) - fz \cdot \frac{x - \mu}{x + \mu}$$

$$\frac{dz}{dt} = x - z \qquad (10)$$

Equations similar to (10) were suggested several years ago by Rössler[34] as a model of the BZ reaction.

Now, since $\epsilon \ll 1$, semiquantitative properties of solutions of (10) can be derived from the two nullclines,

$$\frac{dx}{dt} = 0, \; fz = \frac{x(1 - x)(x - \mu)}{(x + \mu)} \qquad (11)$$

$$\frac{dz}{dt} = 0, \; z = x \qquad (12)$$

These elementary functions are plotted in FIGURE 1. From Equations 10–12 we can derive the information collected in TABLE 2 by the methods used in Reference 26.

These expressions are especially useful in comparing the model with the experimental data reported by Zhabotinskii *et al.* (bromomalonic acid is used instead of malonic acid as a major reactant, which is very convenient from a theo-

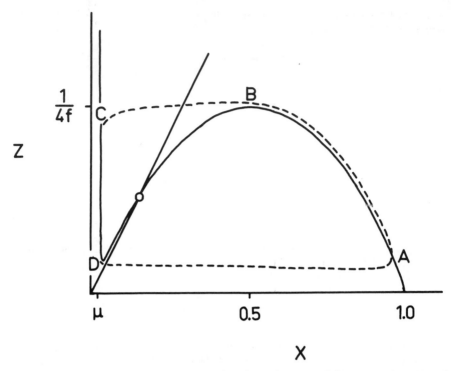

FIGURE 1. Phase plane for Equation 10. Plotted are the two nullclines, Equations 11 and 12, for $\frac{1}{2} < f < 1 + \sqrt{2}$, and a periodic solution (dashed line). Coordinates of the points $A–D$ are:

$$A(x_{max}, z_{min}), \; B(x_{jump \atop down}, z_{max}),$$

$$C(x_{min}, z_{max}), \; D(x_{jump \atop up}, z_{min})$$

where x_{max}, etc. are given in TABLE 2.

retical point of view, since (R9) can be ignored).[28] According to Equations 6 and 8, and the rate constants in TABLE 1,

$$\mu \simeq 10^{-5}, \; k_5 \simeq \frac{0.07 \text{ sec}^{-1}}{0.2 \text{ M} + B}, \; \gamma \simeq (0.2 \text{ M})\frac{B}{A^2} \tag{13}$$

Assuming $f = 1$, we have

$$\log\left(\frac{[Br^-]_{max}}{[Br^-]_{jump \atop down}}\right) \simeq \log\left(\frac{[Ce^{+4}]_{max}}{[Ce^{+4}]_{min}}\right) \simeq -\log(20\mu) \simeq 4 \tag{14}$$

$$\text{Period} \simeq \frac{-\ln(20\mu)}{k_5 B} \simeq \frac{0.2 \text{ M} + B}{B} \; 100 \text{ sec} \tag{15}$$

Neither of these estimates are particularly good. Measurements of Field et al., with a bromide selective electrode, indicate that the maximum bromide ion concentration is only a factor of 10 greater than the switching concentration.[14] Zhabotinskii et al. observed $[Ce^{+4}]$ oscillations of much smaller amplitude (factor of 2–3) and period (10–100 sec for B = 0.1–0.01 M).[28] Thus, though the Oregonator provides a simple qualitative account of the BZ oscillations, its amplitude is several orders of magnitude too large, and its period is too long by a factor of 30. Perhaps it is too much to expect such a drastically simplified model to be quantitatively accurate, but closer examination of its difficulties may suggest ways to improve the model. First, notice that,

$$[Br^-]_{\substack{jump \\ down}} \cong \frac{1.7}{\beta} \cong 10^{-5}[BrO_3^-] \tag{16}$$

is essentially correct. Indeed, development of the model keyed exactly on this point. Thus, since the Oregonator predicts $[Br^-]_{max}$ several orders of magnitude too large, it must be neglecting certain steps which limit bromide ion accumulation when process (A) is predominant. A likely candidate is HOBr, which consumes Br^- according to R1. Secondly,

$$[Ce^{+4}]_{max} \cong \frac{1}{4f\gamma} \cong 10^{-2}\ M \tag{17}$$

cannot be correct, since the total cerium concentration is usually about 10^{-3} M. This is easily corrected by replacing F = 1 with $F \propto [Ce^{+3}]$; that is, the term k_3AX in Equation 3 is replaced by,[10]

$$k_3'AX(C - Z), \quad C = [Ce]_{total} \tag{18}$$

TABLE 2

PROPERTIES OF PERIODIC SOLUTIONS OF EQUATION 7
FOR $1/2 < f < 1 + \sqrt{2}$, $\epsilon' \ll \mu \ll 1$, AND $\epsilon \ll 1$*

$x_{max} = 1 - 6\mu$	$x_{min} = \mu(1 + 8\mu)$
$x_{\substack{jump \\ down}} = 0.5$	$x_{\substack{jump \\ up}} = (1 + \sqrt{2})\mu$
$y_{max} = \dfrac{1}{8\mu}$	$y_{min} = 6\mu$
$y_{\substack{jump \\ down}} = \dfrac{2 + \sqrt{2}}{2} \cong 1.7$	$y_{\substack{jump \\ up}} = 0.5$
$z_{max} = \dfrac{1}{4f}$	$z_{min} = \dfrac{(1 + \sqrt{2})^2}{f}\mu \cong \dfrac{5.8\mu}{f}$
$T_{AB} = \dfrac{1}{f - 1} \ln\left[\dfrac{1}{2}\left(\dfrac{2f}{2f - 1}\right)^{2f-1}\right]$, $T_{AB} \rightarrow 2 \ln 2$ as $f \rightarrow \dfrac{1}{2}$	
$T_{CD} = -\ln[4(5.8 - f)\mu]$	

*Compare with FIGURE 1.

This ensures that $[Ce^{+4}]_{max} < [Ce]_{total}$. Taking this precaution we see that the Oregonator predicts $[Ce^{+4}]_{min}$ several orders of magnitude too low. The period discrepancy is related to this difficulty because,

$$\text{Period} \simeq \frac{-1}{k_5 B} \int_{[Ce^{+4}]_{max}}^{[Ce^{+4}]_{min}} \frac{d[Ce^{+4}]}{[Ce^{+4}]}$$

$$= T_{1/2} \log_2 \left(\frac{[Ce^{+4}]_{max}}{[Ce^{+4}]_{min}} \right) \tag{19}$$

where $T_{1/2} = 0.7/k_5 B$ is the half-life for Ce^{+4} decay. According to Equation 14, Period $\simeq 13 T_{1/2}$. However, if the amplitude of the ceric oscillation is only a factor of 2, then Period $= T_{1/2}$, which is much more reasonable.

In terms of kinetics, the Oregonator requires that $[Ce^{+4}]$ should become very small when process (A) predominates in order to shut off Br^- production (M5), so that $[Br^-]$ can drop to $[Br^-]_{jump\ down}$. If, as we have already indicated, there are important reactions, neglected by the Oregonator, which consume Br^- during (A), then it would not be necessary for $[Ce^{+4}]_{min}$ to be so small. Thus, it seems that all the quantitative difficulties of the Oregonator can be traced to some inaccuracy in its description of (A). Can this be remedied while still retaining the simplicity of only a few essential time-dependent intermediates?

Finally, notice that, if we are willing to choose μ, k_5, and γ "semiempirically," then the Oregonator can fit the observations of Zhabotinskii et al.[28] with,

$$\mu = (2.5 \times 10^{-3}) \sqrt{\frac{A}{B}}, \ k_5 = \frac{0.15 \text{ sec}^{-1}}{0.2 \text{ M} + B}, \ \gamma = (500 \text{ M}^{-1}) \sqrt{\frac{B}{A}} \tag{20}$$

Thus, since the parameters in the model suggested by Zhabotinskii et al. are also chosen semiempirically, there is no reason, on the basis of curve fitting, to prefer their model over the Oregonator, which seems to be a more realistic portrayal of what is known about the mechanism of the BZ reaction.

BISTABILITY

Multiple steady states have been observed in the complete BZ reaction,[3,4] but more convenient from an experimental and theoretical point of view is the appearance of bistability in the bromate-bromide-cerium (III) reaction, recently reported by Geiseler and Föllner.[29] They discuss their observations in terms of the Oregonator equations (3) with $B = 0$ and the additional terms representing material flux in a continuous flow, stirred-tank reactor (in this section only, $Z = [Ce^{+3}]$).

$$\frac{dX}{dT} = k_1 A Y - k_2 X Y + k_3 A X - 2 k_4 X^2 - k_R X$$

$$\frac{dY}{dT} = -k_1 A Y - k_2 X Y + k_R (Y^0 - Y)$$

$$\frac{dZ}{dT} = -2 k_3 A X + k_R (Z^0 - Z) \tag{21}$$

where $k_R = T_R^{-1}$, T_R = residence time of the reactor, and Y^0, Z^0 are concentrations of Br^{-1} and Ce^{+3} in the feed stream. Since the equation for Z can be integrated once $X(T)$ is known, we are dealing with a two component system:

$$\frac{dx}{dt} = \mu y - xy + x - x^2 - kx$$

$$\epsilon \frac{dy}{dt} = -\mu y - xy + \epsilon k(y^0 - y) \tag{22}$$

where

$$x = \frac{2k_4}{k_3 A} X, \quad y = \frac{k_2}{k_3 A} Y, \quad t = k_3 A T \tag{23}$$

$$\mu = \frac{2k_1 k_4}{k_2 k_3}, \quad k = \frac{k_R}{k_3 A}, \quad \epsilon = \frac{2k_4}{k_2} \tag{24}$$

In the limit $|k_2| \to \infty$; i.e., $\epsilon \to 0$ and $\mu \to 0$

$$y \to \frac{\sigma}{x + n\mu}, \quad \text{where} \quad \sigma = \frac{2k_R k_4}{(k_3 A)^2} Y^0, \quad n = 1 + \frac{k_R}{k_1 A} \tag{25}$$

and we are left with a single first-order differential equation,

$$\frac{dx}{dt} = G(x) = x(1 - k - x) - \sigma \frac{x - \mu}{x + n\mu} \tag{26}$$

The function $G(x)$ is plotted in FIGURE 2. Obviously there are three possible steady states for $\sigma_{min} < \sigma < \sigma_{max}$, where σ_{min} and σ_{max} are defined by eliminating x between $G(x) = 0$ and $G'(x) = 0$. If $k \ll 1$, we find that,

$$\sigma_{min} \cong (n + 2 + 2\sqrt{n} + 1)\mu, \quad \sigma_{max} \cong \frac{1}{4} \tag{27}$$

In the experiments of Geiseler and Föllner, $k_R = 4.5 \times 10^{-3}$ sec^{-1} and $A = 2 \times 10^{-3}$ M.[29] Since $k_1 \cong 2$ M^{-1} sec^{-1}, $k_R/k_1 A \cong 1$ and $n \cong 2$. Thus the region of bistability is defined by,

$$7.5 < \frac{[Br^-]^0}{[Br^-]_{jump \atop down}} < \frac{1}{4\mu} \tag{28}$$

where $[Br^-]_{jump\ down} = 2 \times 10^{-8}$ M. Notice as well that in the region of bistable behavior, the two stable steady states are given by,

$$[Br^-] = \begin{cases} \mu[Br^-]^0 \\ \frac{1}{3} [Br^-]^0 \end{cases} \tag{29}$$

For $[Br^-]^0 = 10^{-5}$ M, Geiseler and Föllner have observed one steady state at $[Br^-] = 3 \times 10^{-6}$ M, in excellent agreement with (29). (They report another

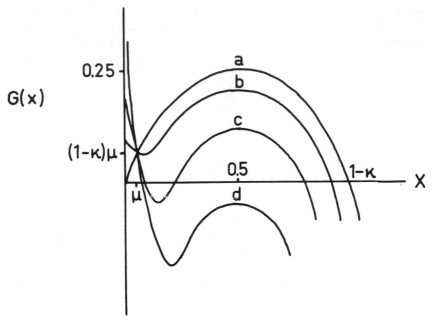

FIGURE 2. $G(x)$ for $k \ll 1$. a) $\sigma = 0$, b) $0 < \sigma < \sigma_{min}$, c) $\sigma_{min} < \sigma < \sigma_{max}$, d) $\sigma_{max} < \sigma < \infty$.

steady state at 2×10^{-7} M, but it is not clear that the bromide ion selective electrode can be trusted at such low concentrations.)

Measurement of the upper and lower limits of $[Br^-]^0$, for which bistability is observed, and of $[Br^-]$, at the two stable steady states, should yield a good estimate of the parameter μ, and thus help clarify quantitative features of the Oregonator.

ECHO WAVES

In an interesting paper Krinskii et al.[30] showed that two adjacent cells, neither of which alone is oscillatory, may excite each other so that a pulse of activity "echoes" back and forth between the cells.[31,32] Though they were primarily interested in waves of excitation in heart or neural tissue, they realized that their ideas applied equally well to chemical waves. In the BZ reaction we observe waves of echo-type at the center of spiral waves.[8,31-33]

From our discussion of the Oregonator it is easy to see that the qualitative arguments of Krinskii et al. apply. Consider two identical, well-stirred compartments separated by a barrier permeable to $HBrO_2$. Using Equation 10 to represent the kinetics within a compartment, we write,

$$\epsilon \frac{dx_1}{dt} = x_1(1 - x_1) - fz_1\left(\frac{x_1 - \mu}{x_1 + \mu}\right) + D(x_2 - x_1), \quad \frac{dz_1}{dt} = x_1 - z_1$$

$$\epsilon \frac{dx_2}{dt} = x_2(1 - x_2) - fz_2\left(\frac{x_2 - \mu}{x_2 + \mu}\right) + D(x_1 - x_2), \quad \frac{dz_2}{dt} = x_2 - z_2 \qquad (30)$$

Suppose that $f \leq 0.5$ so that either compartment in isolation ($D = 0$) is characterized by a stable steady state. Recall as well that, since $\epsilon \ll 1$, either $x = O(1)$ or $x = O(\mu)$. If $x_1 = O(1)$ and $x_2 = O(\mu)$, then

$$\epsilon \frac{dx_1}{dt} \cong x_1(1 - D - x_1) - fz_1 \qquad (31)$$

Thus the $dx_1/dt = 0$ nullcline is shifted down and to the left. If,

$$\frac{1 - D}{2} < f < \frac{1}{2} \qquad (32)$$

then (for ϵ sufficiently small) the nullcline intersects $x = z$ in an "unstable" position (cf. FIGURE 3). Meanwhile, the $dx_2/dt = 0$ nullcline is given by,

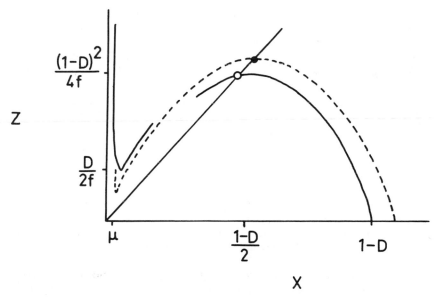

FIGURE 3. Nullclines defined by Equations 31 and 33. The dashed line is the (reference) nullcline defined by Equation 11.

$$fz_2 = \frac{x_2 + \mu}{x_2 - \mu} [x_2(1 - D - x_2) + Dx_1] \tag{33}$$

which has a local minimum at,

$$x_2{}^{min} \simeq \sqrt{\frac{2\mu Dx_1}{1 - D}}, \quad fz_2{}^{min} \simeq Dx_1 + \sqrt{2\mu(1 - D)Dx_1} \tag{34}$$

To get a pulse of excitation echoing between the two cells, we would like to have the situation illustrated in FIGURE 4. At certain times $t_i > 0$, the two cells are in the following positions:

$$\text{Cell } 1—1, 2, 3, 4, A, B, C, D, 1, \ldots$$

$$\text{Cell } 2—A, B, C, D, 1, 2, 3, 4, A, \ldots$$

To achieve this we require that (as $\epsilon \to 0$)

$$T_{23} = T_{BC}, \quad T_{D1} = T_{4A}, \quad z_2{}^{min} < z_C \tag{35}$$

From the analysis in the section on "oscillations" we find that

$$\ln\left(\frac{z_B}{z_C}\right) \simeq 2\ln 2, \quad \left(\frac{fz_B}{1 - D}\right)^2 \simeq 5.3\,\mu \tag{36}$$

Thus, $D < \sqrt{\mu} \ll 1$, and,

$$z_A \simeq z_B \simeq 4.6\sqrt{\mu}, \quad z_C \simeq z_D \simeq 1.15\sqrt{\mu}$$

$$T_{D1} = -\ln(9.2\sqrt{\mu}), \quad T_{BC} = \ln 4 = 1.4 \tag{37}$$

I have checked these results numerically, using $f = 0.505$, $D = 0.004$, $\mu = 0.001$, $\epsilon = 0.01$, $\epsilon' = 0.0001$. Even though f is slightly larger than 0.5, the homogeneous steady state (at $x_1 = x_2 = z_1 = z_2 \simeq 0.497$) is stable. However, starting from proper initial conditions, we find the stable periodic solution illustrated in FIGURE 4. It agrees well with the estimates in Equation 37. The periodic solution exists only for a narrow range of D ($10^{-3} < D < 10^{-2}$). If D is too small, then the steady state in FIGURE 3 near $x = (1 - D)/2$ is stable (cf. Equation 3). If D is too large, then $z_2{}^{min} > z_C$ and, as cell 1 goes from position 2 to 3, cell 2 goes from position B to 2 to 3.

What sort of bifurcation diagram accounts for the simultaneous existence of a stable homogeneous steady state and a stable echo wave? Two facts bear on this question:

1. There exist no inhomogeneous steady states for μ sufficiently small (see PROPOSITION 1 in the appendix).
2. The homogeneous steady state is asymptotically stable for all values of D, if $f < 0.5$, and μ is sufficiently small (see PROPOSITION 2 in the appendix).

Thus it seems that, as D increases at constant $f < 0.5$, echo solutions do not arise by bifurcation from a steady-state branch. I guess that the bifurcation diagram looks like FIGURE 5, but this remains an open question.

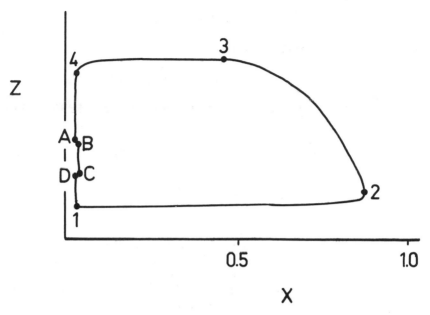

FIGURE 4. Echo wave. For $f = 0.505$, $D = 0.004$, $\mu = 0.001$, $\epsilon = 0.01$, $\epsilon' = 0.0001$. Co-ordinates of the indicated points are:

$A\,(0.0010, 0.16)$	$1\,(0.0095, 0.0095)$
$B\,(0.0011, 0.14)$	$2\,(0.96, 0.06)$
$C\,(0.0015, 0.030)$	$3\,(0.50, 0.50)$
$D\,(0.0012, 0.025)$	$4\,(0.0011, 0.46)$

Elapsed times are: $T_{AB} = 0.11$, $T_{BC} = 1.28$, $T_{CD} = 0.48$, $T_{D1} = 1.11$.

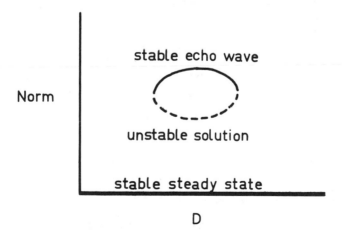

FIGURE 5. Conjectured bifurcation diagram.

Let us compare these results with Winfree's observations of spiral waves.[8] If an echo wave is a reasonable approximation of the core of a spiral wave, then we derive from Equations 13 and 36 that,

$$\text{Period} \simeq \frac{2}{k_5 B}(T_{BC} + T_{D1}) \simeq \frac{0.2\,M + B}{B} \quad (150 \text{ sec})$$

Since certainly $B \leq 0.2\,M$, this estimate is larger than the observed period (15 sec) by at least a factor of 20. Part of this discrepancy is undoubtedly due to quantitative difficulties of the Oregonator (the "kinetic" part of the model). However, if we use the parameter estimates in Equation 20, which fit rather well observed periods of oscillation, then,

$$\text{Period} \simeq \frac{0.2\,M + B}{B} \quad (35 \text{ sec})$$

which is still too large by a factor ≥ 5.

So far we have only considered the case $f \leq \frac{1}{2}$. However if $f \geq 1 + \sqrt{2}$, the steady state is also unstable, but now it is located at $x = O(\mu)$. In this case it is impossible to satisfy the requirement equivalent to $T_{23} = T_{BC}$ of Equation 35. Thus echo waves do not exist for $f \simeq 1 + \sqrt{2}$, even though the medium is still "excitable." Except for the difficulty of demonstrating the nonexistence of a phenomena, it should be possible to check this prediction experimentally.

APPENDIX

PROPOSITION 1. For μ sufficiently small, Equation 30 has no constant solutions for which $x_1 > 0$, $x_2 > 0$ and $x_1 \neq x_2$.

Proof: Let $\sigma = x_1 + x_2$ and $\delta = x_1 - x_2$. The conditions for an inhomogeneous ($\delta \neq 0$) steady-state solution of Equation 30 are,

$$(1 - 2D - \sigma) - f\frac{\sigma^2 - \delta^2 + 4\mu(\sigma - \mu)}{\sigma^2 - \delta^2 + 4\mu(\sigma + \mu)} = 0 \tag{A1}$$

$$\sigma - \left(\frac{\sigma^2 + \delta^2}{2}\right) - f\frac{\sigma(\sigma^2 - \delta^2) + 4\mu(\sigma^2 - \sigma\mu)}{\sigma^2 - \delta^2 + 4\mu(\sigma + \mu)} = 0 \tag{A2}$$

We seek a solution (an "admissable" inhomogeneous steady state) for which $\sigma > 0$ and $\delta^2 < \sigma^2$. Equation A1 implies that,

$$\delta^2 = (\sigma + 2\mu)^2 + \frac{8f\mu^2}{1 - f - 2D - \sigma} \tag{A3}$$

Combining (A2) and (A3) we obtain,

$$\sigma^3 - (2 - 2f - 3D - 2\mu)\sigma^2 + [(1 - f - 2D)(1 - f - D) - $$
$$\mu(3 - f - 6D - \mu)]\sigma + \mu(1 + f - 2D)(1 - f - 2D - \mu) = 0 \tag{A4}$$

For $\mu \rightarrow 0$, the roots of (A4) are,

$$\sigma_1 = -\frac{1 + f - 2D}{1 - f - D}\mu + O(\mu^2)$$

$$\sigma_2 = 1 - f - D + O(\mu)$$

$$\sigma_3 = 1 - f - 2D - \frac{2f\mu^2}{D(1 - f - 2D)} + O(\mu^3)$$

and the corresponding values of δ^2 are given by,

$$\delta_1^2 = \sigma_1^2 + \frac{4D\sigma_1\mu}{2D + f - 1} + O(\mu^3)$$

$$\delta_2^2 = \sigma_2^2 + 4\sigma_2\mu + O(\mu^2)$$

$$\delta_3^2 = \sigma_3^2 + 4D\sigma_3 + O(\mu)$$

In each case, if $\sigma_i > 0$, then $\delta_i^2 > \sigma_i^2$; i.e., there are no admissable inhomogeneous steady states for μ sufficiently small. $\square$

PROPOSITION 2. For $f \leq \frac{1}{2}$ and μ sufficiently small, Equation 30 has a steady state solution,

$$x_1 = x_2 = z_1 = z_2 = 1 - f + \frac{2f\mu}{1 - f} + O(\mu^2) \qquad (A5)$$

which is asymptotically stable for all values of D.

Proof: That (A5) satisfies (30) to terms of order μ^2 is an elementary calculation. Stability of the steady state is determined by the roots of the characteristic equation

$$0 = \begin{vmatrix} -\alpha - \lambda & -\beta & \gamma & 0 \\ 1 & -1 - \lambda & 0 & 0 \\ \gamma & 0 & -\alpha - \lambda & -\beta \\ 0 & 0 & 1 & -1 - \lambda \end{vmatrix} = \lambda^4 + a_1\lambda^3 + a_2\lambda^2 + a_3\lambda + a_4$$

where,

$$\alpha = \frac{1}{\epsilon}\left(1 - 2f + D + \frac{6f\mu}{1 - f}\right) > 0, \quad \beta = \frac{f}{\epsilon}\left(1 - \frac{2\mu}{1 - f}\right) > 0, \quad \gamma = \frac{D}{\epsilon} > 0$$

and

$$a_1 = 2(1 + \alpha), \quad a_2 = 2(\alpha + \beta) + (1 + \alpha)^2 + \gamma^2$$

$$a_3 = 2(1 + \alpha)(\alpha + \beta) + 2\gamma^2, \quad a_4 = (\alpha + \beta)^2 + \gamma^2$$

The Routh-Hurwitz criteria for stability reduce to determining the sign of,

$$\Delta_4 = (a_1a_2 - a_3)a_3 - a_1^2a_4 = [(1 + \alpha)^2(\alpha + \beta) + \alpha\gamma^2][(1 + \alpha)^2 + \gamma^2]$$

Since $\Delta_4 > 0$, the (homogeneous) steady state is asymptotically stable. $\square$

REFERENCES

1. BELOUSOV, B. P. 1958. *In* Collections of Abstracts on Radiation Medicine.: 145. Medgiz, Moscow.
2. ZHABOTINSKII, A. M. 1964. Periodic course of oxidation of malonic acid in solution (investigation of the kinetics of the reaction of Belousov). Biophys. 9: 329–335.
3. DE KEPPER, P., A. ROSSI & A. PACAULT. 1976. Etude expérimentale d'une réaction chimique périodique. Diagramme d'état de la réaction de Belousov-Zhabotinskii. C. R. Acad. Sci. Ser. C 283: 371–375.
4. GEISELER, W. 1974. Erregungsphysiologische phänomene an auslösbaren und periodischen chemischen reaktionen. Math.-Naturwiss. Fak., Rh.-Westf. Tech. Hochschule Aachen, F.R.G. Thesis.
5. SCHMITZ, R. A., K. R. GRAZIANI & J. L. HUDSON. 1977. Experimental evidence of chaotic states in the Belousov-Zhabotinskii reaction. J. Chem. Phys. 67: 3040–3044.
6. RÖSSLER, O. E. & K. WEGMANN. 1978. Chaos in the Zhabotinskii reaction. Nature (London) 271: 89 90
7. ZAIKIN, A. N. & A. M. ZHABOTINSKII. 1970. Concentration wave propagation in two-dimensional liquid-phase self-oscillating system. Nature (London) 225: 535–537.
8. WINFREE, A. T. 1972. Spiral waves of chemical activity. Science 175: 634–636.
9. WINFREE, A. T. 1973. Scroll-shaped waves of chemical activity in three dimensions. Science 181: 937–939.
10. VAVILIN, V. A. & A. M. ZHABOTINSKII. 1969. Autocatalytic oxidation of trivalent cerium by bromate ion. Kinet. Catal. 10: 65–69.
11. VAVILIN, V. A. & A. M. ZHABOTINSKII. 1969. Induced oxidation of tribromoacetic and dibromomalonic acids. Kinet. Catal. 10: 538–540.
12. KASPEREK, G. J. & T. C. BRUICE. 1971. Observations on an oscillating reaction. The reaction of potassium bromate, ceric sulfate, and a dicarboxylic acid. Inorg. Chem. 10: 382–386.
13. NOYES, R. M., R. J. FIELD & R. C. THOMPSON. 1971. Mechanism of reaction of bromine(V) with weak one-electron reducing agents. J. Am. Chem. Soc. 93: 7315.
14. FIELD, R. J., E. KÖRÖS & R. M. NOYES. 1972. Oscillations in chemical systems. II. Thorough analysis of temporal oscillations in the bromate-cerium-malonic acid system. J. Am. Chem. Soc. 94: 8649–8664.
15. BORNMANN, L., H. BUSSE & B. HESS. 1973. Oscillatory oxidation of malonic acid by bromate. 3. CO_2 and Br^- titration. Z. Naturforsch. 28c: 514–516.
16. JWO, J.-J. & R. M. NOYES. 1975. Oscillations in chemical systems. IX. Reactions of cerium(IV) with malonic acid and its derivatives. J. Am. Chem. Soc. 97: 5422–5431.
17. HERBO, C., G. SCHMITZ & M. VAN GLABBEKE. 1976. Réaction oscillante de Belousov. 1. Cinétique de la réaction bromate-céreux. Can. J. Chem. 54: 2628–2638.
18. BARKIN, S., M. BIXON, R. M. NOYES & K. BAR-ELI. 1977. The oxidation of cerous ions by bromate ions—comparison of experimental data with computer calculations. Int. J. Chem. Kinet. 9: 841–862.
19. FIELD, R. J. & R. M. NOYES. 1974. Oscillations in chemical systems. IV. Limit cycle behavior in a model of a real chemical system. J. Chem. Phys. 60: 1877–1884.
20. BRAY, W. C. & H. A. LIEBHAFSKY. 1935. The kinetic salt effect on the fourth order reaction $BrO_3^- + Br^- + 2H^+$. J. Am. Chem. Soc. 57: 51–56.
21. WEST, R. W. 1924. The action between bromine and malonic acid in aqueous solution. J. Chem. Soc. 125: 1277–1282.
22. NOYES, R. M. & K. BAR-ELI. 1977. A comparison of mechanisms for the oxidation of cerium(III) by acidic bromate. Can. J. Chem. 55: 3156–3160.
23. NOYES, R. M. & J.-J. JWO. 1975. Oscillations in chemical systems. X. Implications of cerium oxidation mechanisms for the Belousov-Zhabotinskii reaction. J. Am. Chem. Soc. 97: 5431–5433.
24. NOYES, R. M. & R. J. FIELD. 1977. Mechanisms of chemical oscillators: experimental examples. Acc. Chem. Res. 10: 273–279.
25. TYSON, J. J. 1976. The Belousov-Zhabotinskii reaction. Lect. Notes Biomath. Vol. 10. Springer-Verlag, Berlin.

26. FIELD, R. J. 1975. Limit cycle oscillations in the reversible Oregonator. J. Chem. Phys. **63**: 2289–2296.
27. TYSON, J. J. 1977. Analytic representation of oscillations, excitability, and traveling waves in a realistic model of the Belousov-Zhabotinskii reaction. J. Chem. Phys. **66**: 905–915.
28. ZHABOTINSKII, A. M., A. N. ZAIKIN, M. D. LORZUKHIN & G. P. KREITSER. 1971. Mathematical model of a self-oscillating chemical reaction (oxidation of bromomalonic acid with bromate, catalyzed by cerium ions). Kinet. Catal. **12**: 516–521.
29. GEISELER, W. & H. H. FÖLLNER. 1977. Three steady state situation in an open chemical reaction system. Biophys. Chem. **6**: 107–115.
30. KRINSKII, V. I., A. M. PERTSOV & A. N. RESHETILOV. 1972. Investigation of one mechanism of origin of the ectopic focus of excitation in modified Hodgkin-Huxley equations. Biophys. **17**: 282–289.
31. BALAKHOVSKII, I. S. 1965. Several modes of excitation movement in ideal excitable tissue. Biophys. **10**: 1175–1179.
32. GUL'KO, F. & A. A. PETROV. 1972. Mechanism of formation of closed pathways of conduction in excitable media. Biophys. **17**: 271–281.
33. WINFREE, A. T. 1974. Rotating chemical reactions. Sci. Am. **230**: 82–95.
34. RÖSSLER, O. E. 1972. A principle for chemical multivibration. J. Theor. Biol. **36**: 413–417.

THE BIFURCATION DIAGRAM OF MODEL
CHEMICAL REACTIONS

M. Herschkowitz-Kaufman and T. Erneux

Faculté des Sciences de l'Université Libre de Bruxelles
Campus Plaine
Bd du Triomphe,
1050 Brussels, Belgium

INTRODUCTION

In open systems, diffusion and reactions can interact to produce a wide variety of interesting behaviors including, for example, spatially ordered steady states, temporal oscillations, standing or traveling concentration waves. The onset of these organized patterns is generally related to *bifurcation phenomena*. When some physicochemical parameter is changed, the "most disordered" state looses its stability and a new, qualitatively different solution will emerge from the original state. This new solution, in turn, can undergo a stability change, and in this way a succession of transitions can lead to a spontaneous complexification of the state of the system.

It is at present widely accepted that the mechanisms responsible for order in reaction-diffusion systems may also play a basic role in understanding several aspects of biological order.[1,2] It is clear, indeed, that nonlinear kinetics and matter (and/or energy) fluxes are present at various levels in biological systems. Biochemical oscillations[3a,3b] and embryonic development[4] have been successfully modeled using the ideas and techniques of bifurcation theory. The purpose of this paper is to illustrate some general mechanisms and principles underlying pattern formation and rhythmic phenomena in systems of reacting and diffusing substances.

We assume that, to an adequate approximation, the systems are isothermal, there is no convective transport, and external fields are of no consequence. In vector form, the equations of change of the active species within such a system can be written,

$$\frac{\partial X}{\partial t} = F(X) - \nabla J \qquad (1.1)$$

where X is the N-component vector of concentrations, $F(X)$ is the vector of production rates, and J is the vector of diffusive fluxes. We take,

$$J = - D \nabla X \qquad (1.2)$$

with D a diagonal matrix of constant diffusion coefficients. Equation (1.1) is supplemented with appropriate boundary conditions, as the systems we shall deal with are bounded in space. This is an important point in view in modeling experimental and biochemical situations where this factor may actually have a real influence. The results presented here are thus quite different from those described in works dealing with unbounded media or with transient regimes.[5]

296

0077–8923/78/0316–0296 $1.75/1 © 1979, NYAS

We focus principally on the global, long-time solutions available in reaction-diffusion systems. We are interested in the variety and multiplicity of asymptotic solutions. And we want to know how these solutions and, in general, the bifurcation diagram depend on the various parameters or on small disturbances in the system, how sensitive they are to global features of the system, such as size, geometric form, and conditions at the limits. In short, we would like, at least for a model chemical reaction, to construct as far as possible a complete bifurcation diagram. In addition we expect the knowledge of asymptotic states will also provide insight into various possible evolutionary paths corresponding to different initial conditions.

Different situations are illustrated with the trimolecular scheme which involves only two intermediate species, and corresponds to the reaction sequence:[6]

$$A \rightarrow X$$

$$B + X \rightarrow Y + D$$

$$2X + Y \rightarrow 3X$$

$$X \rightarrow E \tag{1.3}$$

This nonlinear model presents a great richness of dissipative structures, and provides detailed insight into various phenomena predicted by the general theory within the framework of relatively simple equations.

The outline of the paper is as follows. The next section is devoted to steady-state spatial patterns. We shall consider the stabilization of the successive primary solutions by a mechanism of secondary bifurcations. This approach is of interest in connection with the observation of the sequential character of certain processes in developmental biology.[7,8] A discrete succession of *stable* patterns could indeed account for the reproducibility of these phenomena. The third section deals with simple spatiotemporal regimes in 1- and 2-dimensional media. The fourth section describes the interactions between time-periodic and time-independent solution branches. However a complete stability analysis of all the small-amplitude solutions available is not presented, but some typical numerical examples of stable concentration patterns are given in each section.

STEADY-STATE SPATIAL PATTERNS IN ONE DIMENSION

In a bounded system there is a discrete set of distinct spatial patterns arising from the homogeneous steady state at discrete values of the bifurcation parameter (FIGURE 1). Bifurcation theory insures the stability of the first bifurcating solution when it is supercritical. On the contrary, the following branches are unstable when they emerge. However, as we move away from the first bifurcation point, computer simulations with the trimolecular scheme reveal the existence of a multiplicity of stable patterns (FIGURE 2). The situation can be understood by a mechanism of stabilization of the successive primary branches at some distance of their primary bifurcation point. Analytical calculations of secondary bifurcation points were performed by Mahar and Matkowski,[9a] and Keener[9b] for the two first consecutive branches. They apply in the limit, where the distance $\delta = |\lambda_m - \lambda_{m'}|$,

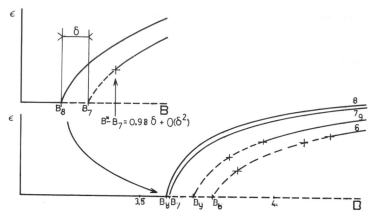

FIGURE 1. Bifurcation diagram for the first four primary steady-state branches when $A = 2$, $D_1 = 0.0016$, $D_2 = 0.008$, and $l = 1$; $\mathcal{E}$ is a characteristic function of amplitude of the different solutions ($\mathcal{E} = 0$ corresponds to the homogeneous steady state X_{st}). The secondary bifurcation points were evaluated numerically, but the secondary branches have not been drawn. Full and broken lines denote stable and unstable solutions, respectively.

$m' = m \pm 1$ between the two bifurcation points, and the critical amplitude ϵ^* (where stability change occurs) is small. We use the same methods, relating the occurrence of secondary points to splitting multiple primary bifurcation points,[10] to determine the stabilization of *all* successive primary branches. In this procedure we assume that no flux of the intermediate reactants occurs at the edges of the system; we also determine, by a convergent perturbative method, the analytical

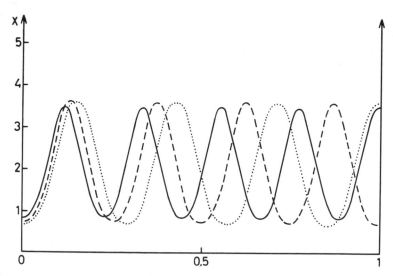

FIGURE 2. Three different steady-state patterns obtained asymptotically for the same value of $B = 4.6$ but different initial conditions. The abscissa indicates space in arbitrary units.

form for the primary solutions of a general reaction system near their bifurcation point[11]; that is,

$$x \equiv X_P - X_{st} = \epsilon p_m \cos \frac{m\pi r}{l} + \epsilon^2 \left(p_{2m} \cos \frac{2m\pi}{l} r + p_0 \right) + O(\epsilon^3) \quad (2.1)$$

with
$$\lambda - \lambda_m = \epsilon^2 \gamma_2 + O(\epsilon^3) \quad (2.2)$$

where γ_2 and the coefficients p are determined by the particular dynamics considered.

The critical points on the primary solutions are found by solving the eigenvalue problem,

$$Mu + D\Delta u = 0 \quad (2.3)$$

where the elements $M_{ij} \equiv (\partial F_i / \partial X_j)$ of the Jacobian M are evaluated at the reference state X_P and the boundary conditions read,

$$\left(\frac{\partial u}{\partial r} \right)_{r=0,l} = 0, \quad 0 \leq r \leq l \quad (2.4)$$

In (2.3) is is assumed that stability changes occur through real eigenvalues (complex eigenvalues will be considered later).

In the limit of $\delta = |\lambda_m - \lambda_{m'}| \to 0$ and $\epsilon_j^* \to 0$ ($j = m$ or m'), we introduce asymptotic expansions in terms of a new, small parameter η:

$$\delta = \eta \delta_0 + \eta^2 \delta_1 + \cdots$$
$$\mathcal{E}_j^* = \eta b_0 + \eta^2 b_1 + \cdots \qquad (j = m \text{ or } m') \quad (2.5)$$
$$x^* = \eta x_0 + \eta^2 x_1 + \cdots$$
$$u = \eta u_0 + \eta^2 u_1 + \cdots$$

The first relation in (2.5) determines the way in which ϵ_j^* depends on δ rather than imposing it a priori. The secondary bifurcation points are then obtained by satisfying solvability conditions which are possible only for some values of the unknown parameters $b_0, b_1, \delta_0, \ldots$. Application to the interactions of the first two steady-state branches, shown in FIGURE 1, for the trimolecular scheme yields stabilization of the solution with dominant spatial dependence $\sim \cos 7\pi r/l$. Indeed, for $m = 7$, $\delta_0 = 0$ and $\epsilon_7^* = O(\delta^{1/2})$, while the branch corresponding to $m = 8$ remains stable. It should be emphasized that in some cases, where, for example, $m' = m/2$ or $2m$, the critical amplitude is of the first order in δ; i.e., $\delta_0 \neq 0$ and $\epsilon^* = O(\delta)$.

In a next step, the asymptotic construction of the secondary bifurcating branches is approached as a perturbation problem of a "degenerate" bifurcation diagram (FIGURE 3).[12] Taking, for example, the case with $\epsilon^* = O(\delta^{1/2})$, we introduce the asymptotic expansions in η; thus,

$$x = \eta x_0 + \eta^2 x_1 + \cdots$$
$$\lambda = \bar{\lambda} + \eta^2 \lambda_1 + \cdots$$
$$\delta = \eta^2 \delta_1 \quad (2.6)$$

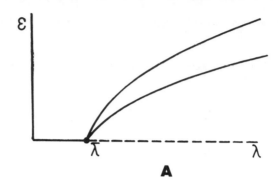

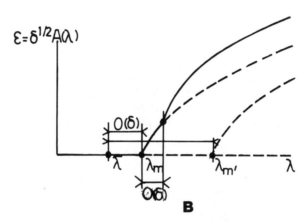

FIGURE 3. Secondary bifurcation considered as a perturbation of a degenerate bifurcation diagram. A) Degenerate bifurcation diagram: $\lambda_m = \lambda_{m'} = \lambda$. B) Perturbated bifurcation diagram: for $\delta = \lambda_m - \lambda_{m'}$ small, the secondary branches can be calculated asymptotically in the limit of $\delta \to 0$.

where $\overline{\lambda}$ is a point of multiple bifurcation. This reduces the determination of the secondary branches to the discussion of two algebraic equations which have in general the form,

$$d_1\{(\lambda - \lambda_m) - \gamma_2(m)d_1^2 + d_2^2 F_1\} = 0$$

$$d_2\{(\lambda - \lambda_{m'}) - \gamma_2(m')d_2^2 + d_1^2 F_2\} = 0 \qquad (2.7)$$

where $\gamma_2(m)$ and $\gamma_2(m')$ are defined by (2.1) for the two primary branches bifurcating at λ_m, $\lambda_{m'}$, respectively. They already appear, together with the expressions F_1, F_2 in the determination of the bifurcating branches for the degenerate problem: $\overline{\lambda} = \lambda_m = \lambda_{m'}$. The ratio d_1/d_2 determines $x_0 = d_1 p_m \cos(m\pi r/l) + d_2 p_{m'} \cos(m'\pi r/l)$, the first approximation to the secondary branch.

The same results are obtained with a convergent perturbation procedure introducing a power series expansion in a small parameter μ ($\mu \ll \epsilon^*$) characterizing the deviation from the secondary point.

The stabilization of the next primary branches needs more than one secondary bifurcation point. Indeed, the stability of these new solutions may be related to the linearized problem for the trivial steady state X_{st} which, for these values of λ, has already at least two eigenvalues with positive real parts. In the situation illustrated in FIGURE 1 we need, respectively, 2 and 3 secondary points to stabilize the primary branches with spatial dependence $\sim \cos 9\pi r/l$ and $\approx \cos 6\pi r/l$, respectively. The preceding calculations thus have to be extended for interactions between nonconsecutive branches. The values of $\delta = |\lambda_m - \lambda_{m'}|$, $m \neq m'$ may then become quite large as is seen in TABLES 1 and 2 in our example, together with the results from the asymptotic procedure (2.5). For a number of interacting branches the latter provides only an indication about the existence or not of secondary points. The divergence, for growing δ from the exact numerical resolution of problem (2.3) by a Galerkin technique is shown in FIGURE 4.

More accurate values for the critical amplitude $\mathcal{E}_m{}^*$ valid whatever δ were determined analytically by expanding the eigenfunctions u in (2.3) in a finite Fourier series[13]; that is,

$$u = \sum_{k:1}^{N} p_k \cos \frac{k\pi r}{l} \qquad (2.8)$$

with $k = jm \pm m'$, $m' \neq m$, $j = 0, 1, 2, \ldots$, retaining those terms which, taking into account the structure of M and of the power serie (2.1) for the steady-state solution, will give nonzero contributions. This procedure leads to a complex eigenvalue problem for the $\{p_k\}$ which at the critical point $\sigma = 0$ and for ϵ^* sufficiently small, gives in first approximation.

$$\epsilon^{*2} = \sum_{j:1}^{n} \psi_j \left(\prod_{k:1}^{N} \delta_k \right)^j \qquad (2.9)$$

where $\delta_k = \lambda_k - \lambda_m$ and ψ_j is a complicated expression depending on δ, m, m' and the other physicochemical parameters.

Clearly (2.9) justifies the perturbation expansions proposed in (2.4) and corroborates the genericity of the mechanism of appearance of secondary bifurcation by splitting of a multiple bifurcation point when $\delta \equiv \delta_m$ deviates from zero.

TABLE 1*

m / m'	8	7	9	6
5	−0.619	−0.614	−0.518	−0.446
6	−0.174	−0.168	−0.071	
7	−0.006		0.097	0.168
8		0.006	0.102	0.174
9	−0.102	−0.097		0.071
10	−0.284	−0.278	−0.181	−0.110
11	−0.527	−0.521	−0.425	−0.354
12	−0.824	0.818	−0.721	−0.650
13	−1.16	−1.16		−0.993

*For the same parameters values as in FIGURE 1, the distances $\delta = (B_m - B_{m'})$ are computed for the first four primary steady states.

TABLE 2*

m' \ m	8	7	9	6
5	—	—	—	—
6	—	—	—	—
7	—	—	1.05	0.917
8	—	0.978	1.03	0.904
9	—	—	—	0.938
10	—	—	—	—
11	—	—	—	—
12	—	—	—	—
13	—	—	—	0.046

*The existence of secondary bifurcation is analyzed for the different possible interactions between primary branches by evaluating the first approximation of the critical amplitude given by the pertubation procedure (2.5); i.e., $B^* - B_m = c \mid \delta \mid + 0 \ (\delta^2)$. Since all the solutions considered in our example are supercritical, secondary stability change only occurs for positive c. The latter is indicated in this table when positive. There is a qualitative agreement with FIGURE 1.

Exact numerical evaluation for the conditions considered here gives the secondary points and stabilizing effects indicated in FIGURE 1. Here again, the secondary branches may be calculated as indicated above. Moreover, interactions between the different secondary branches or between secondary and primary branches are susceptible to generate tertiary bifurcations.[13]

To conclude this section let us mention that the existence of small perturbations resulting, for example, from impurities, imperfections, or small initial

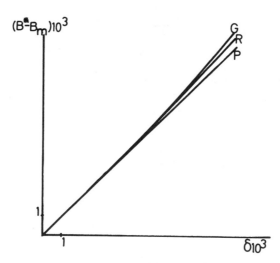

FIGURE 4. The exact locus of the secondary bifurcation point on branch 7 versus the distance $\delta = B_7 - B_8$ is represented by curve R. This numerical result is compared with the evaluation of the first approximations obtained analytically from the pertubation procedure for small δ (straight line P) and the Galerkin-type procedure (curve G), respectively. The conditions are the same as in FIGURE 1.

inhomogeneities in the system, may sometimes modify or even compromise the existence of bifurcations. Using a singular perturbation technique,[12] we have determined for model (1.3) the conditions in which the bifurcations either remain or disappear when a weak diffusion of the initial product A is allowed within the system.

SPATIOTEMPORAL SOLUTIONS

No Flux Boundary Conditions

One-Dimensional Systems

Let us first consider the case of no-flux boundary conditions which are frequently encountered in laboratory experiments and in biological *in vivo* situations. A typical bifurcation diagram is described in FIGURE 5A. Distinct pairs of complex eigenvalues cross the imaginary axis at the successive bifurcation values $\lambda = \lambda_j$, $j = 0, 1, 2, \ldots$. The first time-periodic solution bifurcating from X_{st} at $\lambda = \lambda_0$ is uniform in space, while the next branches, appearing for growing λ are characterized by an increasing basic wave number. This situation is common for two-variable reaction-diffusion systems which are the most studied models in the context of developmental biology. From bifurcation theory we know that only the first supercritically bifurcating solution is stable when λ increases. However the succeeding branches bifurcate in an unstable way. Stable, wavelike activity in the above-enumerated conditions can thus only result from a stability change of one of the oscillatory regimes. Let us therefore discuss the stability of the first two time-periodic solutions X_0 and X_1.[15] For $(\lambda - \lambda_j)$ sufficiently small the time-periodic solutions X_j, $j = 0, 1, 2, \ldots$, can be expanded in convergent analytical series,[19] which yields:

a. For the uniform limit-cycle solution near λ_0,

$$X_0 - X_{st} = \epsilon \, \mathrm{Re}(a_1 e^{i\omega_0 t}) + \epsilon^2 [\mathrm{Re}(a_2 e^{2i\omega_0 t}) + a_0] + O(\epsilon^3)$$

with period $T_0 = 2\pi/\omega_0$.

b. For the first space-dependent periodic solution with basic wave number 1,

$$X_1 - X_{st} = \epsilon \, \mathrm{Re}(b_1 e^{i\omega_1 t}) \cos \frac{\pi r}{l}$$

$$+ \epsilon^2 \sum_{k=0,2} [\mathrm{Re}(b_{k2} e^{2i\omega_1 t}) + b_{k0}] \cos \frac{k\pi r}{l} + O(\epsilon^3) \quad (3.2)$$

with period $T_1 = 2\pi/\omega_1$, and ϵ and ω_j defined by,

$$\lambda - \lambda_j = \epsilon^2 \gamma_{j_2} + O(\epsilon^4) \quad (3.3)$$

$$\omega_j - \omega_{j_0} = \epsilon^2 \omega_{j_2} + O(\epsilon^4) \quad (3.4)$$

where $j = 0$ and $j = 1$, respectively.

To investigate the stability changes of these 2π-periodic regimes, we must consider the linearized equations of motions around X_j:

$$\sigma u + \omega_j \frac{\partial}{\partial \tau} u = Mu + D\Delta u \tag{3.5}$$

with $\tau = \omega_j t$ and $\frac{\partial}{\partial r} u = 0 \bigg|_{r=0,l}$

The basic idea for solving this problem is again to relate the occurrence of secondary stability changes to the coalescence of the primary bifurcation points when some parameter tends to zero. In our problem the only way to relate the different λ_j's in the same bifurcation point is to take $D_1 = 0$; $\theta_j = D_j/D_1 = O(1)$, $j = 2, 3, \ldots, N$. In a previous work,[15] the Floquet exponents σ which determine the

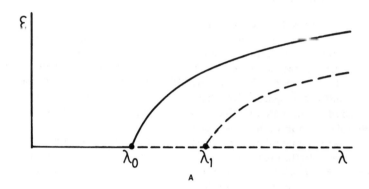

A

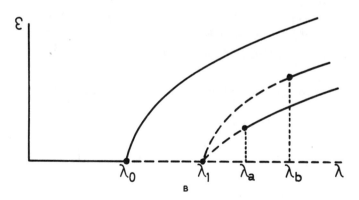

B

FIGURE 5. A) Bifurcation diagram illustrating successive branching of time-periodic solutions as a function of λ, for a 1-dimensional system subject to zero-flux boundary conditions. At $\lambda = \lambda_0$, the homogeneous limit-cycle solution bifurcates from the steady state X_{st}; at $\lambda = \lambda_1$ an inhomogeneous time-periodic regime with basic wave number 1 appears as an unstable branch.

B) Bifurcation diagram for the time-periodic solutions on the circle. The stabilization of the two different branches of solutions emerging from the same bifurcation point $\lambda = \lambda_1$ is conjectural. The secondary branches have not been drawn.

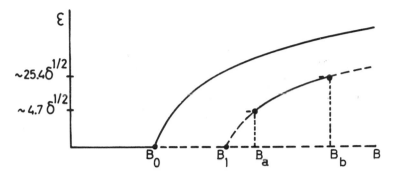

FIGURE 6. For $A = 2$, $\theta = D_2/D_1 = 0.5$, and $\delta = D_1\pi^2 \rightarrow 0$, the first bifurcating solution remains stable while the second bifurcating solution presents two stability changes. The secondary branch(es) emerging from B_a, B_b has not been calculated.

stability of X_j were evaluated when X_j begins to grow using the method of Joseph and Sattinger.[16] We know from this that, in the present case of time-periodic interacting branches, the stability changes occur most probably through complex conjugate exponents. Thus, assuming that a stability change occurs at $\epsilon = \epsilon^*$, $\lambda_j = \lambda_j^*$, and that the limit of small diffusion coefficients is valid, we introduce in (3.5) the convergent perturbation expansion in the small parameter $\delta = D_1\pi^2(\theta_j = O(1))^{14}$ such that,

$$\epsilon_j^* = \delta^{1/2}b_1 + \delta b_2 + \cdots$$

$$\sigma = c_0 + \delta c_2 + \cdots$$

$$u = u_0 + \delta^{1/2}u_1 + \cdots \quad (3.6)$$

In the limit $\delta \rightarrow 0$ one has $u_0 = \beta_1(r)\rho e^{i\tau} + \beta_2(r)\bar{\rho}e^{-i\tau}$. The fundamental corrections $b_1 c_2$, as well as $\beta_1(r)$, $\beta_2(r)$, ..., are completely determined by looking for the solutions of the eigenvalue problem,

$$[\alpha_1 + b_1^2\alpha_2 + \alpha_3\Delta + b_1^2\phi^2 2\alpha_4]\beta_1 - ic_2\beta_1 + b_1^2\phi^2\alpha_4\beta_2 = 0$$

$$b_1^2\phi^2\bar{\alpha}_4\beta_1 + [\bar{\alpha}_1 + b_1^2\bar{\alpha}_2 + \bar{\alpha}_3\Delta + b_1^2\phi^2 2\bar{\alpha}_4]\beta_2 - ic_2\beta_2 = 0, \quad (3.7)$$

which is obtained by expressing the solvability conditions; α_i $(i = 1, 4)$ are complex coefficients depending on all the physicochemical parameters except the bifurcating parameter λ and D_1. For zero-flux boundary conditions we have $\phi^2(r) = \cos^2 \pi r/l$.

For our model reaction scheme (1.3) this analysis confirms the stability of the uniform limit cycle X_0 in the entire range of parameters considered. For X_1 the procedure yields two critical points B_a^* and B_b^* at which a stability change occurs. The situation, together with numerical values, is depicted on FIGURE 6. In some interval of the bifurcation parameter we get the coexistence of *two stable time-periodic solutions*, one of which corresponds to a wavelike activity.

Numerical simulations confirm these analytical results. An inhomogeneous solution with basic wave number of 1 is indeed found for $5.32 < B < 5.66$, $\theta = \frac{1}{2}$,

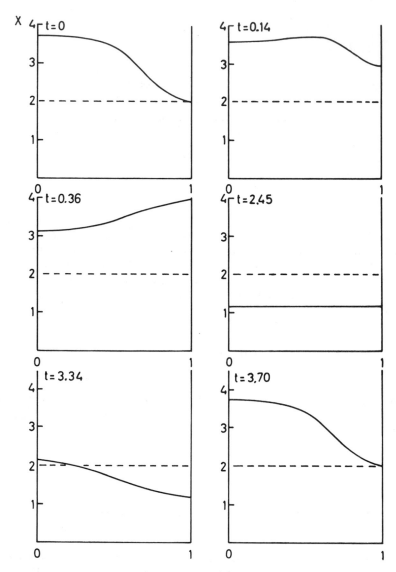

FIGURE 7. Spatiotemporal regime for a 1-dimensional system subject to zero-flux boundary conditions. The characteristic stages of the spatial distribution of X during one period are shown. We observe a short progressive stage from the left to the right. Time $t = 0$ is taken after the system has presented a regular periodicity; $A = 2$, $D_1 = 0.008$, $D_2 = 0.004$, $B = 5.4$, $l = 1$. The abscissa indicates space in arbitrary units.

$D_1 = 8.10^{-3}$, and $A = 2$. Otherwise, the system evolves to a uniform limit-cycle solution whatever the initial conditions. The wave presented in FIGURE 7 is, in the first approximation, a standing wave, however, during a complete period we observe the presence of a slow, progressive phase which from the point of view of the analytical calculations are introduced by the second-order corrections. It should be noted that, even though linear stability calculations indicate that X_0 remains stable, the uniform limit cycle has never been reached numerically in the range of stability of X_1, even starting from entirely homogeneous situations. This raises the problem of global stability when several locally stable solutions coexist.

Two-Dimensional Systems[17]

Let us now consider a system in a 2-dimensional space with circular boundaries. In this case the bifurcation point λ_1 is one of even multiplicity 4 and we prove, for a general reaction scheme, that two *different* solutions with basic spatial dependence $\sim J_1(k_1 r)$ bifurcate from the same point (FIGURE 5B). One of them is a rotation around the center that remains stationary, and is related to the symmetry of the circle. In the first approximation we have,

$$X_a - X_{st} = \epsilon \, \mathrm{Re}(ce^{i(\omega_a t + \phi)}) J_1(k_1 r) + O(\epsilon^2) \tag{3.8}$$

with period $T = 2\pi/\omega_a$ and $\lambda - \lambda_1 = \epsilon^2 \gamma_{a2} + O(\epsilon^4)$.

The second solution is a standing wave with a symmetry plane across the center and recalls the properties of the 1-dimensional wave in FIGURE 7; thus,

$$X_b - X_{st} = \epsilon \, \mathrm{Re}(ce^{i\omega_b t}) \cos \phi J_1(k_1 r) + O(\epsilon^2) \tag{3.9}$$

with period $T = 2\pi/\omega_b$ and $\lambda - \lambda_1 = \epsilon^2 \gamma_{b2} + O(\epsilon^4)$. Rotating waves are however, not only related to circles. Solutions showing the general characteristics of rotating waves also exist for ellipses, but appear as secondary branches bifurcating from nonrotating primary branches. This has been shown by an asymptotic analysis.[17] Upon slightly deforming the circular boundary into an elliptic shape, the problem can be handled as the perturbation of a known bifurcation diagram. In this way we obtain conditions for the occurrence of secondary bifurcations that lead to "imperfect" rotations, since the center remains oscillatory, contrary to what happens in (3.8).

Similar conclusions may also be drawn from the study of other 2-dimensional systems corresponding to a cylindrical envelope or a spherical membrane; i.e., closed surfaces in a 3-dimensional space; again, two classes of solutions are found.

The stability analysis of all these small-amplitude solutions has not yet been completed, however, the existence of stable solutions with the characteristics presented above has been tested numerically by direct integration of kinetic Equations 1.1 and 1.2 for the trimolecular model (1.3). Two typical examples of inhomogeneous time-periodic solutions obtained for circular boundaries are shown on FIGURES 8 and 9. They confirm the coexistence for the same reaction-diffusion equations of two families of solutions: fast, traveling waves and standing (or slow progressive) waves.[17,18]

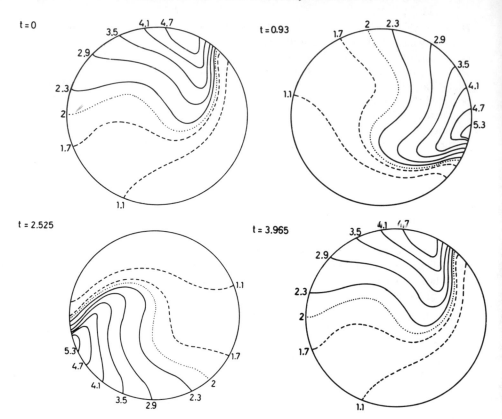

FIGURE 8. Rotating solution: Curves of equal concentration for X are represented by full or broken lines when the concentration is larger or smaller, respectively, than the value of the unstable steady state $X_0 = A$; $A = 2$, $D_1 = 0.008$, $D_2 = 0.004$, $B = 5.8$, $r_0 = 0.5861$. The $r \times \phi$ grid used is 12×24. Time $t = 0$ is chosen after the system has settled down to a periodic regime. The center's oscillations remain less than 10^{-4} of X_0. The various curves are separated by a concentration difference of 0.6, except for the curve $X = 2$, which is indicated by a dotted line. The initial condition used to excite this rotating behavior was: $X = X_{st} + 1.8(U(B, \phi) - X_{st})J_1(k_1 r)$, where $U(B, \phi)$ is the uniform solution 2π-periodic in τ, bifurcating at B_0: $U(B, \phi) = X_0(B, \tau)$.

Fixed Boundary Conditions in One Dimension

The bifurcation diagram is like that in FIGURE 5A but now the first bifurcating, time-periodic solution is space dependent. It has a dominant contribution $X_1 - X_{st} = \epsilon \operatorname{Re}(ae^{i\omega t}) \sin \pi r/l$ and is in a first-approximation symmetric around the middle point. In contrast, the second, unstable branch is asymmetric and converges to $X_2 - X_{st} = \epsilon \operatorname{Re}(be^{i\omega t}) \sin 2\pi r/l$ when $\epsilon \to 0$. Again, a stabilization of this solution may be expected through a mechanism of secondary bifurcation. Computer simulations confirm the surprising possibility of obtaining an asym-

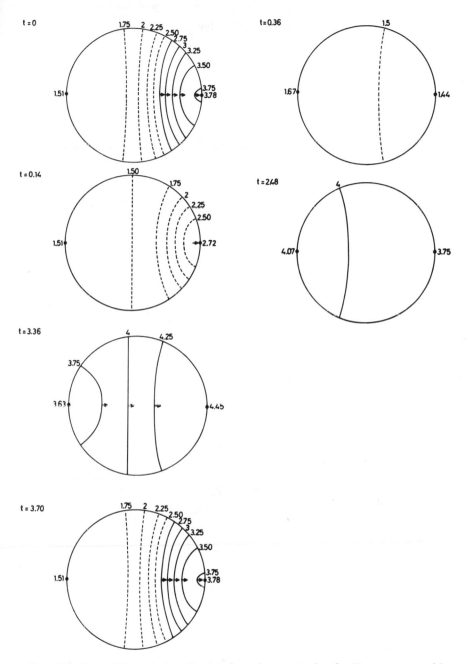

FIGURE 9. Nonrotating solution: Curves of equal concentration for Y are represented by full or broken lines when the concentration is, respectively, larger or smaller than the value of the unstable steady state $Y_0 = B/A$. The curves are separated by the same concentration difference of 0.25. Time $t = 0$ is chosen after the system has settled down to a periodic regime; $A = 2$, $D_1 = 0.008$, $D_2 = 0.004$, $B = 5.4$, $r_0 = 0.5861$. The principal stages of this evolution recall many characteristics of the 1-dimensional, time-periodic solution shown in FIGURE 7.

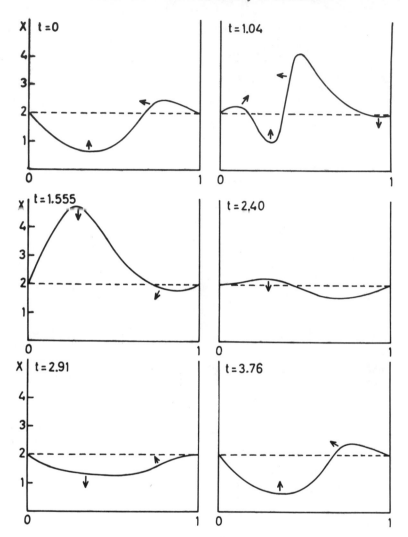

FIGURE 10. Stable time-periodic regime corresponding to the second bifurcating primary branch in a 1-dimensional system subject to fixed boundary conditions. As in the corresponding case with null fluxes, a secondary stability change is needed to observe a stable regime; $A = 2, D_1 = 0.008, D_2 = 0.004, B = 6, l = 1$. The abscissa indicates space in arbitrary units.

metric concentration wave in a system subject to fixed and symmetric boundary conditions (FIGURE 10). The existence of this kind of asymmetrical behavior could be tested experimentally for an artificial membrane reactor for which fixed boundary conditions could be realized. Time oscillations and spatiotemporal structures have indeed been already described for membrane systems with pH-dependent enzyme activity, e.g., for the membrane-bound papain reaction.[20,21]

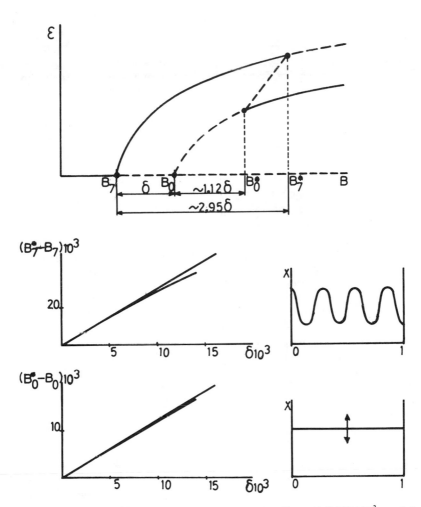

FIGURE 11. For $A = 1$, $D_2 = 0.0055$, $l = 1$, $D_1 = (0.937802 - \delta 1.50279)10^{-3}$, and $\delta = B_0 - B_7 > 0$ the bifurcation analysis in a) predicts stability changes for the steady state pattern of basic wave number 7 and for the homogeneous limit cycle solution; b) and c) test the validity of the perturbation procedure in small δ (straight line) to a better approximation obtained numerically from a Galerkin-type algorithm. By investigating the bifurcation diagram the asymptotic behavior of the system can be discussed more precisely. For $B < B_0^*$, the system evolves to the steady-state pattern. For $B > B_7^*$, the uniform time-periodic regime is the unique stable solution. For $B_0^* < B < B_7^*$, there are two stable patterns. The final state will depend on the initial conditions, but transient evolution will probably be slow resulting from the low stability of the two solutions in this region.

Interaction between Time-Periodic and Steady-State Solutions[22]

We have focused attention on the particular case in which the two consecutive interacting branches are the homogeneous limit-cycle solution described by Equation 3.1, and a steady-state pattern of basic wave number m given by Equation 2.1. The secondary branches are "mixed" states with spatiotemporal dependence. Using the perturbation method proposed in the second and third sections,[9b] which is valid for $\delta = |\lambda_m - \lambda_0| \to 0$ and small ϵ^*, we have determined conditions for the occurrence of secondary bifurcation points on the two branches. These produce either a destabilization of the limit cycle and/or a stabilization of the steady-state pattern or vice versa, depending on which branch bifurcates first. Explicit examples in which different conditions are satisfied have been found for the trimolecular scheme.

The stability change of the limit-cycle solution through real Floquet exponents was analyzed for small ϵ^* and any δ by an analytical Galerkin procedure. This yields in first approximation $\epsilon^{*2} = O(\delta)$ which justifies the analytical expansion in $\delta^{1/2}$ for ϵ^*.

Figure 11 presents the numerically computed values for the secondary bifurcation point on the limit-cycle solution in the function of $\delta = B_0 - B_7$. We observe the convergence towards the curve $\epsilon^{*2} = \delta b_0$ when $\delta \to 0$. For this set of numerical values there exists a stability change for the two successive primary branches simultaneously and, as seen in Figure 11a, a slight change in the value of the bifurcation parameter induces a transition between strongly differentiated behaviors as a stationary pattern and time-oscillating regime. An example of stable, wavelike solution emerging as a secondary branch from the limit-cycle solution through interaction with a steady-state branch has not yet been obtained numerically, although appropriate conditions may be determined theoretically.

Conclusion

We have seen that reaction-diffusion systems can display a multitude of patterns. These patterns may be simultaneously stable and are reached through different initial conditions. Or, they may be obtained by slight modifications of, for example, boundary conditions, size parameters, the symmetry of the reaction vessel, or other macroscopic parameters relating the system to its environment. Of special interest is the possible complexification of the patterns via interactions between the successive bifurcations. Important questions which remain open are: How do these factors, in general, govern the direction and stability properties of bifurcations? And how far can the results obtained for reaction-diffusion systems be extended to other transport mechanisms frequently encountered in biological systems? In our opinion the knowledge of the full bifurcation diagram of model systems could help to gain insight into these questions and progress in the modeling of various biological processes, which involve pattern formation, chemical wave propagation, or information transmission mechanisms.

REFERENCES

1. NICOLIS, G. & R. LEFEVER. 1974. Membranes, dissipative structures and evolution. Adv. Chem. Phys. **29**.
2. WINFREE, A. T. *In* Lect. Notes Biomathem. P. van den Driessche, Ed., Springer-Verlag, Berlin.
3a. SEL'KOV, E. E. 1968. Self-oscillations in glycolysis. Eur. J. Biochem. **4:** 79–86.
3b. GOLDBETER, A. & S. R. CAPLAN. 1976. Oscillatory enzymes. Annu. Rev. Biophys. Bioeng. **5:** 449–476.
4. BABLOYANTZ, A. & J. HIERNAUX. 1975. Bull. Math. Biol. **37:** 637–657.
5. KOPELL, N. & L. N. HOWARD. 1973. Stud. Appl. Math. **52:** 291–328, 1977.
6. LEFEVER, R. 1968. Stabilité des structures dissipatives. Acad. R. Belg., Cl. Sci. **54:** 712–719.
7. KAUFFMAN, S. A., R. M. SHYMKO & K. TRABERT. 1978. Control of sequential compartment formation in drosophila. Science. **199:** 259–270.
8. MEINHARDT, H. J. 1977. Cell. Sci. **23:** 117–139.
9a. MAHAR, T. J. & B. J. MATKOWSKI. 1977. SIAM J. Appl. Math. **32:** 394–404.
9b. KEENER, J. P. 1976. Stud. Appl. Math. **55:** 187–211.
10. BAUER, L., H. B. KELLER & E. L. REISS. 1975. SIAM Rev. **17:** 101–122.
11. HERSCHKOWITZ-KAUFMAN, M. 1975. Bull. Math. Biol. **37:** 589–636.
12. MATKOWSKI, B. J. & E. L. REISS. 1977. SIAM J. Appl. Math. **33:** 230.
13. ERNEUX, T. & M. HERSCHKOWITZ-KAUFMAN. In preparation.
14. ERNEUX, T. & M. HERSCHKOWITZ-KAUFMAN. 1978. Bull. Math. Biol. In press.
15. NICOLIS, G., ERNEUX, T. & M. HERSCHKOWITZ-KAUFMAN. 1978. Adv. Chem. Phys. In press.
16. JOSEPH, D. D. & D. H. SATTINGER. 1972. Arch. Rat. Mech. Anal. **45:** 79–109.
17. ERNEUX, T. & M. HERSCHKOWITZ-KAUFMAN. Submitted for publication.
18. ERNEUX, T. & M. HERSCHKOWITZ-KAUFMAN. 1977. J. Chem. Phys. **66:** 248–250.
19. AUCHMUTY, J. F. G. & G. NICOLIS. 1976. Bull. Math. Biol. **38:** 325–350.
20. THOMAS, D. 1974. *In* Membranes, dissipative structures and evolution. Adv. Chem. Phys. **29**.
21. CAPLAN, S. R., A. NAPARSTEK & N. J. ZABUSKY. 1973. Nature **245:** 364–366.
22. BLUMENTHAL, R. 1975. J. Theory Biol. **49:** 219–239.

CHEMICAL REACTORS AND
SOME BIFURCATION PHENOMENA

Rutherford Aris

Department of Chemical Engineering and Materials Science
University of Minnesota
Minneapolis, Minnesota 55455

INTRODUCTION

The chemical reactor is the heart of any chemical process, as it is here that the chemical transformation takes place for which the whole plant was designed. Like the biological heart, it does not work in isolation; a great deal of ancillary equipment may be needed to prepare and deliver the raw materials and to separate or purify the products. Though the reactor cannot be considered in isolation from its surroundings, it still remains the locus of the transformation from less valuable inputs to more valuable outputs. In crass economic terms it is where the profit is made, and this alone would command a certain attention from the engineer. But from a scientific, rather than technological, point of view its claims are no less; for chemical reactor theory is the source of some of the most varied and interesting models in the whole of engineering science while the analytical understanding of the natural chemical reactors of the biological sciences is still in its infancy.

Chemical reactors come in all shapes and sizes. Sulfur dioxide for sulfuric acid is sometimes made by the reaction of anhydrite, shale, and coke in a rotating kiln 230 feet long and 11 feet in diameter, built to withstand a firing temperature of 1600°C near the inlet. On the other hand, a superphosphate fertilizer is ammoniated in a 3- by 3-foot drum, which granulates the product at the same time. The Fischer-Tropsch synthesis of fuel requires a vertical cylinder 30 feet high and 3 feet in diameter containing 7 tons of catalyst at some 500°F and 350 psi. This bed of catalyst is kept in suspension by the mixture of synthesis gas and cooling oil that is the feed. In other processes, such as the catalytic cracking of crude oil, the catalyst bed may be fully fluidized: a bubbling bed of fine particles with many fluid characteristics. An important part of the reaction may take place in the transfer pipes by which the reactor is fed—a surprise perhaps to the first designers, but something they were quick to take advantage of. There are self-operating reactors in which changes are not intended or perhaps cannot be controlled. No doubt certain reactions will take place in the Alaska pipeline, making it the largest man-made reactor of all; but these will be incidental and so slow as to be unimportant. One challenge for the biochemical engineer might be to breed a bug that would improve crude oil, say by desulfurizing it, so that the long journey from Alaska could be turned to positive advantage.

These are continuous processes and require continuous reactors. There are also batch reactors for discontinuous processes, ranging from the beaker or flask used in the lab to 30,000-gallon fermentors for producing lactic acid from cornstarch. Specialty soaps are made in kettles ranging from 5 to 15 feet in height and 4 to 10 feet in diameter, fitted with steam coils or jackets and stirred with paddles

0077-8923/78/0316–0314 $1.25/1 © 1979, NYAS

or a "crutcher." This is an Archimedean screw mounted on the axis of the cylindrical reactor within a steam-heated jacket. When the screw is rotated it draws the reaction mixture up through the jacket and circulates downward through the annulus between the jacket and the reactor walls. In the food industry the design of the stirring equipment is often of great importance, for badly stirred pockets in the reactor may ruin the quality of the food.

Although pollution control devices are of comparatively recent origin, they are already the commonest of reactors. General Motors alone makes 30,000 catalytic converters a day, each containing over 100,000 pellets. Such reactors may take the form of a shallow bed of spherical pellets or a monolithic ceramic cylinder through which 2000 or more parallel passages pass. The spheres are impregnated or the passage walls coated with a catalyst that promotes the final removal of unburned hydrocarbons, carbon monoxide, and the oxides of nitrogen. The challenge in the design of these reactors is to ensure they work well under highly transient conditions of operation and maintain proper performance for 50,000 miles or more.

In the biological realm, chemical reactors are much smaller but vastly more complex; the wonder is that so many highly specific reactions can take place simultaneously with such efficiency. The cell itself—often called "a bag of enzymes"— contains a number of different sites of reaction and types of catalyst, with the reactants and products passing into, out of, or within the cell by various processes of active or passive transport. The processes of growth and differentiation bespeak an underlying fabric of chemical reactors organized in space and time to produce branching limbs or patterned structures.

At the other end of the physical scale, the environment itself provides examples. The air trapped by an inversion over Los Angeles is a badly stirred reactor, some 500 cubic miles in volume, which is fed from below by automobile exhaust and whose photochemical, smog-forming reactions are stimulated from above by the sun.

REACTOR MODELS

But if chemical reactors are of all sorts and conditions, what of their mathematical models? These too take on very varied forms, ranging from simple algebraic equations to large systems of partial differential equations. The aim of a mathematical modeler is to devise a system of mathematical equations that will reproduce certain features of the reactor. Needless to say there will be a considerable degree of simplification even in the most sophisticated model of the least sophisticated reactor, but this is not necessarily a bad thing. Certain features may not be important in the model and hence must obviously be thrown out. For example, the placing of the feed pipes to a stirred reactor may be of importance in designing the physical plant, but if the reactor is indeed well mixed this feature would find no echo in the equations. On the other hand, if the reactor is broken up into stages and cold feed is mixed with the reaction stream between stages to keep the temperature from running away, then the volume or length attributed to the several stages is important. Other features may be important but impossible to

TABLE 1

DIAGRAM OF MODEL*

Feature to be Modeled	Degree of Sophistication of Model Least ⟶ Most			
Mixing	Composition and temperature both uniform throughout		Turbulent fluctuations of concentration and temperature	
Reactor wall	Ignore wall	Wall temperature same as contents	Wall temperature different but uniform	Distribution of wall temperature given by heat conduction equation
Cooling coil	Cooling rate adjusts to changes instantly		Transient behavior of cooling coil considered	
Purpose	Steady-state design ⟶ Control and start-up			

*Models come in different degrees of sophistication to serve various purposes. For example, ignoring the wall temperature is obviously the simplest thing to do. Treating it the same as the contents only adds to the heat capacity; if it is different but uniform, a new ordinary differential equation is added; but if the variation of temperature through the thickness of the wall is to be considered, a partial differential equation is needed.

incorporate in the less sophisticated models, or would make the more sophisticated models so complicated as to be impossible to solve. For models do not exist in isolation; part of the art of model building is to choose the right level of model for the purpose at hand.

In steady-state design the time factor is, by definition, absent, and one may want to do many calculations with simple algebraic models, perhaps even optimizing the design from an economic point of view. But it is not always safe to leave the matter there. Unless the controls are overpoweringly dominant (and that would make them outrageously expensive), it may not be possible to hold the system at the desired steady state—and it is important to consider the controllability of the reactor. It is also useful to work out the regimen for the start-up of a reactor. Both these purposes call for a model that will describe the transient behavior of the system. This brings in time as an independent variable and transforms algebraic into differential equations or ordinary differential equations into partials. The art of modeling is to pick a system of equations simple enough to work with but sophisticated enough to serve the purpose at hand. The idea of a hierarchy of models and their relationship is illustrated in TABLE 1.

Even more valuable than its use in practical engineering design is the use of the model to enlarge theoretical vision and elucidate new concepts. For models have a life of their own and are not wholly to be judged by the relation to their prototype. A mathematical model is necessarily more abstract than its original and incurs in the engineer's mind the suspicion of being less "real" than the reactor itself. From a philosophical point of view, however, there is a reciprocity in the relationship, and, as with Menander and life, each imitates the other; indeed in logic, model theory is concerned with building less abstract structures that model systems of axioms. Above all the model, by bringing out the essentials of the process, allows the mind to grasp its features more completely and synoptically. Let me try to defend this claim with a more detailed example.

THE STIRRED TANK REACTOR

One of the simplest forms of chemical reactor is the so-called continuous-flow stirred tank reactor, or CFSTR, which at an early stage lost its "F" and, by the addition of a vowel, acquired one of the more imaginative acronymic sobriquets— "C*." This type of reactor, as its name indicates, is a vessel into which the reactants flow and from which the product stream (containing some of the unreacted feed as well as the products) is withdrawn.[2,4] It is sufficiently well stirred that its composition and temperature are uniform, and it may be equipped with a temperature control. For example, if the reaction is exothermic and generates heat, then the vessel may have an immersed cooling coil or an external jacket through which cold water flows; on the other hand an endothermic reaction might call for an immersion heater.

The reaction is supposed to take place only within the reactor, either because the reactants come together there for the first time or because a needed catalyst is present only inside. In Carberry's "spinning basket" reactor, for example, the catalyst pellets are held in a cruciform basket whose spinning serves to stir the feed.[3] In the interesterification of lard, the reactor is in stages and is stirred with an almost scraping action to prevent pockets of lard from hardening.

Because it is well stirred, the composition inside the reactor is that of the product stream—and, if, as is often the case, the reaction rate decreases as the product is formed, then the C* is relatively inefficient. Even so, its simplicity of design may make it attractive; while for certain reactions where kinetic or thermal effects give faster reaction rates within the reactor than at feed conditions, the C* has positive advantages.

As befits a simple reactor the stirred tank has a simple model, yet with even the simplest of reactions its behavior is surprisingly rich and interesting. Such a reaction is the first-order, exothermic, irreversible one in which a reactant forms products at a rate kc, where c is the concentration and k a proportionality constant known as the "rate constant." The rate constant is a function of temperature of the form $k = A \exp -E/RT$, where T is the temperature, R the gas constant, E the activation energy, and A the frequency or preexponential factor. The volume, V, in the reactor is constant because the flow in and the flow out are both the same—say, q. Then mass and heat balances can be written down expressing the fact that the rate of accumulation (whether of mass or heat) equals the rate of inflow minus the rate of outflow plus the rate of generation. In the mass balance for the disappearing reactant the "generation" is negative; in the heat balance it has two terms, a positive generation by the exothermic reaction and a negative generation, or removal, by the cooling coil. The resulting model is a pair of ordinary differential equations:

$$\frac{dc}{dt} = \frac{c_f - c}{\theta} - k(T)c = f(c, T)$$

$$\frac{dT}{dt} = \frac{T_f - T}{\theta} + Jk(T)c - L(T - T_c) = g(c, T)$$

expressing the rate of change of each variable in terms of the current values of those variables.

These equations are nonlinear, since the right-hand sides involve more complicated functions than the first powers of c and T, and are not patient of simple, explicit solutions. But they can be solved numerically, and $c(t)$ and $T(t)$ can be calculated once c_o and T_o, their values at $t = 0$, are given. Very often the values of c and T approach constant, or steady-state, values. FIGURE 1a shows a case where the concentration and temperature in the reactor approach steady-state values by damped oscillations. FIGURE 1b shows how these two graphs may be combined into one by plotting a succession of points $T(t), c(t)$ for increasing t in the "phase plane" whose coordinates are T and c. The time becomes a parameter along the curve and the damped oscillations of c and T give a trajectory that spirals into the steady state (T_s, c_s). The beauty of the phase plane presentation is that many trajectories can be shown in the same "phase portrait."[1] Thus the curve in FIGURE 1b might be just one of the many in FIGURE 1c. This figure shows that when the reactor starts up from a high value of c_o and moderate T (point P) there is a significant swing to high temperatures before it settles down to the steady state. In fact if safety considerations forbade temperatures in excess of T_M, the

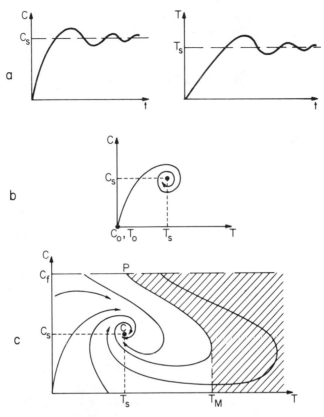

FIGURE 1. (a) Concentration and temperature in a reactor approach steady-state values by damped oscillations. (b) Combination of graphs in (a) into a so-called phase plane. (c) Graph of many possible trajectories in the phase plane.

phase portrait would show that it would be unsafe to start up from any condition represented by a point in the shaded area.

This steady state is also known as a critical point, and c_s and T_s can be found by setting up the time derivatives equal to zero, that is, solving simultaneously

$$qc_f - qc_s - VAe^{-E/RT_s}c_s = 0$$

and

$$qC_pT_f - qC_pT_s + (-\Delta H)VAe^{-E/RT_s}c_s - h(T_s - T_c) = 0.$$

In this case the first equation can be solved rather easily for c_s in terms of T_s and then substituted in the second equation to give a single equation for T_s,

$$qC_p(T_s - T_f) + h(T_s - T_c) = (-\Delta H)qc_f\frac{VAe^{-E/RT_s}}{q + VAe^{-E/RT_s}}.$$

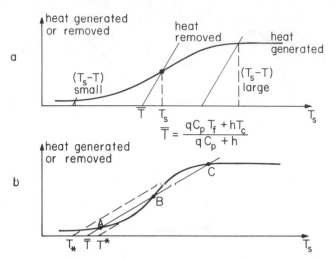

FIGURE 2. (a) The steady state shown is an intersection of a straight line and an S-shaped curve. (b) Same as (a) but with a line giving three steady states.

This is simply the equality of heat generated and heat removed at steady state.[8] The two sides of the equation are plotted in FIGURE 2a, one being a straight line and the other an S-shaped curve, and their intersection giving the steady state. The intercept of the heat removal line, $\overline{T}$, is a weighted mean of the feed and coolant temperatures. At lower values of $\overline{T}$ the steady-state temperature T_s is not much larger than $\overline{T}$ itself (i.e., $T_s - \overline{T}$ is small, as the triangle on the left shows), but when $\overline{T}$ is large $T_s - \overline{T}$ approaches this asymptote $(-\Delta H)c_f q/(qC_p + h)$, as shown by the right-hand triangle. Physically, this corresponds to complete reaction with the greatest possible heat release.

THE CUSP CATASTROPHE

But what if the straight line of heat removal is so sloped that the diagram looks like that in FIGURE 2b? If the greatest slope of the heat generation curve is greater than the slope of the heat removal line, then this figure shows that there will always be two temperatures T_* and T^* such that if $\overline{T}$ is between T_* and T^* there will be three intersections (shown here as A, B, and C). This means that there are three possible steady states for any such value of $\overline{T}$. Clearly the values of T_* and T^* depend on the slope of the heat removal line; if this is smaller than the maximum slope of the heat generation curve, as in FIGURE 3a, then by varying $\overline{T}$ we can plot the rate of heat generation at steady state as a function of $\overline{T}$ (in the lower part of the figure), getting a curve that folds back on itself between T_* and T^*. If the line is steep, as in FIGURE 3c, then the steady-state heat generation as a function of $\overline{T}$ is very like the curve of heat generation; in fact if h is infinite they are the same, for this simply means that the cooling is so efficient that the reactor temperature equals the coolant temperature. There is a critical value of $(qC_p + h)$

where it is just equal to the greatest slope of the heat generation curve and the interval of multiplicity (T_*, T^*) shrinks to a point T_c (FIGURE 3b). The heat generation curve as a function of $\overline{T}$ gives a unique steady state but has a vertical tangent at T_c.

This can be represented very neatly by one of Thom's elementary catastrophes, the cusp.[6,9] For simplicity, let feed and coolant temperatures be the same (so that $T_c = T_f = \overline{T}$) and we will simply refer to $\overline{T}$ as the "feed temperature." Also let the ratio $\lambda = h/(h + qCp)$ be the "relative cooling," for when $\lambda = 0$ there is no deliberate cooling (i.e., all the heat of reaction is taken up in bringing the incoming reactants up from the feed temperature, $\overline{T}$, to the steady-state temperature, T_s), while when $\lambda = 1$ all the burden is taken by the cooling system and the heating of the feed is negligible. Then the "control variables" in the cusp catastrophe are the feed temperature, $\overline{T}$, and the relative cooling, λ; the "behavior variable" is the reaction rate at steady state, which is proportional to the rate of heat generation.

The cusp catastrophe is shown in FIGURE 4. The behavior surface can be seen to have two regions, a high plateau where the reaction rate is relatively high and a low plain where it is very low; we might call them states of ignition and extinction. For near absolute cooling (at the back of the figure) the transition between the two states is a smooth one, but for little or no deliberate cooling it is catastrophic. Thus, if the feed temperature is increased from such a point as A, the reaction rate remains low until C is reached; when sudden ignition to D takes place. Further increase of feed temperature to E increases the reaction rate only slightly; but when $\overline{T}$ is decreased again we can go past D without extinguishing the reaction, for the extinction catastrophe takes place at F, when the reaction rate suddenly drops to B. Something akin to this happens when a gas flame is turned

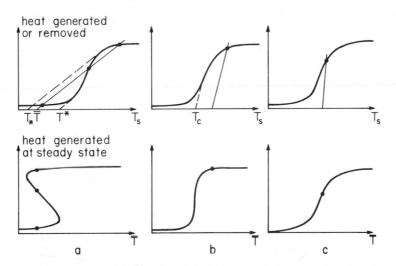

FIGURE 3. (a) A case where the slope of the heat removal line is smaller than the maximum slope of the heat generation curve. (b) A case where the value of $(qC_p + h)$ is just equal to the greatest slope of the heat generation curve; the interval of multiplicity (T_*, T^*) shrinks to a point T_c. (c) A case where the slope of the heat removal line is steep.

up too high—it "blows itself out." If there is a pilot light the flame will be re-kindled when the flow rate is decreased but not necessarily at the same point of extinction. I say "something akin," for we shall find that the flow rate generates a more interesting variety of surface. But first we should say something about the stability of the steady state.

There is no natural way of getting to any point represented by the underfold of the catastrophe surface, and it should not surprise us therefore if it corresponded to unstable steady states. An unstable steady state is one that is theoretically possible but cannot be realized in practice, since the slightest deviation from it would grow until the system finished in a quite remote state. By contrast, a steady state is locally stable if any sufficiently small deviation dies away and the system returns to the original state from which it was displaced. The unique steady state in FIGURE 1c is stable; in fact it is globally stable since, no matter how large a displacement is made, the state always returns to the critical point C.

The classical argument to demonstrate that the underfold is unstable is shown in FIGURE 5. The intermediate steady state B is such that if the temperature were increased slightly we would find that the rate of heat generation exceeded the rate of heat removal, a situation that obviously leads to even higher temperatures. Conversely, if the temperature falls, the heat removal is greater than the heat generation and the temperature continues to fall. This is clearly an unstable situation and is the result of the slope of the heat generation curve being greater than that of the heat removal line. Unfortunately it is not safe to conclude that if the heat removal line has a greater slope than the generation curve at their intersection, then the steady state is stable. Two conditions are necessary to show that any possible deviation dies away. FIGURE 5 was drawn from an equation that already embodied the solution of another, for we solved $q(c_f - c_s) - Vkc_s = 0$ for c_s and substituted in the heat balance. Thus any deviation of T_s that can be

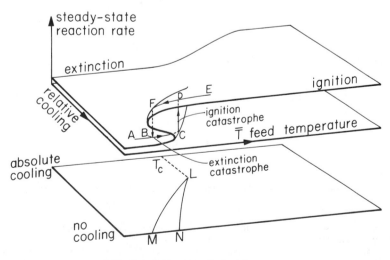

FIGURE 4. A cusp catastrophe.

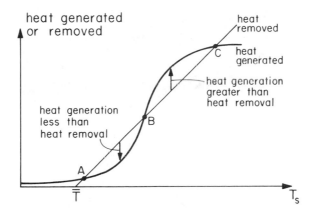

FIGURE 5. Demonstration that the underfold is unstable.

shown in FIGURE 5 implies a particular deviation of c_s that satisfies the last equation. This is too special to ensure that an arbitrary deviation of c_s and T_s will die away. Another way of saying this is that stability is really a property of the steady state which involves the dynamics, and cannot be studied from the steady-state equations alone.

PHASE PORTRAITS

The phase portrait of the system when there are three steady states is illuminating. It is shown in FIGURE 6. The state A, where the reaction rate is low, corresponds to a high value of c_s (since little of the reactant has reacted), so the three steady states are disposed in the T, c plane as shown. Suppose A and C are both stable; B, as we know, must be unstable. The behavior of the system is completely shown by the phase portrait. If the reactor starts in a state such as P, the concentration will rapidly fall and the temperature rise until the first surge of the reaction is spent by using up the reactant and the system goes through a maximum temperature at Q before the cooling takes over and brings it back in a damped oscillation to the steady state C. If the state starts at R, it goes with much more directness to the other stable steady state at A. In fact there is a line, DBE, which divides the plane into two parts; initial states to the left of this so-called separatrix go to A, while those to the right lead to C.

This is immensely important for start-up since A is an unprofitable steady state and C is the one with the healthy reaction rate; the phase portrait thus shows the right initial combination of c_o and T_o. At the same time it warns that a starting point like P may lead to a dangerous excursion of temperature. Indeed it suggests that a safe starting condition may be a point like S, obtained by filling the reactor with an inert substance and raising the temperature to the desired steady-state level. Thus when the feed is introduced, the inert substance serves as both a diluent and a heat sink and the temperature falls before rising again and coming to the steady-state value, with only a modest overshoot. We notice that the separatrix

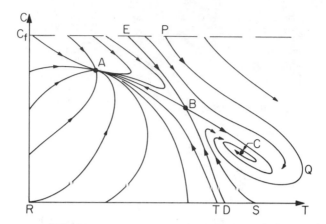

FIGURE 6. Phase portrait of a system where there are three steady states.

consists of two trajectories that start at the boundaries and actually lead to the unstable state B. Thus, the trajectory from S tries to go toward B but veers off to C, while the one starting at T, one the other side of D, veers off to A. So delicate is this knife edge that it is not even possible to integrate the equations from D or E to B on the computer, for the round-off error would provide a deviation that would grow and lead the trajectory either to A or to C. As drawn in FIGURE 6, C is a stable focus, because the trajectories spiral inward to the steady state; A is a stable node, since the values of C and T come in without oscillation; and B is a saddle point. It is also possible to have unstable nodes and foci from which the trajectories flee away.

We can now see in some detail what is meant by structural stability. Associated with each point in the plane of the control variables, λ and $\overline{T}$, of FIGURE 4 is a phase portrait. If this point lies outside the shaded region L, M, N, the phase portrait looks like FIGURE 1c, distorted in some way perhaps but essentially the same, with one steady state to which all trajectories go—for this purpose we need not distinguish between nodes and foci. But if the point lies between the arms LM and LN, the phase portrait is as in FIGURE 6, which is structurally different. MLN is called the "bifurcation set" since when the control point passes across it the structure of the system changes, whereas within either of the two regions separated by the bifurcation set the system is structurally stable.

Steady states however are not the only critical features of the solution of the model equations. There may also be limit cycles, such as the curve Γ in FIGURE 7a. Inside Γ is the unique steady state C, but this is unstable, so all trajectories wind outward from it and approach the limit cycle. Similarly any trajectory from outside winds inward to Γ; while if a trajectory started on Γ it would stay on Γ for all time. Thus Γ is a stable limit cycle, since any perturbation from it dies away as time goes on.

FIGURE 7b shows that there can also be unstable limit cycles. Here the unique steady state is stable and lies within an unstable limit cycle, Γ', which itself lies within Γ, a stable limit cycle. Thus C is approached from any starting point within

Γ' and Γ from any starting point outside Γ'. If one could start from a starting point exactly on Γ', then one would stay on it, but, as with the saddle point and separatrix, even computer rounding error is sufficient to drive the numerical solution away. However, in this case it is easy to reverse the sign of time in the equations. This is the same as reversing the direction of the arrows on all the trajectories, and makes stable things unstable and vice versa. Then Γ' would be the only stable set, and a computation with reversed time starting within Γ would lead to Γ'.

Nor is this situation without practical importance, for if the system is designed to operate at the steady state C, which is unique and stable, it is important to know whether it is globally stable (as in FIGURE 1c) or only stable to perturbations within a limited region (the inside of Γ' in FIGURE 7b). Twenty-five years ago when the modeling of reactors was in its infancy, pioneers, like N. R. Amundson of Minnesota, would often find their expositions to industry greeted with thinly veiled derision, only to be accosted later with the remark, "You know, we have such and such a reactor and find it pretty difficult to control"! The situation depicted in FIGURE 7b is just such a difficult-to-control reactor (though a very simple case of one), for a comparatively slight disturbance might send it into an oscillatory mode of operation. Recent investigations have shown that it is sometimes beneficial to operate in a limit cycle, but this is rare and would probably still meet with great resistance if it were suggested to a plant manager.

To display the more esoteric forms of reactor behavior, we need to digress and discuss the number of parameters in the system. An equation like

$$V\frac{dc}{dt} = qc_f - qc - VAe^{-E/RT}c$$

has all terms of the same dimensions, in this case amount per unit time. By choice of units, say gram moles per hour, these will have certain magnitudes, but there are also certain natural magnitudes that are characteristic of the problem. Thus the feed concentration of the reactant is characteristic of the concentration and, if we put $c = c_f(1 - x)$, x would have no dimensions but would be the fractional conversion of the reactant to products. As such it would always be between 0 and

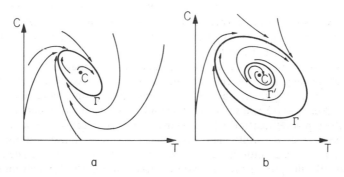

FIGURE 7. Limit cycles. Γ is stable; Γ' unstable.

1. A similar way of making the temperature dimensionless would be to take the y deviation $(T - T_f)$ as a fraction of T_f; but it turns out to be better to multiply this by $\gamma = E/RT_f$ and to write $T = T_f(1 + y/\gamma)$, for then

$$k(T) = Ae^{-E/RT} = Ae^{-E/RT_f}e^{y/(1+y/\gamma)}.$$

The first two factors give $k(T_f)$, while, since T is an absolute temperature and $T - T_f$ small compared with T_f, y/γ can often be neglected in comparison with 1 to give $k(T) = k(T_f)e^y$. Actually there is a lot more to this approximation than meets the eye, but A. B. Poore has established that it is a valid one and this lies at the foundation of the beautifully complete picture that he, A. Uppal, and W. H. Ray have given of the behavior of the reactor.[7]

There are a number of quantities that have the dimensions of time and could be used to make time dimensionless. The most obvious is V/q, the residence time of the reactor, so called because it is indeed the average residence time of any molecule in the tank. If we want to study the effect of this flow rate, it is not good to take residence time as characteristic, since then all parameters will change with the changing flow rate. The other candidates are $1/k(T_f)$ and VC_p/h, and we choose the latter. Then, with $\tau = ht/VC_p$ and $T_c = T_f$, equations become:

$$\frac{dx}{d\tau} = -\frac{x}{\theta} + D(1 - x)e^y,$$

$$\frac{dy}{d\tau} = -\left(1 + \frac{1}{\theta}\right)y + BD(1 - x)e^y.$$

These equations contain three parameters: the dimensionless residence time or reciprocal flow rate, h/qC_p; the so-called Damköhler number, $D = Vk(T_f) \cdot C_p/h$, which is a measure of the intensity of the reaction since it is proportional to $k(T_f)$; and $B = (-\Delta H)c_fE/RC_pT_f^2$, a measure of the exothermicity of the reaction. We have suppressed a fourth, namely $\gamma = E/RT_f$, by using the approximation $y \ll \gamma$ in the exponential. When we consider that one behavior variable, the steady-state reaction rate, depends on four control variables θ, γ, B, and D, we might ask whether a C^* is really a butterfly—the elementary catastrophe with one behavior and four control variables. This is perhaps not an intelligent question, for, though there might be some nonlinear transformation that would reduce the singularity to canonical form, such an effort would be unlikely to be rewarding. However, some of the known results can be exhibited by suppressing γ and giving a fixed value to B, for then the steady-state reaction rate can be shown as a surface over the plane of θ and D.

FIGURE 8 illustrates one feature that the analysis of Uppal, Ray, and Poore has discovered, namely, the existence of fingers in the surface. The sections of the surface for increasing D are shown below the figure proper, and it can be seen how as D decreases (sections F to A) the fold tucks itself up underneath until it touches the top sheet (section E); an island breaks off (section D), and decreases size (sections D, C, B), while the overhand from which it parted diminishes (sections D, C) to give a smooth curve (section B), above which the island simply disappears like the grin of the Cheshire cat (section A). The bifurcation set is the distorted W shown on the base plane. FIGURE 9 shows another way in which a finger

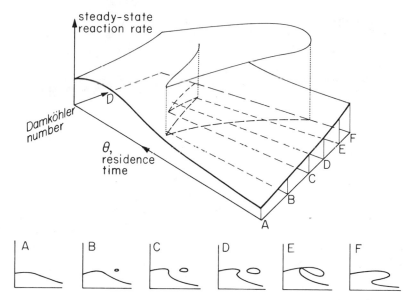

FIGURE 8. Finger in the surface of reaction rate as a function of D and θ.

is formed on the surface (in this case for a higher, but still fixed, value of B). The mushroom that develops from the fold (sections G to D) becomes a pebble resting on the surface (section C), from which it levitates (section B) and vanishes (section A). This time the bifurcation set is hook-shaped; in both cases there are three steady states for each pair of values θ and D.

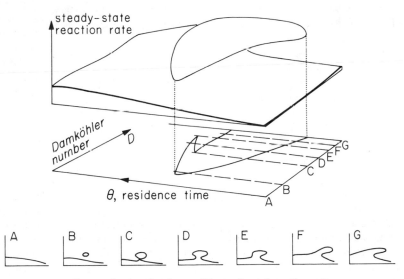

FIGURE 9. Another form of the surface of reaction rate.

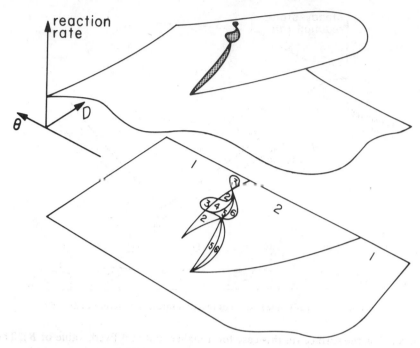

FIGURE 10. Patches of different qualitative behavior.

But this is far from being the complete story, for we have spoken as yet only of steady states. It is clearly unnatural in moving about on such a surface as is shown in FIGURE 8 to be able to reach the underside of any surface, for we have seen that ignition and extinction catastrophes keep one on the uppersides. It is not surprising that the steady states represented by underside points are unstable, but it is also true that not all the upperside points correspond to stable steady states. In fact there are patches of unstable points, as shown in FIGURE 10; one patch is an 8-shape over the fold and the other lies along the rim of the finger. If we project these regions onto the base plane, we have a complicated configuration of six kinds of region in each of which the system has a structurally different behavior. In region 1 there is a unique steady state, and the phase portrait is as shown in FIGURE 1c; 2 is as in FIGURE 6; 3 as in FIGURE 7a; 4, 5, and 6 are modifications of 2, as shown in FIGURE 11.

CONCLUSION

This does not exhaust the tale of the stirred tank behaviors, but it suffices to show how fecund even the simplest of cases is. The catalyst particle is, in some ways, analogous to a stirred tank—or rather to an imperfectly stirred tank. The transport of reactant, products, and heat to and from its external surface corresponds to the feed and take-off pipes to a C^*. Their diffusion within the particle is a process of mixing, though not always a very effective one. The mathematical model now consists of a set of parabolic quasilinear partial differential equations, and it goes without saying that it is much more difficult to obtain a comprehensive picture of its behavior.

For example, if the external conditions are fixed—i.e., if the same irreversible first-order reaction is considered with the limiting case of $\gamma \to \infty$—there are again only two remaining parameters. One is analogous to the Damköhler number, D, though in this context it is called a Thiele modulus, in honor of E. W. Thiele, one of the pioneers in this subject; the other is analogous to B, the heat-of-reaction parameter. In this case the steady-state reaction rate surface shown in FIGURE 12 has a nest of cusps, as if one had reached into the single cusp and given it further twists. It is not surprising that none of these inner folds seems to correspond to a stable steady state, for, just as with the underside of the cusp fold, there is no natural way of getting to them. This, of course, does not constitute a proof, though it corresponds to the results in a simple case that has been proved by R. Jackson. The stability picture is also more complicated, and there are distributed limit cycles in which the concentration and temperature at each point in the particle

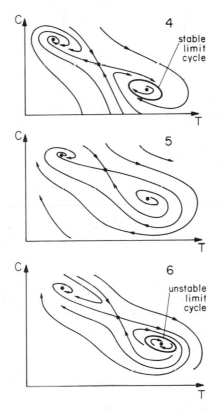

FIGURE 11. Phase portraits for Regions 4–6 of Figure 10.

oscillate in time and the distribution of temperature in the particle has peaks and valleys that move back and forth, like water sloshing about in a container. Even less well understood is the effect of the shape of the particle. For example, FIGURE 12 seems to apply to a sphere, but a very long cylinder or thin, flat plate would give a surface like FIGURE 4.

The final question that is often asked is whether these phenomena of mathematical models, such as fingers and convoluted cusps, have any practical importance. In many cases that have been explored the interesting action takes place within very narrow ranges of parameter values, and it is argued that this means

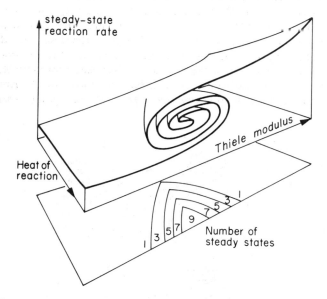

FIGURE 12. A steady-state reaction rate surface shown as a nest of cusps.

that esoteric behavior can be ignored. But this may not always be the case; it is quite possible that some future plant may be called upon to operate under conditions where esoterica loom much larger. The phenomenon of the isolation of steady states on the top of a finger has been remarked on by such a hard-headed practical engineer as C. van Heerden and has been proved experimentally by R. A. Schmitz and his colleagues.[5] But in the long run it is theory, in its original meaning of "vision," that is the most practical of all. For it is by theory that new concepts arise and new understanding is achieved, on the basis of which new progress can be made.

REFERENCES

1. AMUNDSON, N. R. & O. BILOUS. 1955. Amer. Inst. Chem. Eng. J. **1**: 513; 1956, *ibid.* **2**: 117.

2. ARIS, R. 1965. Introduction to the Analysis of Chemical Reactors. Prentice-Hall. Englewood Cliffs, N.J.
3. CARBERRY, J. J. 1976. Chemical and Catalytic Reaction Engineering. McGraw-Hill Book Company, Inc. New York, N.Y.
4. LEVENSPIEL, O. 1972. Chemical Reaction Engineering. 2nd edit. John Wiley & Sons, Inc. New York, N.Y.
5. SCHMITZ, R. A. 1974. Rev. 3rd Int. Cong. on Chem. Reac. Eng., Evanston, Ill. 1975. Adv. Chem. **148.**
6. THOM, R. 1975. Structural Stability and Morphogenesis. W. A. Benjamin, Reading, Pa.
7. UPPAL, A., W. H. RAY & A. POORE. 1974. Chem. Eng. Sci. **29:** 967. 1976. Chem. Eng. Sci. **31:** 205.
8. VAN HEERDEN, C. 1953. Ind. & Eng. Chem. **45:** 1242.
9. ZEEMAN, E. C. 1976. Sci. Amer. **234**(4): 65.

INTERACTING OSCILLATORY
CHEMICAL REACTORS*

Donald S. Cohen and John C. Neu

*Department of Applied Mathematics
California Institute of Technology
Pasadena, California 91125*

INTRODUCTION

We shall study the interaction of two oscillating systems of chemical reactions. In addition to the fundamental importance of these systems in chemical reactor theory, such systems often serve as models in biology, biochemistry, and ecology for the study of rhythms, synchronization, spatiotemporal control of various developmental processes, pattern formation, and population genetics.

Experiments carried out by M. Marek and I. Stuchl[1] and M. Marek and E. Svobodova[2] on coupled chemical reactors clearly show that starting from a system of uncoupled nonlinear chemical oscillators there occur many different types of oscillatory solutions as the coupling and various other parameters change. In the next section we shall account for these observations via various perturbation and bifurcation techniques that clearly reveal the mechanisms and quantities that control the various phenomena and their bifurcations. The purpose of this paper is to present the results giving various physical and heuristic reasons but without detailed mathematical derivations. The full analysis can be found in the paper of J. C. Neu.[3] Here we shall concentrate on the phenomena of synchronization and the bifurcation from this state to that known as rhythm splitting.

Later we shall present and analyze a simple model for subharmonic response between coupled reactors with diffusion. In our simplest situation the coupling is assumed to occur as forcing on the boundary of one reactor as a result of oscillations in the other. We shall present the equations and the results only; the complete mathematical analysis and other results are presented in the paper of D. S. Cohen.[4]

COUPLED CONTINUOUS STIRRED TANK REACTORS

Marek and Stuchl[1] and Marek and Svobodova[2] have observed many different types of phenomena as a result of coupling two continuous stirred tank reactors in each of which a Belousov-Zhabotinsky reaction with different parameters was taking place. The reactors were separated by three removable plates, the middle one being perforated with a variable number of holes. We have found that the observed phenomena do not significantly depend on the special Belousov-Zhabotinsky kinetics, and thus, we present our analysis for general kinetics. Therefore, we study the system

*This work was supported in part by the U.S. Army Research Office under Contract DAHC-04-68-C-0006 and the National Science Foundation under Grant GP-32157X2.

0077-8923/78/0316-0332 $1.75/1 © 1979, NYAS

$$\dot{x}_1 = F(x_1, y_1) + \epsilon[\lambda f(x_1, y_1) + k(x_2 - x_1)],$$

$$\dot{y}_1 = G(x_1, y_1) + \epsilon[\lambda g(x_1, y_1) + k(y_2 - y_1)],$$

$$\dot{x}_2 = F(x_2, y_2) + \epsilon k(x_1 - x_2),$$

$$\dot{y}_2 = G(x_2, y_2) + \epsilon k(y_1 - y_2).$$

(1)

The parameter k is a positive coupling constant. When $\epsilon = \lambda = 0$, we have two identical uncoupled oscillators described by

$$\dot{x}_i = F(x_i, y_i),$$

$$\dot{y}_i = G(x_i, y_i), \quad i = 1, 2,$$

(2)

where we assume that F and G are such that (2) possesses a stable, T-periodic limit cycle given by

$$x_i = X(t + \psi_i), \quad y_i = Y(t + \psi_i), \quad i = 1, 2.$$

(3)

Here the ψ_i are arbitrary constants. When $\lambda = 0$ and $\epsilon \neq 0$, the two identical oscillators are coupled; and when $\lambda \neq 0$ and $\epsilon \neq 0$, two different oscillators are coupled.

It is convenient to change from the variables x_i, y_i ($i = 1, 2$) to the variables A_i, θ_i ($i = 1, 2$) by means of the transformation

$$x_i = X(\theta_i) + \epsilon A_i Y'(\theta_i),$$

$$y_i = Y(\theta_i) - \epsilon A_i X'(\theta_i), \quad i = 1, 2.$$

(4)

Here θ_i parametrizes points on the limit cycle, and A_i measures displacements perpendicular to the limit cycle. We seek a multiscale (two-variable) asymptotic expansion of the solution in the form

$$A_i = A_i^0(t, \tau) + \epsilon A_i^1(t, \tau) + \epsilon^2 A_i^2(t, \tau) + \cdots,$$

$$\theta_i = \theta_i^0(t, \tau) + \epsilon \theta_i^1(t, \tau) + \epsilon^2 \theta_i^2(t, \tau) + \cdots, \quad i = 1, 2,$$

(5)

$$\tau = \epsilon t.$$

Upon substituting (4) and (5) into (1), using the fact that (3) solves (2), and proceeding with the techniques of multiscale asymptotics, we find (after some difficult manipulations) a hierarchy of equations for the A_i^k and θ_i^k, $i = 1, 2$, $k = 0, 1, 2, \ldots$. The detailed analysis is given by J. C. Neu.[3] To the lowest order it is easy to show that

$$\theta_i^0(t, \tau) = t + \psi_i(\tau), \quad i = 1, 2,$$

(6)

where the $\psi_i(\tau)$ are "slow time" phase shifts. In an elegant derivation Neu[3] has shown that the phase shift

$$\Psi(\tau) \equiv \psi_2(\tau) - \psi_1(\tau)$$

satisfies

$$\frac{d\Psi}{d\tau} = kM(\Psi) - \lambda\beta,$$

(7)

where β is a constant whose value depends on certain integrals involving the nonlinearities F, G, f, and g, and where $M(\Psi)$ is a function involving certain averages of the nonlinearities and the periodic functions.[3] Neu has shown that the function $M(\Psi)$ looks as sketched in FIGURE 1, the main features being that $M(\Psi)$ is T-periodic with slope -2 at $\Psi = 0, \pm T, \pm 2T, \ldots$.

Therefore, we see from (4), (6), and (7) that to leading order in ϵ all questions of synchronization and rhythm splitting are controlled by the slow time evolution of $\Psi(\tau)$. Although the quantities $M(\Psi)$ and β are complicated quantities to compute, the computations can, in fact, be carried out, but more importantly $M(\Psi)$ has qualitatively simple features which allow a simple and revealing analysis of both synchronization and rhythm splitting. We now present this analysis.

Clearly, synchronization (or phase-locking) takes place at values of Ψ which are zeroes of the right-hand side of (7). Thus, if Ψ_j satisfies $kM(\Psi_j) - \lambda\beta = 0$, then phase locking occurs at $\Psi = \Psi_j$. Since (7) is a simple first-order differential equation, the stability of the zeroes is easy to resolve. As is well known, values of Ψ_j at which the derivative of the right-hand side is negative are stable, and values of Ψ_j at which the derivative is positive are unstable. Thus, for $\lambda = 0$ (coupled identical oscillators) we see from FIGURE 1 that $\Psi \equiv 0$ is a stable solution as we would expect. For $\lambda \neq 0$ and $\epsilon \neq 0$ we see that $KM(\Psi_j) - \lambda\beta = 0$ has two roots, Ψ_j ($j = 1, 2$), provided that $|\lambda\beta/k| < \max|M(\Psi)|$. The root with negative slope, say Ψ_2, is stable, and the system will evolve to stable oscillations with a constant phase shift Ψ_2.

For $|\lambda\beta/k| < \max|M(\Psi)|$ the zeroes Ψ_j of $kM(\Psi) - \lambda\beta$ clearly depend continuously on λ, β, and k. Thus, as $\lambda\beta/k$ is decreased, the phase shift Ψ_2 varies continuously. At $|\lambda\beta/k| = \max|M(\Psi)|$ an interesting bifurcation takes place, namely, the change from synchronization to rhythm splitting. To show this, write (7) as

$$\frac{d\Psi}{d\tau} = kM(\Psi) - kM(\Psi_0) + kM(\Psi_0) - \lambda\beta$$

$$= kM(\Psi) - kM(\Psi_0) + \delta^2,$$

(8)

where $\delta^2 = kM(\Psi_0) - \lambda\beta$, and $|\lambda\beta/k|$ is slightly greater than $\max|M(\Psi)| \equiv M(\Psi_0)$ so that $0 < \delta^2 \ll 1$ (as sketched in FIGURE 1). Neu[3] has shown that (8) constitutes a singularly perturbed problem in the small parameter δ, the structure of the solution $\Psi(\tau)$ being that Ψ changes very slowly (from constant values) except in boundary layers as illustrated in FIGURE 2a. The detailed analysis is pre-

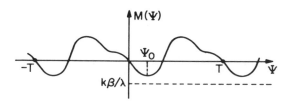

FIGURE 1. Sketch of function $M(\Psi)$.

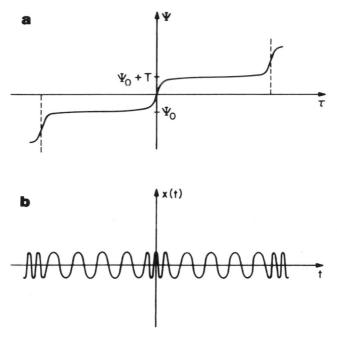

FIGURE 2. (a) Sketch of Equation 8 showing changes at boundary layers. (b) A typical concentration $x(t)$ showing the (relatively) constant phase difference Ψ with the corresponding fast changes.

sented by Neu[3]; however, we can obtain a feeling for the reasons for this behavior as follows: Away from Ψ_0, Equation 8 becomes $d\Psi/d\tau \approx kM (\Psi) - kM(\Psi_0)$ indicating that Ψ is changing more rapidly than near Ψ_0 where the whole right-hand side of (8) is small. The singular perturbation analysis confirms that $\Psi(\tau)$ is nearly constant at the values $\Psi = \Psi_0 + nT, n = 0, \pm 1, \pm 2, \ldots$, with boundary layers joining these values as shown in FIGURE 2a. A typical concentration $x(t)$ is sketched in FIGURE 2b showing the (relatively) constant phase difference Ψ with the corresponding fast changes in the layers. Thus, the boundary layers in $\Psi(\tau)$ as $|\lambda\beta/k|$ crosses the value max $|M(\Psi)| \equiv M(\Psi_0)$ give rise to the pattern of rhythm splitting as it bifurcates from the state of synchronization of oscillations.

SUBHARMONIC RESPONSE IN REACTORS WITH DIFFUSION

We consider a simple reaction-diffusion system

$$u_t = \theta_1 \Delta u + F(u, v, \lambda),$$

$$v_t = \theta_2 \Delta v + G(u, v, \lambda), \tag{9}$$

where λ represents some dimensionless physical parameter we shall vary. These equations are to hold in some domain D. We assume that this reactor is coupled to another reactor whose effect to lowest order (in asymptotics in large equal diffusivities θ_1 and θ_2) is felt through the boundary as

$$\frac{\partial u}{\partial n} = k_1[u - T(t)],$$

$$\frac{\partial v}{\partial n} = k_2[v - C(t)]. \tag{10}$$

That is, oscillations in the temperature and concentration fields in the other reactor drive the system (9) according to boundary conditions (10) which are to hold on the boundary of D. In chemical reactor theory the dimensionless constants k_1 and k_2 vary inversely as the (assumed large) diffusivities θ_1 and θ_2, and we make this assumption here. It is further assumed that in the absence of diffusion ($\theta_1 = \theta_2 = 0$) the system (9) possesses a limit cycle which comes about in the usual way by Hopf bifurcation as λ passes through a bifurcation value $\lambda = \lambda_0$. The addition of arbitrary initial conditions then makes (9) and (10) a meaningful problem.

For large (assumed equal) diffusivities ($\theta_1 = \theta_2 = \theta$), system (9) subject to (10) and arbitrary initial conditions constitutes a singularly perturbed problem. D. S. Cohen[4] has carried out the analysis, and he finds that to leading order near $\lambda = \lambda_0$ we have $u(x, t) \sim a(t)$ and $v(x, t) \sim b(t)$, where $a(t)$ and $b(t)$ satisfy the forced ordinary differential system

$$\frac{da}{dt} = \alpha(\lambda)a - \beta(\lambda)b + f(\lambda, a, b) - T(t),$$

$$\frac{db}{dt} = \beta(\lambda)a + \alpha(\lambda)b + g(\lambda, a, b) - C(t). \tag{11}$$

Here $\alpha(\lambda) \pm i\beta(\lambda)$ are the complex conjugate pair of eigenvalues that give rise to the assumed limit cycle in the Hopf bifurcation, and f and g are nonlinearities containing no linear terms (i.e., f and g contain only quadratic and higher terms in a and b). Cohen[4] has analyzed (11) in several different situations; we shall confine our investigation here to reporting on his study of subharmonic response.

In one case where $T(t) = B \cos \omega t$, $C(t) \equiv 0$, Cohen[4] defines a small parameter $\epsilon = \lambda - \lambda_0$ and employs a multiscale perturbation method by assuming that

$$a(t) \equiv a(t^*, \tau) = \epsilon a_1(t^*, \tau) + \epsilon^2 a_2(t^*, \tau) + \cdots,$$

$$b(t) \equiv b(t^*, \tau) = \epsilon b_1(t^*, \tau) + \epsilon^2 b_2(t^*, \tau) + \cdots, \tag{12}$$

where $\tau = \epsilon^2 t$ and $t^* = \omega t$. In the simplest situation where $B = O(\epsilon)$ it is easy to show that to lowest order

$$a_1(t^*, \tau) = R(\tau)\cos\mu - \frac{\omega}{\beta_0{}^2 - \omega^2} \sin t^*,$$

$$b_1(t^*, \tau) = R(\tau)\sin\mu + \frac{\omega}{\beta_0{}^2 - \omega^2} \cos t^*, \tag{13}$$

where

$$\beta_0 = \beta(\lambda_0), \quad \mu = \frac{\beta_0}{\omega} [(t^* + \phi(\tau)],$$

and where the slowly varying amplitude satisfies an equation of the form

$$\frac{dR}{d\tau} = \eta R - \gamma R^3. \tag{14}$$

The constants η and γ involve lengthy computations and asymptotics which Cohen[4] has carried out. For our presentation here it is necessary to know only that in certain situations the actual physical parameters are such that η and γ are positive so that (14) has the solution

$$R^2(\tau) = \frac{\eta}{\gamma} \frac{1}{1 + Ke^{-\eta\tau}}, \quad (K = \text{constant}). \tag{15}$$

Thus, $R^2(\tau) \to \eta/\gamma$ as $\tau \to \infty$. Hence, from (13) we see that the response has a component at angular frequency of β_0/ω times that of the forcing. So, for example, if $\beta_0/\omega = \frac{1}{3}$, then the terms $R(\tau)\cos \mu$ and $R(\tau)\sin \mu$ would represent subharmonics of order $\frac{1}{3}$. In this simplest case it is also easy to show that $\phi(\tau) \to$ constant as $\tau \to \infty$. Far more complicated relationships between $R(\tau)$ and $\phi(\tau)$ can occur in other situations as is shown in Reference 4.

REFERENCES

1. MAREK, M. & I. STUCHL. 1975. Synchronization in two interacting oscillatory systems. Biophys. Chem. 3: 241–248.
2. MAREK, M. & E. SVOBODOVA. 1975. Nonlinear phenomena in oscillatory systems of homogeneous reactions—Experimental observations. Biophys. Chem. 3: 263–273.
3. NEU, J. C. 1978. Interacting nonlinear chemical oscillators. SIAM J. Appl. Math.
4. COHEN, D. S. In press. Forced and synchronized nonlinear oscillations in reaction-diffusion systems.

NONTHERMAL INSTABILITIES OF "OPTIMIZED" CHEMICAL REACTORS—HYSTERESIS JUMPS IN BINARY REACTIONS WITH HOMOGENEOUS CATALYSIS NEAR THE STATE OF MAXIMAL YIELD EVEN UNDER STRICTLY ISOTHERMAL CONDITIONS*

Friedrich Franz Seelig

Institut für Physikalische und Theoretische Chemie
Universität Tübingen (Lehrstuhl für Theoretische Chemie)
Tübingen, Federal Republic of Germany

The binary reaction A + B → P, if catalyzed by (e.g., a metal) M via the formation of a complex MAB, can show discontinuous hysteresis jumps in the concentrations of A and B and thus in the reaction rate in an open system like the continuous-flow stirred tank reactor (CSTR) if the formation of the complexes MA_2 and MB_2 that subtract active catalyst is substantial. The phenomenon occurs even under ideally isothermal conditions because it is caused by internal chemical reasons and not at all by heat effects. This feature seems to be of technical importance because the jump from a highly productive state to one of considerably less yield occurs usually very near the state of maximal yield and can thus be caused by transient fluctuations in the input fluxes of the formally optimized reactor.

INTRODUCTION

The continuous-flow stirred tank reactor (CSTR) is one of the most common chemical reactor types and can be described by a very simple model. It is also the most prominent prototype of a technical open chemical reaction system, and thus—if kept sufficiently far from thermodynamical equilibrium—it is submitted to all features of instability, flipping phenomena, oscillations, and so on, that are abundantly known from biological systems.

Chemical reactions that are to proceed voluntarily need a decrease of free enthalpy as the driving force, which is ultimately delivered as heat. So exothermal reactions play an important role. Contrary to small laboratory systems, the considerable heat production in large technical systems is a decisive state quantity and, because of the autocatalytic feedback loop of heat production, temperature rise, and rate acceleration, can give rise to phenomena of instability or oscillations even if the chemistry of the process is extremely simple.[1] This effect is considered nearly trivial here and will not be treated. This kind of *external* instabilities can, at least in principle, be removed by appropriate devices of heat exchangers and control circuits that render the plant nearly isothermal.

*This work was supported by the Deutsche Forschungsgemeinschaft.

338

But even under ideally isothermal conditions, certain classes of chemical reaction systems, which of course are of a somewhat higher complexity, show internal instabilities that arise only from the circuitry of the reaction network itself.

It is well known that rather strange conditions, like autocatalysis of the second order, as in the "brusselator" of Prigogine and co-workers,[2,3] can lead to those effects, but much more ubiquitous causes can yield the same effects. So, if the role of some homogeneous catalyst M, say, a (transition) metal, consists of bringing two reactants A and B together because they repel each other due to equal charges, or if the function of M is such that only in a complex MAB the correct arrangement of A and B is attained for a sufficiently long time until reaction occurs, if whatsoever the formation of the mixed ternary complex MAB is a necessary transient step, the competing symmetrical complexes MA_2 and MB_2 will have to be considered, too, an effect that is known as substrate inhibition. Since these parasitic complexes subtract active catalyst, a decrease of the rate of the reaction MAB $\rightarrow$ M + P appears to be quite natural. But what cannot be foreseen without rigorous mathematical analysis is the fact that discontinuous jumps occur because the whole system shows hysteresis. Hysteresis caused by inhibition of one substrate is known from enzyme reactors at least since 1948 from a paper by Denbigh, Hicks, and Page[4] and has been studied in our group in more detail since 1972[5]. The considerations were applied to the complicated oxo synthesis and it was shown that hysteresis[6] as well as oscillatory behavior[7] is possible there. In a recent paper I showed for a two-substrate system that substrate inhibition by one reactant can yield a parametrically universal oscillator type,[8] but not show hysteresis in that reaction scheme. In that model only the reactant performing the inhibition had an overflow so that it cannot be applied to a CSTR. Funny enough, the same model, but now with both reactants having the same overflow as is typical for the CSTR, can show hysteresis, but not oscillations. In the following sections, I shall go over to the more general and more natural but more complicated case in which both reactants form inhibiting complexes.

At first sight this work may seem to be another extract of the same idea, just a little bit more complicated. But the astonishing result was that the instability and transition to an unfavorable state does not occur under pathological conditions, but almost coincides with the state of formally optimal performance of the reactor instead. Together with the ambiguity of a hysteresis loop it turned out that it is nearly impossible to reach that optimal state on a random path of approach without a detailed knowledge of the existence, position, and kind of instability. One can wonder whether those existing reactors to which this model applies are working optimally at all.

THE MODEL SYSTEM IN ACTUAL VARIABLES AND PARAMETERS

The CSTR (FIGURE 1) is externally characterized by a volume V (as measured in m^3, e.g.), a turnover v (as measured in $m^3 h^{-1}$, e.g.), and the concentrations of the reactants R_i at the entrance, $[R_i]_0$ (as measured in kmol m^{-3}, e.g.). It is convenient to formulate the rate equations in intensive rather than extensive quantities, leading to a formal first-order rate constant k_0, that characterizes the

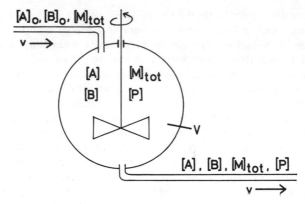

FIGURE 1. Continuous-flow stirred tank reactor (CSTR) with concentrations of substrates and catalyst inside, at entrance and exit, turnover v and volume V.

overflow and constant influxes j_i for the reactants. The relations of these quantities are

$$k_0 = v/V \qquad (1)$$

and

$$j_i = k_0[R_i]_0. \qquad (2)$$

The internal reactions are formulated as usual according to the law of mass action.
    The system of homogeneous catalysis

$$A + B \xrightarrow{\quad M \quad} P$$

via the complex MAB, but in the presence of M, MA, MB, MA$_2$, and MB$_2$ as well,

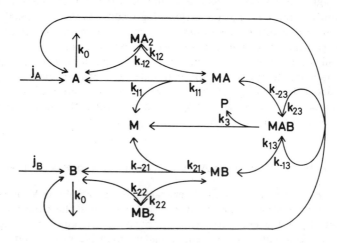

FIGURE 2. Reaction system and assignment of rate constants.

whose concentrations add to the constant sum concentration $[M]_{tot}$, is given in FIGURE 2 and yields the rate equations

$$\frac{d[A]}{dt} = j_A - k_0[A] - k_{11}[A][M] + k_{-11}[MA] - k_{12}[A][MA]$$

$$+ k_{-12}[MA_2] - k_{13}[A][MB] + k_{-13}[MAB], \tag{3}$$

$$\frac{d[B]}{dt} = j_B - k_0[B] - k_{21}[B][M] + k_{-21}[MB] - k_{22}[B][MB]$$

$$+ k_{-22}[MB_2] - k_{23}[B][MA] + k_{-23}[MAB], \tag{4}$$

$$\frac{d[M]}{dt} = -k_{11}[A][M] + k_{-11}[MA] - k_{21}[B][M] + k_{-21}[MB] + k_3[MAB], \tag{5}$$

$$\frac{d[MA]}{dt} = k_{11}[A][M] - k_{-11}[MA] - k_{12}[A][MA] + k_{-12}[MA_2]$$

$$- k_{23}[B][MA] + k_{-23}[MAB], \tag{6}$$

$$\frac{d[MB]}{dt} = k_{21}[B][M] - k_{-21}[MB] - k_{22}[B][MB] + k_{-22}[MB_2]$$

$$- k_{13}[A][MB] + k_{-13}[MAB], \tag{7}$$

$$\frac{d[MA_2]}{dt} = k_{12}[A][MA] \quad k_{12}[MA_2], \tag{8}$$

$$\frac{d[MB_2]}{dt} = k_{22}[B][MB] - k_{-22}[MB_2], \tag{9}$$

$$\frac{d[MAB]}{dt} = k_{13}[A][MB] - k_{-13}[MAB] + k_{23}[B][MA]$$

$$- k_{-23}[MAB] - k_3[MAB]. \tag{10}$$

These equations express the fact that MAB may be attained on two different paths: $M + A \rightarrow MA$, $MA + B \rightarrow MAB$, or $M + B \rightarrow MB$, $MB + A \rightarrow MAB$. Furthermore, (5) through (10) are redundant because they add up to $d[M]_{tot}/dt = 0$, which is consistent with

$$[M]_{tot} = [M] + [MA] + [MB] + [MA_2] + [MB_2] + [MAB] = \text{const.} \tag{11}$$

All complex-forming reactions are considered to be reversible because they are usually much faster than the desired reaction $MAB \rightarrow M + P$. From the same argument, all species containing M can be assumed to be quasistationary, which is equivalent to setting (5) through (10) equal to zero. This yields:

$$[MA_2] = \frac{k_{12}[A]}{k_{-12}} [MA], \tag{12}$$

$$[MB_2] = \frac{k_{22}[B]}{k_{-22}} [MB], \tag{13}$$

$$[MAB] = \frac{k_{13}[A][MB] + k_{23}[B][MA]}{k_{-13} + k_{-23} + k_3}, \tag{14}$$

$$[MA]\left(1 + \frac{k_{23}[B](k_{-13} + k_3)}{k_{-11}(k_{-13} + k_{-23} + k_3)}\right) - [MB]\frac{k_{-23}k_{13}[A]}{k_{-11}(k_{-13} + k_{-23} + k_3)}$$
$$= \frac{k_{11}[A][M]}{k_{-11}}, \tag{15}$$

$$-[MA]\frac{k_{-13}k_{23}[B]}{k_{-21}(k_{-13} + k_{-23} + k_3)} + [MB]\left(1 + \frac{k_{13}[A](k_{-23} + k_3)}{k_{-21}(k_{-13} + k_{-23} + k_3)}\right)$$
$$= \frac{k_{21}[B][M]}{k_{-21}}, \tag{16}$$

which renders (11) as

$$[M]_{\text{tot}} = [M] + [MA]\left(1 + \frac{k_{12}[A]}{k_{-12}} + \frac{k_{23}[B]}{k_{-13} + k_{-23} + k_3}\right)$$
$$+ [MB]\left(1 + \frac{k_{22}[B]}{k_{-22}} + \frac{k_{13}[A]}{k_{-13} + k_{-23} + k_3}\right), \tag{17}$$

whereas (3) and (4) are reduced to

$$\frac{d[A]}{dt} = j_A - k_0[A] - k_3[MAB], \tag{18}$$

$$\frac{d[B]}{dt} = j_B + k_0[B] - k_3[MAB]. \tag{19}$$

Equations 15 and 16 allow an expression for [MA] and [MB] each as a function of [A], [B], and [M], namely,

$$[MA] = [M]\frac{k_{11}[A]}{k_{-11}} \cdot \frac{N_1}{D} \tag{20}$$

$$[MB] = [M]\frac{k_{21}[B]}{k_{-21}} \cdot \frac{N_2}{D} \tag{21}$$

with

$$N_1 = 1 + \frac{k_{13}[A]}{k_{-21}} \cdot \frac{k_{-23} + k_3}{k_{-13} + k_{-23} + k_3} + \frac{k_{13}}{k_{11}} \cdot \frac{k_{21}[B]}{k_{-21}} \cdot \frac{k_{-23}}{k_{-13} + k_{-23} + k_3}, \tag{22}$$

$$N_2 = 1 + \frac{k_{23}[B]}{k_{-11}} \cdot \frac{k_{-13} + k_3}{k_{-13} + k_{-23} + k_3} + \frac{k_{11}[A]}{k_{-11}} \cdot \frac{k_{23}}{k_{21}} \cdot \frac{k_{-13}}{k_{-13} + k_{-23} + k_3}, \tag{23}$$

$$D = 1 + \frac{k_{13}[A]}{k_{-21}} \cdot \frac{k_{-23} + k_3}{k_{-13} + k_{-23} + k_3} + \frac{k_{23}[B]}{k_{-11}} \cdot \frac{k_{-13} + k_3}{k_{-13} + k_{-23} + k_3}$$
$$+ \frac{k_{13}[A]}{k_{-21}} \cdot \frac{k_{23}[B]}{k_{-11}} \cdot \frac{k_3}{k_{-13} + k_{-23} + k_3}, \tag{24}$$

so that with (17) [M] as a function of [A] and [B] is now given as

$$M = [M]_{tot} \cfrac{D}{D + \cfrac{k_{11}[A]}{k_{-11}} N_1 \left(1 + \cfrac{k_{12}[A]}{k_{-12}} + \cfrac{k_{23}[B]}{k_{-13} + k_{-23} + k_3}\right) \\ + \cfrac{k_{21}[B]}{k_{-21}} N_2 \left(1 + \cfrac{k_{13}[A]}{k_{-13} + k_{-23} + k_3} + \cfrac{k_{22}[B]}{k_{-22}}\right)}. \tag{25}$$

Reversely, the rate $k_3[MAB]$ is given by

$$k_3[MAB] = k_3[M]_{tot} \cfrac{\cfrac{k_{11}[A]}{k_{-11}} \cfrac{k_{21}[B]}{k_{-21}} \left(\cfrac{k_{13} \cfrac{k_{-11}}{k_{11}}}{k_{-13} + k_{-23} + k_3} N_2 + \cfrac{k_{23} \cfrac{k_{-21}}{k_{21}}}{k_{-13} + k_{-23} + k_3} N_1\right)}{D + \cfrac{k_{11}[A]}{k_{-11}} \cdot N_1 \left(1 + \cfrac{k_{12}[A]}{k_{-12}} + \cfrac{k_{23}[B]}{k_{-13} + k_{-23} + k_3}\right) \\ + \cfrac{k_{21}[B]}{k_{-21}} \cdot N_2 \left(1 + \cfrac{k_{13}[A]}{k_{-13} + k_{-23} + k_3} + \cfrac{k_{22}[B]}{k_{-22}}\right)}. \tag{26}$$

Here $k_3[M]_{tot}$ is the maximally attainable reaction rate and the complicated residual factor represents the fraction of active catalyst in the form of MAB. In all, (26) together with (22), (23), and (24) makes up a rather unwieldy function of the two variables left, [A] and [B], with not less than 17 parameters (note that $[M]_{tot}$ is a parameter, too!). But these parameters repeat themselves in certain combinations, most of which being, in connection with [A] and [B], dimensionless. Indeed, the introduction of new dimensionless parameters allows for a considerable reduction in complexity without further loss of exactness.

REDUCTION TO DIMENSIONLESS VARIABLES AND PARAMETERS

This procedure, which is exact but not unique, has two purposes: to introduce convenient abbreviations for complicated factors and terms, and to reduce the number of parameters.

The actual concentrations of A and B in the described open reactor may and indeed will differ from those in true equilibrium, but the individual rate constants must be the same in both cases. In equilibrium, whose concentrations are subscripted "e", we get

$$[MA]_e = \frac{k_{11}[A]_e[M]_e}{k_{-11}}, \tag{27}$$

$$[MB]_e = \frac{k_{21}[B]_e[M]_e}{k_{-21}}, \tag{28}$$

$$[MAB]_e = \frac{k_{13}[A]_e[MB]_e}{k_{-13}} = \frac{k_{23}[B]_e[MA]_e}{k_{-23}}, \tag{29}$$

because otherwise the two paths leading to MAB would allow the construction of a *perpetuum mobile* of the second kind.

Therefore, the constants involved are not all independent, but

$$\frac{k_{13}}{k_{-13}} \frac{k_{-11}}{k_{11}} = \frac{k_{23}}{k_{-23}} \frac{k_{-21}}{k_{21}}. \tag{30}$$

We define

$$\tau = k_0 t, \tag{31}$$

$$\alpha = \frac{k_{11}}{k_{-11}} [A], \tag{32}$$

$$\beta = \frac{k_{11}}{k_{-11}} [B], \tag{33}$$

which establish the variables, and

$$\phi_A = \frac{k_{11}}{k_{-11} k_0} j_A = \frac{k_{11}}{k_{-11}} [A]_0, \tag{34}$$

$$\phi_B = \frac{k_{11}}{k_{-11} k_0} j_B = \frac{k_{11}}{k_{-11}} [B]_0, \tag{35}$$

$$\mu = \frac{k_{11} k_3}{k_{-11} k_0} [M]_{tot}, \tag{36}$$

which are easily adjustable parameters (ϕ_A and ϕ_B by means of the adjustable quantities $[A]_0$ and $[B]_0$, μ doubly flexible by $[M]_{tot}$ and k_0), and finally,

$$\kappa_1 = \frac{k_{12}}{k_{-12}} \frac{k_{-11}}{k_{11}}, \tag{37}$$

$$\kappa_2 = \frac{k_{22}}{k_{-22}} \frac{k_{-21}}{k_{21}}, \tag{38}$$

$$\kappa_3 = \frac{k_{13}}{k_{-13}} \frac{k_{-11}}{k_{11}} = \frac{k_{23}}{k_{-23}} \frac{k_{-21}}{k_{21}}, \tag{39}$$

$$\kappa_4 = \frac{k_{-13}}{k_{-13} + k_{-23} + k_3}, \tag{40}$$

$$\kappa_5 = \frac{k_{-23}}{k_{-13} + k_{-23} + k_3}, \tag{41}$$

$$\kappa_6 = \frac{k_{13} k_{-11}}{k_{-21} k_{11}}, \tag{42}$$

$$\kappa_7 = \frac{k_{23} k_{-21}}{k_{-11} k_{21}}, \tag{43}$$

$$\kappa_8 = \frac{k_{21}}{k_{-21}} \frac{k_{-11}}{k_{11}}, \tag{44}$$

which are fixed parameters for the particular reaction system at a fixed specified temperature. If each equilibrium constant K_i for the formation of some complex is defined by k_i/k_{-i}, then κ_1 through κ_3 and κ_8 represent the ratios of the equilibrium constants K_{12}/K_{11}, K_{22}/K_{21}, $K_{13}/K_{11} = K_{23}/K_{21}$, and K_{21}/K_{11}, respectively.

With these 11 instead of 17 parameters, the system is given by

$$\frac{d\alpha}{d\tau} = \phi_A - \alpha - \rho, \tag{45}$$

$$\frac{d\beta}{d\tau} = \phi_B - \beta - \rho, \tag{46}$$

$$\rho = \mu \frac{\kappa_3 \alpha \beta' (\kappa_4 N_2 + \kappa_5 N_1)}{D + N_1(1 + \kappa_1 \alpha + \kappa_3 \kappa_5 \beta') + \beta' N_2(1 + \kappa_3 \kappa_4 \alpha + \kappa_2 \beta')}, \tag{47}$$

with

$$N_1 = 1 + (1 - \kappa_4)\kappa_6 \alpha + \kappa_4 \kappa_7 \beta', \tag{48}$$

$$N_2 = 1 + \kappa_5 \kappa_6 \alpha + (1 - \kappa_5)\kappa_7 \beta', \tag{49}$$

$$D = 1 + (1 - \kappa_4)\kappa_6 \alpha + (1 - \kappa_5)\kappa_7 \beta' + (1 - \kappa_4 - \kappa_5)\kappa_6 \kappa_7 \alpha \beta', \tag{50}$$

where

$$\beta' = \kappa_8 \beta \tag{51}$$

is used in (47) through (50) instead of β itself.

The total system shows a certain structural symmetry which can be utilized by the rotation of the axes caused by the transformation

$$\gamma = \frac{1}{2}(\alpha + \beta), \tag{52}$$

$$\delta = \frac{1}{2}(\alpha - \beta), \tag{53}$$

with $\gamma > 0$ and $-\gamma < \delta < \gamma$ for $\alpha, \beta > 0$ and

$$\phi = \frac{1}{2}(\phi_A + \phi_B), \tag{54}$$

$$\chi = \frac{1}{2}(\phi_A - \phi_B), \tag{55}$$

with $\phi > 0$ and $-\phi < \chi < \phi$ for $\phi_A, \phi_B > 0$, which renders (45) and (46) as

$$\frac{d\gamma}{d\tau} = \phi - \gamma - \rho, \tag{56}$$

$$\frac{d\delta}{d\tau} = \chi - \delta. \tag{57}$$

(γ and δ replace α and β, and ϕ and χ replace ϕ_A and ϕ_B from now on; δ and χ may, contrary to all other quantities, be negative). This shows that the system is virtually one-dimensional because δ is independent from γ, but γ depends on δ.

The steady state is given by

$$\delta_{ss} = \chi, \tag{58}$$

and the roots γ_{ss} of the complicated implicit function

$$\phi - \gamma_{ss} - \rho_{ss} = 0, \tag{59}$$

which leads to a polynomial of 4th degree in γ_{ss}. The stability of the steady states is given by the eigenvalues λ_1 and λ_2 of the Jacobian matrix in the steady states

$$\mathbf{J}_{ss} = \begin{pmatrix} \dfrac{\partial}{\partial\gamma}\dfrac{d\gamma}{d\tau} & \dfrac{\partial}{\partial\delta}\dfrac{d\gamma}{d\tau} \\[2mm] \dfrac{\partial}{\partial\gamma}\dfrac{d\delta}{d\tau} & \dfrac{\partial}{\partial\delta}\dfrac{d\delta}{d\tau} \end{pmatrix}_{ss} = \begin{pmatrix} -1 - \left(\dfrac{\partial\rho}{\partial\gamma}\right)_{ss} & -\left(\dfrac{\partial\rho}{\partial\delta}\right)_{ss} \\[2mm] 0 & -1 \end{pmatrix}, \tag{60}$$

namely,

$$\lambda_1 = -1, \tag{61}$$

$$\lambda_2 = -1 - \left(\frac{\partial\rho}{\partial\gamma}\right)_{ss}. \tag{62}$$

Since one eigenvalue is always negative, oscillations are impossible, a result which is in agreement with the statement that the system is virtually one-dimensional. The steady states are either stable nodes for $\lambda_2 < 0$ or saddle points for $\lambda_2 < 0$; i.e., their stability depends on whether $d\gamma/d\tau$ decreases or increases with increasing γ, so the necessary conditions for hysteresis are fulfilled. The occurrence of hysteresis depends on the multiplicity of the roots of (59).

Since the nullcline $d\delta/d\tau = 0$ is $\delta = \chi$, i.e., a parallel to the γ-axis, the possibility of hysteresis can easily be seen from the nullcline $d\gamma/d\tau = 0$, yielding the implicit function $\phi - \gamma - \rho = 0$: if the latter, viewed δ as a function of γ has a minimum and a maximum and a point of inflection in between for some set of the parameters save χ, an appropriate choice of χ will yield three critical points, two of which might be stable, as a stability analysis can decide. The implicit function $f(\gamma, \delta) = \phi - \gamma - \rho = 0$ can be established numerically by a computer with $\gamma = \delta = \phi$ as a starting point.

Before the results attained so far are depicted for particular examples, the problems of the design of an optimal reactor shall be treated in the next section. Only one typical situation showing the relation between steady state, nullclines, and hysteresis is depicted in FIGURE 3 for the moment.

OPTIMIZATION OF THE REACTOR

As stated before, the parameters κ_1 through κ_8 (comprised in the following in the vector $\underline{\kappa}$), are fixed for a given reaction and a specified temperature, but ϕ, χ, and μ can be freely altered in a wide range.

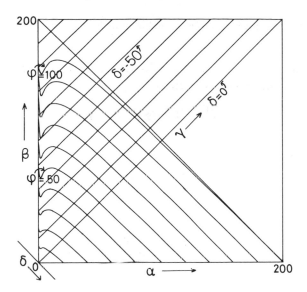

FIGURE 3. Plot of dimensionless steady-state concentration of sum of A and B, γ, vs their difference, δ, for different values of sum of influxes to A and B, ϕ, with all other parameters fixed ($\kappa_1 = \kappa_2 = \kappa_3 = 1$, $\kappa_4 = 0.09$, $\kappa_5 = 0.9$, $\kappa_6 = 0.1$, $\kappa_7 = 10$, $\kappa_8 = 0.01$, $\mu = 100$). Parallel lines $\delta = \chi = $ const. intersect steady-state curves $d\gamma/d\tau = 0$ once, and for sufficiently great values of ϕ thrice, giving rise to hysteresis.

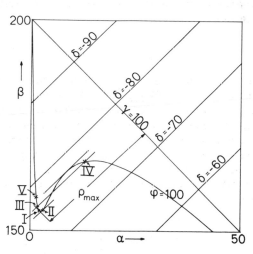

FIGURE 4. Upper left part of FIGURE 3 enlarged: steady-state curve for $\phi = 100$. Points: (I) optimum of yield, (II) jump to inhibited state, (III) suboptimal, but relatively safe against fluctuations, (IV) same χ as (III) but operation in the inhibited region, (V) absolutely safe starting point.

The production rate in a stable stationary state is given by

$$\rho_{ss} = \phi - \gamma_{ss}, \tag{63}$$

which is at the same time the dimensionless concentration of the product P, namely, π, in the steady state according to

$$\frac{d[P]}{dt} = k_3[MAB] - k_0[P] \tag{64}$$

or

$$\frac{d\pi}{d\tau} = \rho - \pi, \tag{65}$$

if

$$\pi = \frac{k_{11}}{k_{-11}}[P]. \tag{66}$$

Here

$$\gamma_{ss} = \gamma_{ss,\kappa}(\phi, \chi, \mu). \tag{67}$$

General optimization is possible for ρ_{ss} with respect to ϕ, χ, and μ if each corresponding partial derivative is zero and if certain conditions concerning the second derivatives are fulfilled.

As only steady-state conditions will be discussed in the following, we drop the subscript "ss." Preliminary investigations suggest that ρ increases monotonously with ϕ and μ so that an optimization with respect to these quantities would not be possible. But a special optimization with respect to χ can be attained.

Starting from (63) we get

$$\frac{\partial\rho}{\partial\chi} = -\frac{\partial\gamma}{\partial\chi} \tag{68}$$

which means that a maximum of ρ corresponds to a minimum of γ. A typical situation for a special set of parameters save χ is given in FIGURE 3 and the essential part of it enlarged in FIGURE 4. It is seen that the optimum, point (I), is in dangerous proximity to point (II), where the jump to the unproductive branch of the hysteresis curve occurs, so that the optimal state of operation is *practically* unstable. Therefore, the reactor should be run slightly suboptimally, say at point (III). Since this point is still in the region of χ where another far less favorable state exists at point (IV), the reactor should be started at (V) and then shifted to (III) by increasing χ.

DISCUSSION

The very general case of a two-substrate reaction with substrate inhibition treated here, where both substrates can enter into metal complexes with up to two ligands in any order, leads to complicated formulas for the steady-state concentrations in a continuous-flow stirred tank reactor (SCTR), but does not differ basically from the simpler case treated in former papers. If both substrates have exactly the same complex formation constants, a case which is only likely to occur if both substrates are optical antipodes, no hysteresis can arise. Normally one substrate is a stronger ligand and is then the one whose second-order complex is essential for the phenomenon of substrate inhibition. Hysteresis appears only in such regions where the influx of this strong ligand is smaller than that of the other substrate. In the opposite case the system is always more or less inhibited causing only poor yields of product. This result can qualitatively be seen from FIGURES 3 and 4 where substance A is assumed to be the stronger ligand, and consequently only for $\chi < 0$ or $\phi_A < \phi_B$ can hysteresis occur. Oscillations could be excluded for the case that the overflows of both substrates are equal, which applies to the case of the CSTR without side reactions. If side reactions are present and differ for A and B in rate (which will normally be the case) oscillations can arise, too. It was shown that the discontinuous jump from a productive to an inhibited state occurs in the immediate vicinity of the state of theoretically optimal performance with respect to yield, thus practically preventing optimization. Insofar as the hysteresis effect is an unwanted one, the goal of this treatment is to give quantitative criteria for its occurrence so that such states may be avoided.

One can hope that this normally negative effect can be used positively to design a pulse reactor, e.g., a tube with a chromatography-like charge of substrate just under a threshold concentration, which is triggered by a small stimulus giving rise to the formation of a spatially concentrated, but wandering and increasing avalanche of product very similar to the operation of a laser. Investigations in this direction are underway in our group.

Computations and plots were performed on the calculator HP 9820 with peripherals.

REFERENCES

1. PERLMUTTER, D. D. 1972. Stability of Chemical Reactors. Prentice-Hall, Inc. Englewood Cliffs, N.J.
2. PRIGOGINE, I. & G. NICOLIS. 1967. J. Chem. Phys. **46:** 3542.
3. PRIGOGINE, I. & R. LEFEVER. 1968. J. Chem. Phys. **48:** 1695.
4. DENBIGH, K. G., M. HICKS & F. M. PAGE. 1948. Trans. Faraday Soc. **44:** 479.
5. SEELIG, F. F. & B. DENZEL. 1972. FEBS Lett. **24:** 283.
6. SEELIG, F. F. 1976. Z. Naturforsch. **31b:** 336.
7. SEELIG, F. F. 1976. Z. Naturforsch. **31b:** 929.
8. SEELIG, F. F. 1976. Z. Naturforsch. **31a:** 731.

DISCUSSION PAPER:

BIFURCATIONS AND PERTURBED ATTRACTORS IN PHYSICOCHEMICAL SYSTEMS

Peter Ortoleva

Department of Chemistry
Indiana University
Bloomington, Indiana 47401

THE SEARCH FOR AND EXTENSION OF ELEMENTARY ATTRACTORS

As a concrete example for discussion let us consider a reacting diffusing system. We introduce an N dimensional column vector: Ψ of concentrations which is taken to obey the continuity equation

$$\frac{\partial \Psi}{\partial t} = \mathfrak{D} \nabla^2 \Psi + \mathfrak{F}[\Psi] \tag{1}$$

where $\mathfrak{D}$ and $\mathfrak{F}$ are the matrix of diffusion coefficients and column vector of chemical rates, respectively. One approach that has been used to analyze this prototype equation is to seek various attracting subspaces in the N-dimensional concentration space embedded in the rate law $\mathfrak{F}[\Psi]$.

Consider the associated ordinary differential equation to (1) corresponding to homogeneous evolution $\Psi_h(t)$,

$$\frac{d\Psi_h}{dt} = \mathfrak{F}[\Psi_h]. \tag{2}$$

Let M be the dimension of the attracting subspace. Examples of attracting subspaces are the stable steady state ($M = 0$), the limit cycle ($M = 1$) and the invarient torus ($M = 2$). Orbits within the M dimensional subspace may by definition be determined by M characteristic parameters $\zeta = \{\zeta_1, \zeta_2, \ldots, \zeta_M\}$. In the absence of diffusion, $\mathfrak{D} = 0$, it is clear that there are inhomogeneous solutions with the parameters ζ_i varying from point to point. The question then arises regarding the possibility of new phenomena due to the presence of "weak diffusion" such that at each point $\underline{r}$, Ψ is close to some point on the attractor $\mathfrak{A}$ embedded in (2). This concept has been applied for limit cycles using the method of constrained coordinates,[1-3] singular perturbation theory for plane waves,[3] frequency renormalization for periodic solutions[3] and multiple time scales[4] and integral equations.[5] We adopt a multiple time scale procedure here since it appears to be the most elegant and contains the flexibility afforded by most of the other procedures.

A complete weak diffusion picture involves the consideration of appropriate lengths to scale the effect of diffusion. We introduce a characteristic diffusion coefficient D_c and a reaction time T_c and let $\mathfrak{D} = D_c D$ and $\mathfrak{F} = T_c^{-1} F$. With this we may construct the characteristic reaction-diffusion length L_c such that $L_c^2 \equiv D_c T_c$. Now we may rewrite (1) in scaled form by using dimensionless time

0077-8923/78/0316-0350 $1.75/1 © 1979, NYAS

$t'(t = T_c t')$ and position $\underline{r}'(\underline{r} = L\underline{r}')$ for a class of phenomena varying in space on a scale L. With this we obtain (letting ∇^2 denote the Laplacian with respect to $\underline{r}'$)

$$\frac{\partial \Psi}{\partial t'} = \epsilon D \nabla^2 \Psi + F[\Psi], \tag{3}$$

where the length scale ratio ϵ is defined by

$$\epsilon = (L_c/L)^2. \tag{4}$$

With such a scaling argument the quantity ϵ presents itself as a natural smallness parameter, which may be used to develop asymptotic solutions for weakly inhomogeneous phenomena, $\epsilon \ll 1$. The key assumption is that the *only* length scale in the solution is L (an ansatz that is known to breakdown in certain wave and multiple scale phenomena[7]).

To properly carry out such a procedure we must recognize that even within the context of the assumption that the class of phenomena is on one (long) length scale, we must be able to account for many possible time scales. This manifests itself, for example, in the wavevector dependence of the frequency (dispersion relation) for plane chemical waves.[1-5] Thus we introduce a sequence of time scales τ_n such that $\tau_n = \epsilon^n t'$. If the attractor $\mathcal{C}$ is to characterize the solutions as $\epsilon \to 0$ then the spatio-temporal extensions of the characteristic parameters ζ_i must only vary on the slower times τ_n, $n \geq 1$. Otherwise the leading term $\Psi_{(0)}$ in Ψ in an expansion of the form

$$\Psi = \sum_{n=0}^{\omega} \Psi_{(n)} \epsilon^n \tag{5}$$

will not obey the scaled homogeneous equation

$$\partial \Psi_{(0)}/\partial \tau_0 = F[\Psi_0], \tag{6}$$

guaranteeing that $\Psi_{(0)}$ is on $\mathcal{C}$. Denoting an arbitrary orbit on the attractor as $\phi(\tau, \zeta)$, where again ζ are the parameters fixing the orbit, we have $\Psi_{(0)} = \phi(\tau_0, \zeta_0)$ to lowest order.

To first order in ϵ one obtains (making developments for ζ similar to that in (5)

$$L\Psi_{(1)} = D\nabla^2 \Psi_{(0)} - \partial \Psi_{(0)}/\partial \tau_1, \tag{7}$$

where the linear operator L is $\partial/\partial \tau_0 - \Omega(\Psi_{(0)})$ and $\Omega(\Psi)$ is $\partial F/\partial \Psi$. Both terms on the RHS of (7) are not zero since $\Psi_{(0)}$ implicitly depends on $\tau_{n \geq 1}$ and $\underline{r}'$ via the characteristic parameters $\zeta(\underline{r}', \tau_1, \tau_2, \ldots)$, e.g.,

$$\partial \Psi_{(0)}/\partial \tau_1 = \sum_{i=1}^{M} \left(\frac{\partial \phi}{\partial \zeta_{(0)i}} \right) \frac{\partial \zeta_{(0)i}}{\partial \tau_1}, \tag{8}$$

and similarly for $D\nabla^2 \Psi_{(0)}$. A key observation that one now must make is that the quantities $\partial \Psi_{(0)}/\partial \zeta_{(0)j}$ are in the null space of the operator L (as can be easily seen by taking the derivative of (6) with respect to $\zeta_{(0)j}$). Thus in order that (7) may be solved the RHS must be orthogonal to this null space. This introduces M

orthogonality conditions, which are indeed partial differential equations for $\zeta_{(0)}$. This program has been carried out successfully for the case of the limit cycle[4] (a one dimensional attractor). For the limit cycle the single parameter ζ is the phase of the oscillation. The theory has been used to study a variety of wave phenomena in systems with a homogeneous oscillation.[1-5] The theory has also been applied to multiply periodic attractors such as the invarient torus ($M = 2$).[6]

The Polarator—A Soluble Model of Multiply Periodic Spatio-Temporal Evolution

A class of three variable model systems has been introduced to which exact multiply periodic spatio-temporal solutions have been found.[6] It demonstrates some of the general features of certain multiple dimension ($M \geq 2$) attractors, i.e., the existence of multiparameter families of solutions. Consider the polar variables R, θ, ϕ to evolve according to the homogeneous dynamics $\dot{R} = RB(R)$, $\dot{\theta} = T(R)$, $\dot{\phi} = P(R)$ ($\cdot \equiv d/dt$). A model reacting-diffusing system has been constructed by transforming to Cartesian variables ($X = R \sin \theta \cos \phi$, $Y = R \sin \theta \sin \phi$, $Z = R \cos \theta$) and letting the "chemical species" X, Y, Z also diffuse via Fick's law. Exact solutions of these reaction diffusion equations have been presented. For example in the case where the diffusion coefficients of X, Y, and Z are all equal one may show that solutions of the form of two copropagating waves with different frequencies. In this case $X(r, t)$ (r being the spatial coordinate for a one dimensional infinite system) takes the form

$$X(r, t) = \tfrac{1}{2} R_k [\sin (\alpha_+ + w_+ t + kr) + \sin (\alpha_- + w_- t + kr)],$$

where $\alpha_\pm$ are constant phases, $w_\pm(k^2) = T(R_k) \pm P(R_k)$ and R_k is the solution of $B(R_k) = k^2 D$, D being the diffusion coefficient. This phenomena represents a two parameter family $(k, \alpha_+ - \alpha_-)$ of solutions, corresponding to the fact that the system has a 2-dimensional multiply periodic attractor (the sphere at $B(R) = 0$) in the homogeneous kinetics. (There is, of course, a third trivial parameter corresponding to the translational invarience of space.)

Chaotic Attractors

For a chaotic attractor (see several articles in this volume) trajectories for homogeneous evolution on the attractor which initially differ by an arbitrarily small amount separate to a distance of order of the dimensions of the attractor as $t \to \infty$. Hence the derivatives such as $\partial\phi/\partial\zeta_{(0)i}$ for some ζ_i (the "secular characteristic parameters") that appear in the theory become unbounded as $\tau_0 \to \infty$. Thus if we are to have weakly inhomogeneous extensions then these ζ_{0i} must be constant since the slow times $\tau_{n \geq 1}$ cannot keep up with τ_0. Alternatively, the weakly distorted attractor pictures must be modified for chaotic attractors. For "weakly chaotic" attractors, where the separation of nearby orbits occurs on a slow time scale, then weakly inhomogeneous solutions might be constructed as bifurcations. Thus consider systems such that as a parameter of the homogeneous kinetics F is varied multiply periodic orbits become chaotic with the time scale for the separation of initially close orbits that diverges as this parameter passes through a critical value, multiple periodicity bifurcating to weak chaos. In this case a bifurcation analysis for inhomogeneous evolution may be possible. This approach is presently being investigated.

For chaotic attractors which do not arise as a continuous bifurcation from multiple periodicity to weak chaos, a weakly perturbed attractor class of solutions for persistent spatio-temporal evolution with nonconstant secular characteristic parameters does not seem possible. In this case steep spatial gradients will build up and jumping from segment to segment of the chaotic attractor seems iminent. This would lead to the possibility of rapidly propagating transition layers for systems such as the Lorenz attractor where the time scale to relax to the attractor is much shorter than that of the evolution within it for certain ranges of parameters. Clearly inhomogeneous phenomena in systems with chaotic attractors are yet to be well understood and present themselves as interesting and challenging problems.

CATASTROPHE THEORY AND JUMPING BETWEEN KINETIC BEHAVIOR SURFACES

The characteristic length and time scales embedded in a phenomonological continuity equation such as (1) may vary over several orders of magnitude. For example let the first f species in the column vector Ψ participate in a sequence of reactions that occur on a time scale t_f that is much shorter than the characteristic time $\overline{T}$ of all other reactions. We may make this fact explicit by writing

$$\mathfrak{F} = \epsilon^{-H}F_s, \quad \epsilon \equiv t_f/\overline{T} \ll 1, \tag{9}$$

where F_s contains only slow time scale processes (i.e., rate parameters of order $\epsilon^0 = 1$ in the time scale ratio ϵ. The diagonal matrix H has all elements zero except the first f elements, which are for simplicity all unity. With this (9) becomes

$$\frac{\partial \Psi}{\partial t} = \mathfrak{D}\nabla^2\Psi + \epsilon^{-H}F_s[\Psi]. \tag{10}$$

For small ϵ it is clear that Ψ must either (1) vary rapidly in space or time or (2) Ψ must lie near one of the attracting intersections of the "behavior surfaces"

$$F_{s,i}[\Psi] = 0, \quad i = 1, 2, \ldots f$$

in concentration (Ψ) phase space. Thus the spatio-temporal evolution of the system consists of a finite or infinite sequence of rapid transitions between attracting branches of the intersection of the surfaces $F_{s,i} = 0$ separating smooth space-time variations on these intersections.

The question immediately arises as to how we may make general classifications of the phenomena that can occur in such systems. First we must be able to categorize the general topologies of these behavior surfaces and a start in this direction has been made using catastrophe theory.[7] Next we must be able to construct solutions to the problem in both the rapidly varying jumps between branches of the behavior surface and the constrained evolution lying on points in phase (Ψ) space near the attracting branches of the behavior surface. The technique of matched asymptotic expansions has been used to study these phenomena in reaction diffusion systems by a number of authors (see the citations in Reference 7).

Preliminary work combining catastrophe theory and matched asymptotic techniques to classify and predict phenomena appears to be a very promising direction in the future. From the preliminary studies[7] it is clear that for one fast variable ($f = 1$) and four or fewer slow variables ($N \leq 5$) that the four cuspoid catastrophes of Thom characterize all the basic features that can arise on the behavior surface. Some results for multiple fast variables, $f \geq 2$, have been obtained using the umbilic catastrophies. The approach thus far has led to the classification or discovery of a variety of propagating chemical wave phenomena such as the pulse, the finite train of pulses, the single jump pulse with smooth return, front multiplicity, wave train encroachment, and a variety of other phenomena (details and references are to be found in Reference 7).

DIFFUSIONAL BEHAVIOR SURFACES

Multiple scales in diffusion are not uncommon. Let this be emphasized, for example, by introducing a factor $\epsilon^{\tilde{H}}$ ($\tilde{H}$ is similar in structure to H of the previous section) in the diffusion matrix, $\mathcal{D} = \epsilon^{\tilde{H}} \tilde{D}$, where all the elements in $\tilde{D}$ are of order unity in ϵ and by proper choice of Ψ we can take $\tilde{D}$ to be diagonal. With this (1) becomes

$$\frac{\partial \Psi}{\partial t} = \epsilon^{\tilde{H}} \tilde{D} \nabla^2 \Psi + \mathcal{F}[\Psi]. \tag{11}$$

It is assumed that there is only one time scale in $\mathcal{F}$.

A consideration of steady states $\Psi^*(\underline{r})$ in such systems shows how multiple diffusion scales ($\epsilon \to 0$) bring out "diffusional behavior surfaces" in $\mathcal{F}$. Multiplying (11) by $\epsilon^{-\tilde{H}}$ and noting that Ψ^* is independent of time we obtain

$$\tilde{D} \nabla^2 \Psi^* + \epsilon^{-\tilde{H}} \mathcal{F} = 0. \tag{12}$$

Clearly as $\epsilon \to 0$ the spatial profile $\Psi^*(\underline{r})$ either varies rapidly in space or lies on attracting branches of the diffusional behavior surfaces

$$\mathcal{F}_i[\Psi] = 0, \quad i = 1, 2, \ldots d, \tag{13}$$

for the d species $i = 1, 2, \ldots d$ with small diffusion coefficients $\epsilon \tilde{D}_i$. Here, as in the case of multiple time scales discussed in the previous section, a combined program of catastrophe theory and matched asymptotic expansions has been suggested.[7] A very complete and elegant treatment of the multiple diffusion approach for two species ($N = 2$) systems has been carried out by Fife who also considered the possibility of simultaneous multiple scaling of diffusion and rate processes.[7] The combined program of catastrophe theory and matched asymptotic expansions for classification or prediction of qualitatively new of phenomena is certain to unfold a great richness of possibilities.

BIFURCATION FROM UNSTABLE SUBSPACES

Bifurcation theory has been used in the study of the onset of new states that arise when one state of the system loses its stability. Indeed many beautiful examples of this have been given in this conference. The possibility of bifurcation of new states from unstable states of physicochemical systems brings about a variety of interesting phenomena. For example the loss of stability of an oscillatory system has been considered.[3] When one or more of the Floquet exponents (that determine the stability of the limit cycle to small inhomogeneous perturbations) has a real part that transverses the origin, new inhomogeneous and possibly aperiodic states of evolution may arise. In this context it would be interesting to develop these ideas for the bifurcation from more complex states such as similarly weakly unstable aperiodic or chaotic subspaces.

PADÉ APPROXIMANTS AND CENTER WAVES

Recently it has been shown that Padé approximants may be used to construct center waves (circular and spiral) that are either periodic or *aperiodic*.[13] The scheme involves a well-defined ordering scheme for coupling the wave center (core) to the plane-wave-like outer regions. The parameters of the Padé's are generated as solutions of simple differential equations.

SELECTED PHYSICOCHEMICAL BIFURCATIONS

Perhaps it would be of interest to point out a number of physicochemical phenomena which have not received a great deal of attention in the context of bifurcation theories.

Insect Flight. Many insects (for example flies) have a wing beat rate that exceeds the maximum nerve repetitive firing rate. Thus the individual wing beats cannot be triggered by individual nervous signals. However, it is known that the wing-thorax system is to a good approximation, a damped oscillator with a frequency of the correct order of magnitude. Furthermore flight muscle has the interesting property that when stretched it tends to contract (in excess of the elastic force!) presumably due to the contraction chemical kinetics. Using a simple model of a damped oscillator coupled to a simple contractile chemical kinetics it has been shown that such a system can enter a state of auto-oscillation.[4]

Fucus Egg Symmetry Breaking. In the very beautiful experiments of Jaffe and others it has been shown that the state of spherically symmetric membrane potential makes a transition to an asymmetric state with a net north-south polarity. This is believed to be the essence of the asymmetric cell differentiation responsible for the root/leaf differentiation that characterizes the subsequent biomorphogenesis. Introducing a set of electrophysiological equations and a simple model it has been shown that the polarized states may bifurcate from the spherically symmetric solutions.[8]

Spontaneous Pattern Formation in Precipitating Systems. It has been shown experimentally that the state of uniform precipitation from a supersaturated phase may spontaneously make a symmetry breaking (pattern forming) transition.[9,10] A simple theory based on diffusion and the competition of small and large particles for the growth material has been presented and describes many of the features of this phenomena.[10]

Nonlinear Phenomena at Local Sites of Reaction. It has become clear that a great variety of phenomena (including multiple states, oscillations, propagating waves and chaotic evolution) may occur in reacting diffusing systems with bulk kinetics. Studies on systems where reactions are localized to membranes or catalytic walls have indicated that this variety also exists in systems with heterogeneous kinetics.[11] When many local sites are present cooperative phenomena, in analogy to equilibrium phase transitions, may arise.[11] The description of these systems may often be reduced to sets of coupled nonlinear integral equations. Many of the mathematical methods used to describe nonlinear phenomena in systems with homogeneous kinetics, including bifurcation and attractor perturbation theory, have been applied to the analysis of these systems.[4]

Electrochemical Waves. In the best known example of chemical wave propagation, the Belousov-Zaikin-Zhaboutinsky reaction, most important chemical species are ionic yet the strong tendency toward charge neutrality that couples ionic motions has been neglected in most theories of these waves. Recently full account of the electrochemical nature of chemical waves in ionic media has been taken.[12] It has been shown that, in the presence of imposed fields, chemical waves in such media can be forced to remain stationary and even breakdown, making a transition to qualitatively new modes of wave propagation.[12]

References

1. ORTOLEVA, P. & J. ROSS. 1973. J. Chem. Phys. **58**: 5673.
2. ORTOLEVA, P. 1976. J. Chem. Phys. **64**: 1395.
3. ORTOLEVA, P. & J. ROSS. 1974. J. Chem. Phys. **60**: 5090.
4. ORTOLEVA, P. 1978. Selected topics from the theory of nonlinear phenomena in physico-chemical systems. Adv. Theor. Chem.
5. KOPELL, N. & L. HOWARD. 1973. Stud. Appl. Math. **52**: 291.
6. DELLEDONNE, M. & P. ORTOLEVA. In press. Turbulent spatio-temporal dynamics in reacting-diffusing systems: Results for a soluble model. J. Chem. Phys.
7. FEINN, D. & P. ORTOLEVA. 1977. J. Chem. Phys. **67**: 2119. (See also the work of Fife and Stanshine and Howard cited therein.)
8. ORTOLEVA, P. 1977. J. Theoret. Biol.
9. FLICKER, M. R. & J. ROSS. 1974. J. Chem. Phys. **60**: 3458.
10. FEINN, D., P. ORTOLEVA, W. SCALF, S. SCHMIDT & M. WOLFF. In Press. Spontaneous pattern formation in precipitating systems. (J. Chem. Phys.).
11. BIMPONG-BOTA, E. K., A. NITZAN, P. ORTOLEVA & J. ROSS. 1977. J. Chem. Phys. **66**: 3650; and references cited.
12. SCHMIDT, S. & P. ORTOLEVA. 1977. J. Chem. Phys.
13. ORTOLEVA, P. 1978. J. Chem. Phys. In press.

SYNERGETICS AND BIFURCATION THEORY*

H. Haken

Institut für Theoretische Physik
Universität Stuttgart
7 Stuttgart 80
Federal Republic of Germany

SYNERGETICS, A NEW FIELD OF INTERDISCIPLINARY RESEARCH

Synergetics[1-6] is a rather new field of interdisciplinary research related to mathematics, physics, astrophysics, electrical and mechanical engineering, chemistry, biology, ecology, and possibly to other disciplines. It studies the self-organized behavior of complex systems (composed of many subsystems) and focuses its attention to those phenomena where dramatic changes of macroscopic patterns or functions occur owing to the cooperation of subsystems. Some examples are exhibited in FIGURE 1. In spite of this rather general scope, synergetics has been able to unearth astounding analogies between entirely different systems. In the course of this research program it more and more transpired that bifurcation theory plays a crucial role.

As long as the systems we are encountered with can be described by mathematical models we very often deal with the following problem. A system is represented by a set of time-dependent variables

$$(q_1, q_2, \ldots, q_N) = \mathbf{q} \tag{1}$$

which in many cases can be interpreted as describing the properties of certain subsystems. If the subsystems are continuously distributed, for instance, molecules in liquids or atoms in a laser, $\mathbf{q}$ will depend not only on time t but also on space coordinate $\mathbf{x}$,

$$\mathbf{q} = \mathbf{q}(\mathbf{x}, t). \tag{2}$$

As we further know from many explicit examples, a system is controlled by external parameters, for instance temperature, energy flux, matter flux, information input, and so forth. We symbolize a set of control parameters by α. A large class of problems is then described by the following set of stochastic partial nonlinear differential equations

$$\dot{\mathbf{q}} = \underset{\text{nonlinear}}{\mathbf{N}(\mathbf{q}, \nabla, \alpha, t)} + \underset{\text{fluctuating forces}}{\mathbf{F}(\mathbf{q}, \nabla, t)} \tag{3}$$

In it, $\mathbf{N}$ is a nonlinear function of the variables $\mathbf{q}$ and may contain spatial derivatives ∇. $\mathbf{N}$ may depend on the parameters α as well as on t. As is well known from physical, chemical, and other systems, random *fluctuations,* caused by internal motion or external influences, play an important role. These fluctuations are taken care of by the fluctuating forces in (3). A general solution of equation 3 seems

*This work was supported by the Volkswagenwerk Foundation, Hannover.

357

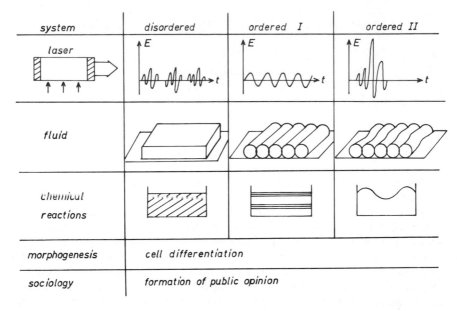

system	disordered	ordered I	ordered II
laser			
fluid			
chemical reactions			
morphogenesis	cell differentiation		
sociology	formation of public opinion		

FIGURE 1. Examples for the spontaneous formation of structures: The laser light-source emits random wave-tracks, a coherent wave, an ultrashort light pulse. A fluid layer heated from below shows no macroscopic motion, motion in form of rolls, oscillating rolls. Sustained chemical reactions show no structure, a layer structure, waves. In morphogenesis, cell differentiate leading to patterns or functions. Public opinion may be indifferent or show pronounced structures.

hopeless. However, by methods that have emerged in the past and whose relation to bifurcation theory will transpire below, it has become possible to classify a number of important cases of Equation 3. Very often, these classes correspond to phenomena, where the above-mentioned dramatic changes of macroscopic structures (in the widest sense of this word) occur.

ORDER PARAMETERS AND SLAVING: REDUCTION OF THE DEGREES OF FREEDOM

To elaborate our basic ideas we first choose the following special case of (3),

$$\mathbf{F} \equiv 0, \quad \mathbf{N} = \mathbf{N}(\mathbf{q}, \alpha). \tag{4}$$

Furthermore we assume that we have found for a certain parameter α a time-independent solution. To check its stability we perform conventional stability analysis by putting

$$\mathbf{q} = \mathbf{q}_0 + \mathbf{u} \tag{5}$$

Inserting it in (3) and linearizing the resulting equation with respect to $\mathbf{u}$ yields

$$\dot{\mathbf{u}} = A\,\mathbf{u} \tag{6}$$

In it A is a constant matrix. The solutions of (6) read

$$\mathbf{u}_\mu = \exp(\lambda_\mu t)\mathbf{v}_\mu. \tag{7}$$

If

$$\mathrm{Re}\,\lambda_\mu > 0, \tag{8}$$

the "mode" $\mathbf{v}_\mu$ is unstable whereas

$$\mathrm{Re}\,\lambda_\mu < 0 \tag{9}$$

is related to a stable mode. For an illustration of our approach let us assume that (6) yields one unstable and one stable mode. We then introduce the time-dependent amplitudes ξ_u (u = unstable) and ξ_s (s = stable) and put

$$\mathbf{q} = \mathbf{q}_0 + \xi_u \mathbf{v}_u + \xi_s \mathbf{v}_s. \tag{10}$$

Inserting (10) into (3) yields, after some straightforward manipulations, equations for ξ_u and ξ_s which are, for example, of the form

$$\dot{\xi}_u = \lambda_u \xi_u - \xi_u \xi_s \tag{11}$$

$$\dot{\xi}_s = -|\lambda_s|\,\xi_s + \xi_u^{\,2} \tag{12}$$

Let us first assume that λ_u and λ_s are both real. When we change the control parameter α so that ξ_u passes from the stable to the unstable region, λ_u must pass through zero. This implies (at least if we neglect nonlinearities) that ξ_u changes very slowly. On the other hand, it transpires from (12) that ξ_s is driven by ξ_u. If λ_s is unequal to zero, and negative, we can neglect the time-derivative on the left hand side and express ξ_s directly by

$$\xi_s(t) = \xi_u^{\,2}(t)/|\lambda_s|. \tag{13}$$

In the following we shall call the unstable amplitudes "order parameters". Because ξ_s obeys *immediately* ξ_u according to (13), we shall say that the stable modes are *slaved* by the order parameters. Inserting (13) into (11) we obtain

$$\dot{\xi}_u = \lambda_u \xi_u - \xi_u^{\,3}/|\lambda_s|. \tag{14}$$

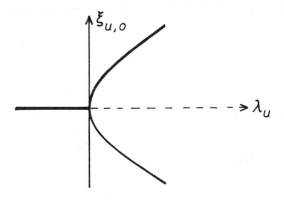

FIGURE 2. Bifurcation of order parameter ξ_u.

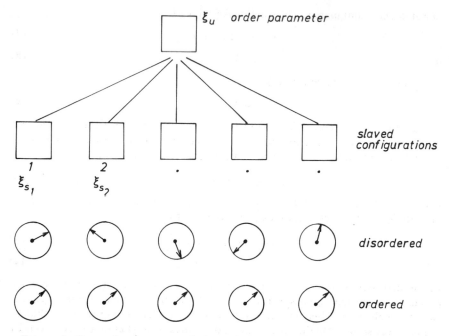

FIGURE 3. One order parameter slaving subsystems (modes). (a) $\xi_u = 0$, subsystems driven by fluctuations; (b) $\xi_u \neq 0$, subsystems are well ordered.

When we put

$$\dot{\xi}_u = 0, \tag{15}$$

we may immediately make contact with the conventional Ljapunov-Schmidt† bifurcation theory, which leads to the well known bifurcation diagram of FIGURE 2. The slaving principle has two important consequences. It can be applied equally well to *many slaved modes* (FIGURE 3) and it contains the *dynamics*. Indeed we can solve (14) in the time-dependent case. The corresponding solution is presented in FIGURE 4. Once we have solved this problem the dynamics of all slaved modes is determined by (13). The slaving principle implies an enormous reduction of the degrees of freedom: In many cases of practical interest it turns out that at the instability point only one or very few modes become unstable and act as order parameters while all the other modes remain stable and can be eliminated. This implies that the dynamics of the whole system is governed, not by many degrees of freedom, but only by very few of them (FIGURE 3). In the special case of Equation 14, it is even a *single* order parameter. In the following part of my article I want to study how far the concept of order parameters and slaving can be extended to more general cases. In this analysis the only crucial requirement will turn out to be that the slaved modes are uniquely determined by

†See, for instance, Vainberg & Trenogin.[7] For more recent papers on bifurcation theory see References 38–43 and 45. For further references see Joseph[9] and Sattinger.[10]

the order parameters. We shall now start to discuss generalizations (which, in each case, also hold for many slaved modes and several order parameters).

If λ_u and λ_s are complex, the elimination scheme (13) might fail to be valid. In this case, however, we solve (12) by

$$\xi_s = \left(\frac{d}{dt} - \lambda_s\right)^{-1} \xi_u^{\,2}. \tag{16}$$

The operator on the right-hand side of (16) depends on initial conditions. Therefore we write (16) more explicitly as

$$\xi_s(t) = \int_{t_0}^{t} \exp[\lambda_s(t - \tau)]\xi_u^{\,2}(\tau)d\tau$$
$$+ \exp[\lambda_s(t - t_0)]\xi_s(t_0), \quad t \geq t_0. \tag{17}$$

For further evolution, we invoke the following self-consistency requirement. We assume that the resulting equation for ξ_u alone allows (only) for small solutions which change much more slowly than $\exp[\lambda_s(t - t_0)]$. This permits us to drop the second term in (17). For the same reason, we may extend the lower limit of the integral to $-\infty$. We thus obtain

$$\xi_s(t) = \int_{-\infty}^{t} \exp[\lambda_s(t - \tau)]\xi_u^{\,2}(\tau)d\tau. \tag{18}$$

As it has been discussed elsewhere, $\infty > \xi_u^{\,2} \geq a > 0$ in ordered phases.[4] Therefore (18) exists if and only if Re $\lambda_s < 0$. In our above example we have chosen a particularly simple case in which we can resolve (12) in a single step with respect to ξ_s. In the general case the right-hand side of (12) contains bilinear functions in ξ_s and ξ_u or still higher powers. All essential steps can be seen if (12) is replaced by the more general equation:

$$\dot{\xi}_s = \lambda_s\xi_s + \xi_u^{\,2} + \xi_s\xi_u + \xi_s^{\,2}. \tag{19}$$

For simplicity, we have put all constant coefficients equal to unity. When we first

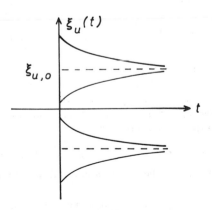

FIGURE 4. Time-dependent solution of Equation 14 for different initial conditions.

neglect the last term, ξ_s^2, the solution of (19) can be written as

$$\xi_s(t) = \left[\frac{d}{dt} - \lambda_s - \xi_u(t)\right]^{-1} \xi_u^2(t). \qquad (20)$$

The operator $[\cdots]^{-1}$ exists if $|\xi_u^2| < \infty$ and

$$|\xi_u| < |\operatorname{Re}\lambda_s|. \qquad (21)$$

In this case, $\xi_s(t)$ is still slaved by $\xi_u^2(t)$. In the example of the Lorenz model we have studied what happens if (21) is violated. We found, that we obtain chaotic motion.

We thus may state: If the order parameter is not able to slave the subsystems, we may obtain chaos.

The last term in (19), ξ_s^2 (and other terms of higher powers) can be taken care of by a rapidly converging procedure in the form of continued fractions, provided (21) is fulfilled. By the above procedure the more conventional bifurcation analysis is decomposed into two steps: (1) determining the unstable modes and eliminating the slaved variables, and (2) solving the order parameter equations.

BIFURCATION THEORY OF CONTINUOUS MEDIA: GENERALIZED GINZBURG-LANDAU EQUATIONS‡

Here we treat space-dependent variables $\mathbf{q}(\mathbf{x}, t)$ and

$$\dot{\mathbf{q}} = \mathbf{N}(\mathbf{q}, \nabla, \alpha) + \mathbf{F}(\mathbf{q}, \nabla). \qquad (22)$$

We start from a quiescent and homogeneous solution $\mathbf{q} = 0$ and assume the non-linear expression $\mathbf{N}$ in the form

$$(H_{\mu\nu} + \delta_{\mu\nu}D_\mu\nabla^2)q_\mu + \text{bilinear} + \text{cubic}, \quad H_{\mu\nu} = \text{const}. \qquad (23)$$

where we keep terms up to third power. We admit a *continuous spectrum* of the linear operator in (23). To determine the bifurcating solutions we write $\mathbf{q}(\mathbf{x}, t)$ as a superposition of wave packets

$$\mathbf{q}(\mathbf{x}, t) = \sum_{\mathbf{k},j} \mathbf{O}^{(j)}(\nabla) \underset{\substack{\text{slowly}\\\text{varying}}}{\xi_{\mathbf{k},j}(\mathbf{x}, t)} \underset{\substack{\text{solution of}\\\text{wave eq.}}}{\chi_{\mathbf{k}}(\mathbf{x})} \qquad (24)$$

where $\chi_{\mathbf{k}}$ values are solutions of the wave equation, $\xi_{\mathbf{k},j}$ are slowly varying space and time dependent functions which play the role of order parameters and slaved mode amplitudes. $\mathbf{O}_j$ is an operator solution of the linearized wave equations with constant coefficients (for more details see Reference 5). Inserting (24) into (22) leads after some analysis to equations of the form

$$\dot{\xi}_{\mathbf{k},j} = \hat{\lambda}_j(\nabla, \mathbf{k})\xi_{\mathbf{k},j} + \Sigma \cdots \xi_{\mathbf{k}'j'}\xi_{\mathbf{k}''j''} + \text{cubic terms} \qquad (25)$$

The stable and unstable modes belong to a continuum (compare FIGURE 5). Tak-

---

‡We present here our own approach.[5,8] For other bifurcation schemes, with emphasis on fluid dynamics, see References 9 and 10.

ing into account small band excitations we allow for a whole continuum of unstable modes. Eliminating the stable modes by a formula which is a straightforward generalization of (16) and making the assumption of small band
excitation we find the following set of coupled differential equations

$$\frac{d}{dt} \xi_{[k,u]} = (\alpha_k + \beta_k \nabla^2)\xi_{[k,u]}$$

$$+ \Sigma \ldots \xi_{[k',u']}\xi_{[k'',u'']}$$

$$+ \Sigma \ldots \xi_{[k',u']}\xi_{[k'',u'']}\xi_{[k''',u''']} + F \qquad (26)$$

where the dots indicate coefficients.

When we specialize these equations to a single-order parameter and drop the

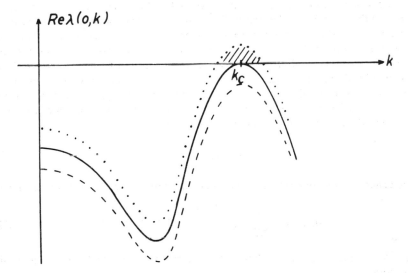

FIGURE 5. Example of unstable mode continuum at $|k| \neq 0$.

second, bilinear, term we obtain the Ginzburg-Landau equation being used in the
phase transition theory of superconductors, ferromagnets, and other materials.[4]
The same type of equations could be derived for lasers [11,4] and the soft mode
instability of a chemical reaction model[12] (Brusselator). In the latter cases we again
deal with nonequilibrium phase transitions. Because of these reasons I have called
the coupled set (26) generalized Ginzburg-Landau equations. In the above mentioned case (single-order parameter) the stationary solution of the corresponding
Fokker-Planck equation reads[11]

$$f(\{\xi(x)\}) = \mathfrak{N} \exp\{-\int dx [\hat{\alpha}|\xi(x)|^2 + \hat{\beta}|\xi(x)|^4 + \hat{\gamma}|\nabla\xi|^2]\}. \qquad (27)$$

It has played a fundamental role in the theory of superconductivity and now applies to lasers, chemical reactions, and so forth. There are important classes of

the general equations (26) to which we can explicitly construct the stationary distribution function corresponding to (27), namely, when the principle of detailed balance holds.[4] In any case the stationary distribution function can be written in the form

$$f(\{\xi_u\}) = \exp[-V(\{\xi_u\})], \tag{28}$$

which introduces again a new aspect into bifurcation theory: The minima of the potential function V allow us to find those configurations of the order parameters which are globally or locally stable. It thus introduces a new criterium for realizations of bifurcating solutions. See below, "Stochastic Approach."

BIFURCATION OF LIMIT CYCLES AND MULTIPERIODIC FLOWS[13]

The problem of bifurcation of limit cycles has been treated by a number of authors in the mathematical literature[14-20] and is also dealt with by other speakers at this meeting. The following procedure seems to be more general in some respects. We do not confine ourselves to the bifurcation to tori, but allow for still more complex motions. Furthermore we treat bifurcation of multiperiodic flows as defined by Equation 30 below. We assume that for a certain set of control parameters we have found a stable limit cycle described by $\mathbf{q}_0(t)$. We now assume that this limit cycle becomes unstable due to a change of control parameters. Since $\mathbf{q}_0$ is a periodic function, it can be written in the form

$$\mathbf{q}_0(t) = \sum_m \mathbf{q}_m e^{im\omega t} \tag{29}$$

More generally we can consider a multiperiodic flow in which case we assume $\mathbf{q}_0$ in the form

$$\mathbf{q}_0(t) = \sum_{m_1, m_2 \ldots} \mathbf{q}_{m_1, m_2} \cdots e^{i(m_1\omega_1 + m_2\omega_2 + \ldots)t} \tag{30}$$

where the number of ω values is finite. We first perform the usual stability analysis by putting

$$\mathbf{q}(t) = \mathbf{q}_0(t) + \mathbf{u}(t), \tag{31}$$

inserting it into (3) with (4) and linearizing (3), which yields

$$\dot{\mathbf{u}}(t) = L(t)\mathbf{u}(t). \tag{32}$$

Since N (Equation 4) does not depend explicitly on t (autonomous system), $L(t)$ has the same periodicity as (29) or (30). In the case of a periodic $L(t)$, according to Floquet's theory the solution of (32) reads

$$\mathbf{u}_\mu(t) = e^{\lambda_\mu t}\mathbf{v}_\mu(t), \tag{33}$$

where $\mathbf{v}_\mu$ has the same periodicity as $L(t)$, provided there is no degeneracy (otherwise, finite powers of t can occur in $\mathbf{v}$). More recently it occurred to me that this theorem can be generalized to $L(t)$ in a form analogous to (30) provided certain analyticity conditions for L and $\mathbf{v}$ are fulfilled.[21] In the following we assume that the bifurcating solution stays close to the original trajectory (in the sense of orbital stability) but that it need not be stable in the usual sense. That means the point on the bifurcating trajectory can move away from the original point arbitrarily far. To take this into account we introduce a phase $\Phi(t)$ and make the general hypothesis

$$\mathbf{q} = \mathbf{q}_0(t) + \Sigma \xi_\mu(t) \mathbf{v}_\mu[t + \Phi(t)]. \tag{34}$$

Since the discussion of this hypothesis and the further treatment are rather lengthy we must refer the reader to the original publication.[13] We just mention an essential point. When we insert (34) into the original nonlinear equation we can derive exact equations for ξ_μ terms, which can again be grouped into two classes Re $\lambda_\mu \gtrless 0$ with $\mu = u$ (unstable) and $\mu = s$ (stable). A typical example, analogous to (12), is

$$\dot{\xi}_s = \lambda_u \xi_s - g(t)\xi_u{}^2 \tag{35}$$

The important result is that (35) can be solved explicitly and uniquely by

$$\xi_s = -\left(\frac{\mathrm{d}}{\mathrm{d}t} - \lambda_\mu\right)^{-1}[g(t)\xi_u{}^2], \tag{36}$$

where the inverse exists, for instance, if ξ contains only oscillatory terms or belongs to a still broader class of slowly increasing or decreasing periodic functions. In general we obtain a whole set of equations for the order parameters and the slaved modes, and Equations 35 become more complicated. However, one can also devise a rapidly converging iteration scheme to express ξ_s as a functional of ξ_u.

In conclusion of the above we can state the following: We have developed general schemes to eliminate the slaved modes and to derive equations for the order parameters alone. Since in general the order parameter equations are much fewer than the original equations, an enormous reduction of the degrees of freedom has been obtained. Now let us go from mathematics over to applications to complex systems belonging to various disciplines. We now recognize that even very complicated temporal and spatial patterns or functions can be described by very few variables. Since the resulting order parameter equations can be grouped into "universality" classes we now understand that seemingly completely different complex systems behave in an entirely analogous fashion. Space does not allow us here to list the numerous examples belonging to these classes.

A note should be added. It is transparent from what has been said above that our procedure implies a self-consistency requirement for the unstable modes, namely we imply that the resulting order parameter equations do not give rise to "run-away solutions." This assumption must be checked in each case (one can in fact construct counter examples where the self-consistency requirement is not fulfilled, at least not for all times. A simple example is the Lorenz attractor which we will discuss below). Self-consistency methods are widespread in physics and chemistry and many results of the present day are based on them. This should be particularly stressed at the present interdisciplinary meeting because usually mathematicians first seek existence theorems for certain classes of solutions (a procedure well known since Gauss) and then devise construction methods. In a number of cases this requires one to make very precise assumptions which are then often not met by systems found in nature. When mathematicians and other scientists talk to each other and develop their methods they should be fully aware of these difficulties.

STOCHASTIC APPROACH: NONEQUILIBRIUM PHASE TRANSITIONS

In many cases, especially at instability points, fluctuations are crucial because they drive the system into new configurations. There are essentially three methods at our disposal (see Haken[5] for a review with further references):

(a) Langevin equations of the type

$$\dot{\mathbf{q}} = \mathbf{N}(\mathbf{q}, \alpha) + \mathbf{F}(\mathbf{q}, t) \tag{37}$$

(or still more general equations of the type of Equation 3). If the fluctuating forces depend on the system coordinates usually Îto's formulation is applied.

(b) If the process is Markovian, the time-dependent distribution function of $\mathbf{q}$ obeys the Chapman-Kolmogorov equation:

$$P(\mathbf{q}, t) = \int p_{t,t'}(\mathbf{q}, \mathbf{q}')P(\mathbf{q}', t')d^n q', \quad t \geq t'. \tag{38}$$

(c) If the process is a continuous Markov process the Fokker-Planck equation holds:

$$\dot{f}(\mathbf{q}, t) = \left\{ -\sum_j \frac{\partial}{\partial q_j} K_j(\mathbf{q}) + \frac{1}{2} \sum_{jk} \frac{\partial^2}{\partial q_j \partial q_k} Q_{jk}(\mathbf{q}) \right\} f(\mathbf{q}, t). \tag{39}$$

It is often useful to proceed from (38) to the so-called master equation by taking an appropriate limit $t' \rightarrow t$. As we have seen above, an essential generalization of bifurcation theory seems to consist in the slaving principle by which we could reduce the total system of equations to a system of order parameter equations of much smaller number. Within the Langevin equations the elimination procedure described above can be immediately repeated. The master equation or Fokker-Planck equation can also be subjected to an elimination procedure,[5] which we describe now by means of a simple example referring to the Fokker-Planck equation. We transform the Fokker-Planck equation to unstable and stable modes, which have been determined, for instance, by linearized equations of motion attached to (39). We then put

$$f(\mathbf{q}, t) \rightarrow f(\xi_u, \xi_s, t) = h(\xi_s/\xi_u)g(\xi_u), \tag{40}$$

and require that the conditional probability h obeys the equation

$$\left(-\frac{\partial}{\partial \xi_s} K_s + \frac{1}{2} Q_{ss} \frac{\partial^2}{\partial \xi_s^2} \right) h = 0, \tag{41}$$

which implies an adiabatic principle. The distribution function of the order parameter ξ_u is then determined by

$$\dot{g}(\xi_u) = -\frac{\partial}{\partial \xi_u} \int K_u h d\xi_s \cdot g(\xi_u) + \frac{1}{2} Q_u \frac{\partial^2 g}{\partial \xi_u^2}. \tag{42}$$

Using this method directly (or using the slaving principle with respect to Langevin equations) we find for example the reduced Fokker-Planck equation for ξ_u in the case of Equations 11 and 12 to which suitable fluctuating forces (Gaussian and Markovian) have been added. The stationary solution reads[22,23]

$$f(\xi_u) = \mathfrak{N} \exp[(\lambda_u \xi_u^2 + \xi_u^4/(2\lambda_s)](4Q)^{-1}. \tag{43}$$

The shape of this distribution changes qualitatively if λ_u (λ_u real) changes its sign (compare FIGURE 6). This distribution function has played an enormous role in physics, where Equation 28 appears in the Landau theory of phase transitions, of systems in thermal equilibrium, for instance, structural or magnetic phase transitions.§ At that time it was derived only by means of plausibility arguments, whereas nowadays it can be derived as a solution of the Fokker-Planck equation. As we now know, (28) describes also transitions of physical systems far from thermal equilibrium ("nonequilibrium phase transitions," e.g., the laser) and even distributions in sociology (e.g., formation of public opinion). Once $f(\mathbf{q}, t)$ is de-

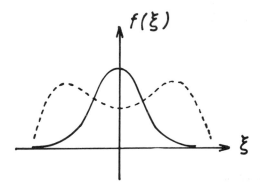

$f(\xi)$

FIGURE 6. The function (28) for $\lambda_u < 0$ (solid curve) and $\lambda_u > 0$ (dashed curve).

termined one can calculate moments and correlation functions. Lack of space does not allow me to discuss here the general methods including renormalization group techniques.

AN EXAMPLE FROM MODERN PHYSICS: THE LASER AND ITS INSTABILITY HIERARCHIES

The left part of FIGURE 7 shows a typical experimental set up of a laser. It consists of a rod of active material, for instance a ruby crystal whose two end faces carry mirrors, which select a certain light wave. As indicated in that figure the atoms of the active material are energetically pumped from the outside. The atoms emit light and thus create the laser field E. The laser is a truly synergetic sys-

§Landau & Lifshitz.[24] For a detailed discussion of analogies between equilibrium and nonequilibrium systems including fluctuations see Landauer.[44]

tem. In it the cooperation of the atoms determines the properties of the emerging laser light field. We have found that there is an instability hierarchy. When the laser is pumped only weakly, the electric field consists of randomly emitted wave tracks representing noise (actually all emitted light from the usual type of lamp is just noise). At a certain critical pump strength, quite suddenly a coherent (sinusoidal) wave appears.[25] With further increased pump strength, again suddenly the coherent wave breaks into pulses.[26–28] The connection with bifurcation theory becomes immediately evident if we treat the field strength E as order parameter[28] and plot it against the pump strength. As the left hand side of FIGURE 8 reveals first there is no coherent emission at all. The order parameter is zero. At the pump strength D_1 the coherent wave solution emerges. When we increase the pump strength further this bifurcated solution becomes unstable again. A new order

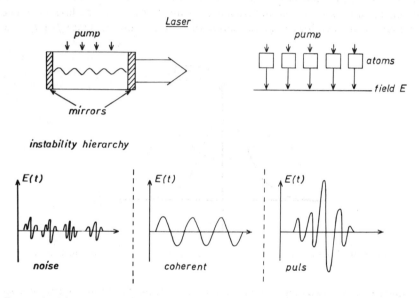

FIGURE 7. Scheme of experimental set-up of a laser (upper left) and instability hierarchy.

parameter $\tilde{\xi}_u$ of pulses occurs (right-hand side). While Ohno and I[28] have first found inverted bifurcation (right-hand side, lower part), Kirchgäßner and Renardy[29] found usual bifurcation. We could recently clarify this result,[30] showing that the kind of bifurcation depends on the length of the laser. The order parameter equation of the oscillating mode reads

$$\dot{\xi}_u = \beta\xi_u + \alpha\xi_u^3 + \beta\xi_u^5 \tag{44}$$

The coefficients β and α depend strongly on the laser length as shown in FIGURE 9.

Quite amazingly we could recently show[31] that under certain other conditions the new occurring instability of the laser just leads to the "strange" Lorenz attractor[33] (see below, "New Interpretation of the Lorenz Model of Turbulence").

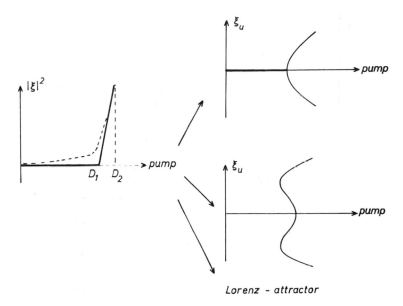

FIGURE 8. Consult text.

For readers more interested in the laser problem we represent the basic equations used in laser theory.[32]

$$\frac{\partial E}{\partial t} + c\,\frac{\partial E}{\partial x} + \kappa E = \alpha P,$$ (45)

$$\frac{\partial P}{\partial t} + \gamma P = \beta ED,$$ (46)

$$\frac{\partial D}{\partial t} = \gamma''(D_0 - D) + \delta EP,$$ (47)

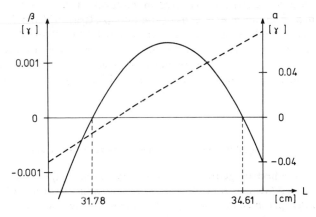

FIGURE 9. The coefficients β (solid curve) and α (dashed line).

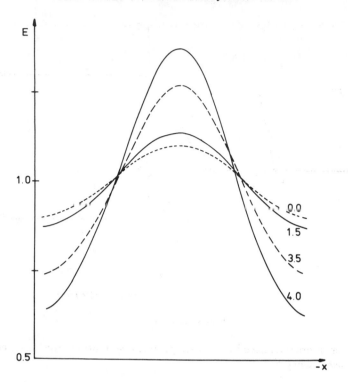

FIGURE 10. Build-up of ultrashort laser pulse (after Ohno and Haken[30]).

where the field strength E, the polarization P, and the so-called atomic inversion D of the two-level systems are space- and time-dependent functions; c is the light velocity, γ and γ'' the damping constants, D_0 depends on the prescribed pump intensity and represents the former control parameter α. The kind of the first bifurcation does not depend on relative magnitudes of κ, γ, and γ''. However, whether bifurcation is normal or leads to the Lorenz attractor depends on the relative size of κ, γ, and γ'': bifurcation or inverted bifurcation if $\kappa < \gamma + \gamma''$, Lorenz attractor if $\kappa > \gamma + \gamma''$. FIGURE 10 represents the increase of pulses determined by higher order approximations to generalized Ginzburg-Landau[5] equations (compare with above, "Bifurcation of Limit Cycles and Multiperiodic Flows").

AN EXAMPLE FROM BIOLOGY: MORPHOGENESIS

Using the method of generalized Ginzburg-Landau equations, we recently treated the Gierer-Meinhardt model[34] of morphogenesis. It turns out[35] that there are three order parameters belonging to modes whose wave vectors form an equilateral triangle. These modes stabilize each other and give rise to hexagonal

patterns. Taking into account the new concept of time-dependent order parameters slaving the other modes, we could determine the evolving structure as shown
in FIGURE 11. These pictures clearly demonstrate the applicability of the order
parameter concept.

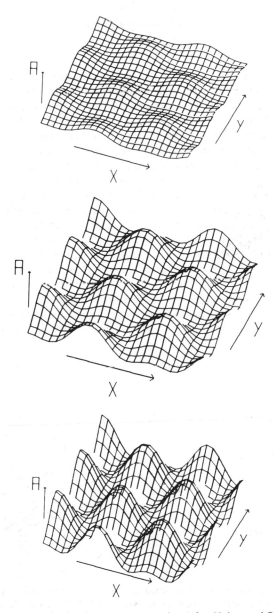

FIGURE 11. Growth of activator-concentration (after Haken and Olbrich[35]).

NEW INTERPRETATION[36] OF THE LORENZ MODEL OF TURBULENCE[33]

When we neglect the spatial derivative in Equation 45 and scale the quantities E, P, and D correspondingly, we find the equations

$$\dot{X} = \boxed{- X} + Y \tag{48}$$

$$\dot{Y} = \boxed{- Y} + XZ \tag{49}$$

$$\dot{Z} = \boxed{Z_0 - Z} - XY, \tag{50}$$

which are, aside from coordinate transformations, equivalent to the by now well-known Lorenz equations of turbulence. Knowing the laser case, however, we can give a new interpretation to these equations. From the laser we know that the terms in the dashed box represent damping whereas the residual terms on the right-hand side stem from a Hamiltonian. In other words those latter terms describe a conservative system allowing for conservation laws. These conservation laws describe energy conservation and the length of the so-called pseudo-spin of the two-level atoms. In our context it is good to know that these conservation laws can be given a form in which one law means that the representative point must move on a sphere; the other one, that the representative point must lie on a cylinder in X, Y, Z-space. Since both laws are valid simultaneously, the representative point must move on the cross-section between sphere and cylinder. Now comes the essential point. When the radius of the cylinder is small, the motion is restricted to one region of space (FIGURE 12a). If, however, that radius is bigger than a critical radius the motion covers two regions of space (FIGURE 12b). When we switch on damping terms (dashed box), both sphere and cylinder start breathing, thus giving rise to jumps from one region to the other one, well known from the Lorenz attractor. Since the jumps depend very sensitively on the rela-

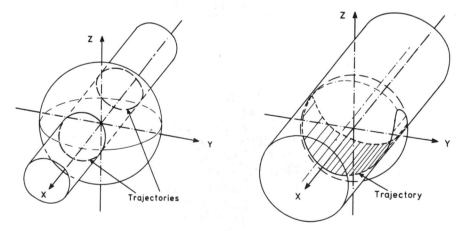

FIGURE 12. The Lorenz model of turbulence. (Left) When the radius of the cylinder is small, the motion is restricted to one region. (Right) If radius is larger the motion includes two regions of space.

TABLE 1

THE GENERAL SCHEME

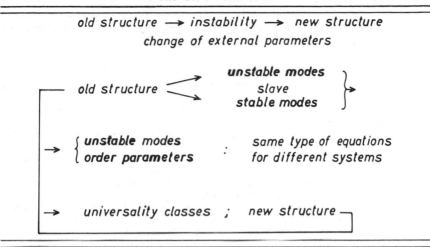

---

tive size of the radii and the actual position of the representative point of the trajectory, it becomes at least qualitatively clear that a seemingly statistical motion evolved.¶

SOME EPISTEMOLOGY

With our above remarks we have indicated only a few ideas and mathematical tools employed in the field of synergetics. However, we hope that we have indicated that the order parameter concept is a very powerful tool which allows us to classify quite different systems and to understand their common features. This result is pointed out in the scheme of TABLE 1. So far I have been talking only about hard sciences in which processes can be described or modeled by mathematical equations. Let me now make a big jump over to a soft science, namely, to epistemology. It appears to me that the above concepts apply also to epistemology in the following sense. In the developments of science we very often find a situation in which a good deal of individual facts are known and more or less incoherent hypotheses have been established to account for these facts. But suddenly an entirely new coherent concept evolves that is capable of explaining all hitherto only poorly understood phenomena. When the new idea (in our context, the order parameter) is born, from then on it governs the thinking of a scientific discipline. In science we are usually accustomed to assume that these new ideas are uniquely determined. However, ecology or politics teach us that new evolving structures, which are again governed by order parameters, need not be uniquely

---

¶Numerous examples of chaotic motion have been studied by O. E. Roessler. For a list of references consult his article in Reference 6. An interesting class of "complicated dynamics" has been studied by May.[37]

determined. There are several political systems or economical systems possible. These different possibilities are no surprise to us here. They just represent the bifurcating solutions. It would certainly be interesting to discuss these ideas in other soft sciences, for instance with respect to languages, political developments, and so forth, but this is far beyond the scope of this article.

REFERENCES

1. HAKEN, H. & R. GRAHAM. 1971. Synergetik—die Lehre vom Zusammenwirken. Umschau **6**: 191–195.
2. HAKEN, H., Ed. 1973. Synergetics—Cooperative Effects in Multicomponent Systems. B.G. Teubner. Stuttgart.
3. HAKEN, H., Ed. 1974. Cooperative Effects, Progress in Synergetics. North-Holland. Amsterdam.
4. HAKEN, H. 1975. Cooperative Effects in Systems far from thermal equilibrium and in nonphysical systems. Rev. Mod. Phys. **47**: 67–121.
5. HAKEN, H. 1977. Synergetics. An Introduction. Nonequilibrium Phase Transitions and Self-Organization in Physics, Chemistry and Biology. Springer. New York.
6. HAKEN, H., Ed. 1977. Synergetics. A Workshop. Springer. New York.
7. VAINBERG, M. M. & V. A. TRENOGIN. 1974. Theory of Branching of Solutions of Nonlinear Equations. Noordhoff International Publishers. Leyden.
8. HAKEN, H. 1975. Z. Phys. B **22**: 69–72, 73–77.
9. JOSEPH, D. D. 1976. Stability of Fluid Motions, I, II. Springer. New York.
10. SATTINGER, D. H. 1973. Topics in Stability and Bifurcation Theory. Lecture Notes in Mathematics.: 309. Springer. New York.
11. GRAHAM, R. & H. HAKEN. 1970. Z. Phys. **237**: 31–46.
12. HAKEN, H. 1975. Z. Phys. B **20**: 413–420.
13. HAKEN, H. 1978. Z. Phys. To be published.
14. RUELLE, D. & F. TAKENS. 1971. Commun. Math. Physics **20**: 167–192; **23**: 343–344.
15. LANFORD, O. E., III. 1973. *In* Lecture Notes in Mathematics. Vol. 322. Springer. New York.
16. JOSEPH, D. D. 1977. *In* Synergetics, A Workshop. Springer. New York.
17. GUREL, O. & O. E. ROESSLER. 1977. Preprint.
18. ROESSLER, O. E. Private communication.
19. CRANDALL, M. G. & P. H. RABINOWITZ. 1976. The Hopf Bifurcation Theorem. MRC Technical Summary Report No. 1604.
20. WEINBERGER, H. 1978. On the Stability of Bifurcating Solutions. *In* Nonlinear Analysis. Academic Press. New York, N.Y.
21. HAKEN, H. 1978. To be published.
22. RISKEN, H. 1965. Z. Phys. **186**: 85–98.
23. HEMPSTEAD, R. D. & M. LAX. 1967. Phys. Rev. **161**: 350–366.
24. LANDAU, L. D. & E. M. LIFSHITZ. 1952. Course of Theoret. Physics. Vol. **5**: 430–456.
25. HAKEN, H. 1964. Z. Phys. **181**: 96.
26. GRAHAM, R. & H. HAKEN. 1968. Z. Phys. **213**: 420–450.
27. RISKEN, H. & K. NUMMEDAL. 1968. J. Appl. Phys. **39**: 4662–4672.
28. HAKEN, H. & H. OHNO. 1976. Opt. Commun. **16**: 205–208.
29. KIRCHGÄSSNER, K. & RENARDY. Private communications.
30. OHNO, H. & H. HAKEN. 1978. To be published.
31. HAKEN, H. 1975. Phys. Lett. **53A**: 77–78.
32. HAKEN, H. 1970. Laser theory. *In* Encyclopedia of Physics. Vol. **25**: 2c. Springer. New York.
33. LORENZ, E. N. 1963. J. Atmosph. Sci. **20**: 130–141.
34. GIERER, A. & H. MEINHARDT. 1974. *In* Lectures on Mathematics in the Life Sciences. Vol. **7**: 163–183. Springer. New York, N.Y.
35. HAKEN, H. & H. OLBRICH. 1978. To be published.

36. HAKEN, H. & A. WUNDERLIN. 1977. Phys. Lett. **62A** (3): 133–134.
37. MAY, R. M. 1976. Nature **261**: 459–467.
38. GUREL, O. 1975. Collective Phenomena **2**: 89–97.
39. GUREL, O. 1976. Simulation of Systems. L. Dekker, Ed.: 53–58. North Holland Publishing Company. Amsterdam.
40. GUREL, O. 1976. Dynamical Systems. Vol. 2: 255–259. Academic Press, New York.
41. GUREL, O. 1977. Phys. Lett. **61A**: 219–223.
42. GUREL, O. 1978. Poincaré's Bifurcation Analysis. Ann. N.Y. Acad. Sci. This volume.
43. RABINOWITZ, P. H. 1977. J. Tract. Anal. **25**: 412.
44. LANDAUER, R. 1977. The role of fluctuations in multistable systems and in the transition to multistability. Ann. N.Y. Acad. Sci. This volume.
45. KIRCHGASSNER, K. 1977. Bifurcation of a Continuum of Unstable Modes. *In* Synergetics. A Workshop. H. Haken, Ed. Springer. New York, N.Y.

CONTINUOUS CHAOS—FOUR
PROTOTYPE EQUATIONS

Otto E. Rössler

Institute for Physical and Theoretical Chemistry
University of Tübingen
7400 Tübingen, Federal Republic of Germany

Institute for Theoretical Physics
University of Stuttgart
Federal Republic of Germany

INTRODUCTION

If oscillation is *the* typical behavior of 2-dimensional dynamical systems (Euclidean and on manifolds), then chaos, in the same way, characterizes 3-dimensional continuous systems. First a method to obtain chaos in degenerate (relaxation type) dynamical systems in two variables is outlined whereby five basic flow patterns emerge. Second, following a piecewise linear degenerate equation, four prototypically simple quadratic differential equations in three variables that realize nondegenerate analogs of those five flows are presented. Finally a possible equation for an even higher type of qualitative behavior beyond chaos is proposed.

THE ASYMPTOTIC CASE

Liénard's "building-block principle"[2] allows not only for relaxation oscillations, but also for higher types of relaxation behavior. The five simplest next-higher possibilities are shown in FIGURE 1. These illustrations are implications of Khaikin's universal circuit principle.[3,4] FIGURE 1c was independently seen in a similar form by Takens as an example of a "constrained" differential system with nontrivial behavior.[5] The examples in FIGURE 1a, e, and b have been described elsewhere.[4,6,7] The circuit in FIGURE 1d is new. Combinations of these flows yielding more complicated 3-dimensional networks are also possible.[7]

When looking at the flows of FIGURE 1, two basic questions come to mind: 1) How can the observed "entangled" trajectorial behavior be described? 2) Are there analogous "nondegenerate" flows? The first question is easy to answer: The five cross sections through the flows of FIGURE 1 (indicated by the arrow marked "P") apply if a ray of recurrent reentry is assumed to be drawn through each flow (FIGURE 2).

These maps belong to the class of Fatou-Julia-Myrberg-Ulam-Sharkovsky-Lorenz-Li-Yorke maps, and hence, under certain quantitative conditions, imply "chaos," that is, an infinite number of unstable periodic trajectories and an uncountable number of nonperiodic recurrent ones.[8,9,10,11]

0077-8923/78/0316-0376 $1.75/1 © 1979, NYAS

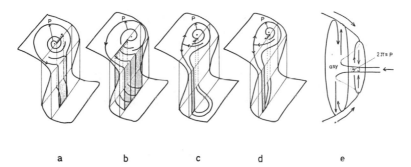

a   b   c   d   e

FIGURE 1. The five simplest possibilities to obtain chaos in 3-dimensional relaxation systems: a) spiral type chaos, b) screw type chaos, c) "inverted" spiral-plus-saddle-type chaos, d) "noninverted" spiral-plus-saddle-type chaos, e) toroidal chaos (P = Poincaré arrows; asy = asymmetry).

THE NONDEGENERATE CASE

By relaxing the constraint of "infinite velocity" of fast transitions at the cliffs of the slow manifold (see Ref. 12 for notation), analogous flows of "finite width" are generated. The former "paper models" are now replaced by "flintstone models." Assuming that the deviation from the idealized case is very small, the cross sections in FIGURE 3 are obtained. They are only approximate because the (thin) second dimension has been neglected. The main message of this figure is that some of the sharp edges in FIGURE 2 (namely, in a, b, e) have been rounded off. This is due to the fact that a continuous gradation between "switched" and "nonswitched" trajectories exists in these "saddle-free" flows. The maps in FIGURE 3 are still Li-Yorke maps, thus generating chaos. Three of them (a, b, e), however, now generically possess at least one attracting orbit among the infinite set of periodic trajectories determined. This is due to the presence of a smooth maximum (see Refs. 9 and 13, where a special case is considered in detail).

FIGURE 4 shows how the same cross sections look like when the second dimension is considered explicitly. The "nonidealized" maps in this figure now belong into two different classes: "bent" and "cut" maps. The bent maps (a, b, e) can be

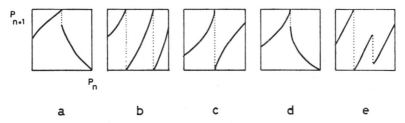

a   b   c   d   e

FIGURE 2. Poincaré maps corresponding to the different types of chaos shown in a–e of FIGURE 1, respectively. Only the attracting submaps are depicted (cf. Refs. 4, 7).

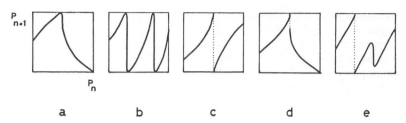

FIGURE 3. Poincaré maps, like those of FIGURE 2, obtained if the idealizing assumptions of "infinitely fast" relaxation are dropped in FIGURE 1 (approximate 1-dimensional maps).

classified as generalized horseshoe maps.[4,7] The simplest case is the "walking-stick diffeomorphism" in FIGURE 4a. The second iterate of such maps contains a horseshoe map (nonlinear, double, attracting) if certain natural simplicity assumptions are fulfilled.[7] Horseshoe maps, as introduced by Smale,[15,16] are known to imply an infinite number of periodic trajectories and an uncountable set of non-periodic ones.[15] Thus, the theory of horseshoe diffeomorphisms generalizes and explains the results that hold true for related 1-dimensional endomorphisms.

The result mentioned above that smooth-maximum 1-dimensional maps

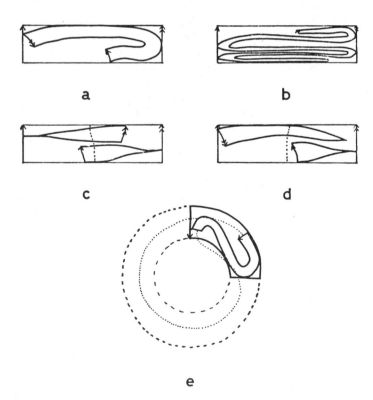

FIGURE 4. Two-dimensional Poincaré cross sections (nonapproximate case of FIGURE 3).

generically contain a periodic attractor[13] suggests that bent 2-dimensional diffeo-morphisms also generically contain an attracting periodic solution (Smale, personal communication). This is indeed the case for the maps in FIGURE 4a, b, and e (due to absence of a property of lateral elongation everywhere—see below). The remaining two maps (FIGURE 4c, d) are of the cut type and hence not diffeomorphic. They become diffeomorphisms only if a certain line segment in the original cross section is deleted; thus they represent an especially "benign" type of 2-dimensional endomorphisms. For these maps it is not hard to show that all the infinitely many periodic orbits are laterally unstable, that is, nonattracting. Their existence follows from Smale's horseshoe proofs.[16] As a consequence, these maps describe "strange attractors" (in the sense of Ruelle and Takens[17]) just as the corresponding maps in FIGURE 2 (and all of the maps in FIGURE 1) do.

Of the two nondiffeomorphic maps in FIGURE 4 (called sandwich maps[18]), FIGURE 4c is well investigated.[19,20,21] It yields the so-called Lorenz attractor which, in fact, corresponds to an infinite family of related attractors.[20] The other (folded) sandwich map in FIGURE 4d has apparently not been described before. The structure of the new strange attractor formed in this map is related to that of the Lorenz attractor.

To sum up, Liénard's principle of coupling through very slowly changing "pseudo-parameters" proves well suited for the design of higher-order dynamical systems of prescribed behavior. This result was to be expected from the 2-dimensional case.[22,23]

AN IDEAL EXAMPLE

The design principle underlying the flows in FIGURE 1a and b is realized by the following relaxation system,

$$\dot{x} = -y + ax - bz$$
$$\dot{y} = x + 1.1 \tag{1}$$
$$\epsilon\dot{z} = (1 - z^2)(x + z) - \epsilon z.$$

This equation has the advantage that, in the limit of ϵ approaching zero from above, a piecewise linear system in x and y is obtained. The parameter z then has the constant value $+1$ or -1 depending on which of the two hysteresis thresholds of x ($x = 1$ or $x = -1$) has been passed last. (For techniques showing how to deal with piecewise linear systems, see e.g., Ref. 24.) Moreover, singular perturbation techniques can be applied to Equation 1 in a neighborhood of the limit.[25,26] Thus Equation 1 is an "ideal" example insofar as it lends itself to an analytical approach.

The numerical simulations displayed in FIGURE 5 show that the behavior of Equation 1 is still rather close to asymptotic when ϵ is as large as 0.03. The difference, which causes an unusual type of nonuniform convergence under singular perturbation, is also visible: one of the two "holes" of the loop is covered by a continuous curtain. In FIGURE 5B, a trajectory in this region is clearly visible.

An analogous ideal equation can be set up for the flow of FIGURE 1e, the

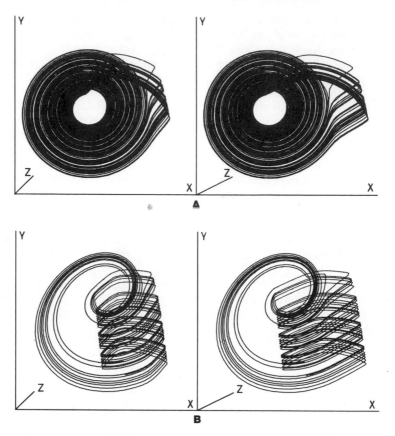

FIGURE 5. Spiral-type (A) and screw-type (B) chaos in Equation 1. Both are stereoscopic plottings. Numerical simulation is done by using a standard Runge-Kutta-Merson integration routine. The left-hand picture is for the right eye and vice versa. (Try to fixate on a pencil about 14 cm from your eyes so that, of the four blurred pictures behind the pencil, the two innermost merge; then just wait for them to come into focus.) Parameters: (A) $\epsilon = 0.03$, $a = 0.1$, $b = 1$; (B) $\epsilon = 0.03$, $\alpha = 0.5$, $b = 5$. Initial conditions: (A) $x(0) = y(0) = 0$, $z(0) = -1$, $t_{end} = 630.3$; (B) the same, $t_{end} = 11.6$ Axes: (A) $-3.5 \ldots 2.5$ for x, $-2 \ldots 4$ for y, $-1 \ldots 1$ for z; (B) $-6 \ldots 4$ for x, $-2 \ldots 8$ for y, $-1 \ldots 1$ for z.

rotation-symmetrical case.[27] Equations that are more or less accessible analytically may also be found for the other flows in FIGURE 1. Such equations, however, are not prototypic in the sense of maximum possible simplicity. Therefore, four simplified equations will be presented in the following section which seem to retain the qualitative behavior of corresponding more complicated equations.

FOUR PROTOTYPIC EQUATIONS

First equation. The following simple quadratic differential equation contains only one nonlinear term

$$\dot{x} = -y - z$$
$$\dot{y} = x + ay \qquad\qquad (2)$$
$$\dot{z} = bx - cz + xz$$

Equation 2 has been simplified from an equation,[4] which like (1), realizes the flow of FIGURE 1a and b in the limit of an infinitely weakly coupled third variable. Note the absence of a cubic (or second quadratic) nonlinearity in the third line as would be required for bistability of the "switching variable" z.

The flows generated by Equation 2 (FIGURE 6) are closely related to those of Equation 1 (FIGURE 5). If the flows generated by (1) do have 2-dimensional cross sections of the types shown in FIGURE 4a and b not only if ϵ is very small (so that an asymptotic approximation can be carried out), but also for larger ϵ (as in FIGURE 5), then the same must be true also for the flows in FIGURE 6.

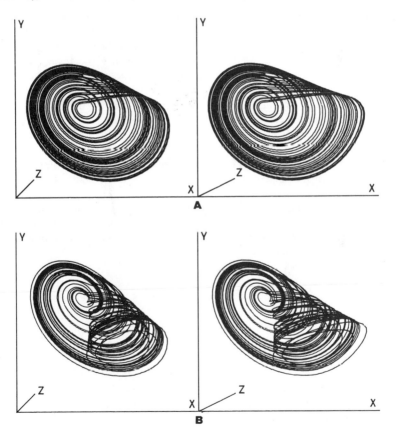

FIGURE 6. Spiral-type (A) and screw-type (B) chaos in Equation 1. Numerical simulation as in FIGURE 5. Parameters: (A) $a = 0.36$, $b = 0.4$, $c = 4.5$; (B) $a = 0.5$, $b = 0.4$, $c = 4.5$. Initial conditions: (A) $x(0) = z(0) = 0$, $y(0) = 3$, $t_{end} = 269.4$; (B) the same for $x(0)$, $y(0)$, $z(0)$, $t_{end} = 244$. Axes: (A) $-6 \ldots 10$ for x, $-8 \ldots 8$ for y, $0 \ldots 8$ for z; (B) $-8 \ldots 14$ for x, $-14 \ldots 8$ for y, $0 \ldots 18$ for z.

Of course, a proof that the chaotic regimes in FIGURE 6 have exactly the same properties as those which can be derived for more complicated analogs that are sufficiently close to their asymptotic limits cannot be made. This is because the Poincaré maps governing the flows in FIGURE 6 might possess additional properties that are not predictable from the limiting cases. The existence of equations like (2) makes it necessary, therefore, to distinguish between "essential" and "accidental" properties of walking-stick-like maps. The robustness of their main properties toward distortions seems to be rather great. This problem is taken up below (following FIGURE 11).

A film displaying the trajectorial behavior of almost the same equation (the term bx replaced by a constant[29]) under variation of a exists.[18] It shows the following bifurcations under a slow increase of a (the parameter that determines instability of the unstable focus). At first, there is a small planar limit cycle around the unstable focus, growing steadily in size. This limit cycle then becomes nonplanar (that is, tilted at one side) beyond a certain size and thereafter unstable while a new limit cycle of double periodicity, split off from the former, replaces it. The new limit cycle looks like a circular rope laid double. After an infinite number of further splittings, spiral-type chaos appears. With further increasing instability of the unstable focus, screw-type chaos develops. FIGURES 6A and B, which both differ in the numerical value of a alone, give a good impression of the underlying increase in the back-bending of the reinjected trajectories. In the film, the transition from spiral-type to screw-type chaos is not completely smooth: before screw-type chaos is fully established, a new period-2 limit cycle appears for a short while. Also, either chaotic regime displays an attracting period-3 solution over a certain small range of a values. The 6-minute super-8 film has a sound track on which $z(t)$ is recorded so that the different regimes and their bifurcations are audible.[18]

Second equation. The scheme underlying FIGURE 1c calls for a combination of a focus-plus-saddle subsystem with a switching variable. If the focus-plus-saddle subsystem contains two foci, the switching variable can be simplified up to linearity, as Equation 3 shows,

$$\dot{x} = x - xy - z$$
$$\dot{y} = x^2 - ay \tag{3}$$
$$\dot{z} = b(cx - z)$$

Here the first two lines determine a double-focus-plus-saddle system.[18] As seen in stereo pictures,[7,18] the behavior of Equation 3 is apparently the same as that of the Lorenz equation.[30] By introducing a cubic term into the third line of (3), a more complicated equation is obtained which, in the limit, provably possesses a Lorenz attractor of the very type postulated by Guckenheimer,[19] Williams,[20] and Kaplan and Yorke.[21]

The simulation presented in FIGURE 7 is slightly different from ordinary Lorenzian behavior: the two spirals have been replaced by two screws. This variant of Lorenzian behavior is, despite the somewhat more complicated structure of the cross section, equivalent to the ordinary (spiral) type up to a critical value of ϵ, beyond which "foldings" appear in the cross section. Note that ϵ ($\equiv b^{-1}$) is rather large in this figure.

In the film,[18] Equation 3 is also simulated (at "ordinary" values of b); thereby, critical emergence of Lorenzian chaos is demonstrated under the variation of an "asymmetry-generating" parameter (namely, a constant added to the third line). Such an additional parameter is also needed if the asymmetrical flow shown in FIGURE 1c (the reinjected portion of the flow entirely crosses the saddle's stable manifold) is to be obtained from Equation 3. The flow of FIGURE 7 obviously lacks this property, and so does Lorenz's original flow.[30] (Asymmetrical versions of both flows have been described in Reference 7.) The addition of a symmetry-generating term is actually unnecessary since the symmetrical flows still have the same cross sections: If the reinjection loop in FIGURE 1c does not cross the saddle's ω-separatrix as a whole, P can be positioned in such a way that, instead of pointing outward, it cuts through the saddle's separatrix in order to

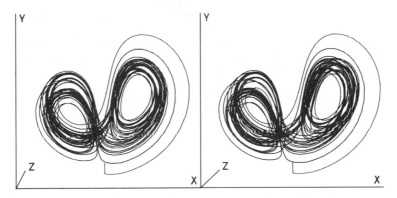

FIGURE 7. Inverted spiral-plus-saddle-type chaos (Lorenzian chaos) in Equation 3; numerical simulation as in FIGURE 5. Parameters: a = 0.1, b = 0.08, c = 0.125. Initial conditions: $x(0)$ = 0, $y(0)$ = 0.2, $z(0)$ = -10^{-6}, t_{end} = 1100. Axes: $-1 \ldots 1.2$ for x, $0 \ldots 2.2$ for y, $-0.02 \ldots 0.02$ for z.

obtain the cross section in FIGURE 2c. The map in FIGURE 4c was, in fact, first proposed as a possible map for ordinary Lorenzian flows.[5,19] Then it was independently introduced to explain an asymmetrical form of chaos observed in Equation 3 with a constant added to the third line.[18] The equivalence of both flows (the Lorenzian one and, supposedly, the simpler asymmetrical one) came as a surprise.[7]

Both the Lorenz equation and Equations 3 produce globally attracting chaotic regimes. Both equations incidentally provide for walking-stick map chaos in a different region of parameter space (see Ref. 18 for Equation 3 and Refs. 31 and 32 for the Lorenz equation). An attracting chaotic regime governed by two nonlinear horseshoe maps in the form of a "cycle"[33] is also possible in the Lorenz equation and Equation 3.[32]

Third equation. The flow in FIGURE 1d is, in a finite thickness version, for example, generated by the (in the third line simplified) Equation 4,

$$\dot{x} = -xy - ax - z$$

$$\dot{y} = -x + by + cz \tag{4}$$

$$\dot{z} = d + exz + fx$$

A simulation result is provided in FIGURE 8. Equation 4 is not more complicated than the Lorenz equation (or Equation 3). Again, there are only two quadratic terms. One difference is that the chaotic attractor is (as in Equation 2) not globally attracting. Equation 4 can probably be simplified further—there is a constant term left, and the parameters are not yet close to unity.

A flow like that of FIGURE 8 has apparently not been observed before. It is almost as simple as the related (but saddle-free) flow in FIGURE 6A. The saddle eliminates the need for a "compression zone" in the region of folding and is, therefore, responsible for turning the chaotic attractor in FIGURE 6A into a closely related strange attractor (see below).

In contrast to FIGURE 1d, the reinjection loop does not completely cross the saddle's stable manifold in FIGURE 8. Both types of flow are again equivalent, however, as a repositioning of P shows (so that it cuts through the saddle's stable manifold). Complete crossing is, however, also possible in Equation 4, this time without the need for an additional term. Incidentally, Equation 4 also produces Lorenzian behavior.

Fourth equation. A simple equation realizing a flow like that of FIGURE 1e is:

$$\dot{x} = -y - z$$

$$\dot{y} = x \tag{5}$$

$$\dot{z} = a(y - y^2) - bz$$

Simulation results for three different sets of parameters are given in FIGURE 9. FIGURE 9A shows an invariant torus. This behavior occurs if the parameter b is set

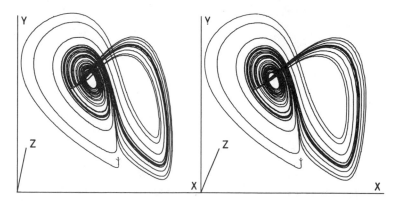

FIGURE 8. Noninverted spiral-plus-saddle-type chaos in Equation 4; numerical simulation as in FIGURE 5. Parameters: $a = 18.5$, $b = 4$, $c = 0.07$, $d = 22$, $e = 2$, $f = 690$. Initial conditions: $x(0) = -19.24$, $y(0) = 1$, $z(0) = -345$, $t_{end} = 37.2$. Axes: $-250\ldots170$ for x, $-5\ldots25$ for y, $-345\ldots6000$ for z.

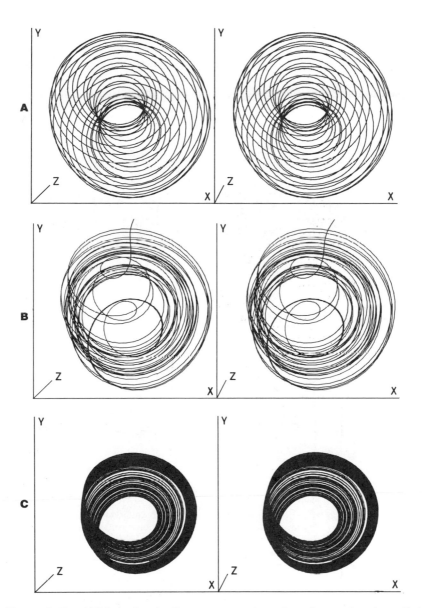

FIGURE 9. Toroidal behavior A, divergence zero chaos B, and attracting chaos C, in Equation 5; numerical simulation as in FIGURE 5. Parameters: (A) $a = 0.2$, $b = 0$; (B) $a = 0.4$, $b = 0$; (C) $a = 0.386$, $b = 0.2$. Initial conditions: (A) $x(0) = 0.3$, $y(0) = 0$, $z(0) = -0.25$, $t_{end} = 174$; (B) the same for $x(0)$, $y(0)$, $z(0)$, $t_{end} = 192$; (C) $x(0) = 0.4$, $y(0) = -0.4$, $z(0) = -0.7$, $t_{end} = 751$; Axes: (A) $-1.1 \ldots 1.1$ for x, $-0.6 \ldots 1.6$ for y, $-0.5 \ldots 0.5$ for z; (B) the same for x, y, $-0.55 \ldots 0.55$ for z; (C) $-1.8 \ldots 1.8$ for x, $-1 \ldots 2.6$ for y, $-0.9 \ldots 0.9$ for z.

equal to zero, and a is sufficiently small. Since the divergence of Equation 5 is zero when $b = 0$, a whole family of invariant tori (that is, quasiperiodic flows) is obtained over a certain range of initial conditions. Beyond that range (and a certain size of the torus) the tori cease to be closed, so that trajectories starting at such large initial conditions do not turn back toward the interior after having completed an outer spiral, but escape to infinity. The illustration in FIGURE 9B

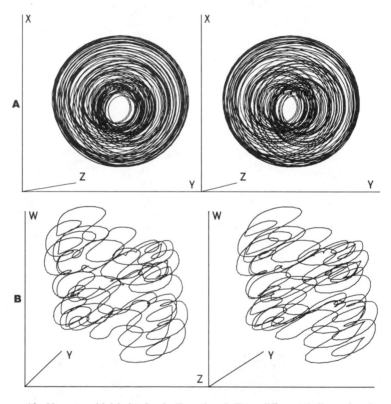

FIGURE 10. Hypertoroidal behavior in Equation 6. Two different 3-dimensional projections of the same flow are shown in A and B; numerical simulation as in FIGURE 5. Parameters: (A) $a = 0.2$, $b = d = 0$, $c = 0.04$. Initial conditions: (A) $x(0) = 0$, $y(0) = 0.75$, $z(0) = 0.2$, $w(0) = -0.75$, $t_{end} = 496$; (B) $t_{end} = 333.7$ ($\simeq 1$ round). Axes: $-1.2 \ldots 1.4$ for x, $-0.6 \ldots 2$ for y, $-0.2 \ldots 0.9$ for z, $-0.81 \ldots -0.69$ for w.

applies when the torus is sufficiently distorted, as occurs if a is increased. Then an intermediary regime that is chaotic appears. Intermediate initial conditions now lead to recurrence for an unpredictable number of rounds before the eventual escape (see FIGURE 9B, on top of the convolute). Thus, there is chaos (for some time), but no chaotic attractor. This situation is analogous to that encountered in more complicated celestial equations and area-preserving maps[34,35,36] as well as (perhaps) in magnetic bottles. The threshold value of a beyond which the core of the invariant tori has disappeared altogether is ~ 0.454. Note that the flow of FIGURE 9B is, like that of FIGURE 9A, time-reversal invariant.

The flow of FIGURE 9C is, unlike that of 9B, determined by a chaotic attractor. Such a flow is obtained if a exceeds the threshold value mentioned while b has an appropriate *nonzero* value. In this way, chaotic regimes of very low nonzero divergence may be realized. Note that the divergence of Equation 5 is equal to b.

While thus giving rise to a chaotic attractor, Equation 5 does not contain a toroidal attractor, however. An attracting torus requires at least two additional quadratic terms or one cubic term. This empirical constraint apparently reflects the presence of a constant of motion in Equation 5 which cannot be destroyed by adding one second-order term. (Note that in 2-dimensional systems, analogously, one quadratic term is not sufficient for a periodic attractor to be possible.) In this situation, the occurrence of chaos is somewhat unexpected. In FIGURE 4e, the transition, toroidal attractor → chaotic attractor, was explained by the occurrence of distortion in an area-contracting annular map (as can indeed be observed with more complicated equations). However, there is no toroidal attractor now. The explanation is that an attracting circular scroll, that is, a limit cycle with toroidal transients (so that the cross section shows an attracting focus), can also be distorted to a sufficient degree to give rise to chaos. This time, the "folded nose" depicted in FIGURE 4e develops out of an attracting torus of zero circumference. A reaction kinetic variant of Equation 5, displaying a toroidal attractor, is described in Reference 37. Another equation, which can be transformed into Equation 5, is proposed in Reference 7. There is some hope that analytical approaches to quadratic differential equations may prove applicable to equations like (5).[28]

HIGHER-ORDER SYSTEMS

Equation 6 is an example of a nontrival higher-order system:

$$\dot{x} = -y - z - w$$
$$\dot{y} = x$$
$$\dot{z} = a(y - y^2) - bz \qquad (6)$$
$$\dot{w} = c\left(\frac{z}{2} - z^2\right) - dw.$$

This equation has a structure similar to that of (5); that is, the position in state space of a suboscillator is slowly displaced by a "parameter variable" which, in turn, is part of a higher-order oscillator whose (apparent) second variable is the numerical value of the displacement of the suboscillator. The difference from Equation 5 is that this principle occurs twice in Equation 6: on a first level, z and y^2 play the roles of parameter and displacement, respectively, as they do in (5); on a second level, w and z^2 also play these roles.

This torus-generating principle is a simplified version of the building-block principle of FIGURE 1e.[27] The fact that it worked once (in Equation 5) was unexpected already. FIGURE 10 now shows that it can be applied repetitively (presumably over a greater number of steps). The hypertoroidal flow generated by Equation 6 is displayed in two different 3-dimensional projections in FIGURE 10. In

FIGURE 10A, the flow looks like a toroidal flow that "runs out of truth" in a periodic manner. That is, the torus rhythmically changes its position while expanding and shrinking (breathing). FIGURE 10B shows a less intuitive projection. Note that the outer layer moves down, while the inner layer moves up; the problem is that both layers cross the middle so that there is no hole in this torus. There is a whole family of invariant hypertori. Again, divergence-zero chaos is possible (not shown).

Three types of *attracting* chaos can be expected this time: 1) the type in Equation 5 (distortion of the subtorus xy/z); 2) an analogous new type (distortion of the subtorus xyz/w); 3) a novel "composed" type (simultaneous distortion in both subtori, that is, the hypertorus $xy/z/w$). The last two types have yet to be found in Equation 6. If not all of them are possible, a more complicated equation may be tried. Thus Equation 6 constitutes but one possible equation for a next-higher type of chaos. Such types of chaos are bound to be found if the building-block principle introduced in FIGURE 1 is not confined to 3 dimensions.[18] The toroidal case (FIGURE 1e) is especially easy to iterate.

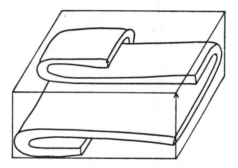

FIGURE 11. Folded-towel map (compare with walking-stick map of FIGURE 4a).

In an equation like (**6**), but somewhat more complicated (so that an attracting hypertorus is formed), the map of FIGURE 11 can be expected to be found if the torus-shaped cross section is distorted (folded over) to a sufficient degree in two independent directions. This map is the next-higher analog to the walking-stick map of FIGURE 4a. It can be obtained through distortion of a hollow solid torus map, just as the walking-stick map could be obtained through distortion of an annular map (FIGURE 4e).

The flow of FIGURE 9c was not obtained from a distorted attracting torus, but rather from a distorted attracting scroll. Similarly, one may expect the distorted map of FIGURE 11 to obtain not only from a distorted attracting hypertorus, but from a distorted attracting hyperscroll (that is, a limit cycle with hypertoroidal transients), as well. Equation 6 produces such a hyperscroll: the hypertorus of FIGURE 10 shrinks to a simple limit cycle (of this type) if both b and $d > 0$ and $d \ll b \ll 1$.

The 3-dimensional map of FIGURE 11 is not more difficult to follow, in its basic properties, than, for example, the map of FIGURE 4a. Actually, it is somewhat easier to understand because of the fact that spatial intuition is 3-dimensional. These typical properties can be illustrated by the preparation of pastry dough. Roll out the dough, fold it over on two adjacent sides, roll it out again, and so forth. If the dough's volume shrinks during the process and the lost volume is subsequently replaced by new material of the same kind, the essential features of the map in FIGURE 11 may be recognized. Asking for an attractor means asking for the geometric locus of all points of the original layer as $n \to \infty$. In the case of the pastry dough, the original layer (which may be colored red) can be easily traced in the final product. If the folding involves only a minor portion of the dough in both directions, the red coloring will be confined to a place near the four sides of the dough so that "scrolls" appear when the dough is sliced. As soon as the folding-over exceeds a certain critical value in either direction, however, the red coloring will be fairly evenly distributed along a 2-dimensional sheet within the dough. As n grows larger in an unbound fashion, this "red sheet" will become infinitely thin, infinitely large, and there will be an infinity of folds. Simultaneously, the volume (measure) of the sheet will go to zero.

Thus, a 2-dimensional manifold with usual properties (*strange manifold*) is formed as an attractor within the map in FIGURE 11. This can be thought of as a higher-order analog of an attracting fixed point of a 1-dimensional diffeomorphism. An apparent counterargument to calling this attracting strange manifold an attractor is that, within the attracting sheet, lumps of red coloring may form in such a way that a lower-dimensional attractor applies. This situation is apparently the rule. Nonetheless, the formation of such a lower-dimensional "genuine" attractor does not interfere with the formation of the attracting sheet itself. The lower-level attractor may therefore be considered an *internal property* of the attracting strange manifold. The higher-level attractor hereby alternates between being minimal and nonminimal.

The distinction between levels of attraction allows one to classify all systems considered above under the common heading of attracting "chaos" as belonging to a single class of systems; namely, to the class characterized by the presence of a 1-dimensional strange manifold (strange line) in the cross section.

Strange lines (continuous or broken) are found in the cross sections of three types of systems: 1) systems governed by walking-stick maps (as in FIGURE 4a, b, e); 2) systems governed by sandwich maps (as in FIGURE 4c, d); 3) systems governed by Smale's folded torus map.[16] The latter is a 3-dimensional diffeomorphism in the form of a solid torus whose first iteration is the same torus, but elongated, shrunk, folded (wrapped), and put back into the original torus. If you think of this as dough, a strange line is formed. If no mistake is made in the preparation of the dough (some part of the torus being forgotten to elongate), all neighboring points of the original red layer will exponentially drift apart under the iteration. The only major difference between this strange line and that found in systems of type 1, is that the quantitative conditions for avoiding an internal attractor are easier to meet in the present case (just as in type 2). Thus, there is some justification to subsume all systems governed by a strange line under a common heading. The term *chaos* (or hyperoscillation[18]) is a possible label for the whole class, in spite of the fact that not all members display the characteristic behavior of the class for an infinite period of time.

Similarly, *strange surfaces* are found in the cross sections of three types of systems: a) systems governed by "folded-towel" maps (as in FIGURE 11); b) systems governed by analogous "cut" maps (as in FIGURE 12); c) systems governed by higher-dimensional maps in which, once more, the higher dimensionality allows one to avoid local compression zones without using cuts. As before, types b and c are easily kept free of lower-dimensional attractors while type 2 generally possesses a lower-level attractor (for example, a strange attractor).

These different systems also form a class of systems. Even though no computer pictures of the corresponding flows and their time projections are available as yet, a possible name for the characteristic behavior of this class is *hyperchaos*.[10]

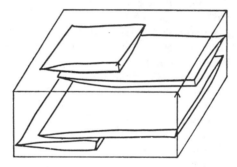

FIGURE 12. Big map (compare with sandwich map of FIGURE 4d).

The proposed distinction between "levels of attraction" should, if real, be reflected in differing bifurcation laws on the respective levels. Higher-level attractors should, whether or not they contain lower-level attractors, always behave in an ordinary (Liapunov) fashion, while lower-level (internal) attractors should follow more complicated "implosion laws."[38] Catastrophe theory[39] and the global stability theorem[40] may both apply to higher-level attractors, but not to their internal dynamics (which in general is structurally unstable[20,33]). In short, the map in FIGURE 11 suggests that external and internal bifurcations of attractors may be distinguished.

CONCLUSION

Higher-dimensional dynamical systems are capable of a whole new world of qualitative types of behavior. To find some of them, a building-block approach can be used. In considering a few examples of quadratic differential equations, there are apparently five new facts to observe:

1) There is a simple strange attractor in three dimensions besides the Lorenz attractor (see Equation 4).
2) Divergence-zero chaos is possible in three-variable Euclidean dynamics (see Equation 5).
3) Higher-order toroidal flows can be found in simple low-dimensional systems (see Equation 6).
4) There is a natural sequence of attractors depending on dimensionality (chaos is almost synonymous to specifically three-variable behavior).
5) *Internal* and *external* bifurcations of attractors may be distinguished.

Chaos (and hyperchaos) may occur as frequently in natural systems as oscillation. The "quantum jump" between oscillation and chaos seems to be rather small, in chemistry, for example.

SUMMARY

A survey of different types of simplest chaos in Euclidean 3-space was presented. Five basic types of chaos were illustrated by four three-variable quadratic differential equation systems containing at most two nonlinear terms. One of these systems showed a new type of strange attractor. A simple four-variable quadratic differential equation producing a hypertoroidal flow, and presumably a higher type of chaos, was also presented.

REFERENCES

1. LIÉNARD, A. 1929. Etude des oscillations entretenues (2 parts), Rev. Gen. Electr. **23**: 901–912, 946–954.
2. RÖSSLER, O. E. 1974. Chemical automata in homogeneous and reaction-diffusion kinetics. Springer Lecture Notes Biomath. **4**: 399–418.
3. KHAIKIN, S. E. 1930. Continuous and discontinuous oscillations. Zh. Prikl. Fiz. 7: 21.
4. RÖSSLER, O. E. 1976. Chaotic behavior in simple reaction systems. Z. Naturforsch. **31a**: 259–264.
5. TAKENS, F. 1976. Implicit differential equations: Some open problems. Springer Lecture Notes Math. **535**: 237–253.
6. RÖSSLER, O. E. 1977. Chaos in abstract kinetics: Two prototypes. Bull. Math. Biol. **39**: 275–289.
7. RÖSSLER, O. E. 1977. Continuous chaos. *In* Synergetics: A Workshop. H. Haken, Ed. Springer-Verlag, New York & Heidelberg. 184–199.
8. ULAM, S. 1963. Some properties of certain nonlinear transformations. *In* Mathematical Models in the Physical Sciences. S. Drobot, Ed. Prentice Hall. 85–95.
9. MYRBERG, P. J. 1963. Iteration of the real polynomials of second degree III (in German). Ann. Acad. Sci. Fenn. Ser. A, 336/3: 1–18.
10. MAY, R. M. 1976. Simple mathematical models with very complicated dynamics. Nature **261**: 459–467.
11. LI, T. Y. & J. A. YORKE. 1975. Period three implies chaos. Am. Math. Mon. **82**: 985–992.
12. ZEEMAN, E. C. 1972. *In* Toward a Theoretical Biology C. H. Waddington, Ed. Edinburgh University Press. **4**: 8–67.
13. SMALE, S. & R. F. WILLIAMS. 1976. The qualitative analysis of a difference equation of population growth. J. Math. Biol. **3**: 1–3.

14. RÖSSLER, O. E. 1977. Syncope implies chaos in walking-stick maps. Z. Naturforsch. **32a:** 607–613.
15. SMALE, S. 1965. Diffeomorphisms with many periodic points. *In* Differential and Combinatorial Topology. S. Cairns, Ed. Princeton University Press. 63–80.
16. SMALE, S. 1967. Differentiable dynamical systems. Bull. Am. Math. Soc. **73:** 747–817.
17. RUELLE, D. & F. TAKENS. 1971. On the nature of turbulence. Commun. Math. Phys. **20:** 167–192.
18. RÖSSLER, O. E. 1976. Different types of chaos in two simple differential equations. Z. Naturforsch. **31a:** 1664–1670.
19. GUCKENHEIMER, J. 1976. A strange, strange attractor. *In* The Hopf Bifurcation and Its Applications. J. E. Marsden & M. McCracken, Eds. Springer-Verlag, New York & Heidelberg. 368–381.
20. WILLIAMS, R. F. 1976. The structure of Lorenz attractors. Preprint.
21. KAPLAN, J. L. & J. A. YORKE. 1977. Preturbulence: A regime observed in a fluid flow model of Lorenz. Preprint.
22. MINORSKI, N. 1974. Nonlinear Oscillations. R. E. Krieger, Huntington, N.Y.
23. RÖSSLER, O. E. 1972. Basic circuits of fluid automata and relaxation systems. Z. Naturforsch. **27b:** 333–343.
24. CRONIN-SCANLON, J. 1974. A mathematical model for catatonic schizophrenia. *In* Ann. N.Y. Acad. Sci. O. Gurel, Ed. **231:** 112–120.
25. O'MALLEY, R. E. 1974. Introduction to Singular Perturbations. Academic Press, New York & London. 86.
26. PLANT, R. E. 1977. Crustacean cardiac pacemaker model: An analysis of the eigenvalue approximation. Math. Biosci. In press.
27. RÖSSLER, O. E. 1977. Quasiperiodic oscillation in an abstract reaction system. Biophys. J. **17:** 281a. Abstract.
28. GERBER, P. D. 1973. Left alternative algebras and quadratic differential equations III. IBM Res. Rep. RC 4440. 1–22.
29. RÖSSLER, O. E. 1976. An equation for continuous chaos. Phys. Lett. **57A:** 397–398.
30. LORENZ, E. N. 1963. Deterministic nonperiodic flow. J. Atmos. Sci. **20:** 130–141.
31. HÉNON, M. & Y. POMEAU. 1976. Two strange attractors with a simple structure. Springer Lecture Notes Math. **565:** 29–68.
32. RÖSSLER, O. E. 1977. Horseshoe-map chaos in the Lorenz equation. Phys. Lett. **60A:** 392–394.
33. NEWHOUSE, S. & J. PALIS. 1976. Cycles and bifurcation theory. Astérisque **31:** 44–140.
34. BIRKHOFF, G. D. 1927. On the periodic motions of dynamical systems. Acta Math. **50:** 359–379.
35. MOSER, J. 1973. Stable and Random Motion in Dynamical Systems, with Special Emphasis on Celestial Mechanics. Princeton University Press.
36. GUMOWSKI, I. 1976. Contribution de l'automatique au problème de réversibilité microscopique-irréversibilité macroscopique. R.A.I.R.O. Automatique **10:** 7–42.
37. RÖSSLER, O. E. 1977. Toroidal oscillation in a 3-variable abstract reaction system. Z. Naturforsch. **32a:** 299–301.
38. THOM, R. 1976. Introduction to qualitative dynamics (in French). Astérisque **31:** 3–13.
39. THOM, R. 1976. Structural Stability and Morphogenesis. Benjamin, Reading, Mass.
40. GUREL, O. 1976. Peeling studies of complex dynamical systems. *In* Simulation of Systems. L. Dekker, Ed. North-Holland, Amsterdam. 53–58.

THE BIFURCATION SPACE OF THE
LORENZ ATTRACTOR

R. F. Williams

Department of Mathematics
Northwestern University
Evanston, Illinois 62201

In a joint paper[3] John Guckenheimer and the present author showed that the space of the title is two-dimensional, in that the pair (k_l, k_r) of kneading sequences is a complete conjugacy class invariant for the flows near the three-dimensional Lorenz system.[4,8] But there are reasons for obtaining more complete knowledge of the space $\mathcal{K}$ of all such pairs, k. Thus, one can use the "Parry coordinates" (λ, c) (see below) but *a priori,* one doesn't know these all correspond to Lorenz models—i.e., to differential equations in $\mathbb{R}^3$. Secondly, a recurring problem in bifurcation theory is just how *do* the periodic orbits change, as one changes a bifurcation parameter? Thus we shall present a formula for computing the periodic orbits in terms of the parameter k, i.e., the kneading sequences.

THE SPACE $\mathcal{K}$ OF KNEADING SEQUENCES

DEFINITION. Let 2^∞ denote the set of all sequences (finite or infinite) $x = x_0 x_1 \ldots$ such that $x_i = 0$ or 1. Define an ordering in 2^∞ by $00 < 0 < 01$—that is, 0's count negatively, the empty symbol is neutral and 1's are positive. Define $s : 2^\infty \to 2^\infty$ by $s(x_0 x_1 \ldots) = x_1 x_2 \ldots$. Finally, say $k = (k_l, k_r) \in \mathcal{K}$ iff

AXIOM 1. $k_l < k_r$.

AXIOM 2. $k_l \leq s^i k_l, s^i k_r \leq k_r, i = 0, 1, 2 \ldots$.

If equality occurs only in the two cases $k_l = s^0 k_l, s^0 k_r = k_r$, we say $k \in \mathcal{K}^0$. Finally, $\mathcal{K}$ inherits an *order* from 2^∞ on each coordinate, so that it becomes a subset of a 2-dimensional disk. In fact, for those points where k_l begins with several 0's and k_r begins with several 1's, it *is* locally a disk.

PROPOSITION. The finite and eventually periodic pairs are dense in $\mathcal{K}$.

PROOF. As I know of no reasonable proof, I will outline one special (in fact generic) case. Let $k = (k_l, k_r) \in \mathcal{K}$ and let n be a positive integer. It will suffice by symmetry to find $\overline{k} \in \mathcal{K}$ $\overline{k} = (k_l', k_r)$, where k_l' agrees with k_l for n terms and is either finite or periodic. We take the special case that k_l begins with m 0's and that it has the syllable 0^m infinitely many times. Then we choose an initial segment $0^m 1 w 1 0^m$ of k_l, which has length greater than n. Let $k_l' = 0^m 1 w 1 (\overline{0^{m-1} 1 w 1})$ where overlining means infinite periodic repetition.

Note that if n is large enough, we still have Axiom $1 : k_l' < k_r$. However, there remains the problem that some terminal segment $s^i k_r$, may be $> k_l'$, so that we must define a k_r' similarly. Finally, by considering three cases, $i \leq m, m < i <$

0077-8923/78/0316-0393 $1.75/1 © 1979, NYAS

length w, length $w \leq i$, one can show that $k_i' \leq s^i k_i'$. One certainly needs a more conceptual proof!

A simple proof has been found.

MODELS FOR THE RETURN MAP

We single out three standard forms for the return map f of a Lorenz attractor. First, we state our basic assumption on such a return map:

$$f : [0, 1] \to [0, 1]$$

(1) f has one point of discontinuity, say $c \in (0, 1)$.

(2) $f(c^+) = 0, f(c^-) = 1$.

(3) f is "Parry" in the sense that given any subinterval J of I, there is an n such that $J \cup f(J) \cup \cdots \cup f^n(J) = I$. Condition (3) can be replaced with (3') $f' > \sqrt{2}$.

REMARK. That $3' \Rightarrow 3$ is quite easy. For example, see Reference 9.

THEOREM (Parry[6]). Any f satisfying (1)–(3) is topologically conjugate to one of *constant slope*.

DEFINITIONS. We call such a map a *Parry-model;* it has the formula (FIGURE 1)

$$f_{\lambda,c}(x) = \begin{cases} \lambda(x - c) + 1, & x < c. \\ \lambda(x - c), & x > c. \end{cases}$$

for some choice of λ, c. We call (λ, c) the *Parry-coordinates.*

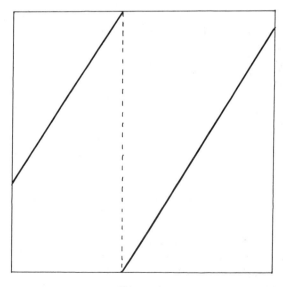

FIGURE 1.

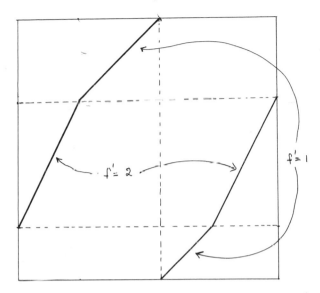

FIGURE 2.

A Lesbesgue-Measure-Preserving Model

Given two real numbers $0 \leq \alpha \leq \beta \leq 1$, we can construct a function as in FIGURE 2.

$$f(x) = \begin{cases} 2x + \alpha, & 0 \leq x \leq \frac{1}{2}(\beta - \alpha) \\ x + 1 - \frac{1}{2}(\alpha + \beta), & \frac{1}{2}(\beta - \alpha) \leq x \leq 1 - \frac{1}{2}(\beta + \alpha) \\ x - 1 + \frac{1}{2}(\beta + \alpha), & 1 - \frac{1}{2}(\beta + \alpha) \leq x \leq 1 - \frac{1}{2}(\beta - \alpha) \\ 2x + \beta - 2, & 1 - \frac{1}{2}(\beta - \alpha) \leq x \leq 1. \end{cases}$$

Note that where f is 2-to-1, $f' = 2$ and where f is 1-to-1, $f' = 1$. Thus f preserves Lebesgue measure.

Lorenz Models

The return map that comes up in the situation of the Lorenz attractor 0 (see References 2,5,7, and 9, and especially Reference 3) have the form (FIGURE 3)

$$f_{s,t}(x) = \begin{array}{ll} \alpha s \sqrt{-x} + (\sqrt{2} + \beta s)x + \frac{1}{2}, & -\frac{1}{2} \leq x < 0 \\ \alpha t \sqrt{x} + (\sqrt{2} + \beta t)x - \frac{1}{2}, & 0 < x \leq \frac{1}{2}. \end{array}$$

Here α, and β are certain constants chosen to make possible the choice

$$0 \leq s, t \leq 1$$

and

$$f_{0,t}(x) = \sqrt{2}\,x + \tfrac{1}{2} \qquad -\tfrac{1}{2} \leq x < 0$$

$$f_{s,0}(x) = \sqrt{2}\,x - \tfrac{1}{2} \qquad 0 < x < \tfrac{1}{2}$$

$$f'_{s,t}(x) > \sqrt{2}, \qquad \text{otherwise,}$$

and

$$f_{1,t}(-\tfrac{1}{2}) = -\tfrac{1}{2}, \qquad f_{s,1}(\tfrac{1}{2}) = \tfrac{1}{2}.$$

These in turn guarantee a "full range" of kneading sequences.

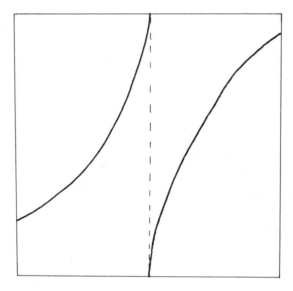

FIGURE 3.

REALIZING KNEADING SEQUENCES

PROPOSITION 1. If each of k_l, k_r is finite or periodic, then there is a Markov partition of $[0, 1]$ and a Parry model f realizing (k_l, k_r).

PROOF. We are given two sequences $k_l = a_0 a_1 a_2 \ldots, k_r$ where each is either finite or periodic. If one is finite, we add to it the letter m (for "middle"). Then the set F of all terminal sequences

$$\{s^i k_l, s^i k_r, \quad i, j = 0, 1, 2, \ldots\}$$

is finite. It contains the letter m if one or both of k_l, k_r is finite. The set F is ordered lexicographically where $0 < m < 1$.

Example. $k = (00\overline{1}, 1\overline{10})$ where overlining means infinite repetition. Then the words of F, in their natural order, are $00\overline{1}$, $0m$, $\overline{01}$, m, $10m$, $\overline{10}$, $110m$.

Then we partition $[0, 1]$ by points P_σ, $\sigma \in F$ in this given order where P_{k_l} and P_{k_r} are the left and right end points. Next, the map s induces a map S: $F \to F$,

$$
S(P_\sigma) = \begin{cases}
P_{s\sigma}, & \sigma \neq m \\
P_{k_l}, & \sigma = m^- \\
P_{k_r}, & \upsilon = m^+.
\end{cases}
$$

The complication $m^\pm$ corresponds to the point of discontinuity of the return map f.

Then F separates I into finitely many line intervals $\{I_i\}$; typically: $[P_\sigma, P_{s\sigma}]$, $[P_i, P_m]$. Then $\{I_i\}$ is a Markov partition and S defines a map on this partition, where typically, an interval I_i will be mapped into the sum of several.

Thus S can be thought of as an $n \times n$ matrix over $\mathbb{Z}^+$, where F contains $n + 1$ points. Hence by the Perron-Frobenius Theorem, there is a positive eigenvalue λ, and corresponding eigenvector $(x_1, x_2, \ldots, x_n)$ where each $x_i > 0$. We normalize to get

$$
x_1 + x_2 + \cdots + x_n = 1.
$$

Thus we may take x_i to be the length of the ith interval in our Markov partition. Then the map f (essentially S) has constant slope λ, and only one point of discontinuity P_m. That is, we have found a first return map corresponding to k.

By passing to limits we can conclude that each pair k of kneading sequences in $\mathcal{K}$ can be realized with a Parry model. Similarly, one can use this approach to show that to each $k \in \mathcal{K}$, can be realized with a Lorenz Model. An alternative approach is to use the known structure of $\mathcal{K}$ (it is a two dimensional disk) to show that each point $k \in \mathcal{K}$ can be realized with return maps of various types. That is, knowing the boundary of $\mathcal{K}$ is realized, with a Lorenz model, and that this "loop" of Lorenz models can be shrunk to a point, one concludes (using continuity) that each point of $\mathcal{K}$ is realized.

PERIODIC ORBITS

In an earlier paper, I used the family of all periodic orbits as a topological invariant for the Lorenz attractor. One writes down a symbolic infinite series, $\eta(x, y)$, to include this family as a "single" invariant. Each term of the series is a monomial in x and y where x and y are the generators in a certain free *non-Abelian* group. Thus one cannot follow the usual practice in periodic orbit theory of exponentiating, essentially because there is no good determinant in a non-Abelian situation. Thus I spent some time trying to show that nothing is lost if one allows x and y to commute. However, it turns out that something is lost:

Example. There exist distinct points $k_i \in \mathcal{K}$, $i = 1, 2$, such that the periodic orbits of the corresponding flows φ_{it}, $i = 1, 2$, have the same structure, when Abelianized.

In order to present the example, we need a formula that gives the collection η_n of all periodic orbits of length n in terms of the kneading sequences. One could proceed as well by evaluating determinants—but the task of evaluating the determinant of a 15×15 symbolic matrix is formidable!

Definition. Given a map f (as above) a word $w = x_0, x_1, \ldots, x_{n-1}$ of 0's and 1's (we use 0's and 1's instead of x's and y's) is a periodic word for f, provided there is an $x \in [0, 1]$ such that

(1) $f_0^i(x)$ is to the left or right of the point of discontinuity according as to whether x_i is 0 or 1.

(2) $f^n(x) = x$.

FORMULA. Given a return map f and its kneading sequences k_l, k_r, a finite word w of 0's and 1's is a periodic word for f, iff for each $i = 0, 1, 2, \ldots$

$$k_l < s^i(\overline{w}) < k_r.$$

PROOF. Let c be the point of discontinuity for f. In fact, we show that the set of all kneading sequences for points of x under f, are exactly the words k such that $k_l \leq s^i k \leq k_r, i = 0, 1, 2, \ldots$.

PROOF. We assume k begins with n 0's, as the other case is similar.

Consider as Case 1 that k_l begins with more than n 0's. Thus there is an interval $[a, b]$, $a < b \leq c$ where $f^{n+1}(a) = c = f^n(b)$. We claim that f^{n+1} maps the interval (a, b) onto the interval $(c, 1)$. This is just the simple fact that $f^n(a, b) = (f^n(a), c)$ together with the fact that $f(c^-) = 1$. Thus we find an interval $[a, b]$ whose points have kneading sequences agreeing with k_0 for the n times that it is 0, and whose $(n + 1)^{\underline{st}}$ images span all of $(c, 1)$. Thus we proceed to find a subinterval along which we can find points whose succeeding iterates under f stay on the right side of c the correct number of times, since by assumption, this is $\leq$ the number of times 1 stays. One finishes by induction.

In Case 2 k_l also begins with n 0's. Then we let $(a, b) = (0, b)$ where $f^n(b) = 0$: We may assume $f^{n+1}(0) > c$, as $f^{n+1}(0) \geq 0$ and $f^{n+1}(0) = 0$ is just Case 1 over again. This means that $f^{n+1}((a, b)) = (f^{n+1}(0), 1)$ and we do not have the full range of $(c, 1)$ to search for the next terms of our sequence k_0. But we know that $k_l \leq k_0$, and after the initial n zeros, k_l has, say, m 1's. Then we can have, for points in (a, b), as few as m 1's by letting $x = a$, or as many as k_r begins with, by taking $x = b$. Again one finishes by induction.

Example. Let $x = 000\,1010$, $y = 110\,0100$, let $k = (x01, y01)$ and $k' = (x10, y10)$. Since k and k' agree for the first seven digits the periodic words for f and f' are the same up to and including those of length seven.

We need only check the words of length ≤ 15. This is because the Abelianized words of length i is shown to be trace B^i, where B is a symbolic matrix (see Reference 9, p. 16). In our case B is 15×15 as there are 16 symbols in the set F (see above). Thus for k we have $000\,101\,001$, $110\,010\,001$, so that the last three symbols, $s^i k_l = s^i k_r$, $i = 6, 7, 8, 9$. These four together with $s^i k_l$, $s^i k_r$, $i = 0, 1, 2, \ldots, 5$ give 16 symbols. (See above.) The computation for k' is similar. Thus we need only verify that they have the same Abelianized words of lengths 8 through 15. Next, by our formula, words w such that

$$x10 < \overline{w} < y10$$

occur for *both* functions f and f'. Hence it suffices to show that those words w and w' such that

$$x01 < \overline{w} \leq x10 \qquad y01 \leq \overline{w}' < y10 \qquad (*)$$

correspond 1-to-1 in such a way that corresponding words have the same number of 0's and 1's. We let U_i be the set of such w's of length $8 + i$ and U_i' the set of such w''s of length $8 + i$. Note that $U_0 = \{x1\}$ and $U_0' = \{y0\}$. Counting, $x1$ and $y0$ have 6 0's and 3 1's each; so far, so good. Next, both $U_1 = \phi = U_1'$; e.g., $x00$ and $x01$ violate the left inequality of (*) whereas $x10$ and $x11$ violate the right inequality (for w, not w'). Next one finds that $U_2 = \{x101\}$ and $U_2' = \{y010\}$. (A sticky point: $x011$ and $y100$ satisfy (*), but they are "saddle connections," not real periodic orbits.) Continuing in this manner, one verifies this correspondence. See Reference 1 for more details.

REFERENCES

1. CHORIN, A., J. E. MARSDEN & S. SMALE. 1977. Turbulence seminar, notes by T. Ratiu and P. Bernard. Lecture Notes No. 615: 108. Springer. New York, N.Y.
2. GUCKENHEIMER, J. 1976. A strange, strange attractor. *In* The Hopf Bifurcation and Applications. Applied Mathematical Sciences.: 19. Springer Verlag. New York.
3. GUCKENHEIMER, J. & R. WILLIAMS. In press. Structural stability of the Lorenz attractor.
4. KAPLAN, J. and J. YORKE. 1978. Preturbulence: A metastable chaotic regime in the system of Lorenz. Ann. N.Y. Acad. Sci. This volume.
5. LORENZ, E. 1963. Deterministic non-periodic flow. J. Atmospheric Sci. 20: 130–141.
6. PARRY, W. 1966 Symbolic dynamics and transformations of the unit interval. Trans. Amer. Math. Soc. 122: 368–378.
7. PARRY, W. In press. The Lorenz attractor and a related population model. Preprint.
8. RUELLE, D. 1978. Bifurcation to turbulent attractors. Ann. N.Y. Acad. Sci. This volume.
9. WILLIAMS, R. 1976. The structure of Lorenz attractors. Preprint, Northwestern University.

THE ONSET OF CHAOS IN A FLUID FLOW
MODEL OF LORENZ*

James L. Kaplan

Department of Mathematics
Boston University
Boston, Massachusetts 02215

James A. Yorke

Institute for Physical Science and Technology
and
Department of Mathematics
University of Maryland
College Park, Maryland 20742

There have been many attempts in the scientific literature to provide a mathematical explanation of the nature of turbulence in fluids. We take the most common approach, arguing that turbulence is a phenomenon of systems of differential equations. The behavior of the fluid is represented as the solution of a system of ordinary differential equations and the system is assumed to depend on a parameter r; hence

$$u' = f(r, u) \tag{E}$$

when $u \in R^n$, $r \in R^1$, and f is highly differentiable. The parameter r usually corresponds to the Rayleigh or Reynold's number.

CHAOS NEAR A REST POINT

The equations of fluid dynamics are so complex that it is necessary to study the dynamics near a rest point; that is, a point u_0 for which $f(r, u_0) = 0$. We now outline briefly and heuristically one line of turbulence research. We assume coordinates are chosen so that 0 is a rest point, and that we can write

$$f(r, u) = A(r)u + g(r, u),$$

where $A(r)$ is an $n \times n$ matrix and $g(r, u)$ consists of terms which are quadratic and higher order in u. Suppose further that for small values of r, the rest point 0 is stable; trajectories starting near 0 may oscillate but damp down to 0. Suppose as r is increased slightly 0 becomes unstable, after a critical value $r = r^*$. It is assumed $A(r^*)$ has a pair of pure imaginary eigenvalues. This line of research assumes that the nonlinear term g produces damping effects so that for r slightly above r^* trajectories remain bounded near 0 while 0 itself is unstable. Landau[12] and E. Hopf[9] investigate such situations. Their efforts at describing turbulence are thus only for fluid systems fitting this picture. They suggest that as r is in-

*This work was supported by Grant MCS76-24432 from the National Science Foundation.

0077–8923/78/0316–0400 $1.75/1 © 1979, NYAS

creased a large number of independent frequencies appear, usually one at a time. This view holds that solutions will be quasiperiodic and the presence of many frequencies in solutions cause the appearance of chaos, which is the nature of turbulence. They work near the equilibrium, where the nonlinear effects are small. Hence we might expect trajectories of the nonlinear system to be like those of linear systems for which trajectories are oscillatory and bounded and undamped. In particular if there is a one-parameter family of periodic solutions for the linear system $x' = A(r^*)x$, Hopf[10] showed we may expect a family of periodic solutions in (r, x) space for the nonlinear system. This family, which starts from $(r^*, 0)$, must persist[1] in some sense, though as we follow the family, if these periodic orbits become unstable, they will not describe the observed behavior of the system. As r increases a number of pairs of nonzero eigenvalues of $A(r)$ may switch from having negative real parts to positive real parts; in this case it has been shown recently[3] that there will still exist at least one family of periodic orbits. In particular harmonic resonance between these frequencies cannot prevent the existence of at least one family of periodic orbits. These orbits certainly need not be stable.

Ruelle and Takens[21] investigated the behavior near 0 when there are $k > 1$ undamped frequencies, that is, when $A(r)$ has k pairs of complex eigenvalues with positive real parts. They argue that there will be a k-dimensional torus, T^k, attracting all nearby orbits. This by itself is consistent with the ideas of Landau and Hopf. Since the torus T^k is attracting, they can now restrict their investigation to the behavior of trajectories of differential equations on T^k. They note that Peixoto studied the set of all smooth systems of differential equations on T^2 and found that an open and dense collection of these systems had at least one stable attracting periodic orbit. Thus for any differential equation on T^2 there is a perturbation, as small as desired, such that the perturbed system has a stable attracting periodic orbit. The presence of such an orbit would mean that the limit set of a trajectory near the torus would be only a part of the torus. If, on the other hand, trajectories on the torus were quasiperiodic (and not periodic), the limit set of a trajectory near the torus would be the entire torus. The phenomenon of systems with two or more frequencies in which small nonlinear terms create a stable attracting periodic orbit is well known and is called "entrainment" of the frequencies. More generally we will use the term entrainment for the tendency of small nonlinearities (or perturbations) to cause orbits on a space to approach a proper subset of that space, that is, to create a "track," a proper invariant subset that nearby trajectories approach. In particular, we allow this track to be a strange attractor, or a periodic orbit, or other invariant sets.

More significant for the understanding of turbulence were the investigations by Ruelle and Takens for $k \geq 4$. For systems on T^k having quasiperiodic trajectories (with each trajectory on T^k coming arbitrarily close to each point of T^k), small perturbations could introduce "strange attractors" into T^k. They defined a strange attractor to be in essence any compact connected attracting set which is neither a rest point nor a periodic orbit nor a surface of any dimension. No definition of turbulence is universally accepted so it is difficult to say whether the existence of a strange attractor implies the existence of turbulence, but it is clear that Ruelle and Takens identified a source of irregular and somewhat chaotic behavior. The orbits of a strange attractor are more chaotic than those Landau described, even though his construction requires an infinite dimensional space. His orbits are almost periodic or at least this is our interpretation of his paper.[12] Ruelle and Takens defined a trajectory to be turbulent if its positive limit set is compact and is neither a point nor a surface of any dimension. (See Reference 13 for a discussion of invariant measures and turbulent trajectories. Lasota related turbulent motions to one dimensional maps in Reference 14.)

To put their result in context, it should also be pointed out that for any quasiperiodic flow on any T^k, there are perturbations as small as desired which change the system into one with an attracting stable periodic orbit (with no strange attractor). Furthermore we have no reason to believe at present that perturbations that change the system into one with a strange attractor are more common than those that change the system into one with a stable attracting periodic orbit.

The case $k = 2$ has been studied in detail by Arnold[2] and he finds the correct interpretation of Peixoto's result is not obvious. While in a topological sense the set of systems on T^2 for which entrainment occurs is large, measured theoretically the set is small. Precise statements cannot be given here; loosely speaking, he assumes that the nonlinearity is small. If we assume the linear system has two frequencies and these are irrationally related, then if two frequencies are chosen at random, he shows the probability is nearly zero that we will see entrainment. The probability is nearly one that each orbit will eventually approach arbitrarily close to every point on a two-dimensional torus. Certainly, Landau was familiar with the phenomena of entrainment. His emphasis on quasiperiodic motions seems to suggest he felt entrainment would not be a significant phenomenon. The findings of Arnold support Landau in that entrainment should not be significant as long as the nonlinearities involved are small. It should be emphasized that the arguments of Ruelle and Takens are perturbation arguments and so require that the nonlinearities are small. It is plausible that the likelihood of strange attractors increases as the size of the nonlinearity increases.

If the corresponding result is true for $k > 2$, then we would not often expect to see strange attractors when the nonlinearities are small. Nonetheless we feel that for systems with no stable rest points and no stable periodic orbits, we must expect to find strange attractors common unless the instability of some rest point is so weak that trajectories can remain quite close. Then the theory of Arnold might be applicable. The present experimental evidence (Swinney and Golub[24]) is not in agreement with any of the above-mentioned mathematical theories. Recent unpublished numerical experiments of J. Curry and J. Yorke indicate the attracting torus T^2 can develop fringes and folds—infinitely many—and thereby become a strange chaotic attractor. This may be the phenomenon seen in Reference 24. There is also a suggestive result of Oxtoby and Ulam[26] on measure-preserving maps that says that ergodic maps are frequent (á la Baire-Category) in the set of all homeomorphisms. This again suggests chaos may occur without invoking bifurcations involving more degrees of freedom. See also numerical experiments in R^2 of Henon[27] and Rannon[28] though these do not relate directly to Hopf bifurcation.

As a fluid becomes turbulent, the trajectories become unstable. We feel that turbulence is the lack of stable rest points and stable periodic orbits. Bounded orbits can then be expected to tend to strange attractors. While this approach seems highly plausible, it is based on inference rather than rigor at this time. This approach to what happens, discarding the approach of staying near an equilibrium, is based on an attempt to fit an example of Lorenz[15] into the turbulence picture, for Lorenz's example gives the first mathematical demonstration of what might be happening in turbulent systems far from equilibrium, far from a rest point.

TURBULENCE FAR FROM A REST POINT

Systems that are far from linear and far from a rest point present even greater difficulties, and no general analysis is available. We are reduced to examination of highly idealized models. One of the most intriguing models was studied by Lorenz.[15]

Lorenz considered the forced dissipative system

$$x' = -\sigma x + \sigma y,$$

$$y' = -xz + rx - y, \tag{1}$$

$$z' = xy - bz.$$

These ordinary differential equations are an approximation to a system of partial differential equations describing finite amplitude convection in a fluid layer heated from below. If the unknown functions in the partial differential equations are expanded in a Fourier series and all the resulting Fourier coefficients are set equal to zero except three, system (1) results. For $\sigma = 10$, $b = \frac{8}{3}$, Lorenz found numerically that the system (1) behaves "chaotically" whenever the Rayleigh number r exceeds a critical value $r_2 \approx 24.74$; that is, all solutions are unstable and almost all of them are aperiodic, though there are an infinite number of periodic solutions of different periods. The chaotic behavior and sensitive dependence upon initial conditions of solutions of differential equations provides a mechanism for understanding turbulence. Guckenheimer[8] gives an example of a dynamical system whose behavior appears topologically identical to that of Lorenz's system; the Guckenheimer system is more easily analyzed. Williams[25] analyzes the pattern of winding and twisting exhibited by these flows. Higher dimensional analogues of this system have been studied in Curry[5] and McLaughlin and Martin.[17]

In a recent paper[11] we were able to show that prior to the onset of chaotic behavior there exists a "preturbulent state" where turbulent orbits exist but represent an exceptional set (measure zero) of initial conditions. Lorenz demonstrated chaotic dynamics; we describe how the chaos arises. Our methodology was to utilize the general short-term behavior of the system, determined numerically, to predict the behavior of particular orbits for all future time. In particular, we show that chaotic behavior actually first occurs when r exceeds $r_0 \approx 13.926$. The value r_0 is the first value for which system (1) possesses a homoclinic orbit (that is, there is a bounded nonperiodic orbit having the same positive and negative limit set). The justification of this claim is based upon arguments similar to Smale's famous horseshoe.[22,23] In some sense, subsequent to the appearance of a homoclinic orbit, system (1) contains a "broken horseshoe." For $r < r_0$ there are no periodic orbits, while for $r = r_0 + \epsilon$ for small $\epsilon > 0$, there are an infinite number of periodic orbits and an infinite number of bounded orbits that do not tend asymptotically to any rest point or periodic orbit.

The definition of the Ruelle and Takens term "strange attractor" is based on the shape or geometry of an invariant set rather than the dynamics within that set. A number of examples have appeared that suggest the need for a detailed taxonomy based on the dynamics rather than shape. In particular, Lorenz argued that his attractor was chaotic by examining functions $\tau:[a, b] \rightarrow [a, b]$. In some of his examples, the entire interval exhibits chaotic dynamics. Nothing about the "shape" of an interval suggests the nature of the dynamics.

DEFINITION 1. Let X be a space with a metric $D(\cdot,\cdot)$, let $E \subset X$, and let $\tau:E \rightarrow X$.

We say a set $C \subset E$ is *invariant* if $\tau(C) = C$.

We say C is (Liapunov) *stable* if for each $\epsilon > 0$ there is a $\delta \subset (0, \epsilon]$ such that $d(x, C) \leq \delta$ implies that for every positive integer n, $d(\tau^n(x), C) \leq \epsilon$, (that is "you stay close if you start sufficiently close to C").

We say C is *attracting* (or is an *attractor*) if for each x sufficiently close to C, $\tau^n(x)$ approaches C; that is, $d(\tau^n(x),C) \rightarrow 0$ as $n \rightarrow \infty$. This careful separation of attraction from stability is standard in the study of topological dynamics and dynamical systems. A well-known example by Denjoy[6] of a differential equation on a torus T^2 emphasizes the need for this distinction. In his example there is a non-empty connected compact invariant set $C \neq T^2$ that is neither a point nor a periodic orbit, (nor in fact does it contain any rest points or periodic orbits) but it is an attractor (as defined above). Hence it is a strange attractor. However, it is not stable. The strange attractors of Ruelle and Takens[21] and Guckenheimer,[8] and presumably Lorenz, are stable; they should be called stable strange attractors.

We say C is *inherently unstable* if every trajectory in C is Liapunov unstable even when the dynamics are restricted to C; more precisely, for each $x \in C$, there is a $\epsilon > 0$ and a sequence $\{ y_i \} \subset C$ with $y_i \rightarrow x$ such that for each y_i there is a positive $n(=n(i))$ for which $\mathrm{dist}(\tau^n(y_i), \tau^n(x)) > \epsilon$.

We say a compact invariant set C is *chaotic* if C is inherently unstable and there is a dense orbit in C; that is, the closure of the set $\{\tau^n(x): n = 1, 2, \ldots\}$ is C. If C has one dense orbit, then "most" points of C have dense orbits, provided "most" is interpreted in the topological sense of Baire Category; (see Reference 7, Theorem 9.20). The horseshoe example of Smale[18,22] contains a chaotic set which is neither stable nor attracting. We show that for certain "preturbulent" parameter values the Lorenz system has a chaotic set that is clearly neither stable nor attracting. Axiom A maps often have attractors that are inherently unstable.

We will say a flow has a chaotic set if some Poincaré "return" map has a chaotic set.

Our principal result for system (1) can now be stated as follows.

THEOREM 1. For $r = r_0 + \epsilon$, ϵ a small, positive number, there are an infinite number of periodic orbits of arbitrarily long period, as well as an uncountable number of points whose trajectories are neither periodic nor asymptotic to any periodic orbit. Further, there exists a chaotic set.

The proof of Theorem 1 can be found in Reference 11. In that paper we carefully demonstrated the existence of the strange set only for r slightly above the critical value $r_0 \approx 13.926$; although we also described what we seem to see over a wider range of r. TABLE 1 summarizes the known dynamic behavior of system (1). The chaotic set is a Cantor set of orbits each orbit being a saddle. As r increases from r_0 to $r_1 \approx 24.06$, the chaotic set grows in size, without any change in its topology. (This is in contrast to the infinitely many topological types of the chaotic stable attractor for $r > r_2$.[25]

THE TRANSITION TO TURBULENCE

We conducted the following experiments in collaboration with E. Yorke. For any initial point p (given r) we define $\sigma(p)$ to be the number of sign changes of $x(t)$, the coordinate of the solution that represents the angular velocity. This in

TABLE 1

A SUMMARY OF THE APPARENT RANGE OF BEHAVIORS AS DETERMINED
BY THEORY AND COMPUTATIONS

For $r < 1$, 0 is globally attracting.
$r = 1$ is a transition value.
For $r > 1$ there are 3 rest points.
For $r < 13.926$ all trajectories tend to one of the rest points.
$r = r_0 \approx 13.926$ is a transition value. There exist two homoclinic orbits, trajectories starting from and going to 0.
For $r > 13.926$ there are infinitely many periodic orbits, and infinitely many "turbulent orbits" that do nothing to any point or periodic orbit.
$r = r_1 \approx 24.06$ is a transition value. The unstable orbits from 0 tend to asymptotically unstable periodic orbits. These two periodic orbits are saddles and the orbits are in the stable manifolds of these periodic orbits.
For $r > 24.06$ there is a chaotic stable attractor. Between 24.06 and 24.74 there exist a chaotic stable attractor and a pair of stable attracting rest points.
$r = r_2 \approx 24.74$ is a transition value.
For $r > 24.74$ there are no stable points.
For some much larger values of $r > 50$ Lorenz has found stable periodic orbits, and for such values no chaotic attractor is observed.

*We list the critical changes as r is varied. This list is calculated for $\sigma = 10$ and $b = \frac{8}{3}$.

essence counts the number of times the orbit switches from an oscillation around one nonzero critical point of system (1) to an oscillation around the other. For various values of r we chose many points at random from the region near the non-attracting chaotic set. There was a striking dichotomy in the results when r was large ($22 < r < r_1$). While $\sigma(p)$ was found to be 0 or 1 for many of the randomly chosen points, many other points produced large values of $\sigma(p)$. In particular, for points near any of the three critical points, $\sigma(p)$ is found to be 0 or 1. Excluding all those points for which $\sigma(p)$ was 0 or 1, the rest were roughly distributed according to a (discrete) exponential distribution, the mean of which appears to go to ∞ as $r \to r_1$. At $r = 23$, the mean of these $\sigma(p)$ values appears to be well over 100, but for r near 24 the computer time required for careful statistics becomes prohibitive, and it is this range that is most interesting.

For r slightly less than 24.06, we thus have "metastable chaos," that is, chaotic behavior that is observed to persist for a long but finite time. Physically, our preturbulent state would appear to be one in which there is an attracting state (two stable attracting points) and a complementary region in which orbits oscillate chaotically. The observed trajectories act as if there were a half-life to their stay in the chaotic region; if a trajectory in the chaotic state is observed for time T, knowledge of T tells us nothing about how much longer it will remain in the chaotic region. As $r \rightarrow r_1$ there is little change in the apparent volume of the metastable chaotic region surrounding the strange set. For $r < r_1$, almost every point in the chaotic region will eventually be sucked toward one of the attracting critical points, but the mean number of preceding oscillations is large. As r passes beyond r_1, the mean time becomes infinite and "suddenly" the ill-defined chaotic region becomes the region of attraction of the strange attractor. Robbins[19] investigates the transition to turbulence and reports she observed (numerically) orbits in the Lorenz system and related systems which oscillate chaotically for quite a while before settling down. See also Robbins.[20] Her parameter values correspond to our situation with r slightly less than r_1. Based on analysis of piecewise linear mappings on the real line to approximate the dynamics of the Lorenz system, Robbins (Reference 19, Section 3) argues that for all $r \in (r_0, r_1)$ (using the notation in our analogous situation) there is an unstable periodic orbit, in agreement with our findings. She argues that for large r in (r_0, r_1) there are others that do not approximate a steady solution, and for larger r in (r_0, r_1), "the set of trajectories may become uncountable." Her method disagrees with our findings of chaos near r_0. This metastable chaos regime may have been observed physically: Creveling et al.[4] reports an experiment involving fluid flow through pipes in which over a hundred oscillations are observed before the oscillations become regular and damp out. The existence of metastable chaotic regimes in physical situations could appear turbulent, for metastable chaos can persist for long durations. It is particularly difficult to determine whether apparent chaos in experiments in fact represents actual turbulence or just metastable chaos.

Lorenz mentioned the transition at $r = 24.74$ since this is the critical point at which the two regions of attraction of the nonzero rest points shrink out of existence. McCracken[16] shows that at $r = r_2$ the nonzero critical points are unstable. McLaughlin and Martin[17] espouse the viewpoint that the nature of this bifurcation causes "an immediate transition to turbulence," which seems to underemphasize the fact that a chaotic set was established earlier at r_1.

References

1. ALEXANDER, J. & J. A. YORKE. In press. Global bifurcation of periodic orbits. Amer. J. Math.
2. ARNOLD, V. I. 1965. Small denominators, I. Translations Amer. Math. Soc. 2nd Series. **46:** 213–284.
3. CHOW, S. N., J. MALLET-PARET & J. A. YORKE. In preparation.
4. CREVELING, H. F., J. F. DePAZ, J. V. BALADI & R. J. SCHOENHALS. 1975. Stability characteristics of a single phase free convection loop. J. Fluid Mech. **67:** 65–84.
5. CURRY, J. H. 1976. Transition to Turbulence in Finite-Dimensional Approximations to the Boussinesq Equations. Ph.D. Thesis, Univ. of Cal., Berkeley.

6. DENJOY, A. 1932. Sur les courbes défines par les équations différentielles à la surface du tore. J. de Math. Ser 9. **11**: 333–375.

7. GOTTSCHALK, W. H. & G. A. HEDLUND. 1968. Topological Dynamics. Amer. Math Soc., Colloquium Pub., Vol. 36. Revised edit. Providence, R.I.

8. GUCKENHEIMER, J. 1976. A strange strange attractor. In the Hopf Bifurcation Theorem and its Applications. J. E. Marsden & M. McCracken, Eds.: 368–381. Springer-Verlag. New York.

9. HOPF, E. 1948. A mathematical example displaying features of turbulence. Commun. Pure Appl. Math. **1**: 303–322.

10. HOPF, E. 1942. Abzweigung einer periodischen Losung von einer stationaren Losung eines Differentialsystems. Ber. Math.-Phys. Sachsische Academie der Wissenschaften Leipzig. **94**: 1–22.

11. KAPLAN, J. L. & J. A. YORKE. In press. Preturbulence: A regime observed in a fluid flow model of Lorenz.

12. LANDAU, L. D. & E. M. LIFSCHITZ. 1959. Fluid Mechanics. Pergamon Press, Oxford. [See also L. Landau. 1944. C. R. Acad. Sci., U.R.S.S. **44**: 311.]

13. LASOTA, A. & J. A. YORKE. In press. On the existence of invariant measures for transformations with strictly turbulent trajectories. Bull. Polish Acad. Sci.

14. LASOTA, A. 1972. Relaxation oscillations and turbulence, in Ordinary Differential Equations. Academic Press. New York.

15. LORENZ, E. N. 1963. Deterministic nonperiodic flow. J. Atmos. Sci. **20**: 130–141.

16. MARSDEN, J. E. & M. MCCRACKEN. 1976. The Hopf Bifurcation and its Applications.: 141–148. Springer Verlag. New York.

17. MCLAUGHLIN, J. B. & P. C. MARTIN. 1975. Transition to turbulence in a statically stressed fluid system. Phys. Rev. A **12**: 186–203.

18. NITECKI, Z. 1971. Differential Dynamics, An Introduction to the Orbit Structure of Diffeomorphisms. M.I.T. Press. Cambridge, Mass.

19. ROBBINS, K. A. In press. A new approach to subcritical instability and turbulent transitions in a simple dynamo. Math. Proc. Cambridge Phil. Soc.

20. ROBBINS, K. A. 1976. A moment equation description of magnetic reversals in the earth. Proc. Nat. Acad. Sci. U.S.A. **73**(12): 4297–4301.

21. RUELLE, D. & F. TAKENS. 1971. On the nature of turbulence. Comm. Math. Phys. **20**: 167–192.

22. SMALE, S. 1961. A structurally stable differentiable homeomorphism with an infinite number of periodic points. Proc. Internat. Symp. Nonlinear Vibrations. Vol. II. Izdat. Akad. Nauk. Ukrain SSR, Kiev.

23. SMALE, S. 1967. Differentiable dynamical systems. Bull. Amer. Math. Soc. **73**: 747–817.

24. SWINNEY, H. L., P. R. FENSTERMACHER & J. B. GOLLUB. 1977. Transition to turbulence in circular couette flow. A preprint for the Symposium on Turbulent Shear Flows.

25. WILLIAMS, R. F. In press. The structure of Lorenz attractors. A preprint.

26. OXTOBY, J. C. & S. M. ULAM. 1941. Measure preserving homeomorphisms and metrical transitivity. Ann. Math. **42**: 87–92.

27. HÉNON, M. 1976. A two-dimensional mapping with a strange attractor. Comm. Math. Phys. **50**: 69–77.

28. RANNON, F. 1974. Numerical study of discrete plane area-preserving mappings. Astron and Astrophys. **31**: 289–301.

SENSITIVE DEPENDENCE ON INITIAL CONDITION
AND TURBULENT BEHAVIOR
OF DYNAMICAL SYSTEMS

David Ruelle

Institut des Hautes Etudes Scientifiques
91440 Bures-sur-Yvette
France

GENERALITIES

The purpose of this talk is to discuss some qualitative features of the time evolution of natural systems. The time evolution is described by an equation

$$x_{t+1} = f(x_t) \qquad \text{(discrete time)} \qquad (1)$$

or

$$\frac{dx_t}{dt} = X(x_t) \qquad \text{(continuous time)}. \qquad (2)$$

The qualitative features that we want to discuss are those associated with sensitive dependence on initial condition. We shall try to analyze sensitive dependence on initial condition, see how it manifests itself as turbulent behavior, and find the simplest examples in which it occurs.

In applications, x_t represents the state (at time t) of the natural system under consideration, and may vary in some infinite dimensional space (as for instance in hydrodynamics). For the convenience of the mathematical discussion we shall, however, suppose that the state space M of our system is finite dimensional.

The time evolution is defined by maps $f^t: M \to M$ (t discrete or continuous). We shall assume that either M is $\mathbb{R}^m$ and there is some bounded open $U \subset \mathbb{R}^m$ such that closure $f^t U \subset U$ for $t > 0$, or that M is a compact differentiable manifold.* We assume that the map f in (1) or the vector field X in (2) are C^r, i.e., r times continuously differentiable with $r \geq 1$.

In what follows I shall try to be mathematically correct, or at least not misleading. It has to be realized, however, that many of the questions to be discussed are poorly understood mathematically. I shall definitely not limit myself to those topics that are completely elucidated. I shall try to give an idea of what lies beyond, at the cost of some conjectures and heuristic considerations.

SENSITIVE DEPENDENCE ON INITIAL CONDITIONS

It will often be convenient to distinguish the following three cases:

Maps: discrete time, f not necessarily invertible.

---

*A compact differentiable manifold may always be thought of as imbedded in $\mathbb{R}^n$ for suitably large N. The "tangent spaces" to M may then be identified with subspaces of $\mathbb{R}^N$ as intuition dictates.

0077-8923/78/0316-0408 $1.75/1 © 1979, NYAS

Diffeomorphisms: discrete time, f has a differentiable inverse, so that f^t is defined for $t = -1, -2, \ldots$.

Flows: continuous time.

Sensitive dependence on initial conditions means that if there is a small change δx_0 in the initial condition x_0, the corresponding change $\delta x_t = f^t(x_0 + \delta x_0) - f^t(x_0)$ of $x_t = f^t(x_0)$ grows and becomes large when t becomes large. More precisely we require δx_t to *grow exponentially with t*. Why this requirement is reasonable will appear in the next section.

The difference $f^t(x_0 + \delta x_0) - f^t(x_0) = \delta x_t$ is meaningful only if our manifold M is $\mathbb{R}^m$. In general, it is preferable to interpret δx_t as a vector tangent to the manifold; we have then

$$\delta x_t = (T_{x_0} f^t) \delta x_0,$$

where $T_{x_0} f^t$ is a linear operator mapping the tangent space $T_{x_0} M$ to M at x_0 to the tangent space $T_{x_t} M$ at x_t. If M is $\mathbb{R}^m$, then the tangent spaces can be identified with $\mathbb{R}^m$ and $T_x f^t$ is just the $m \times m$ matrix of partial derivatives of $f^t(x)$ with respect to x. The discussion of sensitive dependence on initial condition translates thus into the study of $T_x f^t$ for large t. We shall not try to make statements valid for all $x \in M$ (all initial conditions), but rather for almost all x with respect to some probability measure ρ (on M) invariant under f^t (i.e., invariant under time evolution). We shall try to argue later why an ensemble average (with respect to some ρ) corresponds to time average for "most" initial conditions. For the moment we accept as a fact of life the fact that x is distributed according to the f^t-invariant probability measure ρ. The *noncommutative ergodic theorem* of Oseledec describes then the behavior of $T_x f^t$ for large t.

THE NONCOMMUTATIVE ERGODIC THEOREM

We first give the version of the theorem which is appropriate for the study of maps.

THEOREM 1. Let $(M. \Sigma, \rho)$ be a probability space and $\tau : M \to M$ a measurable map preserving ρ. Let also $T : M \to M_n(\mathbb{R})$ be a measurable map into the $m \times m$ matrices, such that†

$$\log^+ \| T(\cdot) \| \in L^1(M, \rho)$$

and write $T_x^n = T(\tau^{n-1} x) \ldots T(\tau x) T(x)$.

There is $\Omega \subset M$ such that $\rho(\Omega) = 1$ and for all $x \in \Omega$

$$\lim_{n \to \infty} (T_x^{n*} T_x^n)^{1/2n} = \Lambda_x$$

exists [* denotes matrix transposition].

Let $\exp \lambda_x^{(1)} < \ldots < \exp \lambda_x^{(s(x))}$ be the eigenvalues of Λ_x [with possibly $\lambda_x^{(1)} = -\infty$], and $U_x^{(1)}, \ldots, U_x^{(s(x))}$ the corresponding eigenspaces. If $V_x^{(r)} = U_x^{(1)} + \cdots + U_x^{(r)}$ we have

---

†We write $\log^+ x = \max\{0, \log x\}$.

$$\lim_{n \to \infty} \frac{1}{n} \log || T_x^n u || = \lambda_x^{(r)} \quad \text{when} \quad u \in V_x^{(r)} \backslash V_x^{(r-1)}$$

for $r = 1, \ldots, s(x)$.

The theorem published by Oseledec[6] assumes τ and T invertible. Its proof has been simplified by Raghunathan.[8] The above result can be obtained by modifying Raghunathan's argument.

Let $m_x^{(r)} = \dim U_x^{(r)} = \dim V_x^{(r)} - \dim V_x^{(r-1)}$. The numbers $\lambda_x^{(1)}, \ldots, \lambda_x^{(s(x))}$, with multiplicities $m_x^{(1)}, \ldots, m_x^{(s(x))}$ constitute the *spectrum* of (ρ, τ, T) at x. The $\lambda_x^{(r)}$ are also called *characteristic exponents*. When n tends to ∞, $(1/n)$ log $|| T_x^n ||$ tends to the maximum characteristic exponent $\lambda_x^{(s(x))}$. The spectrum is τ-invariant; if ρ is τ-ergodic, the spectrum is almost everywhere constant.

To apply Theorem 1 to differentiable maps of M, it suffices, if $M = \mathbb{R}^m$, to take $\tau = f$, $T(x) = T_x f$. If M is a compact manifold, it may be necessary to cut it into a finite number of measurable pieces, each of which is diffeomorphic‡ to a subset of $\mathbb{R}^m$, so that $T_x M$ is identified to $\mathbb{R}^m$ for all x. We see thus that the asymptotic behavior of $\delta x_t = (T_x f^t) \delta x$ is exponential with t for large t. We have sensitive dependence on initial condition if the maximum characteristic exponent is strictly positive. Of course, other characteristic exponents are allowed to be negative.

If f is a *diffeomorphism*, the following version of the noncommutative ergodic theorem gives extra information.

THEOREM 2. Keeping the notation and assumptions of Theorem 1, let τ have a measurable inverse, let T^{-1} exist such that

$$\log^+ || T(\cdot)^{-1} || \in L^1(M, \rho)$$

and write $T_x^{-n} = T(\tau^{-n}x)^{-1} \ldots T(\tau^{-2}x)^{-1}(T(\tau^{-1}x)^{-1}$.

We can assume that for $x \in \Omega$ there is a splitting $\mathbb{R}^m = W_x^{(1)} \oplus W_x^{(2)} \oplus \ldots$ *such that the following limits exists*

$$\lim_{k \to \pm \infty} \frac{1}{k} \log || T_x^k u || = \lambda_x^{(r)} \quad \text{if} \quad u \in W_x^{(r)} \backslash \{0\}.$$

Obviously the $\lambda_x^{(r)}$ are the characteristic exponents, and the $m_x^{(r)} = \dim W_x^{(r)}$ their multiplicities.

The splitting $\mathbb{R}^m = W_x^{(1)} \oplus W_x^{(2)} \oplus \ldots$ depends measurably on x. If M is a manifold and $\tau = f$ a diffeomorphism, the splitting is in general not continuous. One may assume that Ω is a Borel set of measure 1 with respect to every f-invariant measure, and that the dependence of the splitting and the spectrum on $x \in \Omega$ is Borel.

Suppose that one can take for Ω a closed f-invariant set, that there is $\epsilon > 0$ such that the spectrum is disjoint from the interval $(-\epsilon, +\epsilon)$, and that the spaces

$$W_x^+ = \sum_{r:\lambda_x^{(r)} > 0} W_x^{(r)}, W_x^- = \sum_{r:\lambda_x^{(r)} < 0} W_x^{(r)}$$

---

‡We require thus the existence of a differentiable map with differentiable inverse, from an open neighborhood of the closure of the piece of M considered, to an open set in $\mathbb{R}^m$.

depend continuously on $x \in \Omega$. One says then that Ω is *hyperbolic*. This is the main ingredient in the Axiom A of Smale. There exists a detailed theory of Axiom A diffeomorphisms, to which we shall refer to test various ideas. By contrast the ergodic theory of non-Axiom-A diffeomorphisms is in a state close to non-existence.

We shall make no special discussion of the noncommutative ergodic theorem for *flows*. The results are those expected.

ASYMPTOTIC MEASURES

The use of the noncommutative ergodic theorem assumes that x is distributed according to some f-invariant measure ρ (for simplicity we discuss the discrete time case). We have to examine this assumption, and also try to restrict the choice of ρ, the invariant measures being often quite numerous.§

A natural idea is to define ρ as a time average

$$\rho = \lim_{n \to \infty} \frac{1}{n} \sum_{k=0}^{n-1} \delta_{f^k x} \qquad (3)$$

where δ_x is the Dirac measure at x and the limit is in the sense that the integrals of continuous functions converge (vague limit). This procedure may not work: one can find a diffeomorphism f and an open set (nonempty) of x such that (3) fails to exist (R. Bowen, private communication). Nevertheless one can hope that the limit exists in many cases. For C^2 Axiom A diffeomorphisms¶ one can show that there is a set M' such that $M \setminus M'$ has zero Lebesgue measure** and the limit (3) exists for $x \in M$, taking a finite number of values. The *asymptotic measures* determined in this manner for Axiom A diffeomorphisms are precisely those ergodic measures that make maximum the expression

$$p(\rho) = h(\rho) - \int \rho(dx) \sum_{r : \lambda_x^{(r)} > 0} m_x^{(r)} \lambda_x^{(r)} \qquad (4)$$

the maximum being in fact zero. In (4), $h(\rho)$ is the *entropy* (or Kolmogorov-Sinai invariant) of ρ with respect to f.

One can verify[11] that for every differentiable map f of a compact manifold M into itself, and every f-invariant measure ρ, $p(\rho) < 0$, where $p(\rho)$ is defined by (4). This suggests that one should look for the limits (3) among the measures satisfying (4) *provided one discards a set of x with zero Lebesgue measure*. Here is a heuristic argument in favor of that idea.

Let ν be the Lebesgue measure on M, normalized to 1. Suppose the measures

---

§An Axiom A diffeomorphism may have either only finitely many ergodic measures (carried by periodic orbits) or continuously many ergodic measures.

¶There is a similar result for Axiom A flows. See References 2, 9, and 14.

**By "Lebesgue measure" on a compact manifold M we mean the measure associated with any Riemann metric (any two such measures are equivalent).

$$\frac{1}{n} \sum_{k=0}^{n-1} f^k \nu$$

tend to a limit ρ when $n \rightarrow \infty$, then one can expect that the

$$\frac{1}{n} \sum_{0}^{n-1} \delta_{f^k x}$$

tends for ν − almost all x to ρ or to one of its ergodic components. [If it tends to a limit ρ_x for ν − almost all x, then $\rho = \int \nu(dx)\rho_x$]. We assume that for ρ − almost all $\xi \in M$, there is an "unstable manifold" $\mathcal{V}_\xi^-$ tangent to

$$W_\xi = \sum_{r:\lambda_\xi^{(r)}<0} W_\xi^{(r)},$$

such that the family of the $\mathcal{V}_\xi^-$ is invariant under f. (Such a result is known in certain cases, see Pesin.[7]) The manifolds $\mathcal{V}_\xi^-$ are expanded by f^n exponentially fast for large n. Because of this stretching along the manifolds $\mathcal{V}_\xi^-$, the measures $f^n \nu$ for large n will tend to remain smooth in the direction parallel to the V_ξ^- (while their density may acquire a large transverse derivative). Therefore we expect ρ to have, along the $\mathcal{V}_\xi^-$, conditional measures that have continuous density with respect to Lebesgue measures on the $\mathcal{V}_\xi^-$. If that is the case, one can estimate the entropy of ρ to be at least the integral of the logarithm of the expansion coefficient along the unstable manifolds, i.e.,

$$h(\rho) \geq \int \rho(dx) \sum_{r:\lambda_x^{(r)}>0} m_x^{(r)} \lambda_x^{(r)}.$$

Since the reverse inequality also holds, we have $p(\rho) = 0$.

To the above heuristic argument there should correspond a theorem. Its precise formulation and conditions of applicability are not yet known but should be more general than the rather restricted Axiom A class. The important question of the stability of the asymptotic measures under small stochastic perturbations is discussed below in the appendix.

We can try to make use of the equality

$$h(\rho) = \sum_{r:\lambda_x^{(r)}>0} m_x^{(r)} \lambda_x^{(r)}, \tag{5}$$

for asymptotic measures ρ (we assume ρ to be ergodic and the right-hand side is given its ρ-almost-everywhere-constant value). In particular, the right-hand side is quite accessible numerically, while $h(\rho)$ is not. One can thus estimate $h(\rho)$ by taking some initial condition ξ, computing $x = f^N \xi$ for some large N, and then computing

$$\sum_{r:\lambda_x^{(r)}>0} m_x^{(r)} \lambda_x^{(r)}$$

by Theorem 1 above. Notice that if one somehow knows that the topological entropy:

$$\sup \{h(\rho): \rho \text{ invariant}\}$$

vanishes, then (5) implies that there is no sensitive dependence on initial condition. The converse is a theorem: If all characteristic exponents of all invariant measures are ≤ 0, then (because $p(\rho) \leq 0$) the topological entropy is 0.

Since the measures which satisfy (5) are those which maximize $p(\rho)$ [as given by (4)], one can try to approximate them by a variational procedure. It remains to be seen if one can use this method practically to get an idea of the asymptotic measures for a problem such as that of fluid turbulence.

Finally, let us remark (after Benettin *et al.*[1]) that the largest characteristic exponent $\chi = \lambda_x^{(s(x))}$, which characterizes the sensitive dependence on initial condition, satisfies

$$\frac{1}{m} h(\rho) \leq \chi \leq h(\rho)$$

if (5) holds.

NOTE ADDED IN PROOF: The existence almost everywhere of stable manifolds $\mathcal{V}_\xi^-$ has now been established (Ruelle, to appear). It is also known that if ρ has, along the $\mathcal{V}_\xi^-$, conditional measures that have continuous density, then (5) holds. (This follows from work by Pesin as pointed out to me by A. Katok). However R. Bowen and A. Katok have shown me an example where $\sup p(\rho) < 0$.

TURBULENT BEHAVIOR

It is generally impractical to change the initial condition of a dynamical system occurring in nature, and to observe the behavior of δx_t as t increases. There are, however, indirect ways in which the sensitive dependence on initial condition manifests itself, giving rise to what we shall call turbulent behavior.†† In particular, x_t will be neither asymptotically constant nor periodic, but will have an apparently erratic appearance.

If one can measure the position x_t at each time with x high but with only finite precision as is the case for natural phenomena, it is found that systems with sensitive dependence on initial condition *lose information* at the rate

$$\sum_{r:\lambda_x^{(r)} \geq 0} m_x^{(r)} \lambda_x^{(r)}$$

[i.e., $h(\rho)$ according to the relation (5)]. One would like to deduce from this some decay properties of the time correlation functions

$$F_{\varphi\psi}(t) = \int \rho(dx)\, \varphi(x)\, \psi(f_x^t) - [\int \rho\varphi][\int \rho\psi]$$

††The connection with the theory of fluid turbulence is not discussed here. See Lorenz[4] and Ruelle and Takens.[12]

when $| t | \rightarrow \infty$ (φ and ψ are assumed differentiable). At this moment a theorem is known only for Axiom A diffeomorphisms, where it has been proved that $F_{\varphi\psi}$ decreases exponentially at infinity. In particular the positive measure

$$\omega \rightarrow \sum_{t \in \mathbf{Z}} e^{-i\omega t} F_{\varphi\varphi}(t)$$

has continuous density, a property known as *continuous (frequency) spectrum*. One expects continuous spectrum to occur under much more general conditions than Axiom A. Actually continuous spectrum is observed experimentally in fluid turbulence.[3]

SIMPLEST EXAMPLES OF TURBULENT BEHAVIOR

The simplest examples of sensitive dependence on initial condition have been reviewed recently in Reference 10. We recall that the lowest dimension for which sensitive dependence occurs is respectively 1,2,3 for maps, diffeomorphisms, and flows.

On the 3-torus, a flow with sensitive dependence on initial condition can be obtained by an arbitrarily C^2-small perturbation of a quasiperiodic flow. For a m-torus, $m > 3$, C^2 can be replaced by C^∞.[5] This means that if a small coupling is introduced between three or more oscillators, turbulent behavior may result. Using oscillating electric circuits, it should be possible to visualize the transition to continuous spectrum when a suitable coupling is introduced between the oscillators. Alternatively if the frequencies are in the audible range, the transition to continuous spectrum should correspond to a change in the musical nature of the corresponding sound. These experiments—contrasting the coupling of three oscillators with the coupling of two oscillators—have not been performed as far as I know. Since they are easy, I strongly suggest that they should be attempted.‡‡

CONCLUSIONS

The domain of research that has been reviewed here is one where progress is slow, due to great mathematical difficulties. The potential applications are, however, very important, both from the purely theoretical viewpoint (understanding of turbulence), and from a very practical viewpoint (discussion of individual systems with sensitive dependence on initial condition). It seems to me that even in the present very imperfect state of the theory it has started to be possible to interpret in a meaningful way some of the typical aspects of "turbulent" differentiable dynamical systems.

‡‡A computer study of three coupled oscillators has been made by Sherman and McLaughlin.[13] Unfortunately their paper does not make clear exactly what mathematical system is treated.

It has to be remarked that, although the flows on T^2 do not have sensitive dependence on initial condition, they may have only the trivial eigenfunction 1 (see A. N. Kolmogorov[15]). This situation appears to be exceptional (M. Herman, private opinion), and it is not clear what the Fourier transform of $F_{\varphi\varphi}$ then looks like.

APPENDIX

We have justified above the consideration of certain *asymptotic measures* by the fact that they describe the ergodic averages for almost all initial conditions (with respect to Lebesgue measure). Another point of view is that the time evolution (1) or (2) is in practice always perturbed by some noise, being thus replaced by a stochastic process. One can argue heuristically that the same class of measures is obtained in this manner as before, namely, invariant measures that are "continuous along the unstable direction." This is because the smearing corresponding to the noise term preserves the "continuity along the unstable direction." In the Axiom A case this argument has been made rigorous.[14,16,17]

One more remark about the asymptotic measures. The conditional measure on an unstable manifold has a density with respect to the "Lebesgue measure" on that leaf, and, by f-invariance, these densities satisfy proportionality relations involving the Jacobian of f in the unstable direction. These conditions correspond to the "Gibbs state" condition just as the variational principle above corresponds to the definition of equilibrium states in statistical mechanics. Thus, technically, the asymptotic measures or "ensembles" representing turbulence bear a remarkable resemblance with the ensembles of equilibrium statistical mechanics.

REFERENCES

1. BENETTIN, G., I. GALGANI & J.-M. STRELCYN. 1976. Kolmogorov entropy and numerical experiments. Phys. Rev. A **14**: 2338–2345.
2. BOWEN, R. & D. RUELLE. 1975. The ergodic theory of Axiom A flows. Inventiones Math. **29**: 181–202.
3. GOLLUB J.-P. & H. L. SWINNEY. 1975. Onset of turbulence in a rotating fluid. Phys. Rev. Lett. **35**: 927–930.
4. LORENZ, E. N. 1963. Deterministic nonperiodic flow. J. Atmos. Sci. **20**: 130–141.
5. NEWHOUSE, S., D. RUELLE & F. TAKENS. Occurrence of strange Axiom A attractors near quasi periodic flows on T^m, $m \geq 3$. Commun. Math. Phys. To appear.
6. OSELEDEC, V. I. 1968. A multiplicative ergodic theorem. Ljapunov characteristic numbers for dynamical systems. Trudy Moskov. Mat. Obšč. **19**: 179–210. [English translation: 1968. Trans. Moscow Math. Soc. **19**: 197–231.]
7. PESIN, IA. B. 1976. Ljapunov characteristic exponents and ergodic properties of smooth dynamical systems with an invariant measure. Dokl. Akad. Nauk SSSR **226**(4): 774–777. [English translation: 1976. Soviet Math. Dokl. **17**(1): 196–199.]
8. RAGHUNATHAN, M. S. A proof of Oseledec' multiplicative ergodic theorem. Advan. in Math. To appear.
9. RUELLE, D. 1976. A measure associated with Axiom A attractors. Amer. J. Math. **98**: 619–654.
10. RUELLE, D. 1978. Dynamical systems with turbulent behavior. *In* Mathematical Problems in Theoretical Physics, Proceedings, Rome 1977. Lecture Notes in Physics No. 80.: 341–360. Springer. Berlin. 1978
11. RUELLE, D. An inequality for the entropy of differentiable maps. Preprint. Bol. Soc. Bras. Mat. To appear.
12. RUELLE, D & F. TAKENS. 1971. On the nature of turbulence. Commun. Math. Phys. **20**: 167–192; **23**: 343–344.
13. SHERMAN, J. & J. MCLAUGHLIN. 1978. Power spectra of nonlinearly coupled waves. Commun. Math. Phys. **58**: 9–17.
14. SINAI, IA. G. 1972. Gibbs measures in ergodic theory. Uspehi Mat. Nauk **27**(4):21–64. [English translation: 1972. Russian Math. Surveys **27**(4): 21–69.]

15. KOLMOGOROV, A. N. 1953. On dynamical systems with integral invariants on the torus. Dokl. Akad. Nauk SSSR **93**(5): 763–766.
16. KIFER, JU. I. 1974. On the limiting behavior of invariant measures of small random perturbations of some smooth dynamical systems. Dokl. Akad. Nauk. SSSR **216**(5): 979–981. [English translation: 1974. Soviet Math. Dokl. **15**: 918–921.]
17. KIFER, JU. I. 1974. On small random perturbations of some smooth dynamical systems. Izv. Akad. Nauk SSSR. Ser. Mat. **38**(5): 1091–1115. [English translation: 1974. Math. USSR Izvestija. **8**: 1083–1107.]

PHASE TRANSITIONS AS A PROBLEM
IN BIFURCATION THEORY*

John J. Kozak

*Department of Chemistry
and
Radiation Laboratory
University of Notre Dame
Notre Dame, Indiana 46556*

INTRODUCTION

The class of problems with which we shall be concerned in this talk deals with an approach to the study of phase transitions initiated by Kirkwood[1] and Vlasov[2] a generation ago. These authors suggested that changes in the behavior of solutions to certain of the nonlinear integral or integro-differential equations comprising the Bogolyubov-Born-Kirkwood-Yvon-Green (BBKYG) hierarchy might be associated with the onset of a phase transition. In order to implement this suggestion, both authors derived stability criteria, Kirkwood using Fourier-transform methods and Vlasov using bifurcation theory, and, subject to the uncertainty involved in truncating the hierarchy, it was believed that the incidence of a phase transition could be identified with the point(s) at which a well-defined solution exhibited an abrupt change in behavior. Several years ago it was pointed out that the Kirkwood stability criterion, developed originally in a theory of melting, could be derived from a bifurcation analysis of the Kirkwood nonlinear integral equation for the singlet distribution function.[3] In effect, the Kirkwood criterion was shown to be the Fourier transform of the equation obtained via linearization of the nonlinear problem; the points where the criterion was satisfied were just the eigenvalues of the associated linear integral equation. However, inasmuch as bifurcation need not occur unless the eigenvalues of the linear equation are of odd multiplicity, it was noted that, even within the context of the Kirkwood-Vlasov philosophy, satisfaction of the Kirkwood criterion was at most a necessary but not sufficient condition for characterizing onset of a phase transition. More recently, it was found that a similar analysis could be carried thru starting from the Yvon-Born-Green nonlinear integral equation for the pair distribution function,[4] and in fact by taking into account the short-range properties of the kernel of that equation, a bifurcation condition could be formulated that could be applied to the fluid–gas, as well as the fluid–solid transition. This bifurcation criterion was implemented in an extensive numerical study of the statistical mechanics of the square-well fluid,[5] and it was demonstrated that a locus of points could be generated in the temperature–density (T, ρ) plane (FIGURE 1), which differentiated the gas, fluid, and "periodic" regions of the phase diagram. Moreover, in accord with the predictions of bifurcation theory, it was shown numerically that the

*This work was supported by the Division of Basic Energy Sciences of the Department of Energy. This is Document No. NDRL-1820 from the Notre Dame Radiation Laboratory.

417

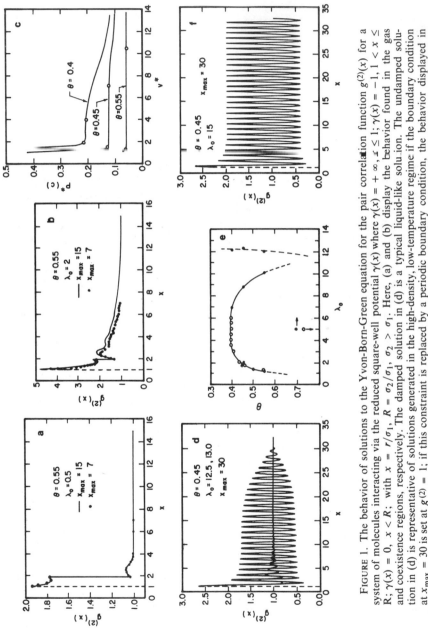

FIGURE 1. The behavior of solutions to the Yvon-Born-Green equation for the pair correlation function $g^{(2)}(x)$ for a system of molecules interacting via the reduced square-well potential $\gamma(x)$ where $\gamma(x) = +\infty, x \leq 1; \gamma(x) = -1, 1 < x \leq R; \gamma(x) = 0, x < R$; with $x = r/\sigma_1$, $R = \sigma_2/\sigma_1$, $\sigma_2 > \sigma_1$. Here, (a) and (b) display the behavior found in the gas and coexistence regions, respectively. The damped solution in (d) is a typical liquid-like solution. The undamped solution in (d) is representative of solutions generated in the high-density, low-temperature regime if the boundary condition at $x_{max} = 30$ is set at $g^{(2)} = 1$; if this constraint is replaced by a periodic boundary condition, the behavior displayed in (f) is realized. The plot (c) shows the isotherms generated in the coexistence region and the plot (e) is the locus of points in the reduced temperature ($\theta = \epsilon/kT$, where ϵ is the depth of the well), reduced density ($\lambda_0 = 4\pi n/\sigma_1^3$, where n is the number density) plane generated via the application of bifurcation criteria.

generated locus of points essentially coincided with those values of (T, ρ) at which solutions to the underlying nonlinear integral equation appeared to experience an abrupt qualitative change in structure.

While very suggestive, the evidence presented in FIGURE 1 should be regarded as "experimental," inasmuch as certain simplifications were introduced in the numerical and theoretical analysis of the problem. Although these approxima- tions were motivated by the underlying physics of the problem, the fact that even *one* approximation was introduced, regardless of how harmless that approxima- tion might seem, means that an *exact* statement regarding the *essential* correct- ness of a bifurcation approach to the problem of phase transitions could not be made. The open question is: can it be proved rigorously that criteria derived from bifurcation theory represent either a necessary or sufficient condition for the onset of a phase transition? It is to this question that the present talk is addressed, and what we shall attempt to do here is to make precise in the formulation of the problem, the statistical mechanical issues that must be taken into account in treating the problem of phase transitions from the standpoint of bifurcation theory.

FORMULATION OF THE PROBLEM

Suppose that the pair potential Φ characterizing the interactions between mole- cules is stable and regular, and let

$$U(x)_n = \sum_{1 \le i < j \le n} \Phi(x_j - x_i) \tag{1}$$

$$\psi(x)_n = \exp[-\beta U(x)_n], \tag{2}$$

where $\beta = 1/kT > 0$. Let us introduce as well the characteristic function χ_Λ of the bounded measurable set $\Lambda \epsilon R^\nu$ (where ν is the dimension) and write

$$\chi_\Lambda(x)_n = \prod_{i=1}^{n} \chi_\Lambda(x_i). \tag{3}$$

Using this notation, the grand partition function for the problem

$$\Xi(\Lambda, z, \beta) = \sum_{n=0}^{\infty} \frac{z^n}{n!} \int_{\Lambda^n} dx_1 \dots dx_n \exp[-\beta U(x)_n] \tag{4}$$

and the m-particle distribution function

$$\rho_\Lambda(x)_m = \Xi(\Lambda, z, \beta)^{-1} \sum_{n=0}^{\infty} \frac{z^{m+n}}{n!} \int_{\Lambda^n} dx_{m+1} \dots dx_{m+n} \exp[-\beta U(x)_{m+n}], \tag{5}$$

(where z is the activity) may be rewritten as:

$$\Xi(\Lambda, z, \beta) = \sum_{n=0}^{\infty} \frac{z^n}{n!} \int d(x)_n \chi_\Lambda(x)_n \psi(x)_n \tag{6}$$

and

$$\rho_\Lambda(x)_m = \Xi(\Lambda, z, \beta)^{-1} \sum_{n=0}^{\infty} \frac{z^{m+n}}{n!} \int dx_{m+1} \dots dx_{m+n} \chi_\Lambda(x)_{m+n} \psi(x)_{m+n}. \qquad (7)$$

The distribution functions $\rho_\Lambda(x)_m$ satisfy various systems of integro-differential or integral equations, among which the more familiar are the Yvon-Born-Green (YGB), the Kirkwood (K), and the Kirkwood-Salsburg (KS) equations. In this section, we focus attention on the Kirkwood-Salsburg equations, given by

$$\rho_\Lambda(x_1) = \chi_\Lambda(x_1) z \left[1 + \sum_{n=1}^{\infty} \frac{1}{n!} \int dy_1 \dots dy_n K[x_1, (y)_n] \rho_\Lambda(y)_n \right] \qquad (8)$$

$$\rho_\Lambda(x)_m = \chi_\Lambda(x)_m z \exp[-\beta W^1(x)_m]$$

$$\cdot \left[\rho_\Lambda(x)_{m-1}^1 + \sum_{n=1}^{\infty} \frac{1}{n!} \int dy_1 \dots dy_n K[x_1, (y)_n] \rho_\Lambda[(x)_{m-1}^1, (y)_n] \right]$$

$$m > 1, \qquad (9)$$

where,

$$W^1(x)_m = \sum_{i=2}^{m} \Phi(x_i - x_1) \qquad (10)$$

$$(x)_{m-1}^1 = (x_2, \dots, x_m) \qquad (11)$$

$$K[x_1, (y)_n] = \prod_{j=1}^{n} \{\exp[-\beta\Phi(y_j - x_1)] - 1\}. \qquad (12)$$

The set of integral equations, Equations 8 and 9, form a *linear* inhomogeneous system for the (generic) distribution functions $\rho_\Lambda(x)_n, n > 1$. Similarly, the set of integro-differential equations generated by Kirkwood, and Yvon, Born, and Green form an inhomogeneous system, *linear* in the m-particle distribution functions.

In a very important paper, Ruelle[6] has taken advantage of the linear structure of the Kirkwood-Salsburg (KS) equations and has shown how these equations may be transformed into a single equation for

$$\rho_\Lambda = [\rho_\Lambda(x)_n]_{n>1} \qquad (13)$$

in the Banach space E_ξ (given $\xi > 0$, E_ξ is the space of sequences

$$\phi = [\phi(x)_n]_{n>1},$$

where $\phi(x)_n$ is a complex Lebesque measurable function on $R^{n\nu}$ such that

$$|| \phi ||_\xi = \sup_{n \geq 1} (\xi^{-n} \operatorname*{ess\,sup}_{(x)_n \epsilon R^{n\nu}} | \phi(x)_n |) < +\infty.$$

In particular, he shows that ρ_Λ satisfies the linear equation

$$\rho_\Lambda = z\chi_\Lambda \alpha + z\chi_\Lambda \Pi K \rho_\Lambda \qquad (14)$$

(where α is such that $\alpha(x)_1 = 1$, $\alpha(x)_n = 0$ for $n > 1$ and Π is a permutation operator), and considers as well the equation

$$\rho = z\alpha + z\Pi K\rho, \tag{15}$$

where ρ is the sequence of "infinite volume correlation functions" $\rho(x)_n$. The theorem proved by Ruelle, which will be referred to in our later discussion, can now be stated.

THEOREM: Let Φ be a stable, regular pair potential, and let z be a complex number satisfying

$$|z| < e^{-2\beta B - 1} C(\beta)^{-1} \tag{16}$$

where

$$C(\beta) = \int dx \, | e^{-\beta\Phi(x)} - 1 | < +\infty \tag{17}$$

$$\Phi(x) = U(0, x) \geq -2B \tag{18}$$

Then, the grand partition function, Equation 6, has no zero in (16). If the distribution functions $\rho_\Lambda(x)_n$ are defined by Equations 8 and 9 for z in (16), there exist "infinite volume correlation functions" $\rho(x)_n$ and a function $\epsilon(\cdot)$ positive and decreasing, such that

$$\lim_{\lambda \to \infty} \epsilon(\lambda) = 0 \tag{19}$$

and

$$| \rho_\Lambda(x)_n - \rho(x)_n | \leq \xi^n \epsilon(\lambda) \tag{20}$$

if ξ and z satisfy

$$|z| < e^{-2\beta B} \xi \exp[-\xi C(\beta)] \tag{21}$$

and λ is the minimum distance from $x_1, \ldots, x_n \in \Lambda$ to the boundary of Λ. The sequence ρ_Λ of the $\rho_\Lambda(x)_n$ and the sequence ρ of the $\rho(x)_n$ are the only solutions of Equations 14 and 15, respectively, in E_ξ when ξ satisfies (21) [in particular, if $\xi = C(\beta)^{-1}$] and ρ_Λ and ρ depend analytically on z.

It should be noted that if $|z|$ actually attains the value given by the right-hand side of Equation 21, this does *not* of necessity mean that a phase transition will occur. Rather, the bound must be interpreted as a lower bound on the limit of stability of a pure phase.

Now, in light of the exact results expressed in this commandment, let us turn our attention to the problem of phase transitions interpreted as a problem in bifurcation theory, as posed within the context of the KS equations. Suppose, for definiteness we consider the first equation in the KS hierarchy (8), and then re-express this equation in terms of the n-particle correlation function $g_\Lambda(y)_n$, where

$$\rho_\Lambda(y)_n = \rho_\Lambda(y_1) \ldots \rho_\Lambda(y_n) g_\Lambda(y)_n. \tag{22}$$

By taking advantage of the symmetry of the functions $K[x_1; (y)_n]$ and $g_\Lambda(y)_n$ with respect to interchange of the $(y)_n$, Equation 8 may be written as a homogeneous integral equation; in particular, if we introduce the transformation

$$\rho_\Lambda(x_1) \to \tilde{\rho}_\Lambda(x_1) + \tilde{\alpha}(x_1) \tag{23}$$

and define

$\tilde{K}[x_1, (y)_n]$

$$= \sum_{m=n}^{\infty} \frac{1}{n!(m-n)!} \, \tilde{\alpha}(x_1)^{m-n} \int_{\Lambda^{m-n}} K[x_1, (y)_m] g_\Lambda(y)_m dy_{n+1} \ldots dy_m, \qquad (24)$$

where here $\tilde{\alpha}(x_1) \equiv \alpha$ is a number that satisfies the equation

$$\frac{\alpha}{z} = 1 + \sum_{n=1}^{\infty} \frac{1}{n!} \, \alpha^n \int_{\Lambda^n} K[x_1, (y)_n] g_\Lambda(y)_n dy_1 \ldots dy_n, \qquad (25)$$

then the first equation in the KS hierarchy can be rewritten as:

$$\tilde{\rho}_\Lambda(x_1) = z \sum_{n=1}^{\infty} \frac{1}{n!} \int_{\Lambda^n} \tilde{K}[x_1, (y)_n] \tilde{\rho}_\Lambda(y_1) \ldots \tilde{\rho}_\Lambda(y_n) dy_1 \ldots dy_n. \qquad (26)$$

It was pointed out several years ago[7] that the structure of Equation 26 is formally the same as an integral power series. Recall that an integral power term of order n relative to the function f is defined as:

$$L(f) = \int K[s, (y)_n] \, f(s)^{\gamma_0} \, f(y_1)^{\gamma_1} \ldots f(y_n)^{\gamma_n} \, dy_1 \ldots dy_n \qquad (27a)$$

if

$$\gamma_0 + \gamma_1 + \ldots + \gamma_n = n, \qquad (27b)$$

where the $\gamma_0, \ldots, \gamma_n$ are nonnegative numbers; an integral power series is a summation from $n = 1$ to infinity of terms of the type, Equation 27. In operator notation, Equation 26 can be written

$$\tilde{\rho}_\Lambda(x_1) = z \mathcal{Q} \tilde{\rho}_\Lambda(x_1) \qquad (28a)$$

where,

$$\mathcal{Q} \tilde{\rho}_\Lambda(x_1) = \sum_{n=1}^{\infty} \frac{1}{n!} \int_{\Lambda^n} \tilde{K}[x_1, (y)_n] \rho_\Lambda(y_1) \ldots \rho_\Lambda(y_n) dy_1 \ldots dy_n. \qquad (28b)$$

Nonlinear operators having the structure of $\mathcal{Q}$ were first studied by Lyapunov, and later by Lichtenstein, Sobolev, and Vainberg, and in the literature such operators are referred to as Lichtenstein-Lyapunov integral power series operators.

We now draw attention to the fact that although the full system of KS equations could be written as a *linear* operator equation, i.e., an operator ΠK was identified by Ruelle which mapped E_ξ into itself with norm

$$\| \, \Pi K \, \|_\xi \leq e^{2\beta B} \xi^{-1} \exp [\xi C(\beta)], \qquad (29)$$

consideration of a *single* equation in the KS hierarchy (viz., the first) leads to a *nonlinear* operator equation, with the operator $\mathcal{Q}$ being of the Lichtenstein-Lyapunov type. This same general feature also characterizes the other, previously-mentioned hierarchies; for example, although the system of integral equations referred to as the Kirkwood coupling parameter hierarchy forms an inhomogeneous system, linear in the m-particle distribution functions, the first equation in this hierarchy can be written *formally* as a nonlinear operator equation of the classic Hammerstein type.[3] In any case, the operative word here is *formal,* since the resultant nonlinear problem can be specified precisely only if certain assumptions are made regarding the m-body correlation problem. For example, within the context of the KS hierarchy, an examination of the kernel in Equation 24 for the case $n = 1$, viz.

$$\tilde{K}[x_1, (y)_1] = \sum_{m=1}^{\infty} \frac{1}{(m-1)!} \alpha^{m-1} \int_{\Lambda^{m-1}} K[x_1, (y)_m] g_\Lambda(y)_m \, dy_2 \ldots dy_m \quad (30)$$

reveals the essential role played by the $g_\Lambda(y)_m$, the m-body correlation functions. If the $g_\Lambda(y)_m$ were completely determined then $\tilde{K}[x; (y_1)]$ would be a perfectly well-defined kernel, and existence and uniqueness properties of the nonlinear operator equation (28) could be studied in a straightforward way. But, unfortunately, the $g_\Lambda(y)_m$ are *not* known; to have the full set of $g_\Lambda(y)_m$ available from the outset would be tantamount to having solved the m-body problem for the system under study. In other words, as it stands Equation 28 is *not* a closed equation for the unknown function $\tilde{\rho}_\Lambda(x_1)$; we have simply swept our ignorance of the m-body problem into the kernel $\tilde{K}$. Therefore, to make any progress whatever in analyzing the properties of any *single* equation of the KS hierarchy, one must make some assumptions regarding the m-body correlation problem; specifically one must introduce a *closure* to restrict the vector $\rho_\Lambda = [\rho_\Lambda(x)_n]_{n \geq 1} \epsilon E_\xi$ to a finite number of elements. The most commonly used closure is the one introduced by Kirkwood,

$$g_\Lambda(x_1, x_2, x_3) = g_\Lambda(x_1, x_2) g_\Lambda(x_1, x_3) g_\Lambda(x_2, x_3), \quad (31)$$

called the superposition approximation. This approximation, when used within the context of one of the hierarchies mentioned previously, effectively decouples the infinite system of integral or integro-differential equations, with the consequence being that one then has the option of studying the properties of a single, closed *nonlinear* equation for the distribution function $\rho_\Lambda(x)_m$, rather than an infinite hierarchy of equations *linear* in the m-body distribution functions. Presuming that there is some advantage to be gained in studying the analytic properties of a single $\rho_\Lambda(x)_n$, the crucial question which arises before this option can be exploited is whether the contraction of the m-body problem specified by the above (or some other) closure makes any sense, either analytically or numerically. The answer here seems to be that the closure (31) isn't that bad, and in fact, in a few simple cases, it is even exact. However, this is by no means a general result, and accordingly one must be sensitive to possible errors introduced into an analysis based on (31) (or any other closure). To illustrate the kind of difficulty that can arise, suppose we examine the consequences of introducing the following closure in each equation of the KS hierarchy:

$$\rho(y)_m = 0 \quad \forall \ n > m. \quad (32)$$

One then obtains

$$\rho(x)_m = z\chi(x)_m \exp[-\beta W^1(x)_m] \cdot$$
$$\cdot \{\rho(x)_{m-1} + \int dy_1 K(x_1, y_1) \rho[(x)_{m-1}^1, y_1]\} \quad (33)$$

Then, for the resultant hierarchy, the Banach spaces E_ξ, and the various norms can be defined in the same manner as Ruelle,[6] and the formal methods he introduced can also be followed to yield the following bound on $|z|$, namely,

$$|z| < e^{-2\beta B} \xi \frac{1}{1 + \xi C(\beta)}. \tag{34}$$

Incidentally, the approximate generic distribution functions generated via the closure (32) can be related to corresponding, approximate "infinite volume correlation functions," provided the conditions specified by Ruelle in the previously-quoted theorem are satisfied. Now if we embed ρ_Λ in the same Banach space as Ruelle, i.e., the one for which

$$\xi = C(\beta)^{-1} \tag{35}$$

the bound (34) obtained with the closure (32) is

$$|z| < e^{-2\beta B} C(\beta)^{-1} 2^{-1} \tag{36}$$

whereas the largest bound that can be constructed for $|z|$ is

$$|z| < e^{-2\beta B} C(\beta)^{-1} \tag{37}$$

Hence, it would appear that one consequence of introducing an approximate closure is that the bound which determines a range of analyticity of the low-density (gas phase) correlation functions tends to move around. To specialize these remarks to a specific case, recall the results obtained for the case of a one-dimensional gas of hard rods, each of diameter a. The bound on $|z|$ determined via the Ruelle theorem is:

$$|z| \leq \frac{1}{2e} a^{-1}, \tag{38}$$

whereas the maximum bound on $|z|$ determined via the approximate closure (32) is

$$|z| \leq \frac{1}{2} a^{-1} \tag{39}$$

Now the last result is interesting inasmuch as Penrose[8] determined that the corresponding Mayer series

$$\beta p = \sum_{n=1}^{\infty} b_n z^n \tag{40}$$

for the hard-rod problem had a radius $\mathcal{R}$ of convergence

$$\frac{1}{2e} a^{-1} \leq \mathcal{R} \leq \frac{1}{2} a^{-1}. \tag{41}$$

It can be speculated that the consequence of introducing the closure (32) has been to shift the Ruelle bound (which corresponds to a lower bound on the radius $\mathcal{R}$ of convergence of the Mayer expansion) to a point which corresponds to the upper bound on $\mathcal{R}$ determined by Penrose for the hard-rod problem. Whether or not we take seriously this speculation, the point to be stressed is that the quantitative estimate of the radius $|z|$ differs from that of Ruelle, this despite the fact that the closure (32) investigated here would *not* appear to be that serious an approximation for the hard-rod problem, since the properties of the resultant kernel $K[x_1, (y)_n]$ are such that only *one* term in each of the equations of the full hierarchy is neglected via this closure. Yet the bound on $|z|$ does shift, indicating that the introduction of an approximate closure may lead to results that are in significant quantitative disagreement with those obtained in a rigorous analysis.

From the above considerations, two rather unpleasant factors have emerged. First, to achieve a contracted representation of the m-body problem, an approximation (i.e., a closure) must be introduced, and the results obtained via a poor choice of closure may be quantitatively in error. Secondly, upon introducing a closure, although a major simplification would appear to have been realized in contracting the full hierarchy of linear integral equations to a single nonlinear equation for $\rho_\Lambda(x)_n$, it must be stressed that the resulting equation is nonlinear, perhaps even highly nonlinear (e.g., the nonlinearity may be exponential). Hence, an investigation of the analyticity properties of the correlation functions defined via such a contracted representation requires the availability of general theorems on the existence and uniqueness of solutions to nonlinear equations, and as is known, progress in obtaining such theorems, especially global ones, has been painfully slow. Given these remarks, one may wonder at this point what advantages have been gained in introducing a closure within the KS (or any other) hierarchy.

A partial answer to this question can be provided by recalling the formal expression for the equation of state of a system characterized by an intermolecular pair potential $\Phi(x_{12})$:

$$\beta \overline{p} = \frac{\overline{N}}{\Lambda} - \frac{\beta}{6\Lambda} \int_{\Lambda^2} x_{12} \Phi'(x_{12}) \rho_\Lambda(x_1, x_2) \mathrm{d}(x)_2 \qquad (42)$$

From this expression, it is evident that once the pair distribution function $\rho_\Lambda(x_1, x_2)$ for a system is known, the pressure may be determined explicitly. In effect, then, although the concept of a distribution function becomes precise only within the context of the full m-body problem, from a practical point of view, detailed information about the spatial distribution of two particles is sufficient to generate the thermodynamics. Naturally, considerable advantage has been taken of this observation, and in the last three decades a rather extensive bank of data, both experimental and computer-generated, has been compiled on the structure of the pair correlation function in various regions of the (T, ρ)-plane. From these data, one can monitor the apparent radial and angular correlations between particles; it is found (FIGURE 1) that the pair correlation function changes in a smooth and regular way in certain regions of the thermodynamic parameter space, but as certain critical T-ρ loci are crossed, rather dramatic changes in the structure of this function can occur, and these dramatic changes have been associated with the onset of a phase transition. From a mathematical point of view, it would be of interest to investigate how it is that the correlation functions can be apparently analytic in certain regions of the thermodynamic parameter space, but suffer discontinuities as the thermodynamic parameters are varied. This, in turn, requires that one provide a mathematical structure for the problem which is rich enough to describe the complex behavior noted above. Ruelle[9] has pointed out that one possible way of attacking this problem is to try first to define clearly the notion of a phase; this led to the description of the states of a physical system as positive linear functions on a B^*-algebra, and the characterization of the group invariance properties of physical states. An alternative approach to the problem, one that would be guided by the available structural information on the pair correlation function, would be to utilize that mathematical theory which has proved successful in describing structural changes at the macroscopic, deterministic level, e.g., such changes as the bending of a rod as certain critical loads are achieved, the onset of convective motion as a certain thermal stress is realized, or the generation of dissipative structures as a chemical network is driven away from equilibrium via chemical reactions. In other words, we may opt for a description of phase transitions as a problem in bifurcation theory.

A SIMPLE EXAMPLE

Before implementing the grand strategy of using the full apparatus of nonlinear functional analysis to attack the problem of phase transitions, it would be of interest to show that an exact analysis carried out using bifurcation theory gave a result which coincided with a previously known, exact result of statistical mechanics. Since it is often the case in mathematics that the main features of a problem can be identified via a carefully chosen counterexample, we consider here, once again, the one-dimensional gas of hard rods. For this system we know that there exists *no* phase transition. Specifically, the equation of state of the hard-rod gas was determined by Tonks[10] to be

$$\frac{pal}{NkT} = \frac{l}{l-1}, \tag{43}$$

where

$$l = \frac{L}{Na}, \tag{44}$$

and Hauge and Hemmer[11] have shown that the zeros of the grand partition function $\Xi(\Lambda, z, \beta)$ of the problem fill the *negative* real axis from $-1/e$ to $-\infty$. Therefore, either from the formal result of Tonks, or via application of the Yang-Lee theorems, there should exist *no* phase transition in a gas of hard rods. Suppose, however, that we didn't have this information; suppose all we knew was that the first equation of the KS hierarchy under a suitable closure was of the form of a Lichtenstein-Lyapunov nonlinear operator equation. In accordance with the Kirkwood-Vlasov philosophy, we might then seek to determine the bifurcation point(s) of this nonlinear operator equation and, should bifurcation occur, to associate this mathematical behavior with the onset of an instability, i.e., a phase transition.

Before one can carry through a traditional bifurcation analysis of the nonlinear operator equation, Equation 28, and in particular, before one can use the celebrated Theorem 2.1 in Krasnosel'skii,[12] it is necessary to establish first of all the conditions under which the operator $\mathcal{C}$ is completely continuous. This matter is a technical one and for the case that $e^{-\beta\Phi}$ is assumed continuous on $\mathcal{R}^\nu$, the proof has already been presented[7] using Banach spaces of bounded continuous functions. For all intents and purposes, we may regard the hard-rod problem to be sensibly covered by this analysis, say, by considering the hard-rod potential to be a limiting case of the potential

$$\Phi(r) = m^m \qquad\qquad 0 < r < a + 1/m - 1 \qquad (45a)$$

$$= \frac{1}{(r - a + 1)^m} \qquad a + 1/m - 1 < r \qquad (45b)$$

when $m \to \infty$; officially, of course, one should use the Banach space of bounded measurable functions for the hard-rod problem.[9]

An examination of the homogeneous integral equation (26) for $\tilde{\rho}_\Lambda(x_1)$ reveals that $\tilde{\rho}_\Lambda(x) = 0$ is a solution of that equation; this solution will be referred to as the trivial solution, and it corresponds to the choice $\rho_\Lambda(x_1) = \alpha$. Following the notation of Krasnosel'skii[12] here, one identifies the trivial solution with the null vector θ, so that in operator form, equation 26 is written

$$z\mathcal{C}\theta = \theta, \qquad (46)$$

where now z_0 is understood to be a variable parameter (in the present context, it is just the activity). For small values of the parameter z, it is straightforward to show that the null solution is unique. However, it is of interest to see if one can go beyond this minimal result, and ask whether for increasing values of the parameter z, starting at some z_0, a nonzero solution makes its appearance in the neighborhood of θ. One says that μ_0 is a bifurcation point of Equation 46 if, for every $\epsilon > 0$, $\delta > 0$, there exists a characteristic value μ of the operator $\mathcal{C}$ such that $|\mu - \mu_0| < \epsilon$, and such that this characteristic value has at least one eigenfunction ϕ,

$$\phi = \mu\mathcal{C}\phi \qquad (47)$$

with norm $\|\phi\| < \delta$. Stated explicitly, our hypothesis is that such z_0, if they exist, have something to do with the onset of a phase transition. To determine the possible existence of a bifurcation point(s) for the problem at hand, we may use the results of Krasnosel'skii.[12] In particular, Krasnosel'skii shows that for the Lichtenstein-Lyapunov operator $\mathcal{C}$, the Fréchet derivative $\mathcal{B}$ at the origin of the space C is the linear operator (in our notation)

$$\mathcal{B}\tilde{\rho}_\Lambda(x_1) = \int_\Lambda \tilde{K}(x_1; y_1)\tilde{\rho}_\Lambda(y_1)dy_1. \qquad (48)$$

Therefore, according to the Krasnosel'skii Theorem 2.1, the bifurcation points of the full nonlinear operator $\mathcal{C}$ are determined by the eigenvalues of *odd* multiplicity of the linear equation

$$z_0 \int_\Lambda \tilde{K}(x_1, y_1)\phi(y_1)dy_1 = \phi(x_1). \qquad (49)$$

In this last statement, we have a concrete equation from which the bifurcation points of the first equation in the KS hierarchy can be determined, at least in principle.

Our earlier discussion of the kernel $\tilde{K}$ stressed the crucial role played by the $g_\Lambda(y)_n$. There we pointed out that before the nonlinear problem could be posed correctly, a closure had to be introduced. However, even if an *exact* closure is introduced, one further aspect of the problem must be addressed. In the distribution function theories of statistical mechanics, the ρ_Λ are functionals of the density; however, the activity z, which plays the role of a strength parameter in the nonlinear problem (47), *also* depends on the density. Therefore, in seeking possible eigenvalues z_0 of the linear equation (49), we may consider only those activities consistent with a given specification of the density. Keeping this additional self-consistency condition in mind, we now investigate the possible existence of a bifurcation point for the hard-rod problem in two limiting cases. In the first case, we shall introduce a very naive approximation, since in exploring the consequences, we shall be able to isolate some features of the overall problem, and then in the second calculation, we shall do the problem exactly.

In the first calculation, our naive approximation consists in neglecting all many-body correlation effects, and in addition, suppressing the previously noted self-consistency condition. Mathematically, this amounts to setting $\forall\, g_\Lambda(y)_n = 1$. Since we impose no device to break the translation invariance of the problem, a nontrivial solution of Equation 46, should one exist, must be a constant (different from α). Simple manipulations then allow calculation of the bifurcation point z_0 as the nonzero eigenvalue of Equation 49; the result is

$$z_0 = e^{-1}\tilde{C}(\beta) \tag{50}$$

where,

$$\tilde{C}(\beta) = \int (e^{-\beta\Phi(r)} - 1)\,dr. \tag{51}$$

Thus, for the particular case of a one-dimensional gas of hard rods, the bifurcation point calculated using the extreme closure introduced above, is just:

$$z_0 = -\frac{1}{2e}\,a^{-1} \tag{52}$$

You may recognize this number; the absolute value of the right-hand side is the same bound on $|z|$ determined by Ruelle in his exact analysis of the full KS hierarchy. Is this correspondence with Ruelle's result coincidental? Or, more to the point, what does the existence of a bifurcation point mean in a problem for which we know that *no* phase transition exists? Well, first of all, Pastur[13] has noted that the spectrum of the operator K of Equation 14 (looked upon as a bounded linear operator with domain and range in a Banach space) is an eigenvalue spectrum consisting of the inverses of the complex values of z for which $\Xi(z) = 0$ (his proof has been generalized recently by Moraal[14]). Now, as noted previously, Hauge and Hemmer[11] have shown that the zeros of $\Xi(z)$ are distributed along that part of the negative real axis stretching from $-1/e$ to $-\infty$. Hence, the bifurcation point of the operator equation (46) calculated using the closure $\forall\, g_\Lambda(x)_n = 1$ would appear to be a point in the spectrum of the operator K, and accordingly might be interpreted in light of the Ruelle theorem as representing a lower bound on the limit of stability of pure phase. Given this interpretation, the existence of a bifurcation point in the present problem might then be regarded as a necessary but not sufficient condition for the onset of a phase transition. This interpretation of a bifurcation point is obviously in contrast to the interpretation given in such *deterministic* theories as hydrodynamics, elasticity, and reaction-diffusion theory; e.g., if one achieves a critical load in the buckling problem, the rod bends. Are we led to this relaxed interpretation of the bifurcation point because of the underlying *probabilistic* nature of statistical mechanics, or has the interpretation arisen because of the introduction of approximations in the calculation of the bifurcation point? Stated explicitly, what would be the result in a bifurcation analysis of the hard-rod problem *if* the exact closure were introduced, and *if* self-consistency were preserved?

To answer this question, let us recall the work of Salsburg, Zwanzig, and Kirkwood.[15] These authors showed that if one labels the particles along the one dimension of the problem so that $y_1 < y_2 < \ldots < y_i < \ldots$, the superposition law

$$\frac{g^{(n+1)}(1, 2, \ldots, n + 1)}{g^{(n)}(1, 2, \ldots, n)} = \frac{g^{(2)}(n, n + 1)}{g^{(1)}(n)} \tag{53}$$

is *exact* for a one-dimensional system in which the interactions are restricted to nearest neighbors. Using this *exact* closure and the fact that the Mayer f functions satisfy the relations

$$f_{1\sigma} = 0 \quad r_{1\sigma} > a, \tag{54a}$$

$$f_{1\sigma} = -1 \quad r_{1\sigma} < a, \tag{54b}$$

Salsburg *et al.* computed an *exact* expression for the pair correlation

$$g^{(2)}(x) = l \sum_{k=1}^{\infty} A(x - k) \frac{1}{(l - 1)^k} \frac{(x - k)^{k-1}}{(k - 1)!} \exp\left(-\frac{(x - k)}{(l - 1)}\right) \tag{55}$$

[where x is the reduced distance r/a, and $A(x - k)$ is the step function], from which the thermodynamics of the system can be calculated exactly, the result being the Tonks equation of state. Notice that the functional dependence of $g^{(2)}(x)$ on the (reduced) interparticle separation variable x is parametrized by the mean density $\rho = 1/la$. Corresponding to this one-parameter family of functions $g^{(2)}(x)$, there will exist a one-parameter family of kernels $\tilde{K}$, and in turn, a *possible* one-parameter family of points z_0. However, as stressed earlier, there also exists a functional relationship between the activity z and the density ρ; for the hard-rod gas, Penrose[8] has specified this relationship as

$$z = \frac{p}{kT} \exp\left(\frac{ap}{kT}\right) = \frac{1}{a(l - 1)} \exp\left(\frac{1}{l - 1}\right). \tag{56}$$

Hence, only those activities z_0 that are consistent with the particular choice of density $\rho = 1/la$ specified in the prior identification of the kernel $\tilde{K}$ are acceptable solutions of the linearized equation. By using Equations 49, 30, and 56 in conjunction with the properties (54) of the Mayer functions for the hard-rod gas, and the fact that the shifting parameter α may be identified explicitly as $1/la$, we determine that a nontrivial solution of the original nonlinear operator equation can arise when

$$z_0 = \{+2 + 2l[\exp(-\beta\mu^E) - 1]\}^{-1}a^{-1}, \tag{57}$$

where the term involving excess chemical potential μ^E is given (exactly)[15] by

$$\exp(-\beta\mu^E) = \frac{l-1}{l} \exp\left(\frac{1}{1-l}\right). \tag{58}$$

However, when this result is analyzed it is found that there is, in fact, *no* density $\rho = 1/la$ for which self-consistency between the bifurcation result (57) and the thermodynamic constraint (56) can be realized. Hence, we may conclude that there is no density at which new solutions of the nonlinear operator equation appear, and from this last remark we argue that there is *no* phase transition in a one-dimensional gas of hard rods predicted by the bifurcation theoretical analysis.

I now point out that in some previous studies of the bifurcation properties of nonlinear equations derived from the BBKYG hierarchy (as well as in the preceding calculation where we set $\forall\ g_\Lambda (x)_m = 1$), the existence of a bifurcation point was predicted when the analysis was specialized to the case of a one-dimensional line of hard rods. The characterizing feature of these previous studies is that approximations were introduced at one point or other in the analysis, and so we may now try to identify the feature or features of the exact analysis which have led to a conclusion which is in qualitative disagreement with these earlier studies (but in accord with the exact statistical-mechanical result). From a qualitative point of view, the closure itself would not appear to be the crucial factor, since the introduction of the exact closure led to a (possible) one-parameter family of bifurcation points. In retrospect, it was the self-consistency condition (56) that was crucial, and consequently it is important to understand how the particular relationship between the activity and the density specified by (56) arises. First of all, the relationship between the activity z and the pressure p, as defined by (56), is a formal result; the explicit expression of z in terms of the density $\rho = 1/la$ follows from the incorporation of the exact equation for the pressure of a hard-rod system, viz., (43). Now, as it happens, Salsburg, Zwanzig, and Kirkwood[15] studied the asymptotic behavior of the partition function for the hard-rod problem by using the method of steepest descent. The saddle-point in this determination was located at a point c on the positive real axis. For a system of hard rods, one can show that $c = 1/[a(l - 1)]$, and in fact Gürsey[16] had noted earlier that the constant c introduced by the method of steepest descents is identically equal to βp. In other words, the Tonks equation of state, used in specifying the consistency relationship for the hard-rod problem, gives that pressure which follows from an asymptotic analysis of the partition function. However, an analysis of the asymptotic behavior of the partition function is one way of realizing the importance of the thermodynamic limit, which in statistical mechanics is a necessary precondition that must be satisfied before one is assured that the thermodynamics can follow from an evaluation of the partition function. Hence, the conclusion that follows from the present analysis is that it is only by taking into account in an *explicit* way the mathematical conditions specified by the Van Hove theorem[17] that phase transitions interpreted as a problem in bifurcation theory can make any statistical-mechanical sense. Phrased simply, the bifurcation point for the problem must lie on the path of steepest descent in the asymptotic analysis of the problem; more generally, it would appear that the thermodynamic limit plays the role of a statistical-mechanical stability condition which must be incorporated in a formal bifurcation analysis so that the stable states of the system may be properly identified. Exactly the same feature was uncovered in the one other statistical-mechanical study in which bifurcation theory was used in conjunction with an asymptotic analysis of the partition function. That study, though approximate in the sense that it was carried out only to order ϵ^2 in a bifurcation parameter ϵ, also showed that though bifurcation could be realized in all dimensions, no phase transition occurred in the one-dimensional problem because the bifurcation point did not lie on the path of steepest descent. The study referred to was a study of the asymptotic behavior of the D-dimensional Ising model,[18] carried out by R. Goldstein and the author. Incidentally, to demonstrate that *positive* results can be obtained when bifurcation theory is coupled with asymptotics, it was found in this study that for $D = 2$, the phase transition was characterized by a logarithmic singularity in the specific heat (in accordance with the exact behavior found by Onsager). For $D = 3$, the transition was characterized by an algebraic singularity in the specific heat (which seems to be the behavior suggested by computer studies). The estimates of the transition temperature in 2 and 3 dimensions were somewhat displaced from the accepted values, but again it should be emphasized that the bifurcation analysis was carried through only to order ϵ^2.

A further general feature emerged in the lattice study· by posing the Ising problem as a problem in bifurcation theory, a theorem could be proved on the asymptotic behavior of the continuum representation of the D-dimensional Ising model. In the limit of vanishing $H (\equiv J/kT$, where J is the interaction energy) and $N \rightarrow \infty$, it was shown that the free energy of the system behaves like the free energy of the corresponding D-dimensional Gaussian model. Then, when the strength parameter H exceeded a certain critical value (determined in the bifurcation analysis) a model resembling the spherical model bifurcated from the Gaussian model. Given the seminal role played by fixed point theorems in developing theorems on bifurcation, the theorem cited above suggests a possible connection between the existence of fixed points and phase transitions. We mention this explicitly because of the results obtained using the renormalization-group approach, first advanced by K. Wilson.[19] Here it was found that the thermodynamic properties of a lattice system may have a character determined by one fixed point in one range of parameters, and a quite different character in another range of parameters. In particular, for a system with a scalar order parameter and short-range forces, there exists a trivial Gaussian fixed point in which order parameter fluctuations are completely uncorrelated, a second Gaussian fixed point which is related to the spherical model, and an Ising fixed point which is assumed to describe the liquid-vapor critical point. Starting from the Wilson picture, several authors already[20] have begun to exploit the connection between the ideas of the renormalization group, and the underlying relationship between fixed point theorems and bifurcation theory. In terms of this picture, the lattice study mentioned previously differs from studies based on the ideas of the renormalization group in that the order in which asymptotics and bifurcation theory are introduced into the analysis of the partition function is essentially inverted.

Conclusion

Perhaps the main point we have wanted to stress in this talk is the importance of weighing carefully the statistical-mechanical consequences of introducing approximations in a bifurcation theoretic approach to the problem of phase transitions. The two approximations of crucial importance are: (1) the closure introduced to decouple the hierarchy of integral or integro-differential equations, and (2) the manner in which the asymptotic nature of the underlying statistical-mechanical theory is taken into account. These factors have been elucidated by examining in detail a certain physically uninteresting but mathematically well-posed problem in the equilibrium theory of phase transitions—the rectilinear, hard-rod gas. In particular, by correlating the results obtained with exact results obtained previously by Ruelle, Penrose, and Hauge and Hemmer, the closure introduced was found to be of crucial importance in affecting quantitative estimates on the location of the (possible) bifurcation point. Furthermore, from this study and from an earlier one on the asymptotic behavior of the D-dimensional Ising model, we noted that failure to account properly for the thermodynamic limit may affect the qualitative results obtained; viz., one may predict the existence of a phase transition when, in fact, none exists.

References

1. KIRKWOOD, J. G. & E. MONROE. 1941. J. Chem. Phys. 9: 514.
2. VLASOV, A. A. 1961. Many-Particle Theory and its Application to Plasma. Gordon and Breach. New York.
3. WEEKS, J. D., S. A. RICE & J. J. KOZAK. 1970. J. Chem. Phys. 52: 2416.
4. LINCOLN, W. W., J. J. KOZAK & K. D. LUKS. 1975. J. Chem. Phys. 62: 2171.
5. LUKS, K. D. & J. J. KOZAK. 1978. The statistical mechanics of square-well fluids. Adv. Chem. Phys. 37: 139–201.
6. RUELLE, D. 1963. Ann. Phys. 25: 109–120.
7. CHENG, I. Y. & J. J. KOZAK. 1973. J. Math. Phys. 14: 632.
8. PENROSE, O. 1962. J. Math. Phys. 4: 1312.
9. RUELLE, D. 1969. Statistical Mechanics. W. A. Benjamin, Inc. New York.
10. TONKS, L. 1936. Phys. Rev. 50: 955.
11. HAUGE E. H. & P. C. HEMMER. 1963. Physica 29: 1338.
12. KRASNOSEL'SKII, M. A. 1964. Topological Methods in the Theory of Nonlinear Integral Equations. Pergammon Press, Ltd. London.
13. PASTUR, L. A. 1973. Preprint of the Academy of Sciences of the Ukrainian SSR (Kiev).
14. MORAAL, H. 1975. Physica 81A: 469.
15. SALSBURG, Z. W., R. W. ZWANZIG & J. G. KIRKWOOD. 1953. J. Chem. Phys. 21: 1098.
16. GÜRSEY, F. 1950. Proc. Cambridge Phil. Soc. 46: 182.
17. GOLDSTEIN, R. A. & J. J. KOZAK. 1974. Physica 71: 267.
18. GOLDSTEIN, R. A. & J. J. KOZAK. 1974. *In* Global Analysis and its Applications. Vol. 2: 245–261. International Atomic Energy Agency. Vienna.
19. WILSON, K. G. 1971. Phys. Rev. B 4: 3174, 3184.
20. DOMB, C. & M. S. GREEN, Eds. 1976. Phase Transitions and Critical Phenomena, Vol. 6. Academic Press. New York.

THE ROLE OF FLUCTUATIONS IN MULTISTABLE SYSTEMS AND IN THE TRANSITION TO MULTISTABILITY

Rolf Landauer

IBM T. J. Watson Research Center
Yorktown Heights, New York 10598

INTRODUCTION

Computers are built out of bistable systems and these have been discussed in detail in connection with the attempt to develop a theory of the ultimate limits of the computational process.[1] In this paper we shall not focus on that subject, per se, but instead will discuss some of the more general results coming out of that work which have a bearing on physical systems that are multistable or approach a transition from monostability to multistability. Most of our discussion will emphasize fluctuations, and the role they have in establishing relative probability distributions. We shall also emphasize active kinetic and dissipative systems far from thermal equilibrium. The behavior of systems close to equilibrium seems to be a relatively settled subject. We will, however, draw upon the frequent, though limited, analogies between the behavior of thermodynamic phases on the one hand, and the dissipative systems on the other. Such analogies have been appreciated for a long time, at the deterministic (nonstochastic) level. Thus, for example, Busch[2] in 1921 gave a perceptive description of the analogy between the van der Waals curve and hysteresis phenomena in the ballast resistor. Indeed, as pointed out by B. Ross,[3] the analogy between hysteresis in negative resistance circuits and hysteresis in ferromagnets was understood[4] as early as 1905. The extension of such analogies to include a discussion of the motion of phase boundaries, i.e., domain wall propagation, was given by this author in an unpublished note.[5]

In recent years, we have seen the extension of these analogies to include stochastic phenomena, and the resulting probability distributions. In addition to Haken's definitive reviews,[6,7] we can cite only a few of the many recent publications in this area.[8-11] The importance of fluctuations can be recognized from the simple example shown in FIGURE 1. In classical mechanics it is customary to label B in FIGURE 1 as the state of absolute stability, as a result of the fact that it is the state of lowest energy. In actual fact, however, classical mechanics cannot really describe the relative stability of various competing states of local stability. A particle caught in the left-hand well of FIGURE 1 will simply stay there; it has no chance of making a transition to B. Once, however, we deviate from mechanics and bring in temperature and fluctuations, via the Boltzmann distribution, $\exp[-U/kT]$, it becomes obvious that B is a more likely state than A. Thus, a deterministic theory, e.g., catastrophe theory, which classifies points of local stability and their appearance and disappearance as the system is changed, can tell us nothing about relative stability. Questions about relative stability for particles in potentials, as shown in FIGURE 1, or for thermodynamic phases are, of course, easy to answer in terms of energy, or free energy. Recent work, however, has answered these questions also for many dissipative systems, far from equilibrium.

0077–8923/78/0316–0433 $1.75/1 © 1979, NYAS

Thus, for example, in discussions of computer limitations[12] we have been concerned with the relative stability of *"structures* which are in a steady (time invariant) state, but in a *dissipative* one, while holding on to information." (Italics added here for emphasis.)

In the early sections of this paper we shall discuss stability questions. The final section will treat heat flow in systems whose external constraint parameters, e.g., the degree of excitation, are subject to change. In that section we will focus on a system approaching a bifurcation point.

STABILITY CRITERIA

When we ask questions about stability we may be asking one of several questions:

(1) What is the stability of a time-dependent mode of operation, e.g., an oscillatory mode? We will not address this question.

(2) What is the stability of a given time-independent solution of the macroscopic equations, without regard to distribution functions and fluctuations? If the system is perturbed away from this solution, will it return to the original state?

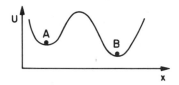

FIGURE 1. Damped potential with metastable state A and state of lowest energy at B.

(3) What is the stability of the distribution function, which describes the relative probability with which various states are occupied?

Let us discuss (3), and put it aside. The time-dependent distribution function for a time-independent Markov process is a solution of the "master" equation.[13] This is a linear equation and has negative eigenvalues, except for one time-independent solution, toward which all other distributions must decay.[13,14] (Admittedly there are, in general, somewhat more complex situations possible if a population in one state cannot leak directly, or via a sequence of intermediate states, to all other states.[13] Even in that case, however, we do not believe that *unstable* steady states can exist.) Thus, the time-independent solution to the master equation is not unstable, and we see no need for stability criteria for that solution.

This leaves us with question (2), the stability of a steady state as described by the macroscopic equations of motion. In that case, local stability can easily be analyzed in the obvious way: Linearize the equations of motion about the time-independent state and see if all the solutions decrease with time. It is not clear that a more sophisticated and powerful approach is important. Since, however, the literature has provided many discussions of stability criteria, we shall consider some of these. The Brussels school, in particular, has drawn attention to this question.[15] In a series of three papers,[16-18] this author has provided a critique of the Brussels work, and we will review here a portion of that. This author's discussions have emphasized (but are not in all cases limited to) electrical networks, whereas the Brussels work emphasizes chemical reactions. It is, undoubtedly, easier to treat one or the other of these than it is to understand their relationship completely. Some of the differences are obvious, but perhaps superficial, others are deeper. Among the obvious differences we find that electrical circuits provide two kinds of energy storage mechanisms, inductive and capacitive. Capacitive energy storage is closely akin to storing chemical energy via a variable number of a given species, deviating from the equilibrium concentration. The equivalent of an inductance *does* exist in the chemical reaction case, but is very unimportant there. There is a contribution to the molecular kinetic energy which is proportional to the rate at which a reaction proceeds. The exactly equivalent quantity in electrical circuits is a contribution to the inductance that is related to electron inertia, rather than magnetic fields, and has been measured.[19] The simultaneous presence of inductances and capacitances allows electrical circuits (just as in the case of resonant mechanical systems) to oscillate in a very simple way, in contrast to chemical systems where oscillations are a sophisticated modern refinement.

A more subtle difference involves the way contact with the thermal reservoir is maintained. In chemical systems the molecules bounce into the walls and thus achieve a temperature. This temperature in turn determines a reaction rate. By contrast, in the electrical case the capacitor, which stores charge, has no direct contact with the reservoir. The resistances, which are equivalent to the reaction pathways, provide the contact with the thermal reservoir, and thus in turn determine the energy of fluctuations present in the capacitors.

One of the key differences between chemical systems and electrical systems is that in electrical systems a circuit description makes it easy to focus on a few essential degrees of freedom that are intimately related to the ongoing dynamic process. Thus charge fluctuations in a capacitor are obviously determined by the noise generated in the circuit and have, at best, a very indirect connection to the ambient temperature. By contrast, in discussing chemical kinetics we are likely to mask the entropy contribution of the fluctuations related to the progress of the reaction by the entropy of the huge number of unimportant degrees of freedom. This leads to a "local equilibrium" assumption, which the author has, for the sake of emphasis, called a "dead entropy." [17] These distinctions were appreciated by van Kampen[20] as early as 1959.

The Brussels literature invokes the fact that in a stable steady state the energy storage coefficients, e.g., differential inductances and capacitances, are positive.[15] We do not believe this is a correct assumption,[18] though undoubtedly it is satisfied in a broad range of cases. One of the simplest counterexamples is provided in the interior of a ferroelectric or ferromagnetic domain wall.[5] A domain wall provides a transition in space between two different locally stable states. In the center of the domain wall, however, we have a state which, by itself, would be unstable. For that state, the differential inductances or capacitances can be negative. If we take a sample and apply a strong bias field in one direction at one end of the sample and in the other direction at the other end, then the state with the domain wall is not only a time-independent state, it is in fact the equilibrium state. We shall not dwell further on this point here; hereafter assume that our discussion is restricted to systems with positive differential inductances and capacitances.

Minimum Entropy Production

An early theorem by Prigogine is described as follows[21]:

"Linear systems, in particular, systems close to equilibrium, evolve always to a disordered regime corresponding to a steady state which is stable with respect to all disturbances. Stability can be expressed in terms of a variational principle of *minimum entropy* production."

However, this statement should not be put forth as a general physical principle. Indeed, it is possible to give counterexamples, for instance, the very simple circuit shown in FIGURE 2.

Assume that a steady state with a time-independent current has been reached. Now increase the battery voltage slightly, and then keep it fixed. (The earlier and lower voltage is invoked only to help define an initial state, which is not a steady state. Our actual detailed concern is with the circuit behavior under the time-independent final battery voltage.) The current will now start rising toward its new and higher steady state value, and will approach this higher current value in

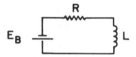

FIGURE 2. Battery across resistance and inductance in series.

an exponential fashion, with a time constant $\tau = L/R$. During this relaxation to the new steady state the value of the current and, as a result, the energy dissipation and entropy production *increase* continually. Note that the battery voltages involved can be arbitrarily small in magnitude; i.e., we can be as close to equilibrium as desired.

Reference 17 states the more restricted conditions, for electrical circuits, which permit the principle to be salvaged; we shall not discuss the restrictions here. Within these restrictions a proximity to equilibrium is necessitated only by the need for linear resistances. Thus it is not necessary for the circuit to have all of its elements at the same temperature. If, however, the circuit elements are at different temperatures the theorem becomes one of minimal energy dissipation rather than entropy production.

Excess Entropy Production

We have pointed out that when we are sufficiently close to equilibrium, and impose additional restrictions,[17] then the steady state is a state of minimal entropy production, when compared to nearby states relaxing toward it. Far from equilibrium, even within these cited restrictions, this statement ceases to be valid. Is there an alternative variational principle? A plausible candidate for such a variational principle must be stationary about the steady state; i.e., it must exhibit no linear variation with deviations from the steady state. Let i_{0k} and V_{0k} be the steady-state current and voltage in the kth dissipative element. Then

$$\delta^2 P = \sum_k (i_k - i_{0k})(V_k - V_{0k}) \tag{1}$$

is clearly stationary, as required. Furthermore

$$\delta^2 P = \sum_k i_k V_k - \sum_k i_{0k} V_{0k} - \sum_k (V_k - V_{k0})i_{0k} - \sum_k (i_k - i_{k0}) V_{0k}. \tag{2}$$

If we are in the linear range of the resistive elements (which will certainly be the case if we are close to equilibrium) then

$$(V_k - V_{k0})/(i_k - i_{k0}) = V_{k0}/i_{k0}. \tag{3}$$

As a result, the last two sums on the right-hand side of Equation 2 cancel and

$$\delta^2 P = \sum_k i_k V_k - \sum_k i_{0k} V_{0k}. \tag{4}$$

In other words, close to equilibrium, $\delta^2 P$ reduces to the entropy production of the earlier principle, except for an additive constant. This, together with the stationary property, makes $\delta^2 P$ a likely candidate for a Liapunov function. The Brussels school calls $\delta^2 P$ "excess entropy production." This is, perhaps, a misleading name, since it suggests the expression given on the right-hand side of Equation 4; $\delta^2 P$ is, however, equal to that only when we are in the linear resistance range.

The stability of a dissipative steady-state solution can be examined through the linearized equations characterizing the departure from the steady state. In the case of electrical circuits the linearized equations characterize a circuit with the same topological structure as the original circuit. Furthermore, the linearized circuit has capacitances, inductances, and resistors in the same places as the original circuit. The linearized circuit, however, contains a short circuit in place of the voltage sources (e.g., batteries) of the original circuit and an open circuit in place of the current sources of the original circuit. The linearized circuit has normal modes, which can either decay with advancing time, or build up. If, as assumed, the differential capacitances and inductances are positive, then growing modes are associated with a build-up in stored energy. As a result of this growth, the power dissipation in the linearized circuit,

$$\sum_k \delta i_k \delta V_k, \tag{5}$$

written as a sum over resistive elements, must be negative. If this power dissipation is guaranteed to be positive for all states in the vicinity of a steady state, then growing (or undiminished) modes are excluded and the steady state must be stable. Thus, a positive excess entropy production throughout the neighborhood of a given steady state is a sufficient condition for stability. Is it also a necessary condition?

To answer that question let us first specialize to circuits that have only resistive and capacitive elements, but no inductances. The argument to follow will apply equally well to the case where inductances are present, but no capacitances. Now consider the capacitors C_i attached to an otherwise purely resistive network. Some of the differential resistances can, of course, be negative; otherwise stability is inevitable and trivial. The capacitive charges will then satisfy an equation

$$\mathrm{d}q_i/\mathrm{d}t = -\sum_j G_{ij} V_j, \qquad (6)$$

where V_j is the voltage across capacitor j. The term G_{ij} will be a real and symmetric matrix, as a consequence of the reciprocity theorem in circuit theory. (Unlike the Onsager relations, it is not based on proximity to equilibrium.) If we now change variables to $u_i = q_i/\sqrt{C_i}$ we find:

$$\mathrm{d}u_i/\mathrm{d}t = -\sum_j (C_i C_j)^{-1/2} G_{ij} u_j. \qquad (7)$$

Thus the matrix A, in $\mathrm{d}\mathbf{u}/\mathrm{d}t = A\mathbf{u}$, is real and symmetric, and its characteristic values are real. The normal modes are then pure exponentials, without oscillatory

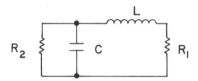

FIGURE 3. Resonant circuit: R_2 is small and positive; R_1 is very large and negative.

behavior. (We ignore the case of degenerate characteristic values. It can be considered as the limiting case of close, but distinct, characteristic values.) The total capacitive energy stored in the circuit then consists of a sum of growing or decaying exponentials, one per mode, and the cross terms between modes. The cross terms, summed over all capacitors, vanish as a result of the orthogonality of the characteristic vectors of A. Thus, if all the modes are decaying modes, the stored energy must decrease monotonically in time and the expression in (5) is positive. Hence if all the possible modes in the vicinity of a steady state are decaying modes, the expression (5) must be positive throughout the vicinity, and that becomes a *necessary* and *sufficient* condition, completely equivalent to stability.

    Let us now return to the more general case where inductive and capacitive components are present simultaneously. Consider a differential circuit as shown in FIGURE 3. Let R_1 be positive but small enough so that we have an oscillatory and under-damped circuit. Let R_2 be a negative resistance, but very large, so that very little current flows through it, and the effect of R_1 is dominant. Thus we deal with a stable circuit. As the damped oscillations proceed there will be instants of time, every half cycle, when the current through L and R_1 vanishes and the capacitive charge q will be close to its maximum amplitude. At that instant the positive dissipation due to R_1 vanishes, and the negative dissipation of R_2 will be uncompensated. Thus, despite stability there will be moments when the expression in Equation 5 is not positive. In this more general case, then, a positive excess entropy production throughout the neighborhood of a steady state is only a sufficient but not a necessary condition for stability. This fact was not properly stated in Reference 17, and the correction here is a result of discussions with C. Gardiner.

The literature contains some very sophisticated discussions and debates related to the Glansdorff-Prigogine excess entropy production theorem, its interpretation and validity.[14,22-27] We shall not attempt to resolve these. After all the extent to which our electrical circuits are really equivalent to the chemical system has been left unsettled. We furthermore stress that our discussion has assumed elements that are purely resistive, purely capacitive, or purely inductive, and do not mix these types within one element. The inclusion of multiport resistive elements, e.g., transistors, also causes a problem, since the reciprocity theorem cannot be applied in that case. Finally we stress that Equation 5 once again relates to energy dissipation. It bears a simple relation to entropy production only if all dissipative elements are kept at the same temperature. To stress the relationship to the Glansdorff-Prigogine theorem, however, we will continue to refer to "excess entropy production."

In the next section we will go on to a consideration of the stochastic case where we cannot just follow a single valued solution in time, but must consider a distribution function. We will restrict ourselves to the case that led to the strongest results in the preceding considerations, where only resistive and capacitive elements are present.

EXCESS ENTROPY PRODUCTION, STOCHASTIC CASE

Up to this point, our discussion of $\delta^2 P$ has been confined to systems in which fluctuations are unimportant, and which can be described by the evolution, in time, of a single-valued macroscopic set of currents and voltages. Is there a sensible version of the excess entropy production theorem applicable to the case where the system must be described by a distribution function, e.g., for a bimodal distribution? Let us, here, first remind the reader of our earlier point: Stability of the steady-state distribution function is assured and not an open question; a criterion for that is unnecessary. We shall return to this point later again, in connection with a subsequent discussion of Schnakenberg's discussion[14] of excess entropy production.

Now if we consider a particular ensemble member, at an instant of time, excess entropy production need not be positive. Even in equilibrium, energy can flow in the "wrong" direction some of the time, i.e., be opposed to the direction of macroscopic relaxation. A time average for a particular circuit, which starts away from the steady-state ensemble average is not much better. The circuit may have started in a stage in which the fluctuations in its energy-storing elements were low, and power then is likely to flow out of the resistive elements into the reactances. Furthermore, if we extend our time averaging over a sufficiently long period we will be effectively taking an ensemble average *for* the steady state, we won't be watching a relaxation *to* the steady state. We are thus led to the conclusion that a generalization must be sought by averaging *over* an ensemble. This leaves us with two obvious choices. Let us define deviations from the steady-state ensemble average

$$i_k - i_{k0} = \delta i_k, \tag{8}$$

$$V_k - V_{k0} = \delta V_k, \tag{9}$$

where the subscript 0 denotes the steady-state ensemble average. We can then consider either

$$\sum_k <\delta i_k \delta V_k> \quad \text{or} \quad \sum_k <\delta i_k> <\delta V_k>,$$

where $< \; >$ denotes ensemble averaging. It turns out that this choice is immaterial. It can be shown that in both cases the defined quantity is not necessarily positive in the relaxation to a nearby steady state. "Nearby" means that the initial distribution function is not far from that for the steady state, a point that has been emphasized by Gardiner.[25] The analysis is very similar for the two cases. We shall, therefore, only describe that for $\sum_k <\delta i_k> <\delta V_k>$ explicitly. Before going into the details, let us sketch, in a general way, the point that is involved. The deviation from the steady-state distribution function can be described through expansion in a set of normal modes,[14] each with its own exponential decay time. We can then choose a combination of modes with very different time constants. In the early portion of the relaxation process the main contribution to $<\delta V_k>$ can come from one of these modes, while that to $<\delta i_k>$ can come from another. Therefore the signs for $<\delta i_k>$ and $<\delta V_k>$ can be chosen independently, and we can achieve either sign for $\sum_k <\delta i_k> <\delta V_k>$, in the early part of the relaxation process.

Up to this point the subscript k denoted a circuit element. Let us now change notation; a subscript will denote a node at which circuit elements come together; i_{ij} will denote the current from node i to node j; and V_i is the potential at node i. From the requirement for continuity of current, we have for each node in a circuit

$$\sum_j <\delta i_{ij}> = 0. \tag{10}$$

Then, multiplying by the node potential deviation δV_i and summing again, we obtain

$$\sum_{i,j} \delta V_i <\delta i_{ij}> = 0. \tag{11}$$

We can then ensemble-average again, leaving

$$\sum_{i,j} <\delta V_i> <\delta i_{ij}> = 0. \tag{12}$$

Equation 12 can be written as a sum over all circuit branches

$$\Sigma <\delta V> <\delta i> = 0, \tag{13}$$

where δV is now a deviation in voltage drop from a steady state. Current and voltage sources drop out of Equation 13 since in each case one of the factors in Equation 13 will vanish. For capacitors $\delta i = i = C(V)dV/dt$. Thus Equation 13 becomes

$$-\sum_C <\delta V> <C(V)dV/dt> = \sum_R <\delta V> <\delta i>, \tag{14}$$

where R, and C denote the type of circuit element in the summation. Let us assume, as discussed earlier, that C is positive. To show that $\sum_R \delta V \delta i$ need not be positive, it will be adequate to consider the simpler case $C(V) = \text{const}$ and to give a counterexample applicable to that case. The left-hand capacitive term in (14) will be positive if $<\delta V>$ and $<dV/dt>$ are opposite in sign. Are they?

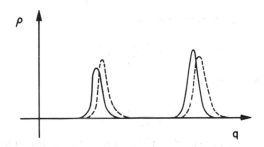

FIGURE 4. Solid line is steady state distribution function. Dotted line is perturbed, nearby state.

Reference 17 discussed this, using a tunnel diode circuit. But it is not really necessary to invoke the detailed properties of that circuit, we only need to know that the distribution function is bimodal. FIGURE 4 shows the steady-state distribution for the charge on the capacitor placed across the tunnel diode. It also shows the perturbed distribution function. In achieving the perturbation each peak is displaced slightly to the right, whereas some of the total intensity is taken out of the right peak and shifted to the left peak. We can easily make this latter shift big enough, so that it dominates $<\delta V_C>$ (the subscript C identifies the capacitor). For the case shown in FIGURE 4 that will be negative. On the other hand $<dV_C/dt>$ will be determined by the relaxation process, which will be very fast for redistribution within each peak, but very slow for the activated transitions between the two peaks. This disparity in time constants can be made as large as we wish. As a result we can easily have the intra-peak relaxation dominate $<dV_C/dt>$ initially. That will be negative for the case shown in FIGURE 4. As a result, the "excess entropy production" given by Equation 14 will be negative during the initial stages of the relaxation process, despite the proximity to the steady-state distribution function (which is inevitably stable).

SCHNAKENBERG'S VERSION OF THE GLANSDORFF-PRIGOGINE STABILITY CRITERION

In this section we will take issue with some concepts presented by Schnakenberg in a review paper,[14] which in all its other aspects is a most exemplary and definitive presentation. Schnakenberg's viewpoint is based upon earlier and related discussions by Schlögl.[26] Schnakenberg defines an excess entropy production not in terms of the macroscopic system, but in terms of the master equation for the system. Let $<i/j>$ represent the transition probability per unit time from j to i, and p_i the occupation probability for state i. Then Schnakenberg defines a current.

$$J_{ij} = <i/j> p_j - <j/i> p_i. \tag{15}$$

He then defines an associated force

$$A_{ij} = \log \frac{<i/j> p_j}{<j/i> p_i}. \tag{16}$$

This then leads to an entropy production:

$$P = \frac{1}{2} \sum_{i,j} J_{ij} A_{ij}. \tag{17}$$

and an excess entropy production

$$\delta^2 P = \frac{1}{2} \sum_{i,j} \delta J_{ij} \delta A_{ij} \geq 0, \tag{18}$$

where δJ_{ij} and δA_{ij} represent deviations from the steady state. Schnakenberg then attempts to identify his $\delta^2 P$ with the macroscopic excess entropy production of Glansdorf and Prigogine. There certainly are cases where this identification applies, but we question its general validity for several reasons:

(1) Schnakenberg's discussion is based entirely on the master equation and introduces no other physics. How can we take a master equation which, to take one example, may describe biological population statistics, and associate an entropy production with it, via Equations 16 and 17?

(2) Many systems are known,[6] in which detailed balance is obeyed, in the dissipative steady state, far from equilibrium. Thus $J_{ij} = 0$ and hence, via Equation 17, Schnakenberg's $P = 0$, whereas there certainly is entropy production in the real system. Schnakenberg's answer: The independent compensating mechanisms that lead to $J_{ij} = 0$ must be treated as separate transitions, each with its own J and driving force. But how, in general, do we know when we have reached an adequate decomposition?

(3) A macroscopic steady state can be metastable, and thus be stable toward *small* deviations, but not large ones. It is then in a region in which the Glansdorff-Prigogine (GP) value of $\delta^2 P$, as discussed in Equation 5, is positive near the metastable state and furthermore is stationary and vanishing at the point of metastability. On the other hand, the steady-state solution of the master equation will be very different; a *distribution* function centered near the metastable state will certainly *not* be stable, and will shift toward the neighborhood of greatest stability.

(4) In the bimodal example that we discussed we have seen that the reasonable generalizations of the GP expression for $\delta^2 P$ can be negative, for distribution functions near a steady state, whereas Equation 18 will yield a positive value for $\delta^2 P$.

C. Gardiner[25] has attempted to rescue the GP criterion for the case of FIGURE 4. He defines three fluxes and associated driving forces. There is a flux describing the relaxation within each peak, and an additional flux to describe the transition between peaks. He then finds $\delta^2 P \geq 0$ for small deviations from the steady-state distribution functions. It seems to us that this is simply a partial step away from the usual macroscopic definition of fluxes, and toward Schnakenberg's master equation fluxes. Gardiner has just gone far enough toward Schnakenberg to make $\delta^2 P \geq 0$ in this case, but in this process has incurred some of our objections to Schnakenberg.

The "excess entropy production" of Schnakenberg and that of Gardiner may be interesting and significant quantities in their own right. We stress here, however, that they are *not* equivalent to the GP excess entropy production.

INADEQUACY OF LOCAL CRITERIA FOR METASTABILITY

Consider the bistable potential shown in FIGURE 5, in which the left-hand well is metastable. Changing the height of the intervening barrier near its peak will

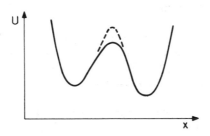

FIGURE 5. Bistable potential well. Relative occupation near minima is unaffected by changes in the barrier.

change the jump rate between the wells, but not the relative distribution between the two wells. The Boltzmann distribution, $\exp(-\beta U)$, gives us information about relative stability, independent of the exact behavior at the unlikely states between the two valleys.

Can we find a similar simple description for dissipative bistable systems, far from equilibrium? Much of the literature on nonequilibrium thermodynamics and statistical mechanics, is concerned with a search for simple criteria for the favored state. The applicability of these criteria to multistable systems is seldom examined. Any really broadly applicable criterion for the favored state should be able to analyze multistable systems. As stressed elsewhere,[1] biological evolution is akin to a series of activated transitions between states of local stability. Thus, a stability criterion that is unable to treat *relative* stability cannot be much help in discussing evolution. We shall demonstrate that, unfortunately, a criterion for relative sta-

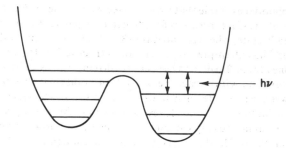

FIGURE 6. Biased bistable well. Transition out of highest right-hand level is pumped with laser light.

bility as simple as the Boltzmann factor cannot be expected for systems far from equilibrium. Equilibrium systems are very special: The fluctuations throughout the transition from one state of relative stability to another are characterized by the single temperature T and are not independently modifiable along different portions of that path. In nonequilibrium systems, on the other hand, we can play around with the kinetics in some of the intermediate transitions, along the path between metastable states. We shall here give an example of this, to supplement other examples given in earlier discussions.

Consider a symmetrical bistable well, in a biasing field, favoring the right-hand well as shown in FIGURE 6. Let the barrier between the wells be sufficient so that the quantized levels are essentially decoupled, and the relative well shifts, due to the bias potential, are large compared to the level-splitting arising from tunneling. Now apply radiation, e.g., from a laser to induce transitions between the uppermost level in the right-hand well and the level above that. If the level spacing is large compared to kT these two levels are very unequally occupied, before the laser light is applied. In the presence of intense radiation the level populations can be brought much closer to each other. Note that the light intensity has to be high, but if the lower of the two levels is many kT above the ground state, and sparsely occupied, the net absorbtion need not be large, and the intensity can be built up by reflection between the mirrors of a cavity wall. (We can invoke more complex optical pumping schemes than the one discussed here, analogous to those used in pumping four-level CW lasers, to achieve approximate level equalization, or even inversion, with still lesser light intensity.)

Now let us assume that the interlevel transitions, other than the one controlled by the laser, occur largely between adjacent levels. The *relative* population of all other levels, unaffected by the laser, will then still be controlled by the Boltzmann factor. Now consider the case where the two levels, whose population has been brought to approximate equality by the laser, are further apart than the relative displacement, in energy, of the two well minima, as set by the bias. The right-hand well, which was originally the favored well, will then become the more sparsely occupied well. Thus, the relative population of the two wells has been controlled by manipulating the kinetics of sparsely occupied states, without changing the kinetics in the bottom of each well. Any attempt to predict the relative occupation of the two wells that uses local properties characterizing the behavior only in the bottom of each well, cannot be successful. This doesn't prevent the existence of variational principles or Lyapunov functions for the steady-state distribution function. Such criteria, however, must give adequate weight to the transitions in sparsely occupied states. The evaluation of the relative stability of two or more states of local stability must, as has been stressed, take into account the fluctuations present in the intervening states. Attempts to discuss this relative stability on the basis of purely macroscopic deterministic equations, e.g., by the equivalent of a Maxwell construction, are likely to be incorrect, or at the very least require subsidiary justification. Such deterministic approaches have been employed by Bedeaux, Mazur, and Pasmanter[28] and by Takeyama and Kitahara,[29] and in both cases applied to negative resistance systems. My own alternative viewpoint has also been expounded[30,31] as a reaction to these papers, but was already spelled out in a much earlier publication,[32] and was recently confirmed by the work of Nicolis and Lefever.[33]

REVERSIBLE HEAT FLOW AT THE APPROACH TO BIFURCATION

The relation $TdS = dQ$ as an equality, rather than an inequality, is normally applied only to systems that are displaced slowly through a series of successive equilibrium states. In a system far from equilibrium, in the presence of continual steady-state dissipation, this equality seems hopelessly inapplicable. We'll construct two simple examples to show that it isn't all that hopeless. Consider first the system in FIGURE 7. This consists of a highly dissipative and open system, e.g., a jet engine, together with a nondissipative system being modulated. The dissipative system will cause a steady-state heat flow. Then if we move the piston, and change

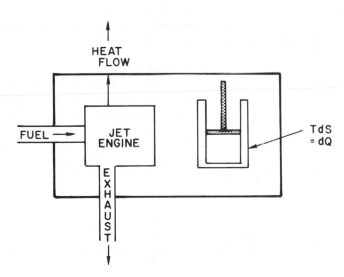

FIGURE 7. Fuel and exhaust flow at a uniform rate and the engine operates at a uniform speed, giving rise to a time independent heat flow. The compressible gas gives an independent contribution to the heat flow.

the volume of the gas, $Td\mathcal{S} = dQ$ will apply to the gas cylinder. If we then take the total heat flow out of the box enclosing both subsystems, and subtract the steady-state dissipation, we are left with $Td\mathcal{S} = dQ$, where $d\mathcal{S}$ is related to the change in gas volume, and dQ represents heat flow above and beyond the steady-state flow. FIGURE 8 shows a somewhat less artificial example. If the resistances are linear and not too far from equilibrium, the dc power flow in the circuit is independent of, and simply superimposed upon, the behavior of the fluctuations. Thus, if the resistances are raised to a higher temperature, the capacitive fluctuations and their entropy will increase. We will again have $Td\mathcal{S} = dQ$, where $d\mathcal{S}$ represents the capacitive entropy change and dQ the heat flow above and beyond the steady-state dissipation. These two examples leave us with the question: How far can we push this? How separate do the dissipative process and the modulated parts of the system have to be to permit this sort of analysis? These questions were discussed in an earlier paper[34] and a subsequent elaboration.[37] We shall here, instead, present an example that avoids most of the subtleties and complexities faced in the more general discussion.

We will consider a symmetrical circuit, which has a monostable symmetric macroscopic solution and is brought toward a point of transition beyond which there are two macroscopic solutions in which the symmetry has been broken. FIGURE 9 illustrates such a circuit and stresses the analogy to the symmetry-breaking events that occur in a uniaxial ferromagnet at its Curie temperature. In the magnetic case the transition out of a monostable symmetric state occurs as the temperature is lowered. In the circuit of FIGURE 9b, the transition takes place as the B+ voltage is increased, with a resulting increase in the gain of the transistors. The circuit of FIGURE 9b illustrates a typical memory circuit. If it is to be a useful circuit, additional circuit elements and connections are needed to interrogate it, and to permit it to be reset. FIGURE 9b does not actually show the two key capacitances in the circuit. We can think of these as being connected between the source and drain of each transistor, though a realistic model will have more capacitances than that.

In view of the complexity of a three terminal device we will find it simpler, instead, to consider the tunnel diode circuit shown in FIGURE 10 and in the remaining discussions will have that circuit in mind. FIGURE 11 illustrates the range of interest in this discussion. The two tunnel diode characteristics intersect near their maxima. Let us assume that the single stable steady state represented by the intersection in FIGURE 11 is disturbed, and thus the charge balance between the two capacitors shown in FIGURE 10a is upset. This situation is indicated by the vertical bar A in FIGURE 11, lying to the left of the dot marking the original intersection.

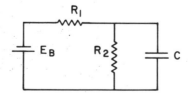

FIGURE 8. Battery with two resistors and one capacitor.

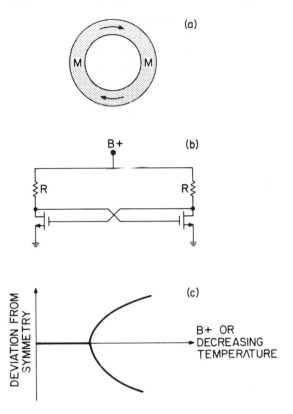

FIGURE 9. (a) Magnet. (b) Bistable field effect transistor circuit in which one transistor conducts and the other is shut off. (c) Both of the above systems exhibit second order phase transitions.

The lower diode has less than half the voltage, the upper diode more, and the diode currents are unequal. The sign of the current inequality is such as to restore the circuit to the symmetrical voltage division. The current differences, however, for the voltage division shown in FIGURE 11 have become very small because the characteristics intersect close to their maxima. Thus, the restoration becomes sluggish and we face the usual lengthening of relaxation times characterizing the approach to a second order transition.[35]

While the circuit of FIGURE 10a shows two capacitors, these are not independent degrees of freedom since the total voltage across the two is fixed by the externally imposed voltage E_B. Let each diode have a capacitance C. In the symmetrical state the total stored capacitive energy, summing over both capacitors, will be $\frac{1}{4} C E_B^2$. If the midpoint voltage between the two capacitors shifts by δV away from the symmetrical voltage division, then the total stored energy increases by $C(\delta V)^2$ above the value $\frac{1}{4} C E_B^2$. Instead of specifying δV we can specify a total charge δq which has been injected into the midpoint junction between the capacitors. In that case the increase in energy due to the unbalance becomes

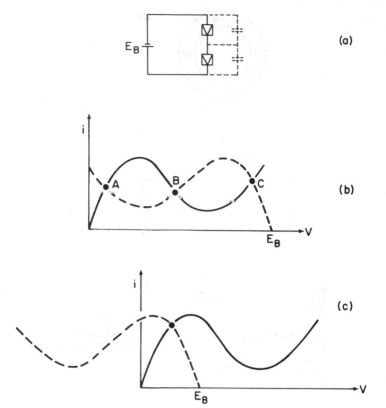

FIGURE 10. (a) Bistable tunnel diode circuit, with associated tunnel diode capacitances. (b) Solid line is the current through the lower tunnel diode as a function of the voltage V at the junction between the two diodes. Dashed line gives current through the upper diode as a function of the same midpoint voltage V. (c) Corresponding situation for a lower battery voltage E_B.

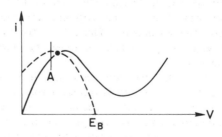

FIGURE 11. Tunnel diode characteristic as in Figure 10c, but closer to the transition to the bistable regime: A indicates a state in which a fluctuation has taken us away from the exact steady state solution. The two diode currents and voltages are now unequal, but the two voltages still add up to give E_B.

$\frac{1}{4}(\delta q)^2/C$. In the remainder of this section we shall use the symbol q to represent δq, the charge injected into the midpoint between the capacitors.

As has been discussed elsewhere,[36] systems of this sort can be described by a flux of probability in q space:

$$j = \rho v(q) - D \partial \rho / \partial q. \tag{19}$$

Here ρ is the distribution function, v is the restoration velocity toward $q = 0$, given by the circuit equations. The final right-hand diffusion term represents the effects of noise. This diffusion term permits ensemble members that start out together to separate subsequently. As discussed in Reference 36, Equation 19 is accurate only near the points of local stability, but that is the only place where we need it. In the steady state $j = 0$ and Equation 19 integrates to

$$\rho = C \exp[\int (v/D)dq]. \tag{20}$$

Near a point of local stability $v = -q/\tau$, where τ is the relaxation time. Thus, Equation 20 becomes

$$\rho = C \exp(-q^2/2D\tau). \tag{21}$$

Let the threshold value of E_B marking the transition between monostability and bistability be denoted by E_T. Let $E_T - E_B$, defining the distance from the transition, be denoted by δE. As we have pointed out above, the relaxation time of this circuit, for the restoration of small deviations from symmetry obeys:

$$\tau = \alpha(\delta E)^{-1}, \tag{22}$$

where α is independent of E_B, for small δE. On the other hand, the noise behavior of this circuit shows no particular anomaly at $E_B = E_T$; the diffusion coefficient D in Equation 19 is well behaved and continuous in that range. Thus Equation 21 yields a distribution

$$\rho = C \exp(-q^2 \delta E/2\alpha D). \tag{23}$$

This distribution becomes broad as $\delta E \to 0$ and bifurcation is approached. The associated entropy, $S = -k \int \rho \log \rho \, dq$, becomes large. In equilibrium such entropy changes can be associated with a reversible heat flow. We shall show that there are similar terms in our dissipative circuit, if the steady state is subject to a slow modulation.

As pointed out in the earlier treatment[34] our circuit exhibits conservation of energy, based on the macroscopic circuit equations, following the ensemble averages:

$$E_B \langle i_B \rangle = \sum_{\substack{\text{circuit} \\ \text{elements}}} \langle V \rangle \langle i \rangle. \tag{24}$$

E_B and i_B are the voltage and current associated with the battery. The circuit also obeys conservation of energy for each ensemble member. Averaging this over the ensemble yields

$$\langle E_B i_B \rangle = \sum_{\substack{\text{circuit} \\ \text{elements}}} \langle V i \rangle. \tag{25}$$

Since E_B does not fluctuate, then $E_B \langle i_B \rangle = \langle E_B i_B \rangle$. Thus, equating the right-hand sides of Equations 24 and 25 we obtain

$$\sum_C (\langle V_C i_C \rangle - \langle V_C \rangle \langle i_C \rangle) = -\sum_R (\langle V_R i_R \rangle - \langle V_R \rangle \langle i_R \rangle), \tag{26}$$

where C and R denote the capacitive and resistive (i.e., lossy) circuit elements, respectively. The right-hand side of Equation 26, after multiplication by dt, can be written as $dU - dQ_0$, where dQ represents the heat flow from the reservoir into the (nonlinear) resistive elements, and dQ_0 represents that quantity calculated from the macroscopic circuit equations. Thus, $dQ - dQ_0$ can be considered as the "interesting" heat flow, above and beyond that associated with the obvious steady-state dissipation. The left-hand side of Equation 26, after multiplication by dt, is entirely analogous to the term $dU - pdV$ in the analysis of a compressible gas. While this quantity will not, in general, be a perfect differential, it will be one if the capacitors are linear. In that case the left-hand side of Equation 26, after multiplication by dt, can be written as $dU - dU_0$, representing changes in stored energy; U_0 is just the energy, $\frac{1}{4} C E_B^2$, of the symmetrical state. The actual energy is higher because the fluctuations given by Equation 23 allow deviations from symmetry. Thus

$$U - U_0 = \tfrac{1}{4} \langle q^2 \rangle / C \sim 1/\delta E \tag{27}$$

and

$$dU - dU_0 \sim - d\delta E/(\delta E)^2. \tag{28}$$

Applying $dU - dU_0 = dQ - dQ_0$ to Equation 28 yields

$$dQ = dQ_0 - \gamma d\delta E/(\delta E)^2, \tag{29}$$

with γ independent of δE. As E_B is allowed to approach E_T, $d\delta E$ is negative. Equation 29 thus represents a divergent heat flow *into* the circuit as the transition is approached. The term dQ_0 has no anomalies at $E_B = E_T$. The divergence we have discussed is a result of Equation 23. If E_B approaches very closely to E_T, $v = (-q/\tau)$ will eventually be inadequate, and higher powers of q must be taken into account in that relationship. Note that the "interesting" heat flow, $dQ - dQ_0$, is reversible. If the changes in E_B are reversed, $dQ - dQ_0$ also reverses sign.

　　Can the divergence in the heat flow be detected, *e.g.,* by cycling the system back and forth near its transition and the use of phase-locked detection? Since we are dealing with a single degree of freedom and a divergence that is limited by the departures from $v = - q/\tau$, it doesn't seem all that likely. Furthermore our equations assume a modulation rate slow compared to the relaxation time τ of the circuit. This relaxation time becomes very long as the transition is approached. Thus, we must have extremely slow modulation, and must detect the contributions to the divergent term dQ in the presence of the nonanomalous, steady-state heat dissipation dQ_0. The only hope, it seems, would be the use of very small systems, perhaps a great many of them. These would have to be locked together in their modulation parameter, but would have to act as independent systems as far as their fluctuations are concerned.

ACKNOWLEDGMENTS

Much of the material related to stability criteria, resulted from discussions, debates, and correspondence with J. Keizer, G. Nicolis, F. Schlögl, J. Schnakenberg, J. Ross, O. Rössler and, most particularly, with C. Gardiner.

REFERENCES

1. LANDAUER, R. 1976. Fundamental limitations in the computational process. Ber. Bunsenges. Phys. Chem. **80**: 1048–1059.
2. BUSCH, H. 1921. Über die Erwärmung von Drähten in verdünnten Gasen durch den elektrischen Strom. Ann. Phys. (Leipzig) **64**: 401–450.
3. ROSS, B. I. 1975. The Ballast Resistor: A Simple Dissipative Structure. Ph.D. thesis. Mass. Inst. of Technology.
4. SIMON, H. T. 1905. Über die Dynamik der Lichtbogenvorgänge und über Lichtbogenhysteresis. Phys. Z. **6**: 297–336.
5. LANDAUER, R. 1976. Moebius strip coupling of bistable elements. IBM Research Report RC 6093. Available from the author.
6. HAKEN, H. 1975. Cooperative phenomena in systems far from thermal equilibrium and in nonphysical systems. Rev. Mod. Phys. **47**: 67–121.
7. HAKEN, H. 1977. Synergetics, an Introduction. Springer. Heidelberg.
8. RISTE, T. 1975. Fluctuations, Instabilities, and Phase Transitions. Plenum Press. New York.
9. ROSS, J. 1976. Temporal and Spatial Structures in Chemical Instabilities. Ber. Bunsenges. Phys. Chem. **80**: 1112–1125.
10. KEIZER, J. 1976. Fluctuations, stability, and generalized state functions at nonequilibrium steady states. J. Chem. Phys. **65**: 4431–4444.
11. MATHESON, I., D. F. WALLS & C. W. GARDINER. 1975. Stochastic models of first-order nonequilibrium phase transitions in chemical reactions. J. Stat. Phys. **12**: 21–34.
12. LANDAUER, R. 1961. Irreversibility and heat generation in the computing process. IBM J. Res. Devel. **5**: 183–191.
13. KEIZER, J. 1972. On the Solutions and the Steady States of a Master Equation. J. Stat. Phys. **6**: 67–72.
14. SCHNAKENBERG, J. 1976. Network theory of microscopic and macroscopic behavior of master equation systems. Rev. Mod. Phys. **48**: 571–585.
15. GLANSDORFF, P. & I. PRIGOGINE. 1971. Thermodynamic Theory of Structure, Stability and Fluctuations. Wiley-Interscience. New York.
16. LANDAUER, R. 1975. Inadequacy of entropy and entropy derivatives in characterizing the steady state. Phys. Rev. A **12**: 636–638.
17. LANDAUER, R. 1975. Stability and entropy production in electrical circuits. J. Stat. Phys. **13**: 1–16.
18. LANDAUER, R. 1976. Can capacitance be negative? Collective Phenomena **2**: 167–170.
19. MESERVEY, R. & P. M. TEDROW. 1969. Measurements of the kinetic inductance of superconducting linear structures. J. Appl. Phys. **40**: 2028–2034.
20. VAN KAMPEN, N. G. 1959. The definition of entropy in non-equilibrium states. Physica **25**: 1294–1302.
21. PRIGOGINE, I., G. NICOLIS & A. BABLOYANTZ. 1974. Nonequilibrium problems in

biological phenomena in mathematical analysis of fundamental biological phenomena. Ann. N.Y. Acad Sci. **231**: 99–100.

22. DE SOBRINO, L. 1975. The Glansdorff-Prigogine thermodynamic stability criterion in the light of Lyapunov's theory. J. Theor. Biol. **54**: 323–333.

23. KEIZER, J. & R. F. FOX. 1974. Qualms Regarding the Range of Validity of the Glansdorff-Prigogine Criterion for Stability of Non-Equilibrium States, Proc. Nat. Acad. Sci. **71**: 192–196.

24. GLANSDORFF, P., G. NICOLIS, & I. PRIGOGINE. 1974. The thermodynamic stability theory of non-equilibrium states. Proc. Nat. Acad. Sci. **71**: 197–199.

25. GARDINER, C. W. 1977. A Stochastic Basis for Isothermal Equilibrium and Non-Equilibrium Chemical Thermodynamics. Preprint. Univ. Waikato, Hamilton, New Zealand.

26. SCHLÖGL, F. 1971. On stability of steady states. Z. Phys. **243**: 303–310.

27. LANDSBERG, P. T. 1972. The fourth law of thermodynamics. Nature **238**: 229–231.

28. BEDEAUX, D., P. MAZUR & R. PASMANTER. 1977. The ballast resistor. Physica **86A**: 355–382.

29. TAKEYAMA, K. & K. KITAHARA. 1975. A theory of domain- and filament-formation due to carrier-density fluctuations in conductors with negative differential resistance. J. Phys. Soc. Japan **39**: 125–131.

30. LANDAUER, R. 1977. The ballast resistor. Phys. Rev. A **15**: 2117–2119.

31. LANDAUER, R. 1976. The role of noise in negative resistance circuits. J. Phys. Soc. Japan **41**: 695–696.

32. LANDAUER, R. 1962. Fluctuations in Bistable Tunnel Diode Circuits. J. Appl. Phys. **33**: 2209–2216.

33. NICOLIS, G. & R. LEFEVER. 1977. Comment on the kinetic potential and the Maxwell construction in non-equilibrium chemical phase transitions. Phys. Lett. **62A**: 469–471. See also Nicolis, G and J. W. Turner, this volume.

34. LANDAUER, R. 1973. Entropy changes for steady state fluctuations. J. Stat. Phys. **9**: 351–361. (Errata in 1974. J. Stat. Phys. **11**: 525.)

35. LANDAUER, R. 1971. Cooperative effects, soft modes, and fluctuations in data processing. Ferroelectrics **2**: 47–55.

36. LANDAUER, R. & J. W. F. WOO. 1973. Cooperative Phenomena in Data Processing. In Synergetics. H. Haken, Ed.: 97–123. B. G. Teubner, Stuttgart, Germany. See Section 5 and the Appendix.

37. LANDAUER, R. 1978. $dQ = TdS$ far from equilibrium. Phys. Rev. A **18**: 255–266.

MASTER EQUATIONS, STOCHASTIC DIFFERENTIAL EQUATIONS, AND BIFURCATIONS

C. W. Gardiner, S. Chaturvedi, and D. F. Walls

Physics Department
University of Waikato
Hamilton, New Zealand

INTRODUCTION

Master equations of the form obtained from stochastic chemical reaction theory exhibit bifurcation behavior related to that of the deterministic differential equation obtained in a large volume limit. A stochastic differential equation can be obtained by expanding the distribution function in Poisson distributions. Certain bifurcations occurring in chemical reaction theory can be investigated using this technique, and the behavior near critical points can be shown to be closely related to the Landau-Ginsburg theory.

Because of the wide variety of phenomena that can occur in nonequilibrium systems, it is of interest to study at first the simplest possible systems of this kind. Schlögl[1] first suggested that constrained chemical reaction systems provided analogues of many phase transition phenomena, and many other workers developed his insight.[2-10,28,29] A stochastic approach to these systems was developed by the present authors,[5-7] who used master equation techniques, as well as by Nitzan et al.[4] and Keizer,[8-10] who used Langevin equation techniques, and Haken,[28,29] who has used techniques similar to those used in laser theory and solid-state physics. A variety of phenomena have been investigated using the techniques, including analogues of second-order phase transitions, systems exhibiting hysteresis, and systems exhibiting spatial and temporal structures. The techniques have hitherto not been applied to non-steady-state systems.

In this paper, we wish to review a particular technique recently introduced by us, the Poisson representation method, in which Master equations can be converted *exactly* into Fokker-Planck equations and, hence, stochastic differential equations. We are then able to use perturbation techniques to develop systematic approximations to various quantities of interest, such as mean values and correlation functions. We show how these techniques can be applied in three typical cases, which demonstrate that the stochastic differential equations are defined in terms of a quasiprobability, which can take on negative and even complex values. In fact, in certain cases the quasiprobability is defined as a function of a complex variable.

A further significant generalization comes in the definition of what we call *higher order noise*. This is required since the Fokker-Planck equations involve third- or higher order noise has recently been given by Hochberg.[11]

The paper concludes with a brief outline of certain other results that have been obtained using the Poisson representation, namely the onset of instability in the Brusselator, a study of two time correlation functions and their relation to

453

0077-8923/78/0316-0453 $1.75/1 © 1979, NYAS

generalized fluctuation dissipation theorems, and a generalized Landau theory for chemically reacting systems.

Stochastic Master Equations for Chemical Instabilities

Master equations for chemical kinetics have been in use for some years, and have been developed and applied by many authors.[2,3,5-7] A system which gives an instability, and which has been well studied by ourselves and others[4,5,7] is the "second-order phase transition" model first studied by Schlögl[1]:

$$A + B \underset{k_1}{\overset{k_2}{\rightleftarrows}} 2X,$$

$$B + X \underset{k_3}{\overset{k_1}{\rightleftarrows}} C,$$

$$(1)$$

for which the deterministic rate equations are

$$\frac{dx}{dt} = k_2 Ax + k_3 C - k_1 Bx - k_4 x^2. \tag{2}$$

Here, it is understood that A, B, C, (the numbers of molecules of compounds A, B, and C) are held fixed, and that x represents the *number* of molecules in the system, which can vary. The stochastic master equation which we use to represent this system is an equation for the time rate of change of $P(x, t)$ the probability of there being x molecules in the system at time t, and is

$$\frac{dP(x, t)}{dt} = [k_1 B(x + 1) + k_4 x(x + 1)] P(x + 1, t)$$

$$+ [k_2 A(x - 1) + k_3 C] P(x - 1, t) \tag{3}$$

$$- [k_1 Bx + k_4 x(x - 1) + k_2 Ax + k_3 C] P(x, t)$$

This model assumes that the probability per unit time of a reaction occurring is proportional to the number of ways of assembling the reactants, and further, it takes no account of the spatial distribution of reactants, and would thus be valid only in the case that the spatial distribution was kept homogeneous.

If we wish to describe a system in which spatial diffusion can occur, we must divide the system into cells of length l, labeled by an index i, so that a diffusion-reaction master equation for the reaction would be

$$\frac{\partial P(\mathbf{x}, t)}{\partial t} = \sum_{ij} \{d_{ij}(x_i + 1) P(x_i + 1, \hat{\mathbf{x}}, t) - d_{ij} P(\mathbf{x}, t) + \sum_i \left(\frac{\partial P}{\partial t}\right)_{i, \text{chem}} \tag{4}$$

(here $\hat{\mathbf{x}}$ is the vector of all $\mathbf{x}$ except x_i) where $(\partial P/\partial t)_{i, \text{chem}}$ is of the form of $\partial P/\partial t$ in Equation 3 with x replaced by x_i.

THE POISSON REPRESENTATION

The distribution function $P(x)$ for a system is a multivariate Poisson, when the system is described by a grand canonical ensemble, which we[12] and van Kampen[13] have emphasized recently, and a natural procedure is then to expand $P(x)$ in terms of equilibrium distributions. For a one variable system, we write

$$P(x) = \int d\alpha \, \frac{e^{-\alpha}\alpha^x}{x!} f(\alpha). \tag{5}$$

The major advantage is the relation between the factorial moments of x and the moments of α, i.e.,

$$\langle x^r \rangle_f \equiv \langle x(x-1) \cdots (x-r+1) \rangle = \langle \alpha^r \rangle \equiv \int d\alpha \, \alpha^r f(\alpha). \tag{6}$$

The range of the α integration depends on the system under study. To illustrate the various possibilities we study three examples, which have been studied in more detail in our more detailed paper.[12]

Second-Order Phase Transition Model

This is the model described by Equations 1, 2, and 3. Assuming a Poisson representation for $P(x, t)$ as in (5) we can, after integrating by parts, obtain the Fokker Planck equation

$$\frac{\partial f(\alpha, t)}{\partial t} = -\frac{\partial}{\partial \alpha} \left\{ [K_3 V + (K_2 - K_1)\alpha - K_4 V^{-1}\alpha^2] f(\alpha, t) \right.$$
$$\left. + \frac{1}{2} \frac{\partial}{\partial \alpha} [2(K_2\alpha - K_4 V^{-1}\alpha^2) f(\alpha, t)] \right\} \tag{7}$$

where we have defined

$$k_3 C = K_3 V, \quad k_2 A = K_2, \quad k_1 B = K_1, \quad k_4 = K_4 V^{-1} \tag{8}$$

and thereby exhibited the volume dependence of the parameters. The steady state solution of (7) is straightforwardly found. In order to carry out the integration by parts it is necessary for $f(\alpha, t)$ and its derivative to vanish on the boundaries of the region of integration, from which it readily follows that in the steady state the term inside the curly brackets vanishes, and hence

$$f_{ss}(\alpha) = e^{\alpha}(K_2 V - \alpha)^{V(K_1 - K_3/K_2) - 1} \alpha^{(K_3 V/K_2 - 1)} \tag{9}$$

and the range of the α integration is $(0, K_2 V)$. (There is another choice of contour which satisfies the boundary condition, but the resulting $P(x)$ can be negative, so it is excluded, as we have demonstrated in our major work.[12] Using (9) we can now calculate the various moments of α by steepest descents techniques, giving an asymptotic expansion in inverse powers of V, the system volume.

An alternative approach is to use the stochastic differential equation equivalent to this Fokker-Planck equation.[14,15] By defining a variable

$$\eta = \alpha/V \tag{10}$$

and for simplicity setting $K_4 = 1$, we obtain the Ito stochastic differential equation[14,15]:

$$d\eta = [K_3 + (K_2 - K_1)\eta - \eta^2]dt + V^{-1/2}[2(K_2\eta - \eta^2)]^{1/2}dw(t), \qquad (11)$$

which may be solved iteratively as an expansion in powers of $V^{-1/2}$. The two methods are equivalent in this one variable case, but when diffusion is introduced as in equation 4, the resulting multivariate Fokker-Planck equation is no longer soluble, whereas the stochastic differential equation technique is easily generalized. If ΔV is the volume of the cell, and the various K terms are defined as in (8) using ΔV instead of V, the stochastic differential equations for $\eta_i = \alpha_i/\Delta V$ are

$$d\eta_i = \left[\sum_j D_{ij}\eta_j + K_3 + (K_2 - K_1)\eta - \eta_i^2\right] dt$$

$$+ (\Delta V)^{-1/2}[2(K_2\eta_i - \eta_i^2)]^{1/2}dw_i(t) \qquad (12)$$

(here $D_{ij} = d_{ij} - \delta_{ij}(\sum_k d_{ik})$).
This model displays a sharp transition in the steady-state solution in the limit $K_3 = 0$; i.e., we have

$$\eta_i = 0 \qquad (K_2 < K_1)$$

$$= K_2 - K_1 \quad (K_2 \geq K_1) \qquad (13)$$

and the fluctuation behavior near this transition is of great interest. The long wavelength behavior can be best studied in a continuum approximation, where we write

$$\eta_i = \eta(\mathbf{r}_i), \qquad (14)$$

$$\sum D_{ij}\eta_j \rightarrow D\nabla^2\eta(\mathbf{r}_i), \qquad (15)$$

and using the physicist's notation,

$$dw_i(t) = (\Delta V)^{1/2}\xi(\mathbf{r}_i, t), \qquad (16)$$

so that

$$\langle\xi(\mathbf{r}, t)\xi(\mathbf{r}'t')\rangle = \delta^3(\mathbf{r} - \mathbf{r}')\delta(t - t'),$$

we obtain the Langevin equation:

$$\frac{\partial\eta(\mathbf{r}, t)}{\partial t} = D\nabla^2\eta + K_3 + (K_2 - K_1)\eta - \eta^2 + [2(K_2\eta - \eta^2)]^{1/2}\xi(\mathbf{r}, t). \qquad (17)$$

This Langevin equation differs from those usually written[4,8-10,16] for reaction-diffusion systems in some significant ways:

(i) The variable $\eta(\mathbf{r}, t)$ is a quasivariable, not the concentration, though its mean value is the mean value of the concentration. The relationship is very much the same as that which occurs in quantum optics, when coherent state variables are used.[17,18]

(ii) The noise term is given exactly from the master equation, and is not merely a Gaussian approximation arising from a central limit theorem.[8-10,19]

(iii) The noise term arises only from the reaction part of the system, not the diffusion, and describes only fluctuations in excess of the equilibrium Poissonian value. In the conventional chemical Langevin equations, the noise involves gradient terms, and is quite complicated.[20,21]

Solutions. We can employ iterative methods to determine spatial correlations. The first-order solutions are

$$<\rho(\mathbf{r}), \rho(\mathbf{r}')> = <\rho> \delta(\mathbf{r} - \mathbf{r}') + \frac{K_1 <\rho> \exp(-|\mathbf{r} - \mathbf{r}'|/l_c^2)}{4\pi D |\mathbf{r} - \mathbf{r}'|}, \quad (18)$$

where

$$l_c^2 = <\rho>/D \quad (19)$$

and $\rho(\mathbf{r})$ is a concentration variable, such that $\rho(\mathbf{r}_i)\Delta V = x_i$. Higher order corrections can be calculated, and these prove to be negligible except near the critical point when $K_1 = K_2$, when they are negligible as long as

$$|K_2 - K_1| \gg \frac{K_1^2}{D^3}. \quad (20)$$

This result is the same as would arise by applying the Ginsburg criterion,[22] in the form that

$$<\rho(\mathbf{r}), \rho(\mathbf{r}')> \Big|_{|\mathbf{r}-\mathbf{r}'| \sim l_c} \ll <\rho(\mathbf{r})>^2. \quad (21)$$

This result has been demonstrated by us in the context of a generalized Landau theory based on the Poisson representation.[23]

A Reaction That Gives $f(\alpha)$ with Negative Variance

The reaction

$$B \xrightarrow{k_1} X, \qquad 2X \xrightarrow{k_2} A \quad (22)$$

has been considered by several authors.[2,12,24,25] Setting

$$K_1 V = k_1 B,$$
$$K_2 V^{-1} = k_2, \quad (23)$$

we obtain the Fokker-Planck equation, by using Poisson methods,

$$\frac{\partial f(\alpha, t)}{\partial t} = -\frac{\partial}{\partial \alpha} \left\{ (K_1 V - 2K_2 V^{-1} \alpha^2) f(\alpha, t) + \frac{\partial}{\partial \alpha} [K_2 V^{-1} \alpha^2 f(\alpha, t)] \right\}. \quad (24)$$

In this equation, the noise term is negative for real α. Remembering that $f(\alpha, t)$ is a quasiprobability, and hence need not be positive, and that α is a quasivariable, which need not be real, we can write the steady-state solution as

$$f(\alpha) = \alpha^{-2} \exp\left(2\alpha + \frac{ar^2}{\alpha}\right), \tag{25}$$

with $a = 2K_1/K_2$. The range of the α integration will now be a closed contour encircling the origin. (In this case the argument on boundary conditions used above is not applicable, because the contour of integration is closed. However, the most general steady-state solution turns out not to be single valued, so that boundary value terms would never vanish. Equation 25 is the unique admissible solution.)

Unlike the previous example, $f(\alpha)$ has a minimum at the deterministic steady-state value, which is a saddle point in the complex plane. Thus, we find, by steepest descent methods,

$$\langle\alpha\rangle \simeq V(a/2)^{1/2}, \tag{26}$$

$$\langle\alpha^2\rangle - \langle\alpha\rangle^2 \simeq -\frac{V}{4}\left(\frac{a}{2}\right)^{1/2}, \tag{27}$$

so that the variance in the α variable is negative. Thus, in the x variable

$$\langle x\rangle \simeq V(a/2)^{1/2}. \tag{28}$$

Thus the negative α variance corresponds to the diminution of the noise in the x variable to less than the Poissonian value. The corresponding stochastic differential equation to (24) is

$$\frac{d\eta}{dt} = K_1 - 2K_2\eta^2 + V^{1/2}(2K_2)^{1/2}\eta\xi(t), \tag{29}$$

where, as before, $\eta = \alpha/V$.

This correspondence must be regarded as only heuristic, although the results predicted are the same as those of the Fokker-Planck equation, as far as we have checked. The Poisson representation appears to open up as an area for realistic study the realm of stochastic processes in the complex plane, with complex probabilities.

A SINGLE VARIABLE REACTION WITH A TRIMOLECULAR STEP

The process

$$A + 2X \underset{k_2}{\overset{k_1}{\rightleftarrows}} 3X,$$

$$A \underset{k_4}{\overset{k_3}{\rightleftarrows}} X, \tag{30}$$

has been studied by many authors previously.[4,6,26] It exhibits, in the steady state a possibility of bistability. The Fokker-Planck equation corresponding to it is

$$\frac{\partial f(\alpha, t)}{\partial t} = -\frac{\partial}{\partial \alpha} [(K_1 V^{-1}\alpha^2 - K_2 V^{-2}\alpha^3 + K_3 V - K_4\alpha)f(\alpha, t)]$$

$$+ \frac{1}{2}\frac{\partial^2}{\partial \alpha^2} [4(K_1 V^{-1}\alpha^2 - K_2 V^{-2}\alpha^3)f(\alpha, t)]$$

$$- \frac{\partial^3}{\partial \alpha^3} [(K_1 V^{-1}\alpha^2 - K_2 V^{-2}\alpha^3)f(\alpha, t)]. \tag{31}$$

When the cubic in the bracket has three roots, we have one unstable solution, and two stable solutions, and this corresponds to a bimodal steady state $f(\alpha)$. A similar situation pertains for $P(x)_{ss}$.

The range of the variable α is determined by the second- and third-order diffusion coefficients, which vanish at

$$\alpha = 0$$

and

$$\alpha = \frac{K_1 V}{K_2}, \tag{32}$$

which is the range of α. The steady state $f(\alpha)$ can be determined, and is given by

$$f(\alpha) = e^\alpha \alpha^{-2}(1 - K_2\alpha/K_1 V)^{-1}\alpha^b F(b - a - 1, b + a, 2b, K_2\alpha/K_1 V), \tag{33}$$

where

$$V^2 K_4/K_2 = -a(a + 1),$$
$$V^2 K_3/K_1 = -b(b - 1). \tag{34}$$

The contour of integration is a Pochhammer contour, as described in Whittaker and Watson.[27] The range of variation of the stochastic variable is thus on a multisheeted surface in the complex plane.

The analysis of the implications of this solution is by no means easy, though it can be carried out by writing an integral representation for the hypergeometric function, and using a double application of the steepest decents methods. A much more interesting method is to introduce a concept of *higher order noise*.

We consider the process of third-order noise, by introducing the stochastic variable $v(t)$, whose probability distribution function obeys the third-order equation

$$\frac{\partial P(v, t)}{\partial t} = -\frac{1}{6}\frac{\partial^3 P(v, t)}{\partial v^3}. \tag{35}$$

We have demonstrated in our more detailed work[12] that if $v(0)$ is a deterministic value, (i.e., $\langle v(0)^n \rangle = \langle v(0) \rangle^n$,) then

$$\langle [v(t) - v(0)]^n \rangle = 0$$

if n is not a multiple of 3, and

$$<[v(t) - v(0)]^{3m}> = \left(\frac{1}{6}\right)^m \frac{3m!}{m!} \tag{36}$$

These relations are not possible unless $P(v, t)$ takes on negative values, as can be shown by direct solution of (33). We can then introduce the differential $dv(t)$. and show quite directly that a stochastic differential equation of the form

$$dy(t) = a(y)dt + b(y)dw(t) + c(y)dv(t) \tag{37}$$

is equivalent to the third-order Fokker-Planck equation:

$$\frac{\partial P(y, t)}{\partial t} = -\frac{\partial}{\partial y} [a(y)P] + \frac{1}{2} \frac{\partial^2}{\partial y^2} [b(y)^2 P] - \frac{1}{6} \frac{\partial^3}{\partial y^3} [c(y)^3 P]. \tag{38}$$

In the same way as a Langevin force $\xi(t)$ is defined, we may define a third-order noise source by

$$dv(t) = \zeta(t)dt, \tag{39}$$

where

$$<\zeta(t)> = <\zeta(t)\zeta(t')> = 0,$$
$$<\zeta(t)\zeta(t')\zeta(t'')> = \delta(t - t')\delta(t' - t''). \tag{40}$$

We now apply this method to our Fokker-Planck Equation (40). Defining

$$\alpha = \eta V,$$
$$\mu = V^{-1/6}, \tag{41}$$

we obtain

$$\frac{d\eta}{dt} = K_1\eta^2 - K_2\eta^3 + K_3 - K_4\eta + \mu^3 [4(K_1\eta - K_2\eta^3)]^{1/2}\xi(t)$$
$$+ \mu^4 [6(K_1\eta^2 - K_2\eta^3)]^{1/3}\zeta(t). \tag{42}$$

We can now expand η as a power series in μ, and solve iteratively, the expansion parameter now being μ. Solving for steady-state moments, we get

$$\left.\begin{array}{l} <x> = V\eta_0 + 2ab/c^2 \\[2mm] <x^2> - <x>^2 = V\left(\frac{2a}{c}\right) + \left(\frac{28}{3}\frac{a^2b^2}{c^4} + \frac{8ab^2\eta_0}{c^3} - \frac{36K_2a^2}{c^3}\right. \\[4mm] \left. + \frac{8ab}{c^3}\right) + \cdots, \\[4mm] <(x - <x>)^3> = V\left(\frac{8a}{c} - \frac{12a^2b}{c^3} + \eta_0\right), \end{array}\right\} \tag{43}$$

where

$$a = K\eta_0^2 - K_2\eta_0^3,$$
$$b = 2K_1 - 3K_2\eta_0, \qquad\qquad \left.\begin{array}{c} \\ \\ \\ \end{array}\right\} \text{(44)}$$
$$c = K - 2K_1\eta_0 + 3K_2\eta_0^2$$

and η_0 is a solution of the steady-state equation

$$K_1\eta_0^2 - K_2\eta_0^3 + K_3 - K_4\eta_0 = 0. \qquad\qquad \text{(45)}$$

Although it is not obvious from Equation 44, the third-order noise contributes to $O(V^{-1})$ in the mean, to $O(1)$ in the variance, and to $O(V)$ in the skewness coefficient. Various kinds of singular behavior occur as $C \to 0$, which occurs when the slope of a graph of η_0 against K_3 becomes infinite, which can occur above a bifurcation point.

Our simple-minded expansion procedure indicates that the singularity will be different from the naive $1/c$ behavior expected in the lowest order approximation.

Of course we can in principle determine the critical exponents as $c \to 0$ directly from the solution (31), or by other means.[30] A generalization to a diffusing system, however, for which the Langevin Equation (42) would be modified by the addition of a term $D\triangledown^2\eta$, is relatively straightforward. The iterative process is then easily applicable.

CONCLUSION

We have presented here only an outline of some of the results on the Poisson representation; a more complete version is being published elsewhere.[12,31,32] We have treated in detail the full theory of two-time correlation functions and fluctuation dissipation theorems,[31] calculated[32] the spatial and temporal correlations in the oscillating chemical reaction introduced by the Brussels[33] group, and formulated a generalized Landau theory[23] based on the Poisson representation.

The methods presented here have been mainly applied to rather simple problems, all of which can be treated by various other methods. When diffusion is introduced, however, we believe that our methods are easily the most powerful. The fact that the noise terms in the stochastic differential equations are not constant makes application of renormalization group and scaling techniques not straightforward. It is clear from the perturbation analysis that the classical theory breaks down near critical points, and that a critical dimension of $d = 4$ exists in these models.

A more interesting question arises from the third-order noise terms, which can turn up in reactions with trimolecular steps. It appears that no dramatic results will arise from this, but we have not yet carried out a full analysis of such a system when diffusion is included, and hence are not certain.

The essential conclusion of our work is that stochastic master equations, with only polynomial transition probabilities and integer steps, are equivalent to appropriate stochastic differential equations, in which the probabilistic aspect is handled by a quasiprobability, whose values may be negative or complex.

ACKNOWLEDGMENTS

One of us (C.W.G.) is grateful to the University of Waikato for travel sponsorship and the IBM Corporation for the award of a visiting scientist fellowship at the IBM Thomas J. Watson Research Center.

REFERENCES

1. SCHLÖGL, F. 1972. Z. Phys. **253:** 147.
2. NICOLIS, G. & I. PRIGOGINE. 1971. Proc. Nat. Acad. Sci. U.S.A. **68:** 2102.
3. PRIGOGINE, I. & NICOLIS, G. 1973. Proc. 3rd Internatl. Conf. From Theoretical Physics to Biology. S. Karger. Basel.
4. NITZAN, A., P. ORTOLEVA, J. DEUTCH & J. ROSS. 1974. J. Chem. Phys. **61:** 1056.
5. MCNEIL, K. J., D. F. WALLS. 1974. J. Stat. Phys. **12:** 21.
6. MATHESON, I. S., D. F. WALLS & C. W. GARDINER. 1975. J. Stat. Phys. **14:** 307.
7. GARDINER, C. W., K. J. MCNEIL, D. F. WALLS & I. S. MATHESON. J. Stat. Phys. **14:** 307.
8. KEIZER, J. 1975. J. Chem. Phys. **63:** 398.
9. KEIZER, J. 1975. J. Chem. Phys. **63:** 5037.
10. KEIZER, J. 1976. J. Chem. Phys. **64:** 1679.
11. HOCHBERG, K. 1977. A Signed Measure on Path Space. Preprint. Courant Institute.
12. GARDINER, C. W. & S. CHATURVEDI. 1977. J. Stat. Phys. **17:** 429–468.
13. VAN KAMPEN, N. G. 1976. Phys. Lett. **59A:** 333.
14. ARNOLD, L. 1974. Stochastic Differential Equations. Wiley-Interscience. New York.
15. HAKEN, H. 1975. Rev. Mod. Phys. **47:** 67.
16. VAN KAMPEN, N. G. 1976. Fluctuations in continuous systems. In Topics in Statistical Mechanics and Biophysics. R. Picirelli, Ed. Am. Inst. Physics. New York.
17. GLAUBER, R. J. 1963. Phys. Rev. **130:** 2529; **131:** 2761.
18. SUDARSHAN, E. C. G. 1966. Phys. Rev. Lett. **16:** 534.
19. KURTZ, T. G. 1972. J. Chem. Phys. **57:** 2976.
20. GROSSMAN, S. 1976. J. Chem. Phys. **65:** 2007.
21. GARDINER, C. W. 1976. J. Stat. Phys. **15:** 451.
22. KADANOFF, L. P., W. GOTZE, D. HAMBLEN, R. HECHT, E. A. S. LEWIS, V. V. PALCAUSKAS, M. RAYL & J. SWIFT. 1967. Rev. Mod. Phys. **39:** 395.
23. GARDINER, C. W. & D. F. WALLS. 1977. A generalised Landau theory for chemical instabilities. J. Phys. A **11:** 161.
24. NICOLIS, G. 1972. J. Stat. Phys. **6:** 195.
25. MAZO, R. M. 1975. J. Chem. Phys. **62:** 4244.
26. JANSSEN, H. K. 1974. Z. Phys. **270:** 67.
27. WHITTAKER, E. T. & G. N. WATSON. 1927. A Course of Modern Analysis. 4th edit. Cambridge University Press. London.
28. HAKEN, H. 1975. Z. Phys. B **22:** 69–72.
29. HAKEN, H. 1975. Z. Phys. B **20:** 413–420.
30. NICOLIS, G. & J. W. TURNER. 1977. Physica. **89A:** 326.
31. CHATURVEDI, S. & C. W. GARDINER. 1978. The Poisson representation II: Two time correlation functions. J. Stat. Phys. **18:** 503.
32. CHATURVEDI, S., C. W. GARDINER, I. S. MATHESON & D. F. WALLS. 1977. J. Stat. Phys. **17:** 469.
33. GLANDSORFF, P. & I. PRIGOGINE. 1971. Thermodynamic Theory of Structure, Stability and Fluctuation. Wiley Interscience. New York.

DISCUSSION PAPER:
FRACTALS, ATTRACTORS,
AND THE FRACTAL DIMENSION

Benoit B. Mandelbrot

IBM Research
Yorktown Heights, New York 10598

The evidence is that among the diagrams shown by the various speakers there is hardly one that is not either a confirmed or a suspected fractal, or at least related to a fractal. In the talks fractals are mostly referred to as "monster sets" or "strange sets," and the like, with many of them being labeled "strange attractors." In other circles, we hear some of them described as "exceptional sets," "weird sets," or "dragons." It is, however, to be hoped that none of these terms is to become entrenched in a technical sense. Since the Latin for "irregular and fragmented" is *fractus,* I coined "fractal" to denote them.* The recent recognition of their role in science has had several different and independent sources, of which the first are in Lorenz's well-known paper of 1963 and in my own papers.†

Students of fractal attractors and the like will have to evaluate their diverse "degrees of monstrosity," and study the relationships, if any, between their fractal and topological properties. Several alternative measures have been debated by mathematicians. Hausdorff and Besicovitch have advanced a notion of continuous dimension D, which became extremely well known to an extremely small number of very abstract-minded scholars. I claim this dimension is immensely intuitive and widely useful.* The impression that D could have no concrete application was due in part of the fact that its definition referred to the *local* properties of certain weird sets. However, the most interesting among these sets possess the property of *scaling,* which expresses a very strong relationship between their local, global, and intermediate characteristics, and it implies that D is also capable of tackling *numerically* a certain important nontopological facet of overall geometric "form." It has been convenient to call D the *fractal dimension.* For any set (in a metric space), it can be shown that $D \geq D_T$, where D is the fractal dimension and D_T the topological dimension. As is fit for a dimension, the topological D_T is always an integer, but "strangely" enough, we find more often than not that the fractal D is *not* an integer. For example, the fractal dimension of a surface is not less than 2, but it need not be exactly 2. A set is fractal if and only if $D > D_T$.* It may be of interest that the fractal dimension of the Lorenz attractor (as estimated by Velarde on the basis of an argument due to Pomeau and Ibanez) is about

*MANDELBROT, B. 1975. Les objects fractals: forme, hasard et dimension. Flammarion. Paris. [2nd ed, in English: 1977. Fractals: Form, Chance, and Dimension. W. H. Freeman and Co. San Francisco.]

†BERGER, J. M. & B. B. MANDELBROT. 1963. A new model for the clustering of errors on telephone circuits. IBM J. Res. Devel. 7: 224–236; MANDELBROT, B. B. 1967. Sporadic random functions and conditional spectral analysis. Proc. 5th Berkeley Symposium on Mathematical Statistics and Probability, 1965. Lucien LeCam & J. Neyman, Eds. University of California Press. Berkeley. 3: 155–179; MANDELBROT, B. B. 1977. Fractals and turbulence: Attractors and dispersion. Turbulence Seminar, Berkeley 1976/77. Lecture Notes in Mathematics. 615: Vol. 83–93. Springer-Verlag. New York.

2.06, meaning it is a *fractal* surface. Since $D - 2$ is small, this surface is "not too strange." This prediction having been made, it should be possible to go back to the data to which the Lorenz equation relates, in order to find out if this value of D is acceptable. The inequality $D \geq D_T$ is but one aspect of the fact that a set's fractal and topological properties are to a large extent mutually independent.

ECOSYSTEM STABILITY AND BIFURCATION IN THE LIGHT OF ADAPTABILITY THEORY

Michael Conrad

Department of Computer and Communication Sciences
University of Michigan
Ann Arbor, Michigan 48109

INTRODUCTION

In the author's laboratory an experiment of the following type is performed. Replicas of "jar microecosystems" are prepared as identically as possible from undefined pond material (water, bottom matter). Ensembles of these replicas are then cultured in incubators under varying degrees of statistical uncertainty and physical stress. Such microecosystems, like all ecosystems, undergo a sequence of changes (called succession) until they reach a climax form of organization. This is a form of organization which appears to be stable in the sense that it persists for a relatively long time. A necessary condition for such persistence is that the system is somehow capable of either dissipating or absorbing without visible effect the perturbations to which it is being subjected; after all, if it were not capable of doing this it would continue to change. By choosing the degree of environmental uncertainty it is thus possible to prepare (through the culturing process) ecosystems capable of absorbing or dissipating at minimum this degree of environmental uncertainty.

The ability to cope with (absorb or dissipate) environmental uncertainty will be called adaptability. Thus, the imposed uncertainty of the culturing environment provides an operational definition of minimum adaptability. Once systems with defined minimum adaptabilities are prepared, many questions might be asked. For example, does actual adaptability tend to decrease in the direction of the minimum possible? What is the relation between adaptability and the stability of succession? What is the relation to the spectrum of particular forms of adaptability (e.g., genetic, organismic, populational) and to patterns of community organization? The intention of this paper, however, is not to consider the experimental aspect of this problem, but rather to describe a formalism which makes it possible to approach complete, self-sustaining biological systems of the above type theoretically, in particular to analyze the complex of adaptabilities underlying their dynamics.

The essence of this formalism (hierarchical adaptability theory) is an entropy theory analysis of the state-to-state behavior of ecosystems in which the state specifications are structured in a way which reflects the hierarchical and compartmental organization of complete, self-sustaining biological systems (including their environment). After briefly reviewing this formalism (cf. also Conrad[1-7]), we shall state and prove the bootstrap principle of hierarchical adaptability theory (that the assumption of structured descriptions is self-justifying). The bootstrap principle underlies the major conclusion, viz., that as the dynamics of any particular level in an ecosystem appears more autonomous (i.e., more amenable to

465

0077–8923/78/0316–0465 $1.75/1 © 1979, NYAS

description without reference to variables associated with other levels), the contribution of adaptabilities at other levels in general becomes more important as support for these dynamics. This makes it possible to cross-correlate the elements of hierarchical adaptability theory with concepts deriving from dynamical styles of analysis and therefore to make some statements about the hidden biology underlying ecosystem stability and bifurcation.

FORMALISM OF HIERARCHICAL ADAPTABILITY THEORY

Structured State Specifications

The ecosystem is supposed to consist of a living (biotic) part and a physical environment, each with a finite set of distinguishable states observed on a discrete time scale. (These assumptions simplify some technical problems, but could be dropped.) The state of both the living part (community) and the environment is furthermore supposed to be adequately specifiable in terms of a finite set of observable properties. In the case of the community these properties include: species composition, foodweb structure, number of organisms in each species, locations of each organism, physiological state of each organism, pattern of gene activation and also genotype (DNA base sequence) of each organism. In the case of the environment the physical space is divided into local regions and the state of each of these regions (excluding that portion occupied by living matter) specified in terms of macroscopic variables (e.g., temperature, pressure, mole numbers, other mechanical forces).

The above state specifications can be thought of as structured descriptions of ecological systems. In the case of the community this structuring corresponds to the apparent organizational structure, i.e., to the levels of organization in biological systems, with more or less identifiable units (or compartments) at the various levels. Thus, the community can be considered as a unit, populations as subunits of the community, organisms as subunits of populations, and genome and phenome (all of the organism aside from the genome) as subunits of the organism. Species composition and foodweb organization are thus partial states (or partial characterizations) of the community, without reference to species, organisms, or other lower level compartments. Number and locations of organisms are partial states of species, again without reference to lower level subunits. For the present purposes, the pattern of gene activation can be thought of as the partial state of the organism in terms of phenome and genome and physiological state and DNA sequence as complete states of the phenome and genome. The procedure could be carried further, by increasing the number of levels and decreasing the number of properties required to characterize compartments at each level (assuming that a complete set of properties has been chosen in the first place).

The above correspondence between structured state descriptions and hierarchical and compartmental structure can be made formal by defining complete states as many-tuples of partial states. Thus,

$$\alpha^a(t) = [\alpha_{13}{}^b(t), \ldots, \alpha_{hk}{}^c(t), \ldots, \alpha_{(2n)0}{}^d(t)] \equiv \bigcap_{i,j} \alpha_{ij}{}^f(t), \tag{1}$$

where the α^u are complete states of the community, the $\alpha_{ij}{}'$ are partial states of compartment i at level j; level 3 can be taken as the community level (and there is only one community), level 2 can be taken as the population level, level 1 as the organism level, level 0 as either the genome or phenome level (in which case the states are complete), and the re-representation is possible for some choice of superscripts. (Actually, since matter cycles in an ecosystem and since organisms are born and die, it is necessary to define a reference structure with a very large number of compartments, some "unborn" or "dead" at any given time.[3]) Also, for the environment,

$$\beta^a(t) = [\beta_{i0}{}^b(t), \ldots, \beta_{m0}{}^c(t)] \equiv \bigcap_h \beta_{h0}{}^g(t), \tag{2}$$

where the b^v are complete states of the environment, the $\beta_{i0}{}^s$ are complete states of region i at level 0, the convention is that all compartments are at level 0, and the re-representation is possible for some choice of superscripts.

Later it is shown that structurability of community descriptions is a self-justifying assumption.

Transition Schemes and Transition Scheme Entropies

Imagine a large number of replicas of microecosystems, all prepared as identically as possible and undergoing state to state transitions in time. One may suppose that for any given time, starting from any given initial state, there would exist in principle a set of transition probabilities

$$\Omega = \{p[\alpha^u(t + 1), \beta^v(t + 1) \mid \alpha^r(t), \beta^s(t)]\}. \tag{3}$$

For the community and environment separately this implies the partial schemes

$$\omega = \{p[\alpha^u(t + 1) \mid \alpha^r(t), \beta^s(t)]\} \tag{4a}$$

$$\omega' = \{p[\beta^v(t + 1) \mid \alpha^r(t), \beta^s(t)]\} \tag{4b}$$

where the prime notation distinguishes the environment scheme. On these partial schemes entropies may be defined, e.g.,

$$H(\omega') = -\Sigma p[\alpha^r(t), \beta^s(t)] \, p[\beta^v(t + 1) \mid \alpha^r(t), \beta^s(t)] \times$$
$$\log p[\beta^v(t + 1) \mid \alpha^r(t), \beta^s(t)] \tag{5}$$

and

$$H(\omega \mid \omega') = -\Sigma p[\alpha^r(t), \beta^s(t), \beta^v(t + 1)] \, p[\alpha^u(t + 1) \mid \alpha^r(t), \beta^s(t), \beta^v(t + 1)] \times$$
$$\log p[\alpha^u(t + 1) \mid \alpha^r(t), \beta^s(t), \beta^v(t + 1)], \tag{6}$$

where indices run over all sums and corresponding entropies may be written for $H(\omega)$ and $H(\omega' \mid \omega)$. In analogy to a well known identity of entropy theory,[8]

$$H(\omega) - H(\omega \mid \omega') + H(\omega' \mid \omega) = H(\omega'), \tag{7}$$

which can easily be proved for this generalized case by substituting the transition scheme definitions. Equation 7 is always true and expresses what is in fact a tautological relation between the observed statistical properties of the community and the observed statistical properties of the environment.

Operational Definition of Adaptability

In order to use the foregoing to develop a definition of adaptability it is necessary for the terms on the left-hand side of Equation 7 to represent potentialities, independent of the particular statistical character of the environment. This can be done by varying over all possible environments, looking for the most uncertain environment compatible with the system remaining alive. If this environment has transition scheme ω', Equation 7 can be expressed as the inequality

$$H(\hat{\omega}) - H(\hat{\omega} \mid \hat{\omega}') + H(\hat{\omega}' \mid \hat{\omega}) \geqq H(\omega') \qquad (8)$$

where $H(\hat{\omega})$ is the potential uncertainty of the community, $H(\hat{\omega} \mid \hat{\omega}')$ reflects the potential ability to anticipate the environment, $H(\hat{\omega}' \mid \hat{\omega})$ reflects the potential indifference to the environment, and ω' is the transition scheme of the actual environment. The entire left-hand side of Equation 8 will be called the *maximal adaptability* of the biological system. Actually the condition that the system undergo complete demise is a rather strong one if the biological system is a complete community (as opposed to, say, a single population or organism) and in general it is more reasonable to replace it by a weaker condition, e.g., that the community not be pushed out of a given edaphic climax (in which case maximal adaptability is defined only for this climax) or that it not exhibit more than some predefined degree of decrement in biomass. The latter condition introduces a certain arbitrariness into the definition, but this can be removed to the extent that the relative adaptabilities (of at least initially identical systems) can be expected to be properly ordered.

It is possible to replace the above definition of (maximal) adaptability by a definition of another quantity (minimal adaptability) which extends the idea of relative ordering in the sense that it requires no measurement of cell death and which at the same time may quite reasonably be expected to provide an index of actual adaptability. This definition, however, requires a postulate, viz., the *culture procedure postulate* (really the fundamental postulate of microbiology since the work of Koch). To express this, first consider an environment with transition scheme $\bar{\omega}'$. Equation 8 can then be written, for this special case, as

$$H(\bar{\omega}) - H(\bar{\omega} \mid \bar{\omega}') + H(\bar{\omega}' \mid \bar{\omega}) = H(\bar{\omega}'), \qquad (9)$$

where $\bar{\omega}$ is the transition scheme of the biota in this environment. If the community is allowed to culture in an environment $\bar{\omega}'$ for a very long time (in practice, long relative to successional time scales) the left-hand side of Equation 9 will be called the minimal adaptability of the community. The culture procedure postulate is: *Minimal adaptability is an index of maximal adaptibility in the sense that a community with higher minimal adaptability will suffer less damage (as measured by biomass decrement) than a community with lower minimal adaptability when both communities are exposed to an environment more uncertain than either culturing environment.* (The condition should also be added that the two systems are initially identical and that the culturing environment is as nearly as identical as possible except for the difference in uncertainty.)

The culture procedure postulate is based on the idea that never used components of adaptability (larger than necessary behavioral repertories, better than necessary ability to anticipate, more than necessary indifference) always tend to atrophy in the succession or evolution process. Thus Equation 8 can be replaced by

$$H(\hat{\omega}) - H(\hat{\omega} \mid \hat{\omega}') + H(\hat{\omega}' \mid \hat{\omega}) \rightarrow H(\omega'), \tag{10}$$

where the arrow indicates a tendency for the adaptability to decrease in the direction of the actual uncertainty of the environment and thus to decrease in the direction of minimal adaptability (if $\omega' = \bar{\omega}'$). Thus, the arrow can be taken to mean that in the course of succession or evolution adaptability decreases in the sense that the amount of biomass decrement following on an increase in the uncertainty of the environment would be greater.

The culture procedure postulate is a local postulate since only systems which are in all respects similar except for the uncertainty of the culturing environment are being compared.

Structure of Adaptability

The components of adaptability may be further decomposed, using the structured state descriptions expressed in Equations 1 and 2. To this the transition scheme of the community is written in terms of the joint occurrence of all possible subcompartment transitions

$$\omega = \left\{ p \left[\bigcap_{i,j} \alpha_{ij}{}^{d}(t + 1) \middle| \bigcap_{i,j} \alpha_{ij}{}^{f}(t), \bigcap_{h} \beta_{h0}{}^{g}(t) \right] \right\}. \tag{11}$$

Defining local transition schemes for each compartment

$$\omega_{pq} = \left\{ p \left[\alpha_{pq}{}^{u}(t + 1) \middle| \bigcap_{i,j} \alpha_{ij}{}^{f}(t), \bigcap_{h} \beta_{h0}{}^{g}(t) \right] \right\}. \tag{12}$$

Equation 11 can be expressed as

$$\omega = \prod_{i,j} \omega_{ij}, \tag{13}$$

where care should be taken to realize that this is not a bona fide product, but rather shorthand for the composition of all (or some set) of local schemes. Similarly, the transition scheme of the environment can be expressed in terms of a "product" of transition schemes for local regions

$$\omega' = \left\{ p \left[\bigcap_{h} \beta_{h}{}^{e}(t + 1) \middle| \bigcap_{i,j} \alpha_{ij}{}^{f}(t), \bigcap_{h} \beta_{h0}{}^{g}(t) \right] \right\} = \prod_{h} \omega'_{h0}, \tag{14}$$

where

$$\omega'_{h0} = \left\{ p \left[\beta_{h0}{}^{v}(t + 1) \middle| \bigcap_{i,j} \alpha_{ij}{}^{f}(t), \bigcap_{r} \beta_{r0}(t) \right] \right\}. \tag{15}$$

Substituting the above compositions into Equation 10, we find

$$H\left(\prod_{i,j}\hat{\omega}_{ij}\right) - H\left(\prod_{i,j}\hat{\omega}_{ij}\middle|\prod_{h}\omega'_{h0}\right) + H\left(\prod_{h}\hat{\omega}'_{h0}\middle|\prod_{i,j}\hat{\omega}_{ij}\right)$$
$$\rightarrow H\left(\prod_{h}\omega'_{h0}\right). \quad (16)$$

Each of the terms in the above equation may be expressed (using identities analogous to Equation 7) in terms of sums of unconditional and conditional entropies. However, since there are many possible expansions of this type, it is convenient to choose a single canonical form which puts all the subcompartments on an equal footing. To do this, define the effective entropy, $H_e(\omega_{ij})$ as the sum of the unconditional and all possible conditional entropies of compartment i at level j, but normalized by normalizing the linear combination of all possible expansions (for an example, see Equations 19a,b,c). With this convention Equation 16 can be written

$$\sum_{i,j} H_e(\hat{\omega}_{ij}) - \sum_{i,j} H_e\left(\hat{\omega}_{ij}\middle|\prod_{h}\omega'_{h0}\right) + \sum_{h} H_e\left(\hat{\omega}'_{h0}\middle|\prod_{i,j}\hat{\omega}_{ij}\right) \rightarrow \sum_{h} H_e(\omega'_{h0}).$$
$$(17)$$

Each term in this equation can be identified with a possible mechanism of adaptability. Thus the unconditional parts of the effective entropies express the indeterminacy of the genetic endowment of an organism, phenotypic plasticity (e.g., morphological, physiological, behavioral), developmental plasticity (i.e., alternate expressibility of the genome), numerical and topographical plasticity of populations, and so forth. The conditional terms express the degree of independence among these different plasticities, the anticipation terms reflect the effectiveness with which they are used, and the indifference term expresses the niche structure (both spatial and physiochemical). The equation itself formally expresses a principle of compensation among adaptabilities, viz., that change in the adaptability of one subsystem tends to be compensated by opposite changes in the adaptability of other subsystems, at the same or different levels, or by opposite changes in the indifference to the environment. The costs and advantages of such compensations depend on factors such as the morphological organization of the system and the time scale of environmental disturbance. In this sense the spectrum of adaptabilities in nature can be regarded as an optimization problem.

Equation 17 could also be written in terms of the $\bar{\omega}_{ij}$, in which case the compensations are among minimum adaptabilities. In what follows the argument could be expressed either in terms of the $\hat{\omega}_{ij}$ or the $\bar{\omega}_{ij}$.

BOOTSTRAP PRINCIPLE OF ADAPTABILITY THEORY

The bootstrap principle of adaptability theory is: *The assumption of the hierarchical and compartmental structurability of the state description of biological systems is self-justifying within the framework of adaptability theory in the sense that such structurability is a necessary condition for efficient adaptability.* By efficiency of adaptability is here meant the relative cost of adaptability in terms of energy or in terms of interference with biological function (which ultimately appears in terms of an energy cost). Thus, suppose the morphological structure of an organism is extremely complex. Developmental and genetic plasticity would be relatively more costly in this case since genetic variations are more likely to be disruptive and alternative modes of development would require the accumulation and maintenance of much more genetic information. As another example, suppose that modifications of organism behavior are coordinated to morphological modifications involving growth processes. This clearly increases the energy that must be expended to maintain behavioral plasticity. More generally stated: *Any increase in the extent of modification required to achieve a given degree of adaptability increases the biological costs (ultimately measurable in terms of energy utilization) associated with that adaptability.* This generally is in fact implicit in the culture procedure postulate.

Given the above, the bootstrap principle is fairly easy to prove. To make things concrete, consider the simple, three compartment decomposition of a single organism (organism one at level one, with genome and phenome compartments one and two at level zero). For the behavioral uncertainty component of adaptability

$$H(\hat{\omega}_{10}\hat{\omega}_{20}\hat{\omega}_{11}) = H_e(\hat{\omega}_{10}) + H_e(\hat{\omega}_{20}) + H_e(\hat{\omega}_{11}), \tag{18}$$

where

$$H_e(\hat{\omega}_{10}) = \tfrac{1}{3}\{H(\hat{\omega}_{10}) + \tfrac{1}{2}H(\hat{\omega}_{10} \mid \hat{\omega}_{20}) + \tfrac{1}{2}H(\hat{\omega}_{10} \mid \hat{\omega}_{11}) + H(\hat{\omega}_{10} \mid \hat{\omega}_{20}\hat{\omega}_{11})\} \tag{19a}$$

$$H_e(\hat{\omega}_{20}) = \tfrac{1}{3}\{H(\hat{\omega}_{20}) + \tfrac{1}{2}H(\hat{\omega}_{20} \mid \hat{\omega}_{10}) + \tfrac{1}{2}H(\hat{\omega}_{20}\mid\hat{\omega}_{11}) + H(\hat{\omega}_{20} \mid \hat{\omega}_{10}\hat{\omega}_{11})\} \tag{19b}$$

$$H_e(\hat{\omega}_{11}) = \tfrac{1}{3}\{H(\hat{\omega}_{11}) + \tfrac{1}{2}H(\hat{\omega}_{11} \mid \hat{\omega}_{10}) + \tfrac{1}{2}H(\hat{\omega}_{11} \mid \hat{\omega}_{20}) + H(\hat{\omega}_{11} \mid \hat{\omega}_{10}\hat{\omega}_{20})\}. \tag{19c}$$

Each effective entropy consists of an unconditioned modifiability term (to be called the modifiability term) and a number of conditioned modifiability terms (to be called the independence terms). For given modifiabilities the effective entropies are largest when each of the conditioned terms are equal and equal to the modifiability term (since they cannot be larger). Thus, the conditions for a maximum uncertainty component of adaptability for given observable modifiabilities are

$$H(\hat{\omega}_{10}) = H(\hat{\omega}_{10} \mid \hat{\omega}_{20}) = H(\hat{\omega}_{10} \mid \hat{\omega}_{11}) = H(\hat{\omega}_{10} \mid \hat{\omega}_{10}\hat{\omega}_{11}), \tag{20a}$$

$$H(\hat{\omega}_{20}) = H(\hat{\omega}_{20} \mid \hat{\omega}_{10}) = H(\hat{\omega}_{20} \mid \hat{\omega}_{11}) = H(\hat{\omega}_{20} \mid \hat{\omega}_{10}\hat{\omega}_{11}), \tag{20b}$$

$$H(\hat{\omega}_{11}) = H(\hat{\omega}_{11} \mid \hat{\omega}_{10}) = H(\hat{\omega}_{11} \mid \hat{\omega}_{20}) = H(\hat{\omega}_{10} \mid \hat{\omega}_{10}\hat{\omega}_{20}), \tag{20c}$$

where $H(\hat{\omega}_{10})$, $H(\hat{\omega}_{20})$, and $H(\hat{\omega}_{11})$ are the observable modifiabilities. To the extent that these conditions are not realized, the observable modifiabilities of different compartments will be correlated (because the independence terms are relatively small) and therefore the extent of modification required to achieve a given degree of adaptability would increase. However, according to the above version of the culture procedure postulate, this increases the costs of adaptability. In short, a necessary condition for efficient adaptability is that the independence terms are not small relative to their modifiability term. To the extent that this is true, however, it is also true that compartmental (and level) structure will be a necessary condition for efficient adaptability. In this sense the initial assumption of hierarchical and compartmental structure is self-justifying from the standpoint of adaptability theory.

The above conclusion clearly generalizes for an arbitrary number of compartments and levels and therefore for the community as a whole. This does not mean that the condition of maximum independence of compartments is ever obtained. In general this is impossible and in many cases strong correlation is inherent in fundamental biological constraints. The most important example is the relation between genome and phenome. Specification of the phenome completely determines the genome, but specification of the genome only partially determines the phenome. This is because a given genome only specifies a repertoire of possible phenome states, but in general any genetic change (other than certain degenerate changes) modifies this repertoire. This constraint is clearly fundamental since it is the basis of evolution. However, to the extent that the effects of environmental disturbance can be prevented from ramifying to more and more compartments, the absorption or dissipation of these disturbances will be less costly.

DYNAMICAL AUTONOMY AND THE ROLE OF UNREPRESENTED ADAPTABILITIES

Degree of Autonomy

A compartment (or level) will be said to be autonomous to the extent that its behavior is independent of other compartments (or levels). From the standpoint of the transition scheme ω_{ij}, a compartment or level is autonomous to the extent that the transition probabilities need not be conditioned on other compartments or levels. From the standpoint of the scheme entropies, a compartment or level is independent to the extent that its independence terms are large relative to the modifiability term. The degree of autonomy can be expressed by the ratio

$$\xi_{ij} = H_e(\hat{\omega}_{ij})/H(\hat{\omega}_{ij}), \qquad (21)$$

where compartment (i, j) is completely autonomous if $\xi_{ij} = 1$ and not autonomous if $\xi_{ij} < 1$ (since the effective entropy is less than the unnormalized modifiability).

Autonomy and Predictability (for Discrete Time, Finite State Dynamical Systems)

If compartment (i, j) is completely autonomous ($\xi_{ij} = 1$), $\hat{\omega}_{ij}$ is as good a predictor of the behavior of this compartment as the global community scheme $\hat{\omega}$. In this case only properties (variables) of the compartment need enter into the predictor. If $\xi_{ij} < 1$, the "law will be broken," with the degree of breaking depending on the degree of autonomy (or degree of dependence). In this case variables from other compartments must be added to the predictor to restore its predictive power. (If the modifiability is high and the anticipation term low, reflecting good anticipation, a still better predictor could of course be constructed by including environmental variables.)

The predictor here has been written in the form of a transition scheme (cf.

Equations 4a & 13). The data provided by this setup are: a set of states of the community, a set of inputs (states of the environment), a set of states of the compartment in question (which may be the community or a subunit of the community), the probability distribution of initial states of the community and environment, and a rule (assumed for the sake of generality to be probabilistic) for determining the state of the compartment at the next instant of time. This is the same information that would be necessary to define the compartment as an automaton (since, according to Equation 1, specification of the state of the community automatically entails specification of the states of all its subcompartments). Thus the predictor can be interpreted as the transition function (or rule) governing the behavior of the compartment, viewed as an automaton (actually semiautomaton, or automaton without output). A necessary condition for the rule to be the best possible rule is that the compartment be completely autonomous. If the compartment is incompletely autonomous, a better rule can always be written, but at the expense of choosing as the automaton an enlarged system, including more compartments. If no such union of compartments is autonomous, a better rule can always be chosen, until finally the automaton is the complete community itself, with all its subcompartments. Thus, the above condition, to be called the *predictivity condition,* can be stated more generally: *A transition scheme incorporating variables associated only with a particular compartment (level) can only approach optimal predictivity to the extent that the compartment is autonomous.* (A level of organization of a compartment is a compartment if it is the union of all the subcompartments at that level and in this case autonomy would be defined for the level and enlargement could correspond to addition of levels.)

Autonomy provides a necessary but not sufficient condition for a good rule because the initial state of the entire community is specified, not just of the compartment itself. If the rule is not good in this case, it could never be good if only the initial state of the compartment were specified. Alternatively, it must be good if the rule for the compartment is to be good. It would be possible to redefine transition schemes solely in terms of compartments (using only the initial states of the compartments). This would make it possible to formulate a sufficient condition, but would also generate many more conditional terms, representing the uncertainty in the behavior of the environment and the rest of the community given the behavior of the compartment.

The Bath of Unrepresented Adaptabilities

In and of itself the predictivity condition is basically trivial—all it says is that if other components give information about the compartment of interest, it is possible to construct a more predictive but less local transition scheme. However, in conjunction with the compensation equation (Equation 17) it is highly nontrivial and leads directly to the following key statement: *As the predictive value of a transition scheme (transition function, rule, law) used to describe the behavior of any given compartment (level) of a global biological system depends less on the incorporation of variables associated with other compartments (levels) the adaptabilities of these other compartments (levels) becomes in general more important for main-*

taining this predictive value. This statement follows directly from the fact that predictivity means high independence, but high independence means that environmental disturbances affecting other compartments (levels) are absorbed and dissipated in those compartments or levels, with minimum ramification to the compartment (or level) of interest. There are four possibilities (assuming an uncertain environment):

(A) Independence high (i.e., comparable to modifiability)

(1) Low modifiability. In this case adaptabilities elsewhere in the system enable the compartment to support arbitrary but determinate and predictable dynamics despite environmental uncertainty, i.e., are unrepresented in these dynamics but are critical for supporting them.

(2) High modifiability. In this case the compartment of interest makes an efficient contribution to total adaptability, providing the anticipation term is relatively low (implying good anticipation). An optimal predictor can be constructed without reference to other compartments, but the best predictions are necessarily probablistic. If the transition scheme of the environment is added to the predictor it will be better, provided that the coordinated anticipation term is smaller than the modifiability, and can be deterministic to the extent that the anticipation term is small. If the adaptability of the compartment of interest is less than the total required adaptability (in general the case), adaptabilities unrepresented in its dynamics are critical for maintaining these dynamics.

(B) Independence low (smaller than the modifiability)

(3) Modifiability low. In this case adaptabilities elsewhere in the system absorb most of the environmental uncertainty, but nevertheless the modifications of these other compartments ramify to the compartment of interest, making its behavior less than optimally predictable without enlargement of the set of variables. Thus fewer unrepresented adaptabilities would play a role in supporting a predictable dynamics.

(4) Modifiability high. In this case the compartment of interest makes a less than maximally efficient contribution to total adaptability, providing the anticipation term is relatively low, but an optimal predictor cannot be constructed without enlargement of the set of variables, thereby implying that fewer unrepresented adaptabilities would play a role in supporting the (in general stochastic) dynamics.

As a somewhat contrived example, imagine a population undergoing phyletic evolution. In the case of low modifiability and high independence for the population level, all the uncertainties would be absorbed at the genetic and organismic levels. More realistically, these modifications would also appear in the population dynamics (the case of low modifiability and low independence). In the case of high modifiability and high independence, the lower level adaptabilities would not appear, but the population dynamics would itself contribute to the adaptability. A more realistic case is the one of high modifiability and low independence, in which case the genetic and organismic modifiabilities would manifest themselves at the population level, but the population dynamics would itself contribute to adaptability.

The conclusion is thus that any apparently autonomous compartment or level of biological organization is really highly controlled by a very much larger system of unviewed compartments or levels that are protecting it from environmental disturbance. It is important to recognize that this is not the same as saying that underlying the dynamics of, for example, a population, there are an enormous number of physiological and genetic processes, involving metabolism, repair, irritability, movement, and so forth. It is only in so far as these processes undergo either adaptive or maladaptive change that they can produce modifications in the dynamics of the population, and it is only in so far as any adaptive changes are unrealistically efficient that they can fail to produce modifications in these dynamics. In general the learning at one level can only rarely be completely hidden from the standpoint of the behavior of another.

The statements made in this section have been made for discrete time, finite state, probablistic dynamical systems. They are clearly also true for the deterministic case and could be generalized for continuous time and differentiable dynamical systems.

FUNCTIONAL SIGNIFICANCE OF STABILITY CONCEPTS

Role of the Represented Adaptabilities

Now it is possible to address the question, what is the relation between the various forms of stability or instability exhibited by biological systems and their adaptability. According to the previous section, as the behavior of any particular level of biological organization appears more autonomous, the contribution of unrepresented adaptabilities at other levels becomes in general more critical for maintaining this behavior. The key to using hierarchical adaptability theory to study the functional significance of stability and bifurcation is to look at this statement from the other side, viz., from the side of the represented adaptabilities. From this standpoint, as the behavior of any particular level of biological organization appears more autonomous, its potential contribution to maintaining the behavior patterns at other levels increases (though it can only make such a contribution if the modifiability term is large).

Recovery of Stability Concepts

Basic stability concepts (e.g., weak stability, asymptotic orbital stability, structural stability) are ordinarily used in biology within the context of differential models, whereas hierarchical adaptability theory has been formulated in a discrete time, finite state representation. Here the most convenient way to make cross-correlation possible is to define analog concepts for the representation used here, basing these on the notion of a tolerance (originally discussed in the context of biological models by Zeeman[9] and most prominently developed for automaton systems by M. Dal Cin[10]). This is a binary, symmetric, reflexive, but nontransitive relation on a set of states, to be denoted by " $\sim$ " and which roughly corresponds to the idea of a (here arbitrarily established) allowable difference between states. Analog definitions are:

(A) *Weak Stability.* A trajectory of a deterministic transition scheme is weakly stable if and only if it is possible to define a nontrivial tolerance such that for all n if $\alpha_{ij}^{s}(t) \sim \alpha_{ij}^{r}(t)$, then $\alpha_{ij}^{u}(t + n) \sim \alpha_{ij}^{v}(t + n)$, where α_{ij}^{u} and α_{ij}^{v} are the states at time $t + n$ and α_{ij}^{r} and α_{ij}^{s} are the states at time t. A scheme all of whose trajectories are weakly stable will be called weakly stable. Such schemes are analogous to differentiable dynamical systems with a constant of the motion (since a constant of the motion means that nearby states always remain nearby but never become identical).

(B) *Strong Stability.* A state α_{ij}^{r} of a (probabilistic or deterministic) transition scheme is strictly stable if and only if it is possible to define a tolerance such that the system returns to α_{ij}^{r} by time $t + n$ (n sufficiently large) if $\alpha_{ij}^{s}(t) \sim \alpha_{ij}^{r}(t)$, and furthermore if for all $k < n$ states at $t + n - k$ are either identical to or in tolerance with α_{ij}^{r}. Strong stability could also be defined for a trajectory if it is only required that by time $t + n$ the states are identical, but not necessarily identical to the initial state. For deterministic systems, strong stability is analogous to asymptotic orbital stability, where nearby states change to more and more nearby states (or where trajectories converge, as for stable limit cycles).

(C) *Structural Stability.* Consider the joint transition scheme $\omega_{pq}\omega_{rs} = \{p(\alpha_{pq}^{a}(t + 1), \alpha_{rs}^{b}(t + 1) \mid \alpha(t), \beta(t)\}$. The term α_{rs}^{b} will be called a parameter (and ω_{rs} a parametric scheme) if it remains the same for all t, while α_{pq}^{a} in general changes (i.e., is the variable). The joint scheme will be called structurally stable to a change from α_{rs}^{a} to α_{rs}^{c} if it is possible to define a tolerance on the states of compartment (p, q) such that $\alpha_{pq}^{u}(t + n) \sim \alpha_{pq}^{v}(t + n)$, where n is arbitrary and the α_{pq}^{u} are states of the compartment with parameter α_{pq}^{a} and the $\alpha_{pq}^{v}(t + n)$ are states of the compartment with parameter α_{pq}^{c}. If the scheme is not structurally stable to this particular change in parameter, the change will be called a bifurcation-producing change and the scheme will be said to bifurcate. The definition of structural stability is analogous to the dynamical notion of qualitative invariance to change in parameter and the notion of bifurcation to the notion of qualitative change in response to slight change in parameter. (If $\alpha_{pq}^{u}(t + n) = \alpha_{pq}^{v}(t + n)$, then rather than calling the joint scheme structurally stable it will be said that the two compartments do not interact.)

Biology of Stability, Instability, and Bifurcation

Stability and bifurcation play a fundamental role in relation to all the basic biological capabilities—metabolism, repair, growth, reproduction, irritability, and movement. Adaptability is also a fundamental biological capability, but in a higher order sense that it involves variation and modulation of these first order capabilities so as to absorb and dissipate disturbance. Here the concern is only with this connection, but in fact it is the decisive connection as regards the particular form stability and bifurcation take in relation to processes involving metabolism, repair, growth, and so forth. *A major conclusion is that instability and bifurcation make a fundamental contribution to the stability of biological systems, or at least to the stability of the most essential processes in such systems.*

To justify this statement, consider the fundamental biology of each of the basic notions of stability (from the standpoint of adaptability theory):

(A) *Weak Stability.* Weakly stable models are frequently criticized (as candidates for descriptions of biological systems) on the grounds that they are not structurally stable to a change in parameter that would represent the addition of even the smallest amount of dissipation. Nevertheless a compartment accurately describable by a weakly stable model would be capable of making a highly efficient contribution to adaptability and could under suitable conditions persist in a form so describable. The suitable condition is that the compartment be completely protected by a bath of adaptabilities provided by other compartments (possibly including its subcompartments or its supercompartments) from all disturbances that could introduce a source of dissipation into its dynamics. The efficiency of the contribution would result directly from this necessarily high autonomy together with the fact that slight perturbations (not introducing sources of dissipation) produce only slight modifications in the system and an enormous number of slight perturbations can therefore be absorbed without significant cost. The likelihood of the necessary conditions being met may not be compelling; nevertheless the above considerations should be taken into account if it appears that a system is accurately describable by a weakly stable model. (Examples of weakly stable models are the Lotka-Volterra equation, Kerner's[11] statistical mechanics of Lotka-Volterra systems, and Goodwin's[12] statistical mechanics of gene-protein interactions in cells.)

(B) *Strong Stability.* The argument generally made in favor of strongly stable models is that they are necessarily dissipative (since they forget perturbation), and therefore not necessarily structurally instable to a perturbation that introduces more dissipation. The fundamental equations of physics are conservative, therefore weakly stable, so that ultimately the dissipation of disturbance is equivalent to the absorption of disturbance in a (weakly stable) heat bath. This heat bath is thus the unrepresented (physical) adaptability that supports the strongly stable dynamics of the compartment in question. However, the particular form of these dynamics is also maintained by other, biological adaptabilities, which are necessary to prevent different types of dissipative perturbation from mixing in, and which play an increasingly important role insofar as the degree of autonomy increases.

The above considerations suggest how strongly stable dynamics (with asymptotically stable states or limit cycle behavior) might be significant from the standpoint of a system performing particular biological functions, how the strong stability contributes to adaptability by representing the absorption of disturbances in a heat bath, and how the appearance of some degree of dynamical autonomy (or predictability in terms of a restricted set of variables) can more or less be supported by unrepresented adaptabilities. Taking the other point of view, however, instabilities of a strongly stable system (involving transitions among multiple steady states or multiple limit cycles) potentially provide adaptabilities that may be unrepresented in the dynamical descriptions of other compartments, but which support these dynamics. The costs of such adaptabilities depends on the biological structures and processes required, the modifications required to move from one stable state (organization) to another, degree of independence of such modifications from modifications of other compartments, and the compatibility of the different states with the efficiency of the compartment and other compartments (which might be expected to increase as the "nearness" of the states increases).

(C) *Structural Stability*. The argument for structurally stable models is the same as above, except that the dynamical structures can be so chosen that more or less wide classes of perturbation can be mixed in also so chosen that the bifurcations (catastrophes) occur in a structurally stable way.[13,14] From the standpoint of adaptability there are three basic considerations, indicating both costs and advantages of structural stability and structural instability:

(1) Relation between independence and structurally stable behavior (stable or unstable). The states assumed by the varying compartment (i.e., the variable) are changed gradually as the states of parametric compartments (represented by the parameter) undergo changes. This means that modifications in different compartments are correlated, thereby apparently decreasing independence and increasing the cost of adaptability. Some caution is necessary, however, since the system is really more structurally stable if the "nearbyness" of the behavior of the variable compartment for different states of the parametric compartment increases, so that independence increases as structural stability increases provided the compartments interact. Indeed, another way of saying that the dynamics of a subcompartment of an interconnected system of compartments has high independence (or high degree of autonomy) is to say that these dynamics are structurally stable relative to changes in the other compartments (whose states may be lumped into parameters). To increase this independence (nearbyness of the behavior of a compartment despite possible change in the behavior of other compartments) the system must be built so that it is capable of supporting more modes of dissipation compatible with a given class of behaviors. This (along with structural constraint or internal decorrelation concomitant to weakened interaction) is the *cost of independence*. Furthermore, since independence has a basis in the construction of the system, it is itself an aspect of organization that must be protected from disturbance, implying that the whole hierarchy of adaptabilities contributes to the maintenance of the independence (i.e., structural stability or noninteraction) which increases their efficiency (self-consistency of the bootstrap principle).

(2) Contribution of structural stability of stable behavior. By structural stability of stable behavior is meant the structural stability away from the bifurcation points. Of two systems structurally stable to the same class of perturbations, the one which undergoes less quantitative change (therefore is more independent) is here called structurally stable. For the system less structurally stable (in this sense) adaptive or maladaptive changes in compartments (generally represented in terms of a single lumped parameter) causes any law written solely in terms of the variable of the compartment of original interest to be broken. However, the consequent quantitative changes in the compartment of interest potentially subserve the absorption of disturbance and therefore may also be adaptive, though in this case at a real increase in the cost of adaptability (because of the real decrease in

independence). This cost can be reduced if the parametric compartment is organized in such a way that the modifications are really inexpensive. It can then be thought of as a sensor or control element and in fact low independence in this case increases adaptability by increasing anticipation (i.e., decreasing terms such as $H(\hat{\omega}_{pq} \mid \hat{\omega}'\hat{\omega}_{rs})$, where $\hat{\omega}_{rs}$ is the transition scheme of the parametric compartment). It might be noted that if the parametric compartment were excluded from consideration, the quantitative variation of behavior of the compartment of interest would appear as a form of unstable behavior.

(3) Contribution of structural stability of unstable behavior. By structural stability of unstable behavior is meant structural stability of the bifurcation process. Insofar as bifurcation allows for different modes of behavior (e.g., different modes of development), it provides a mechanism for absorbing disturbance, in just the same sense that weak stability or multiple steady states potentially absorb disturbance, except that the unstable behavior would appear quite incomprehensible if an attempt were made to describe it solely in terms of variables of the compartment of interest. As with the case for stable behavior, a decrease in independence would in general mean correlated modifications and therefore an increase in the cost of adaptability; but if the parametric compartment is specialized as an inexpensively modifiable sensor or controller, better anticipation is possible and it is more likely that a mode of behavior appropriate to the environment will be assumed.

In sum, the modifiability terms cross-correlate to instabilities, associated with jumps to alternate weakly or strongly stable states (absorption of disturbance) or to stable behavior involving the direct return to an initially strongly stable state subsequent to disturbance (dissipation of disturbance). The independence terms cross-correlate with either structural stability or weakening of interaction. The translations, however, are strictly applicable only to discrete time, finite state dynamical systems, although it is not unreasonable to expect essential carryover to more general situations.

CONCLUSIONS

Adaptability theory has a number of applications which have been described elsewhere, e.g., to homeostasis,[3] patterns of adaptability in populations,[1] routability of matter and energy flow in communities,[15] and evolution of levels of organization.[1] The discussion in the present paper, which deals with the cross-correlation of the adaptability theory framework and dynamic biological models, also has evident implications for the construction and interpretation of dynamical models in biology.

A pertinent example is the long-standing issue as to the relation between complexity and stability of ecological systems. The elegant studies of May[16] have clearly established that the probability of stability decreases as complexity—number and interdependence among species—increases. The basic reason is roughly that as the dimensionality of a space gets larger it becomes less likely to find multidimensional valleys and if such a valley is found it is more likely to be one which a small perturbation will cause the system to escape. Yet as May[17] points out, complex communities in nature are not necessarily less stable than simple ones, and in fact climax communities are often quite complex. From the standpoint of adaptability theory the increase in complexity of the community level is quite compatible with the lowered stability, provided that adaptabilities unrepresented in the population dynamics are increased, either by increasing the modifiability of subcompartments (increasing genetic and phenotypic plasticities) or by increasing the independence of the various forms of modifiability. In this way the community earns the advantages of complexity (e.g., specialization of labor) but without incurring the disadvantage of instability at the population level, essentially because the bath of unrepresented adaptabilities causes the environment it sees to appear quieter. However, it should also be remembered (from the standpoint of constructing models) that instability of the population dynamics may also make a contribution to adaptability, particularly if species can be stored (in spore or seed form) and if organisms are not too expensive to construct (e.g., microorganisms). Indeed adaptability theory suggests that as the environment is made harsher and more uncertain (thus increasing the requirements for adaptability and at the same time making it more difficult to pay for this adaptability) there will at some point be a switch between the two types of adaptability patterns. In either case to the issue of the relation between stability and complexity must be added the consideration of adaptability.

References

1. CONRAD, M. 1972. Statistical and hierarchical aspects of biological organization. *In* Towards a Theoretical Biology. C. H. Waddington, Ed. Vol. **4:** 189–221. Edinburgh University Press. Edinburgh.
2. CONRAD, M. 1972. Can there be a theory of fitness? Intern. J. Neuroscience **3:** 125–134.
3. CONRAD, M. 1975. Analyzing ecosystem adaptability. Math. Biosci. **27:** 213–230.
4. CONRAD, M. 1976. Biological adaptability: the statistical state model. Bioscience **27:** 319–324.
5. CONRAD, M. 1976. Patterns of biological control in ecosystems. *In* Systems Analysis and Simulation in Ecology. B. C. Patten, Ed. Vol. **4:** 431–456. Academic Press. New York.
6. CONRAD, M. 1977. Functional significance of biological variability. Bull. Math. Biol. **39:** 139–156.
7. CONRAD, M. 1977. Biological adaptability and human ecology. *In* Proceedings of the First International Congress on Human Ecology. H. Knötig, Ed.: 467–473. Georgi Publishing Co. Switzerland.
8. KHINCHIN, A. I. 1957. Mathematical Foundations of Information Theory. Dover. New York.
9. ZEEMAN, E. C. & O. P. BUNEMAN. 1968. Tolerance spaces and the brain. *In* Towards a Theoretical Biology. C. H. Waddington, Ed. Vol. **1:** 140–151. Edinburgh University Press. Edinburgh.
10. DAL CIN, M. 1974. Modifiable automata with tolerance: a model of learning. *In* Physics and Mathematics of the Nervous System. M. Conrad, W. Güttinger & M. Dal Cin, Eds.: 442–458. Springer-Verlag. Heidelberg.
11. KERNER, E. 1957. A statistical mechanics of interacting biological species. Bull. Math. Biophys. **19:** 121–146.
12. GOODWIN, B. C. 1963. Temporal Organization in Cells. Academic Press. New York, N.Y.

13. THOM, R. 1970. Topological models in biology. *In* Towards a Theoretical Biology. C. H. Waddington, Ed. Vol. **1**: 95–120. Edinburgh University Press. Edinburgh.
14. GÜTTINGER, W. 1974. Catastrophe geometry in physics and biology. *In* Physics and Mathematics of the Nervous System. M. Conrad, W. Güttinger & M. Dal Cin, Eds.: 2–30. Springer-Verlag. Heidelberg.
15. CONRAD, M. 1972. Stability of foodwebs and its relation to species diversity. J. Theoret. Biol. **34**: 325–335.
16. MAY, R. M. 1975. Stability and Complexity in Model Ecosystems. 2nd edit. Princeton University Press. Princeton, N.J.
17. MAY, R. M. 1976. Patterns in multi-species communities. *In* Theoretical Ecology. R. M. May, Ed.: 136–152. Blackwell Scientific Publications. London.

ON BIFURCATION IN ECOSYSTEM MODELS*

Thomas G. Hallam

Departments of Mathematics and Ecology
University of Tennessee
Knoxville, Tennessee 37916

INTRODUCTION

Recent publications[1-4] have demonstrated a need for ecologists to be aware of the fundamental nature of bifurcation in modeling biological processes. The purpose of this paper is to indicate the importance of bifurcation phenomena in two additional areas of ecosystem analysis: persistence extinction and model linearization.

Ecosystem structure can be drastically affected by extinction of a trophic level component. Paine[5] and Connell[6] have documented the disruption and collapse of food webs resulting from extinction of a biological component. Ecosystem persistence-extinction mechanisms are complex; modeling such phenomena is, at best, a difficult task. We shall examine, in an elementary modeling framework, the structure of the persistence-extinction mechanism for simple food chains, that is, ones in which a single species composes a trophic level and its dynamics are governed by the levels immediately preceding and succeeding. In each of the simple food chain models, the persistence-extinction mode is a bifurcation; this holds independent of the mechanistic character of the model formulations. This property is not valid for general food webs since introduction of complexity into the trophic structure can change the form of the persistence-extinction mechanism.

The basic ecological concern of system response to perturbation and the concept of linearization[7] as related to bifurcation is discussed briefly in the third section of this article. Linear models have been widely employed in ecological settings and have been most useful in simulating climax communities. Mathematically, such conclusions are deemed natural as climax communities can be represented as equilibrium states about which classical perturbation theorems for ordinary differential equations conclude local validity of linearization. We conclude with some speculative interpretations of recent parametric perturbation results as they are applicable to model linearization. These indicate that in an appropriate setting, catastrophic behavior can ensue from perturbed linear models.

THE PERSISTENCE-EXTINCTION MECHANISM AS A BIFURCATION

Extinction of a species component of an ecosystem often results in "grave" consequences for the ecosystem as well as the unfortunate species. Extinction is a topical and prominent event. Levins[8] has suggested that 10^8 or 10^9 species extinctions have occurred in the 10^8–10^9 years since the onset of the Cambrian geological era. In some classical work on island biogeography, Simberloff and Wilson[9] found some population extinction times of a few days. In this section, the

*This work was partially supported by the Office of Naval Research.

0077-8923/78/0316-0482 $1.75/1 © 1979, NYAS

persistence-extinction mechanism in many long-termed simple food chain models is illustrated by several examples to be a bifurcation.

The continuous time models considered here are usually of the form

$$\frac{dx_i}{dt} = x_i f_i(x_1, x_2, \ldots, x_n), \quad i = 1, 2, \ldots, n \tag{1}$$

where each f_i is continuous from $R_+^n = \{x \in R^n : x = (x_1, x_2, \ldots, x_n), x_i \geq 0, i = 1, 2, \ldots, n\}$ to $R = (-\infty, \infty)$. System (1) is said to be *persistent* if each solution $\phi = \phi(t)$ of (1) satisfies $\lim \sup_{t \to \tau} \phi(t) > 0$ for all $\tau \in (0, T_\phi]$ where $[0, T_\phi)$ is the maximal interval of existence of ϕ. If (1) is not persistent, (1) has a species that goes *extinct*.

The main theme, persistence-extinction in simple food chains is bifurcation, is now illustrated with a sequence of examples. The developmental perspective is the assumption of a definitive model with different allowable ecologies. The persistence-extinction mechanism is regarded as the system behavioral response as the ecology transects from persistence to extinction.

Logistic Equation

The continuous time logistic equation

$$\frac{dx}{dt} = ax(b - x) \quad a > 0, b > 0$$

has been utilized as a model for many biological and social phenomena. The persistence-extinction parameter is $\rho = b$ with value $\rho = 0$ as the extinction threshold. The persistence-extinction mechanism may be viewed as a sequence of populations, each modeled by the logistic equation, each with successively smaller carrying capacities, and the sequential limit of the carrying capacities being equal to zero. As the ecology transects in this fashion from persistence to extinction, $\rho = b$ goes from a positive value to zero and the carrying capacity state $x = b$ approaches (in the limit, coincides with) the zero state. The persistence-extinction mechanism is bifurcation at persistence parameter value $\rho = 0$.

Lotka-Volterra Predator-Prey Model

The two dimensional Lotka-Volterra model

$$\frac{dx}{dt} = x(a - bx - cy),$$

$$\frac{dy}{dt} = y(-e + fx), \tag{2}$$

wherein a, b, c, e, and f are positive constants, has a single parameter $\rho = af - be$ which determines persistence. When $\rho > 0$, the model (2) predicts ecosystem persistence; otherwise, it is not persistent. The value $\rho = 0$ is the persistence threshold; again the persistence extinction mechanism is bifurcation at the threshold. As will occur in most of the subsequent examples, a positive value of the persistence parameter is equivalent to a nontrivial equilibrium; here, $\rho > 0$ guarantees that (2) has an equilibrium with both components positive.

Conservative Three-Level Food Chain

In this example we consider an aquatic food chain consisting of a nutrient (n), a phytoplankton population (p), and a zooplankton population (z).[10-12] The model formulation traces nutrient flow through the system and can be written in a nondimensional setting as

$$\frac{dp}{d\tau} = \frac{np}{\alpha + n} - \beta p - \epsilon z f(p),$$

$$\frac{dn}{d\tau} = \frac{-np}{\alpha + n} + \beta p + \delta z,$$

$$n(\tau) + p(\tau) + z(\tau) = 1. \tag{3}$$

In (3) the ecological parameters α, β, ϵ, δ are positive constants, $f(p)$ is a function representing zooplankton grazing, which satisfies $f \in C^1[(0, 1), (0, 1)]$, $0 < f'(p)$, $\lim_{p\to 0+} f(p)/p \neq 0$. The traditional grazing formulations: mass action, $f(p) = p$; Ivlev, $f(p) = 1 - e^{-\lambda p}$, $\lambda > 0$; and Michealis-Menten-Monod, $f(p) = p/(\gamma + p)$ are included in this representation class. It is known[11] that the model is persistent if and only if the parameters satisfy

$$\rho = 1 - \frac{\alpha\beta}{1 - \beta} - f^{-1}(\delta/\epsilon) > 0,$$

where f^{-1} denotes the inverse of f.

This inequality corresponds geometrically to an equilibrium point component ordering. There are, in general, three equilibriums of the form I $(1, 0, 0)$, II $(n_{II}, p_{II}, 0)$ and III $(n_{III}, p_{III}, z_{III})$. The inequality $\rho > 0$ is equivalent to $p_{II} > p_{III}$. The persistence-extinction mechanism is again a bifurcation at $\rho = 0$, with equilibrium III coalescing with II. Persistence is independent of the stability characteristics of the interior (nontrivial) equilibrium point III.

Nonconservative Three-Level Food Chain

Freedman and Waltman[13] consider the following model of a food chain wherein $x = x(t)$ represents the resource species density, $y = y(t)$ represents the density of the first predator and $z = z(t)$ represents the density of the top predator, all at time t:

$$\frac{dx}{dt} = xg(x) - yp(x);$$

$$\frac{dy}{dt} = y[-r + cp(x)] - zq(y);$$

$$\frac{dz}{dt} = z[-s + dq(y)]; \tag{4}$$

where r, s, c, d are positive constants; $g(x)$ satisfies $g(0) = \alpha > 0$, $dg(x)/dx \leq 0$ for $x \geq 0$, and there exists a $K > 0$ with $g(K) = 0$; $p(x)$ satisfies $p(0) = 0$, $dp(x)/dx > 0$ for $x \geq 0$; $q(y)$ satisfies $q(0) = 0$, $dq(y)/dy > 0$ for $y \geq 0$.

With these hypotheses, the points I $(0, 0, 0)$ and II $(K, 0, 0)$ are equilibriums. If r/c is in the range of p and $p(x_{III}) = r/c$ where $x_{III} < K = x_{II}$, then equilibrium III $(x_{III}, y_{III}, 0)$ exists with $y_{III} = x_{III} g(x_{III})/p(x_{III})$. The nontrivial (interior) equilibrium IV (x_{IV}, y_{IV}, z_{IV}) exists if s/d is in the range of $q(y)$ so $q(y_{IV}) = s/d$; y_{IV} in the range of the operator $xg(x)/p(x)$ yields x_{IV}. The last component of the equilibrium is given by

$$z_{IV} = \frac{y_{IV}[-r + c\,p(x_{IV})]}{q(y_{IV})}$$

with $z_{IV} > 0$ whenever $x_{IV} > x_{III}$.

Freedman and Waltman's persistence theorem is that a necessary condition for persistence is $\rho = -s + dq(y_{III}) \geq 0$ and a sufficient condition for persistence is $\rho > 0$. This statement deletes a phrase concerning limit cycles from Freedman and Waltman's wording of the result; in this model, a limit cycle in the xy-plane cannot be an attractor from the first octant.

Rewriting the persistence inequality $\rho > 0$ as $q(y_{III}) > s/d = q(y_{IV})$ and utilizing the monotone characteristic of q, we find that $y_{III} > y_{IV}$. The persistence-extinction mechanism is again one of bifurcation with $\rho = 0$ being the bifurcation value.

An n-Dimensional Lotka-Volterra Food Chain

Gard and Hallam[14] have obtained a classification of persistence in simple food chains as modeled by Lotka-Volterra dynamics. For such systems, the persistence-extinction mechanism is bifurcation; there is a single bifurcation parameter that depends upon the interspecific and intraspecific interaction coefficients as well as the length of the food chain. The simple food chain with dynamics modeled by Lotka-Volterra kinetics is

$$\frac{dx_1}{dt} = x_1(a_{10} - a_{11}x_1 - a_{12}x_2)$$

$$\frac{dx_2}{dt} = x_2(-a_{20} + a_{21}x_1 - a_{23}x_3)$$

$$\vdots \qquad \vdots$$

$$\frac{dx_{n-1}}{dt} = x_{n-1}(-a_{n-1,0} + a_{n-1,n-2}x_{n-2} - a_{n-1,n}x_n)$$

$$\frac{dx_n}{dt} = x_n(-a_{n0} + a_{n,n-1}x_{n-1}) \qquad (5)$$

with all parameters a_{ij} positive excepting a_{11}, which is nonnegative.

If the resource level in (5) has a positive carrying capacity ($a_{11} > 0$), then (5) predicts persistence whenever

$$\rho = a_{10} - \sum_{j=1}^{m} a_{2j,0} \prod_{i=1}^{j} \frac{a_{2i-2,2i-1}}{a_{2i,2i-1}} - \sum_{k=1}^{m} a_{2k+1,0} \prod_{i=1}^{k} \frac{a_{2i-1,2i}}{a_{2i+1,2i}} > 0;$$

the model (5) predicts nonpersistence if $\rho < 0$. Here m is the integer such that n is written as $2m + 1$ if n is odd or $2m$ if n is even.

The condition $\rho > 0$ is equivalent to an ordering of equilibrium coordinates. Let the coordinates of equilibriums associated with a given hierarchical trophic level K (a generic Roman numeral) be denoted by $x_j{}^K$; for example, I $(0, 0, \ldots, 0)$, II $(x_1{}^{II}, 0, 0, \ldots, 0)$, III $(x_1{}^{III}, x_2{}^{III}, 0, \ldots, 0)$, etc. The ordering equivalent to persistence is

$$0 < x_1{}^{II}, x_1{}^{II} > x_1{}^{III}, x_1{}^{IV} > x_1{}^{III}, x_1{}^{IV} > x_1{}^{V}, \ldots$$

or equivalently,

$$0 < x_2{}^{III}, x_2{}^{III} > x_2{}^{IV}, x_2{}^{IV} < x_2{}^{V}, x_2{}^{V} > x_2{}^{VI}, \ldots$$

or equivalently,

$$0 < x_3{}^{IV}, x_3{}^{IV} > x_3{}^{V}, x_3{}^{V} < x_3{}^{VI}, x_3{}^{VI} > x_3{}^{VII}, \ldots$$

or equivalently,

$$0 < x_4{}^{V}, x_4{}^{V} > x_4{}^{VI}, x_4{}^{VI} < x_4{}^{VII}, x_4{}^{VII} > x_4{}^{VIII}, \ldots$$

... or equivalently,

$$0 < x_{n-1}{}^{N}, x_{n-1}{}^{N} > x_{n-1}{}^{N+I}$$

or equivalently,

$$0 < x_n{}^{N+I}.$$

Here, N represents the Roman numeral corresponding to n.

This ordering mandates (once again!) that ecosystems need to be studied on the complete level; static field measurements reflect not only species location in the food chain but the length of the food chain as well. The persistence-extinction mechanism is again one of bifurcation at the threshold $\rho = 0$.

Analogous persistence results have been obtained for (5) in the absence of a resource carrying capacity ($a_{11} = 0$). The specific theorem will not be stated here but an interesting related situation in which persistence is not intricately related to equilibriums is worth mentioning. For (5), with $a_{11} = 0$ and n odd, persistence occurs even though there is no nontrivial positive equilibrium.

The next example demonstrates that complexity in a trophic level can change the character of the persistence-extinction mechanism.

Competition in Two Dimensions

The classical quadratic competition model is

$$\frac{dx}{dt} = x(a - bx - cy),$$

$$\frac{dy}{dt} = y(e - fx - gy),$$

where a, b, c, e, f, and g are positive parameters representing a fixed ecology. The persistence-extinction mechanism is determined by location of the equilibrium point IV $[(ag - ce)/(bg - cf), (be - fa)/(bg - cf)]$ with respect to the equilibrium density line $L: bx/a + gy/e = 1$. If IV is above L, persistence results; if IV is below L, extinction occurs. As the ecology transects from persistence to extinction, this system has four equilibriums transecting to the equilibrium density line segment then contracting to four equilibriums again. This mechanism is different than those above; in particular, the number of persistence parameters has increased as the complexity increased with $\rho = be - af$ and $\sigma = ag - ce$ determining persistence.

BIFURCATION AND LINEARIZATION

Linear models of many physical, mechanical, and electrical phenomena have been quite successful. In many of these situations, it is recognized through inherent difficulties in model validation that these linear models are idealizations of nonlinear systems. For ecological systems, Patten[7] has supported linearization as an evolutionary tendency of ecosystems. In this section we discuss some recent results on linearization and bifurcation as they relate to perturbation effects.

Basic hypotheses for the homogeneous linear equation include admissibility of the pair of Banach spaces (B_J, D_J), where (B_J, D_J) is *admissible* for $dy/dt = A(t)y(t \in J, y \in R^n)$ if for each $g \in B_J$, the nonhomogeneous linear equation $dw/dt = A(t)w + g(t)$ $(t \in J, w \in R^n)$ has a solution $w \in D_J$. Tacitly assuming the admissibility of (B_J, D_J) for the linear part of

$$\frac{dx}{dt} = A(t)x + f(t, x, \lambda) \quad t \in J, x \in R^n, \lambda \in \Lambda \subset R^m, \tag{6}$$

we discuss the asymptotic behavior of solutions $x(t, \lambda)$ for small λ. The character of the results differ according to choice of $J = R_+ = [0, \infty)$ or $J = R = (-\infty, \infty)$; on $J = R_+$ bifurcation conditions are trivial while on $J = R$ they play an essential role. Only the results on R are indicated here; the interested reader can refer to Hallam[15] for a complete development.

Denote by $M[R, R^n]$ the space of all equivalence classes of locally integrable functions f from R to R^n with the property that

$$\sup_{t \in R} \int_t^{t+1} |f(s)| \, ds < \infty.$$

$M[R, R^n]$ is a Banach space with respect to the norm

$$|f|_M = \sup_{t \in R} \int_t^{t+1} |f(s)| \, ds.$$

For illustration purposes we shall employ the hypothesis

(H) Let there exist positive constants K_j, α_j and supplementary projections $P_{jL\infty}$, $j = 0, \pm 1, \infty$ (i.e., $P_{jL\infty}{}^2 = P_{jL\infty}$ and $\sum_j P_{jL\infty} = I$) such that the fundamental matrix Y of the linear system associated with (6) satisfies the inequalities

$$|Y(t)P_{0L\infty}Y^{-1}(s)| \leq K_0 e^{-\alpha_0|t-s|}, t, s \in R,$$

$$| \, Y(t) P_{\infty L \infty} Y^{-1}(s) \, | \; \leq \begin{cases} K_\infty e^{-\alpha_0 (s-t)}, & 0 \leq t \leq s, \\[2mm] K_\infty e^{-\alpha_0 (t-s)}, & t \geq s \geq 0, \end{cases}$$

$$| \, Y(t) P_{-1 L \infty} Y^{-1}(s) \, | \; \leq \begin{cases} K_{-1} e^{-\alpha_{-1}(t-s)}, & t \geq s \geq 0, \\[2mm] K_{-1} e^{-\alpha_1 (t-s)}, & s \leq t \leq 0, \end{cases}$$

$$| \, Y(t) P_{1 L \infty} Y^{-1}(s) \, | \; \leq \begin{cases} K_1 e^{-\alpha_1 (t-s)}, & t \leq s \leq 0, \\[2mm] K_1 e^{-\alpha_1 (s-t)}, & 0 \leq t \leq s. \end{cases}$$

The hypothesis (II) is necessary and sufficient for each of the pairs of Banach spaces $(M[R_+, R^n], L^\infty[R_+, R^n])$ and $(M[R, R^n], L^\infty[R_-, R^n])$ to be admissible for $dy/dt = A(t) y$ on R_+ and R_-, respectively.[16] The decomposition of the solution space of $dy/dt = A(t) y$ as structured by (H) may be regarded as a partitioning of solutions into two exponential dichotomies. This leads to four sets which are composed of those solutions bounded on R_+ but unbounded on R_-, bounded on R_- but unbounded on R_+, bounded on R, and unbounded on both R_+ and R_-.

THEOREM. Suppose (H) is satisfied. Let $\Phi(\cdot)$ denote the $n \times r$ matrix valued function whose columns form a basis for span $\{Y(\cdot) P_{0L\infty}\}$. Let $\Psi(\cdot)$ denote the $k \times n$ matrix whose rows form a basis for span $\{P_{\infty L\infty} Y^{-1}(\cdot)\}$. Let $f:R \times R^n \times \Lambda \to R^n$ be continuous together with its second partial derivatives and suppose that $f(t, x, 0) = 0$ for (t, x) in $R \times R^n$. Define

$$H(a_0, \lambda) = \int_{-\infty}^{\infty} \Psi(t) \frac{\partial f}{\partial \lambda} (t, \Phi(t) a_0, 0) \lambda dt.$$

Suppose a continuous map $\alpha : \Gamma \to R^p$ (Γ denotes a compact set on the surface of the unit sphere in Λ) can be found such that $H(\alpha(\lambda), \lambda) = 0$ and rank $(\partial H/\partial a_0)(\alpha(\lambda), \lambda) = k$ for $\lambda \in \Gamma$. Then, there exists a positive constant $\sigma = \sigma(\Gamma)$ such that (6) has an $L^\infty(R)$-solution $x^* = x^*(\cdot, \lambda)$ that is uniformly bounded on $C = \{\lambda \in \Lambda : |\lambda| \leq \sigma, \lambda/|\lambda| \in \Gamma$ if $\lambda \neq 0\}$, continuous on $C - \{0\}$, and satisfies

$$x^*(\cdot, \lambda) = \Phi(\cdot)\alpha(\lambda/|\lambda|) + v(\cdot, \lambda),$$

where $\lim_{\lambda \to 0} v(\cdot, \lambda) = 0$.

The above result demonstrates that when the parameters in a perturbed linear model are vector valued, the solution need not be continuous as a function of that parameter. A system perturbation, applied to a linear system then completely relaxed, could result in an asymptotic return to a solution of the linear equation other than the solution mode present before perturbation.

Mechanisms, such as exhibited in the above theorem, might be model formulations of catastrophic biological events. Events, once regarded as a stable situation but in which significant change has occurred, such as the destruction by Crown-of-thorns starfish (*Acanthaster planci*), of coral reefs in the Pacific, or the emergence of the wood-boring isopod *Sphaeroma terebrans* as a dominant factor in the red mangrove (*Rhizophora mangle*) swamps and island ecosystems of south-western Florida could possibly be viewed in this setting.

Summary

After examining several examples containing many model formulations, it is concluded that the persistence-extinction mechanism in simple food chains seems to be a bifurcation. Within more complex trophic level structures, the mechanism is more complicated.

Possible ramifications, in present form quite speculative, of a recent theoretical result on bifurcation and parametric perturbations of linear systems are considered from the perspective of linearly modeling ecosystems in which catastrophic biological events have occurred.

References

1. Rosen, R. 1970. Dynamical System Theory in Biology. Vol. 1. Wiley and Co. New York, N. Y.
2. Gurel, O. 1973. Global Analysis of Ecological Systems. *In* Working Proceedings, NATO Conference: Mathematical Analysis of Decision Problems in Ecology, Istanbul, Turkey July 9–13. 293–302.
3. May, R. M. & G. F. Oster. 1976. Bifurcations and dynamic complexity in simple ecological models. Amer. Natur. 110: 537–599.
4. Hoppensteadt, F. & J. M. Hyman. 1977. Periodic solutions of a logistic difference equation. SIAM J. Appl. Math. 32: 78–81.
5. Paine, R. T. 1966. Food web complexity and species diversity. Amer. Natur. 100: 65–75.
6. Connell, J. H. 1975. Some mechanisms producing structure in natural communities. *In* Ecology and Evolution of Communities. M. L. Cody & J. M. Diamond, Eds.: 460–490.
7. Patten, B. C. 1975. Ecosystem linearization: An evolutionary design problem. *In* Ecosystem Analysis and Prediction. S. A. Levin, Ed.: 182–197.
8. Levins, R. 1970. Extinction. *In* Some Mathematical Questions in Biology. A.M.S. 2: 75–107.
9. Simberloff, D. S. & E. O. Wilson. 1969. Experimental zoogeography of islands. I, II. Ecology. 50: 278–314.
10. Canale, R. P. 1969. Predator-prey relationships in a model for the activated process. Biotech. Bioeng. 11: 887–907.
11. Hallam, T. G. 1978. Structural sensitivity of grazing formulation in nutrient controlled plankton models, J. Math. Biology. 5: 269–280.
12. Quinlan, A. V. & H. M. Paynter. 1976. Some simple nonlinear dynamic models of interacting element cycles in aquatic ecosystems. Trans. ASME J. Dyn. Syst. Meas. Control. 98: 6–19.
13. Freedman, H. I. & P. Waltman. 1977. Mathematical analysis of some three species food-chain models. Math. BioSci. 33: 257–276.
14. Gard, T. C. & T. G. Hallam. 1977. Persistence in food webs. I: Lotka-Volterra food chains. To appear.
15. Hallam, T. G. 1976. Parametric perturbation problems in ordinary differential equations. Trans. Amer. Math. Soc. 224: 43–59.
16. Coppel, W. A. 1971. Dichotomies and stability theory. Lecture Notes in Math. 206: 160–162.

SOME DIFFUSIVE PREY AND PREDATOR SYSTEMS AND THEIR BIFURCATION PROBLEMS

Masayasu Mimura

Department of Applied Mathematics
Konan University
Kobe, Japan

Yasumasa Nishiura and Masaya Yamaguti

Department of Mathematics
Kyoto University
Kyoto, Japan

INTRODUCTION

One of the classical problems in mathematical ecology is that of two interacting populations. In the prey-predator situation, a general spatial homogeneous model is described by

$$N_t = h(N, P)N$$
$$P_t = k(N, P)P, \tag{1}$$

where N and P represent the population densities of a prey species and its predator. Functional responses h and k were studied in many ecological literatures.[1] For a simple explicit form, we have

$$h(N, P) = a - \alpha P \quad \text{and} \quad k(N, P) = -b + \beta N,$$

which implies the well-known Lotka-Volterra system. Another form is

$$h(N, P) = r\left(1 - \frac{N}{K}\right) - \frac{RP}{N + D} \quad \text{and} \quad k(N, P) = -s\left(1 - \frac{\gamma P}{N}\right).$$

The ecological interpretation lying behind these forms and their mathematical studies are stated in Reference 1. On the other hand, reasonable modelings of migrational effect for a population have been studied by many authors. Among them, Skellam,[2] Kerner,[3] and Montroll and West[4] discussed a population that both grows and diffuses. Following their theories, we combine the diffusion process with the system (1). The resulting system takes the form

$$N_t = d_1 N_{xx} + h(N, P)N$$
$$P_t = d_2 P_{xx} + k(N, P)P, \tag{2}$$

where d_1 and d_2 are both nonnegative constants.

Our concern here is to discuss the parametric dependency of (d_1, d_2) on spatial structures of $(N(t, x), P(t, x))$ of (2) subject to zero flux boundary conditions. The boundary condition can be interpreted as an assumption that there is no migration across the boundaries (for example, a lake, an island, etc.). This study is motivated by the papers of Cassie[5] and Steele[6] who reported that plankton

490

0077-8923/78/0316-0490 $1.75/1 © 1979, NYAS

displayed spatial heterogeneity in a homogeneous environment in spite of the diffusion processes. Such a phenomenon is called *patchiness* in ecology. Segal and Jackson[7] showed that the combination of prey autocatalysis (which is called Allee effect) with high dispersal ability of predator leads to the destabilization of uniform spatial patterns and the resultant inhomogeneous patterns appear. Their theory is based on Turing's idea of diffusion driven instability.[8] Steele[6] and Okubo[9] also suggested the development of patchiness when the homogeneous steady state is unstable with diffusion. Such an idea has been pursued by various authors. In chemistry, biology, and other fields, they proposed interesting models which are described by the class of (2).[10-12] Their mathematical study has been performed considerably. From an ecological point of view, Segel and Levin[13] discussed (2) in a particular case when $h(N, P) = a + bN - \alpha P$ and $k(N, P) = -c - dP + \beta N$ and obtained small amplitude spatial patterns by the use of multiple time scale perturbation analysis. Meanwhile, in the case of $h(N, P) = h_0(N) - p$ and $k(N, P) = -k_0(P) + N$, Mimura[14] showed that the solution tends to be homogeneous asymptotically when $h_0'(N) \leq 0$ and $k_0'(P) \geq 0$ for $N \geq 0$ and $P \geq 0$. Therefore he concluded that the existence of patchiness necessitates the introduction of the Allee effect in the nonlinear term $h_0(N)$ from both mathematical and ecological points of view.

This paper is intended to discuss spatial structures of the solution of (2) within the framework of the papers mentioned above.[5-14] Our results are that, roughly speaking, patchiness is due to the "hump" effect in the nonlinear term $h(N, P)$. In the next section we show the existence of small-amplitude steady-state solutions using the Lyapunov-Schmidt bifurcation analysis with two parameters d_1 and d_2, and discuss their stability. Following that, we shall explore the spatial structures of the steady-state solutions when d_1 is zero or sufficiently small, which can arise in certain plant-herbivore systems. In the last section, we present some numerical results and discuss their relevance to the results for heterogeneous steady-state solutions with small d_1.

FORMULATION OF THE PROBLEM

For simplicity, we write (2) as the vector form

$$U_t = DU_{xx} + F(U), \quad (t, x) \in (0, +\infty) \times I = (0, l), \tag{3a}$$

where $U = {}^t(u, v)$, D is a diagonal matrix whose elements are nonnegative constants d_1 and d_2, and $F(U)$ can be described by a more explicit form than the nonlinear term in (2) as:

$${}^t(f_0(u, v)\{f(u) - v\}u, -g_0(u, v)\{g(v) - u\}v).$$

Of course it concludes the realistic examples in the previous section. We assume that the functions f_0, f, g_0, and g are appropriately smooth in the quadrant $(u, v) \geq 0$. We also assume that U satisfies zero flux boundary conditions

$$U_x(t, 0) = U_x(t, l) = 0 \tag{3b}$$

and the initial condition

$$U(0, x) = U_0(x), \quad x \in \bar{I}. \tag{3c}$$

Remark 1. In planktonic interaction models, $0 < d_1 \leq d_2$ is observed from the characteristics of phytoplankton (prey) and zooplankton (predator).[9] For plant-herbivore systems, $d_1 = 0$ or is sufficiently small and $d_2 > 0$ are reasonable assumptions.

Remark 2. If $u_0(x) = 0$ or $v_0(x) = 0$ for all $x \in I$, then (3) reduces to a single diffusion equation. In this case it is shown that nonnegative heterogeneous steady state solutions are unstable even when they exist. Consequently we assume that $u_0(x)$ and $v_0(x)$ are not identically zero.

In most realistic models, f_0, f, g_0, and g satisfy the following conditions:

(H1) $f_0(u, v) > 0$ and $g_0(u, v) > 0$ for $(u, v) \geq 0$.
(H2) There is a constant $c_1 > 0$ such that $f(u) < 0$ for $u > c_1$.
(H3) There is a constant $c_2 \geq 0$ such that $g(v) > c_2$ for $v \geq 0$.
(H4) There exists at least one positive constant solution $\bar{U} = {}'(\bar{u}, \bar{v})$ satisfying $f(\bar{u}) = \bar{v}$ and $g(\bar{v}) = \bar{u}$.

Remark 3. In many ecological models, the $f(u)$ and $g(v)$ curves are represented by two types as shown in FIGURES 1 and 2, respectively. The Type (2) response in FIGURE 1 is usually called the "hump" effect.

Our attention is restricted to a particular case of (H2) and (H3), as follows:

(H5) $f(u)$ satisfies Type (2) and $g(v)$ is a strictly monotone increasing function such that $g(v) = c_2 + c_3 v^m$ for some positive constants c_3 and m, which is illustrated in Type (2) in FIGURE 2.

Other cases were partially discussed in Reference 14. It is worth noting that the diffusive Lotka-Volterra system never exhibits patchiness asymptotically for large time (see, for example, Murray[15]). This result leads to assumption (H5).

Remark 4. It is known that the assumptions (H1) and (H5) yield uniform boundedness of the solution $U(t, x)$ of the problem (3). Therefore the global existence theorem can be proved.[14]

The assumptions (H4) and (H5) imply that the intersecting point $(\bar{u}, \bar{v})$ lies in the right hand side or in the left hand side of the summit of $v = f(u)$. Concerning

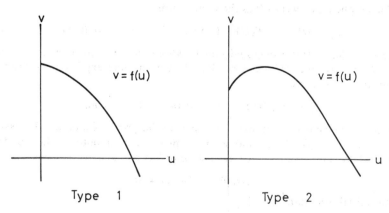

FIGURE 1.

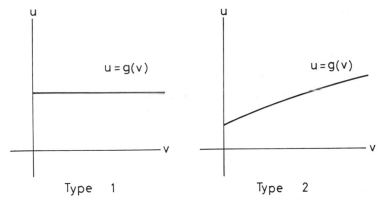

FIGURE 2.

the former case, we can prove nonlinear stability of $(\bar{u}, \bar{v})$. This result implies that small perturbations with respect to $(\bar{u}, \bar{v})$ do not give rise to patchiness. Hence, in this paper, we are only interested in the latter case.

HETEROGENEOUS STEADY-STATE SOLUTIONS I

This section demonstrates the stable heterogeneous steady-state solutions of (3). We employ here the Lyapunov-Schmidt method developed by Sattinger,[16] Sather,[17] and Crandall-Rabinowitz.[18] A good deal of work has been done on particular cases of (3) using the technique of Poincaré-Lindstedt, to product explicit heterogeneous bifurcating solutions. Since these studies did not refer to the proof of the existence of the one-parameter family of heterogeneous bifurcating solutions, we show it in this section and appendices. Moreover, we present a stability criterion that is useful in applications.

It is more convenient to introduce a perturbation vector $V(t, x) = U(t, x) - \bar{U}$. The resulting problem for V is

$$V_t = DV_{xx} + G(V), \quad (t, x) \in (0, +\infty) \times I \tag{4a}$$

$$V_x(t, 0) = V_x(t, l) = 0, \tag{4b}$$

$$V(0, x) = U_0(x) - \bar{U} = V_0(x), \quad x \in \bar{I}, \tag{4c}$$

where $G(V) = F(V + \bar{U})$. Using the Taylor expansion, we may write $G(V)$ as

$$G(V) = BV + H(V),$$

where $B = \{b_{ij}\}$ is the Jacobi matrix of G at $V = 0$ and $H(V)$ is the smooth nonlinear term such that $H(0) = H_V(0) = 0$. From the above discussion, our concern is restricted to the case when B satisfies

$$\det B > 0, \text{ and } b_{11} > 0, b_{12} < 0, b_{21} > 0 \text{ and } b_{22} < 0.$$

Here we assume that

(H6) $\operatorname{tr} B < 0$,

which implies that $V = 0$ is stable to small homogeneous perturbations.
    The stationary problem for (4) is as follows:

$$DV_{xx}^{s} + BV^{s} + H(V^{s}) = 0, \quad x \in I \tag{5a}$$

$$V_{x}^{s}(0) = V_{x}^{s}(l) = 0. \tag{5b}$$

We now consider the bifurcation problem from the trivial branch $V^{s} = 0$ as D varies in the quadrant $R_{+}^{2} = \{(d_1, d_2) \mid d_i > 0, i = 1, 2\}$. Hereafter, we use the notation D for the diagonal matrix whose elements are $\{d_i\}$ or a point (d_1, d_2) in R_{+}^{2}, as the case may be.
    We begin by studying the linear eigenvalue problem to find possible bifurcation points of D in R_{+}^{2}:

$$D\Psi_{xx} + B\Psi = \lambda\Psi, \quad x \in I, \tag{6a}$$

$$\Psi_{x}(0) = \Psi_{x}(l) = 0. \tag{6b}$$

The Fourier series expansion

$$\Psi = \sum_{n=0}^{\infty} \Psi_{n} \cos \frac{n\pi x}{l}$$

implies that the problem (6) is equivalent to the following system:

$$(B_n - \lambda)\Psi_n = 0, \quad n = 0, 1, 2, \ldots, \tag{7}$$

where $B_n = B - \gamma n^2 D$ for $\gamma = (\pi/l)^2$. The characteristic equation of (7) is as follows:

$$\lambda^2 - \{(b_{11} + b_{22}) - (d_1 + d_2)\gamma n^2\}\lambda$$
$$+ (b_{11} - d_1\gamma n^2)(b_{22} - d_2\gamma n^2) - b_{12}b_{21} = 0. \tag{8}$$

It is found that (8) has a zero root if and only if D satisfies

$$(b_{11} - d_1\gamma n^2)(b_{22} - d_2\gamma n^2) - b_{12}b_{21} = 0. \tag{9}$$

Noting that (6) has a zero eigenvalue if and only if D satisfies (9) for some n (≥ 1), the set of bifurcation points with respect to D is represented by the following hyperbolic curves $\{C_n\}$ in R_{+}^{2}:

$$C_n: d_2 = \frac{\dfrac{b_{12}b_{21}}{(\gamma n^2)^2}}{d_1 - \dfrac{b_{11}}{\gamma n^2}} + \frac{b_{22}}{\gamma n^2}, \quad n = 1, 2, \ldots.$$

    Remark 5. It is seen that all curves C_n are tangent to the line $d_2 = sd_1$ for $n = 1, 2, \ldots,$ where s is given by

$$s = \frac{(\det B - b_{12}b_{21}) + 2\sqrt{-b_{12}b_{21}\det B}}{b_{11}^{2}}.$$

We call the domain Ω_s *the stable region* with respect to the trivial solution $V^s = 0$, where Ω_s is the set $\{D = (d_1, d_2) \mid D \in R_+^2$ and $\text{Re}(\lambda) < 0$ for all $n = 1, 2, \ldots\}$, where λ is a root of (8) (FIGURE 3). We call the boundary of the stable region and its upper side *the bifurcation curve* Γ and *the unstable region,* respectively. Here Γ is represented by

$$\Gamma = \bigcup_{n=1}^{\infty} \hat{C}_n, \text{ where } \hat{C}_n = \{(d_1, d_2) \in C_n \mid P_n \leq d_1 < P_{n-1}\},$$

where $P_0 = b_{11}/\gamma$ and $P_n (n \geq 1)$ is an abscissa of the intersecting point of C_n and C_{n+1}. We note that $\hat{C}_n \neq \phi$ and $\hat{C}_n \cap \hat{C}_m = \phi$ when $n \neq m$.

 Remark 6. If d_1 and d_2 are both large enough, it is seen that the solution of (3) tends to be homogeneous asymptotically (see, for example, Conway-Hoff-Smoller[19]). Consequently, as the sufficient condition that patchiness appears, we have that either d_1 or d_2 is not large. FIGURE 3 helps us to understand this result.

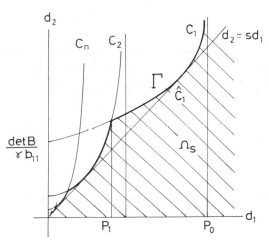

FIGURE 3.

 Our bifurcation problem is considered when D varies along any fixed path $D = D(\sigma)$ starting from the stable region and going into the unstable region. Here $D = D(\sigma)$ is defined as follows:

(H7a) $D: I_0 \to R_+^2$ is a smooth mapping, where I_0 is an open interval in R^1 which contains 0, and $D(0) = (d_1^0, d_2^0)$.

(H7b) $D(0)$ lies on some $\hat{C}_n$ and D intersects transversally with $\hat{C}_n$ at $D = D(0)$, i.e., $D'(0)$ is not parallel to the tangent of $\hat{C}_n$ at $D(0)$.

(H7c) $D(0)$ is not an intersection point of two curves of $\{C_n\}$, i.e., the zero eigenvalue of (6) with $D = D(0)$ is simple.

Thus our problem will be discussed within the framework of "bifurcation theory at simple eigenvalues".

 Remark 7. If (H7c) does not hold, i.e., the zero eigenvalue is double, we can discuss the bifurcation problem by using the techniques of Graves[20] and McLeod-Sattinger.[21] The details will appear in a forthcoming paper.

Now we recast the system (5) in the form of an operator equation in $E = (L^2(I))^2$, which has a real parameter σ,

$$L(\sigma)V + H(V) = 0, \quad \sigma \in I_0, \tag{10}$$

where

$$L(\sigma) = D(\sigma)\frac{d^2}{dx^2} + B$$

and $D(\sigma)$ satisfies (H7). The domain of definition of $L(\sigma)$ is defined by

$$D(L(\sigma)) = D(L_0) = (H_N^2(I))^2,$$

where $L_0 = L(0)$ and $H_N^2(I) =$ closure of

$$\left\{\cos\frac{n\pi x}{l}\right\}_{n=0}^{\infty} \quad \text{in } H^2(I).$$

$H(V) \colon D(L_0) \cap \hat{O} \to E$ is a smooth nonlinear operator with $H(0) = H_V(0) = 0$, where $\hat{O}$ is an open set in E with $0 \in \hat{O}$. (Here $D(L_0)$ is topologized by the graph norm of L_0.) We denote the norm of E, and the inner product by $\|\cdot\|$ and $(,)$, respectively. Thus we have the following theorems.

THEOREM 1 (Existence). There exists some constant $\epsilon_0 > 0$ such that (10) has a unique one parameter family of solutions $(\sigma(\epsilon), V(\epsilon)) \in R^1 \times \{(H_N^2(I))^2 \cap \hat{O}\}$ for $|\epsilon| < \epsilon_0$. Here $\sigma(\epsilon)$ is a smooth function of with $\sigma(0) = 0$ and

$$V(\epsilon) = \epsilon\Phi_n + \hat{V}(\sigma(\epsilon), \epsilon),$$

where Φ_n is the normalized eigenvector corresponding to the zero eigenvalue of

$$L_0\Psi^0 = \lambda\Psi^0, \quad x \in I$$

and

$$\Psi_x^0(0) = \Psi_x^0(l) = 0,$$

and $\hat{V}(\sigma(\epsilon), \epsilon)$ satisfies the estimate

$$\|\|\hat{V}(\sigma(\epsilon), \epsilon)\|\| \leq C|\epsilon|(|\sigma(\epsilon)| + |\epsilon|)$$

for some positive constant C independent of σ and ϵ, where $\|\|\cdot\|\|$ denotes the graph norm of L_0, i.e.,

$$\|\|V\|\| = \|L_0V\| + \|V\| \quad \text{for} \quad V \in D(L_0).$$

We expand the nonlinear term $H(V)$ as $H(V) = H_Q(V, V) + H_C(V) + H_R(V)$, where H_Q, H_C and H_R denote the quadratic part, the cubic part and the remainder, respectively.

THEOREM 2 (Bifurcation Diagram). The relation between σ and ϵ, that is, *the bifurcation diagram,* is determined by the following scalar equation:

$$\alpha\sigma + \beta\epsilon^2 + \eta(\sigma, \epsilon) = 0 \quad \text{(bifurcation equation)}.$$

Here α is some positive constant, β is given by

$$\beta = (H_C(\Phi_n) - H_Q(\Phi_n, KQH_Q(\Phi_n, \Phi_n)) - H_Q(KQH_Q(\Phi_n, \Phi_n), \Phi_n), \Phi_n^*)$$

and $\eta(\sigma, \epsilon)$ is a smooth function with respect to σ and ϵ, satisfying $|\eta(\sigma, \epsilon)| = O(|\sigma|^2 + |\epsilon|^2)$ with $|\eta(0, \epsilon)| = O(|\epsilon|^3)$. (For the definitions of K, Q and Φ_n^*, see Lemmas A1 and A2.) Moreover, if $\beta \neq 0$, the bifurcation is one-sided (FIGURE 4).

Remark 8. If the boundary condition is of Dirichlet type, the diagram is different from the one above.[22]

The form of $\sigma = \sigma(\epsilon)$ determines the structure of the spectrum of the linearized operator of (10) at $V(\epsilon)$ as follows:

THEOREM 3 (Linearized Stability). If $\beta < 0$, then the bifurcating solutions $V(\epsilon)$ are stable in the linearized sense, that is, all the eigenvalues of the problem

$$L(\sigma(\epsilon))\,\hat{\Psi} + H_V(V(\epsilon))\,\hat{\Psi} = \lambda\hat{\Psi}, \quad x \in I$$

and

$$\hat{\Psi}_x(0) = \hat{\Psi}_x(l) = 0$$

have negative real parts. If $\beta > 0$, then they are unstable in the linearized sense.

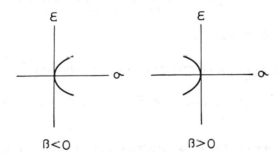

FIGURE 4.

In order to show nonlinear stability, we consider the evolution system of equations of (4) for $W = V - V(\epsilon)$,

$$W_t = D(\sigma(\epsilon))\,W_{xx} + BW + H_V(V(\epsilon))\,W + R(W; V(\epsilon)), \qquad \textbf{(11a)}$$

$$W_x(t, 0) = W_x(t, l) = 0 \qquad \textbf{(11b)}$$

$$W(0, x) = W_0, \qquad \textbf{(11c)}$$

where $R(W; V(\epsilon)) = H(V(\epsilon) + W) - H(V(\epsilon)) - H_V(V(\epsilon))\,W$.

Let us define the operators A, M and $\tilde{A}$ by

$$A = -D(\sigma(\epsilon))\,\frac{d^2}{dx^2} + \hat{I},$$

$$M = -B - \hat{I} - H_V(V(\epsilon)),$$

$$\tilde{A} = A + M,$$

respectively, where $D(A) = (H_N{}^2(I))^2$ and $\hat{I}$ denotes the identity operator. Using the results of Kielhöfer[23] and Theorem 3, we can prove the following:

THEOREM 4. (Nonlinear Stability). If $\beta < 0$, the bifurcating solutions $V(\epsilon)$ in Theorem 1 are stable in the sense that the solution $W = 0$ of (11) is asymptotically stable in the topology of $D(A^\alpha)$, where α is a constant such that $\frac{1}{2} \leq \alpha < 1$. More precisely, for any $\kappa > 0$ there exists a constant $\delta(\kappa) > 0$ such that if $\| A^\alpha W_0 \| < \delta(\kappa)$, (11) has a global strict solution $W(t, x)$,[23] and we have the estimate

$$\| A^\alpha W(t) \| \leq \kappa \exp(-bt), \quad t \in [0, +\infty).$$

The value b is determined by the spectrum of $\tilde{A}$, namely $0 < b < \mathrm{Re}\,\sigma(\tilde{A})$. If $\beta > 0$, the bifurcating solutions are unstable in the topology of E.

Remark 9. Convergence in the topology of $D(A^\alpha)$ implies uniform convergence for $\alpha \geq \frac{1}{2}$.

From the above results, we see that the important thing is to know the sign of β. In general, however, this requires fairly intricate calculations. For this reason, we present here a useful criterion which only requires simple calculations. Before giving it, we introduce some notation:

$$\Gamma_k = \left\{ D = (d_1, d_2) \in \Gamma \;\; 0 < k \leq \frac{d_1}{d_2} \leq \frac{1}{s} \right\} \text{ for a constant } k \text{ with } 0 < k \leq \frac{1}{s}.$$

$$H_Q(V, V) = {}^t({}^tVQ_1V, {}^tVQ_2V),$$

where Q_i is a real symmetric matrix with components $\begin{pmatrix} a_i & b_i \\ b_i & c_i \end{pmatrix}$ for $i = 1, 2$.

$$H_C(V) = {}^t(e_1u^3 + f_1u^2v + g_1uv^2 + h_1v^3, e_2u^3 + f_2u^2v + g_2uv^2 + h_2v^3)$$

$$\text{for} \quad V = {}^t(u, v).$$

Remark 10. We note that for an arbitrary k with $0 < k \leq 1/s$, there exists N such that $\Gamma_k \cap \hat{C}_n \neq \phi$ for $n \geq N$.

THEOREM 5 (Criterion of Stability). Since β depends on n, d_1, and d_2, we write $\beta = \beta(n, D)$.

(1) (High Frequency Modes) If $D \in \Gamma_k$ for some k with $0 < k \leq 1/s$ and

$$3e_1 - \frac{4}{\det B} \{a_1(a_1b_{22} - a_2b_{12}) + b_1(a_2b_{11} - a_1b_{21})\} < 0(\text{resp.} > 0),$$

then it follows $\beta(n, D) < 0$ (resp. > 0) for sufficiently large n.

(2) (Fundamental Mode) If

$$9b_{11}e_1 + 2a_1{}^2 - \frac{16b_{11}}{\det B} \{a_1(a_1b_{22} - a_2b_{12}) + b_1(a_2b_{11} - a_1b_{21})\} < 0(\text{resp.} > 0),$$

then $\beta(1, D) < 0$ (resp. > 0) for sufficiently large d_2.

We indicate here briefly two model examples to which Theorem 5 can apply.

The Simplified Prey-Predator Model. This model was suggested by Segal and Jackson[7];

$$N_t = d_1 N_{xx} + (1 + aN)N - bNP,$$

$$P_t = d_2 P_{xx} - c(P^2 - NP),$$

where a, b, and c are constants. When $0 < a < c$ and $a < b$, the constant steady-state solution $\bar{N} = \bar{P} = 1/(b - a)$ is stable to small homogeneous perturbations. Using the criterion (1) of Theorem 5, we find that the bifurcating solutions with high frequency modes are stable when $0 < a < c$ and $2a < b$.

The Model of Morphogenesis. Gierer and Meinhardt[11] proposed the following system:

$$a_t = d_1 a_{xx} + \rho \rho_0 + \frac{\rho a^2}{h} - a,$$

$$h_t = d_2 h_{xx} + k(\rho a^2 - h),$$

where, ρ, ρ_0 and k are all positive constants. The constant steady state solution $\bar{a} = 1 + \rho \rho_0$ and $\bar{h} = (1 + \rho \rho_0)^2 \rho$ is stable to small homogeneous perturbations when $1 < (2/\bar{a}) < k + 1$. From a biological point of view, the bifurcation solutions with fundamental mode are important. Thus, the criterion (2) can be applied. After some computation, we obtain the stable solutions when $1 < (2/\bar{a}) < k + 1$, $(1/\bar{a}) < 8((2/\bar{a}) - 1)k$, and d_1/d_2 is sufficiently small. We remark that when ρ_0 is sufficiently small (which is a reasonable assumption in their model), the condition of the criterion (2) is necessarily satisfied.

The theorems stated above are proved in the appendices.

HETEROGENEOUS STEADY-STATE SOLUTIONS II

The previous section concerned the existence of small-amplitude steady-state solutions and their stability. We next construct large amplitude solutions of the problem (5) when d_1 is zero or sufficiently small, and show the existence of striking spatial patterns. This study is motivated by diffusive plant-herbivore systems or host-parasite systems where plant or host diffuses scarcely. Large-amplitude solutions depend on the global nonlinearity of $F(U)$ in (3), so we recall that $f(u)$ satisfies the "hump" effect. This means that there exists a constant $c_4 > 0$ such that

$$f'(u) > 0 \quad \text{for} \quad 0 \leq u < c_4$$

$$f'(u) < 0 \quad \text{for} \quad c_4 < u.$$

We assume here that the constant steady-state solution $\bar{U}$ is unique.

We first consider the stationary problem of (3) in the limiting case $d_1 = 0$. The resulting system has the form

$$f_0(\hat{u}, \hat{v})\{f(\hat{u}) - \hat{v}\}\hat{u} = 0, \qquad (12a')$$
$$d_2 \hat{v}_{xx} - g_0(\hat{u}, \hat{v})\{g(\hat{v}) - \hat{u}\}\hat{v} = 0, \qquad x \in I. \qquad (12a'')$$

The boundary conditions are

$$\hat{u}_x(x) = \hat{v}_x(x) = 0 \quad \text{at} \quad x = 0 \text{ and } l. \qquad (12b)$$

From (12a'), it follows that

$$\hat{u} = 0 \quad \text{or} \quad f(\hat{u}) = \hat{v}. \tag{13}$$

Since $f^{-1}(\hat{v})$ is a double-valued function of $\hat{v}$ because of the hump effect, (13) implies that $\hat{u}$ generally takes three different values for $\hat{v}$, say $h_1(\hat{v})(= 0)$, $h_2(\hat{v})$ or $h_3(\hat{v})$ ($h_2 < h_3$). Thus three single equations are derived from (12a):

$$d_2 \hat{v}_{xx}^i + G_i(\hat{v}^i) = 0 \quad \text{for} \quad i = 1, 2 \text{ and } 3, \tag{14}$$

where $G_i(\hat{v}) = -g_0(h_i(\hat{v}), \hat{v})\{g(\hat{v}) - h_i(\hat{v})\}\hat{v}$. Now consider the boundary value problem (14) in the whole domain I subject to (12b). Then it is seen that for $i = 1$ or 3 the only solution is $v^i = 0$. This result follows from the nonlinearity of G_i. On the other hand, for $i = 2$, there may exist heterogeneous solutions $\hat{v}^2(x)$. In fact, if $G_2(\hat{v})$ is a cubic polynomial of v, the solutions can be constructed explicitly. However, it is worth noting that the eigenvalue problem

$$d_2 \phi_{xx} + G_2'(\hat{v}^2(x))\phi = \lambda\phi, \quad x \in I$$

$$\phi_x(0) = \phi_x(l) = 0$$

has $\text{Re}(\lambda_1) > 0$, where λ_1 is the principal eigenvalue. We infer that the solution $(\hat{u}^2(x), \hat{v}^2(x))$ of the problem (12) is unstable. Here $\hat{u}^2(x)$ is constructed from $u = h_2(v)$. Numerical evidence also confirms the instability of the solution. We next consider the case where I consists of at least two different parts $\{l_i\}$ for $i = 1, 2,$ and 3, and (14) is satisfied on each domain l_i, that is, $I = l_i \cup l_2 \cup l_3$ (one of these regions may be empty). It is obvious that patched solutions strongly depend on the combination of l_1, l_2, and l_3. We see that they can be composed of $\{\hat{v}^i(x)\}$ in an infinite number of ways. The discussions are very lengthy so we only show one example. Supposing that $I = l_1 \cup l_3$ and there exists only one patching point v^* in $(f(0), f(u_{max}))$, then the following problems are well formulated:

$$d_2 \hat{v}_{xx} + G(\hat{v}) = 0, \quad x \in I, \tag{15}$$

where $G(v)$ is given by

$$G(v) = \begin{cases} G_1(v) & \text{for} \quad v > v^* \\ G_3(v) & \text{for} \quad 0 \leq v < v^*, \end{cases} \tag{16a}$$

or

$$G(v) = \begin{cases} G_1(v) & \text{for} \quad 0 \leq v < v^* \\ G_3(v) & \text{for} \quad v^* < v \leq f(u_{max}). \end{cases} \tag{16b}$$

Here we note that $G(v)$ has a discontinuity of the first kind at the point $v = v^*$. Recently this kind of equation has been studied.[24,25] Now, considering the problem (15) with (16a) when the boundary conditions (12b) are imposed, we see that there exists no solution other than $\hat{v} = 0$. However, for the problem (15) with (16b), we can see that there exist multiple solutions $\{\hat{v}(x)\}$. Here the solution $\hat{v}(x)$ is defined by $\hat{v}(x) \in C^1(\bar{I})$ and

$$d_2 \hat{v}(x)_{xx} + G(\hat{v}(x)) = 0 \quad \text{for almost all } x \in I.$$

The next problem is to determine the discontinuity point v^*. It may not be unique in general. If one can find v^* explicitly, the spatial pattern can be discussed precisely, by solving (15) subject to (12b). In order to study this problem, we use the singular perturbation technique.[26,27] Consider the stationary problem of (3) with nonzero but sufficiently small d_1,

$$d_1 u_{xx} + f_0(u, v)\{f(u) - v\}u = 0, \tag{17a}$$

$$d_2 v_{xx} - g_0(u, v)\{g(v) - u\}v = 0. \tag{17b}$$

From some numerical evidences, we find the following fact (FIGURE 5): There exists a solution $(u(x), v(x))$ of the problem (17) under the zero-flux boundary conditions, such that

$$\lim_{d_1 \to 0} u(x) = \hat{u}(x) \quad \text{for almost all } x \in \bar{I}$$
$$\lim_{d_1 \to 0} v(x) = \hat{v}(x) \quad \text{for all } x \in \bar{I}. \tag{18}$$

If the boundary condition is of Dirichlet type, Fife[28] studied the limiting processes (18). Following Fife's argument, we get a rigorous justification of our problem. Therefore we continue our discussion under the assumption that (18) holds. By transforming from x to

$$y = \frac{x - x^*}{\sqrt{d_1}},$$

where x^* is an arbitrarily fixed seperating point between l_i and l_j $(i \neq j)$, (17a) is reduced to

$$u_{yy} + f_0(u, v)\{f(u) - v\}u = 0. \tag{19}$$

Noting the continuity of $v(x)$ at the point x^*, we have

$$u_{yy} + f_0(u, \hat{v}(x^*))\{f(u) - \hat{v}(x^*)\}u = 0, \quad x \in R^1, \tag{20a}$$

from (18) and (19). The boundary conditions for (20a) can be reasonably assumed as follows:

$$\lim_{y \to -\infty} u(y) = u^i(x^*)$$
$$\lim_{y \to +\infty} u(y) = u^j(x^*). \tag{20b}$$

Here we must note that (20) is value of $O(1)$. Concerning the problem (20), it is known that there exists a solution if and only if

$$\int_{\hat{u}_i(x^*)}^{\hat{u}_j(x^*)} f_0(s, \hat{v}(x^*))\{f(s) - \hat{v}(x^*)\}s\, ds = 0 \tag{21}$$

(see, for example, Aronson-Weinberger.[29]) From (21), we find the following results:

(1) $I = l_1 \cup l_3$ and
(2) $v(x^*)$ is determined uniquely, say v_c^*, which is independent of the separating point x^*.

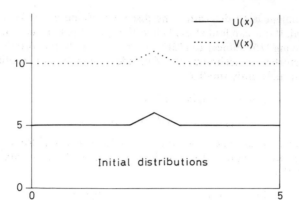

FIGURE 5a.

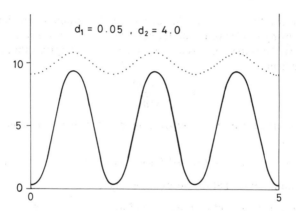

FIGURE 5b.

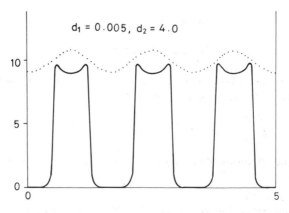

FIGURE 5c.

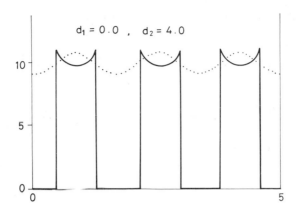

FIGURE 5d.

Thus, we can formulate the well-defined boundary value problem in the whole domain I,

$$d_2 \hat{v}_{xx} + G(\hat{v}) = 0, \quad x \in I, \tag{22a}$$

$$\hat{v}_x(0) = \hat{v}_x(l) = 0, \tag{22b}$$

where $G(v)$ is given by

$$G(v) = \begin{cases} G_1(v) & \text{for} \quad 0 \leq v < v_c^* \\ G_3(v) & \text{for} \quad v_c^* < v \leq f(u_{\max}). \end{cases}$$

We conclude that there exists remarkable heterogeneity in the solution $\hat{u}(x)$, where $\hat{u}(x)$ is determined by $u = f(v)$. From an ecological point of view, l_1 is a dead region of the prey and l_3 is its living region, although the predator is living in the whole domain. The boundaries between l_1 and l_3 can be determined by solving (22). Shapes of the solutions are illustrated in FIGURE 5.

NUMERICAL RESULTS

In this section we consider one simple model system as an illustrative example. The model takes the following form:

$$\begin{aligned} u_t &= d_1 \Delta u + \{f(u) - v\}u \\ v_t &= d_2 \Delta v - \{g(v) - u\}v, \end{aligned} \tag{23}$$

where Δ is the Laplacian in R^r, $d_2 = 4$, and f and g have the explicit forms

$$f(u) = \tfrac{1}{9}(35 + 16u - u^2) \quad \text{and} \quad g(v) = 1 + \tfrac{2}{5}v,$$

respectively; d_1 is only one adjustable parameter. First we present the case $r = 1$

Initial distribution of $u_o(x)$ Initial distribution of $v_o(x)$

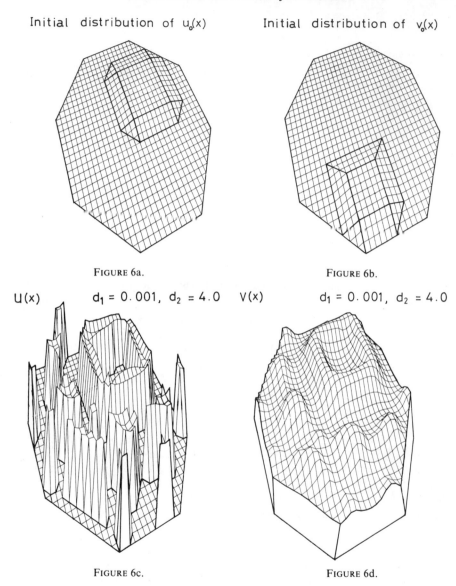

FIGURE 6a.   FIGURE 6b.

$U(x)$ $d_1 = 0.001$, $d_2 = 4.0$ $V(x)$ $d_1 = 0.001$, $d_2 = 4.0$

FIGURE 6c.   FIGURE 6d.

and the length $l = 5$. After some calculation, it is verified that the bifurcating solutions of (23) are stable. We study spatial structures of the heterogeneous steady-state solutions when d_1 varies from the bifurcation point to zero. The numerical method used here is the finite difference method.[30] All of the results were given by solving the evolution problem (23) when the same initial distribution is imposed. FIGURES 5b–d imply that the limiting processes (18) are valid. Furthermore, FIGURE 5d shows that the value of the patching point v_c^* given by (21) is in good agreement with the computational result.

We next consider (23) in a two-dimensional bounded domain, whose shape is octagonal as shown in FIGURE 6a. We note that the discussion given in the previous section is also valid for $r = 2$. FIGURES 6c and d show that the final stable pattern for a long time exhibits striking heterogeneity in the population density of the prey.

CONCLUDING REMARK

In this paper we have constructed heterogeneous steady-state solutions of a diffusive prey-predator system in one-dimensional space. The results are the proof of the existence of bifurcating solutions with arbitrary frequency modes when d_1 and d_2 are chosen appropriately, and the presentation of a useful criterion of stability in applications. Thus, it is found that the spatial structure strongly depends not only on the ecological factors such as growth rate, predation rate and so on, but also on diffusivities. For higher dimensional space, it seems difficult to discuss the bifurcation problem in a manner similar to that of one-dimensional space, although general theories have nearly been found. One of the reasons is that the bifurcation points are not found easily.

In the case where the diffusivity of a prey is zero or sufficiently small, we have shown that there appear interesting phenomena from the ecological point of view. The whole region is divided into two quite different ones. One is the region where prey and predator coexist, the other is where prey is dead and only predator exists. The form of the subregions is determined subject to initial distributions. The stability problem of the above solution is still unsolved.

Although the analysis has been performed for two interacting populations, one might discuss the case of three interacting populations consisting of two prey and one predator with a switching effect.

ACKNOWLEDGMENTS

We would like to thank Miss Keiko Boku and Mr. Masahisa Tabata for so willingly carrying out the numerical calculations in the section on numerical results.

REFERENCES

1a. GUREL, O. 1973. Global analysis of ecological systems. *In* Working Proceedings, NATO Conference, Mathematical Analysis of Decision Problems in Ecology, Istanbul, Turkey, July.: 293–302.
1b. MAY, R. M. 1976. Models for two interacting populations. *In* Theoretical Ecology, Principles and Applications. R. M. May, Ed.: 49–70. Blackwell Scientific Publications, Oxford, England.
2. SKELLAM, J. G. 1951. Random dispersal in theoretical populations. Biometrika **38:** 196–218.
3. KERNER, E. H. 1959. Further considerations on the statistical mechanics of biological associations. Bull. Math. Biophys. **21:** 217–255.
4. MONTROLL, E. W. & B. WEST. 1973. Models of population growth, diffusion, competition and rearrangement. *In* Synergetics, Cooperative Phenomena in Multi-Component Systems. H. Haken, Ed.: 143–156. B. G. Teubner. Stuttgart, West Germany.
5. CASSIE, R. M. 1963. Microdistribution of plankton. Oceanogr. Mar. Biol. Ann. Rev. **1:** 223–252.

6. STEELE, J. 1973. Stability of plankton ecosystems. *In* Ecological Stability. M. B. Usher & M. H. Williamson, Eds.: 179–191. Chapman and Hall. London.
7. SEGEL, L. & J. JACKSON. 1972. Dissipative Structure: An explanation and an ecological example. J. Theor. Biol. **37**: 545–559.
8. TURING, A. M. 1952. The chemical basis of morphogenesis. Phil. Trans. R. Soc. B **237**(37): 37–72.
9. OKUBO, A. 1974. Diffusion-induced instability in model ecosystems. Tech. Rep. 86, Cheasapeake Bay Inst., Johns-Hopkins Univ. Baltimore, Md.
10. PRIGOGINE, I. & R. LEFEVER. 1973. Theory of dissipative structures. *In* Synergetics, Cooperative Phenomena in Multi-Component Systems. H. Haken, Ed.: 124–136. B. G. Teubner. Stuttgart, West Germany.
11. GIERER, A. & H. MEINHARDT. 1972. A theory of biological pattern formation. Kybernetik **12**: 30–39.
12. NOVAK, B. & F. E. SEELIG. 1976. Phase-shift model for the aggregation of amoebae: A computer study. J. Theor. Biol. **56**: 301–327.
13. SEGEL, L, A, & S, A. LEVIN. 1976. Application of nonlinear stability theory to the study of the effects of diffusion on predator-prey interaction. *In* AIP Conference Proceedings **27**: 123–152.
14. MIMURA, M. 1977. Asymptotic behavior of a parabolic system related to a planktonic prey and predator model. SIAM J. Appl. Math. In press.
15. MURRAY, J. D. 1975. Nonexistence of wave solutions for the class of reaction-diffusion equations given by the Volterra interacting population equations with diffusion. J. Theor. Biol. **52**: 459–469.
16. SATTINGER, D. H. 1973. Topics in stability and bifurcation theory. Lecture Notes in Mathematics.: 309. Springer-Verlag. New York, N.Y.
17. SATHER, D. 1973. Branching of solutions of nonlinear equations. Rocky Mountain J. Math. **3**(2): 209–250.
18. CRANDALL, M. G. & P. H. RABINOWITZ. 1971. Bifurcation from simple eigenvalues. J. Functional Anal. **8**: 321–340.
19. CONWAY, E., D. HOFF & J. SMOLLER. 1977. Large time behavior of solutions of systems of nonlinear reaction-diffusion equations. Preprint.
20. GRAVES, L. 1955. Remarks on singular points of functional equations. Trans. Amer. Math. Soc. **79**: 150–157.
21. MCLEOD, J. B. & D. H. SATTINGER. 1973. Loss of stability and bifurcation at double eigenvalues. j. Functional Anal. **14**: 62–84.
22. NISHIURA, Y. 1977. Bifurcation of stable stationary solutions from symmetric modes. Proc. Japan Acad. **53**A(2): 41–45.
23. KIELHÖFER, H. 1974. Stability and semilinear evolution equations in Hilbert space. Arch. Rat. Mech. Anal. **57**: 150–165.
24. STUART, C. A. 1976. Differential equations with discontinuous non-linearities. Arch. Rat. Mech. Anal. **63**(1): 59–75.
25. TABATA, M. Two-point boundary value problems with a discontinuous semilinear terms. Preprint.
26. FIFE, P. C. 1976. Pattern formation in reacting and diffusing systems. J. Chem. Phys. **64**(2): 554–564.
27. MIMURA, M. & J. D. MURRAY. In press. On a planktonic prey-predator model which exhibits patchiness. J. Theor. Biol.
28. FIFE, P. C. 1976. Boundary and interior transition layer phenomena for pairs of second-order differential equations. J. Math. Anal. Appl. **54**: 497–521.
29. ARONSON, D. G. & H. F. WEINBERGER. 1975. Lecture note in mathmatics. J. A. Goldstein, Ed. Springer. New York, N.Y.
30. MIMURA, M. & M. TABATA. On a finite element approximation for a class of semilinear parabolic systems. In preparation.
31. CRANDALL, M. G. & P. H. RABINOWITZ. 1973. Bifurcation, perturbation of simple eigenvalues, and linearized stability. Arch. Rat. Mech. Anal. **52**: 161–180.
32. NISHIURA, Y. On a bifurcation problem of some semilinear parabolic systems and its stability. In preparation.

Here we only prove Theorems 1, 2, 4, and 5. For the proof of Theorem 3, see the results of Sattinger[16] and Crandall-Rabinowitz.[31]

Appendix 1. Proofs of Theorems 1 and 2

We first state two lemmas without proofs.[17]

Lemma A1. (1) $L_0: D(L_0) \to E$ is a linear Fredholm operator. (2) The null space $N(L_0)$ of L_0 and the null space $N(L_0^*)$ of its adjoint operator L_0^* are both one-dimensional and they are spanned by the vector $\Phi_n = \kappa_n \phi_n \cos(n\pi x/l)$ and $\Phi_n^* = \kappa_n^* \phi_n^* \cos(n\pi x/l)$, respectively, where $\phi_n = {}^t(d_2{}^0 \gamma n^2 - b_{22}, b_{21})$, $\phi_n^* = {}^t(d_2{}^0 \gamma n^2 - b_{22}, b_{12})$ and κ_n, κ_n^* are normalized constants such that $\| \Phi_n \| = \| \Phi_n^* \| = 1$. (3) L_0 is a one to one mapping of $D(L_0) \cap N(L_0)^\perp$ onto $R(L_0)$ (range of L_0) so that

$$K = (L_0 |_{D(L_0) \cap N(L_0)^\perp})^{-1}$$

is well defined, and KQ is a linear compact operator defined on E, where Q is defined in the next lemma.

Lemma A2. Let P be the orthogonal projection operator of E onto $N(L_0)$ and Q be that of E onto $R(L_0)$. Then we obtain

(1) $\qquad\qquad Pw = (w, \Phi_n)\Phi_n \quad \text{and} \quad (\hat{I} - Q)w = (w, \Phi_n^*)\Phi_n^*,$

(2) $\qquad\qquad KL_0 w - (\hat{I} - P)w \quad \text{for all} \quad w \in D(L_0),$

(3) $\qquad\qquad L_0 KQw = Qw \quad \text{for all} \quad w \in E.$

Using the projection P, we rewrite V as

$$V = PV + (\hat{I} - P)V$$
$$= \epsilon \Phi_n + \hat{V} \qquad\qquad \text{(A1)}$$

where ϵ is a real parameter. Substituting (A1) into (10), we obtain

$$L_0 \hat{V} + \Lambda(\sigma)(\epsilon \Phi_n + \hat{V})_{xx} + H(\epsilon \Phi_n + \hat{V}) = 0, \qquad\qquad \text{(A2)}$$

where $\Lambda(\sigma) = \{D(\sigma) - D(0)\}$. Using the projection Q, we obtain the equivalent system:

$$L_0 \hat{V} + Q\Lambda(\sigma)(\epsilon \Phi_n + \hat{V})_{xx} + QH(\epsilon \Phi_n + \hat{V}) = 0, \qquad \text{(A3a)}$$

$$(\hat{I} - Q)\{\Lambda(\sigma)(\epsilon \Phi_n + \hat{V})_{xx} + H(\epsilon \Phi_n + \hat{V})\} = 0. \qquad \text{(A3b)}$$

The property (2) of Lemma A2 reduces (A3a) to

$$\hat{V} + KQ\Lambda(\sigma)(\epsilon \Phi_n + \hat{V})_{xx} + KQH(\epsilon \Phi_n + \hat{V}) = 0. \qquad \text{(A4)}$$

From the assumption (H7a), $\Lambda(\sigma)$ can be represented by

$$\Lambda(\sigma) = \sigma\chi + \Theta(\sigma), \qquad\qquad \text{(A5)}$$

where $\chi = \{\chi_i\}$ is diagonal, (χ_1, χ_2) is not parallel to the tangent of $\hat{C}_n$ at $D(0)$, and $\Theta(\sigma)$ is a higher order term. Then (**A4**) is rewritten as

$$\hat{V} + \sigma K Q \chi (\epsilon \Phi_n + \hat{V})_{xx} + K Q \Theta(\sigma)(\epsilon \Phi_n + \hat{V})_{xx} + K Q H(\epsilon \Phi_n + \hat{V}) = 0. \tag{A6}$$

Lemma A3. There exist positive constants σ_1, ϵ_1 such that for any σ, ϵ with $|\sigma| < \sigma_1$, $|\epsilon| < \epsilon_1$, (**A6**) has a unique solution $\hat{V} = \hat{V}(\sigma, \epsilon)$ in $D(L_0) \cap N(L_0)^{\perp} \cap O$. Here the following estimate holds

$$||| \hat{V}(\sigma, \epsilon) ||| \leq C |\epsilon|(|\sigma| + |\epsilon|) \tag{A7}$$

for some positive constant C independent of σ and ϵ. Moreover $\hat{V}(\sigma, \epsilon)$ depends smoothly on σ and ϵ.

Remark A4. From (**A6**), it is easily seen that the principal part of $\hat{V}(\sigma, \epsilon)$, when σ and ϵ are sufficiently small, is as follows

$$\hat{V}(\sigma, \epsilon) = \sigma \epsilon \gamma n^2 K Q \chi \Phi_n - \epsilon^2 K Q H_Q(\Phi_n, \Phi_n) + \epsilon \overline{V}(\sigma, \epsilon), \tag{A8}$$

where $\overline{V}(\sigma, \epsilon)$ is of at least second degree with respect to σ and ϵ.

Proof of Lemma A3. We can get the unique solution $\hat{V}(\sigma, \epsilon)$ of (**A6**) with the aid of the contraction mapping principle in the Banach space $D(L_0) \cap N(L_0)^{\perp}$ with graph topology. Using the standard techniques (see Sather[17]), we can prove (**A7**) and the smooth dependency on σ and ϵ. For the precise discussion, see Nishiura.[32]

Inserting the solution $\hat{V}(\sigma, \epsilon)$ of (**A6**) into (**A3a**), we obtain the scalar equation, what we call *the bifurcation equation,* with respect to σ and ϵ,

$$(\hat{I} - Q)\{\Lambda(\sigma)(\epsilon \Phi_n + \hat{V}(\sigma, \epsilon))_{xx} + H(\epsilon \Phi_n + \hat{V}(\sigma, \epsilon))\} = 0. \tag{A9}$$

Using the relation (1) of Lemma A2 and (**A5**), (**A9**) is reduced to

$$(\sigma \chi(\epsilon \Phi_n + \hat{V}(\sigma, \epsilon))_{xx} + \Theta(\sigma)(\epsilon \Phi_n + \hat{V}(\sigma, \epsilon))_{xx} + H(\epsilon \Phi_n + \hat{V}(\sigma, \epsilon)), \Phi_n^*) = 0. \tag{A10}$$

Using the expansion of $H(V)$ and (**A8**), (**A10**) becomes

$$-\sigma \epsilon \gamma n^2 (\chi \Phi_n, \Phi_n^*) + \epsilon^2 (H_Q(\Phi_n, \Phi_n), \Phi_n^*)$$
$$+ \epsilon^3 (H_C(\Phi_n) - H_Q(\Phi_n, KQH_Q(\Phi_n, \Phi_n)) - H_Q(KQH_Q(\Phi_n, \Phi_n), \Phi_n), \Phi_n^*)$$
$$+ \epsilon \eta(\sigma, \epsilon) = 0, \tag{A11}$$

where $\eta(\sigma, \epsilon)$ is a smooth function that satisfies the properties stated in Theorem 2. Dividing (**A11**) by ϵ, and noting that $(H_Q(\Phi_n, \Phi_n), \Phi_n^*) = 0$ and $-\gamma n^2 (\chi \Phi_n, \Phi_n^*)(= \alpha) > 0$, we have, from (**A11**), the bifurcation equation in Theorem 2

$$\alpha \sigma + \beta \epsilon^2 + \eta(\sigma, \epsilon) = 0. \tag{A12}$$

Using the implicit function theorem, there exists a positive constant ϵ_2 such that (**A12**) has a unique smooth solution $\sigma = \sigma(\epsilon)$ for $|\epsilon| < \epsilon_2$ satisfying $\sigma(0) = 0$. Inserting this solution into $\hat{V}(\sigma, \epsilon)$, we obtain the unique one-parameter family of solutions $V(\epsilon)$ in Theorem 1.

From now on we assume that

$$\beta \neq 0. \tag{A13}$$

This is a generic condition, i.e., (A13) holds for almost all nonlinearities H. There hardly appears the case $\beta = 0$ in applications. Thus applying a suitable change of coordinates to (A12), we know the form of $\sigma = \sigma(\epsilon)$ as shown in FIGURE 4, i.e., bifurcation is one-sided. This completes the proofs of Theorems 1 and 2.

Appendix 2. Proof of Theorem 3

We examine whether the operators A, M and $\tilde{A}$ satisfy the conditions of Theorems 1.4. and 1.5. in Reference 23. Operator A is a self-adjoint and positive definite operator in E with A^{-1} being compact. Noting that $H_N^2(I)$ is continuously imbedded in $C^1(\bar{I})$, we easily see that M has the estimate

$$\| MV \| \leq C \| V \| \quad \text{for } V \in E,$$

where C is a positive constant independent of V. Since the nonlinear term H is smooth and $A^{-1/2}$ (fractional power of A) is a bounded operator from E into $(H^1(I))^2$, the composite operator $T = R \circ A^{-1/2}$ is locally Lipschitz-continuous in E, i.e.,

$$\| T(U) - T(V) \| \leq C(d) \| U - V \| \quad \text{for } \| U \|, \| V \| \leq d,$$

where $C(d)$ is a positive constant which only depends on d, and T also satisfies $\| T(V) \| = o(\| V \|)$. Thus Theorems 1.4 and 1.5 in Reference 23 can be applied. Hence we obtain Theorem 4.

Appendix 3. Proof of Theorem 5

The normalized factors κ_n and κ_n^* are both positive, so it is sufficient to determine the sign of the following quantity

$$\omega(n, D) = (H_C(\Psi_n) - H_Q(\Psi_n, KQH_Q(\Psi_n, \Psi_n))) - H_Q(KQH_Q(\Psi_n, \Psi_n), \Psi_n), \Psi_n^*),$$

where $\Psi_n = \phi_n \cos(n\pi x/l)$ and $\Psi_n^* = \phi_n^* \cos(n\pi x/l)$. We introduce some notations

$$D_m = (b_{11} - \gamma m^2 d_1{}^0)(b_{22} - \gamma m^2 d_2{}^0) - b_{12} b_{21},$$

$$M_m = \begin{pmatrix} b_{22} - \gamma m^2 d_2{}^0 & -b_{12} \\ -b_{21} & b_{11} - \gamma m^2 d_1{}^0 \end{pmatrix},$$

$$K_m = \frac{1}{D_m} M_m.$$

In what follows we write (d_1, d_2) in place of $(d_1{}^0, d_2{}^0)$ for simplicity.

Lemma A4.

(1) $(H_C(\Psi_n), \Psi_n^*) = \frac{3}{8} l(H_C(\phi_n), \phi_n^*)$

(2) $(H_Q(\Psi_n, KQH_Q(\Psi_n, \Psi_n)), \Psi_n^*)$

$$= \frac{l}{8} (H_Q(\phi_n, 2K_0 H_Q(\phi_n, \phi_n) + K_{2n} H_Q(\phi_n, \phi_n)), \phi_n^*)$$

Proof. The proofs are elementary, so we leave them to the reader.

Using Lemma A4, we have

$$\omega(n, D) = \frac{l}{8} (3H_C(\phi_n) - 2H_Q(\phi_n, 2K_0 H_Q(\phi_n, \phi_n) + K_{2n} H_Q(\phi_n, \phi_n)), \phi_n^*).$$

Noting that $D_{2n} > 0$, it is sufficient to know the sign of

$$\hat{\omega}(n, D) = (3D_{2n} H_C(\phi_n) - 2H_Q(\phi_n, 2D_{2n} K_0 H_Q(\phi_n, \phi_n) + M_{2n} H_Q(\phi_n, \phi_n)), \phi_n^*).$$

Recalling that $\phi_n = {}'(d_2\gamma n^2 - b_{22}, b_{21})$ and $\phi_n^* = {}'(d_2\gamma n^2 - b_{22}, b_{12})$, we may consider $\hat{\omega}(n, D)$ to be a polynomial of $d_2\gamma n^2$. Here we note the following properties:

Lemma A5.

(1)
$$d_2\gamma n^2 \to +\infty \text{ as } n \to +\infty.$$

(2)
$$\text{For } n = 1, d_2\gamma \to +\infty \text{ as } d_1 \to \frac{b_{11}}{\gamma}.$$

Proof. Recalling the definitions of C_n and Γ, we can prove this lemma easily. The sign of $\hat{\omega}(n, D)$ for large $d_2\gamma n^2$ is determined by that of the coefficient of the highest degree term of $\hat{\omega}(n, D)$ when it is considered to be a polynomial of $d_2\gamma n^2$. After some tedious computations, we obtain that this coefficient is

$$16\frac{d_1}{d_2}\left(3e_1 - \frac{4}{\det B}\{a_1(a_1 b_{22} - a_2 b_{12}) + b_1(a_2 b_{11} - a_1 b_{21})\}\right).$$

Using the assumption $D \in \Gamma_k$, (1) of Theorem 5 follows. We can prove the latter part of Theorem 5 in a similar fashion by using the property (2) of Lemma A5, so we omit the details.

NONUNIQUE STABLE RESPONSES OF EXTERNALLY FORCED OSCILLATORY SYSTEMS

F. C. Hoppensteadt

Department of Mathematics
University of Utah
Salt Lake City, Utah 84112

J. E. Flaherty

Department of Mathematics
Rensselaer Polytechnic Institute
Troy, New York 12181

Changes in a population's size over many generations are frequently described in terms of reproduction curves in which the next generation's size is plotted as a function of the present generation's size.[1] FIGURE 1a shows one such curve having two stable (δ, L) and two unstable (0, u) equilibria. The value u acts as a threshold of population size: A population initially smaller than u will approach size δ over several generations and those greater than u will approach size L. It may happen that the threshold u changes in response to the population size; for example, it may represent a level of predator satiation.[2] We assume that u increases each generation that the population is at level L and decreases each generation that the population is at level δ. FIGURE 1b gives a simplified version of this model that is particularly well suited for numerical computation.

The model in FIGURE 1b is described by the following system of equations

$$x_{n+1} = \min\{\max[\delta, (s + 1)x_n - su_n], L\}$$

$$u_{n+1} = \{u_n + A(2x_n - L - \delta)\}_+$$

where A is a measure of the response of u to the population's state and $\{\ \ \}_+$ denotes the positive part of this expression.

Solutions of this system rapidly approach the periodic solution shown in FIGURE 2. The period of this relaxation oscillation is listed in TABLE I for some values of A.

Next, we consider the response of this system to external periodic forcing: In particular, we consider the system

$$x_{n+1} = \min\{\max[\delta, (s + 1)x_n - su_n], L\}$$

$$u_{n+1} = \{u_n + A(2x_n - L - \delta) + B\cos(2\pi n/P)\}_+$$

where B gives the amplitude and P the period of the forcing (P is an integer).

A technique for analyzing such problems by direct calculation was developed in Reference 3. This was based on numerical calculation of rotation number of stable oscillations. The rotation number is defined for this problem in the following way.

0077-8923/78/0316-0511 $1.75/1 © 1979, NYAS

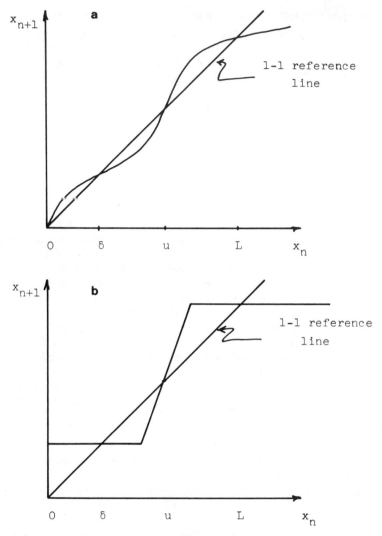

FIGURE 1. (a) A two-level reproduction curve is described. Here x_n = population size in the nth generation. An approximation to it is given in (b). We take $x_{n+1} = \min\{\max[\delta, (s + 1)x_n - su_n], L\}$ where $s + 1 \gg 1$ is the slope of the vertical segment of the reproduction curve.

First, the Poincaré mapping $\mathcal{P}$ is defined by

$$\mathcal{P}(x_0, u_0) = (x_P, u_P)$$

where P is the forcing period. Fixed points of this mapping correspond to periodic solutions having period P or some fraction of P, and finite invariant sets might correspond to subharmonics: An N point set corresponding to a solution

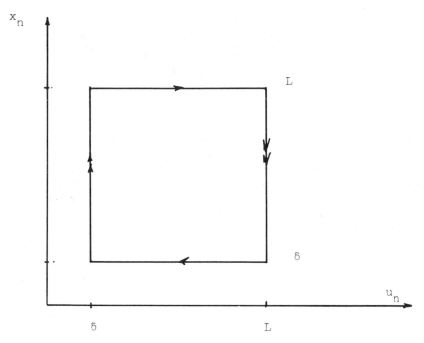

FIGURE 2. Free relaxation oscillation. The periodic solution described here is for $s \gg 1$. The double arrow indicates that the branch is passed over very rapidly. The two horizontal segments are traversed at a rate dependent on A.

having period NP. The free problem ($B = 0$) has a static state $x^* = u^* = (L + \delta)/2$, and we use this as an origin for calculating the rotation number. The rotation number of a stable subharmonic is defined to be the number of revolutions the solution vector emanating from (x^*, u^*) makes per period of the subharmonic, all divided by the period.

This system has several interesting features, some of which are illustrated by the sample data presented in TABLE 2.

The most striking feature of this system is illustrated by the first data. For $0.01 < B < 0.05$, there are two stable responses observed: One having rotation number $\frac{1}{3}$ (period 30) and the other rotation number $\frac{5}{16}$ (period 160). Both of these solutions lie essentially on the free relaxation oscillation, so their domains of attraction are intertwined in a very complicated way. Next, note that the rota-

TABLE 1

PERIOD OF RELAXATION OSCILLATION SHOWN IN FIGURE 2

($s = 100$, $L = 10$, $\delta = 2$)

A	0.02	0.04	0.06	0.08	0.1	1
Period	107	56	40	30	26	7

TABLE 2

ROTATION NUMBERS OF STABLE OSCILLATIONS $(A = 0.08)$

B	$0.01 < B < 0.05$	$0.055 < B < 2.1$	$2.2 < B$
Rotation No.	$\frac{1}{3}$ and $\frac{5}{16}$	$\frac{1}{3}$	$\frac{1}{1}$

tion numbers are constant over intervals. This illustrates the phenomenon of phase locking in which the frequency of the response is constant for nearby parameter values. The data suggest that the rotation number is a monotone function of B where it is single valued. As B decreases, solutions having successively lower rotation numbers appear. These arise from bifurcations of subharmonics. Presumably the solutions persist for decreasing values of B, but as unstable oscillations: For $B < 2.1$, the solution having rotation number $\frac{1}{1}$ is still present but unstable.

We will present a detailed analysis of this system elsewhere.

The model for (x, u) has many features in common with the forced van der Pol

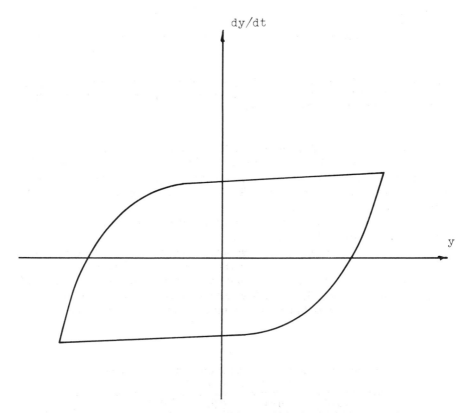

FIGURE 3. Free (relaxation) oscillation of the van der Pol Equation $\epsilon \ll 1$ and $B = 0$.

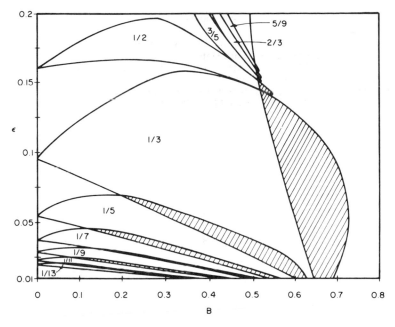

FIGURE 4. Rotation numbers of stable responses to the forced van der Pol relaxation oscillator. Overlap regions where two stable subharmonics coexist are indicated by diagonal lines. These were obtained by the numerical computations described in Reference 3. (From Flaherty & Hoppenstaedt.[3] By permission of The Massachusetts Institute of Technology.)

equation

$$d^2y/dt^2 + (1/\epsilon)(y^2 - 1)(dy/dt) + y = (B/\epsilon)\cos t$$

where $\epsilon \ll 1$. When $B = 0$, the solution of this equation are attracted to the relaxation oscillation in FIGURE 3. The rotation numbers of stable responses to periodic forcing were calculated in Reference 3. The results of these calculations are shown in FIGURE 4.

DISCUSSION

The elementary model introduced here exhibits several important properties of forced relaxation oscillators: Nonunique response, phase locking, and successive bifurcation of subharmonics. In this it shares many features with the van der Pol Equation[3] while avoiding the difficulties of having to solve a stiff differential equation. It is quite suitable as an educational device for introducing these phenomena and their analysis by computers.

REFERENCES

1. HOPPENSTEADT, F. C. 1977. Mathematical Methods in Population Biology. Courant Institute Lecture Notes. New York University.
2. HOPPENSTEADT, F. C. & J. B. KELLER. 1976. Synchronization of periodical Cicada emergences. Science 194: 335–337.
3. FLAHERTY, J. E. & F. C. HOPPENSTEADT. 1978. Frequency entrainment of a forced van der Pol oscillator. Stud. Appl. Math. 58:5–15.

BIFURCATIONS AND DYNAMIC COMPLEXITY
IN ECOLOGICAL SYSTEMS*

Robert M. May

Biology Department
Princeton University
Princeton, New Jersey 08540

INTRODUCTION

As is fitting in a volume devoted to bifurcations, this paper itself undergoes bifurcations that divide it into four distinct sections.

The major bifurcation divides the presentation into a part that reviews some well-established results, and a part that gives a preliminary and unfinished account of some interesting problems. These two parts arc each further subdivided into two unconnected sections. The first section reviews "thresholds" and "breakpoints" in ecological systems which can have two alternative stable states. The second reviews the complicated dynamical behavior that can arise in fully deterministic systems described by simple first-order difference equations possessing one critical point. The third section begins to extend this analysis into the yet more complicated circumstances that can arise in systems described by first-order difference equations with two critical points. The last section notes some bizarre aspects of the global stability behavior of a simple second-order difference equations that occurs in a host-parasite context.

ECOLOGICAL SYSTEMS WITH ALTERNATIVE STABLE STATES: A REVIEW

There are many circumstances in which one is interested in the dynamics of a natural plant or animal population that is being harvested. Examples for which quantitative data are available include vegetation being grazed by herbivores, fish or whale populations being harvested by man, and insect populations being "harvested" by predators. In these situations, the net population growth rate (dN/dt) for the unharvested population (N) tends to be an increasing function of N at low population densities, and a decreasing function of N at high densities, becoming negative beyond some value $N = K$, which may be thought of as the natural "carrying capacity" of the environment. The loss rate due to harvesting is likely to be some nonlinearly increasing function of N at low population densities, and to saturate to some constant rate at high densities. The overall result will often be that at low harvesting rates there is a unique stable state (set primarily by K), and at high harvesting rates there is also a unique stable state (set primarily by the harvester), but that at intermediate harvesting rates there can be two alternative stable states.

Such bifurcation phenomena are trivial from the mathematical point of view,

*This work was supported in part by the National Science Foundation under Grant DEB77-01565.

517

but can have very important implications for management of the natural population, be it pasture, fish, whales, insect pests, or human parasitic disease. I have recently attempted to give a review[1] that draws together the available mathematical and biological material pertaining to such situations, and I will not recapitulate it in this written version of my presentation.

First-Order Difference Equations with One Critical Point: A Review

Elsewhere in this volume, Guckenheimer[2] has outlined the two basic kinds of bifurcations that can arise in simple mappings of the kind

$$x_{t+1} = F(x_t), \tag{1}$$

when $F(x)$ has "one hump," or one critical point. Hoppensteadt[3] and Yorke[4] have also touched on this topic. The upshot is that the dynamics can manifest a stable point, or stable cycles, or apparently "chaotic" solutions.

First-order difference equations of this kind arise in a natural way in many contexts in population biology; the dynamics of temperate zone insect populations, and of fish, are two concrete examples. The extremely complicated dynamical behavior of these deceptively simple first-order difference equations has disturbing implications for the analysis of data about natural populations, and for the hope of long-term predictions in ecology.

In short, this general subject is one which is rich in mathematical interest, and also one which possibly holds serious biological implications (depending on whether or not the nonlinearities in natural situations are typically severe enough to carry the populations into the regime of chaotic behavior). Again, in this written version of my presentation, I shall not repeat material that has been published elsewhere. For an attempt at a thorough account of the mathematical aspects of this subject, see the review by May[5] (and the references therein, and also the work of Sharkovsky[6]). A review oriented towards the biological implications is given by May and Oster.[7]

First-Order Difference Equations with Two Critical Points: Some Problems

General Remarks about Maps with Two Critical Points

In population genetics, an important class of problems center on one genetic locus with two alleles (the familiar A and a, with respective "gene frequencies" p_t and q_t in the t th generation), and are therefore concerned with the mapping that relates p_{t+1} to p_t. In the case where the forces of natural selection are "frequency dependent,"[8-10] it can be that the allele A enjoys a selective advantage when rare, and a disadvantage when common. The result will be that the corresponding gene frequency p will tend to increase at low values, and to decrease at high values. This situation is depicted in FIGURE 1, where, for subsequent notational simplicity, the variable p (defined on the interval $[0, 1]$) is replaced by

$$x = 2p - 1 \tag{2}$$

(whence the mapping is on the interval $[-1, 1]$).

The map in FIGURE 1 has two critical points. In contrast with maps with one critical point, which (as discussed in the previous section) have recently been the subject of exhaustive analysis, such maps have received relatively little attention. Although frequency-dependent natural selection was the motivation for the present work, these maps are likely to be of interest in many other biological contexts.

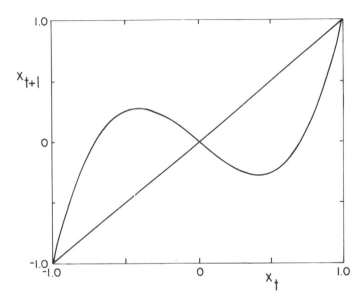

FIGURE 1. First-order difference equation, of the form of Equation 1, with two critical points. Here x is defined on the interval $[-1, 1]$, and the map is specifically the cubic Equation 3, with $a = 2$. The fixed points of period 1 (at $x = 1, 0, -1$) are at the intersections of the map with the 45° line.

Quite apart from possible biological applications, the dynamical behavior generated by maps with two critical points is of intrinsic mathematical interest, as the next step up from that generated by "one hump" maps.

For first-order difference equations with one critical point, the maximum number of fixed points with period k is 2^k: each new iteration of the map doubles the number of possible intersections with the 45° line, so that the k-times-composed mapping can have a maximum of 2^k fixed points. For a first-order difference equation with two critical points, as illustrated in FIGURE 1, each new folding can triple the number of intersections with the 45° line, and there can consequently be a maximum of 3^k fixed points of period k. For the map with one critical point, various techniques have been used[2,5,7] to elucidate the generic order in which the different cycles of period k appear, and in general to classify how the totality of 2^k fixed points of period k are organized. The task of giving a generic classification of how the 3^k fixed points of period k are organized, for maps with two critical points, is more messy. A beginning is made below, but, to the best of my knowledge, much remains to be done.

For maps with one critical point, there can be at most one stable attractor; even in the chaotic regime, where there are infinitely many different periodic orbits and an uncountable number of asymptotically aperiodic orbits, there is in general a unique cycle that attracts almost all initial points. For the general first-order difference equation, Guckenheimer et al.[11] have shown that, for each parameter value, the maximum number of periodic attractors is equal to the number of critical points. Thus, for maps with two critical points, there can be domains of parameter space in which the system possesses two distinct periodic attractors. We shall see, below, how this phenomenon emerges from the bifurcation structure. Maps with two critical points consequently deserve special attention as the simplest first-order difference equations to exhibit the feature of alternative stable states.

In what follows, there is first a generic discussion of the bifurcations that occur in second, and higher, iterations of an (antisymmetric) map with two critical points, as the "hill" and "valley" steepen. Second, these processes are illustrated by the concrete example of the cubic mapping

$$x_{t+1} = ax_t^3 + (1 - a)x_t. \tag{3}$$

Equation 3 is the canonical exemplar of a map with two critical points, in the sense that $x_{t+1} = ax_t(1 - x_t)$ is the standard example of the one-hump map. Finally, frequency-dependent natural selection is discussed in the light of these results.

Generic Aspects of the Bifurcation Structure

Consider a first-order difference equation, as defined by Equation 1, with x defined on the interval $[-1, 1]$. In discussing the case where $F(x)$ has two critical points, it is convenient further to assume that $F(x)$ is antisymmetric, so that

$$F(-x) = -F(x). \tag{4}$$

Such a situation is illustrated in FIGURE 1. The extent to which the subsequent conclusions are specific to these antisymmetric maps is an open question.

We further assume the map is "anchored" to (unstable) fixed points at $x = 1$ and $x = -1$: $F(1) = 1$. Interest now focuses on the fixed point at $x = 0$ ($F(0) = 0$ from Equation 4). The stability analysis follows the lines laid down for one-hump maps,[2,5,7] and depends on the slope, $\lambda^{(1)}(0)$, of the map at the fixed point.

$$\lambda^{(1)}(0) = [dF/dx]_{x=0}. \tag{5}$$

So long as $|\lambda^{(1)}| < 1$, the fixed point at $x = 0$ is an attractor. However, as the hill and valley in FIGURE 1 steepen, $\lambda^{(1)}$ will steepen toward -1, and the fixed point at $x = 0$ will become unstable once $\lambda^{(1)}$ steepens beyond -1.

To see what happens as this fixed point of period 1 ($x = 0$) becomes unstable, we turn, as in the analysis of the one-hump map,[2,5,7] to the map for the second iterate:

$$x_{t+2} = F^{(2)}(x_t). \tag{6}$$

Here $F^{(k)}$ denotes the k-times-composed mapping of F. The period-1 point at $x = 0$ is obviously a degenerate period-2 point, and the slope of the $F^{(2)}$ map at this point is

$$\lambda^{(2)}(0) = [dF^{(2)}/dx]_{x=0}. \tag{7}$$

That is, explicitly writing $F^{(2)}(x) = F(F(x))$,

$$\lambda^{(2)}(0) = [dF/dx]^2_{x=0} \tag{8}$$

$$= [\lambda^{(1)}(0)]^2. \tag{9}$$

The observation that $\lambda^{(2)}(0)$ is the square on $\lambda^{(1)}(0)$ is the key[5,7] to the cascading bifurcations that arise in maps with one critical point: so long as the period-1 point at $x = 0$ is stable, which implies $|\lambda^{(1)}| < 1$, then $|\lambda^{(2)}| < 1$ and the $F^{(2)}$ map intersects the 45° line only once in the neighborhood of $x = 0$; as the period-1 point at $x = 0$ becomes unstable, which implies $|\lambda^{(1)}| > 1$, then $|\lambda^{(2)}| > 1$ and the $F^{(2)}$ map steepens to make a loop which intersects the 45° line three times in the neighborhood of $x = 0$. In this way, two new and initially stable fixed points of period 2 are born by a "pitchfork" bifurcation, at the same time as the period-1 fixed point becomes unstable. For the antisymmetric map under consideration here, we can label these two fixed points of period 2 as $x = \pm\Delta$:

$$\pm\Delta = F(\mp\Delta). \tag{10}$$

The stability of this new orbit of period 2 depends, in turn, on the slope of $F^{(2)}$ at these fixed points $x = \pm\Delta$:

$$\lambda^{(2)}(\Delta) = [dF^{(2)}/dx]_{x=\Delta}. \tag{11}$$

Again, writing $F^{(2)}$ explicitly,

$$\lambda^{(2)}(\Delta) = \left[\frac{dF}{dx}\right]_{x=-\Delta}\left[\frac{dF}{dx}\right]_{x=\Delta} \tag{12}$$

That is,

$$\lambda^{(2)}(\pm\Delta) = \lambda^{(1)}(\Delta)\lambda^{(1)}(-\Delta). \tag{13}$$

Of course, the slope $\lambda^{(2)}$ is the same at each of the two points $x = \Delta$ and $x = -\Delta$.

The antisymmetry of $F(x)$ necessarily implies that $\lambda^{(1)}(\Delta) = \lambda^{(1)}(-\Delta)$, whence

$$\lambda^{(2)}(\pm\Delta) = [\lambda^{(1)}(\Delta)]^2. \tag{14}$$

This is a second key relationship. The stability-setting slope of the $F^{(2)}$ map at these fixed points of period 2 is *necessarily positive*. This is in complete contrast to what happens for maps with one critical point, where the period-2 cycle is born with $\lambda^{(2)} = +1$, and then as the hump continues to steepen $\lambda^{(2)}$ decreases to zero (as one of the period-2 points coincides with the critical point), beyond which $\lambda^{(2)}$ becomes negative (with the period-2 points on opposite sides of the critical point, and consequently dF/dx having opposite signs at the two points), so that eventually $\lambda^{(2)}$ steepens beyond -1 with the period-2 orbit becoming unstable and bifurcating to a stable period-4 orbit.[5,7] For the antisymmetric map with two critical points, however, the stability-setting slope at $x = \pm\Delta$ is born with $\lambda^{(2)} = +1$, decreases to $\lambda^{(2)} = 0$ as the hill and valley steepen, and then increases back toward $\lambda^{(2)} = +1$. As $\lambda^{(2)}$ increases beyond $+1$, the $F^{(2)}$ map again loops to cut the 45° line three times in the neighborhood of each of the period-2 points at $x = \pm\Delta$

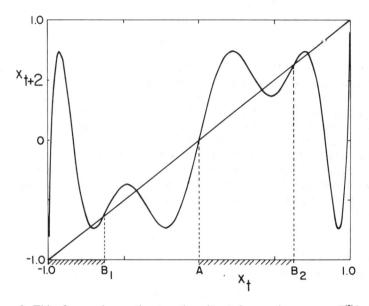

FIGURE 2. This figure shows the two-times-iterated mapping $x_{t+2} = F^{(2)}(x_t)$, for the cubic mapping of Equation 3, with $a = 3.3$. The map intersects the 45° line at 9 points, which are the fixed points of period 2. Here the symmetric period-2 cycle between the points B_1 and B_2 is unstable, and there are two distinct periodic attractors: one attracts points originating in the hatched regions between -1 and B_1, and between A and B_2; the other attracts points originating in the two unhatched regions. For a full discussion, see the text.

(in exactly the manner that initially gave rise to these two fixed points by bifurcation in the neighborhood of $x = 0$). Thus as the first cycle of period 2 becomes unstable, it bifurcates to produce two new and distinct cycles, each of period 2.

   This situation is illustrated in FIGURE 2. The system now has two alternative stable states, each a stable cycle of period 2. Points originating in the intervals $[-1, -\Delta]$ or $[0, \Delta]$ (denoted by the hatched regions along the x-axis in FIGURE 2) will be attracted to one cycle; points in the intervals $[-\Delta, 0]$ or $[\Delta, 1]$ to the other cycle. This now accounts for all $3^2 = 9$ fixed points of period 2 (three of which are, of course, the period-1 points).

For each of these "second generation" cycles of period 2, the stability-setting slope $\lambda^{(2)}$ at the two fixed points will evolve in the manner made familiar by the map with one critical point: $\lambda^{(2)}$ will decrease from $+1$, through zero, toward -1, beyond which point the period-2 cycle will become unstable, giving rise to an initially stable cycle with period 4. Thus each of the two domains of attraction will exhibit the cascading bifurcation process,[2,5,7] generating successively stable cycles of periods $2, 4, 8, 16, \ldots, 2^n$.

In summary, we see that for first-order difference equations with one critical point (as discussed in the previous section) the basic process is a bifurcating hierarchy of cycles with periods $1 \rightarrow 2 \rightarrow 4 \rightarrow 8 \cdots 2^n$; other higher-order cycles echo this theme, with stable cycles of basic period k cascading down through "harmonics" of period $k2^n$. For first-order difference equations with two anti-symmetric critical points (as discussed above), the basic process is cycles with periods $1 \rightarrow 2 \rightarrow$ two distinct 2s, each with its own domain of attraction, and each of which then goes $2 \rightarrow 4 \rightarrow 8 \rightarrow \cdots 2^n$.

Higher-order cycles arise by "tangent" bifurcation,[2,5,7] and are similarly complicated. Again, in general, there will be two stable periodic attractors for each parameter value, for the reasons alluded to above.[11] As an example, consider the fixed points of period 3. For maps with one critical point, there are eight (2^3) such points: two period-1 points, and a stable and unstable pair of period-3 cycles. For maps with two critical points, there are potentially $3^3 = 27$ such points. Subtracting the three period-1 points, there remain two constellations of 12 points. The first set of 12 points originates as two pairs of 3-cycles, one initially stable and one unstable; that is, there simultaneously arise two distinct stable 3-cycles, each with its own distinct domain of attraction. The other set of 12 points behaves similarly.

The task of cataloging the way in which the 3^k fixed points of period k are organized, and the order in which the various cycles of period k originate, remains an interesting problem.

A Specific Example: The Cubic Map

The canonical example used in discussions[2,5,12] of the one-hump map is the quadratic difference equation $x_{t+1} = ax_t(1 - x_t)$. The above discussion of anti-symmetric maps with two critical points may be similarly made concrete by considering the cubic difference Equation 3. This equation has nontrivial behavior for $0 < a < 4$; for $a < 0$, the end-points at $x = \pm 1$ are attractors, and for $a > 4$ the hill top and valley bottom lie outside $[1, -1]$.

The stability of the fixed point at $x = 0$ hinges on the slope $\lambda^{(1)}(0)$, which here is

$$\lambda^{(1)}(0) = 1 - a. \tag{15}$$

Thus the period-1 orbit is stable for $0 < a < 2$ (with exponential damping for $0 < a < 1$, and oscillatory damping for $1 < a < 2$).

For $a > 2$, the $F^{(2)}$ map bifurcates to give the symmetric fixed points of period 2 at $x = \pm \Delta$. These points are obtained from Equation 10,

$$-\Delta = a\Delta^3 + (1 - a)\Delta. \tag{16}$$

This gives

$$\Delta = \pm[(a - 2)/a]^{1/2}. \qquad (17)$$

The slope of the $F^{(2)}$ map at these two points follows from Equation 14:

$$\lambda^{(2)}(\pm\Delta) = (2a - 5)^2. \qquad (18)$$

Initially, at $a = 2$, this slope is $\lambda^{(2)} = +1$. It decreases to $\lambda^{(2)} = 0$ at $a = 2.5$, and then increases back to $\lambda^{(2)} = +1$ at $a = 3$. For $a > 3$, this period-2 cycle is unstable.

The subsequent nonsymmetric cycles of period 2 are obtained by finding all the fixed points of Equation 6. For the cubic $F(x)$ of Equation 3 this gives

$$x = x[ax^2 + 1 - a][ax^2(ax^2 + 1 - a)^2 + 1 - a]. \qquad (19)$$

Writing $y = ax^2 + 1 - a$, this can, after some manipulation, be brought to the form

$$[y^2 - 1][y^2 + (a - 1)y + 1] = 0. \qquad (20)$$

The solution $x = 0$ has been discarded. Similarly, the pair of solutions $y = \pm 1$

TABLE 1

DYNAMICAL BEHAVIOR OF THE MAPPING $x_{t+1} = ax_t^3 + (1 - a)x_t$

Value of a	Dynamical behavior
$2 > a > 0$	stable point
$3 > a > 2$	stable cycle of period 2
$1 + \sqrt{5} > a > 3$	two distinct (unsymmetrical) cycles, each with period 2
$4 > a > 1 + \sqrt{5}$	two distinct periodic attractors, with various periods; usually the dynamics is apparently chaotic

lead back to the 4 period-2 fixed points that have already been found: $x = \pm 1$ and the $x = \pm\Delta$ of Equation 17. The remaining quadratic in y in Equation 20 leads routinely to four other period-2 fixed points at $x = \pm\alpha, \pm\beta$, with α and β defined by

$$\alpha, \beta = \left\{ \frac{(a - 1) \pm [(a + 1)(a - 3)]^{1/2}}{2a} \right\}^{1/2} \qquad (21)$$

It is easy to verify that these four fixed points correspond to two distinct period-2 cycles, with $\alpha \to -\beta \to \alpha$ and $\beta \to -\alpha \to \beta$, as illustrated in FIGURE 2. The stability of each of these cycles depends on the slope of the $F^{(2)}$ map at these fixed points, namely,

$$\lambda^{(2)}(\pm\alpha, \pm\beta) = \lambda^{(1)}(\pm\alpha)\lambda^{(1)}(\mp\beta), \qquad (22)$$

$$= 7 + 4a - 2a^2. \qquad (23)$$

The cycles first appear at $a = 3$ (see Equation 21), with $\lambda^{(2)} = +1$. They become unstable, with $\lambda^{(2)} < -1$, for $a > 1 + \sqrt{5}$.

These results are collected in TABLE 1. Beyond $a = 1 + \sqrt{5}$, there is first (for each of the two domains of attraction) the bifurcating hierarchy of cycles of periods 4, 8, and so forth. As for the one-hump maps,[5,7] this then gives way to an apparently chaotic regime of dynamical behavior, the details of which remain to be elucidated.

There are two distinct periodic attractors for all values of a in the range $3 < a < 4$.

Relevance to Frequency-Dependent Natural Selection

It will be recalled that the above analysis was motivated by genetic problems in which the selective forces depend on the gene frequencies.

Specifically, suppose we have one locus with two alleles, A (with frequency p) and a (with frequency $q = 1 - p$). For a diploid population, assume the genotypes AA, Aa, aa have fitnesses $w_{AA} = f(p)$, $w_{Aa} = 1$, $w_{aa} = 1/f(p)$, respectively. The change in gene frequency between one generation and the next will be

$$p_{t+1} = \frac{w_{AA}p_t^2 + w_{Aa}p_t q_t}{w_{AA}p_t^2 + 2w_{Aa}p_t q_t + w_{aa}q_t^2}. \tag{24}$$

Under the above assumptions, this reduces to

$$p_{t+1} = \frac{p_t f(p_t)}{p_t f(p_t) + q_t}. \tag{25}$$

This is now in the standard form of Equation 1,

$$p_{t+1} = F(p_t), \tag{26}$$

with p_t defined on the interval $[0, 1]$, and $F(p)$ related to $f(p)$ by

$$F(p) = \frac{pf(p)}{pf(p) + 1 - p}. \tag{27}$$

Alternatively, it may be more convenient to work with the variable $x = 2p - 1$ (see Equation 2).

Frequency-dependent natural selection typically implies $f(p) > 1$ for small p, and $f(p) < 1$ for large p. The result, via Equation 27, is a mapping $F(p)$ with two critical points, and the above analysis applies directly.

Fortunately or unfortunately, depending on whether you are a biologist or a mathematician, the full complications listed in TABLE 1 will not usually unfold for maps $F(p)$ that arise in this way. It can be shown that if the selection function $f(p)$ is a monotonic function of p, as it will be for the majority of biologically plausible circumstances, then the symmetric 2-point cycle cannot become unstable ($\lambda^{(2)}(\Delta)$ cannot increase beyond $+1$). Thus one gets either a stable point, or a (symmetric) stable cycle of period 2. This result is obtained and discussed elsewhere.[13]

Conversely, the cubic mapping of Equation 3 corresponds to a frequency-dependent selection function

$$f(p) = \frac{1 + 2a(1 - 2p)(1 - p)}{1 - 2a(1 - 2p)p}. \tag{28}$$

This has the biologically strange feature that the maximum selective advantage accrues at gene frequencies corresponding to intermediate rarity (rather than to $p \to 0$), once $a > 1$.

In short, the complexities that can arise in first-order difference equations with two critical points may be of only marginal interest in frequency-dependent natural selection. But the mathematical intricacies are interesting in their own right, and warrant further exploration.

GLOBAL VERSUS LOCAL STABILITY IN SECOND ORDER DIFFERENCE EQUATIONS: SOME PROBLEMS

Biological Motivation for This Work

Studies of temperate zone insects and their predators or parasites, and of host-parasite systems in general, often lead to two coupled first-order difference equations,[14] one for the changes in the host population and the other for the predator or parasite population. Such systems are equivalent to a single second-order difference equation.

Simple such models (often taken directly from the empirically biological literature) can exhibit very complicated dynamical behavior, including stable cycles and strange attractors. Pioneering studies are those by Beddington et al.,[15,16] and by Oster and his collaborators.[11,17,18] The situation is often further complicated by the presence of two or more different attractors.

If such a system has an equilibrium point (a fixed point of period 1) which is locally stable, an important question is whether it is globally stable. If the point is not globally stable, there arises the further question of mapping out its domain of attraction, for any specified set of parameter values. This question is pursued explicitly by Beddington et al.,[16] who make detailed numerical studies of a particular arthropod prey-predator model. Implicit in this study, and in other similar studies, is the notion that the domain of attraction is some tidily closed region surrounding the attracting point. This intuition derives, at least in part, from the corresponding models with differential equations (where locally stable equilibrium points tend to have their domains of attraction defined nicely by some unstable limit cycle).

Here I present a counter-example, where a very simple host-parasite model can have a locally, but not globally, stable equilibrium point, and where the domain of attraction can be exceedingly messy. Nor is this a contrived example; it is a slightly simplified version[19] of the "Crofton model,"[20] which is widely cited in the parasitological literature as *the* simplest model of a host-parasite system with discreet generations.

A Parasitological Example

The pair of coupled first-order difference equations in question can be written in dimensionless form as

$$X_{t+1} = \lambda X_t [1 + Y_t]^{-k}, \tag{29}$$

$$Y_{t+1} = \frac{X_{t+1}}{\lambda} \cdot \frac{Y_t}{1 + Y_t}. \tag{30}$$

Here X_t and Y_t are essentially the number of hosts and of parasites in generation t, respectively. The parameter λ represents the intrinsic growth factor for the host population, and the (negative binomial) parameter k measures the degree of parasite aggregation.

A linearized analysis of the dynamical behavior of this system is presented in

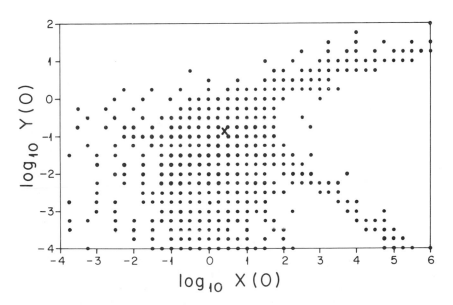

FIGURE 3. The figure indicates the global stability behavior of the pair of first-order, coupled difference Equations 29 and 30, with $\lambda = 2$ and $k = 5$. The locally stable equilibrium values of X and Y are marked by the cross, and those initial points $X(0)$, $Y(0)$ which are marked with dots on the grid do indeed give trajectories that converge to this equilibrium point. All other initial points on the grid lead to oscillations that diverge until Y is less than 10^{-99}. The point at $X(0) = 10^{0.75}$, $Y(0) = 10^{-3.5}$ is exceptional, being attracted to a stable 11-point cycle. For a more full discussion, see the text.

full elsewhere.[19] The equilibrium point is locally stable for all λ if $k < 2$, and for

$$\lambda < \left(\frac{k-2}{k+2}\right)^k, \tag{31}$$

if $k > 2$.

FIGURE 3 is devoted to the specific case $\lambda = 2$, $k = 5$. The cross at $X = 2.30$, $Y = 0.15$ marks the (locally stable) equilibrium point. The figure shows the eventual dynamical fate of the system for various values of the initial populations, $X(0)$ and $Y(0)$: $X(0)$ ranges from 10^{-6} to 10^4 inclusive, and $Y(0)$ from 10^{-4} to 10^2 inclusive (both in steps of 0.25 on a logarithmic scale). The solid dots denote grid points that are attracted to the equilibrium point. All other initial grid points (except one) lead to diverging oscillations that carry Y below 10^{-99}. The initial point at $X(0) = 10^{0.75}$, $Y(0) = 10^{-3.5}$ is attracted into a stable 11-point cycle.

FIGURE 3 is not devoid of pattern. All points within an order of magnitude or so of the equilibrium point are attracted to it, and for large $X(0)$ there is a tendency for points lying along two "spiral arms" to converge to the equilibrium point. But, particularly for relatively small values of $X(0)$, some parts of the grid look rather like a collection of go-stones thrown randomly on a go-board.

An alternative approach to mapping out the domain of attraction of the equilibrium point is to run the equation backward in time, starting from points in the neighborhood of the equilibrium point. This is an effective technique for corresponding differential equations. The pair of Equations 29 and 30 are, moreover, easily inverted, to give

$$Y_t = \frac{\lambda Y_{t+1}}{X_{t+1} - \lambda Y_{t+1}} \tag{32}$$

$$X_t = \frac{X_{t+1}}{\lambda} \left(\frac{X_{t+1}}{X_{t+1} - \lambda Y_{t+1}} \right)^k . \tag{33}$$

When run backward in time, the trajectories are not confined to the positive quadrant. The procedure does not indicate any more coherent a domain than is shown in FIGURE 3.

Implications of This Example

Elsewhere in this volume, Yorke[4] has shown that dynamical systems can have "pre-turbulent" regions in which trajectories wander around for arbitrarily varying periods of time, before suddenly converging in on an equilibrium point. Conceivably, the above system is an example of this kind, where I stopped following trajectories when Y went below 10^{-99}, when in fact such trajectories may eventually have been captured in to the equilibrium point. But I strongly doubt it.

This example warrants further analytical study. It raises disturbing questions about the relation between local and global stability in simple host-parasite and arthropod prey-predator models. The pair of Equations 29 and 30 come from the biological literature, and provide a sensible (if overly simple) description of some of the essential features of host-parasite interactions. Yet their dynamical behavior is horribly complicated, in a way that has not previously been contemplated in these contexts.

ACKNOWLEDGMENTS

I am much indebted to J. R. Beddington and G. F. Oster for helpful discussions.

REFERENCES

1. MAY, R. M. 1977. Thresholds and breakpoints in ecosystems with a multiplicity of stable states. Nature **269**: 471–477.

2. GUCKENHEIMER, J. 1978. Global bifurcation theory of dynamical systems. Ann. N.Y. Acad. Sci. This volume.

3. HOPPENSTEADT, F. C. 1978. Nonunique stable responses of externally forced oscillatory systems. Ann. N.Y. Acad. Sci. This volume.

4. YORKE, J. A. Private communication. Also see KAPLAN, J. & J. A. YORKE. 1978. Preturbulence: A metastable chaotic regime in the system of Lorenz. Ann. N.Y. Acad. Sci. This volume.

5. MAY, R. M. 1976. Simple mathematical models with very complicated dynamics. Nature 261: 459–467.

6. SHARKOVSKY, A. N. 1964. Ukr. Matem. Zh. 16: 61; 1966. ibid. 17: 104. Cited in KLOEDEN, P., M. A. B. DEAKIN & A. Z. TIRKEL. 1976. A precise definition of chaos. Nature 264: 295.

7. MAY, R. M. & G. F. OSTER. 1976. Bifurcations and dynamic complexity in simple ecological models. Amer. Natur. 110: 573–599.

8. CLARKE, B. C. 1969. The evidence for apostatic selection. Heredity 24: 347–352.

9. WRIGHT, S. 1969. Evolution and the Genetics of Populations. Vol. 2. Chicago University Press. Chicago, Ill.

10. ENDLER, J. A. 1977. Geographic Variation, Speciation and Clines. Princeton University Press. Princeton, N.J.

11. GUCKENHEIMER, J., G. F. OSTER & A. IPAKTCHI. 1976. The dynamics of density dependent population models. J. Math. Biol. 4: 101–147.

12. LI, T-Y., & J. A. YORKE. 1975. Period three implies chaos. Amer. Math. Monthly 82: 985–992.

13. OSTER, G. F., R. M. MAY & R. C. LEWONTIN. 1978. Frequency dependent natural selection: stable points, stable cycles, but no chaos. Theor. Pop. Biol.

14. HASSELL, M. P. 1978. Arthropod Predator-Prey Systems. Princeton University Press. Princeton, N.J.

15. BEDDINGTON, J. R., C. A. FREE & J. H. LAWTON. 1975. Dynamic complexity in predator-prey models framed in difference equations. Nature 255: 58–60.

16. BEDDINGTON, J. R., C. A. FREE & J. H. LAWTON. 1976. Concepts of stability and resilience in predator-prey models. J. Anim. Ecol. 45: 791–816.

17. AUSLANDER, D., J. GUCKENHEIMER & G. F. OSTER. 1978. Random evolutionarily stable strategies. Theor. Pop. Biol. 13: 276–293.

18. OSTER, G. F., A. IPAKTCHI & S. ROCKLIN. 1976. Pheontypic structure and bifurcation behavior of population models. Theor. Pop. Biol. 10: 365–382.

19. MAY, R. M. 1977. Dynamical aspects of host-parasite associations: Crofton's model revisited. Parasitology 75: 259–276.

20. CROFTON, H. D. 1971. A model of host-parasite relationships. Parasitology 63: 343–364.

CAUSALLY INDETERMINATE MODELS VIA MULTI-VALUED DIFFERENTIAL EQUATIONS

Henry Y. Wan, Jr.

Economics Department
Cornell University
Ithaca, New York 14850

INTRODUCTION

I present my views as an economist about the possible use of global dynamic analysis in economics. My discussion is neither rigorous nor general. It does not reflect the consensus of the economic discipline. The conceptual issues I raise have rarely surfaced in the economic literature and from personal contacts, it appears that most of us economists are not aware of such problems. I believe that economics can fulfill its central mission in the future if and only if its analysis is tempered by the global dynamic point of view. Both economic theory and econometric practices need fundamental modifications by taking advantage of the recent advances in mathematics.

Economics *is* for prediction. But even with full information and in the absence of randomness, can one *always* predict a *unique* outcome? My answer is no. Uniqueness can be guaranteed only by adopting very strong assumptions on technology and individual behavior. Such assumptions are so strong that no one defends them on their realism, *a priori*. They are usually postulated as "simplifying" assumptions: their precise role in avoiding indeterminacy is extremely rarely mentioned, and perhaps not even recognized by many researchers. It should be imperative to estimate empirically how well such assumptions fit reality.

Econometrics *is* for estimation. Its separate existence from general statistics rests upon its special relationship to economic theory. But has econometrics verified the capability of economic theory to predict a unique outcome? My answer is again no. Current econometric practice precludes such opportunity to verify such predictability.

In this paper our focus is placed on dynamic, deterministic theories. But what holds for nonunique evolutionary path in an intertemporal model corresponds to the nonuniqueness of equilibria in an atemporal equilibrium model, to which comparative static analysis is addressed. In fact, the nonuniqueness of the momentary equilibrium, which resembles the static general equilibrium on one hand and affects the evolution of a dynamic model on the other hand, is the source of nonuniqueness of economic evolution. Whatever is true for a deterministic model, also holds for a realized event chain in a stochastic model. If nonuniqueness prevails under one contingent situation, no probabilistic prediction can be made for the stochastic prospect.

Moreover, what we are concerned with should be relevant for all social sciences, as well. There, *repeated* games would affect future evolutions through the *current* game theoretic solution in which all individuals interact. In game theory, descriptive models often allow for nonunique solutions.

530

0077-8923/78/0316-0530 $1.75/1 © 1979, NYAS

Historically, Thornton was probably the first economist to note the multiplicity of trade equilibria in a critique to J.S. Mill in the mid-19th century.[1] Kaldor subsequently discussed its implications to some extent.[2] Today, the non-uniqueness is conceded by most workers in a temporal general equilibria. The supposedly reasonable macroscopic conditions for uniqueness ("dominant diagonals," etc.) are not anchored in microscopic conditions on individual decision makers.

So far, the dynamic economic models obtain unique outcome through one of the following four approaches (for exact references, see Reference 3):

(1) Modeling the economic system with no explicit reference to individual actions. This dates back to the Malthusian theory of population.

(2) Assuming that one individual is in control of all events in total isolation. The Ramsey model of optimal saving is a good illustration.

(3) Supposing that all individuals are such and act in such a manner as if there were only one individual. Harrod's one sector growth model is perhaps the earliest example.

(4) Postulating a complex of assumptions which manage to rule out multiplicity. The classical example is Uzawa's assumption that under all input price configurations, it requires more capital per worker to produce goods for consumption than for investment.

An alternative exists. All is not lost in the face of multiplicity. The totality of all admissible paths, the set of states reachable from an initial position after a time interval, and the asymptotic behavior of the reachable set can all be studied. This was done in Reference 4, on a rough-and-ready basis. The elegant and rigorous theory arising in the stability analysis for general equilibrium proceeds on similar lines.[5,6] The approach in Reference 5 suits superbly their context. For other problems, the less formal method in Reference 4 appears more intuitive and still adequate to the task on hand.

Science avoids dogmatism. Should one recognize explicitly the bifurcation possibility and tackle it via the multi-valence dynamic processes? Should one side-step it as it is conventionally done? This must be judged case by case. In physical science, a most elaborate model need not be the most appropriate model. In presenting an alternative analysis here, I do not advocate its universal adoption. It should be useful for economists, however, to have an analytic instrument ready for bifurcation, in case this possibility cannot be consciensciously ignored.

We shall present in the following, examples from the economic literature which yield bifurcation. We shall next introduce concepts and analyze one example to greater length. Some methodological remarks will be offered in conclusion.

Four Examples

The Industrialized Center and Its Periphery

Almost two decades and a half ago, Prebisch claimed that by Engel's Law, the industrialized center would gain against the nonindustrialized periphery through the ever deteriorating terms of trade against the less developed countries.[7] Rival camps of economists debate about the validity of this thesis ever since.[8] We shall demonstrate that conclusive answers may be theoretically impossible. Our example adopts simplifying assumptions to facilitate discussions. However, these assumptions are inessential to the presence of epistemological limits for our economic analysis.

Let regions 1 and 2 represent the pheriphery and the center, respectively, and goods 1 and 2 denote primary and manufactured goods. Each region contains $N(t)$ identical, price-taking traders, at time t, so that we may conduct our analysis in a two-person, two-good box diagram. Moreover, individuals in both regions share the same utility index: $U(x_{i1}, x_{i2}) = \min(x_{i1}^2, x_{i2})$, where x_{ij} is the per capita consumption of goods j in region i. The initial endowment vector for an individual in region i is te_i where e_i is the ith unit vector. In other words, at time t, a person in region k has t units of good k and nothing else. As time goes on, a proportional expansion of endowment goes on in both regions.

Our choice for the particular utility index is motivated by its simplicity. It illustrates Engel's Law, that if an individual maximizes his utility $U(x_{i1}, x_{i2})$ subject to the budgetary and nonnegative constraints

$$p_1 x_{i1} + p_2 x_{i2} \leq y_i \quad x_{i1} \geq 0 \quad x_{i2} \geq 0, \tag{1}$$

where y_i is his income and p_i is the price for good i, then given the prices (p_1, p_2), as a person gets richer, the consumption of primary (manufactured) goods rises proportionally slower (faster) than income, i.e.,

$$\partial \log x_{i1} / \partial \log y_i < 1 < \partial \log x_{i2} / \partial \log y_i.$$

For the particular utility index chosen, it is clear that under whatever positive price pair (p_1, p_2), the following equality always holds:

$$x_{i2} = x_{i1}^2. \tag{2}$$

If we take into consideration the fact that the income for region i equals the value of endowment,

$$y_i(t) = p_i t. \tag{3}$$

On the other hand, (p_1, p_2) at instant t must be such to clear the market; i.e., if we equate the aggregate consumption of any good among the two regions to the total supply,

$$\sum_{i=1}^{2} x_{ij}(t) = t \quad j = 1, 2.$$

Solving Equations 1–4 for p_1, p_2, and x_{ij}, one can determine all the possible equilibrium configurations for each t. Since p_1 and p_2 are determined only up to a multiplicative scaler, one may set $p_1 + p_2 \equiv 1$.

In FIGURE 1, the initial holding of region 1, t units of primary good (good 1) and 0 unit of industrial good (good 2), is represented by the "endowment point" $(t, 0)$. The budget line $p_1 x_{11} + p_2 x_{12} = p_1 t$ passes through that point with a slope equal to the price ratio $p_1/p_2 = p_1/(1 - p_1)$. The budget set, i.e., the totality of all affordable consumption bundles, is the shaded triangle below the budget line.

Three niveaux lines of the utility function are shown. Utility is maximized by consuming bundle $x^{(1)}$, where the budget line touches the highest reachable niveaux line. Given the form of the utility index, the intersection between the budget line and the $x_{12} = x_{11}^2$ locus marks the optimal consumption bundle.

FIGURE 2 is the box diagram, frequently used in economics. The $t \times t$ square represents the total available goods, t units of each good. By marking the upper-right corner as 0_2, the origin for region 2, we can represent any allocation of the available goods $(x_{11}, x_{12}, x_{21}, x_{22})$ as a point in the square. Point E denotes the

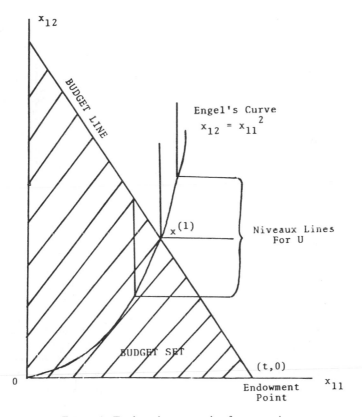

FIGURE 1. Trade and consumption for one region.

"endowment pair," vector $0_1E = (t, 0)$ for region 1, and vector $0_2E = (0, t)$ for region 2. The point e_1 represents an equilibrium allocation. The line $Ee_1L''L'$ represents a budget line for region 1, under which vector 0_1e_1 is its optimal consumption choice. The segment EeL'' of that line represents a budget line for region 2, under which vector 0_2e_1 is its optimal choice. The slope of that line is an equilibrium price ratio, $p_1^{(1)}/p_2^{(1)}$, in that vector 0_1e_1 + vector $0_2e_1 = (t, t)$ so that total consumption of both goods is equated to their respective available supplies. Because of the special form of the utility index, the equilibrium allocation e_1 is identifi-

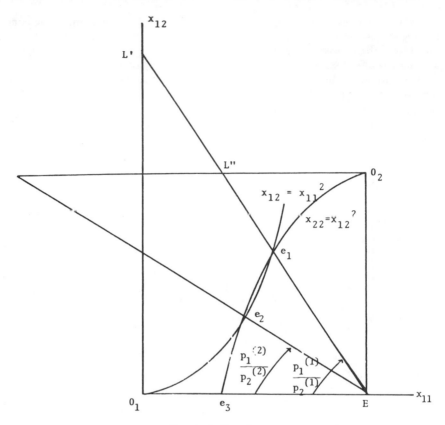

FIGURE 2. Box diagram.

able with the intersection between the locus $0_1 e_2 e_1 : x_{12} = x_{11}{}^2$ and the locus $0_2 e_1 e_2$:
$x_{22} = x_{21}{}^2$. But by the same token, e_2 also qualifies as an equilibrium where
$p_1^{(2)}/p_2^{(2)}$ is the corresponding equilibrium price ratio which can be read off from
the slope of line Ee_2. There is a further equilibrium with $p_1 = 0$ and $p_2 = 1$,
and an equilibrium allocation representable by any point in the segment $[0, e_3]$.
The above situation holds for all t, $1 < t < 2$. Plotting the equilibrium price
p_1 for all positive t yields FIGURE 3. One is tempted to interpret FIGURE 3 as
an example of Riemann-Hugoniot catastrophe.[9] The set of catastrophe points is
evidently: $a = (1, 1)$, $b = (\frac{1}{2}, 2)$, and $c = (0, 1)$ in the $B \times T$ space where $B =$
$[0, 1]$, $T = [0, \infty)$ and (p_1, t) is the generic element.

In FIGURE 3, if we denote $P_1(t)$ as the set of equilibrium p_1-values at t,
then any function $h(0)$ such that $h(t)$ belongs to $P_1(t)$, for all $t \geq 0$, is a "possible
history" for $p_1(t)$. FIGURE 3 shows that initially, the manufactured good is free
($p_1 = 1$, $p_2 = 0$, for $t < 1$) and ultimately, the primary good is free ($p_1 = 0$,
$p_2 = 1$, for $t > 2$). Prebisch may be regarded as vindicated in that limited sense.
But since Prebisch is interested in a world where no good is free, we must limit
our attention to the period where $1 < t < 2$, in which three branches of the graph

for equilibrium price coexist side by side. Again, ignoring the lower branch where the primary good is free, we still have two branches to contend with.

We note that, in general, if $U = \min(x_{i1}{}^q, x_{i2}{}^r)$, $0 < q < r$, the consumption bundle for i must satisfy the equation, $x_{i1} = x_{i2}{}^{r/q}$ where $r/q > 1$. The curvilinear form of such a locus gives rise to possible multiplicity of solutions. Moreover, consider the "direct addilog" utility function:

$$(x_{i1}{}^{-\beta} + x_{i2}{}^{-\gamma})^{-1} = [x_{i1}{}^{-\beta} + (x_{i2}{}^{\gamma/\beta})^{-\beta}]^{-1/\beta} \qquad 0 < \beta < \gamma$$

$$= (\phi[x_{i1}, z_{i2}; \beta])^{\beta}$$

say, where $z_{i2} = x_{i2}{}^{\gamma/\beta}$. It is known[10] that

$$\lim_{\beta \to \infty} \phi[x_{i1}, z_{i2}; \beta] = \min[x_{i1}, z_{i2}].$$

Consequently, for all x_{i1}, x_{i2} belonging to $[0, t]$, the niveaux lines of $\min[x_{i1}, z_{i2}]$ in the space of $x_{i1} \sim z_{i2}$ approximates those of $\phi[x_{i1}, z_{i2}; \beta]$, for large β, and the niveaux lines of the latter coincides with those of the original "direct addilog" function in the $x_{i1} \sim z_{i2}$ space. If we transform back to the $x_{i1} \sim x_{i2}$ space, then for large enough values of β and γ, the niveaux lines approximate those of $\min(x_{i1}, x_{i2}{}^{\gamma/\beta})$. Consequently, multiple solutions are again likely under certain values of t.

Consider now the Houthakker model[11] where at any instant, two regions each produces a single good, both sharing an identical addilog utility index. The trade and consumption patterns of both regions are so determined to maximize utility with respect to the market-clearing equilibrium prices. Prices so determined are then used to evaluate the per capita incomes of both regions and the speed and direction of migration depend upon such income comparisons. Populations are modified by migration and natural growth. Production levels are determined by population and the technological progress. The model is now formally complete.

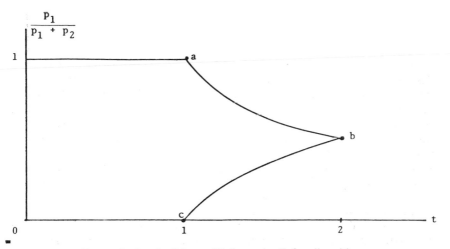

FIGURE 3. Graph of the equilibrium price P_1 for all positive t.

But as we have seen that given the output levels, alternative prices can achieve equilibrium, alternative relative incomes may prevail for the two regions, and hence alternative migration patterns may follow. Hence, the prediction of a unique outcome may not be possible. To resolve this difficulty, Houthakker assumes that each region may consume at the value of the average income between two regions. In other words, budget constraints are no longer binding. But such a model is hardly satisfactory. After all, if irrespective of relative income figures, every person can consume the same value of goods anywhere, why should one ever bother about migration?

Resource Reallocation at Optimal Speed

In Reference 4 a single input with constant supply—labor—is used in two production processes, each producing a different good. Given the outputs, the market demand is such that the output price ratio would vary in inverse proportion to the output quantity ratio. The level of output in each sector is a concave parabolic function of the effective labor force in that sector. The effective labor force differs from the total sectoral labor force in that hiring and firing divert manpower to training and reorganizing activities. The speed of firing or hiring is proportional to such diversions. Given the wage rate and the output prices, firms will solve the variational problem to maximize the profit stream by deciding the optimal speed for hiring and firing. In such decisions, current prices and wages are believed to last forever. In fact, however, wage is decided each instant so that hiring matches firing. Constant plan revision always takes place.

FIGURE 4 summarizes the situation. Panel (a) plots the relationship between L, the proportion of total labor input in sector 1, and its rate of change, $\dot{L}$. Panels (b) and (c) illustrate the cases where the initial value of L is very small and very large, respectively. The shaded "solution funnels" represent the totality of the graphs of all possible evolution path, starting from the known initial position.

Two-Sector Neoclassical Growth

In the early sixties, Uzawa and Inada[12-14] studied the two-sector model of neoclassical growth, where investment and consumption goods are produced with capital and labor according to two production functions that are smooth, concave, and first-order homogeneous. Given the input prices, both sectors strive to produce at minimum unit cost by varying input proportions. Proportions of capital and labor incomes are saved to buy investment goods. Labor increases at a constant proportional rate. The state of the system is signified by the overall capital-per-worker ratio. Given each equilibrium input price ratio, the speed of change for the overall capital-per-worker ratio can be determined. However, there may be more than one equilibrium input price ratio. This is the earliest instance that "causal indeterminacy" is explicitly acknowledged. However, subsequently, all research work done is devoted to the finding of assumptions to avoid such indeterminacy rather than analyze it as such. More will be said in the next section.

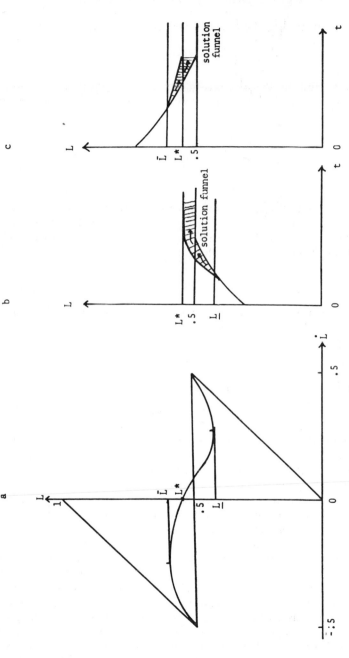

FIGURE 4. Summary. (a) Relationship between L, the proportion of total labor input in section 1, and its rate of change $\dot{L}$. (b) Case where the initial value of L is very small. (c) Case where the initial value of L is very large.

Macroeconomic Stabilization under Uncertainty

The economic system is described by the dynamic equation:

$$\frac{dx}{dt} = f(x, u, v),$$

where x, u, v stands for the state vector, the vector of government controls, and the vector of disturbances caused by speculators, special interest groups, or Nature. It is assumed that $v(t) = V[x(t)]$, where $V(\cdot)$ is not completely known to the researcher. The target of the government is to minimize the integral of

$$\int_t^\infty \{C[x(t)] - \bar{C}\} dt,$$

where $C(\cdot)$ is the welfare loss function and $\bar{C}$ is the level of unavoidable cost. In Reference 15, a special case of this problem is studied where $\bar{C}$ is zero and $C[x(t)] = [x(t) - \bar{x}(t)]^2$ with $\bar{x}(\cdot)$ being the target path. It is then shown that under certain conditions one can construct a policy $u = U^*(x)$ such that for all possible $V(x)$, the solution of the differential equation

$$\frac{d}{dt}(x - \bar{x}) = f[x, U^*(x), V(x)]$$

yields a path convergent to $\bar{x}(t)$. The basic approach is to find a Lyapunov function L, consider its time derivative

$$\frac{dL}{dt} = g(x, u, v),$$

and for each x, determine the saddle value:

$$s(x) = \min_u \max_v g(x, u, v).$$

If $s(x) < 0$ for all x, then asymptotic stability is attained.

A "solution funnel" containing all possible time-paths starting from a given initial position is shown in FIGURE 5. Under the prescribed policy for stabilization, all such time-paths converge to the desired path.

The first three examples were "descriptive" models, predicting what the future may be. In all three of these, price formation plays the key role in causing multiple time paths. At each instant, individuals act according to both the historically given state variables and the market-determined equilibrium prices. Such actions decide the levels of migration, firing/hiring, and capital accumulation in these examples, respectively. Hence, the course of future events will be modified by such individual activities. In the first two examples, individuals exhibit rationality, either traders maximize their instantaneous utility, or firms maximize their expected profit streams. In the third example, individuals save according to given behavior rules. But in all these, their collective actions are coordinated by the "equilibrium" price signals so that the aggregate demand on resources match their respective aggregate supply. The co-existence of alternative equilibrium prices imply alternative time paths.

The problem is ubiquitous. In all dynamic economic models appearing in literature, further disaggregation is likely to cause multiplicity in time path. The problem is usually neglected. The "aggregation" assumptions (usually adopted as an expository expedient) hide away the conceptual problem of what can one really predict. The problem is also relevant. Its policy implications are illustrated by the debates about the Prebisch view (above). Its analytical ramifications are bound to affect the current studies of temporary equilibrium models.[16,17] The crucial issue is not so much a matter of unique time path or a multiplicity of such time paths. Nor is it an issue whether at almost every state, a finite number of nondenumerably many equilibrium price vectors may coexist. This latter has been studied recently through an elegant differential topologic approach.[18] The question is how "far

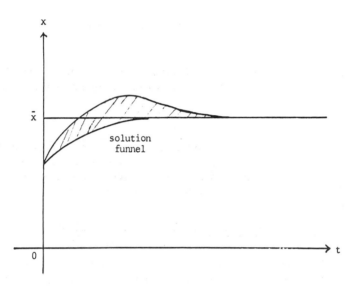

FIGURE 5. A stabilized "solution funnel" containing all possible time-paths starting from a given initial position.

apart" are different evolutionary paths likely to be. In realistic models, a direct evaluation of *all* equilibria is not likely to be appropriate. Existing procedures only aim at the finding of one equilibrium.[19] What I suggest is a three-pronged attack:

(1) A conceptual analysis to relate some bounds for the equilibrium price set to some easily estimatible magnitudes.[20]

(2) An econometric approach that emphasizes the estimation of structural equations. More on this later.

(3) A control procedure along the lines of the fourth example given here to influence the "width" of the solution funnel, if this is deemed desirable for public interest.

Suffice to say, not all is lost in the face of multiple evolutionary paths. We shall now investigate this matter in depth in the context of the third example. This is the most well known among our four examples.

THE TWO-SECTOR NEOCLASSIC MODEL REVISITED

We define k as the overall capital-labor ratio, ω as the wage-rent ratio (i.e., $\omega = w/r$ where w and r are wage and rent rates, respectively), and $f_1(k_1)$ and $f_2(k_2)$ are output per worker ratios in the first (investment) and second (consumption) sectors, which are functions of the capital per worker ratios, k_1 and k_2, of these respective sectors.

We assume the Inada conditions, i.e.,

$$f_i(0) = 0 = f'_i(\infty), f'_i(0) = \omega = f_i(\infty),$$

$$f'_i > 0 > f''_i, \quad i = 1, 2.$$

We shall use the expressions $k_1(\omega)$ and $k_2(\omega)$ to indicate the functional dependence between the least-unit-cost capital per worker ratios and the wage-rent ratio.

It can be shown that

$$\{f_i[k_i(\omega)]/f'_i[f_i(\omega)]\} - k_i(\omega) = 0, \quad i = 1, 2.$$

Again, we define s_r and s_w as the saving ratios out of capital income and labor income, respectively, $(0 < s_w, s_r < 1)$, n and δ as the labor growth and capital depreciation rates, respectively, and $\psi(\omega)$ as that over-all capital-labor ratio consistent with the wage-rent ratio ω, and it can be shown that for each ω, $\psi(\omega)$ takes the unique value:

$$\psi(\omega) = \frac{k_1(\omega)k_2(\omega) + \omega s_w k_1(\omega) + \omega(1 - s_w)k_2(\omega)}{\omega + s_r k_2(\omega) + (1 - s_r)k_1(\omega)}.$$

It can be shown that $\psi(0) = 0$ and $\psi(\infty) = \infty$, and that $\psi^{-1}(k)$ is nonempty over $(0, \infty)$ but it need not be a singleton for all k. Let $\Delta(\omega)$ represent the rate of increase of k, i.e., dk/dt, implied by the wage-rent ratio. It can be shown that $\Delta(\omega)$ takes the form of:

$$[s_w \omega + s_r \psi(\omega)]f'_1(\omega) - (n + \delta)\psi(\omega).$$

We define $F(k) = \{\Delta(\omega):\omega\epsilon\psi^{-1}(k)\}$ as the set of admissible "velocities" of change for k, corresponding to the value, k. $F = \Delta \circ \psi^{-1}$ is a point-to-set correspondence, not always single-valued.

These facts are well-known in the theory of economic growth and summarized in Reference 3. Hitherto, economists strived to obtain weaker and weaker set of conditions under which F is always single-valued, so that one has the differential equation:

$$\frac{dk}{dt} = F(k).$$

The alternative option is to consider the differential correspondence:

$$\frac{dk}{dt} \in F(k).$$

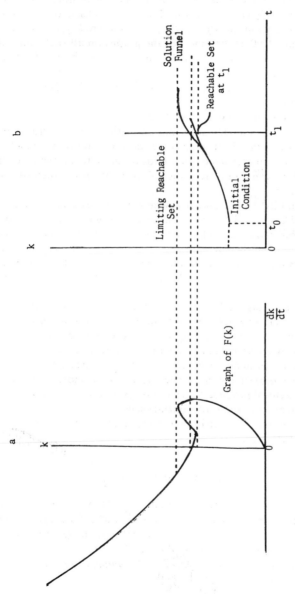

FIGURE 6. Solution funnel and reachable set.

The graph of $F(k)$ can be studied as follows. In the space of ω, k, and dk/dt, the graph of $\Gamma(\omega) = [\omega, \psi(\omega), \Delta(\omega)]$ is a simple, connected curve. The projection of graph Γ on the $(k, dk/dt)$ space is the graph for F.

Define now a function $k(\cdot)$ as a *solution* to the above differential correspondence if $k(t)$ is absolutely continuous on all compact subsets of $[t_0, \infty)$, $dk/dt \in F[k(t)]$ for almost every t in $[t_0, \infty)$ and $k(t_0) = k_0$ where (k_0, t_0) is the given initial condition. Define further the union of the graphs on all such solutions as the *solution funnel* and the cross-section of this funnel at t as the *reachable set* of the equation from (t_0, k_0) at t. Such concepts are either taken from or defined analogously to those in the dynamic process literature.[21-24] These concepts are illustrated in FIGURE 6.

The advantage of such a theory is two-fold:

(1) Quantitatively, given adequate information on the form of sectoral production functions, one can numerically determine the solution funnel, which, after all, supplies all the information one can ask for in such a theoretical model.

(2) Qualitatively, one can still deduce the following types of conclusions, as it is shown elsewhere[25]:

   (a) System stability. In other words, whether there exists a *compact asymptotically stable set* of k-values such that eventually the value of over-all capital–labor ratio must belong or stay close to it. For the two-sector neoclassical model studied here, this question can be answered in the affirmative, by the simple observation that there exists $\underline{k}$ and $\bar{k}$, $0 < \underline{k} > \bar{k} < \infty$, such that: For all $k < \underline{k}$, and $\Delta \epsilon F(k)$, $\Delta > 0$; for all $k > \bar{k}$ and $\Delta \epsilon F(k)$, $\Delta < 0$.

   (b) Sensitivity analysis. Perhaps the most interesting "comparative dynamics" question is that whether the "solution funnel" continuously deform to a limiting time path in an appropriate sense if the structure of a sequence of causally indeterminate model converges toward a limiting model which is causally determinate. Two cases have been considered:

   (i) if the factoral saving ratios become closer and closer to each other, and

   (ii) If the sectoral production technologies become more and more similar to each other in some appropriate sense.

   In both cases, one can conclude that the solution funnel "shrinks" toward a limiting time-path.

CONCLUDING REMARKS

Three questions will be briefly addressed here: (1) What are the econometric evidences of the possible nonuniqueness of time path? (2) Can economic models be reformulated so that unique prediction can be assured? (3) Are there alternative approaches in analyzing path nonuniqueness?

In econometrics, the fundamental *structural equation* can be written as:

$$F(y, z, u) = 0,$$

where y is the jointly distributed vector, z is the predetermined vector, and u is the random vector. The evolution of the system is predictable by a sequence of y_t, over time. "Causal indeterminacy," or nonunique evolution arises when given one particular realization of u and one particular known vector z, there exist more than one vector y satisfying the structural equation. Large scale econometric studies extant today (e.g., as surveyed in Reference 26) usually assume away the problem and start with a reduced form:

$$y = f(z, u).$$

The traditional econometric analysis for "simultaneous equation estimates" postulate that $F(\cdot)$ is linear.[27] Given appropriate rank conditions, causal indeterminacy cannot happen. Recent studies of nonlinear regression maintain that with the addition of nonnegative constraints, there probably can be only one "reduced form" solution $y = f(z, u)$ for the equation, $F(y, z, u) = 0$.[28,29] It seems this is only an opinion and not a conclusion deduced from either assumptions or observations. It is not known, however, that extensive efforts have been made to track down all alternative "reduced form" solutions for one structural equation.

In economic theory, very little work has been done where many price-making individuals coexist in an analytic model. I doubt though, that once we assume that prices are set in some "rational" way, we can still obtain unique time-paths under rather general assumptions.

Finally, within the global dynamic framework, the bifurcation theory usually deals with the rapid changes of system behavior corresponding to some parametric variations. In our discussions above, the parametric changes are included within the full-fledged model. The theory of catastrophe allows a dynamic system to make sudden jumps in the state space at certain catastrophe points, e.g., point b in FIGURE 3 of our first example. Nonetheless, causality is usually preserved in such models. This corresponds to what we termed the "branch continuity" view in Reference 4. Currently, such concepts have been applied for the Keynes-Kaldor nonlinear cycles.[30] Here we eschew such a view. In the first place, "branch continuity" is not anchored in any of the known economic assumptions on individual behavior or institutional framework. One might try to relate such branch continuity to the stability of a tâtonnement process. But the behavioral implications of such stability concepts in a world without an auctioneer is not entirely clear. In the second place, at some points of catastrophe, the direction of "jumps" is presumably still nonunique. Hence, causality is not preserved in any case. These matters deserve further analysis and I will be ready to accept a superior alternative theory. However, in any case, dynamic economic theory should take account of these issues raised above, in one manner or another.

REFERENCES

1. PRIESTLEY, F. E. L., et al., Eds. 1965. Collected Works of John Stuart Mill. Vol. 3. Toronto University Press. Toronto.
2. KALDOR, N. 1934. A classificatory note on the determinateness of equilibrium. Rev. Economic Studies 1: 122–36.
3. WAN, H. Y., JR. 1971. Economic Growth. Harcourt, Brace Jovanovich. New York.
4. KEMP, M. C. & H. Y. WAN, JR.: Hysteresis of long-run equilibrium from realistic adjustment costs. In Trade, Stability and Macro-Economics. G. Horwich & P. A. Samuelson, Eds. Academic Press. New York.
5. CHAMPSAUR, P., J. DREZE & C. HENRY. 1977. Dynamic processes in economic theory. Econometrica 45: 273–294.

6. YUN, K. K. 1977. Quasi-stability in a system of generalized differential equation and its economic application. Proc. 1st Internat. Conf. on Mathematical Modeling. University of Missouri at Rolla. Vol. 5: 2331–2337.
7. PREBISCH, R. 1959. Commercial policy in the underdeveloped countries. Amer. Economic Rev. (Suppl.) 49: 251–273.
8. PINTO, A. & J. KÑÁKAL. 1972. The center–peripheral system, 20 years later. In International Economics and Economic Development. L. E. Dimarco, Ed. Academic Press. New York.
9. THOM, R. 1975. Structural Stability and Morphogenesis. D. H. Fowler, Translator. W. A. Benjamin, Reading, Mass.
10. HARDY, G. H., J. E. LITTLEWOOD & G. POLYA. 1952. Inequalities. 2nd edit. Cambridge University Press. London.
11. HAUTHAKKER, H. S. 1976. Disproportionate growth and the intersectoral distribution of income. In Relevance and Precision, from Quantitative Analysis to Economic Policy. J. S. Cramer et al., Eds. North-Holland. Amsterdam.
12. UZAWA, H. 1961. On a two-sector model of economic growth, I. Rev. Economic Studies 29: 40–47.
13. UZAWA, H. 1963. On a two-sector model of economic growth, II. Rev. Economic Studies 30: 105–118.
14. INADA, K. 1963. On a two-sector model of economic growth, comments and generalization. Rev. Economic Studies 30:119–127.
15. LEITMANN, G. & H. Y. WAN, JR. 1977. Macro-economic stabilization for an uncertain dynamic economy. In New Trends in Dynamic Systems Theory and Economics. A. Mazzolo, Ed. Springer Verlag. New York.
16. GRANDMONT, J. M. 1977. Temporary general equilibrium theory. Econometrica 45: 535–572.
17. FUCHS, G. & G. LAROQUE. 1976. Dynamics of temporary equilibria and expectations. Econometrica 44: 1157–1178.
18. DEBREU, G. 1970. Economies with a finite set of equilibria. Econometrica 38: 387–392.
19. SCARF, H. 1973. The Computation of Economic Equilibria, Yale University Press. New Haven, Conn.
20. WAN, H. Y., JR. 1975. Multiplicity and localizability of exchange equilibria. Cornell Working Papers in Economics, No. 94. Cornell University.
21. BHATIA, N. & G. P. SZEGÖ. 1967. Dynamic Systems: Stability and Applications. Springer-Verlag. New York.
22. CASTAING, C. 1969. Some theorems in measure theory and generalized systems defined by contingency equations. In Mathematical Systems Theory and Economics. H. W. Kuhn & G. P. Szegö, Eds. Springer-Verlag. New York.
23. YORKE, J. 1969. Space of solutions. In Systems Theory and Economics. H. W. Kuhn & G. P. Szegö, Eds. Springer-Verlag. New York.
24. ROXIN, E. 1965. On generalized dynamic systems defined by contingent equations. J. Differential Equations 1: 188–205.
25. WAN, H. Y., JR. 1973. Causally indeterminate models via multi-valued differential equations, two-sector models revisited. Cornell Working Paper in Economics, No. 73. Cornell University.
26. FROMM, G. & L. R. KLEIN. 1973. A comparison of eleven econometric models of the United States. Amer. Economic Rev. (Suppl.) 62: 385–393.
27. HOOD, W. C. & T. C. KOOPMANS. 1953. Studies in econometric methods. Wiley. New York.
28. FISHER, F. M. 1966. The Identification Problem in Econometrics. McGraw-Hill. New York.
29. GOLDFELD, S. M. & R. E. QUANDT. 1972. Nonlinear Methods in Econometrics. North-Holland. Amsterdam.
30. TORRE, V. 1977. Existence of limit cycles and control in complete keynesian system by theory of bifurcation. Econometrica 45: 1457–1466.

ON COMPARATIVE STATICS AND BIFURCATION
IN ECONOMIC EQUILIBRIUM THEORY

Steve Smale

Department of Economics
University of California
Berkeley, California 94750

It is worthwhile to consider what Paul Samuelson had to say about this subject in the early 1940s in *Foundations of Economic Analysis*.[1]

"It was an achievement of the first magnitude for the older mathematical economists to have shown that the number of independent and consistent economic relations was, in a wide variety of cases, sufficient to determine the equilibrium values of unknown economic prices and quantities. . . .

"It is the task of comparative statics to show the determination of the equilibrium values of given variables (unknowns) under postulated conditions (functional relationships) with various data (parameters) being specified. . . . In order for the analysis to be useful, it must provide information concerning the way in which our equilibrium quantities will change as a result of changes in the parameter taken as independent data.

"For few commodities have we detailed quantitative empirical information concerning the exact form of the supply and demand curves even in the neighborhood of the equilibrium point.

"This is a typical problem confronting the economist: in the absence of precise quantitative data he must infer analytically the qualitative direction of movement of a complex system."

Thus the problem is posed: in an economy to study solutions in x of $f_\alpha(x) = 0$ as the parameter α changes, where f is a system of n equations in n unknowns. This is the problem of comparative statics. A bifurcation is a pair (x, α) where the Jacobian of f_α at x is zero. Roughly speaking f_α is the excess demand of an economy and the parameter α could involve either endowments, tastes, or technology. We will make this picture precise in the case of a pure exchange economy where x is an allocation-price equilibrium and α is the endowment vector. For example, changes in this parameter could be caused by new taxation or oil discoveries. The situation in explicit economic models is more subtle because of Walras' Law. But let us specify the model in a simple but basic form.

Suppose there are l commodities so that a commodity bundle is a point in $R_+^l = \{(x^1, \ldots, x^l) = x \mid x^i > 0\}$. Suppose next that in our simple economic model, there are m traders, each having a preference represented by a "utility function" $u_i : R_+^l \to R$ of differentiability class C^2 without critical points, $i = 1, \ldots, m$. Thus agent i prefers x to x' in R_+^l exactly when $u_i(x) > u_i(x')$. To each agent is associated an endowment, e_i in R_+^l, $i = 1, \ldots, m$.

A price system is a vector $p \in R^l$ so that $p = (p^1, \ldots, p^l)$ and p^j is the price of one unit of the jth commodity. The *value* of a commodity bundle $x \in R_+^l$ at

0077 8923/78/0316–0545 $1.75/1 © 1979, NYAS

price system p is the dot product

$$p \cdot x = \sum_{i=1}^{l} p^i x^i.$$

We will suppose that price systems are normalized, $||p||^2 = \Sigma(p^i)^2 = 1$, so that the space of price systems is the unit sphere S^{l-1}. (Usually also the price would be positive.

A *state* of the economy is a pair (x, p) in $(R_+{}^l)^m \times S^{l-1}$ where $x = (x_1, \ldots, x_m)$, $x_i \in R_+{}^l$ is an allocation of the goods to the m traders and p is a price system. The state is *feasible* if

$$\sum_{i=1}^{m} x_i = \sum_{i=1}^{m} e_i. \tag{1}$$

Here the vector of total resources of the economy is just the sum of the endowments. The parameter of the economy is the endowment vector $e = (e_1, \ldots, e_m) \in (R_+{}^l)^m$.

An *equilibrium* state is a feasible state (x, p) where the satisfaction condition is met: For each i, x_i is a maximum of u_i on the budget set $B = \{\hat{x} \in R_+{}^l \mid p \cdot \hat{x} = p \cdot e_i\}$. It follows directly that an equilibrium (x, p) for the economy with endowments $e_1, \ldots, e_m$ is a solution of this system of equations:

$$\Sigma x_i = \Sigma e_i \tag{1}$$

$$p \cdot x_i = p \cdot e_i \quad i = 1, \ldots, m \tag{2}$$

(note the mth equation of Equation 2 is redundant in view of Equation 1).

$$g_i(x_i) = p \quad i = 1, \ldots, m. \tag{3}$$

In condition 3 $g_i(x_i)$ is the normalized gradient of u_i at x_i and Equation 3 is the ordinary first derivative condition for a maximum. For a much more detailed exposition of these matters with further reference.[2,3] The main goal of this note is to give a theorem that assists in presenting a global setting for comparative statics and bifurcation. This theorem states that under a certain convexity condition on the u_i, the set of equilibria in (x, p, e) space is a submanifold, the "equilibria manifold." The key condition may be expressed in the following manner: For $x \in R_+{}^l$, let $K = K_{x,i}$ be the set of v which are mapped into zero by the (total) derivative $DU_i(x): R^n \to R$. Thus K is the tangent to the level surface $u_i^{-1}(c)$ at x. Then

$$\text{for each } i = 1, \ldots, m \quad \text{and} \quad x \in R_+{}^l, \tag{C}$$

the second derivative $D^2 U_i(x)$ when restricted to K is negative definite.

THEOREM. Given the above setting with condition C satisfied, the set Σ of solutions of (1), (2), and (3) (the equilibria of the pure exchange economy) is a submanifold of (x, p, e) space, $(R_+{}^l)^m \times S^{l-1} \times (R_+{}^l)^m$.

Our sketch of the proof follows References 2 or 4, but a closely related theorem can be found in the work of Balasko or Delbaen. See also Balasko[5] for a somewhat related approach to comparative statics.

If Σ were indeed a manifold then the tangent $T_{x,p,e}(\Sigma)$ of Σ at x, p, e would have the following description: T = the set of $\bar{x} = (\bar{x}_1, \ldots, \bar{x}_m)$, $\bar{p}, \bar{e} = (\bar{e}_1, \ldots, \bar{e}_m)$ with $\bar{x}_i \in R^l$, $\bar{p} \in R^l$ with $\bar{p} \cdot p = 0$ and $\bar{e}_i \in R^l$ satisfying the linear equations:

$$\Sigma \bar{x}_i = \Sigma \bar{e}_i \tag{1'}$$

$$\bar{p} \cdot (x_i - e_i) + p \cdot (\bar{x}_i - \bar{e}_i) = 0, \quad i = 1, \ldots, m \tag{2'}$$

$$Dg_i(x_i)\bar{x}_i = \bar{p}, \quad i = 1, \ldots, m. \tag{3'}$$

These equations are simply obtained by differentiating Equations 1, 2, and 3.

Now conversely it can be shown with a short argument using the inverse function theorem that if the space of solutions of (1'), (2'), and (3') has the right dimension, ml, then Σ is a submanifold of dimension ml (or empty!).

PROPOSITION. The linear space T of solutions of (1'), (2'), and (3') has dimension ml.

For the proof of the proposition, consider the linear subspace K^* of $(R^l)^m \times T_p(S^{l-1}) \times (R^l)^m$ of $(\bar{x}, \bar{p}, \bar{e})$ satisfying $\Sigma \bar{e}_j = 0$, $\bar{x}_i \cdot p = 0$, $i \leq m - 1$, and $\pi_p \bar{e}_i = 0$ for $i \leq m - 1$ where $\pi_p : R^l \to p^\perp$ is the orthogonal projection onto the space perpendicular to p. By simple counting one can see that the codimension of K^* is $l + m - 1 + (m - 1)(l - 1) = ml$. The proposition now follows easily from the assertion that $K^* \cap T = 0$. To see the truth of the assertion, let $(\bar{x}, \bar{p}, \bar{e}) \in K^* \cap T$. Let $\gamma_i : p^\perp \to p^\perp$ be restriction of $Dg_i(x)$. Then γ_i has all its eigenvalues negative by virtue of (C) and hence is nondegenerate. Then $\Sigma \bar{x}_i = \Sigma e_i = 0$. Since $\bar{x}_i \cdot p = 0$, $\gamma_i^{-1}(\bar{p}) = \bar{x}_i$ for each i. Then $\Sigma \gamma_i^{-1}(\bar{p}) = \Sigma x_i - 0$ and $\Sigma \gamma_i^{-1}$ is an isomorphism. The last uses the fact that γ_i comes ultimately from a symmetric form. Thus $\bar{p} = 0$ and the rest follows easily.

What does this theorem have to do with comparative statics and bifurcation theory? In fact both of these problems revolve around the study of the map π: $\Sigma \to (R_+^l)^m$, which sends (x, p, e) of Σ into e. To apply comparative statics at (x, p, e) this map must be a local diffeomorphism at (x, p, e) or the derivative $D\pi : T \to (R^l)^m$ must be an isomorphism. This derivative sends $(\bar{x}, \bar{p}, \bar{e}) \in T$ into $\bar{e}$, so that a study of the linear equations defined by (1'), (2'), and (3') enters. The comparative statics is described by the inverse of this derivative.

Finally, the bifurcation of equilibria occur exactly when $D\pi$ at (x, p, r) becomes singular.

The situation is on the surface similar to catastrophe theory. However, because economics does not give us a potential function, (elementary) catastrophe theory does not apply and the situation is closer to the general study of singularities of maps.

Observe in the above situation that changes in the endowment parameters don't necessarily affect a price equilibrium. For example if (x^*, p^*) is a price equilibrium at an endowment vector e^*, then (x^*, p^*) is also a price equilibrium for e where $p^* \cdot (e - e_i^*) = 0$ and $\Sigma e_i^* = \Sigma e_i$. In fact there is an $ml - (m + l)$ parameter family of such e. It is reasonable to ask for an *effective family* of *parameters* for equilibria of a pure exchange economy. This can be done as follows.

Consider a point (x, p, w_i, s) in the space $(R_+^l)^m \times S^{l-1} \times (R_+^l)^m \times R^l$. Here (x, p) is a state as before, w_i is some kind of income parameter, s is the total resources or a supply parameter. The "effective" equilibrium manifold $\hat{\Sigma}$ will be the subset of (x, p, w_i, s) in $(R_+{}^l)^m \times S^{l-1} \times (R_+{}^l)^m \times R^l$ satisfying:

$$\Sigma x_i = s$$

$$p \cdot x_i = w_i, \quad i = 1, \ldots, m$$

$$g_i(x_i) = p, \quad i = 1, \ldots, m.$$

THEOREM. $\hat{\Sigma}$ is a submanifold of dimension $m + l$.

The proof is similar to that of the preceding theorem. The comparative statics situation is similar to that before except that the w_i, and s are *effective* parameters.

Further, if

$$\alpha: (R_+{}^l)^m \times S^{l-1} \times (R_+{}^l)^m \rightarrow (R_1{}^l)^m \times S^{l-1} \times (R_+{}^l)^m \times R^l$$

is the map defined by

$$\alpha(x, p, e) = (x, p, w_i = p \cdot e_i |_{i=1}^m, s = \Sigma e_i)$$

then $\alpha^{-1}(\hat{\Sigma}) = \Sigma$.

REFERENCES

1. SAMUELSON, P. 1971. Foundations of Economic Analysis, Atheneum. New York.
2. SMALE, S. 1978. Global analysis and economics. *In* Handbook of Mathematical Economics. Arrow-Intrilligator. In press.
3. QUIRK, J. & R. SAPOSNIK. 1968. Introduction to General Equilibrium Theory and Welfare Economics. McGraw-Hill. New York.
4. SMALE, S. 1976. Global Analysis and Economics. VI. J. Math. Econ. **3:** 1–14.
5. BALASKO, Y. 1976. Equilibrium Analysis and Envelope Theory. Preprint Ecole Normale Superieure. Paris.

DISCUSSION PAPER

H. Y. Wan, Jr.

Economics Department
Cornell University
Ithaca, New York 14850

This paper shows the relevance of bifurcation theory to economics. As I am an economist, I would like to pose a problem for further research for the bifurcation theorists.

In economics, the dynamics of the system may be described by:

$$z(\alpha, x) = 0 \quad \text{(Smale)}$$

$$\frac{d\alpha}{dt} = \psi(x) \quad \text{(parametric evolution).} \tag{A1}$$

The fact that the implicit function $z(\alpha, x) = 0$ may be not reducible globally to a reduced form $x = \eta(\alpha)$ implies that (**A1**) is a generalized dynamic system studied by Castaing, Roxin, Bhatia, Yorke, and others. A solution to (**A1**) may be defined as any function $\xi(\cdot)$ absolutely continuous over any compact subinterval of R such that:

$$\frac{d\xi}{dt} \epsilon \{\psi(\xi) : z\alpha(t), \xi] = 0\} \text{ almost everywhere} \tag{A2}$$

and that satisfies certain boundary conditions. My paper explores that possibility.

From a more "catastrophe theoretic" point of view, the qualification, almost everywhere, may be replaced by the qualification, "except at catastrophe points." This, however, need not restore causality as FIGURE 1 shows. The generalized dynamic system view regards causality is lost at $\alpha = a$. The catastrophe theory view still must concede that causality may be lost at b.

In the Uzawa two-sector growth model, the Equation A1 takes the following form*

$$\alpha = (K, L)$$

$$x = (K_1, L_1, K_2, L_2, r, w, p)$$

$$z_1(\alpha, x) = K - K_1 - K_2$$

$$z_2(\alpha, x) = L - L_1 - L_2$$

$$z_3(\alpha, x) = \frac{\partial F_1}{\partial K_1} - r$$

$$z_4(\alpha, x) = \frac{\partial F_2}{\partial L_1} - w$$

*UZAWA, H. 1961. Rev. Economic Stud. **29**.

0077-8923/78/0316-0549 $1.75/1 © 1979, NYAS

$$z_5(\alpha, x) = \frac{\partial F_2}{\partial K_2} p - r$$

$$z_6(\alpha, x) = \frac{\partial F_2}{\partial L_2} p - w$$

$$z_7(\alpha, x) = s_r rK + s_w wL - F_1(K_1, L_1)$$

$$z_8(\alpha, x) = (1 - s_r)rK + (1 - s_w)wL - pF_2(K_2, L_2), \quad 0 \le s_w < s_r \le 1$$

$$\frac{dK}{dt} = F_1(K_1, L_1) - \delta K$$

$$\frac{dL}{dt} = nL,$$

where $F_i(\cdot)$ is increasing, concave, nonnegative, first-order homogenous, and twice differentiable.

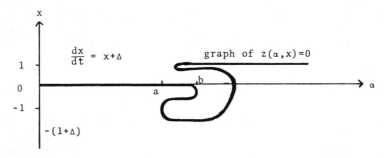

FIGURE 1. Graph of $z(\alpha, x) = 0$, showing that causality is not necessarily implied.

This system is equivalent to

$$\hat{z}(\hat{\alpha}, \hat{x}) = 0$$

$$\frac{d\hat{\alpha}}{dt} = \hat{\psi}(\hat{x}) \tag{A3}$$

where $\hat{\alpha} = k = K/L$, $\hat{x} = \omega = w/r$. Then

$$\hat{z}(\hat{\alpha}, \hat{x}) = \frac{k_1(\omega)k_2(\omega) + \omega[s_w k_1(\omega) + \omega(1 - s_w)k_2(\omega)]}{\omega + s_r k_2(\omega) + (1 - s_r)k_1(\omega)} - k$$

$$= \rho(\omega) - k, \text{ say}$$

where

$$k_i(\cdot) \text{ is the inverse of } \zeta_i(\cdot) \text{ and } \zeta_i(k_i) = \frac{f_i(k_i)}{f_i'(k_i)} - k_i,$$

$f_i(\cdot)$ is so defined that $f_i(k_i) = F_i(k_i, 1)$.

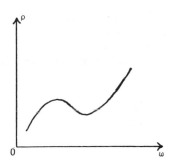
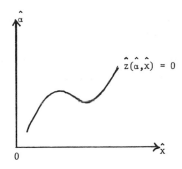

FIGURE 2. Graphs of $\rho(\cdot)$ and $\hat{z}(\cdot)$.

The only restriction is that $k_i'(\omega) > 0$ for all ω.

$$\frac{d\hat{\alpha}}{dt} = [s_w\omega + s_r\rho(\omega)] f_1'(k_1(\omega)) - (n + \delta)\rho(\omega).$$

Noncausality occurs if $\rho(\cdot)$ is not everywhere invertible. Set $s_w = 0$, $s_r = 1$ ("worker spends all, capitalist saves all") then

$$\rho(\omega) = \frac{k_1(\omega)k_2(\omega) + \omega k_2(\omega)}{\omega + k_2(\omega)} = \frac{k_2(\omega)}{\omega + k_2(\omega)}(k_1(\omega) + \omega).$$

Singularity occurs when

$$\rho'(\omega) = 0, \text{ or } \frac{d \log \rho}{d\omega} = 0 = \frac{(\omega k_2'/k_2) - 1}{k_2 + \omega} + \frac{k_1' + 1}{k_1 + \omega}$$

One can easily find a pair of increasing functions $[k_1(\cdot), k_2(\cdot)]$ to achieve that.

In such a case, the graphs of $\rho(\cdot)$ and $\hat{z}(\cdot)$ may be as shown in FIGURE 2 with

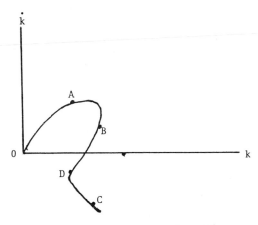

FIGURE 3. Phase diagram of k and $\dot{k}$.

a phase diagram as shown in FIGURE 3. Inada[†] first proposed to interpret the above diagrams as "chaos" rather than nonlinear cycle: ABCD → ABCD → .... The above model postulates behavior rules (i.e., workers' saving/income ratio is s_w, capitalists' is s_r) rather than utility maximization. But one may assume utility functions for workers and capitalists like

$$u_w = (1 - s_w) \log C_w + s_w \log I_w$$

$$u_r = (1 - s_r) \log C_r + s_r \log I_r,$$

with

$C_w + C_r = F_2(K_2, L_2)$ representing consumption and

$I_w + I_r = F_1(K_1, L_1)$, representing gross investment.

My interpretation of Uzawa via generalized dynamic systems of Roxin, Castaing, Yorke, Bhatia, and others has been presented here at this conference[‡] and elsewhere.[§]

†INADA. 1963. Rev. Economic Stud. **30.**

‡WAN, H. Y., JR. 1978. Ann. N.Y. Acad. Sci. This volume.

§WAN, H. Y., JR. 1975. 3rd World Congress Econometric Society, Toronto, August 1975; and Cornell Working Paper No. 73, Cornell University.

BIFURCATIONAL ASPECTS OF CATASTROPHE THEORY

J. M. T. Thompson

Department of Civil Engineering
University College
London, England WCIE 6BT

INTRODUCTION

Catastrophe theory, as sketched by Thom, and powerfully developed by Zeeman and others, has aroused some controversy in scientific circles.[1-3] It is clear that application of this theory to the social and inexact science, where *any* mathematical formulation might be suspect, should be pursued with caution. The controversial aspect should not however blind us to the very real insights that this theory can offer in, for example, the physical sciences, particularly in the study of light caustics and elastic stability.[4-7]

Indeed, catastrophe theory is intimately related to theories delineating bifurcation of equilibrium paths of conservative systems. It is the detailed interrelationships between these theories that are explored and illustrated in this paper, with particular emphasis on the crucial topological concept of structural stability. Special attention is given to the *imperfection-sensitivity* arising in post-buckling of elastic structures. Illustrative examples include the compound instability of stiffened plates and a mechanically stressed atomic lattice.

GENERAL FORMULATION

Catastrophe theory, in its most basic form, relates to the instabilities of discrete systems governed by a potential function. Thus it covers *exactly* the same ground as a sizable body of work in elastic stability.[8-10] Its new and powerful contribution, as we shall see, is the topological concept of structural stability and the associated unfolding rules.[11]

We formulate briefly the two theories by introducing a potential function $V[Q_i, \Lambda^j]$, where the Q_i are a set of n internal generalized coordinates, or state variables, and the Λ^j are a set of h external "control" parameters. A stationary value of V, with respect to the coordinates Q_i, is assumed to be necessary and sufficient for equilibrium, while a *local* minimum of V, with respect to the Q_i, is assumed to be necessary and sufficient for stability.[8] Here the word *local* refers to the domain of the so-called *delay* systems. We shall ignore the less important *Maxwell* systems that require an absolute global minimum of V for stability.

The external variables Λ^j are often called *control parameters*, although they are understood to include all external influences on our system, including small essentially uncontrollable perturbations, or disturbances, of the environment. These may be lengths, gravitational constants, moduli, etc. In elastic structures they are often associated with geometric imperfections within the fabric of the system. If they

0077–8923/78/0316–0553 $1.75/1 © 1979, NYAS

vary with real time, their variation is assumed to be slow in comparison with the fast dynamic changes in the Q_i generated by the minimum-seeking process.

We are interested in equilibrium surfaces in the $(n + h)$-dimensional (Q_i, Λ^j) space. We wish to explore singularities of these surfaces which are intimately related to critical equilibrium states C at which the quadratic form,

$$\delta^2 V = \frac{1}{2} V_{ij}{}^C q_i q_j = \frac{1}{2} \left. \frac{\partial^2 V}{\partial Q_i \partial Q_j} \right|^C \Delta Q_i \Delta Q_j$$

has one or more zero eigenvalues. If we introduce a set of local principal coordinates u_i, by, for example, a nonsingular linear coordinate transformation $q_i = \alpha_{ij} u_j$, we can reduce this second variation to a sum of squares; thus,

$$\delta^2 V = \tfrac{1}{2} C_1 u_1{}^2 + \tfrac{1}{2} C_2 u_2{}^2 + \cdots + \tfrac{1}{2} C_n u_n{}^2$$

Here the C_i represent a set of n stability coefficients.

ELIMINATION OF PASSIVE COORDINATES

Following Koiter,[12] it is a standard procedure in problems of this type to eliminate suitably defined passive coordinates that do not participate in a particular instability under consideration.[8] For example, if we have an m-fold critical point at which $C_1 = C_2 = \cdots = C_m = 0$, with the remaining stability coefficients nonzero, we could chose the first m of the u_i as *active* coordinates and the rest as *passive*.

To achieve the elimination of passive coordinates, we must not simply delete them from the energy function, but we must also solve the nonsingular passive equilbrium equations, $V_\alpha = 0$, $\alpha = m + 1, m + 2, \ldots, n$, to obtain passive coordinates u_α as single-valued functions of the active coordinates and the Λ^j. Passive coordinates can then be substituted back into the original function V to give us a new transformed potential function that contains only active coordinates and control parameters.[6,8]

This solution and back substitution need not be done globally, but can be done locally with the aid of a perturbation procedure.[13] It can be shown that the new transformed energy function, with only the active coordinates, contains all necessary and sufficient information for establishing both equilibrium *and* stability.[12] This elimination of passive coordinates corresponds to the fact that all essential features of developing m-fold instability can be adequately viewed in a suitably defined m-dimensional subspace of the full n-dimensional coordinate space.

The new transformed energy function, whether established by employing the original q_i or principal u_i coordinates, and written here still as V, will now have a completely *null* quadratic form with all coefficients V_{ij} equal to zero. So a Taylor expansion of V about the m-fold critical equilibrium state will start at the cubic term as,

$$V - V^C = \tfrac{1}{6} V_{ijk} q_i q_j q_k + \tfrac{1}{24} V_{ijkl} q_i q_j q_k q_l + \cdots$$

where subscripts on V denote differentiation of the transformed energy function with respect to the active coordinates alone, and the dummy-suffix summation convention implies summation over these active coordinates.

The vanishing of the quadratic form can be thought of as representing a singularity in V with an associated singularity in the equilibrium surface. Catastrophe theory gives us rules to ensure that these singularities are fully analyzed.[11]

ELIMINATION OF PASSIVE CONTROL PARAMETERS

Structural stability, in this context, requires that the singularity of the equilibrium surface be invariant, in topological form, under small random perturbations of the potential function. We can illustrate the associated argument for a system with a single active coordinate Q_1.

Taking the transformed energy function $V(Q_1)$, the equilibrium condition is $V_1 = 0$ and the critical condition is $V_{11} = 0$. Now if we were given a function V, we could *typically* expect a solution of these two nonlinear algebraic equations, provided there were at least two unknowns. This means that to observe a simple, critical equilibrium state under nonpathological conditions we must introduce at least one control parameter Λ (a second coordinate would simply bring with it another equilibrium equation).

Catastrophe theory emphasizes that, not only is this single Λ *necessary* for observing the phenomenon, but it is *sufficient* to understand fully the singularity, provided Λ is contained in V, in a prescribed manner. This argument leads us to the fold catastrophe $q^3 + \lambda q$ (FIGURE 1, top), which is the limit point of elastic stability where a unique equilibrium path loses its stability at a maximum (or minimum) value of Λ.

In a similar manner, a one-coordinate singularity ($m = 1$), with $V_1 = V_{11} = V_{111} = 0$, needs two control parameters for its full necessary and sufficient description. This leads to the cusp catastrophe (FIGURE 1), which can be identified as the distinct symmetric point of bifurcation common in the buckling of engineering structures.[6,8]

A singularity is observable in the real world only if a scan is made through a sufficient number of external control parameters to render the situation non-pathological. The unfolding rules tell us how many external control parameters are required, exactly how they should enter the energy function, and that any control parameters beyond the required number have no affect on the basic topological form of the singularity. Thus we may ignore them. The original problem in $(n + h)$-dimensional space is now reduced to a simpler problem in $(m + k)$-dimensional space, where m is the number of *active* coordinates and k is the number of unfolding parameters. This reduced space is really an *activity subspace* in which all the essential topology is contained.

SEVEN ELEMENTARY CATASTROPHES

A major contribution of catastrophe theory has been to provide a more complete and universal classification of stability phenomena than has previously been available. In particular, for gradient systems, Thom[1] has identified *all* the structurally stable unfoldings that can arise in a control space of up to 4 dimensions; this classification is shown in FIGURE 1. This list has been extended to control

Name	Equation	Classification
FOLD	$q^3 + \lambda q$	LIMIT POINT / ASYMMETRIC
CUSP	$q^4 + \lambda^2 q^2 + \lambda q$	STABLE-SYM / UNSTABLE-SYM
SWALLOW-TAIL	$q^5 + \lambda^3 q^3 + \lambda^2 q^2 + \lambda q$	
BUTTERFLY	$q^6 + \lambda^4 q^4 + \lambda^3 q^3 + \lambda^2 q^2 + \lambda^1 q$	
HYPERBOLIC UMBILIC	$q_2^3 + q_1^3 + \lambda^1 q_2 q_1 - \lambda^2 q_2 - \lambda^3 q_1$	MONOCLINAL / HOMEOCLINAL
ELLIPTIC UMBILIC	$q_2^3 - 3q_2 q_1^2 + \lambda(q_2^2 + q_1^2) - \lambda^2 q_2 - \lambda^3 q_1$	ANTICLINAL
PARABOLIC UMBILIC	$q_2^2 q_1 + q_1^4 + \lambda^1 q_2^2 + \lambda^2 q_1^2 - \lambda^3 q_2 - \lambda^4 q_1$	

FIGURE 1. Seven elementary catastrophes and their equivalents in bifurcation theory. These are the seven modes of instability that can arise in a simple gradient system that has no more than four control parameters. The name of the catastrophe is followed by the simplest potential polynomial that can generate the form; the q_i are incremental coordinates and the λ^j, incremental controls. The right-hand column shows the author's bifurcational classification, which is often finer than Thom's.[1]

spaces of dimension greater than 4, but Thom had a particular interest in developmental biology, where a control space can frequently be related to the 4 dimensions of space and time.

In this list, the first four catastrophes have only a single active coordinate ($m = 1$), written simply as q, while the remainder have two active coordinates $m = 2$ written as q_1 and q_2. The single control parameter of the fold is written simply as λ, while the two control parameters of the cusp are written as λ^1 and λ^2, etc. The control parameters always appear linearly in these polynomials, so λ^2 should always be read as "lambda two," rather than "lambda squared."

BIFURCATIONAL FORMALISM

In many branches of physics and engineering one particular parameter of a system, say Λ, is of special interest. This has given rise to an extensive study, not of equilibrium surfaces, as in catastrophe theory, but of equilibrium paths in the $(n + 1)$-dimensional space of Λ and the Q_i. This is the case for the buckling and post-buckling of engineering structures, where Λ is usually a load applied to the structure.

Interest thus focuses on bifurcations of a primary equilibrium path. Since the fold is the only structurally stable singularity that can be observed with the use of a single control parameter, these bifurcations will be topologically unstable. It is indeed well-known that small imperfections or perturbations of the function V will destroy a bifurcation by rounding off its corners (the number of significant independent imperfections are predicted by the unfolding rules).

Our interest in structurally *unstable* bifurcations arises from the symmetry inherent in nature and in Man's designs.[13] Optimization of the latter can generate high-order singularities with a multiplicity of instability modes (active coordinates) with an associated high degree of imperfection-sensitivity, as observed in the buckling of thin elastic shells.[8,12,14,15]

Noteworthy instability studies within this bifurcation viewpoint have been collected by Keller and Antman,[16] and Leipholz.[17] We mention also the recent contributions of Keener and Keller.[18,19]

NEW BIFURCATION THEOREMS

A basic theorem has been proposed by us in 1970.[20] This theorem has recently been proved rigorously by Kuiper as reported by Chillingworth.[21] We state this theorem formally, following our monograph[8] (see FIGURE 2):

THEOREM 1. An initially stable (primary) equilibrium path rising monotonically with the loading parameter Λ cannot become unstable without intersecting a further distinct (secondary) equilibrium path.

Moreover, by a second theorem (which may not be completely covered by Kuiper's work), we state the conditions under which such a path can reach an *unstable* critical equilibrium state[8,20]:

THEOREM 2. An initially stable equilibrium path rising with the loading parameter Λ cannot approach an unstable equilibrium state, from which the sys-

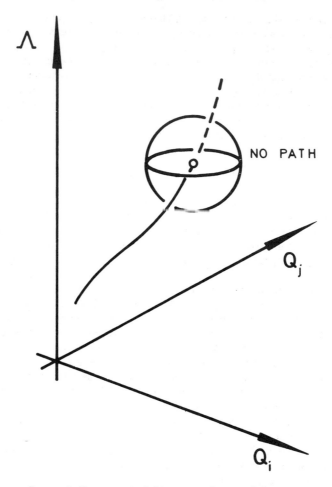

FIGURE 2. Response denied by a recently proved theorem.

tem exhibits a finite dynamic snap, without the approach of an equilibrium path (which may or may not be an extension of the original path) at values of the loading parameter less than that of the unstable state.

This theorem is written to embrace the limit point, as well as unstable bifurcation points.

These nonlinear theorems are of course concerned with real secondary equilibrium paths of finite length; they are not simply statements about the existance of linearized eigenvalues. Some relevant work has been recently reviewed by Stakgold.[22]

THE FOLD CATASTROPHE

The fold, as we have seen, represents the straightforward loss of stability of a unique equilibrium path at an extremum of Λ. It is the only catastrophe that can be observed by a one-parameter control scan. This confirms the experimenter's knowledge that a straightforward loading of any structure (even one that theoretically ought to exhibit a point of bifurcation) will only yield a loss of stability at a limit point.

Due to the fact that, in physics and engineering, there might be an *a priori* difference in parameters, say between a primary control parameter (such as a load) and a secondary control parameter ϵ (such as an imperfection), the fold can take on the appearance of an asymmetric bifurcation point, as illustrated in FIGURE 3. Therefore, with given *distinctive* control parameters, physical scientists may require

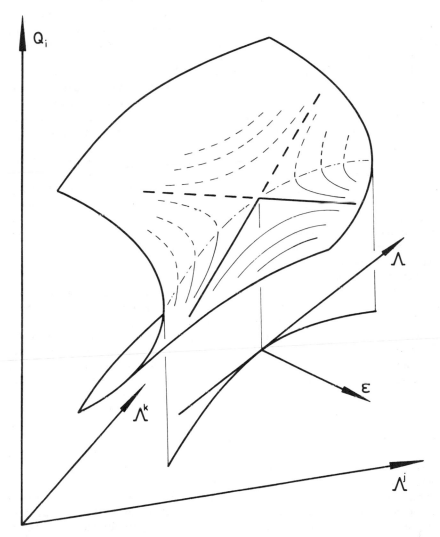

FIGURE 3. Fold catastrophe as an asymmetric point of bifurcation—common in the buckling of frames.[8]

a finer classification than that of Thom (as indicated in FIGURE 1). Such a refined classification has been pursued in a topological context by Wassermann.[23] It is interesting to observe the fold catastrophe and its associated hysteresis cycle in a recent review of ecological instabilities.[24]

THE CUSP CATASTROPHE

We see in FIGURE 1 that the cusp is the only new structurally stable singularity that arises as we increase the number of controls from one to two. It is a distinct critical point with $m = 1$. The energy function yields the folded equilibrium surface of FIGURE 4. The equivalence of this with the well-known unstable symmetric bifurcation point is displayed by the sections at constant values of ϵ.[8] The stability boundary projected into the control space generates the sharp two-thirds power-law cusp, familiar in elastic stability.

Here again the engineer will be interested in a finer classification than that of Thom, since a stable symmetric bifurcation point, with an upwards curving stable post-buckling equilibrium path is for him quite different from an unstable symmetric bifurcation point with a downwards curving unstable post-buckling equilibrium path. To the topologist, however, these points are identical since one is derived from the other by simple inversion.

The stable symmetric point is well-known in the buckling of an Euler column, while the unstable symmetric point arises in the buckling of a shallow arch, as studied by Roorda.[25] His experimental results are shown in FIGURE 5. The lower graph shows the two-thirds power-law imperfection-sensitivity on a plot of the load-carrying capacity against the offset f. Of particular interest is the value f_0 of the controled "imperfection" f that was found necessary to balance all other unavoidable manufacturing tolerances. It was a two-parameter scan of control space, using both the load P and the offset f that was needed in the experiment to locate the peak of the imperfection-sensitivity cusp.

A second illustration of the concept of structural stability arises in this problem if we assume that a small error in manufacture results in the initial post-buckling path being slightly *inclined*. This would, for example, be generated by the introduction of a small perturbing q^3 term in the energy polynomial. Now because the cusp with two controls (a load and an imperfection) is structurally stable, we can be sure that this perturbation will not alter the essential topology of the cusp, although it replaces a symmetric, with an asymmetric, bifurcation point. This is illustrated in FIGURE 6, where the perturbation has tilted the cusp but not destroyed it. An experimental study restricted to *naturally occurring* maxima of the original load deflection curves alone would not pick up the overhanging region (represented by the solid circles and solid triangles).

A small asymmetry of this type must of necessity be present in the experimental models tested by Roorda, and we have reexamined all of his illustrations in the light of this prediction. Nowhere does there seem to be any greater evidence of tilting than the case shown in FIGURE 5, where, close to the cusp point, a certain lack of symmetry is noticeable. It seems, that, while the method of manufacture provides a very noticeable shift f_0, it does not leave a very noticeable tilt in the imperfection-sensitivity diagram. Studies of strut and arch buckling in the context of catastrophe theory have been presented by Chillingworth[26] and Zeeman.[27]

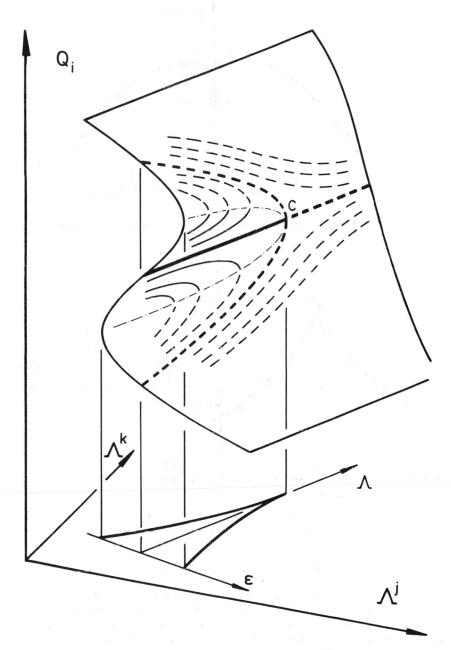

FIGURE 4. Cusp catastrophe as an unstable symmetric, bifurcation point.

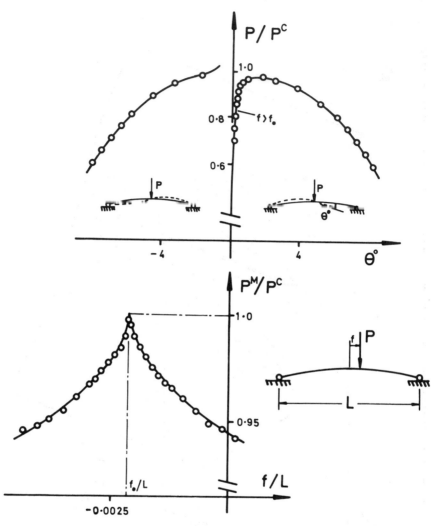

FIGURE 5. Roorda's experimental observation of a cusp by the scan of two control parameters; namely, load P and symmetry-breaking offset f. [25]

COALESCENCE OF BRANCHING POINTS

As two or more distinct bifurcations are brought into proximity and coalescence, for example, by varying a geometric parameter in a buckling problem, secondary bifurcations will often be observed on one or more of the secondary equilibrium paths as illustrated in FIGURE 7. [10,28] This gives rise to the possible stabilization of an unstable secondary path (as illustrated), a feature that may be of interest in thermodynamics for studying oscillating and self-organizing chemical reactions. [29]

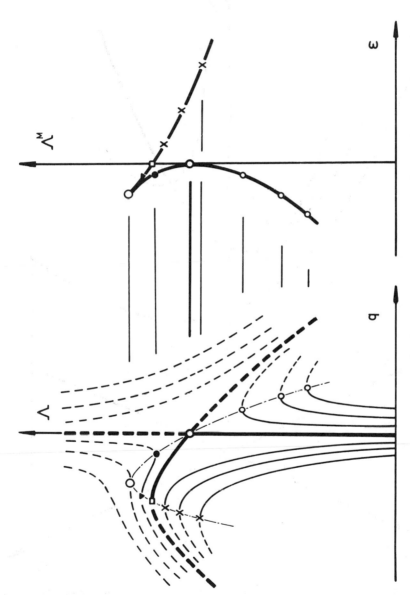

FIGURE 6. Tilting of a cusp by a parameter that destroys the symmetry of a distinct bifurcation.

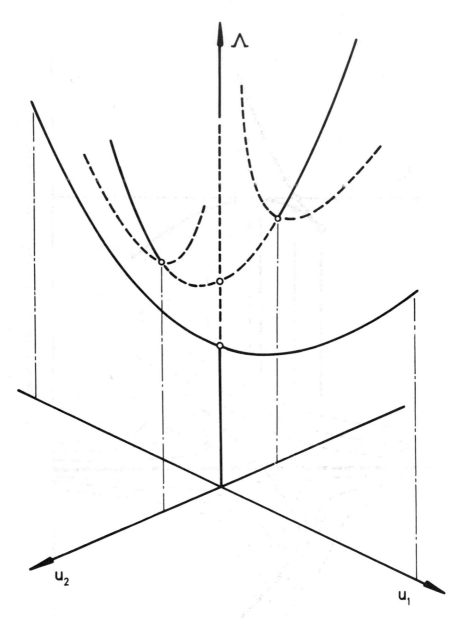

FIGURE 7. Approach of two distinct branching points can generate a secondary bifurcation which may herald a stabilization of a secondary equilibrium path. Here stable paths are solid lines and unstable paths, broken lines.

Secondary bifurcations, involving the destabilization of a secondary equilibrium path, are of course well-known in the progressive loss of symmetry and stability of a rotating self-gravitating liquid mass,[4] such as may arise in astrophysics.[30,31]

SEMISYMMETRIC COMPOUND BRANCHING

We have studied in detail the coalescence of distinct symmetric and asymmetric bifurcation points which generates a compound instability ($m = 2$) with two active coordinates.[4,6] Such compound instabilities are constantly generated by optimization procedures of engineering structures, and bring with them an increasingly severe imperfection-sensitivity.[13] The energy polynomial characterizing such a compound branching point will be symmetric in one active coordinate u_1 but nonsymmetric in the other, u_2.

We have shown that the compound semisymmetric branching point will have either:

(a) A single uncoupled secondary equilibrium path giving the *monoclinal* bifurcation point;

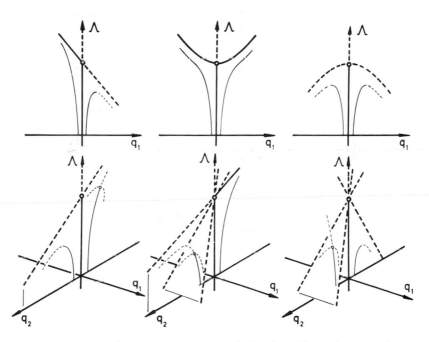

FIGURE 8. Top three diagrams show common distinct branching points; namely, asymmetric, stable symmetric, and the unstable symmetric. The three lower diagrams show semisymmetric branching points; namely the monoclinal, the homeoclinal, and the anticlinal. Light lines are paths of imperfect systems; broken curves are unstable equilibrium states.

    (b) A single uncoupled and two coupled secondary equilibrium paths falling in the same direction when projected into the plane of symmetry, giving the *homeoclinal* bifurcation point;

    (c) A single uncoupled and two coupled secondary equilibrium paths falling in opposite directions when projected into the plane of symmetry, giving the *anticlinal* bifurcation point.

These are illustrated, together with the three simple distinct branching points, in FIGURE 8. Equilibrium paths of imperfect systems are shown as light lines. We see that for certain imperfections, these can exhibit *distinct* bifurcation points when $m = 2$.

More detailed load-deflection illustrations and 3-dimensional imperfection-sensitivity surfaces are shown in our earlier work and in the analysis of Hunt.[4,6,32]

The form of compound branching developed at coalescence depends on the cubic energy coefficients $V_{222}{}^C$ and $V_{112}{}^C$ (FIGURE 9). The triple root line corres-

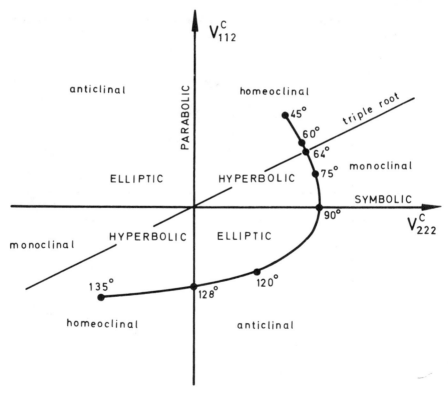

FIGURE 9. Classification of semisymmetric branching points according to values of two nonvanishing cubic energy coefficients. The curved path from 45° to 135° represents branching points of a simple buckling model during the scan of a fundamental parameter of the system. (After Thompson and Gaspar.[34])

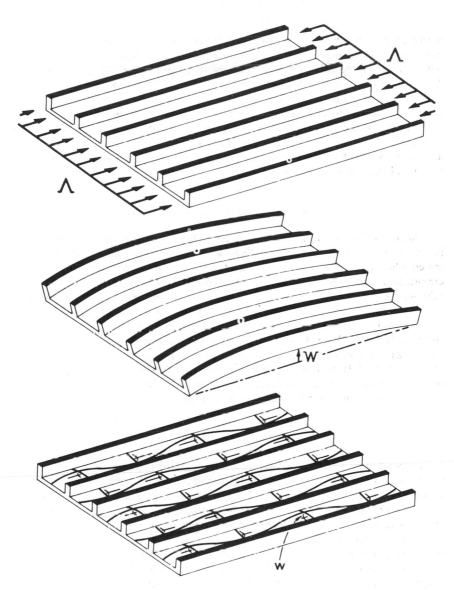

FIGURE 10. Compound instability of a stiffened plate, showing the undeflected com-
pressed state, buckling in the overall Euler mode, and buckling in the local-plate mode.

ponds to the condition,

$$\frac{V_{222}{}^{C}}{V_{112}{}^{C}} = \frac{2V_{22}{}^{0C}}{V_{11}{}^{0C}}$$

where a superscript zero denotes partial differentiation with respect to the loading parameter Λ. Also shown is the catastrophe theory classification of the umbilics, which hinges on the cubic form and not its interaction with the quadratic form $V_{ij}{}^{0C}$. This makes our classification finer than Thom's (see FIGURE 1).

The monoclinal point arises in the buckling of a strut on an elastic foundation,[33] and the homeoclinal point, in the buckling of stiffened plates, as we shall see in the following section. A simple two-degree-of-freedom buckling model, due to Thompson and Gaspar,[34] illustrates all three semisymmetric branching points in an elegant manner by the scan of a fundamental angle of the system, as shown by the curved trajectory in FIGURE 9.

THE BUCKLING OF A STIFFENED PLATE

The interactive buckling of an optimized stiffened plate, similar to that shown in FIGURE 10, has been analyzed by Tvergaard.[35] The compound bifurcation generated when Euler buckling of amplitude W coincides with plate buckling of amplitude w is of the homeoclinal type. The energy function of Tvergaard has been fully unfolded by Hunt[32] to obtain a 3-dimensional imperfection-sensitivity surface. This is an interesting manifestation of the hyperbolic umbilic catastrophe.

This imperfection-sensitivity surface is guaranteed by catastrophe theory to be structurally stable. Hunt has also examined the structural stability of semisymmetric imperfection-sensitivity surfaces under a perturbation that splits primary critical points.[36] Hunt's work yields the perturbed form of the imperfection-sensitivity diagram of the stiffened plate. This provides some interesting new features despite *topological* similarity to the coincident case.

HILL-TOP BRANCHING POINTS

A second, and in some ways rather different, manifestation of the hyperbolic, umbilic catastrophe arises when a limit point of a primary path coincides with a distinct symmetric bifunction point for the response of a mechanically stressed atomic lattice (FIGURE 11).[37-39] Here the top diagram in the figure shows the critical point C in the space of the primary control parameter, the direct stress σ_{11}, and the two active coordinates, the strains ϵ_{11} and ϵ_{12}. The lower left-hand diagram shows the projection of the equilibrium paths onto the $(\sigma_{11}, \epsilon_{12})$-plane, while the lower right-hand diagram shows the bilinear imperfection-sensitivity curve, which replaces the familiar two-thirds power law. The fully unfolded failure locus in the 3-dimensional control space, spanned by the stresses σ_{11}, σ_{12}, and σ_{22}, is shown in our review article,[4] where it can be compared with a cardboard model of the hyperbolic, umbilic stability boundary. This hill-top branching phenomenon can be nicely illustrated by the snapping and buckling of a tied elastic arch.[4]

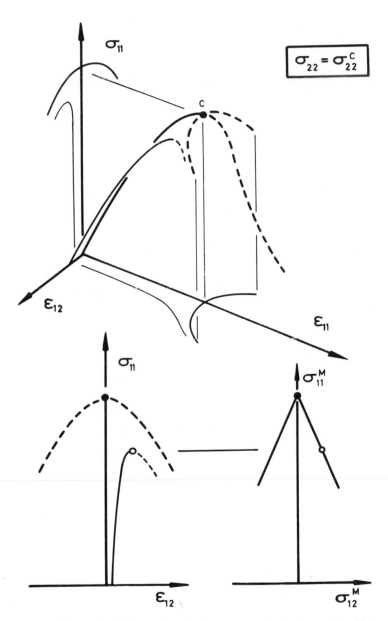

FIGURE 11. Hill-top branching point in the stress-strain response of a mechanically loaded atomic lattice—a second manifestation of the hyperbolic, umbilic catastrophe.

Conclusion

In this brief article attention has been focused on the role of the topological concept of structural stability in the analysis of imperfection-sensitivity. The strength of this concept in guaranteeing *topological* similarity, and its corresponding weakness have been seen, remembering topologically similar situations may have quite different physical manifestations. The analytical consequence of structural stability in semisymmetric branching is examined more fully in a companion paper by Hunt.[36] Finally, we would like to draw attention to an interesting example of elliptic, umbilic catastrophe in the unfolding of unstable fluid flow.[40]

References

1. THOM, R. 1975. Structural Stability and Morphogenesis. D. H. Fowler, Trans. Benjamin, Reading.
2. ZEEMAN, E. C. Catastrophe Theory: Selected Papers. Addison-Wesley, London. In press.
3. POSTON, T. & I. STEWART. Catastrophe Theory and its Applications. Pitman, London. In press.
4. THOMPSON, J. M. T. & G. W. HUNT. 1977. The instability of evolving systems. Interdiscip. Sci. Rev. 2: 240–62.
5. BERRY, M. V. 1976. Waves and Thom's theorem. Adv. Phys. 25: 1–26.
6. THOMPSON, J. M. T. & G. W. HUNT. 1975. Towards a unified bifurcation theory. J. Appl. Math. Phys. 26: 581–604.
7. THOMPSON, J. M. T. 1977. Catastrophe theory and its role in applied mechanics. In Theoretical and Applied Mechanics. Proc. 14th IUTAM Congr, Delft. W. T. Koiter, Ed. North-Holland, Amsterdam. 2: 451–8.
8. THOMPSON, J. M. T. & G. W. HUNT. 1973. A General Theory of Elastic Stability. Wiley, London.
9. HUSEYIN, K. 1974. Nonlinear Theory of Elastic Stability. Noordhoff, Leyden, the Netherlands.
10. SUPPLE, W. J., ED. 1973. Structural Instability. IPC Science and Technology Press, Guildford, England.
11. POSTON, T. & I. STEWART. 1976. Taylor Expansions and catastrophes. Res. Notes Math. 7. Pitman, London.
12. KOITER, W. T. 1945. On the Stability of Elastic Equilibrium. Delft, Holland. NASA, Tech. Trans., F10, 833 (1967). Dissertation.
13. THOMPSON, J. M. T. & G. W. HUNT. 1977. A bifurcation theory for the instabilities of optimization and design. In Mathematical Methods of the Social Sciences. D. Berlinski, Ed. Reidel, Dordrecht, the Netherlands. Synthese 36: 315–351.
14. HUTCHINSON, J. W. & W. T. KOITER. 1970. Postbuckling theory. Appl. Mech. Rev. 23: 1353–1366.
15. BUDIANSKY, B. 1974. Theory of buckling and post-buckling behavior of elastic structures. Adv. Appl. Mech.
16. KELLER, J. B. & S. ANTMAN, EDS. 1969. Bifurcation Theory and Nonlinear Eigenvalue Problems. Benjamin, New York.
17. LEIPHOLZ, H., ED. 1971. Instability of Continuous Systems. Springer-Verlag, Berlin.
18. KEENER, J. P. & H. B. KELLER. 1973. Perturbed bifurcation theory. Arch. Ration. Mech. Anal. 50: 159–175.
19. KEENER, J. P. 1974. Perturbed bifurcation theory at multiple eigenvalues. Arch. Ration. Mech. Anal. 56: 348–366.
20. THOMPSON, J. M. T. 1970. Basic theorems of elastic stability. Int. J. Eng. Sci. 8: 307–313.

21. CHILLINGWORTH, D. R. J. 1976. A problem from singularity theory in engineering. Lecture to Symposium on Nonlinear Mathematical Modelling (29 August). University of Southampton.
22. STAKGOLD, I. 1971. Branching of solutions of nonlinear equations. SIAM Rev. 13:289–332.
23. WASSERMANN, G. 1976. (r,s)-stable unfoldings and catastrophe theory. In Structural Stability, the Theory of Catastrophes, and Applications in the Sciences (Seattle 1975). P. Hilton, Ed. Springer Lect. Notes Math. 525.
24. MAY, R. M. 1977. Thresholds and breaking points in ecosystems with a multiplicity of stable states. Nature. 269: 471–477.
25. ROORDA, J. 1965. Stability of structures with small imperfections. J. Eng. Mech. Div. Am. Soc. Civ. Eng. 91: 87–106.
26. CHILLINGWORTH, D. R. J. 1975. The catastrophe of a buckling beam. In Dynamical Systems (Warwick 1974). A. Manning, Ed. Springer Lect. Notes Math. 468.
27. ZEEMAN, E. C. 1976. Euler buckling. In Structural Stability, the Theory of Catastrophes, and Applications in the Sciences (Seattle 1975). P. Hilton, Ed. Springer Lect. Notes Math. 525.
28. CHILVER, A. H. 1967. Coupled modes of elastic buckling. J. Mech. Phys. Solids. 15: 15–28.
29. NICOLIS, G. & I. PRIGOGINE. 1977. Self-organization in Nonequilibrium Systems: From Dissipative Structures to Order through Fluctuations. Wiley, New York.
30. LYTTLETON, R. A. 1953. The Stability of Rotating Liquid Masses. Cambridge University Press.
31. LEDOUX, P. 1958. Stellar stability. In Handbuch der Physik LI. S. Flugge, Ed. Springer-Verlag, Berlin.
32. HUNT, G. W. 1977. Imperfection-sensitivity of semi-symmetric branching. Proc. R. Soc. London A 357:193–211.
33. HANSEN, J. S. 1973. Buckling of Imperfection-sensitive Structures: A Probabilistic Approach. University of Waterloo. Dissertation.
34. THOMPSON, J. M. T. & Z. GASPAR. 1977 A buckling model for the set of umbilic catastrophes. Math. Proc. Camb. Phil. Soc. 82: 497 507.
35. TVERGAARD, V. 1973. Imperfection-sensitivity of a wide integrally stiffened panel under compression. Int. J. Solids Struct. 9: 177–192.
36. HUNT, G. W. 1978. Imperfections and Near-coincidence for semisymmetric bifurcations. Proc. Conf. Bifurcation Theory Appl. Sci. Disciplines (Oct. 1977). Ann N.Y. Acad. Sci. In this volume.
37. THOMPSON, J. M. T. & P. A. SHORROCK. 1975. Bifurcational instability of an atomic lattice. J. Mech. Phys. Solids. 23: 21–37.
38. THOMPSON, J. M. T. 1975. Experiments in catastrophe. Nature. 254: 392–395.
39. THOMPSON, J. M. T. & P. A. SHORROCK. 1976. Hyperbolic umbilic catastrophe in crystal fracture. Nature. 260 (5552): 598–599.
40. BERRY, M. V. & M. R. MACKLEY. 1977. The six roll mill: unfolding an unstable persistently extensional flow. Phil. Trans. R. Soc. London 287A (1337): 1–16.

IMPERFECTIONS AND NEAR-COINCIDENCE FOR
SEMISYMMETRIC BIFURCATIONS*

Giles W. Hunt

Department of Civil Engineering
University College
London, England WC1E 6BT

INTRODUCTION

In recent years the very rich interaction between Thom's catastrophe theory and the general bifurcation theories, which were developed to cope with sometimes severe nonlinear buckling problems of engineering structures, has been explored in some detail.[1-6] Thompson has presented in these proceedings a most lucid account of the developments including detailed lists of references.[5,7] This work reveals that the philosophy and theoretical base of catastrophe theory can be used to give greater insight into the phenomena of elastic stability than had previously been available.

This new, unified approach has proved valuable in interpreting the *distinct* bifurcations, those with merely a single critical coordinate, but is perhaps even more rewarding when we come to consider *compound* bifurcations, where two or more such points coincide.[8,9] For these problems it is well known that equilibrium paths can suffer severe contortions,[10] and it is here that the concept of *structural stability* (in the catastrophe theory sense) can provide the analyst with a significant extra string to his bow. Such contortions can usually be associated with the choice of structurally *unstable* representations of the phenomena of interest—structurally stable forms, by their very nature, would be unlikely to contort in this manner. Thus, taking this into account analytically, we could concentrate on solving for structurally stable forms. Here, without the convergence problems associated with contortion, the results might well have a greater range of validity than otherwise, and the equations might be easier to solve. We note that this would tend to move us away from the study of equilibrium paths and towards the study of imperfection-sensitivity.

This paper explores these ideas for the semisymmetric points of bifurcation: two-fold compound phenomena with symmetry of the potential energy in one of the contributing modes. These are the simplest of all compound bifurcations and arise in a number of different structural problems, including, for example, the buckling of a spherical shell under a uniform external pressure, and the optimal buckling of a stiffened plate loaded longitudinally. The analysis of their imperfection-sensitivity generates the universal unfoldings in control space of Thom's *hyperbolic* and *elliptic umbilic catastrophes*.[8] The structural stability of these forms ensures that with the addition of an extra control parameter—one which has the effect of separating the two contributing bifurcations—the deformation of the

*This paper was supported by the Science Research Council of Great Britain.

0077-8923/78/0316-0572 $1.75/1 © 1979, NYAS

surfaces is fairly mild. This is illustrated for one of the three subclassifications of semisymmetric branching, the homeoclinal point of bifurcation, which generates a hyperbolic umbilic catastrophe.

Also discussed briefly is the transition from homeoclinal to anticlinal branching, a somewhat more complex form of compound bifurcation, which corresponds to Thom's *parabolic umbilic catastrophe*. Here the extra control parameter is essential for structural stability, so we can predict that the usual imperfection-sensitivity plots would exhibit changes in topological form as this parameter is varied. This is borne out by the illustrations of Thom.[1] We note that this form of branching seems to be of prime significance in the interactive buckling of stiffened plates, as developed in general terms by Koiter,[11] and we term it the *paraclinal point of bifurcation*.

FORMULATION

For a formulation in the most general terms, and a detailed account of the interrelationships between bifurcation theory and catastrophe theory, we refer the reader to Thompson's companion paper (in this volume).[7] In this paper we are obliged to be a little more specific, so we start by considering a discrete (with n degrees of freedom) conservative structural system, which at a certain *critical* load experiences two-fold *semisymmetric* branching behavior.[3,8] As outlined in Thompson's paper, we shall suppose that the system can be described by a total potential energy function $V(Q_i, \Lambda^j)$, where the Q_i are a set of n internal generalized coordinates or state variables and the Λ^j are a set of external (controled) parameters. The latter can include loads, or indeed essentially uncontrolable imperfections, as explained in the earlier work.[5] Normally one of the Λ^j is selected as a distinctive *loading parameter*, considered to be of special significance; we henceforth designate this simply Λ.[12]

We now introduce a refined form of potential function that can be used to describe the system and write it as $\mathcal{Q}(u_i, \Lambda^j)$; this can be obtained from the more general formulation by a series of three energy transformations,[3,7,12] which introduce, in turn, incremental and diagonalized coordinates, and finally eliminate $n - 2$ passive (or nonessential) coordinates from the analysis. Here the u_i represent a diagonalized set of two incremental coordinates measured from a fundamental equilibrium solution F lying along the load axis in u_i-Λ space.

Writing out the $\mathcal{Q}$-function in full, we shall here concentrate on the particular form,

$$\mathcal{Q} = \tfrac{1}{6}\mathcal{Q}_{222}{}^C u_2{}^3 + \tfrac{1}{2}\mathcal{Q}_{112}{}^C u_1{}^2 u_2 + \tfrac{1}{2}\lambda(\mathcal{Q}_{11}{}^{0C} u_1{}^2 + \mathcal{Q}_{22}{}^{0C} u_2{}^2) + \tfrac{1}{2}\mathcal{Q}_{11}{}^{0C}\sigma u_1{}^2$$
$$+ \mathcal{Q}_1{}^{1C}\epsilon^1 u_1 + \mathcal{Q}_2{}^{2C}\epsilon^2 u_2 + \text{higher-order terms} \quad (1)$$

where $\lambda = \Lambda - \Lambda^C$. We see that this constitutes a Taylor series expansion of the potential function about the *critical* equilibrium state C lying on F. Here subscripts denote partial differentiation with respect to the corresponding u_i, a superscript zero denotes partial differentiation with respect to the loading parameter Λ, and further superscripts denote the partial differentiation with respect to the *imperfection parameters* ϵ^i. The new *splitting parameter* σ is introduced to allow for the separation of the two contributing bifurcations on the fundamental path of the *perfect system*, where $\epsilon^i = 0$. The low-order terms which are missing from this Taylor series expansion have zero coefficients, which arise from equilibrium and critical conditions, and the special properties of the $\mathcal{Q}$-function.[3]

We see that when $\sigma = 0$, the compound critical point C of the perfect system is *semisymmetric* in the sense that equal and opposite values of u_1 give identical energy levels. Thus the coefficients $\mathcal{Q}_{111}{}^C$ and $\mathcal{Q}_{122}{}^C$ are zero, these being odd powers of u_1.[8] We note that physical symmetries of engineering structures frequently give rise to symmetries of the potential function,[12] and indeed semisymmetric points of bifurcation can be found in the response of many structural systems, including stiffened plates,[8,11,13] frames,[14] a strut on an elastic foundation,[3,15] and a spherical shell under uniform external pressure.[3]

COMPLETE COINCIDENCE

Post-Buckling Paths of the Perfect System

Let us first set $\epsilon^i = \sigma = 0$, to study the post-buckling equilibrium paths of the perfect system under conditions of complete coincidence. These are well understood.[3] The solution of the equilibrium equations via a perturbation scheme (or perhaps by some other technique[16]) gives the three results for the post-buckling paths in rate space $(u_i^{(1)} = \Lambda^{(1)})$,

$$\left.\frac{u_1^{(1)}}{u_2^{(1)}}\right|^C = 0, \quad \left.\frac{\Lambda^{(1)}}{u_2^{(1)}}\right|^C = -\left.\frac{\mathcal{Q}_{222}}{2\mathcal{Q}_{22}{}^0}\right|^C,$$

$$\left.\frac{u_1^{(1)}}{u_2^{(1)}}\right|^C = \pm\sqrt{\left.\left(\frac{2\mathcal{Q}_{22}{}^0}{\mathcal{Q}_{11}{}^0} - \frac{\mathcal{Q}_{222}}{\mathcal{Q}_{112}}\right)\right|^C}, \quad \left.\frac{\Lambda^{(1)}}{u_2^{(1)}}\right|^C = -\left.\frac{\mathcal{Q}_{112}}{\mathcal{Q}_{11}{}^0}\right|^C. \tag{2}$$

Here a superscript in parentheses denotes differentiation with respect to some independent (or perturbation) parameter s, which for this analysis can remain undefined. We note that these rays in rate space become the post-buckling equilibrium path tangents when mapped directly into coordinate space u_i-Λ.[3]

The first of these solutions is identical to the slope of the single post-buckling path for the distinct asymmetric point of bifurcation critical with respect to u_2. It is thus clear that this path is unaffected by variations of σ, as was first recognized by Chilver[10]; we term it an *uncoupled* post-buckling path. The other two *coupled* paths may or may not exist. We next introduce a subclassification of semisymmetric branching based on the relations between coupled and uncoupled paths.

The first class of systems arises when the coupled paths are imaginary and the condition,

$$\left.\frac{2\mathcal{Q}_{22}{}^0}{\mathcal{Q}_{11}{}^0}\right|^C < \left.\frac{\mathcal{Q}_{222}}{\mathcal{Q}_{112}}\right|^C \tag{3}$$

is satisfied. We term this the *monoclinal point of bifurcation* and note that it arises in the response of a strut on an elastic foundation[15] and in a simple propped cantilever model due to Thompson and Gaspar.[17]

Secondly we have the case when the coupled paths exist, but coupled and uncoupled paths fall in *opposite* directions with respect to u_2 in u_i-Λ space, so $\mathcal{A}_{112}{}^C/\mathcal{A}_{11}{}^{0C}$ and $\mathcal{A}_{222}{}^C/\mathcal{A}_{22}{}^{0C}$ are *opposite* in sign. We term this the *anticlinal point of bifurcation* and note that it also arises in Thompson and Gaspar's propped cantilever.[17] We might also expect it in the interactive buckling between Euler and local stiffener modes of failure of a longitudinally stiffened plate loaded in its plane.[18]

Thirdly we have the class of problems with coupled and uncoupled paths falling in the *same* direction with respect to u_2 in u_i-Λ space, so that $\mathcal{A}_{112}{}^C/\mathcal{A}_{11}{}^{0C}$ and $\mathcal{A}_{222}{}^C/\mathcal{A}_{22}{}^{0C}$ are of the *same* sign. We term this the *homeoclinal point of bifurcation*. This seems to arise more frequently in structural systems than the earlier cases, and can be found in the response of a spherical shell under uniform external pressure,[3] frame buckling,[14] the simple propped cantilever model,[17] and the interactive buckling between Euler and local plate-buckling modes of failure in a stiffened plate.[8,13] Three-dimensional representations in u_i-Λ space of these three forms of branching, showing the perfect equilibrium paths and the paths of some imperfect systems, are given in FIGURE 8 of Reference 7.

Finally we draw the reader's attention to one other important form of branching that in fact represents the transition between anticlinal and homeoclinal branching. Here $\mathcal{A}_{222}{}^C = 0$ and the uncoupled path has a zero slope. Solution of a higher-order equilibrium equation now gives the uncoupled path curvature as

$$\left.\frac{\Lambda^{(2)}}{(u_2^{(1)})^2}\right|^C = -\left.\frac{\mathcal{A}_{2222}}{3\mathcal{A}_{22}{}^0}\right|^C, \tag{4}$$

which is identical to the curvature of the single post-buckling path for distinct symmetric points of bifurcation critical with respect to u_2. The uncoupled path is thus once again unaffected by the approach and coincidence of the two contribution bifurcations. We term this the *paraclinal point of bifurcation*.

This form of branching seems more important in elastic stability than had previously been supposed.[4] We note that Koiter develops a corresponding potential function in his general theory of the elastic buckling of plate assemblages.[11] Clearly the fourth-order coefficient $\mathcal{A}_{2222}{}^C$ must now play a significant role. This can be rigorously confirmed by applying the rules for determinacy and unfolding presented by Poston and Stewart.[19] We shall see later that this form of bifurcation corresponds to Thom's *parabolic umbilic catastrophe*.

Equilibrium Paths of Imperfect Systems

To study the forms of *imperfect* equilibrium paths under conditions of complete coincidence we continue with σ set to zero, but allow the ϵ^i to vary. It is most convenient to adopt a polar coordinate transformation on the imperfection parameters.[8] If this is done we can schematically represent the equilibrium paths of imperfect systems by the solid trees in FIGURES 1, 2, and 3. In each case, the equilibrium paths of the perfect system provide a skeleton for the tree and the

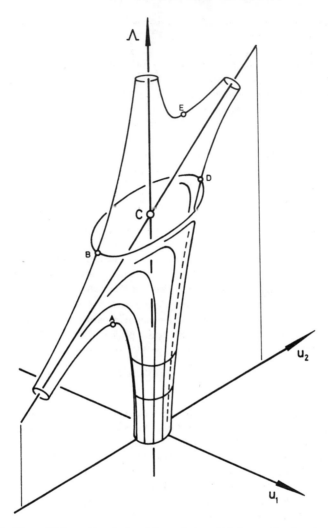

FIGURE 1. The solid tree of perfect and imperfect equilibrium paths for the monoclinal point of bifurcation. The perfect paths provide a skeleton which the imperfect paths fully envelop.

imperfect paths put meat on the skeleton—the whole tree expands with increasing *radial imperfection* ϵ. We see that each perfect path is fully enclosed by imperfect paths.

Catastrophe theory focuses attention on structurally stable topological forms, and the full structurally stable equilibrium surfaces would for these forms of branching be five-dimensional, involving the full u_i-Λ-ϵ^j space. These solid trees represent *crushed-down* versions of the full surface,[5] and a *shell* generated at

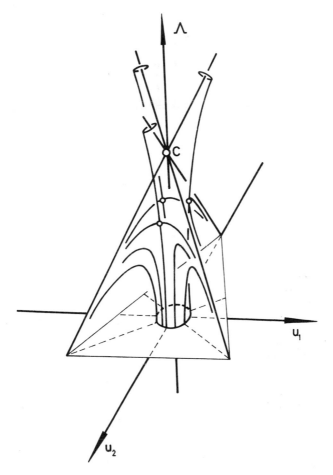

FIGURE 2. The solid tree of perfect and imperfect equilibrium paths for the anticlinal point of bifurcation. Again a skeleton of perfect paths is enveloped by imperfect paths.

a constant radial imperfection denotes a *contour* on the surface. Thus we have here perhaps the most convenient visual representation of these five-dimensional forms.

Imperfection-Sensitivity Surfaces

Continuing with $\sigma = 0$, we next pinpoint *critical* equilibria of perfect and imperfect systems using a perturbation scheme[8] and plot the loci of these points in the control space $\Lambda\text{-}\epsilon^i$ to obtain the *imperfection-sensitivity surfaces*. We note that for semisymmetric branching these three-dimensional failure loci are structurally stable topological forms.

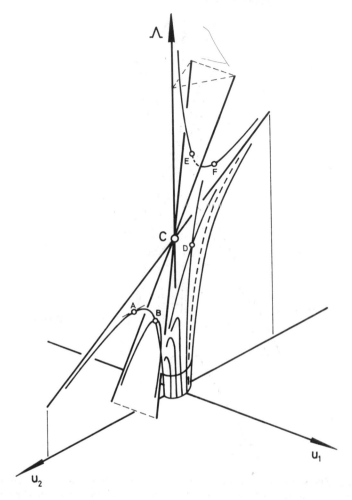

FIGURE 3. The solid tree of perfect and imperfect equilibrium paths for the homeoclinal point of bifurcation. Again a skeleton of perfect paths is enveloped by imperfect paths.

FIGURE 4 shows this surface for the monoclinal point of bifurcation. This can be identified as the universal unfolding of Thom's *hyperbolic umbilic catastrophe*.

FIGURE 5 shows the surface for the anticlinal point of bifurcation. This can be identified as the universal unfolding of Thom's *elliptic umbilic catastrophe*.

In FIGURE 6 the imperfection-sensitivity surface is plotted for a specific structural problem, an axially loaded stiffened plate. (The potential function for this was initially derived by Tvergaard.[13]) This can be identified once again as the universal unfolding of Thom's *hyperbolic umbilic catastrophe*. We note that only the lowest (cusped) sheet of this surface is associated with the *initial* loss of stability of a system loaded from zero, and thus would be of interest to structural engi-

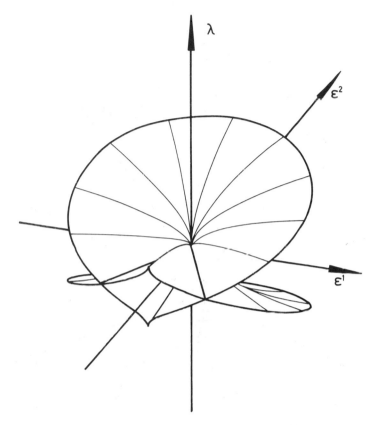

FIGURE 4. Imperfection-sensitivity for the monoclinal point of bifurcation giving the universal unfolding of hyperbolic umbilic catastrophe.

neers; however, for some systems the higher (smooth) sheet will apply.[8] We shall use this specific example later in this paper to illustrate the deformation of the surface with varying σ.

We note that certain specific imperfect equilibrium paths may bifurcate and these *secondary bifurcations* can be determined explicitly from the perturbation scheme.[8] They may be symmetric bifurcations, which generate cusp lines on the imperfection-sensitivity surface or they may be asymmetric bifurcations, where the surface folds back on itself, as can be seen in FIGURE 6.

It is interesting to observe the form of imperfection-sensitivity on the *symmetric section*. If we slice each surface on the Λ-ϵ^2 plane, where $\epsilon^1 = 0$, we obtain two parabolas, one a line of cusps,

$$\Lambda = \Lambda^C + \alpha\sqrt{\epsilon^2}$$

where
$$\alpha = \pm \frac{\mathcal{Q}_{112}}{\mathcal{Q}_{11}{}^0} \sqrt{\left(\frac{2\mathcal{Q}_2{}^2\mathcal{Q}_{11}{}^0}{2\mathcal{Q}_{112}\mathcal{Q}_{22}{}^0 - \mathcal{Q}_{222}\mathcal{Q}_{11}{}^0} \right)}\Bigg|^C \tag{5}$$

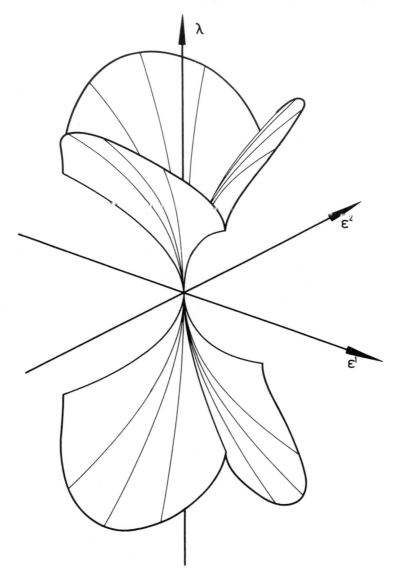

FIGURE 5. Imperfection-sensitivity for the anticlinal point of bifurcation giving universal unfolding of the elliptic umbilic catastrophe.

And the second a line of folds,

$$\Lambda = \Lambda^C + \beta\sqrt{\epsilon^2}$$

where
$$\beta = \pm \frac{\alpha_{222}}{\alpha_{22}{}^0} \sqrt{\left.\left(\frac{2\alpha_2{}^2}{\alpha_{222}}\right)\right|^C} \tag{6}$$

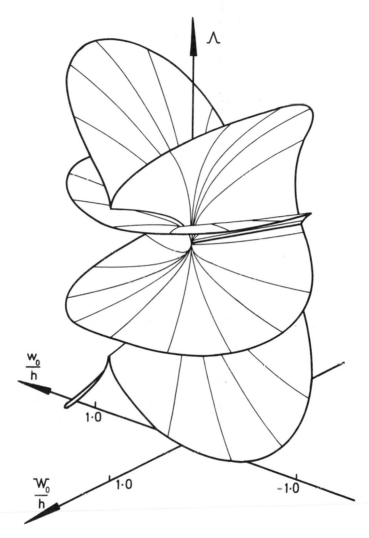

FIGURE 6. Imperfection-sensitivity for the homeoclinal point of bifurcation of a stiffened plate example. Here w_0/h implies local buckling and W_0/h overall buckling. This is again the universal unfolding of hyperbolic umbilic catastrophe, but there are significant differences between this and the form in FIGURE 4.

These two curves will be of particular interest when we come to study the deformation of the surfaces with varying σ.

Finally, we see that these pictures illustrate the need in bifurcational studies for a somewhat finer classification of phenomena than the topological approach of catastrophe theory provides. We see here two quite different forms of the hyperbolic umbilic catastrophe; the difference is of a very real significance in, for example, an interactive buckling problem. We note that the hyperbolic umbilic arises in yet another capacity, in the imperfection-sensitivity of a hill-top branching point where a limit point and an unstable-symmetric point of bifurcation coincide.

NEAR-COINCIDENCE

We now allow variations of the parameter σ; this will have the effect of separating the two contributing bifurcations on the fundamental path of the perfect system. Since this work is largely new we shall present it in a little more analytical detail. Adopting the polar coordinate representation for the imperfections, so $\epsilon^1 = \epsilon \sin \theta$ and $\epsilon^2 = \epsilon \cos \theta$, the governing equations for any *imperfection ray* (with constant θ) may be written,

$$\alpha_{112}x_2 u_1^{M(1)} + \alpha_{112}x_1 u_2^{M(1)} + \alpha_{11}{}^0 x_1 \Lambda^{M(1)} + \alpha_{11}{}^0 x_1 \sigma^{M(1)} \mid^C = 0$$

$$\alpha_{112}x_1 u_1^{M(1)} + \alpha_{222}x_2 u_2^{M(1)} + \alpha_{22}{}^0 x_2 \Lambda^{M(1)} \mid^C = 0$$

$$\alpha_{112}u_1^{M(1)}u_2^{M(1)} + \alpha_{11}{}^0 u_1^{M(1)}\Lambda^{M(1)} + \alpha_{11}{}^0 u_1^{M(1)}\sigma^{M(1)} + \tfrac{1}{2}\dot{\alpha}_1 \epsilon^{M(2)} \mid^C = 0$$

$$\alpha_{112}(u_1^{M(1)})^2 + \alpha_{222}[u_2^{M(1)}]^2 + 2\alpha_{22}{}^0 u_2^{M(1)}\Lambda^{M(1)} + \dot{\alpha}_2 \epsilon^{M(2)} \mid^C = 0 \tag{7}$$

where a dot denotes partial differentiation with respect to the radial imperfection ϵ.

The last two of these equations are the lowest-order, nontrivial *equilibrium* equations derived from a perturbation scheme, and the first two equations are the lowest-order, nontrivial *critical state* equations derived from the same scheme. Here x_j denotes a local eigenvector, obtained by considering a critical state of an imperfect system as a local linear eigenvalue problem.[8] Also a superscript in parentheses denotes full differentiation with respect to the perturbation parameter of the scheme, which here can remain undefined. These equations define critical behavior, and so adopting a convention originally used in the study of distinct critical points, we have introduced a superscript M (maximum or minimum) to distinguish the parametric representations of this imperfection-sensitivity study from those of a pure equilibrium study.[3]

Bifurcations of the Perfect System

Setting $\epsilon = 0$, we can also set $\epsilon^{M(2)C} = 0$ in the above equations and solve to determine the loci of the *bifurcations* that arise in the response of the perfect system with varying σ. We first obtain the trivial solution,

$$u_1^{A(1)C} = u_2^{A(1)C} = \Lambda^{A(1)C} = x_1^C = 0,$$

$$x_2^C \neq 0, \quad \sigma^{A(1)C} \neq 0 \tag{8}$$

This represents an *asymmetric bifurcation A*, critical with respect to u_2 which remains fixed at $\Lambda = \Lambda^C$ on the fundamental path of the perfect system as σ varies. It could, of course, be found by a linear eigenvalue analysis.

Secondly we have the solution,

$$u_1^{S(1)C} = u_2^{S(1)C} = x_2^C = 0,$$
$$x_1^C \neq 0, \quad \Lambda^{S(1)C} = -\sigma^{S(1)C} \tag{9}$$

which denotes a *symmetric bifurcation S*, critical with respect to u_1; this moves along the fundamental path of the perfect system as σ varies. This could likewise be predicted by a linear eigenvalue analysis.

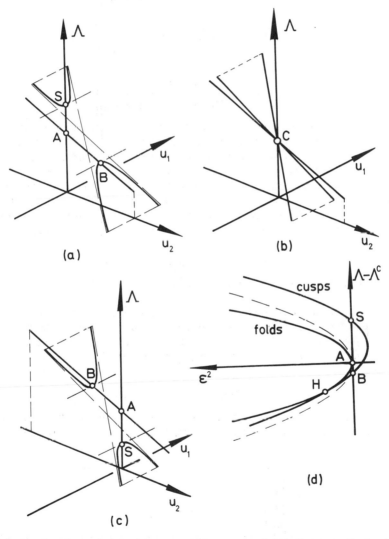

FIGURE 7. (a)–(c) Equilibrium paths of the perfect system for the homeoclinal point of bifurcation with variations of the splitting parameter σ. In (a) $\sigma < 0$, (b) $\sigma = 0$, (c) $\sigma > 0$. (d) Deformation of symmetric section of FIGURE 6 with $\sigma < 0$.

To a structural engineer these two bifurcations on the fundamental path would represent two contributing instabilities of an interactive buckling problem; either one may be the first to be encountered in a natural loading sequence from zero, or they may both be met coincidently. Finally we have the third solution,

$$u_1^{B(1)C} = x_2^C = 0, \quad x_1^C \neq 0$$

$$\left.\frac{\Lambda^{B(1)}}{u_2^{B(1)}}\right|^C = -\left.\frac{\mathcal{Q}_{222}}{2\mathcal{Q}_{22}{}^0}\right|^C$$

$$\left.\frac{\sigma^{B(1)}}{u_2^{B(1)}}\right|^C = \left.\frac{\mathcal{Q}_{222}}{2\mathcal{Q}_{22}{}^0} - \frac{\mathcal{Q}_{112}}{\mathcal{Q}_{11}{}^0}\right|^C \tag{10}$$

which gives the relationship,

$$\left.\frac{\Lambda^{B(1)}}{\sigma^{B(1)}}\right|^C = \left.\frac{\mathcal{Q}_{222}\mathcal{Q}_{11}{}^0}{2\mathcal{Q}_{112}\mathcal{Q}_{22}{}^0 - \mathcal{Q}_{222}\mathcal{Q}_{11}{}^0}\right|^C \tag{11}$$

This represents a symmetric bifurcation B critical with respect to u_1, which lies on the post-buckling path of the asymmetric bifurcation A. Since B lies away from the fundamental path we refer to it as a *secondary bifurcation*.

The equilibrium paths of the perfect system can now be sketched as σ varies.[10] Schematic representations of these paths in u_i-Λ space for the homeoclinal point of bifurcation are given in FIGURE 7 (a)–(c), for $\sigma < 0$, $\sigma = 0$, and $\sigma > 0$, respectively. These illustrations show how the bifurcations A, S, and B arise. We note that very similar results can be obtained for the monoclinal and anticlinal points of bifurcation.

Locus of the Umbilic Point and Deformation of the Symmetric Section

As σ varies, the structurally stable topology of the imperfection-sensitivity ensures that the umbilic point (represented by C at $\sigma = 0$) persists as the surface deforms. We can trace the locus of this point by a new perturbation scheme.

Defining this umbilic point U, a two-fold compound critical point, by the identity,

$$\mathcal{Q}_{ij}[u_k^U(s), \Lambda^U(s), \epsilon^U(s)] \equiv 0 \tag{12}$$

where s is the perturbation parameter, as before; we can use the lowest-order nontrivial full derivative of this with respect to s to obtain the governing equations,

$$\mathcal{Q}_{112}u_2^{U(1)} + \mathcal{Q}_{11}{}^0\Lambda^{U(1)} + \mathcal{Q}_{11}{}^0\sigma^{U(1)} \mid^C = 0$$

$$\mathcal{Q}_{112}u_1^{U(1)} \mid^C = 0$$

$$\mathcal{Q}_{222}u_2^{U(1)} + \mathcal{Q}_{22}{}^0\Lambda^{U(1)} \mid^C = 0 \tag{13}$$

These three equations replace the first two (critical state) equations of (7), which of course apply only to *distinct* critical behavior away from C.

Solving now these equations along with the two equilibrium equations we find that $\dot{\alpha}_1$ must equal zero, so U always arises on the symmetric section of the imperfection-sensitivity surface where $\epsilon^1 = 0$. The complete solution is,

$$u_1^{U(1)C} = 0, \quad \left.\frac{\Lambda^{U(1)}}{u_2^{U(1)}}\right|^C = -\left.\frac{\alpha_{222}}{\alpha_{22}^0}\right|^C$$

$$\left.\frac{\sigma^{U(1)}}{u_2^{U(1)}}\right|^C = \left.\frac{\alpha_{222}}{\alpha_{22}^0} - \frac{\alpha_{112}}{\alpha_{11}^0}\right|^C, \quad \left.\frac{\epsilon^{U(2)}}{(u_2^{U(1)})^2}\right|^C = \left.\frac{\alpha_{222}}{\dot{\alpha}_2}\right|^C \tag{14}$$

which gives the relationship,

$$\left.\frac{\Lambda^{U(1)}}{\sigma^{U(1)}}\right|^C = \left.\frac{\alpha_{222}\alpha_{11}^0}{\alpha_{112}\alpha_{22}^0 - \alpha_{222}\alpha_{11}^0}\right|^C \tag{15}$$

It is interesting to compare this result with the secondary bifurcation B, given by Equation 11.

This is the same locus in Λ-ϵ^2 space as the line of folds for the surface at complete coincidence, given by Equation 6, which is itself the same as the imperfection-sensitivity of a *distinct* asymmetric point of bifurcation critical with respect to u_2.[3] It is clearly associated with the fact that the asymmetric bifurcation A, and its corresponding uncoupled post-buckling path, remain fixed as σ varies.

Combining the above information with the results for the perfect system, we find that the only plausible deformation on the symmetric section of the imperfection-sensitivity is as illustrated in FIGURE 7(d), here shown for the homeoclinal point of bifurcation. We see that the line of folds remains fixed, but the line of cusps of Equation 5, while retaining the same shape, rolls around the fold line as shown;[9] the hyperbolic umbilic point H remains the point of contact but shifts away from the Λ-axis. Entirely similar conclusions can be drawn for the monoclinal and anticlinal points of bifurcation.

Imperfection-Sensitivity at Near-Coincidence

To determine the full imperfection-sensitivity for a nonzero value of σ, we introduced a polar coordinate transformation on λ and σ as follows,

$$\lambda = \bar{\lambda} \sin \phi,$$

$$\sigma = \bar{\lambda} \cos \phi. \tag{16}$$

Substituting these into the potential function of Equation 1 we obtain,

$$\alpha = \tfrac{1}{6}\alpha_{222}^C u_2^3 + \tfrac{1}{2}\alpha_{112}^C u_1^2 u_2 + \tfrac{1}{2}\bar{\lambda}[\alpha_{11}^{0C}(\sin \phi + \cos \phi)u_1^2 + \alpha_{22}^{0C} \sin \phi u_2^2]$$
$$+ \alpha_1^{1C}\epsilon^1 u_1 + \alpha_2^{2C}\epsilon^2 u_2 + \text{higher-order terms.} \tag{17}$$

We see now that by setting,

$$\bar{\alpha}_{11}^{0C} = \alpha_{11}^{0C}(\sin \phi + \cos \phi),$$

$$\bar{\alpha}_{22}^{0C} = \alpha_{22}^{0C} \sin \phi, \tag{18}$$

we can generate a family of associated problems, each of which corresponds to choosing a particular value for ϕ between 0 and 2π. The governing equations of imperfection-sensitivity for each case are reduced to the form of Equations 7, but with $\sigma = 0$, $\overline{\lambda}$ replacing λ, and $\overline{\alpha}_{ii}^{OC}$ replacing α_{ii}^{OC}. These equations have already been solved to generate the undeformed pictures of FIGURES 4, 5, and 6.

Thus the introduction of the polar transformation means that we have, at least implicitly, all the necessary information needed to plot imperfection-sensitivity surfaces with $\sigma \neq 0$. Of course we would like to draw these at constant values of σ. The complete solution must therefore involve some kind of search routine; λ and ϕ both vary over the surface, but the method of solution above demands that ϕ be held constant. However, since these surfaces are most conveniently plotted with the aid of a computer, this fact presents no real difficulty, although an extra degree of complexity is nevertheless introduced.

To illustrate the deformation of a full imperfection-sensitivity surface with varying σ, we select the optimal stiffened plate example discussed earlier,[13] which at a homeoclinal point of bifurcation generates the hyperbolic umbilic form of FIGURE 6. We note here that one possible physical interpretation of a *splitting parameter* σ, which affects the critical load of just one of the contributing bifurcations, could be provided in this example by a small change in the length of the plate. This would raise or lower the Euler buckling load, but leave the local buckling load unaltered.

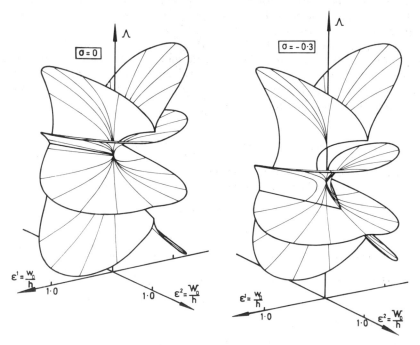

FIGURE 8. Distortion of the surface of FIGURE 6 for $\sigma < 0$. Structural stability ensures a smooth deformation and continuing existence of hyperbolic umbilic point in some imperfect system. One distinctive feature of the hyperbolic umbilic, the bilinear corner at the surface intersection, is here clearly illustrated.

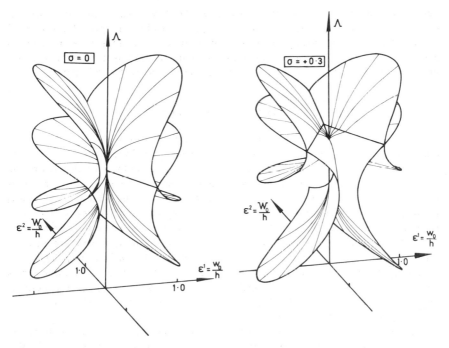

FIGURE 9. Distortion of the surface of FIGURE 6 for $\sigma > 0$. This second view of a distorted surface shows, in particular, the cusp associated with the secondary bifurcation B, which arises on the Λ-axis as shown in FIGURE 7.

The deformed surfaces are shown in FIGURES 8 and 9 together with corresponding views of the surface when it is not deformed. These pictures illustrate very clearly the effect of perturbing a structurally stable topology—deformation takes place slowly and nonviolently. However, if we slice through the surface, the structurally unstable locus obtained on the section could very easily exhibit some severe contortion or change in topological form as it is perturbed. Analytically this could give rise to considerable convergence difficulties.

In conclusion, let us look briefly at the situation that arises when $\alpha_{222}{}^C = 0$ and we have the transitional case of the paraclinal point of bifurcation. This problem is introduced in the earlier discussion of the equilibrium paths of the perfect system at complete coincidence, but we can now enlarge somewhat on these remarks.

Examining the form of the potential function, Thom's theory[1] predicts a *parabolic umbilic catastrophe* at C. We have seen earlier that, in this case, the coefficient $\alpha_{2222}{}^C$ plays a significant role, and this is confirmed by Thom's expression for the *germ* of the catastrophe, which is presented explicitly as underlined terms in FIGURE 1 of Thompson's paper.[7]

As might be expected, this catastrophe requires four control parameters for a universal unfolding, and it can be easily verified that the four parameters considered here, Λ, σ, and ϵ^i would suffice. We can thus predict that an imperfection-sensitivity surface drawn at different, but constant values of σ would, unlike the present example of FIGURES 8 and 9, exhibit changes of topological form. However, although this form of potential function does arise in Koiter's general treatment of interactive buckling of plate assemblages,[11] the exact significance of this change in topology still remains to be investigated in detail.

CONCLUDING REMARKS

The analysis of near-coincidence represents an extension of the discrete general theory of elastic stability,[3] seen from a recently developed viewpoint[4] which takes into account the philosophical base and classificational power of catastrophe theory. Interest is almost exclusively centered on semisymmetric points of bifurcation, which are seen to relate to the umbilic catastrophes. These have received less attention in the mathematics literature than the cuspoid catastrophes, which, with their single active internal variable, corresponds to the distinct points of bifurcation.[4] Two contributions, however, are of immediate interest and are directly relevant to the present approach. First the concept of the *umbilic bracelet,* introduced by Zeeman,[20] has proved very valuable in the study of general cubic forms and the transitions between umbilics.[17] Second, the *differential geometry* study of umbilics due to Porteous[21] develops the same somewhat refined classification of phenomena as our bifurcational view.[22]

Structural stability has been seen to be of key significance, and it can be expected to play an increasingly important role as we turn to more complex phenomena. It is by direct application of this principle that we are able to predict the deformations of the symmetric section for semisymmetric points.[9] The parabolic umbilic catastrophe introduces a new fundamental control parameter of the system; the splitting parameter σ. Normally an imperfection-sensitivity study would not include such a parameter, and a developed surface could then suffer changes in topological form with random fluctuations in σ. However, the precise physical interpretation of these changes remains to be seen.

REFERENCES

1. THOM, R. 1975. Structural Stability and Morphogenesis. D. H. Fowler, Trans. Benjamin, Reading.
2. KOITER, W. T. 1945. On the stability of elastic equilibrium. Dissertation. Delft University of Technology, Holland. [1967. NASA, Tech. Trans. F 10: 833].
3. THOMPSON, J. M. T. & G. W. HUNT. 1973. A General Theory of Elastic Stability. Wiley. London.
4. THOMPSON, J. M. T. & G. W. HUNT. 1975. Towards a unified bifurcation theory. J. Appl. Math. Phys. 26: 581–604.
5. THOMPSON, J. M. T. & G. W. HUNT. 1977. The instability of evolving systems. Interdiscip. Sci. Rev. 2: 240–262.
6. THOMPSON, J. M. T. 1975. Experiments in catastrophe. Nature 254: 392–395.
7. THOMPSON, J. M. T. 1978. Bifurcational aspects of catastrophe theory. Ann. N. Y. Acad. Sci. This volume.
8. HUNT, G. W. 1977. Imperfection-sensitivity of semi-symmetric branching. Proc. R. Soc. London A 357: 193–211.

9. HUNT, G. W. 1978. Catastrophe theory predictions of structural imperfection-sensitivity. To be published.

10. CHILVER, A. H. 1967. Coupled modes of elastic buckling. J. Mech. Phys. Solids. **15:** 15–28.

11. KOITER, W. T. 1976. General theory of mode interaction in stiffened plate and shell structures. Rep. WTHD 91. Delft University of Technology. Holland.

12. THOMPSON, J. M. T. & G. W. HUNT. 1978. A bifurcation theory for the instabilities of optimization and design. *In* Mathematical Methods of the Social Sciences. D. Berlinski, Ed. Synthese **36:** 315–351.

13. TVERGAARD, V. 1973. Imperfection-sensitivity of a wide integrally stiffened panel under compression. Int. J. Solids Struct. **9:** 177–192.

14. LEWIS, G. M. 1973. Numerical analysis of initial post-buckling. Thesis. University College, London.

15. HANSEN, J. S. 1973. Buckling of imperfection-sensitive structures: a probabilistic approach. Thesis. University of Waterloo.

16. SUPPLE, W. J., Ed. 1973. Structural Instability. IPC Science and Technology Press. Guildford, England.

17. THOMPSON, J. M. T. & Z. GASPAR. 1977. A buckling model for the set of umbilic catastrophes. Math. Proc. Camb. Phil. Soc. **82:** 497–507.

18. FOX, W. C., J. RHODES & A. C. WALKER. 1976. Local buckling of outstands in stiffened plates. Aeronaut. Q. **27:** 277–291.

19. POSTON, T. & I. STEWART. 1976. Taylor expansions and catastrophes. Res. Notes Math. 7.

20. ZEEMAN, E. C. 1976. The umbilic bracelet and the double cusp catastrophe. *In* Structural Stability, the Theory of Catastrophes, and Applications in the Sciences (Seattle 1975). P. Hilton, Ed. Springer Lect. Notes Math. 525. Springer-Verlag. Berlin.

21. PORTEOUS, I. R. 1971. The normal singularities of a submanifold. J. Diff. Geom. **5:** 543–564.

22. PORTEOUS, I. R. 1976. Open letter to J. M. T. Thompson and G. W. Hunt.

SOME APPLICABLE HOPF BIFURCATION FORMULAS AND AN APPLICATION IN WIND ENGINEERING

A. B. Poore and A. Al-Rawi

Department of Mathematics
Colorado State University
Fort Collins, Colorado 80523

INTRODUCTION

The application of the theory for the bifurcation of periodic orbits from certain steady states in autonomous ordinary differential equations gives a fundamentally important technique in the overall investigations of the dynamics associated with the problem. To facilitate the application of this theory, we shall give algebraic expressions that resolve the existence, direction, and stability for the bifurcation of a single and a pair of periodic orbits. A problem in wind engineering, which motivated much of this work, is given to illustrate how this bifurcation theory can be used to obtain significant information about the dynamics in the problem.

The bifurcation of periodic orbits from certain critical points of a real, n-dimensional ($n \geq 2$), first-order system of autonomous ordinary differential equations was treated by E. Hopf[7] in 1942. To briefly explain Hopf's work, let the differential equation be denoted by

$$\frac{dx}{dt} = F(x, \epsilon) \tag{1}$$

where ϵ is a real parameter, and let a^ϵ be a critical point. It is assumed that F is analytic in a neighborhood of $(x, \epsilon) = (a^0, 0)$ and that the matrix $F_x(a^0, 0)$ has exactly two, nonzero, purely imaginary eigenvalues, say $\pm i\omega_0$, and no zero eigenvalue. Hopf proved that a nonconstant periodic orbit bifurcates from $(x, \epsilon) = (a^0, 0)$ under the sole additional assumption that $\alpha'(0) \neq 0$ where $\alpha(\epsilon) + i\omega(\epsilon)$ denotes that eigenvalue of $F_x(a^\epsilon, \epsilon)$, which is a continuous extension of $+i\omega_0$. Hopf also supplied a uniqueness theorem and the essential information regarding stability.

Since a discussion of the subsequent literature on the further development of theory can be found in the works of Marsden and McCracken[10] and Poore,[12] we forgo this discussion here. It is, instead, our goal here to give algebraic formulas for direction of bifurcation, stability, and initial conditions of the bifurcated periodic orbit. We treat not only the case of a single pair of complex conjugate purely imaginary eigenvalues but also the case of two pairs. This latter situation arose in our investigations of the wake-oscillator model of Hartlen and Currie[6] in the field of wind engineering, and thus motivated our development of the methods of application for this two-pair case. In addition, D. S. Cohen[3] has given

*This work was supported in part by Grant ENG 76-03135 from the National Science Foundation.

590

examples of problems in biology in which the usual case is the bifurcation of periodic orbits associated with two pairs of complex conjugate eigenvalues crossing the imaginary axis instead of one pair crossing the imaginary axis.

It has been our experience that the Hopf bifurcation can play a fundamental role in the analysis of the dynamics of problems modeled by autonomous ordinary differential equations in at least three ways. First, it is an excellent method for establishing the existence and stability of periodic orbits. Secondly, since these bifurcations are often associated with exchanges in the stability of a steady state (critical point) as some parameter is varied, one can extract the dynamics of the system associated with this exchange. The theory shows that one either has the appearance of a stable oscillatory state or a jump phenomenon associated with the bifurcation of an unstable periodic orbit. Finally, this theory furnishes a basis for sytematically locating the periodic orbits in the "large." By starting near the bifurcation point one can trace these periodic orbits into the large. This technique has been used successfully in chemical reactor problems[13,14] and is the technique which we use on the wake-oscillator model.

Our plan then is to give a review of the theory and algebraic formulas for the single pair case. We shall then proceed to give the theory and algebraic formulas for the direction of bifurcation, stability, and initial conditions for the periodic orbits in the two-pairs case. Following that, our application to the wake-oscillator model is given, where we show the bifurcation of a stable and unstable model periodic orbits from a steady state that is unstable on both sides of the bifurcation point. Other bifurcation diagrams are given to help explain this phenomenon.

THE SINGLE-PAIR CASE

In this section we discuss the case in which a single pair of complex conjugate eigenvalues crosses the imaginary axis. This case is presented here not only for completeness but also to help draw an analogy between the cases of a single pair and a double pair of complex conjugate, purely imaginary eigenvalues. The presentation here is in a much abbreviated form since an in-depth discussion along the present lines has been given by one of the authors.[12]

Starting with the real, n-dimensional ($n \geq 2$), first-order system of autonomous ordinary differential equations of the form

$$\frac{dx}{dt} = F(x, \epsilon), \qquad (2)$$

we assume that there is a critical point a^0 such that the Jacobian matrix $F_x(a^0, 0)$ has a pair of complex conjugate purely imaginary eigenvalues $\pm i\omega_0$ with $\omega_0 > 0$, and no zero eigenvalues. In addition, if there is a purely imaginary eigenvalue, say $i\lambda$, then $\lambda \neq \pm m\omega_0$, where m is a positive integer. The smoothness assumption that we make on F is $F(x, \epsilon) \in C^k[Dx(-\epsilon_0, \epsilon_0)]$ where $\epsilon_0 > 0$, $k \geq 3$, and D is a domain in $\mathbb{R}^n$ containing a^0. By the eigenvalue requirement on the Jacobian matrix $F_x(a^0, 0)$, the implicit function theorem guarantees the existence of a critical point a^ϵ which is k-times continuously differentiable in ϵ and satisfies $F(a^\epsilon, \epsilon) = 0$ for ϵ in a sufficiently small neighborhood of $\epsilon = 0$. Using this definition of a^ϵ, we introduce the change of variables

$$
\begin{cases}
x = a^\epsilon + \mu y, \quad (1 + \mu\eta)s = t, \\[2mm]
A^\epsilon = F_x(a^\epsilon, \epsilon), \quad \epsilon B^\epsilon = A^\epsilon - A^0, \quad B^0 = \left.\dfrac{dA^\epsilon}{d\epsilon}\right|_{\epsilon=0}, \\[2mm]
\mu^2 Q(y, \mu, \epsilon) = F(a^\epsilon + \mu y, \epsilon) - \mu A^\epsilon y, \quad \epsilon = \mu\delta,
\end{cases}
\tag{3}
$$

where μ, δ, and η are auxiliary parameters that are to be defined. This change of variables reduces the differential equation (2) to

$$
\frac{dy}{ds} = A^0 y + \mu G(y, \mu, \delta, \eta),
\tag{4}
$$

where

$$
G(y, \mu, \delta, \eta) = \delta B^{\mu\delta} y + \eta A^{\mu\delta} y + (1 + \mu\eta)Q(y, \mu, \mu\delta).
\tag{5}
$$

The independent small parameter here is μ, which essentially represents the amplitude of the bifurcating periodic orbit. The parameter δ is introduced so that the relation between ϵ and μ can be determined in the course of the development of the theory and not be predetermined. The remaining parameter, η, is introduced to allow for a change in the period of oscillation as μ changes. We now continue with some further definitions that are needed in the explanations of the existence, stability, and algebraic criteria for stability and direction of bifurcation.

Let $T = 2\pi/\omega_0$ and define $\mathcal{P}_T$ to be the Banach space of all T-periodic continuous vector functions which map $\mathbb{R}^1$ into $\mathbb{R}^n$ with the norm $\| y \| = \sup\{| y(s) | : 0 \le s \le T\}$ where $| \cdot |$ denotes a norm on $\mathbb{R}^n$. Our next objective is to define a continuous projection on $\mathcal{P}_T$. From the eigenvalue requirements on the real matrix $A^0 = F_x(a^0, 0)$, it follows that there exists a real non-singular matrix P such that

$$
PA^0 P^{-1} = \text{diag} \left(\begin{pmatrix} 0 & \omega_0 \\ -\omega_0 & 0 \end{pmatrix}, \ C_0 \right),
\tag{6}
$$

where C_0 is a real $(n - 2) \times (n - 2)$ matrix. Using such a matrix P, we define an $n \times 2$ matrix function $\Phi(s)$ by

$$
\Phi(s) = \exp(A^0 s)P^{-1}(e_1, e_2),
\tag{7}
$$

where (e_1, e_2) is an $n \times 2$ matrix whose kth column is e_k. (Here, e_k denotes the n-dimensional column vector with a one in the kth row and zeros elsewhere.) By the special assumptions on A^0, it then follows from J. Hale[5] that the operator U defined for $v \in \mathcal{P}_T$ by

$$
Uv = \Phi(s) \left(\int_0^T {}^t\Phi(s)\Phi(s)ds \right)^{-1} \int_0^T {}^t\Phi(s)v(s)ds,
\tag{8}
$$

is a well-defined continuous projection on $\mathcal{P}_T$; U is a projection of $\mathcal{P}_T$ onto the T-periodic solutions of $dy/ds - A^0 y = 0$. With these definitions we now state the existence theorem in

THEOREM 1. Let $F(x, \epsilon) \in C^k[Dx(-\epsilon_0, \epsilon_0)]$ where $\epsilon_0 > 0$, $k \geq 3$ and D is a domain in $\mathbb{R}^n$ containing a critical point a^ϵ. We assume that the Jacobian matrix $F_x(a^0, 0)$ has a pair of complex conjugate purely imaginary eigenvalues $\pm i\omega_0$ with $\omega_0 > 0$, and if λ is any other eigenvalue, then $\lambda \neq 0$ and $\lambda \neq mi\omega_0$ for any integer m. Let $\alpha(\epsilon) + i\omega(\epsilon)$ denote that eigenvalue of $F_x(a^\epsilon, \epsilon)$ which is a continuous extension of $+i\omega_0$ and let $T = 2\pi/\omega_0$.

If $\alpha'(0) \neq 0$, then for some sufficiently small $\mu_1 > 0$ there exist real-valued functions $\delta(\mu)$ and $\eta(\mu) \in C^{k-2}[-\mu_1, \mu_1]$ and a $y(s, \mu) \in \mathcal{P}_T$ such that $\delta(0) = \eta(0) = 0$, and

$$x(t, \mu) = a^{\epsilon(\mu)} + \mu y\left(\frac{t}{1 + \mu\eta(\mu)}, \mu\right) \tag{9}$$

is a $[1 + \mu\eta(\mu)] \cdot T$-periodic solution of

$$\frac{dx}{dt} = F[x, \epsilon(\mu)] \tag{10}$$

for $\epsilon(\mu) = \mu\delta(\mu)$. If $\delta = \delta(\mu)$ and $\eta = \eta(\mu)$ in (4), then $y = y(s, \mu)$ is a T-periodic solution of (4) with $y(s, \mu) \in C^{k-2}[-\mu_1, \mu_1]$, uniformly in s. Furthermore, it may be assumed that $Uy = \Phi(s)b(\mu)$ where $\Phi(s)$ and U are defined by (7) and (8) and $b(\mu)$ is any arbitrary real, two-dimensional vector function which satisfies $b(\mu) \in C^{k-2}[-\mu_1, \mu_1]$ and $b(0) \cdot b(0) \neq 0$.

REMARKS. Besides the smoothness assumptions on $F(x, \epsilon)$, the essential requirements in this theorem are those on the eigenvalues of the matrix $A^0 = F_x \cdot (a^0, 0)$ and the nonzero derivative of the real part of the eigenvalue $\alpha(\epsilon) + i\omega(\epsilon)$ at $\epsilon = 0$. In applications, these requirements are often satisfied when there is an exchange in the stability of a critical point as two complex conjugate eigenvalues cross the imaginary axis. We also wish to point out the significance of the functions $\eta(\mu)$ and $\delta(\mu)$. Since $\epsilon = \epsilon(\mu) = \mu\delta(\mu) = \mu^2\delta'(0) + o(\mu^2)$ as $\mu \to 0$, the bifurcated periodic orbit (11) exists for (x, ϵ) in a sufficiently small neighborhood of $(a^\epsilon, 0)$ only for $\epsilon > 0$ if $\delta'(0) > 0$ or only for $\epsilon < 0$ if $\delta'(0) < 0$. In this sense we shall say that $\delta'(0)$ determines the direction of bifurcation provided $\delta'(0) \neq 0$. The period of oscillation of the periodic solution $x(t, \mu)$ in (11) is

$$T[1 + \mu\eta(\mu)] = \frac{2\pi}{\omega_0}[1 + \mu^2\eta'(0) + o(\mu^2)]$$

as $\mu \to 0$. Thus the period increases or decreases from $T = 2\pi/\omega_0 > 0$ according to the sign of $\eta'(0)$. Also, the stability of the periodic solution is reduced to an algebraic problem (Theorem 2) once the sign of $\alpha'(0)\delta'(0)$ is known.

By the assumptions of Theorem 1 the differential equation in (4) is continuously differentiable in the parameter μ when $\delta = \delta(\mu)$ and $\eta = \eta(\mu)$ and in the function y in a neighborhood of the periodic orbit. Thus the existing periodic orbit will be asymptotically orbitally stable with asymptotic phase if $n - 1$ of the characteristic multipliers of the variational equation have moduli less than one. (One of the characteristic multipliers—the one corresponding to the periodic orbit—is identically equal to one). If $n - 1$ of the characteristic multipliers have moduli different from one with some larger than one and the remaining less than one, then it is reasonable to speak of stable and unstable manifolds of the periodic orbit. Since a full discussion of these manifolds is given by Hale[5] we shall develop information about the modulus of each of the characteristic multipliers. This is the content of

THEOREM 2. (Stability) Suppose the Jacobian matrix $A^0 = F_x(a^0, 0)$ has a single pair of complex conjugate purely imaginary eigenvalues $\pm i\omega_0$ with $\omega_0 > 0$, and that the remaining eigenvalues have nonzero real parts. Additionally, assume that the remaining hypotheses of Theorem 1 are satisfied. Then the n-characteristic multipliers, denoted by $\rho_l(\mu)$ for $1 \leq l \leq n$, are continuous functions of μ at $\mu = 0$ and satisfy the relations

$$\rho_1(\mu) = \exp[\hat{\beta}(\mu)T], \quad \rho_2(\mu) \equiv 1,$$

$$\rho_l(\mu) = \exp[\lambda_l^0 T + o(1)] \text{ as } \mu \to 0 \ (3 \leq l \leq n), \tag{11}$$

where $\hat{\beta}(\mu)$ is continuous at $\mu = 0$ and $\hat{\beta}(0) = 0$. There, T is the period of oscillation of the periodic solution $y(s, \mu)$ of (4), and the λ_l^0 values denote the $n - 2$ eigenvalues (counting multiplicities) of $A^0 = F_x(a^0, 0)$ with nonzero real parts.

In addition, if $k \geq 4$ in the smoothness assumption $F(x, \epsilon) \in C^k[Dx(-\epsilon_0, \epsilon_0)]$ and $\delta'(0) \neq 0$, then

$$\rho_1(\mu) = \exp[\mu^2\beta(\mu)T] = \exp[-2\mu^2\alpha'(0)\delta'(0)T + o(\mu^2)] \text{ as } \mu \to 0, \tag{12}$$

where $\beta(\mu)$ is a real-valued continuous function with $k - 4$ continuous derivatives for μ in a sufficiently small neighborhood of $\mu = 0$.

REMARKS. To obtain the information in (11), it is sufficient to have $k = 3$ in the smoothness assumption $F(x, \epsilon) \in C^k[Dx(-\epsilon_0, \epsilon_0)]$; however, to establish (12) our proof requires $k \geq 4$. If $k \geq 4$, $\alpha'(0)\delta'(0) > 0$ and $\Re e\lambda_l^0 < 0$ for $l = 3, \ldots, n$, the bifurcated periodic will be asymptotically orbitally stable with asymptotic phase for μ sufficiently small and $|\mu| > 0$. On the other hand, if $\alpha'(0)\delta'(0) < 0$ or $\Re e\lambda_l^0 > 0$ for some $l = 3, \ldots, n$, the periodic orbit will be unstable. Since an algebraic expression for $\alpha'(0)\delta'(0)$ is given in Theorem 3, the stability of the periodic orbit is thus reduced to an algebraic problem whenever $\delta'(0) \neq 0$.

THEOREM 3. (Algebraic Expressions.) Let $F(x, \epsilon)$ satisfy the hypotheses in Theorem 1 and let u and v denote left and right eigenvectors, respectively, for the eigenvalue $+i\omega_0 (\omega_0 > 0)$ of the matrix A^0. If u and v are normalized by the requirement $uv = 1$ (u is a row vector and v, a column vector), then

$$\alpha'(0) + i\omega'(0) = uB^0v \tag{13}$$

and

$$8\alpha'(0)\delta'(0) + i8[\omega'(0)\delta'(0) + \omega_0\eta'(0)] = b(0) \cdot b(0) \cdot$$

$$\{-uF_{xxx}vv\bar{v} + 2uF_{xx}vA^{0-1}F_{xx}v\bar{v} + uF_{xx}\bar{v}(A^0 - 2i\omega_0I)^{-1}F_{xx}vv\}, \tag{14}$$

where $\alpha'(0) + i\omega'(0)$ denotes the derivative of the complex eigenvalue $\alpha(\epsilon) + i\omega(\epsilon)$ of A^ϵ at $\epsilon = 0$, $B^0 = dA^\epsilon/d\epsilon|_{\epsilon=0}$, $A^0 = F_x(a^0, 0)$, $F_{xx} = F_{xx}(a^0, 0)$, $F_{xxx} = F_{xxx}(a^0, 0)$, and $b(0)$ is the $\mu = 0$ value of the two-dimensional vector $b(\mu)$ which occurs in $Uy(s, \mu) = \Phi(s)b(\mu)$.

REMARKS. Written out in component form, (14) can be expressed as

$$8\alpha'(0)\delta'(0) + i8[\omega'(0)\delta'(0) + \omega_0\eta'(0)]$$

$$= b(0) \cdot b(0) \left\{ - u_l \frac{\partial^3 F^l}{\partial x_j \partial x_k \partial x_p} v_j v_k \bar{v}_p + 2u_l \frac{\partial^2 F^l}{\partial x_j \partial x_k} v_j (A^{0-1})_{kr} \frac{\partial^2 F^r}{\partial x_p \partial x_q} v_p \bar{v}_q \right.$$

$$\left. + u_l \frac{\partial^2 F^l}{\partial x_j \partial x_k} \bar{v}_j [(A^0 - 2i\omega_0 I)^{-1}]_{kr} \frac{\partial^2 F^r}{\partial x_p \partial x_q} v_p v_q \right\}, \quad (15)$$

where $(A^{0-1})_{kr}$ denotes the element in the kth row and the rth column of A^{0-1} and repeated indices within each term imply a sum from one to n. We also wish to point out that there is an arbitrary multiplicative positive constant in (14). Let a and b be complex scalars and replace the eigenvectors u and v in (14) by au and bv. Then we can factor out the product $ab b\bar{b}$. By the normalization requirement $uv = (au)(bv) = 1$, $ab = 1$, so that we are left with the positive number $b\bar{b}\, b(0) \cdot b(0)$ multiplying the expression enclosed in parentheses in (14). In particular, this shows that the sign of the real and imaginary parts of the right hand side of (14) are independent of the choice of $h(\mu)$ in $Uy = \Phi(s)h(\mu)$ and the eigenvectors u and v so long as $b(0) \cdot b(0) > 0$ and $uv = 1$.

Finally, from the information in Theorem 1 one can establish the asymptotic relation

$$x(0, \epsilon) \sim a^0 + \sqrt{\frac{\epsilon}{\delta'(0)}} \left(b_1(0) \frac{(\bar{v} - v)}{2i} + b_2(0) \frac{(v + \bar{v})}{2} \right) \text{ as } \epsilon \to 0. \quad (15)$$

It should be stressed that $b(0) = [b_1(0), b_2(0)]$ and v are the same as that used in Theorem 3. Of course, Equation 15 is not a very profound observation; however, the use of (15) in choosing the appropriate initial conditions for numerical computation of the periodic orbit can save a great deal of computer time, especially for higher dimensional problems.

THE TWO-PAIRS CASE

In this section we consider the case in which the matrix $A^0 = F_x(a^0, 0)$ has two pairs of complex conjugate purely imaginary eigenvalues $\pm i\omega_{10}$ and $\pm i\omega_{20}$ where $\omega_{20} > \omega_{10} > 0$. If the dimension n of the vector function F is greater than four, we shall assume that the remaining eigenvalues have nonzero real parts. Let $\alpha_1(\epsilon) \pm i\omega_1(\epsilon)$ and $\alpha_2(\epsilon) \pm i\omega_2(\epsilon)$ denote the eigenvalues of $A^\epsilon = F_x(a^\epsilon, \epsilon)$, which are the continuous extensions of $\pm i\omega_{10}$ and $\pm i\omega_{20}$, respectively. If $\alpha_2'(0) \neq 0$, then Theorem 1 of the previous section guarantees the existence of a non-constant bifurcated periodic orbit with asymptotic period $2\pi/\omega_{20}$. To delineate this situation, we consider three cases. If ω_{20}/ω_{10} is not a positive integer, we can give a relatively complete and tractable set of algebraic expressions which yield information about the existence of two distinct periodic orbits, their directions of bifurcation, and their stabilities. The second case is $\omega_{20}/\omega_{10} = k$, a positive integer greater than one. In this second case we still have bifurcation of a periodic orbit with asymptotic period $2\pi/\omega_{20}$ if $\alpha_2'(0) \neq 0$. However, the unperturbed problem $dv/ds - A^0 y = 0$ now has four linearly independent periodic solutions with period $2\pi/\omega_{10}$ since $k(2\pi/\omega_{20}) = 2\pi/\omega_{10}$. The existence problem reduces to the solution of four equations with the two free parameters δ and η in (4) and four free parameters in the initial conditions. Several existence theorems which utilize various combinations of four of the six free parameters are given in Reference 1. Finally, the case $\omega_{20} = \omega_{10} > 0$ is even more difficult because one has to consider the various canonical forms of the matrix A^0 and the problem of the existence of $\alpha'(0)$. Our attention here is confined to the first case since it appears that in applications the first case is rather common whereas the second two cases seldom occur. This is certainly the case in the application in wind engineering that we give in the next section.

To explain the situation in which $\omega_{20} \neq k\omega_{10}$ for any positive integer k, we begin with some definitions. For $l = 1$ and 2 let $T_l = 2\pi/\omega_{l0}$, and let $\mathcal{P}_{T_l}$ denote the Banach space of continuous T_l-periodic functions which map $\mathbb{R}^1$ into $\mathbb{R}^n$. Next, there is a real nonsingular matrix P for which

$$PA^0P^{-1} = \operatorname{diag}\left(\begin{pmatrix} 0 & \omega_{10} \\ -\omega_{10} & 0 \end{pmatrix}, \begin{pmatrix} 0 & \omega_{20} \\ -\omega_{20} & 0 \end{pmatrix}, C_0\right) \qquad (16)$$

where C_0 is an $(n - 4) \times (n - 4)$ real matrix whose eigenvalues have nonzero real parts. For such a matrix P we define a matrix function $\Phi^l(s)$ for $l = 1, 2$ by

$$\Phi^l(s) = \exp(A^0 s)P^{-1}(e_{2l-1}, e_{2l}) \qquad (17)$$

where e_k denotes the n-dimensional column vector with a one in the kth row and zeros elsewhere. We define projection operators U^l that act on $\mathcal{P}_{T_l}$ by

$$U^l v(s) = \Phi^l(s)\left(\int_0^{T_l} {}^t\Phi^l(s)\Phi^l(s)ds\right)^{-1}\int_0^{T_l} {}^t\Phi^l(s)v(s)ds, \qquad (18)$$

where $v(s) \in \mathcal{P}_{T_l}$. It follows from the work of Hale[5] that U^l is a well-defined continuous projection of $\mathcal{P}_{T_l}$ onto the T_l-periodic solutions of $(dy/ds) - A^0 y = 0$. Next, we replace y, η, and δ by y_l, η_l and δ_l for $l = 1$ and 2 in the change of variables (3). With these definitions we state the existence of two distinct bifurcating periodic orbits in

THEOREM 4. (Existence.) Suppose that $x = a^\epsilon$ is a constant solution of the differential equations $dx/dt = F(x, \epsilon)$ and that $F(x, \epsilon) \in C^k[Dx(-\epsilon_0, \epsilon_0)]$ where $\epsilon_0 > 0$, $k \geq 3$ and D is a domain in $\mathbb{R}^n$ containing a^ϵ. Assume that the Jacobian matrix $A^0 = F_x(a^0, 0)$ has two pairs of complex conjugate purely imaginary eigenvalues $\pm i\omega_{20}$ and $\pm i\omega_{10}$ with $\omega_{20} > \omega_{10} > 0$; and if the dimension n of F is greater than four, we require the remaining eigenvalues to have nonzero real parts. Let $\alpha_l(\epsilon) \pm i\omega_l(\epsilon)$ denote those eigenvalues of $A^\epsilon = F_x(a^\epsilon, \epsilon)$ which are continuous extensions of $\pm i\omega_{l0}$ for $l = 1$ and 2.

If $\alpha_l'(0) \neq 0$ for $l = 1$ and 2, then for $l = 1$ and 2 and for some sufficiently small $\mu_l > 0$ there exist real functions $\eta_l(\mu)$ and $\delta_l(\mu) \in C^{k-2}[-\mu_l, \mu_l]$ and a $y^l \cdot$

$(s, \mu) \in \mathcal{P}_{T_l}$ such that $\delta_l(0) = \eta_l(0) = 0$ and

$$x(t, \mu) = a^{\epsilon(\mu)} + \mu y^l \left(\frac{t}{1 + \mu \eta_l(\mu)}, \mu \right) \qquad (19)$$

is a $[1 + \mu \eta_l(\mu)] \cdot T_l$-periodic solution of

$$\frac{dx}{dt} = F[x, \epsilon(\mu)] \qquad (20)$$

for $\epsilon(\mu) = \mu \delta_l(\mu)$. If $\delta = \delta_l(\mu)$ and $\eta = \eta_l(\mu)$ in (4), then $y^l(s, \mu)$ is a T_l-periodic solution of (4) with $y^l(s, \mu) \in C^{k-2}[-\mu_l, \mu_l]$, uniformly in s. Furthermore, it may be assumed that $U^l y = \Phi^l(s) b^l(\mu)$ where $\Phi^l(s)$ and U^l are defined by (17) and (18) and $b^l(\mu)$ is any arbitrary real, two-dimensional vector function which satisfies $b^l(\mu) \in C^{k-2}[-\mu_l, \mu_l]$ and $b^l(0) \cdot b^l(0) \neq 0$ for $l = 1, 2$.

We first observe that this theorem states the existence of two distinct bifurcating periodic orbits. As before, the sign of $\delta_l'(0)$ indicates the direction of bifurcation and $\eta_l'(0)$ indicates whether the period of oscillation increases or decreases from T_l. The proof of this existence theorem follows along the same lines as that given for the single pair case and consequently will not be given here.

Algebraic expressions for $\alpha_l'(0) \pm i\omega_l'(0)$, $\delta_l'(0)$, and $\eta_l'(0)$ for $l = 1, 2$ are given in

THEOREM 5. (Algebraic Expressions.) Assume that the assumptions of Theorem 4 are satisfied and let u^l and v^l denote left and right eigenvectors of A^0 corresponding to $+i\omega_{l0}$, normalized by $u^l v^l = 1$ for $l = 1$ and 2. Then

$$\alpha_l'(0) + i\omega_l'(0) = u^l B^0 v^l \quad (l = 1 \text{ and } 2) \qquad (21)$$

and

$$8\alpha_l'(0)\delta_l'(0) + i8[\omega_l'(0)\delta_l'(0) + \omega_{l0}\eta_l'(0)] = b^l(0) \cdot b^l(0) \cdot$$
$$[-u^l F_{xxx} v^l v^l \bar{v}^l + 2u^l F_{xx} v^l A^{0-1} F_{xx} v^l \bar{v}^l + u^l F_{xx} \bar{v}^l (A^0 - 2i\omega_{l0} I)^{-1} F_{xx} v^l v^l] \qquad (22)$$

where A^0, B^0, F_{xx}, and F_{xxx} have the same meaning as in Theorem 3.

Having discussed existence of two distinct bifurcating periodic orbits and given the algebraic expressions for $\delta_l'(0)$, $\eta_l'(0)$, $\omega_l'(0)$, and $\alpha_l'(0)$ for $l = 1$ and 2, we turn to the much more complex problem of stability. This is the content of

THEOREM 6 (Stability). Assume that the hypotheses of Theorem 4 are satisfied so that there are two distinct bifurcating periodic orbits. Let $\rho_{lm}(\mu)$ ($l = 1, 2$; $m = 1, \ldots, n$) denote the characteristic multipliers for the periodic solution $y^l(s, \mu)$ which has an asymptotic period of $2\pi/\omega_{l0}$. These multipliers are continuous at $\mu = 0$ and, for $l = 1$ and 2, satisfy the relations:

$$\rho_{11}(\mu) = \exp(\hat{\beta}_{11}(\mu)T_l), \quad \rho_{12}(\mu) \equiv 1$$
$$\rho_{13}(\mu) = \exp\{[i\omega_{20} + \hat{\beta}_{13}(\mu)]T_1\}, \quad \rho_{14}(\mu) = \overline{\rho_{13}(\mu)},$$
$$\rho_{23}(\mu) = \exp\{[i\omega_{10} + \hat{\beta}_{23}(\mu)]T_2\}, \quad \rho_{24}(\mu) = \overline{\rho_{23}(\mu)},$$
$$\rho_{lm}(\mu) = \exp[\lambda_m T_l + o(1)] \quad \text{as} \quad \mu \to 0 \, (5 \leq m \leq n), \qquad (23)$$

where the $\hat{\beta}_{lm}(\mu)$ are continuous at $\mu = 0$ and the λ_m values ($5 \leq m \leq n$) are the $n - 4$ eigenvalues (counting multiplicities) of the matrix A^0 with nonzero real parts.

In addition, if $k \geq 4$ in the smoothness assumption $F(x, \epsilon) \in C^k[Dx(-\epsilon_0, \epsilon_0)]$ and if $\delta_l'(0) \neq 0$, then for $l = 1$ and 2

$$\rho_{l1}(\mu) = \exp(-2\mu^2\alpha_l'(0)\delta_l'(0)T_l + o(\mu^2)) \quad as \quad \mu \to 0.$$

Let

$$\beta_{13}{}^0 = \delta_1'(0)[\alpha_2'(0) + i\omega_2'(0)] + i\omega_{20}\eta_1'(0)$$

$$+ \frac{b'(0) \cdot b'(0)}{4} [u^2 F_{xxx} v^1 \bar{v}^1 v^2$$

$$- u^2 F_{xx} v^2 A^{0-1} F_{xx} v^1 \bar{v}^1$$

$$- u^2 F_{xx} v^1 (A^0 - i\omega_{20} I + i\omega_{10} I)^{-1} F_{xx} \bar{v}^1 v^2$$

$$- u^2 F_{xx} \bar{v}^1 (A^0 - i\omega_{20} I - i\omega_{10} I)^{-1} F_{xx} v^1 v^2]. \tag{24}$$

and

$$\beta_{23}{}^0 = \delta_2'(0)[\alpha_1'(0) + i\omega_1'(0)] + i\omega_{10}\eta_2'(0)$$

$$+ \frac{b^2(0) \cdot b^2(0)}{4} [u^1 F_{xxx} v^2 \bar{v}^2 v^1$$

$$- u^1 F_{xx} v^1 A^{0-1} F_{xx} v^2 \bar{v}^2$$

$$- u^1 F_{xx} v^2 (A^0 - i\omega_{10} I + i\omega_{20} I)^{-1} F_{xx} \bar{v}^2 v^1$$

$$- u^1 F_{xx} \bar{v}^2 (A^0 - i\omega_{10} I - i\omega_{20} I)^{-1} F_{xx} v^2 v^1]. \tag{25}$$

Then

$$\rho_{13}(\mu) = \exp[i\omega_{20} T_1 + \beta_{13}{}^0 \mu^2 T_1 + o(\mu^2)] \quad as \quad \mu \to 0 \tag{26}$$

and

$$\rho_{23}(\mu) = \exp[i\omega_{10} T_2 + \beta_{23}{}^0 \mu^2 T_2 + o(\mu^2)] \quad as \quad \mu \to 0. \tag{27}$$

It is required in (26) that the imaginary part of $\beta_{13}{}^0$ be nonzero whenever $2\omega_{20} = k\omega_{10}$ for a positive integer k.

REMARKS. The periodic orbit y^l is thus stable for μ sufficiently small whenever Re $\lambda_m < 0$ ($m = 5, \ldots, n$), $\alpha_l'(0)\delta_l'(0) > 0$, and Re $\beta_{13}{}^0 < 0$. If any of these inequalities are reversed, the periodic orbit is unstable. We also wish to point out that there is an interaction between the stabilities of the two periodic orbits. This is evident in the algebraic expressions for $\beta_{13}{}^0$ and $\beta_{23}{}^0$.

The proof of Theorem 6 is rather long and we shall not give it here; it can be found, however, in the work of A. R. Al-Rawi.[1] We close this section with a word about appropriate initial conditions for the two distinct periodic orbits. In analogy with the initial conditions given in the previous section, one should take

$$x^l(0, \epsilon) \sim a^0 + \left(\left|\frac{\epsilon}{\delta_l'(0)}\right|\right)^{1/2} \left(b_1{}^l(0) \frac{(\bar{v}^l - v^l)}{2i} + b_2{}^l(0) \frac{(v^l + v^l)}{2}\right), \tag{28}$$

for "small" ϵ and $l = 1$ or 2. Here, ϵ and $\delta'(0)$ have the same sign. These initial conditions are particularly important here because we now have two periodic orbits and it is crucial that we know which orbit we are finding numerically.

AN APPLICATION IN WIND ENGINEERING

One of the more intensely studied problems in the field of wind engineering is the vibration of bluff bodies, e.g., a tall building, under a steady incident flow of air.[2,4,6,8,9,11] In this problem the two shear layers, which separate from the body and bound a broad wake, tend to roll up and form discrete vortices alternately from the two layers, so that an oscillatory pressure loading is produced on what is normally an elastically supported bluff body. The resulting tendency of the bluff body is to oscillate in a plane that is perpendicular to the unidirectional incident flow.

Mathematically, the governing equations are the dynamic equations of motion for the bluff body and the Navier-Stokes equations. Because of the intractibility of these equations, considerable efforts have been extended to the development of simpler models.[11] One of these models, the wake-oscillator model of Hartlen and Currie,[6] is relatively simple and has given good agreement with experimental observations. In this model, the bluff body is a circular cylinder and the air flow is assumed to be normal to the cylinder. The motion of the cylinder, which is due to the vortex shedding, is assumed to be in a plane perpendicular to the flow. The modeling equations are[6]:

$$\ddot{x}_r + \beta \dot{x}_r + x_r = aW^2 c_L. \tag{29}$$

$$\ddot{c}_L - \alpha W \dot{c}_L + \frac{\gamma}{W} \dot{c}_L^3 + W^2 c_L = b\dot{x}_r \tag{30}$$

where x_r is a dimensionless displacement of the cylinder and c_L defines the instantaneous lift coefficient ($x_r = x/D$ where x is the actual displacement and D is the diameter of the cylinder). Equation 29 is the dynamic equation of the circular cylinder in which $-\beta \dot{x}_r$ represents the damping force; $-x_r$, the spring force; and $aW^2 c_L$, the pressure force due to vortex shedding. The second equation (30) for c_L represents the Van der Pol equation in which the lift coefficient, c_L, is linearly affected by the cylinder velocity. The parameters in (29) and (30) are the damping factor β, the dimensionless wind speed W, the Van der Pol coefficients α and γ, and the interaction parameter b. Modifications of these basic equations have been made by several authors[2,4,6,8,9,11] and perhaps one of the more promising models is that of Iwan and Blevins.[8,9]

We begin our investigations of the wake-oscillator by first rewriting the governing equations as a first-order system. If $x_1 = c_L$, $x_2 = \dot{c}_L$, $x_3 = x_r$, and $x_4 = \dot{x}_r$, then (29) and (30) can be written as $dx/dt = F(x, \epsilon)$ where

$$F(x, \epsilon) = {}^t\Bigg(x_2, -(W_0 + \epsilon)^2 x_1 + \alpha(W_0 + \epsilon)x_2$$

$$- \frac{\gamma}{W_0 + \epsilon} x_2^3 + bx_4, x_4, a(W_0 + \epsilon)^2 x_1 - \beta x_4 - x_3\Bigg),$$

where α, β, a, and b are to be held fixed with a variable wind speed $W = W_0 + \epsilon$. The only steady state associated with this differential equation is $x = a^{\epsilon} \equiv 0$ for all ϵ. The stability of this steady state can be determined by finding the eigenvalues of the Jacobian matrix

$$
A^{\epsilon} = \begin{pmatrix} 0 & 1 & 0 & 0 \\ -(W_0 + \epsilon)^2 & \alpha(W_0 + \epsilon) & 0 & b \\ 0 & 0 & 0 & 1 \\ a(W_0 + \epsilon)^2 & 0 & -1 & -\beta \end{pmatrix}.
$$

For a fixed set of values of a, b, α, and β, the eigenvalues of A^{ϵ} vary as the wind speed $W = W_0 + \epsilon$ varies. The stability is determined by finding those values of the wind speed W_0 at which either an eigenvalue passes through zero or one or two pairs of complex conjugate eigenvalues crosses the imaginary axis. For the present problem, it is easily shown that there are no zero eigenvalues for positive wind speeds. Hence, we concentrate on the case of one or two pairs of complex conjugate eigenvalues crossing the imaginary axis as W passes through W_0. From the sign of the derivative of the real part of these eigenvalues at $W = W_0$, we can determine the stability of the steady state to the left and right of $W = W_0$. Since this information is obtained in the course of the application of the bifurcation theory, we proceed to the application of this theory.

For fixed values of a, b, α, and β one can derive a polynomial in the wind speed W such that the positive real roots, denoted by W_0, give the wind speed at which A^0 has one or two pairs of complex conjugate purely imaginary eigenvalues. At these parameter values we can apply the theory of the previous two sections. This has been done and the various types of bifurcation phenomena have been classified by Al-Rawi.[1] We present here four of the types of bifurcation diagrams found. One of these four diagrams is for the case of two pairs and the other three diagrams are given to explain how it develops as the parameter varies. The relevant bifurcation information is contained in the table with four bifurcation diagrams in FIGURE 1.

The information near the bifurcation points was obtained by applying the theory of the previous two sections, and the completed bifurcation diagrams were obtained numerically. For the branches of completely stable periodic orbits in FIGURES 1b–d, we used the initial conditions in Equations 15 and 28. It turned out that they were very good not only for $\epsilon = 0.001, 0.01$, and 0.1 but also for $\epsilon = 1$ and 2. For the remaining branches of periodic orbits, the unstable part near the bifurcation point follows from the application of the theory. The stable part was obtained numerically and the remaining unstable part is simply conjecture. To obtain the stable part of the branch that bifurcates from $W = 1.026$ in FIGURE 1c, we started with a stable orbit in FIGURE 1d for a fixed $W < 1.019$ and gradually decreased β from $\beta = 0.0233$ to $\beta = 0.0161$ and then varied W. This technique also worked well for the cases in FIGURES 1a and b.

In each of the four cases (FIGURES 1a–d) our numerical solutions seem to indicate that the period of oscillation of the stable periodic orbit tends to infinity as W tends to zero. In addition, the amplitude of x, tends to zero whereas the amplitude of c_L appears to remain finite and bounded away from zero as w tends to

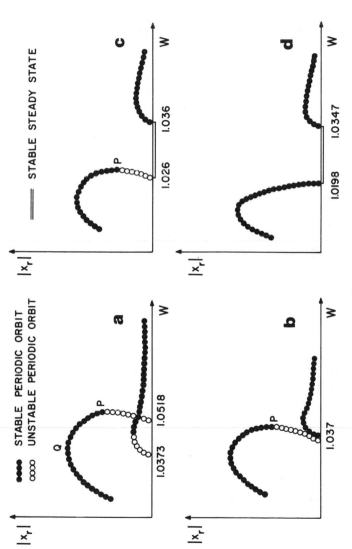

FIGURE 1. Bifurcation diagrams.

TABLE 1
BIFURCATION INFORMATION*

Parameters	Case 1 Single Pair		Case 2 Two Pairs		Case 3 Single Pair		Case 4 Single Pair	
β		0.0083		0.0114		0.0161	0.0198	0.0233
Wind speed, W_0	1.0518	1.0372	1.0370	1.0370	1.0264	1.0365	1.0198	1.0348
Eigenvalues	$\pm i1.0005,$ 0.0016 $\pm i1.0513$	$\pm i1.0366,$ 0.00155 $\pm i1.001$	$\pm i1.00088$	$\pm i1.03611$	$\pm i1.0018,$ -0.0024 $\pm i1.0245$	$\pm i1.03513,$ -0.00235 $\pm i1.0013$	$\pm i1.00415,$ -0.00604 $\pm i1.016$	$\pm i1.0327,$ -0.006 $\pm i1.002$
$\alpha'(0) + i\omega'(0)$	$-0.08 - i0.01$	$0.16 + i1.01$	$\alpha_1'(0) + i\omega_1'(0) =$ $-0.15 - i0.025$	$\alpha_2'(0) + i\omega_2'(0) =$ $0.16 + i1.02$	$-0.35 - i0.04$	$0.16 + i1.05$	$-0.93 + i0.13$	$0.16 + i1.1$
$\delta'(0)$	0.5	2028.7	$\delta_1'(0) = 1.1$	$\delta_2'(0) = 1999.7$	1.8	1956.8	-5.1	1900.2
$\eta'(0)$	0.3	-2026.8	$\eta_1'(0) = 1.1$	$\eta_2'(0) = -2027.0$	5.4	-2031.3	33.7	-2061.9
$\alpha'(0)\delta'(0)$	-0.038	317.3	$\alpha_1'(0)\delta_1'(0) =$ -0.172	$\alpha_2'(0)\delta_2'(0) =$ 317.8	-0.64	36.5	4.7	308.7
$\beta_{13}^{\,0}, \beta_{23}^{\,0}$	—		$\beta_{13}^{\,0} =$ $-14.1 + i3.4$	$\beta_{23}^{\,0} =$ -291.5 $-i2176.8$	—		—	
Stability of orbit	unstable	unstable	unstable	stable	unstable	stable	stable	stable

*Other parameters: $\alpha = 0.011$, $\gamma = 2/3$, $a = 0.02$, $b = 0.4$.

zero. The maximum amplitude of the oscillation of the cylinder, x_r, in the figure represents a displacement of between 3% and 7% of the diameter of the cylinder. As w tends to infinity the amplitude of x_r appears to tend to zero whereas that of c_L remains finite.

We now point out that the steady state is always unstable in FIGURES 1a and b and stable only between the bifurcation points in FIGURES 1c and d. In FIGURE 1b and Case 2 in TABLE 1 we have the bifurcation of two periodic orbits—one stable and one unstable—from the same bifurcation point. Thus we have an example of the bifurcation of a stable periodic orbit from a steady state which is unstable on both sides of the bifurcation point. This is quite different from the usual Hopf bifurcation situation. To explain how this comes about we start with $\beta = 0.0083$ in FIGURE 1a. As β increases to $\beta = 0.0114$, the two bifurcation points come together with the bifurcation points coinciding in FIGURE 1b. Increasing β further to $\beta = 0.0161$ causes the two bifurcation points to pass through one another with the result given in FIGURE 1c. Finally, as β is increased to 0.0233, the direction of bifurcation changes for the bifurcation point on the left, leaving two completely stable branches of periodic orbits in FIGURE 1d.

Next, we wish to remark on the dynamics of the structure as obtained in these bifurcation diagrams. If we start in FIGURE 1a with the wind speed to increase, then the period of oscillation decreases with a peak amplitude at Q. At P, an increase in the wind speed causes a jump in both the amplitude and period of oscillation. Similar phenomena occur in FIGURE 1b. In FIGURE 1c, there is a jump at P, but to a stable steady state, while in FIGURE 1d, there is no jump phenomena at all. Also in FIGURES 1a and b, note that there exist two stable periodic orbits for some values of the wind speed. This is experimentally observed in what is commonly called double amplitude response.

ACKNOWLEDGMENTS

We wish to express our appreciation to Professor J. E. Cermak and to A. Kareem for having suggested the problem in wind engineering and for their many helpful discussions.

REFERENCES

1. AL-RAWI, A. R. 1978. Some Applicable Hopf-Type Bifurcation Formulas and Applications in Wind Engineering. Ph.D. Thesis, Colorado State University.
2. BISHOP, R. E. & A. Y. HASSAN. 1964. The lift and drag forces on a circular cylinder oscillating in a flowing fluid. Proc. R. Soc. Series A 277: 51–75.
3. COHEN, D. S. To appear. Bifurcation from complex eigenvalues.
4. CURRIE, I. G., R. T. HARLEN & W. W. MARTIN. 1972. The response of Circular Cylinders to Vortex Shedding. Symposium on Flow-Induced Vibrations: 128–142. Spring-Verlag. Karlsrube, Germany.
5. HALE, J. 1969. Ordinary Differential Equations. Wiley-Interscience. New York, N.Y.
6. HARTLEN, R. T. & I. G. CURRIE. 1970. Lift-oscillator model of vortex-induced vibration. J. Eng. Mech. 96: 577–591.
7. HOPF, E. 1942. Abzweigung einer periodischer lösung von einer stationären lösung eines differentialsystem. Ber. Verh. Sächs. Akad. Wiss. Liepzig Math. Nat. KL 95: 3–22.

8. IWAN, W. D. & R. D. BLEVINS. 1974. A model for vortex-induced oscillation of structures. J. Appl. Mech. **41**: 581–586.
9. IWAN, W. D. 1975. The vortex induced oscillation of elastic structual elements. J. Eng. Ind. **97** B: 1378–1382.
10. MARSDEN, J. E. & M. McCRAKEN. 1976. The Hopf Bifurcation and Its Applications. Springer-Verlag. New York, N.Y.
11. PARKINSON, G. V. 1972. Mathematical models of flow-induced vibrations of bluff bodies. Symposium on Flow-Induced Structural Vibrations.: 81–127. Springer-Verlag. Karlsrube, Germany.
12. POORE, A. B. 1976. On the theory and application of the Hopf-Friedrichs bifurcation theory. Arch. Rational Mech. Anal. **60**: 371–393.
13. UPPAL, A., W. H. RAY & A. B. POORE. 1974. On the dynamic behavior of continuous stirred tank reactors. Chem. Eng. Sci. **29**: 967–985.
14. UPPAL, A., W. H. RAY & A. B. POORE. 1976. The classification of the dynamic behavior of continuous stirred tank reactors—Influence of reactor residence time. Chem. Eng. Sci. **31**: 205–214.

TOPICS IN LOCAL BIFURCATION THEORY

Jack K. Hale*

Lefschetz Center for Dynamical Systems
Division of Applied Mathematics
Brown University
Providence, Rhode Island 02912

Suppose Λ, X, Z are Banach spaces and $M: \Lambda \times X \to Z$ is a smooth function of $(\lambda, x) \in \Lambda \times X$. Consider,

$$M(\lambda, x) = 0 \tag{1}$$

for $\lambda \in \Lambda$, $x \in X$. A pair (λ, x) satisfying (1) is called a solution, the set of solutions is denoted by S, and $S_\lambda = \{x \in X: (\lambda, x) \in S\}$ is the section of solution set at λ.

The basic problem in bifurcation theory is to determine how the set S_λ varies with the parameter λ. Any point λ for which the structure of S_λ changes is called a *bifurcation point*.

If λ is a scalar parameter, very general results on the existence of bifurcation points can be obtained without imposing too many specific properties about the manner in which the function M depends on (λ, x). If λ is a vector parameter, we must generally assume that more complete knowledge is available.

In the past few years, we have been attacking this problem under the following premises. First, we assume that the parameter is a vector parameter of dimension generally greater than 1. (The dimension we choose is not too large because we wish to discuss the interaction of only a few physical parameters at a time.) It is well-known that many parameters are needed to discuss a complicated bifurcation point. However, it is also known that some parameters have a more drastic effect on the qualitative nature of bifurcation than others. It is, therefore, of interest to understand well the bifurcations in low dimensional parameter space Λ. Second, we wish to devise methods which are applicable to equations which may not be the gradient of some function. Such methods will be applicable to nonconservative physical systems. Third, we want the methods to be extremely elementary and require only the calculus, the implicit function theory, and a small amount of geometric intuition.

The purpose of this paper is to survey some of our efforts in this direction. Three types of problems are discussed. For the first problem, suppose that,

$$M(\lambda_0, x) = 0 \tag{2}$$

for a particular value of λ_0 has the isolated solution $x = 0$ and the linear operator $A = \partial M (\lambda_0, 0)/\partial x$ does not have a bounded inverse. Assuming that dim $N(A)$ is 1 or 2 and that there are some generic conditions on the nonlinearities, we can

*This research was supported in part by the Air Force Office of Scientific Research under AF-AFOSR 76-3092, the U.S. Army Research Office under AROD AAG 29-76-G-0294, National Science Foundation, under MCS 76-07247.

give a complete description of the bifurcation set near the point $(\lambda_0, 0) \in \Lambda \times X$. We give the theory and applications especially to the von Kármán equations considered by Chow et al., List, and Mallet-Paret.[1-3] External forces, imperfections, small curvatures, and variations in shape are considered. The effect of symmetry is considered by Vanderbauwhede and Rodrigues.[4,5] Bifurcation of the nodal lines of a rectangular plate is discussed by Mallet-Paret.[6] Hale gives general lecture notes on bifurcation.[7]

The second problem discussed concerns the case in which Equation 2 has a compact family of solutions. More specifically, suppose there is a C^2 function, $p(t) = p(t + 1), t \in \mathbb{R}$, such that,

$$M(\lambda_0, p(t)) = 0, \quad t \in \mathbb{R} \tag{3}$$

For each $t \in \mathbb{R}$, the operator $A(t) = \partial M(\lambda_0, p(t))/\partial x$ does not have a bounded inverse. We are interested in the bifurcation of solutions near the "circle" $\Gamma = \{p(t), 0 \leq t \leq 1\} \subset M$ for λ near λ_0. The complete structure of the bifurcation is given by Hale[8,9] with applications to nonlinear oscillations under the assumption that dim $N[A(t)] = 1$ for all t. Implications in classical perturbation theory are also given. The case where dim $N[A(t)] = 2$ is discussed by Hale.[10]

The third problem concerns bifurcation from a noncompact family of solutions of (3). This can arise in several different ways in the applications. If the function M has the form, $M(\lambda, x) = Ax + N(\lambda, x)$, where $N(0, x) = 0$ for all $x \in X$. This is the classical problem of a small perturbation of a linear operator. If dim $N(A) \geq 1$, then for $\lambda = 0$ Equation 3 has a linear subspace of solutions; that is, a noncompact set of solutions. These problems are not well understood and are extremely difficult. In Hale,[11] we give a complete description of the bifurcation sets for the classical Duffing equation with or without damping, with all parameters treated as independent.

Another way in which a noncompact family can arise is when there is a family of solutions $p(t), t \in \mathbb{R}$, of Equation 3 with the set $\Gamma = \{p(t), t \in \mathbb{R}\} \subset X$ bounded but not compact. For example, in a second-order autonomous ordinary differential equation, Γ could be an orbit whose α- and ω-limit sets are the same critical point. When this system is subjected to a small periodic forcing, it has been known for a long time that homoclinic points may occur near Γ. This problem is discussed in more detail by Chow[12] when the equation is subjected to both damping and forcing which is not necessarily periodic.

SUMMARY

Suppose Λ, X, Z are Banach spaces, $M : \Lambda \times X \to Z$ is a mapping continuous together with derivatives up through some order r. A bifurcation surface $M(\lambda, x) = 0$ for Equation 1 is a surface in parameter space Λ for which the number of solutions x of Equation 1 changes as λ crosses this surface. Under certain generic hypotheses on M, the author and his colleagues have shown that the bifurcation surfaces can be systematically determined by elementary scaling techniques and the implicit function theorem. This paper gave a summary of these results for the case of bifurcation near an isolated solution or families of solutions of $M(\lambda_0, x) = 0$. The results have applications to the buckling theory of plates and shells under the effect of external forces, imperfections, curvature, and variations in shape. The results on bifurcation near families of solutions have applications in nonlinear oscillations and the theory of homoclinic orbits.

REFERENCES

1. CHOW, S., J. K. HALE & J. MALLET-PARET. 1975. Applications of generic bifurcation, I. Arch. Rat. Mech. Anal. **59**: 159–188. II. 1976 **62**: 209–235.
2. LIST, S. 1976. Generic bifurcation with applications to the von Kármán equations. (Ph.D. Thesis, Brown University.) University of Minnesota preprint. J. Diff. Eqs. In press.
3. MALLET-PARET, J. 1977. Buckling of cylindrical plates with small curvature. Q. Appl. Math. **35**: 383–400.
4. VANDERBAUWHEDE, A. L., Generic and nongeneric bifurcation for the von Kármán equations. J. Math. Anal. Appl. In press.
5. RODRIGUES, H. M. & A. L. VANDERBAUWHEDE. Symmetric perturbations of nonlinear equations: symmetry of solutions. J. Nonlinear Anal.: Theory, Methods, Applications. In press.
6. MALLET-PARET, J., Bifurcation in nodal lines. Preprint.
7. HALE, J. K. 1977. Generic bifurcation with applications. Heriot-Watt Lect. Vol. 1. Pitman.
8. HALE, J. K. 1977. Bifurcation near families of solutions. Proc. Int. Conf. Diff. Eqns. Acta Univ. Upsaliensis. Uppsala.
9. HALE, J. K. & P. Z. TÁBOAS. 1978. Interaction of damping and forcing in a second order equation. J. Nonlinear Anal.: Theory, Methods, Applications. **2**: 77–84.
10. HALE, J. K. & P. Z. TÁBOAS. Bifurcation near degenerate families. To be published.
11. HALE, J. K. & H. M. RODRIGUES. 1977. Bifurcation in the Duffing equation with several parameters. I. Proc. R Soc. Edinburgh. Sect. A. **78**; II. **79**.
12. CHOW, S., J. K. HALE & J. MALLET-PARET. Homoclinic orbits with general forcing and damping. To be published.

QUALITATIVE TECHNIQUES FOR BIFURCATION ANALYSIS OF COMPLEX SYSTEMS*

Philip Holmes

Department of Theoretical and Applied Mechanics
Cornell University
Ithaca, New York 14853

Jerrold E. Marsden

Department of Mathematics
University of California
Berkeley, California 94720

INTRODUCTION

In this paper we consider systems whose dynamical behavior may be represented by an autonomous ordinary differential equation (ODE) with parameters,

$$\frac{dx}{dt} = A_\mu x + B(x) \equiv G_\mu(x); \quad x(0) = x_0 \tag{1}$$

Here x is an element of a finite-dimensional vector space (say $\mathbf{R}^n$) or of a suitable Banach space of functions. In the latter case, (1) represents a partial differential equation (PDE). The control parameter $\mu \in \mathbf{R}^m$ is supposed to vary slowly in comparison with the evolution rate of a typical solution $x(t)$ of (1). Thus we treat (1) as an m-parameter family of ODE's. We are primarily interested in studying the qualitative changes that occur in the vector field or (semi) flow defined by (1) as μ varies.

The techniques used in the study of (1) draw on several fields, notably those of functional analysis and differentiable topology. In this brief paper we are only able to sketch general ideas and must therefore refer the reader to texts such as Chillingworth,[5] and Marsden and McCracken[21] for background information and further details. Both texts contain a wealth of additional references.

The general problem of bifurcation of vector fields—the qualitative study of equations, such as (1)—contains as an important subproblem, the study of bifurcations of equilibria, or stationary solutions. Much of the work done so far in bifurcation theory has been addressed specifically to the latter problem. The usual definitions of a bifurcation point are couched with this in mind. Since we wish to study a more general class of problems, and, in particular, to consider the case of *global bifurcations*, we propose an alternative definition, which is a slight modification of the definitions due to Smale and Thom. First we review the usual definition.

Consider a map $F: X \times \Lambda \rightarrow Y$, where X, Λ and Y are Banach spaces, and Λ is the parameter space. Set $F(x, \lambda) = 0$ and seek the solutions. Let $x(\lambda)$ be

*This work was partially supported by the Science Research Council of the U.K. and the National Science Foundation.

0077–8923/78/0316–0608 $1.75/1 © 1979, NYAS

a curve of known solutions; then we say that (x_0, λ_0) is a bifurcation point if, in any neighborhood of (x_0, λ_0), there is another solution $x_1(\lambda) \neq x(\lambda)$. However, consider FIGURE 1, where we have a 2-dimensional sheet Σ of solutions. According to the above definition, every point of Σ is bifurcation point, whereas we would like to distinguish the true bifurcation point (σ_0, λ_0), where a distinct curve of *new* solutions appears.

DEFINITION 1 (local bifurcation). Let $h: M \to N$ be a continuous map between topological spaces. A point $x_0 \in M$ is a bifurcation point for h if for every neighborhood U of $h(x_0)$ and V of x_0, the sets $h^{-1}(y) \cap V$ for $y \cap U$ are not all homeomorphic; i.e., $h^{-1}(y)$ changes topological type at x_0.

There is a corresponding global version, where the "whole space" $\cap h^{-1}(y)$ replaces $h^{-1}(y) \cap V$.

This can be related to the definition for an equation of the form $F(x, \lambda) = 0$, $x \in X$, $\lambda \in \Lambda$, by letting $\Sigma \subset X \times \Lambda$ be the set of solutions and $h: \Sigma \to \Lambda$ the

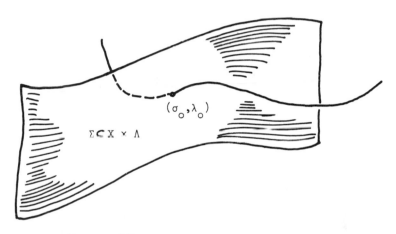

FIGURE 1. Bifurcation at a point on a sheet of solutions.

projection map onto the parameter space. "Parameter-free" DEFINITION 1 has advantages in certain situations, such as in the work of Buchner, Schecter, and Marsden[3] on scalar-curvature equations.

The condition, "same topological type," can be replaced by other relations according to the situation. For instance, if we have a set of vector fields and let the relation be, "have conjugate phase portraits," we recover Thom's definition of a bifurcation point of a family of vector fields as a member of the family which is not *structurally stable,* relative to the family cf. Ref. 5, pp. 228–9). It is noteworthy that Andronov and Pontryagin defined and discussed structural stability as early as 1937.[1]

The next two sections contain a general approach to systems represented by Equation 1. The fourth and fifth sections consider specific examples arising in engineering, which enable us to illustrate some of the relevant concepts.

EXISTENCE AND UNIQUENESS OF SOLUTIONS
AND BIFURCATIONS OF FIXED POINTS

In case Equation 1 is an ODE on $\mathbf{R}^n$, the establishment of local existence and uniqueness theorems is frequently a trivial matter, provided the nonlinear function $B(x)$ has reasonable properties. The proof of *global* existence is generally not so simple, and may require the use of Liapunov functions or related energy methods. However, a proof of local existence suffices for application of the center manifold theorem (THEOREM 2). When (1) is defined on a function space, existence-uniqueness results are often considerably more difficult to obtain, but in many cases may be obtained by use of this theorem originally due to Segal[26] (see Holmes and Marsden[18] for further details and proof).

THEOREM 1.   Let X be a Banach space and let A, with domain $D(A)$ be the generator of a C^0 (linear) semigroup $U_t = e^{tA}$ on X. Let $B:X \to X$ be of class C^k, $k \geq 1$, and let $G = A + B$, $D(G) = D(A)$. Then there is a unique local semiflow $F_t(x) = F(t, x)$ defined on an open set of $[0, \infty) \times X$ containing $\{0\} \times X$ such that for $x_0 \in D(A)$, $F_t(x_0) \in D(A)$ and $F_t(x_0)$ is the unique solution of,

$$\frac{dx}{dt} = G(x)$$

$$x(0) = x_0 \in D(G)$$

REMARK.   In the terminology of Marsden and McCracken, G generates a *smooth semiflow*.[21] The proof shows that if G_μ depends continuously on a parameter μ (with domain fixed), then so does its semiflow $F_t^\mu : X \to X$. From Chernoff and Marsden we note that separate continuity implies joint continuity.[4] As we shall see, a certain amount of smoothness (say $C^k, k \geq 3$) is required for application of the center manifold theorem and for subsequent bifurcation analysis.

Checking the hypotheses of THEOREM 1 in specific cases is generally a lengthy process (for equations of panel and pipe flutter, see Holmes and Marsden.[17,18] We obtain global existence results with a modification of Liapunov's second method:[18]

PROPOSITION 1.   Suppose the conditions of THEOREM 1 hold, and there is a C^1 function $H:X \to \mathbf{R}$ such that:

(i) There is a monotone increasing function, $\phi:[x, \infty) \to [0, \infty)$, where $[a, \infty) \supset$ range of H satisfying $\|x\| < \phi[H(x)]$;

(ii) There is a constant $K \geq 0$ such that if $x(t)$ satisfies Equation 1, then

$$\frac{d}{dt} H[x(t)] \leq KH[x(t)]$$

Then $F_t(x_0)$ is defined for all $t \geq 0$ and $x_0 \in X$. If, in addition, H is bounded on bounded sets and,

(iii) $\dfrac{d}{dt} H[x(t)] \leq 0$,   if $\|x(t)\| \geq B$

then any solution of (1) remains uniformly bounded in X for all time (i.e., given x_0), there is a constant $C = C(x_0)$ such that $\|x(t)\| \leq C$ for all $t \geq 0$.

If H decreases along solution curves of (1) on all of X and $H(0) = 0$ is the global minimum of H, then we can conclude that $x = 0$ is globally stable and that no bifurcations are possible while this condition holds.

In the succeeding discussion we will assume that the bifurcations to be studied occur from a curve of known solutions which, prior to bifurcation, are hyperbolic sinks and thus *locally* stable. We will further assume that the eigenvalues of (1), linearized at such a sink, can be calculated directly or at least estimated by numerical computations based on a finite-dimensional model.[14,17] Specifically, letting $\bar{x}(\mu)$ denote the sink, we consider the system,

$$\frac{dx}{dt} = DG_\mu(\bar{x}(\mu))x \qquad (2)$$

and compute the spectrum $\sigma\{DG_\mu[\bar{x}(\mu)]\}$ of the (Fréchet) derivative of G_μ at $\bar{x}(\mu)$. If the eigenvalues all lie strictly in the left-hand half plane, then $\bar{x}(\mu)$ is a hyperbolic sink. Bifurcation from $\bar{x}(\mu)$ occurs at some parameter value μ_0 when $\sigma\{DG_{\mu_0}[x(\mu_0)]\}$ has at least one eigenvalue on the imaginary axis. If this eigenvalue passes into the right-hand half plane as μ changes, then a bifurcation occurs in which $\bar{x}(\mu)$ becomes a repelling fixed point (either a source or a saddle). In this situation the classical bifurcation theorems, such as that of Hopf[21] can be applied to determine the nature of the secondary solutions bifurcation from $\bar{x}(\mu)$ at $\bar{x}(\mu_0)$. For example, closed orbits and/or additional fixed points may appear near $\bar{x}(\mu)$ for $\mu > \mu_0$.

However, in general, only a relatively small number of eigenvalues cross the imaginary axis simultaneously at $\mu = \mu_0$, thus $\bar{x}(\mu)$ "loses stability" in relatively few directions. In the case of an ODE on a Banach space of functions, for example, suppose a finite number d of eigenvalues crosses the imaginary axis as μ passes through μ_0. For $\mu > \mu_0$, then (say $\mu = \mu_0 + \epsilon$, ϵ small) the *unstable manifold* $W^u[\bar{x}(\mu)]$ of $\bar{x}(\mu)$ is of dimension d and the *stable manifold* $W^s[\bar{x}(\mu)]$, of codimension d. At least in some neighborhood of $\bar{x}(\mu_0)$ and μ_0 in $X \times \Lambda$ the new solutions can be studied by restricting our attention to a d-dimensional submanifold of X. If d is small (say $d = 1, 2$) then this dramatically reduces the complexity of the problem, and, in specific cases, enables us to obtain a complete characterization of local bifurcational behavior. To formalize this notion we use invariant manifold methods.

INVARIANT MANIFOLDS AND THE CENTER MANIFOLD THEOREM

We have already mentioned the stable and unstable manifolds of a fixed point in the second section.

DEFINITION 2. The local stable manifold of a fixed point $\bar{x} \in X$ is the set of points y in some neighborhood U of $\bar{x}$ which approach $\bar{x}$ under the flow F_t as $t \to +\infty$; thus,

$$W_{\text{loc}}^s(\bar{x}) = \{ y \in U \mid F_t(y) \to \bar{x}, \, t \to +\infty \}$$

The unstable manifold is obtained by reversing time in the above definition; hence,

$$W_{\text{loc}}^u(\bar{x}) = \{ y \in U \mid F_t(y) \to \bar{x} \text{ as } t \to -\infty \}$$

These definitions can be globalized by taking the unions of the local manifolds over all time (cf. Ref. 5, pp. 215–221). Clearly $W_{loc}{}^u(\bar{x})$ and $W_{loc}{}^s(\bar{x})$ describe the local splitting of the "phase space" X induced by the flow F_t of (1), or by the semi-flow F_t in the case of the PDE. The definitions generalize to more complex invariant sets, such as closed orbits, etc.[5,21]

When the fixed point $\bar{x}(\mu)$ is structurally unstable and $\mu = \mu_0$ is a bifurcation value then we can define a third local submanifold, the *center manifold* $M(\bar{x})$. Just as the stable and unstable manifolds are associated with (and tangent to) the eigenspaces of those eigenvalues of $DG_{\mu_0}[\bar{x}(\mu_0)]$ with negative and positive real parts, respectively, so $M(\bar{x}(\mu_0))$ is associated with those eigenvalues with zero real parts. The center manifold theorem may be stated for ODE's or for semiflows; here we give the latter version.[21] Without loss of generality we take $\bar{x}(\mu_0) = 0 \in X$

THEOREM 2 (Center manifold theorem for flows). Let X be a Banach space admitting a C^∞ norm away from 0, and let F_t be a C^0 semiflow defined in a neighborhood of 0 for $0 \leq t \leq T$. Assume $F_t(0) = 0$ and that, for $t > 0$, $F_t(x)$ is C^{k+1} jointly in t and x. Assume also that the spectrum of the linear semigroup $DF_t(0):X \to X$ is of the form $\exp[t(\sigma_1 \cup \sigma_2)]$, where $\exp(t\sigma_1)$ lies on the unit circle—i.e., $\mathrm{Re}(\sigma_1) = 0$—and $\exp(t\sigma_2)$ lies inside the unit circle a nonzero distance from it, for $t > 0$; i.e., $\mathrm{Re}(\sigma_2) < 0$. Let Y be the generalized eigenspace corresponding to $\exp(t\sigma_1)$, and assume dim $Y = d < \infty$. Then there exists a neighborhood V of 0 in X and a C^k submanifold $M \subset V$ of dimension d passing through 0 and tangent to Y at 0 such that:

(a) If $x \in M$, $t > 0$ and $F_t(x) \in V$, then $F_t(x) \in M$ (local invariance);
(b) If $t > 0$ and $F_t(x)$ remains defined and in V for all $\mathbf{t}$, then $F_t(x) \to M$ as $t \to \infty$ (local attractivity).

REMARK. If F_t is C^∞ then M can be chosen so as to be C^l for any $l < \infty$. For the semigroup $F_t{}^\mu(x)$ with control parameter $\mu \in \mathbf{R}^m$, if $F_t{}^\mu(x)$ is only assumed to be C^{k+1} in x; and its x-derivatives depend continuously on t and μ, and at $\mu = \mu_0$ part of the spectrum of $DF_t{}^{\mu_0}(0)$ is on the unit circle, as above, then for μ near μ_0 we can choose a family of C^k invariant manifolds M_μ depending continuously on μ. *This family completely captures the bifurcational behavior locally.*

We note that Henry[13] has a version of the theorem to cover the case in which the spectrum of $DF_t(0)$ also has a component $\exp(t\sigma_3)$ comprising a *finite* number of eigenvalues *outside* the unit circle; i.e., $\mathrm{Re}(\sigma_3) > 0$. Thus, in addition to M, we also have invariant stable and unstable manifolds W^s, W^u, the dimensions of which are determined by the number of eigenvalues within and outside the unit circle; thus dim $W^u < \infty$. The theorem now provides a full infinite-dimensional analog of that for ODE's in $\mathbf{R}^n$.[20] However, in this case we need a further result derived from the generalized Böchner-Montgomery theorem:

PROPOSITION 2. Let F_t be a local C^k semiflow on a Banach manifold $\tilde{M}$ and suppose F_t leaves invariant a finite-dimensional submanifold $M \subset M$. Then on M, F_t is locally reversible jointly C^k in t and x, and is generated by a C^{k-1} vector field on M.[21]

THEOREM 2 and **PROPOSITION 2** imply that, under their assumptions, we can find a $(d + m)$-dimensional subsystem $M \times U$, where U is a neighborhood of the critical parameter value $\mu = \mu_0$ such that $M \times U$ provides a local, finite-dimensional *essential model*. More details on the concept of essential models can be found in Holmes and Rand[19], where a general approach to the identification of nonlinear systems is suggested based on the concept of structural stability and the assumption of generic properties.

Since the new fixed points, closed orbits and other invariant sets of F_t^μ created in the bifurcation all lie in the center manifold for μ near μ_0, their structure may be considerably easier to analyze than would otherwise be the case. Of course, the global problem of relating these invariant sets to other invariant sets, perhaps created in other bifurcations, still remains, but, as we shall see in the next section, considerable progress can be made by use of the essential model concept.

AN APPLICATION TO PANEL FLUTTER

The problem of panel flutter has been extensively studied by Dowell[6,7] who has used numerical time-marching methods to solve a finite-dimensional Galerkin approximation to the governing PDE. In this way a *partial* picture of the behavior was obtained, in that only the attracting invariant sets such as sinks and attracting limit cycles were found. Moreover, the convergence of a finite-dimensional Galerkin system to the full PDE is tacitly assumed. The method proposed here is, we feel, complementary to such techniques in that, while it lacks the quantitative accuracy of numerical methods, it provides a fuller description of qualitative behavior and, in particular, of the surprisingly rich bifurcational behavior of the panel. Since the work has been extensively reported elsewhere, we merely outline the main results.[14,17,18]

The equation of motion of a thin panel of length l, fixed at both ends and undergoing cylindrical bending, can be written in terms of lateral deflection $v = v(z, t)$ as,

$$\alpha \dot{v}'''' + v'''' - \{\Gamma + k \,|\, v'\,|^2 + \sigma(v', \dot{v}')^2\} v'' + \rho v' + \sqrt{\rho}\,\delta\dot{v} + \ddot{v} = 0 \qquad (3)$$

Here $(\dot{\ }) = \partial/\partial t$, $(') = \partial/\partial z$ and $|\cdot|$ and $(\cdot, \cdot)$ denote the L_2 norm and inner product, respectively; $\alpha, k, \sigma, \delta > 0$ are fixed mechanical parameters and only the axial load; Γ and the dynamic pressure of the fluid flowing over the panel ρ vary. We collect these in a control parameter $\mu = (\rho, \Gamma)$. Note the strong symmetry of (3) due to the absence of even-order nonlinear terms. We prove the existence, uniqueness, and smoothness of the semiflow generated by (3) in Reference 18.

Working with fixed points and eigenvalues calculated from a finite-dimensional (four mode) approximation,[14] we obtain the bifurcation set of FIGURE 2, in which the bifurcation curves for fixed points only are shown. On B_s, a pair of sinks bifurcate off the trivial solution $x = \{v, \dot{v}\} = \{0, 0\} = \{0\} \in X$ and a Hopf bifurcation occurs on B_{h_1} in which an *attracting* closed orbit is created. The stability and "direction" of the family of orbits created on B_{h_1} can be checked by use of the "$V'''(\cdot)$ algorithm" described by Marsden and McCracken.†[21] The two non-

---

†B. Hassard (SUNY Buffalo) reports that he has carried out computations for 4, 6, 8, and 10 mode models using a version of the stability formula due to himself and Y.-H. Wan. He detected a strong convergence as the order of the approximation increases. The results reported and summarized here are qualitatively correct, although possibly in error by $\approx 5\%$ in quantitative accuracy.[14,17]

trivial fixed points $\bar{x} = \{v, \dot{v}\} = \{\pm\bar{v}, 0\}$ undergo simultaneous Hopf bifurcations on B_{h_2}. Here the bifurcations are *subcritical* so that *repelling* orbits encircle the sinks for parameter values "below" B_{h_2}. We thus have a partial picture of bifurcational behavior near the point $O = \mu_0$, where B_{h_1}, B_{h_2}, and B_s meet.

To complete the picture we note that at O, $\mu_0 = (\Gamma_0, \rho_0)$, $A_\mu : X \to X$ has a zero eigenvalue with multiplicity two, and that near $\{0\} \in X$ and the point $\mu_0 = (\Gamma_0, \rho_0) \in \mathbf{R}^2$ we can define a 2-dimensional center manifold M and a 4-dimensional essential model $M \times U$, where $U \subset \mathbf{R}^2$ is a neighborhood of μ_0. All the nontrivial invariant sets created as μ crosses $B_{h_1} \cup \mu_0 \cup B_s$ from left to right lie on $M \subset X$. Thus the problem reduces to that of completing the bifurcation picture

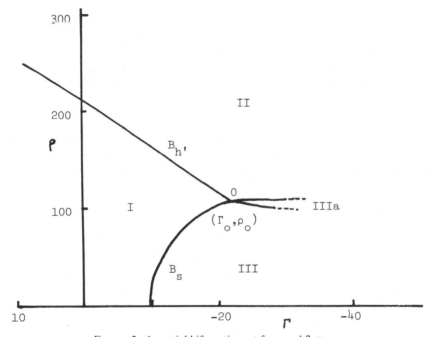

FIGURE 2. A partial bifurcation set for panel flutter.

for a two-parameter, 2-dimensional vector field. In particular, the degenerate singularity occurring at $\{0\} \in M$ for $\mu = \mu_0$ contains our information in its *versal unfolding*.[2]

We now make the key assumption that the bifurcation at $\{0\} \times \mu_0 \in M \times U$ is *generic* in the sense that, under the symmetry group acting in (3) it is the "simplest" degenerate singularity of a 2-dimensional vector field with double zero eigenvalues. We then use the classification of such *codimension-2* singularities, due to Takens,[27] and select the singularity whose unfolding contains the phase portraits we have already detected, specifically, a sink in region I of FIGURE 2, a source and limit cycle in region II, two sinks, and a saddle in region III (near B_s), etc. This is the "$m = 2; -$" normal form. We therefore have:

CONJECTURE 1. Flutter and divergence near $\{0\} \times (\rho_0, \Gamma_0) \in X \times \mathbf{R}^2$ can be modeled by a two-parameter vector field V_μ on a 2-manifold M, where V_μ is diffeomorphic to Takens' "$m = 2; -$" normal form (FIGURE 3). The actual vector fields portrayed in FIGURE 3a belong to the nonlinear oscillator $\ddot{x} + \nu_1\dot{x} + \nu_2x + x^2\dot{x} + x^3 = 0$, which is diffeomorphic to the more complicated form of Takens.[27] It is interesting to note that the essential model is in a sense a *nonlinear normal mode*.

Note that the double *saddle connection* occurring as μ crosses the curve B_{sc} from region IIIb to IIIc is an example of a *global bifurcation* in which the phase portrait changes topological type without local bifurcations of fixed points or closed orbits occurring. The period of the two repelling closed orbits existing in region IIIb tends to infinity and the orbits reappear as a single repelling orbit in region IIIc, whose period decreases from infinity and approaches that of the attracting orbit until the two orbits coalesce and annihilate each other as μ crosses B_{lc}.[17] It is interesting to note that the use of Takens' "$m = 2; -$" normal form as an essential model for panel flutter was proposed by Holmes and Rand[19] purely on the basis of "generic" arguments, without detailed knowledge of the PDE's governing panel motions.

FORCED NONPERIODIC OSCILLATIONS

As a final example we outline some recent work of the first author on the forced oscillations of a system possessing three equilibria, two sinks, and a saddle (the actual system studied is the second-order ODE); hence,

$$\ddot{x} + \delta\dot{x} - \beta x + \alpha x^3 = f \cos \omega t; \quad \alpha, \delta, \beta, \omega > 0 \text{ fixed}, f \geq 0, \text{varies} \quad (4)$$

A detailed preliminary study of Equation 4 has been completed and will appear in due course.[16] Here we merely give the main results.

Although (4) is an ODE, we believe it can be derived from the PDE for the oscillations of a buckled column under transverse sinusoidal loading by invariant manifold techniques. The PDE, in nondimensional form ($v = v(z, t)$ again represents transverse deflections), is

$$v'''' + \Gamma v'' + k |v'|^2 v'' + \delta\dot{v} + \ddot{v} = f(z) \cos \omega t, \quad (5)$$

where k and δ are structural constants and $\Gamma > \pi^2$ is the fixed axial end load. Certainly a study of the behavior of (4) is necessary before the full problem (5) is tackled.

Holmes[16] proves that (4) is globally stable in the sense that after sufficient time all solution curves enter and remain within a bounded set A in the state space, thus (4) always has at least one attractor. We rewrite (4) as an autonomous system on $\mathbf{R}^2 \times S^1$, as follows:

$$\dot{x}_1 = x_2$$

$$\dot{x}_2 = \beta x_1 - \delta x_2 - \alpha x_1^3 + f \cos \theta$$

$$\dot{\theta} = \omega \quad (6)$$

and consider the *Poincaré map* $P_f : \Sigma \to \Sigma$ induced by the flow $\phi_t : \mathbf{R}^2 \times S_1 \to$

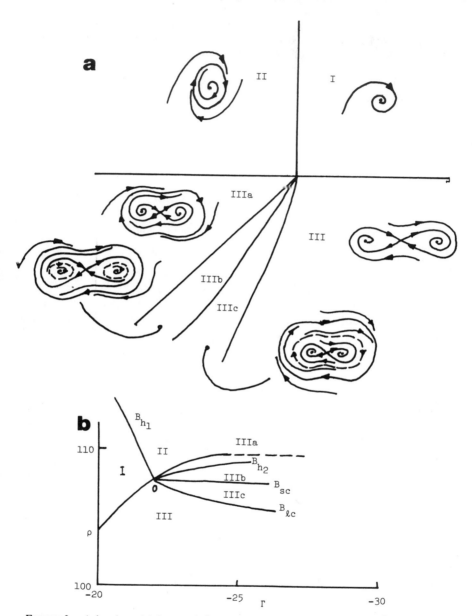

FIGURE 3. A local model for panel flutter near $\{0\} \times (\mu_0, \Lambda_0) \in X \times \mathbf{R}^2$. (a) Takens' "$m = 2; -$" normal form and associated structurally stable vector fields. (b) The completed bifurcation set.

$\mathbf{R}^2 \times S^1$ of (6), where Σ is a global cross section $\Sigma = \{(x, y, \theta) \in \mathbf{R}^2 \times S^1 \mid \theta =$ $0, 2\pi/\omega, \ldots\}$; P_f is the time $2\pi/\omega$ or period-1 Poincaré map.

First consider the trivial system for $f \equiv 0$. Here all cross sections of (6) are identical for all $\theta \in [0, 2\pi/\omega]$. The Poincaré map thus has a structure identical to that of the vector field of the autonomous 2-dimensional system,

$$\dot{x}_1 = x_2, \dot{x}_2 = \beta x_1 - \delta x_2 - \alpha x_1^3 \qquad (7)$$

in the sense that the stable and unstable *manifolds* of the saddle $(0, 0)$ of the Poincaré map are curves identical to the stable and unstable *seperatrices* of $(0, 0)$ for the vector field. However, we must recall that an *orbit* of the map is a sequence of *points* and not a curve, as in the case of vector field.[5] It is easy to check that the vector field of (7), and hence the Poincaré map of (6) for $f = 0$, is (globally) *structurally stable*. We can conclude that for $f \neq 0$, small, the topological type will be identical to that for $f = 0$. This is confirmed by analog computer analysis

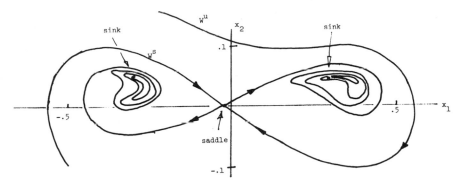

FIGURE 4. Stable and unstable manifolds of the saddle point of P_f for $f = 0.2$ ($\alpha = 1$, $\beta = 10$, $\delta = 100$, $\omega = 3.76$).

(FIGURE 4).[16] Thus, for small f, P_f has a hyperbolic saddle and two hyperbolic sinks, corresponding to the two attracting and one repelling closed orbits of (6).

As f increases it is possible to prove, using the methods of Melnikov[22], that the stable and unstable manifolds of the saddle point approach and ultimately intersect, giving rise to infinitely many homoclinic points. We note that this proof apparently has much in common with recent work of Hale[11] and others. The critical value $f = f_c(\alpha, \delta, \beta, \omega)$— ≈ 0.79 for the case studied—thus computed agrees well with that found in analog computations. FIGURE 5 shows the dispositions of stable and unstable manifolds W^s and W^u just before and after intersection takes place at $f \approx 0.76$. Note that the presence of the period-1 sinks in FIGURE 5b implies that almost all orbits converge to either one of these fixed points of P_f as $t \to \infty$. However, since P_{f_c} has tangencies of W^s and W^u, according to a theorem proved by Newhouse,[23,24] there is an open set of diffeomorphisms near P_{f_c} where each possesses an infinite number of periodic sinks. In addition to the creation of

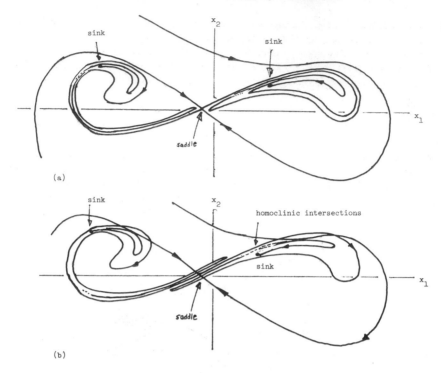

FIGURE 5. Stable and unstable manifolds of P_f for (a) $f = 0.75$ and (b) 0.90.

Smale horseshoes, then, we may have (weakly) attracting sets of periodic points of P_f for $f > f_c$.

As f continues to increase, analog computations show that the fixed points of FIGURE 5b bifurcate to sinks of period 2, and then of period 4. It is possible that further bifurcations, to periods 8, 16, 32, ... also occur but they are difficult to detect with reliability. In any event, for $f \geq 1.08$, successive iterates of P_f are no longer attracted to a clearly periodic orbit, and they appear to wander chaotically back and forth across Σ. FIGURE 6 indicates that this wandering is in fact ordered in the sense that the orbits rapidly converge to and appear to remain on a 1-dimensional curve close to, and perhaps identified with, W^u. Computations of the power spectrum for some 50,000 samples (12,000 cycles of the forcing function) clearly suggest nonperiodic behavior.[15,16] We suggest[16] that for $f \in (1.08, 2.45)$, P_f has a *strange attractor*.[21,25]

In order to study the structure of the attractor more fully, we approximate[16] the true Poincaré map P_f by a simple cubic polynomial mapping P_d given by,

$$(x, y) \mapsto (y, -bx + dy - y^3), \quad b; d > 0 \qquad (8)$$

Fixing $b = 0.2$ and varying d, $d \in (1.2, 2.8)$, we were able to reproduce much of the behavior of P_f for $f \in (0, 1.2)$.

Under a suitable assumption on generic properties, and taking the symmetry of (6) into account, we might expect a Taylor series expansion of the true Poincaré map at $(0, 0)$ to include only odd terms. We *can* explicitly calculate the linear terms by integrating (6), these terms have the form assumed in (8). We thus obtain an approximate map by ignoring all terms of order 5 and higher in the Taylor series. We assume that it is possible to choose coordinates such that the nonlinear term appears in only one component of the map.[12]

The cubic mapping (8) has much in common with the quadratic planar mapping discussed by Hénon, and digital computations strongly suggest that the invariant attracting set S_d, for a large number of values of d in the range $(2.7, 2.8)$, has a local structure isomorphic to the product of a smooth curve and a Cantor set. Successive iterates of $P_d (d \approx 2.77)$ behave much as do successive iterates of the true Poincaré map P_f for $f \in (1.08, 2.45)$ as shown in FIGURE 7. In addition to computing successive iterates, the structure of stable and unstable manifolds of the saddle of P_f at $(0, 0)$ was also studied. As for the true map P_f, the manifolds become tangent at a critical value $d = d_c \approx 2.60$.

Summarizing, then, we prove that (5) has homoclinic orbits for $f > f_c$. $(\alpha, \beta, \delta, \omega) \approx 0.79$ and that (5) has at least one attracting set for all $f < \infty$.[16] Analog computer solutions of (5) indicate (but cannot prove) that the sinks of the Poincaré map P_f of (5) undergo a sequence of "flip" bifurcations for $f \approx f_0 >$

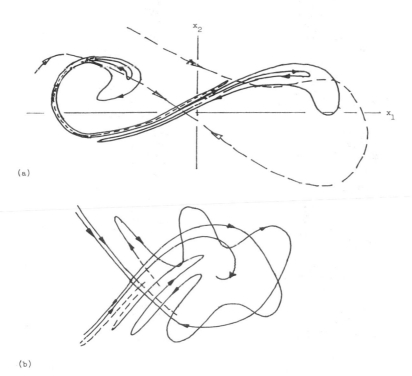

(a)

(b)

FIGURE 6. (a) The attracting invariant set of P_f for $f = 1.10$. (b) Schematic structure of one "lobe."

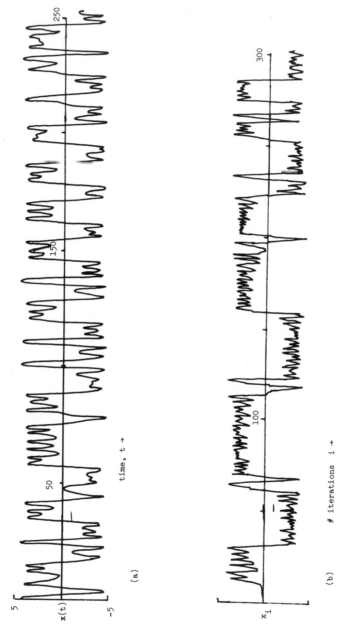

FIGURE 7. Evolutions of the ODE (6) and of the approximate Poincaré map (8).

f_c in which sinks of periods 2, 4, 8, ..., 2^n, ... are created. Ultimately for $f \geq f_d \approx 1.08 (> f_b)$ all orbits apparently approach a nonperiodic attractor as $t \to +\infty$. Working with an approximate Poincaré map P_d, we can prove that P_d indeed undergoes such a succession of flip bifurcations. We then use the digital computer to study the structure of the invariant attracting set S_d of P_d for the appropriate range of values of d in which S_d appears to be topologically conjugate to the strange attractor S of P_f for $f \in (1.08, 2.45)$.

Very little is presently known about the behavior of polynomial mappings of $\mathbf{R}^2$, such as P_d (Equation 8) or the map studied by Hénon.[12] In order to obtain more information in the present case, we intend to study the cubic mapping on $\mathbf{R}$ given by,

$$y \to (dy - y^3) \qquad (9)$$

If we introduce a dummy small parameter $\epsilon < 1$ into (8), and rewrite the equation as

$$(x, y) \mapsto (\epsilon y, -bx + dy - y^3) \qquad (10)$$

we see that (9) is essentially identical to the limiting case of (10) when $\epsilon = 0$. In this way we hope to reduce the complexity of the problem much as Guckenheimer[8,9] and Williams[28] did in their studies of Lorenz equations. Guckenheimer has also reduced the study of the Poincaré map of the forced van der Pol oscillator to that of a map on the circle.[10]

CONCLUSION

We have outlined a general approach to the qualitative analysis of nonlinear dynamical problems and illustrated our methods with two examples taken from engineering science. One of the most interesting features in these examples is the detection of *global bifurcations* in which periodic orbits are created or annihilated in a manner which depends upon the global disposition of stable and unstable manifolds of some invariant set. The "figure-8" loop occurring on B_{sc} (FIGURE 3b) is a simple example, while the creation of infinitely many periodic points when W^u and W^s intersect homoclinically (FIGURE 5) is considerably more complex. In the latter case we have the creation of a countable infinity of periodic points as W^u is "pulled through" W^s, and vice versa. These examples illustrate the need for a more general definition of bifurcation, such as that proposed in the first section.

ACKNOWLEDGMENTS

The authors would like to thank John Ball, David Chillingworth, Brian Hassard, David Rand, and Christopher Zeeman for many helpful comments, criticisms, and computations.

REFERENCES

1. ANDRONOV, A. A. & L. S. PONTRIAGIN. 1937. Systèmes Grossières. Dokl. Akad. Nauk. **14**: 247–251.
2. ARNOLD, V. I. 1972. Lectures on bifurcations in versal families. Russ. Math. Surveys **27**: 54.
3. BUCHNER, M., S. SCHECTER & J. E. MARSDEN. A differential topology approach to bifurcation at multiple eigenvalues. Submitted for publication.
4. CHERNOFF, P. & J. E. MARSDEN. 1970. On continuity and smoothness of group actions. Bull. Am. Math. Soc. **76**: 1044.
5. CHILLINGWORTH, D. R. J. 1976. Differential Topology with a View of Applications. Pitman.
6. DOWELL, E. H. 1977. Nonlinear oscillations of a fluttering plate. AIAA J. **4** (7): 1267.
7. DOWELL, E. H. 1975. Aeroelasticity of Plates and Shells. Noordhoff.
8. GUCKENHEIMER, J. A strange, strange attractor. Chapter 12, Ref. 21.
9. GUCKENHEIMER, J. Structural stability of Lorenz attractors. University of California at Santa Cruz. Preprint.
10. GUCKENHEIMER, J. Symbolic dynamics and relaxation oscillations, University of California at Santa Cruz. Preprint.
11. HALE, J. K. Applications of generic bifurcation. In this volume.
12. HÉNON, M. 1976. A two dimensional mapping with a strange attractor. Commun. Math. Phys. 69–77.
13. HENRY, D. 1975. Geometric theory of parabolic equations. University of Kentucky Lect. Notes. Submitted for publication.
14. HOLMES, P. J. 1977. Bifurcations to divergence and flutter in flow induced oscillations: A finite dimensional analysis. J. Sound Vib. **53**(4): 471–503.
15. HOLMES, P. J. 1977. Strange phenomena in dynamical systems and their physical implications. Appl. Math. Modelling. **1**(7): 362.
16. HOLMES, P. J. A nonlinear oscillator with a strange attractor. Submitted for publication.
17. HOLMES, P. J. & J. E. MARSDEN. 1977. Bifurcations to divergence and flutter in flow-induced oscillations: an infinite dimensional analysis. Proc. Int. Fed. Autom. Control Symp. Distributed Parameter Syst. (Warwick). June–July. Submitted for publication.
18. HOLMES, P. J. & J. E. MARSDEN. 1978. Bifurcations to divergence and flutter in flow-induced oscillations: An infinite dimensional analysis. Automatica (July).
19. HOLMES, P. J. & D. A. RAND. 1975. The identification of vibrating systems by generic modelling. Inst. Sound Vib. Res. Tech. Rept. No. 79. Southampton University.
20. KELLEY, A. 1967. The stable, center stable, center, center-unstable and unstable manifolds. J. Diff. Eqns. **3**: 546.
21. MARSDEN, J. E. & M. McCRACKEN. 1976. The Hopf Bifurcation and its Applications. Springer-Verlag. New York, N.Y.
22. MELNIKOV, V. K. 1963. On the Stability of the Center for time periodic perturbations. Trans. Moscow Math. Soc. **12** (1): 1.
23. NEWHOUSE, S. 1974. Diffeomorphisms with infinitely many sinks. Topology. **13**: 9.
24. NEWHOUSE, S. The abundance of wild hyperbolic sets and non-smooth stable sets for diffeomorphisms. I.H.E.S. Bures-sur-Yvette (Paris). Preprint.
25. RUELLE, D. Bifurcation to turbulent attractors. In this volume.
26. SEGAL, I. Nonlinear semigroups. 1963. An. Math. **78**: 339.
27. TAKENS, F. Forced oscillations and bifurcations. Commun. No. 3. Mathematics Institute, Rijksuniversiteit, Utrecht, the Netherlands.
28. WILLIAMS, R. F. 1977. The structure of Lorenz attractors. Turbulence Seminar, Berkeley 1976/77. Springer Lect. Notes Math. No. 615: 94.

BISTABILITY, OSCILLATION, AND CHAOS
IN AN ENZYME REACTION

Hans Degn and Lars F. Olsen

Institute of Biochemistry

John W. Perram

Institute of Mathematics
Odense University
DK-5230 Odense M, Denmark

INTRODUCTION

Only a few years ago a chemical reaction was supposed to have a decent reaction order and not much more. But now, such a reaction is of low standing if it does not include a few of the following items among its properties: bifurcations, bistability, catastrophe, chaos, dissipative structures, echo waves, multistability, oscillations, super chaos, symmetry break, trigger waves, etc. Our fascination with such phenomena in chemical systems is due to the fact that we believe they are the fundamental elements of dynamics which may integrate to form life.

A few reaction systems have become prominent as models in studies of elements of dynamics. These are the Bray[1] reaction, the Belousov-Zhabotinskii[2,3] reaction, and the peroxidase-oxidase reaction. The latter, incidentally, would have been called the Yamazaki[4,5] reaction except for the fact that biochemists have abolished the tradition of naming reactions after their discoverers. We do not include the glycolytic system[6,7] because it is not a model system.

Bray's reaction, which is purely inorganic, was the first chemical reaction reported to have a periodically changing reaction rate in a homogenous solution. This phenomenon was left alone for nearly fifty years until the demise of a theory denying the existence of oscillating reactions in the homogenous phase. This happened in the sixties when unequivocal biological evidence emerged demonstrating that oscillations in the rate of glycolysis exist in homogeneous yeast cell extracts.[8] At about the same time, oscillations were discovered in the Belousov-Zhabotinskii and peroxidase-oxidase reactions. In the late sixties the Belousov-Zhabotinskii reaction opened up an entirely new field of chemical experimentation when Zhabotinskii[9] found that this reaction can cause the spontaneous formation of spatial concentration patterns in unstirred solutions.

The peroxidase-oxidase reaction system is an open system in which oxygen diffuses from a gas phase into a stirred solution. Therefore it is not suitable for spatial studies. However, the reaction system makes up for this shortcoming because it exhibits all the types of behavior imaginable in a homogeneous chemical reaction system. Beside the trivial monotonic approach to a steady state, it can show bistability, oscillations, and chaos.

We have proposed a kinetic model to explain some of the antics of the peroxidase-oxidase reaction.[10] So far we only know about the behavior of this model from a few computer solutions. However, a general treatment might be worthwhile.

Peroxidases are enzymes, so named because they catalyze the oxidation of a variety of organic substances using hydrogen peroxide. However, some peroxidases also catalyze an alternative reaction using molecular oxygen.[11] The latter type of reaction, which is the subject of this paper shall be called the *peroxidase-oxidase reaction*. NADH, which is one of the main carriers of reducing equivalents in metabolism, can be oxidized in the peroxidase-oxidase reaction[12]:

$$2NADH + 2H^+ + O_2 \rightarrow 2NAD^+ + 2H_2O$$

The analogous substance NADPH is also oxidized. The mechanism of the reaction is not known in detail. However, it involves several different spectroscopically

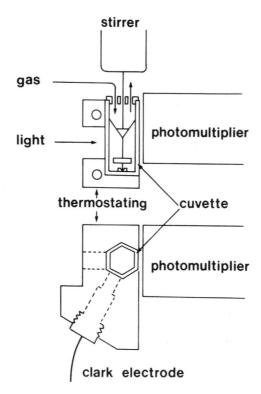

FIGURE 1. Side and top view of the reaction vessel mounted in the dual wavelength spectrophotometer. The reaction vessel is a 5 ml hexagonal glass cuvette inserted into a brass holder containing channels for thermostating water. The stirring shaft is provided with a cone in order to keep the surface of the liquid stable.

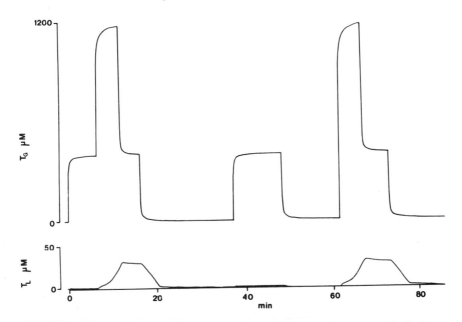

FIGURE 2. Bistability in the peroxidase-oxidase reaction with NADH as the hydrogen donor. *Experimental conditions:* 1.9 μM peroxidase, 3.0 mM NADH in 0.1 M Na acetate buffer pH 6.0; temperature 25°C; T_G and T_L denote oxygen tension in gas and liquid, respectively (units of tension are chosen as concentration units to relate gas and liquid measurements); $T = 1200$ μM corresponds to saturation of liquid with gas containing 100% O_2 at 25°C.

distinguishable forms of the enzyme and free radical intermediates. A commercial preparation of horseradish peroxidase was used in our experiments (Böhringer, Mannheim, Federal Republic of Germany).

FIGURE 1 shows the setup in which the experiments were done. The reaction vessel is a hexagonal glass cuvette fitted with a stirrer and a Clark electrode. The cuvette has a teflon cover with holes in it for the stirrer shaft and inlet and outlet for the flow of a mixture of O_2 and N_2. The ratio of this mixture can be adjusted with the help of a digital gas mixer.[13] At a constant stirring rate, the diffusion v_t of oxygen across the phase boundary is governed by,

$$v_t = k([O_2]_{eq} - [O_2]) \qquad (1)$$

where $[O_2]_{eq}$ is the oxygen concentration in the liquid when it is equilibrated with the gas. The oxygen transfer constant k depends on the volume of the sample and on the rate of stirring. At a steady state, the rate of oxygen consumption by the reaction is equal to the rate of transport of oxygen given by Equation 1. The cell is placed in a Hitachi-Perkin Elmer dual wavelength spectrophotometer. A Harvard Instruments infusion pump was used to pump a solution of reactant into the sample at a rate of 10–20 μl/h.

BISTABILITY

Bistability in a homogeneous chemical reaction system was first observed by Degn[14] in a peroxidase-oxidase reaction. In the original experiments a second enzyme, glucose oxidase, was present to facilitate the switch from a high to a low oxygen steady state. We have repeated the experiment under conditions in which the second, inessential enzyme was ommitted. NADH was added in excess to the sample, and the oxygen partial pressure in the gas was changed at time intervals. FIGURE 2 shows that, when the oxygen partial pressure of the gas is kept constant, the oxygen concentration of the liquid will approach a steady state. However, the steady state is not unique. Two different steady states in the liquid can exist at the same oxygen partial pressure of the gas. One of the two alternative steady states, which is chosen by the system, possesses one bit of memory. A small perturbation of oxygen partial pressure in the gas can cause the system to switch from a low to a high oxygen steady state in liquid. The opposite transition is harder to achieve because the half time of the system is much longer in the high, than in the low, oxygen steady state due to the lower rate of oxygen consumption in the high steady state.

The bistability phenomenon in the peroxidase-oxidase system is explained by the fact that the enzyme is inhibited by high oxygen concentrations. In the case of substrate inhibition of an enzyme reaction, one usually finds the rate law for the reaction to be of the form,[15]

$$v = \frac{Vs}{K + s + Ls^2} \qquad (2)$$

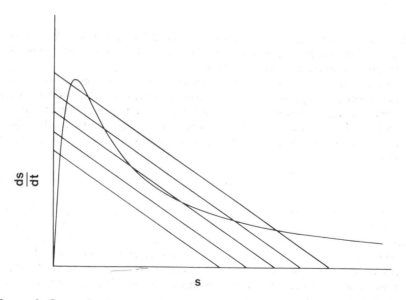

FIGURE 3. Curves showing substrate inhibition according to Equation 2 and simple diffusion rate laws according to Equation 1. The points of intersection of the curves determine the steady states.

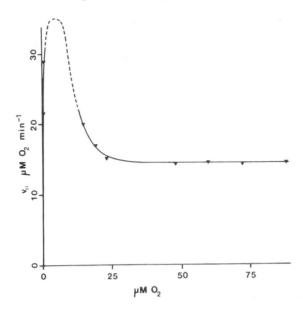

FIGURE 4. Steady state rates of oxygen consumption v_{st} against the oxygen concentration in the liquid for the peroxidase-oxidase reaction. The experiments were performed by equilibrating the liquid phase with the gas phase at different O_2 partial pressures. The reaction was then started by adding NADH to the reaction mixture containing peroxidase. The steady state rate was determined from Equation 1. *Experimental conditions:* 1.9 μM peroxidase, 1.0 mM NADH in 0.1 M Na acetate buffer pH 5.1; temperature 25°C.

where s is the substrate concentration and K, L, V are constants. This function has one maximum, and therefore may have one or three points of intersection with the line given in Equation 1, as shown in FIGURE 3. These three points of intersection correspond with the steady states of the system. Stability analysis shows that only the two extreme steady states are stable, in accordance with experimental findings.

Because bistability is not necessarily caused by forward inhibition, we have done some experiments to determine the steady-state rate of reaction as a function of oxygen concentration. NADH was added to solutions of peroxidase equilibrated with different concentrations of oxygen. The result is shown in FIGURE 4. It was observed that a maximum in the rate-versus-oxygen-concentration curve was found. However, the rate does not fall off to zero at increasing oxygen concentrations as does the rate in the classical substrate inhibition model Equation (2). The following empirical expression describes the experimental findings shown in FIGURE 4:

$$v = \frac{V([O_2] + A[O_2]^2)[NADH]}{K[NADH] + [O_2] + L[O_2]^2} \tag{3}$$

where A, K, V, and L are constants. The bistability phenomenon is explained satisfactorily by Equations 1 and 3. Because the rate law is of the Michaelis-Menten form with respect to NADH, the rate will not be sensitive to the NADH concentration as long as this is in excess. Whether the experiment is performed with a large excess of NADH or with a continuous infusion to maintain a sufficiently high NADH concentration is of no consequence.

OSCILLATIONS

Damped oscillations in the peroxidase-oxidase reaction in a system open to oxygen was first observed by Yamazaki et al.[4] who used NADH as the substrate

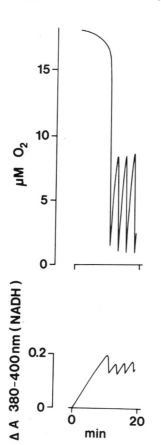

FIGURE 5. Sustained oscillations in the peroxidase-oxidase reaction in the presence of 2,4-dichlorophenol and methylene blue. *Experimental conditions:* 4.5 ml 0.1 M Na acetate buffer containing 1.1 μM peroxidase, 11 μM 2,4-dichlorophenol and 0.1 μM methylene blue were equilibrated with the gas phase containing 1.65% O_2 by volume; experiment was started by infusion of 0.25 M NADH at a rate of 12.5 μl/h; temperature 28°C; wavelengths used for NADH measurement were found optimal under different experimental conditions.

and bubbled air into the solution. This observation was confirmed by Degn[16] who found that other substrates, such as dihydroxyfumaric acid and indoleacetic acid, also gave rise to damped oscillatory kinetics. Subsequently, Nakamura et al.,[5] using NADPH as the substrate and adding a second enzyme, glucose-6-phosphate dehydrogenase, to regenerate NADPH from its oxidation product NADP$^+$, found that the system was capable of sustained stable oscillations when small amounts of methylene blue and 2,4-dichlorophenol were present. We have found that the

NADPH regenerating enzyme can be dispensed with because NAD^+ has no effect on the reaction. We infused NADH at a slow constant rate and obtained sustained oscillations, as shown in FIGURE 5.

In general, oscillations may be caused by any one of the following four different types of "feedback"[17]:

activation

$$A \xrightarrow{\quad} B \to \cdots \cdots \to C \to D$$

inhibition

$$A \xrightarrow{\quad} B \to \cdots \cdots \to C \to D$$

activation

$$A \to B \rightleftharpoons \cdots \cdots \rightleftharpoons C \to D$$

inhibition

$$A \to B \rightleftharpoons \cdots \cdots \rightleftharpoons C \to D$$

As we have seen, the fourth type of feedback, forward inhibition, is a feature of the peroxidase-oxidase reaction which can give rise to bistability. Therefore, it is the first suspect in relation to the oscillations. The system conforms to a simple model which was proposed by Degn[18] to explain electrochemical oscillations, and by Degn and Harrison[19] to explain respiratory oscillations in continuous cultures of microorganisms. Its chemical expression is,

$$A \to Y$$

$$B \rightleftharpoons X$$

inh

$$X + Y \to P$$

The last reaction is assumed to be a two-substrate enzyme reaction which is inhibited by one of the substrates (x) according to the equation,

$$\frac{dp}{dt} = \frac{V \times y}{K + y^2} \tag{4}$$

This model was investigated by Balslev and Degn[20] and by Ibañez *et al.*[21] with respect to spatial properties.

Although this model seems very plausible, we do not believe oscillations in the peroxidase-oxidase reaction are caused by forward inhibition. This is because sustained oscillations are only observed in the presence of 2,4-dichlorophenol, which we have found to prevent bistability because it abolishes the forward inhibition exerted by oxygen. All our attempts to create oscillations by adjusting parameters to values for which the forward-inhibition model predicts oscillations have failed. Another reason for rejecting the forward-inhibition model is that it cannot reproduce the waveform found in the experiments. As observed in FIGURE 5 the waveform of the oscillations in the NADH trace is a sawtooth. Our model[10] is based on explaining the peroxidase-oxidase oscillations using this fact and the assumption that autocatalysis (positive feedback) exists in the reaction.[16,22]

In electronics sawtooth oscillations are produced in circuits which include components with critical voltage limits for "turnon" and "turnoff." An example is the famous glow-lamp oscillator (Kipp-generator). Does the analogous phenomenon—critical concentrations for turnon and turnoff of a chemical reaction—exist? The answer is *yes,* as was shown by Semjonov[23] in his classical studies of branched chain reactions.

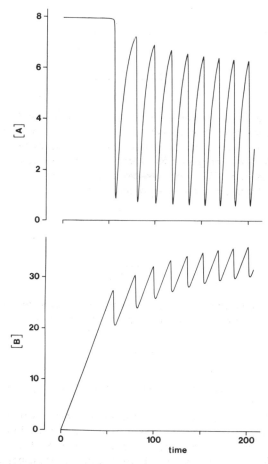

FIGURE 6. Computer simulations of the model described in text. *Rate constants:* $k_1 = 8.5 \ 10^{-2}$, $k_2 = 1.25 \ 10^3$, $k_3 = 4.6875 \ 10^{-2}$, $k_4 = 20.0$, $k_5 = 2.0$, $k_6 x_0 = 10^{-3}$, $k_7 = k_{-7} = 0.1175$, $k_8 b_0 = 0.5$, $a_0 = 8.0$. *Initial conditions:* $a = 8.0$; $b = x = y = 0.0$. The units of time and concentrations are arbitrary. The integration was performed using a standard Runge-Kutta-Merson procedure.

Consider a chain reaction with linear branching and linear termination,

$$A + X \xrightarrow{k_1} 2X$$

$$X \xrightarrow{k_2} B$$

This reaction will not start unless $A > (k_2/k_1)$. Two critical limits energe in models with quadratic branching; that is, models in which autocatalytic molecules (X) interact to form new autocatalytic molecules (Y), which are more efficient. Our simple model incorporating this idea is,

$$A + X \xrightarrow{k_1} 2X$$

$$2X \xrightarrow{k_2} 2Y$$

$$A + Y \xrightarrow{k_3} 2Y$$

$$X \xrightarrow{k_4} B$$

$$Y \xrightarrow{k_5} B$$

In order for the reaction to start, A must exceed the critical limit k_4/k_1. If A decreases below the critical limit k_5/k_3, the reaction stops. Obviously the reaction behaves in a similar way to the glow lamp if $k_5/k_3 < k_4/k_1$. When the above reaction with quadratic branching is supplied with an irreversible input reaction for A, it produces sawtooth oscillations, as revealed in computer solutions by Lindblad and Degn.[24] The model was proposed by Degn[25] to explain the oscillation in the Bray reaction.

The quadratic branching mechanism also can give rise to bistability if a reversible input reaction for A is added. This is readily understood when we consider the ability of a glow lamp in series with a resistor to be either on or off at the same external voltage. For reasons already discussed, we prefer the forward-inhibition model to explain the bistability observed in the peroxidase oxidase reaction.

We have modified the original quadratic branching model to fit the peroxidase-oxidase system by adding a second substrate with reversible input, corresponding to the diffusion of oxygen into the liquid:

$$A + B + X \xrightarrow{k_1} 2X$$

$$2X \xrightarrow{k_2} 2Y$$

$$A + B + Y \xrightarrow{k_3} 2X$$

$$X \xrightarrow{k_4} P$$

$$Y \xrightarrow{k_5} Q$$

$$X_0 \xrightarrow{k_6} X$$

$$A_0 \underset{k_{-7}}{\overset{k_7}{\rightleftharpoons}} A$$

$$B_0 \xrightarrow[k_8]{} B$$

Computer solutions of this model are in very good agreement with the experimental findings, as can be seen by comparing the solutions in FIGURE 6 with the experimental curves in FIGURE 5.

CHAOS

We have repeated the experiment shown in FIGURE 5 at different enzyme concentrations and found that the oscillation depends in an interesting way on enzyme concentration.[26] The system seems to change from a simple, stable through double periodicity, and more complex patterns to a nonmonotonic, nonperiodic

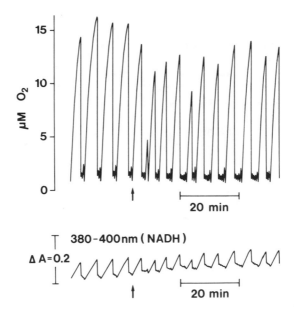

FIGURE 7. Nonperiodic oscillations (chaos) in the peroxidase-oxidase reaction in the presence of 2,4-dichlorophenol and methylene blue. *Experimental conditions:* 4.5 ml 0.1 M Na acetate buffer pH 5.1 containing 1.2 μM peroxidase, 22 μM 2,4-dichlorophenol, and 0.1 μM methylene blue was in contact with gas phase containing 1.65% O_2 by volume, initially 0.25 M NADH was infused at a rate of 12.5 μl/h; at indicated arrow infusion rate was changed into 15 μl/h; temperature 28°C.

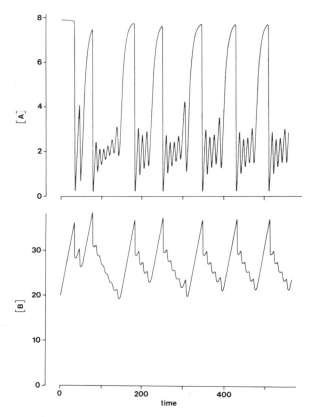

FIGURE 8. Computer simulation of the model described in text. *Rate constants:* as in FIGURE 6 except for k_1 = 6.25 10^{-2} and k_5 = 1.104. *Initial conditions: a* = 8.0, *b* = 20.0, *x = y* = 0.0. Units of concentration and time are arbitrary; standard Runge-Kutta-Merson procedure was used for integration.

behavior, as enzyme concentration is decreased. The same type of behavior can be achieved by increasing the concentration of 2,4-dichlorophenol. An intermediate case is shown in FIGURE 7. Our model can reproduce features of more complex behavior of the experimental system seen by comparing FIGURES 7 and 8. However, under all conditions tried, the computer solution of the model tended to become periodic.

In order to diagnose chaos, Olsen and Degn[26] made the next amplitude plot (A_{n+1} versus A_n) on nonperiodic oscillations obtained in similar experiments and found that the plot was a single-valued bell-shaped curve resembling those obtained from simple, discrete functions, as discussed by May.[27] The shape of this plot was such as to indicate chaos according to the Li and Yorke[28] theorem, similar to Rössler's model.[29]

It is not immediately obvious to us that the successive extrema in the solutions of a nonlinear system of differential equations are related by the same simple next amplitude plot, as the successive states of a difference equation. Nevertheless, there is some indication that this is so. We have therefore tried to find some simple systems in which trajectories may be studied both from the standpoint of difference and differential equations.

The most familiar systems that can yield very complex trajectories are dynamical ones. A simple example are the two coupled, nonlinear springs discussed by May.[30] This system is described by four first-order differential equations, which are reduced to three equations by conservation of energy. A similar system is provided in the case of two particles of masses m_1, m_2 at x_1, x_2, moving along the line $0 < x_1 < x_2 < 1$ under the influence of a potential $V(x)$ between the masses and the ends of the line. The equations of motion are,

$$m_1 \ddot{x}_1 = -V'(x_1) + V'(x_2 - x_1) \tag{5}$$

$$m_2 \ddot{x}_2 = V'(1 - x_2) - V'(x_2 - x_1) \tag{6}$$

Where ($\cdot$) denotes time differentiation and ($'$) differentiation with respect to x. The total energy E is constant, hence,

$$\tfrac{1}{2} m_1 \dot{x}_1^2 + \tfrac{1}{2} m_2 \dot{x}_2^2 + V(x_1) + V(x_2 - x_1) + V(1 - x_2) = E \tag{7}$$

If masses m_1, m_2 and the initial positions and velocities are chosen arbitrarily, the trajectories $x_1(t)$, $x_2(t)$ of the particles will be very complicated functions of time. Further, the complexity of this motion does not depend on the nature of $V(x)$. We have studied the Gaussian potential,

$$V(x) \propto (1/\sqrt{2\pi\sigma^2}) \exp{(-x^2/2\sigma^2)}$$

which is "soft" or "hard" depending on the value of σ. As σ is reduced, $V(x)$ becomes steeper and tends to $\delta(x)$ as $\sigma \to 0$. This is the potential for perfectly elastic collisions, and may, as we shall see, be studied within the context of discrete equations. However, we can readily describe a perfectly elastic model by defining new variables,

$$q_1 = m_1^{1/2} x_1 \tag{8}$$

$$q_2 = m_2^{1/2} x_2 \tag{9}$$

So that,

$$0 < q_1 < (m_1/m_2)^{1/2} q_2 < m_1^{1/2} \tag{10}$$

This region is a right triangle in the (q_1, q_2)-plane, with the hypotenuse making an angle $\beta = \tan^{-1}(k_1^{1/2}/k_2^{1/2})$ with the q_2-axis.[31] Because of the elastic collisions, the trajectories are straight lines in this region. When a trajectory intersects a side of the triangle, it corresponds to a collision in the system. Such a system conserves linear momentum when the particles collide with each other so that,

$$m_1^{1/2} \dot{q}_1 + m_2^{1/2} \dot{q}_2 = m_1^{1/2} \dot{p}_1 + m_2^{1/2} \dot{p}_2 \tag{11}$$

where $\dot{q}$ and $\dot{p}$ are the new and old velocities. A little algebra shows that a trajectory striking the hypotenuse is perfectly reflected there, as well as at the other two sides, corresponding to a collision at a wall. The trajectories in this system thus may by studied by considering the motion of a single particle inside a perfectly elastic right triangle. This is reminiscent of the motion of a single ball on a billiard table, discussed by May[30] as an example of an almost periodic system. We can generate a sequence of states on a $1 \times L$ billiard table by constructing a trajectory and recording the position along one of the unit sides every time the trajectory intersects it. However, in the nth record of this sequence, x_n is not sufficient to predict x_{n+1}. This is because, if α is the initial angle of projection, subsequent trajectories will intersect at angles of either α or $\pi - \alpha$. The trajectories

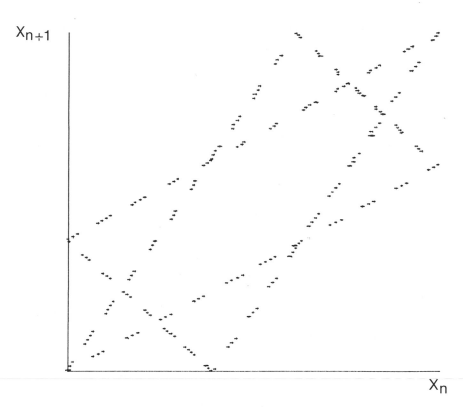

FIGURE 9. Next amplitude plot for the motion of a ball on a billard table with the shape of a right isosceles triangle. The particle starts at the point 0.674 on the hypotenuse and is projected at an angle 1.022398 relative to it.

are calculated by constructing an infinite rectangular lattice by reflecting across each edge of the rectangles. Because of symmetry, this lattice is regular. Then x_{n+1} and x_n satisfy one of the four difference equations,

$$x_{n+1} = (-1)^{i_+(\alpha)} x_n + 2L \cot \alpha - i_+(\alpha) \tag{12}$$

$$x_{n+1} = (-1)^{i_-(\alpha)} x_n - 2L \cot \alpha + i_-(\alpha) \tag{13}$$

where,

$$i_+(\alpha) = \text{int } (x_n + 2L \cot \alpha) \tag{14}$$

$$i_-(\alpha) = \text{int } (1 - x_n + 2L \cot \alpha) \tag{15}$$

and int(y) is the integer part of y. These four lines form a rectangle in the (x_{n+1}, x_n)-plane rotated through 45°. Thus the next amplitude plot is double valued, which corresponds to the uncertainty in α.

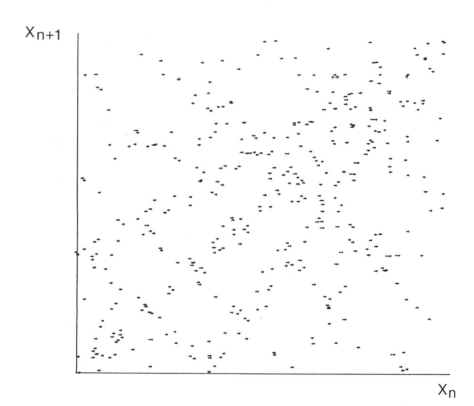

FIGURE 10. Next amplitude plot for the motion of a ball on an irregular right triangular billiard table with $\beta = 0.913561$ radian. The ball is projected from a point 0.781236 along the hypotenuse at an angle 1.021587 radians relative to it.

Let us now consider the *right triangular* billiard table. For arbitrary angle β, the lattice constructed by reflecting across each side of the triangle is not a regular covering of the plane. If $\beta = \pi/4$, corresponding to $m_1 = m_2$, this covering is regular, and it should come as no surprise that this special case gives a well-defined next amplitude plot, as shown in FIGURE 9. It is clear from the nature of the lattice that the derivation of this map in closed form is very complicated. Thus we have proceeded by computer simulation. The map obtained for the isosceles triangle is shown in FIGURE 9, where we have recorded x_n as the distance along the hypotenuse each time the trajectory strikes it.

Because of the failure of the lattice construction for $m_1 \neq m_2$, we are lead to expect that no algebraic relation between x_n and x_{n+1} exists. This is born out by the computer simulation shown in FIGURE 10. The (x_n, x_{n+1}) plot reveals no definite functional relation, although a certain amount of structure can be observed. Nevertheless, this sequence is closely related to the solution of Equations 5 and 6. These equations have a single critical point (x_1, x_2, v_1, v_2) given by $(\frac{1}{3}, \frac{2}{3}, 0, 0)$, which lies on the energy surface only if $E = 3V(\frac{1}{3})$. We thus conclude that a deterministic chaotic system need not be capable of diagnosis as such by the next amplitude plot and the variety of chaos produced by such systems is richer than that possessed by unstable difference equations.

REFERENCES

1. BRAY, W. C. 1921. J. Am. Chem. Soc. **43**: 1262.
2. BELOUSOV, B. P. 1959. Sb. ref. radiats. med. 2a 1958, Medgiz, Moscow.
3. ZHABOTINSKII, A. M. 1964. Biofizika **9**: 306.
4. YAMAZAKI, I., K. YOKOTA & R. NAKAJIMA. 1965. Biochem. Biophys. Res. Comm. **21**: 582.
5. NAKAMURA, S., K. YOKOTA & I. YAMAZAKI. 1969. Nature **222**: 794.
6. GHOSH, A. & B. CHANCE. 1964. Biochem. Biophys. Res. Commun. **16**: 174.
7. PYE, E. K. 1969. Can. J. Bot. **47**: 271.
8. CHANCE, B., B. HESS & A. BETZ. 1964. Biochem. Biophys. Res. Commun. **16**: 182.
9. ZHABOTINSKII, A. M. 1968. *In* Biological and Biochemical Oscillators. B. Chance, E. K. Pye, A. K. Ghosh, and B. Hess, Eds. Academic Press, 1973.
10. OLSEN, L. F. & H. DEGN. 1978. Biochim, Biophys. Acta **523**: 321.
11. SWEDIN, B. & H. THEORELL. 1940. Nature **145**: 71.
12. AKAZAWA, T. & E. E. CONN. 1958. J. Biol. Chem. **232**: 403.
13. LUNDSGAARD, J. & H. DEGN. 1974. IEEE Trans. Biomed. Eng. BME. **20**: 384.
14. DEGN, H. 1968. Nature **217**: 1047.
15. LAIDLER, K. J. & P. S. BUNTING. 1973. The Chemical Kinetics of Enzyme Action. Clarendon Press, Oxford.
16. DEGN, H. 1969. Biochim. Biophys. Acta **180**: 271.
17. HIGGINS, J. 1967. Ind. Eng. Chem. **59**: 19–62; DEGN, H. 1972. J. Chem. Educ. **49**: 302.
18. DEGN, H. 1968. Trans. Faraday Soc. **64**: 1348.
19. DEGN, H. and D. E. F. HARRISON. 1969. J. Theor. Biol. **22**: 238.
20. BALSLEV, I. & H. DEGN. 1975. J. Theor. Biol. **49**: 173.
21. IBAÑEZ, J. L., V. FAIREN & M. G. VELARDE. 1976. Phys. Lett. **59A**: 335.
22. DEGN, H. & D. MAYER. 1969. Biochim. Biophys. Acta **180**: 291.
23. SEMJONOV, N. N. 1959. Some Problems in Chemical Kinetics. Princeton University Press.
24. LINDBLAD, P. & H. DEGN. 1967. Acta Chem. Scand. **21**: 791.
25. DEGN, H. 1967. Acta Chem. Scand. **21**: 1057.
26. OLSEN, L. F. & H. DEGN. 1977. Nature **267**: 177.
27. MAY, R. M. 1976. Nature **261**: 459.
28. LI, T.-Y. & J. A. YORKE. 1975. Am. Math. Mon. **82**: 985.
29. RÖSSLER, O. E. 1976. Z. Naturforsch. **319**: 259.
30. MAY, R. M. 1976. Am. Nat. **110**: 573.
31. HOBSON, A. 1975. J. Math. Phys. **16**: 2210.

OBSERVATIONS OF COMPLEX DYNAMIC BEHAVIOR IN THE H₂—O₂ REACTION ON NICKEL*

R. A. Schmitz, G. T. Renola, and P. C. Garrigan

Department of Chemical Engineering
University of Illinois
Urbana, Illinois 61801

INTRODUCTION

Through several decades of research into the dynamics of chemically reacting systems, the notion prevailed among investigators that any realistic system open to its surroundings would approach either a stable steady state or a time-periodic state. This notion came about, not from any general mathematical arguments, but rather, because all computer simulation studies, as well as published experimental reports, supported it. Furthermore, any time-periodic states observed either through simulations or experiments were "simple" limit cycles, in the sense that all state variables passed through a single peak value per cycle. References to most of the early work on this subject may be found in recent review papers.[1,2,3]

More recently, interest has shifted to the possibility of greater dynamic complexities, such as multipeak periodic states, and to so-called "chaotic" states (i.e., states with sustained nonperiodic time dependency). Theoretical studies of such behavior suggest that the two go hand in hand; that is, systems that exhibit multipeak, cyclic states are also likely to exhibit chaotic states under certain conditions. The most recent interest in this subject was apparently motivated by the publication of several theoretical studies of the dynamics of populations whose histories can be described by a single nonlinear difference equation.[4-8] In essence, these studies show that a surprisingly simple mathematical model can predict a plethora of dynamic complexities, including multipeak periodic states and chaotic behavior. Rössler,[9-11] and May and Leonard[12] have recently presented model systems of nonlinear ordinary differential equations that exhibit similar dynamic complexities. Such equations might also be descriptive of the state of a chemically reacting system.

To our knowledge there have been only a few reports of experimental observations of complex oscillatory states in open chemically reacting systems, and only three of these have cited evidence of chaotic behavior.[13-15] The subject of oscillatory states is particularly interesting for reactions which are catalyzed on solid surfaces. This is because kinetic models, long considered descriptive of the rates of catalytic reactions, and, in fact, generally employed in mathematical models of catalytic reactors, fail to predict oscillatory behavior. Yet laboratory experiments in recent years have demonstrated its existence.[3] It appears that new information regarding catalytic reaction mechanisms and kinetic models might result from research which is aimed at observing and delineating dynamic behavior of catalytic reactions over the space of experimental parameters and coupling such observations with theoretical predictions based on candidate kinetic models.

*This work was supported by grants from the National Science Foundation and the Gulf Oil Corporation.

0077-8923/78/0316-0638 $1.75/1 © 1979, NYAS

Some of the early stages of our work aimed in this general direction are described in this paper. This work consisted of observations from a simple experimental system, which include evidence of chaotic states in nonisothermal oxidation of hydrogen on nickel. There apparently have been no previous reports on observations of chaotic behavior in catalytic reactions, but some prior work does report observations of complex periodic states.[3]

THE EXPERIMENTAL SYSTEM

A diagram of the reactor employed in our experiments is shown in FIGURE 1. Oxygen and hydrogen were continuously fed into the bottom of the Pyrex tube,

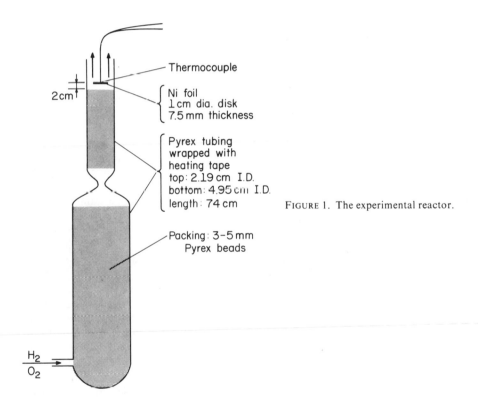

FIGURE 1. The experimental reactor.

which was packed with Pyrex beads. The purpose of the "beads" was to facilitate mixing of reactants and attaining a uniform stream temperature. For this purpose, the reactor tube was necked down to a diameter of about 0.65 cm at a distance of 43 cm from the reactor inlet.

As shown in the figure, the catalyst specimen, a disk-shaped piece of nickel foil, is 2 cm above the top layer of beads, and perpendicular to the direction of flow (99.95% pure foil was obtained from A. B. Mackay, Inc., New York City). An iron-constantan thermocouple is welded to the downstream side of the foil. Its signal was recorded continuously during experimental testing. Both the downstream surface of the foil and the exposed portion of the thermocouple lead wires were coated with aluminum paint to render them catalytically inactive. The insulated lead wires, coated with a layer of ceramic cement to within 1 cm of the foil, were sufficient for supporting the catalyst.

Pretreatment of the catalyst amounted to heating the foil (after it had been electropolished) to a red glow in a butane flame in air for several seconds (we found that an electropolished foil not so pretreated was almost completely inactive under experimental conditions). Heating in the butane flame probably produced an active oxide grain of large surface area, and possibly also caused some thermal faceting. The treated metal was not analyzed, but on visual inspection, it appeared to be considerably rougher than the intial electropolished material.

In our experiments, the superficial hydrogen velocity in the top portion of the Pyrex tube was held constant at 9.2 cm/sec (at this velocity, gas flow was well in the laminar regime; the Reynolds number based on the top tube diameter was about 8).

The heat input was held constant (from heating tapes wrapped around the Pyrex tube). This input was set at a value sufficient to heat a flowing stream of pure hydrogen to 231°C, as measured by the thermocouple attached to the catalyst foil. We thus refer in later figures to the *gas stream temperature* T_g (i.e., steady catalyst temperature at oxygen concentration = 0%). The primary experimental parameter was the concentration of oxygen in the feed stream, symbolized by $[O_2, g]$. We manipulated this parameter between values of 0% and 7% by volume. Notice that changes in the feed composition cause the total gas flow rate to change as well, because the hydrogen flow rate was held constant. This change, however, was slight.

We also tested, in a cursory manner, the effect of other conditions, including hydrogen velocity, the vertical location of the nickel foil, and gas temperature. Though the first two of these conditions had a quantitative effect on observations—the qualitative features of the behavior were not changed. The gas temperature had a large effect on the behavioral characteristics, as expected (this will require further study).

With this simple experimental system, we planned to make preliminary tests and observations in which various reactions and catalysts could be tested easily. Important parameters and interesting dynamics could be sorted out and used as a basis for a more rational design of an experimental study. Clearly one could raise unanswerable questions about observations of this experimental system—for example, the effect of physical transport processes and of the possible variation in their rates over the catalyst surface, or possible nonuniformity of the surface on a microscopic or macroscopic scale. Because of such intricacies, reliable mathematical modeling of the experimental system would be difficult, so we have not strongly pursued it. Nevertheless, we feel that observations at this stage are worth reporting because they are new, interesting, and potentially important.

It should be mentioned that other nickel foils were used—platinum foil was used in some experiments with a hydrogen-lean mixture. Furthermore, platinum wires, rings, and other geometries were explored. In none of these cases, was the dynamic behavior studied as thoroughly and systematically as in the work reported here. We intend to publish a summary of observations of oscillatory states from all

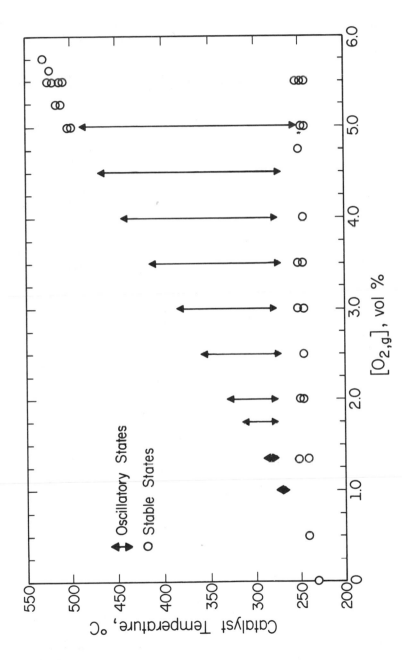

FIGURE 2. Stable and oscillatory experimental states.

of these experiments elsewhere. Suffice it to say, different catalyst samples invariably produced different results, probably because of the difficulty in reproducing the exact pretreatment process. In all cases, however, oscillatory behavior was observed over some ranges of experimental conditions, and the behavior appeared to be chaotic in a number of instances.

RESULTS

Results of experiments employing the apparatus described above are summarized in FIGURE 2. The figure shows catalyst temperatures at observed stable and oscillatory states—the extreme temperatures reached during the course of oscillations are indicated by arrowheads. The oscillatory states were observed for oxygen feed concentrations between 1% and 5%, and stable nonoscillatory states were found over the entire range of oxygen concentrations. The information in the figure represents the results of a large number of tests conducted over 6 weeks.

Tests of the stable states and all of the oscillatory states were repeated several times. The endpoint values of the oscillatory states, (represented by the arrowheads) are average values, varying by as much as 5°C to 10°C. Variations were also noticed in the oscillations themselves. However these variations were slight and could be attributed to the imperfect setting of parameters or to an insufficient time allowed for the oscillations to become fully developed.

The most interesting features of the experimental results were the hysteresis characteristics and the nature of the oscillatory states. To describe the hysteresis, we reconstruct a typical sequence of tests starting with an oscillatory state at an oxygen concentration of 3%. Oscillatory states persisted until the concentration was increased beyond about 5%. In almost all cases, an increase beyond 5% led to a high-temperature nonoscillatory state. (These states, which were at a catalyst temperature of about 500°C, appear to be mass-transfer-controled states.) In all cases a subsequent decrease in oxygen concentration below about 5% caused a sudden transition to a stable low-temperature state at about 250°C (probably a kinetically controled state). Having reached that low state, we were unable to reach either the oscillatory or the stable high-temperature states by changing the oxygen concentration in the range 0% to 7%. Such changes yielded only the low-temperature stable states (FIGURE 2). We found that perturbations in the gas temperature to about 500°C created by temporarily increasing the heat input to the flowing gas, would cause a transition back from the low states to oscillatory states if oxygen concentration was below 5%.

In one test, the cessation of oscillations at the end of the oscillatory range led to a stable low-temperature state. There appeared to be a small overlapping region near the high end of the oscillatory range. In some cases an oscillatory state was observed at a concentration of 5% as the concentration was being increased, but a stable high-temperature state existed at that same concentration when the concentration was being decreased. The existence of oscillatory states near the endpoints of the region was difficult to establish, because of the very long characteristic time (1 hour or more) associated with oscillation. We were never certain, in such cases, if the oscillations would not eventually die away to a high or low stable state.

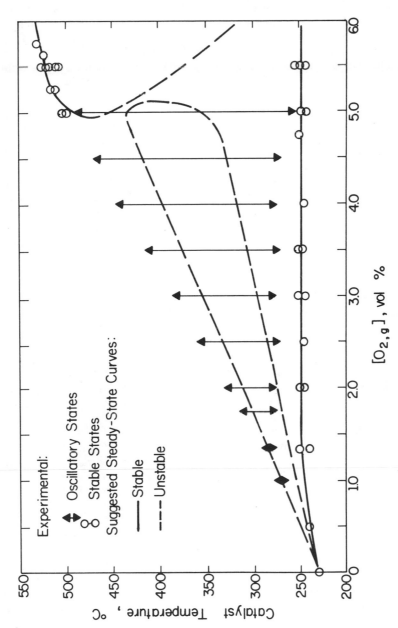

FIGURE 3. Experimental data and presumed steady-state curves.

Changing to a sequence of tests in which the oxygen concentration was decreased, starting at an oscillatory state, we were able to observe oscillations for oxygen concentrations as low as 1.0%, although they were very small at that value and seemed to fade to negligible amplitudes at lower values. The surprising observation was that turning the oxygen flow completely off for long periods of time (9 hours in some cases) would not obliterate the catalyst's "memory" of the oscillatory path. That is to say, increases in the oxygen concentration from 0% would invariably lead to oscillatory states if such states existed before the oxygen flow was turned off. On the other hand, if stable low-temperature states existed as the concentration was decreased to zero, they would also exist when it was again increased. Oscillatory states are interesting, not only because of their nature, but also the manner in which they can be made to appear and disappear.

As described above, the oscillations could be rather easily "turned on" by means of thermal perturbations, but not by perturbations in gas composition   at least in the experimental range. We found that once the oscillatory states were attained, they could not be "turned-off" by moderate decreases in the heat input (decreases that would cause a catalyst temperature change of about 100°C) or by decreasing the oxygen concentration. However a circuitous sequence of changes in the concentration, amounting first to increasing its value beyond 5% then decreasing it to a lower value, would remove them. We also found that if the catalyst was removed from the system and allowed to reach equilibrium with ambient room conditions for several days, stable low-temperature states were reached upon startup regardless of the earlier course of events.

The steady-state picture suggested by these observations is shown in FIGURE 3. The presumed steady-state curves are constructed so as to give an odd number of steady-state temperatures at each value of $[O_{2,g}]$ on the abscissa. Notice that according to the experimental observations, the steady-state temperature curves must converge as $[O_{2,g}]$ approaches zero. It should be kept in mind, however, that only one state variable, the catalyst temperature, was measured. Others, including concentrations of chemisorbed species, gas composition near the interface, and the nature and extent of metal oxidation, should all be represented, if they could be measured, and that the resulting steady-state curve should be shown in higher-dimensional state space. Thus, the convergence of three steady-state curves to a single temperature as $[O_{2,g}]$ approaches zero, as shown in the figure, does not imply that the complete state is identical at that parameter value. In fact we strongly suspect that the nature of the metal surface, or most likely of an oxide layer, is characteristically different along the path of oscillatory states. It would indeed be interesting to study the state of the metal surface *in situ* and also to explore the steady state picture and oscillatory patterns in the 2-dimensional parameter space of $[O_{2,g}]$, T_g. Work along these lines is now in progress.

The nature of the oscillatory states is shown in FIGURE 4 through 7, which present tracings of strip-chart recordings of the catalyst temperature. In all such tracings, the abscissa gives the time elapsed from the beginning of the test at the specified value of $[O_{2,g}]$. We found that if a sufficient length of time was allowed at these states, they eventually became chaotic (exceptions will be described later). Again, no such firm conclusion could be made for states very near the ends of the oscillatory regime.

FIGURES 4 and 5, which show results for oxygen feed concentrations of 1.75% and 2.5%, respectively, typify much of the development and nature of the observed oscillations.

In almost all cases, we found that an initial period of adjustment followed any change in the oxygen concentration. The nature and length of that period depended on the magnitude of the change and on the initial state.

As shown in FIGURES 4 and 5, this initial period was usually followed by

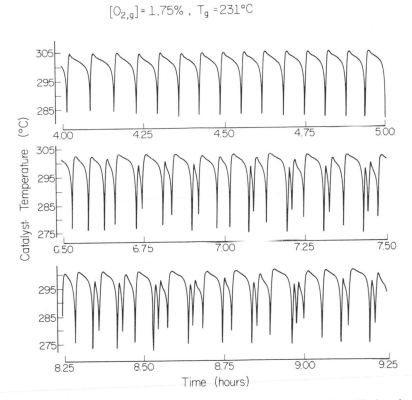

$[O_{2,g}] = 1.75\%$, $T_g = 231°C$

FIGURE 4. Temperature recordings showing the development of chaotic oscillations for an oxygen feed concentration of 1.75%.

single-peak oscillations that were very nearly periodic. After some time, a transition from single- to double-peak sycles occurred. In some instances (FIGURE 5), for an oxygen concentration of 2.5%, double-peak cycles were followed by nearly periodic cycles containing four peaks per cycle. In all cases, eventually a three-peak cycle appeared, and, with rare exceptions, this appearance was followed by chaotic oscillations. A three-peak cycle was evident in FIGURE 4 shortly after the seventh-hour mark and in FIGURE 5 at about the thirteenth hour. The final hour of recordings in FIGURE 4 and the final 2 hours in FIGURE 5 are chaotic, in our

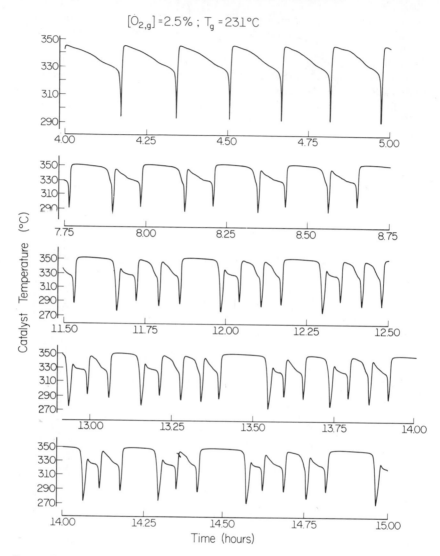

FIGURE 5. Temperature recordings showing the development of chaotic oscillations for an oxygen feed concentration of 2.5%.

judgement. As yet, we have not applied any statistical analysis to the time-dependent temperatures. Chaotic states are easily distinguished by visual inspection.

Specific mention was made in the preceding discussion that a three-peak cycle was usually evident just preceding the onset of chaos. Almost invariably, such a cycle seemed to separate nearly periodic behavior from that which was clearly chaotic. This fact relates nicely to the theoretical result of Li and Yorke, who proved that if the solution of a difference equation yielded a three-point cycle under given conditions, then a chaotic state also existed under those same conditions.[6] Their result cannot be generally extended to the time history of a continuous variable, such as the catalyst temperature trace, in a multivariable system governed by differential equations, but still the correspondence between that theoretical result and our experimental observations is interesting. It is also noteworthy that published computer simulations, again based on difference equations[4,5,7,8] suggest that we could expect bifurcations of periodic states to proceed through cycles with an even number of peaks per cycle before cycles having an odd number of peaks per cycle appear. The development of oscillations shown in FIGURE 6 is interesting in this regard. In this case a single foor-peak cycle started shortly after the two-hour point, and it was followed by several five-point cycles. Three-peak cycles eventually developed and appeared to be stable and nearly periodic over the few remaining hours of that particular test (notice the indicated cycle lengths for the time span between 6.75 and 8.0 hours). Nearly all other tests at this oxygen concentration of 2.0% led to chaotic states. The chaotic nature of the catalyst temperature was much more evident and more quickly established at higher oxygen concentrations. FIGURE 7 shows about 4.5 hours of chaotic catalyst temperature variations for an oxygen concentration of 3%.

The development of the oscillatory patterns described above in connection with FIGURES 4, 5, and 6 call for further comment. The development takes place as if some parameter were being slowly changed. Our opinion, which cannot be

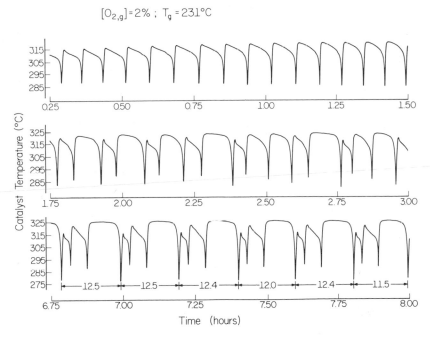

FIGURE 6. Results of a test at an oxygen concentration of 2% which led to a nearly periodic three-peak cycle.

staunchly defended at this point, is that during the development the nature of an oxide layer or other catalyst surface characteristic is adjusting to the change in environment. The adjustment is sufficiently slow that the transitory states observed are "pseudo steady" states. If a quick scan over values of the oxygen concentration could be made at a "frozen" condition, say at 4 hours into the oscillatory pattern, it is very likely that we would observe single-peak cycles over some range, double-peak cycles over others, and so on, presumably into a range of chaotic states. Such an experiment would hardly be feasible, or conclusive, however, because of the initial effects that follow each change.

It should be kept in mind that experimental studies can probably never prove the existence of chaotic states because there is always the possibility that slow changes occur within the system, such as with the catalyst material in our case, and simply prevent the attainment of a truly periodic state. Small extrinsic disturbances may have the same effect. In some of our tests, we allowed the chaotic state to run for days (6 days in the longest run) in order to examine the "ultimate" behavior. Though some changes in the characteristic time scale of the chaotic temperature changes and in some fine points of the tracings were evident, there was no indication that a periodic, or stable state was imminent in these long runs.

A few isolated observations, perhaps spurious, but not necessarily so, should also be mentioned. On one occasion, thermal perturbations from a stable, relatively inactive state led to simple periodic states that showed no tendency to

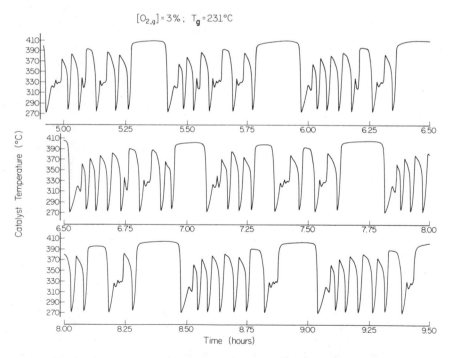

FIGURE 7. Chaotic oscillations for an oxygen feed concentration of 3.0%.

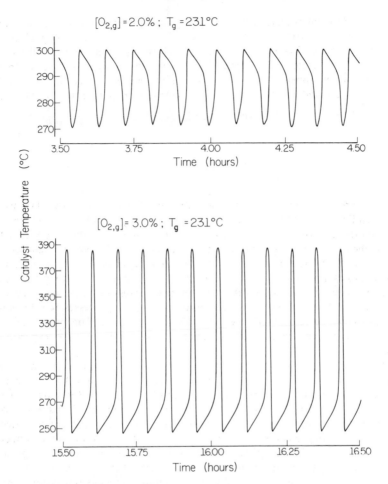

FIGURE 8. Periodic states resulting from tests at oxygen concentrations of 2.0% (top diagram) and 3.0% (bottom diagram).

change to multipeak cycles or to chaotic states, even when subjected to small deliberate perturbations. Two such states are shown in FIGURE 8. The state shown at the bottom of the figure for an oxygen feed concentration of 3.0% is particularly interesting because of its duration and nature. Notice that the slow portion of the cycle for this particular case is the low-temperature portion—completely opposite to that of all others. It is possible that an insufficient length of time was allowed in these experiments or that more than one stable oscillatory state exists under a given set of conditions. In such cases the eventual oscillatory state reached would depend on the history of perturbations. During the course of our experiments we made numerous perturbations in all manipulatable parameters in trying to locate multiple oscillatory states, but in no case did the perturbations have a permanent effect on the oscillations.

On two different occasions, with an oxygen composition of 2.0%, an apparently stable steady state with a catalyst temperature lying between the peak-to-peak values indicated in FIGURE 2 was reached following a perturbation in the heat input. The initial state in these tests was a steady low-temperature state. In both cases, oscillations appeared when the composition was increased to 3.5%, and the oscillations persisted when the oxygen concentration was returned to its original value of 2.0%.

CONCLUSION

If we were to attempt to construct a theoretical model to aid in explaining and understanding our experimental observations, we would incorporate such physical rate processes as heat and mass transfer to and from the catalyst surface, and the chemical processes of oxidation and reduction of the metal, chemisorption of reactants, and chemical reaction on the catalytic surface. The oxidation and reduction steps, and probably the thermal response, are slow relative to the surface processes. Depending mainly on the form of rate expressions, it is very likely that a system of differential equations accounting for these processes would yield a variety of dynamic complexities. Work along these lines is in progress and hopefully will yield some theoretical guidance for future laboratory experimental work.

More attention, however, should probably be given at this stage to a refinement of the experimental system and procedures—a refinement that would result in an accurate description of the catalyst condition and of other state variables. We are concerned, for example, that nonuniformities on the catalyst surface might be very important in the dynamic behavior. The isolated vicinity of the thermocouple may exhibit certain periodic characteristics, while the outer regions may exhibit different ones. The coupling of two such oscillators by thermal conduction along the foil may be an important factor in the onset of chaotic states.

Despite the shortcomings embodied in our simple experimental system, the observations reported here suggest a variety of interesting and important studies with catalyst particles and with assemblies of them. Certainly new information regarding catalytic reactions, their mechanisms and kinetics, which govern their rates, may be extracted if experiments are carefully designed and planned, and if experimental results are linked to a mathematical model.

In addition, analogies to biological systems are evident. These analogies include, not only the oscillatory patterns, but also hysteresis phenomena and the apparent memory of "differentiated" oscillatory states. An assembly of interacting catalyst particles might be viewed as a nonbiological analog of a living system. Exploring the behavior of such a system, for which a confrontation between theoretical and experimental studies seems feasible, might be useful in improving our knowledge and understanding of interacting populations, developmental processes, patterns rhythms, etc. in living systems, for which such a confrontation is extremely difficult if not impossible.

Summary

Measurements of the catalyst temperature in experiments involving the flow of oxygen in excess hydrogen past a suspended nickel catalyst foil at atmospheric pressure give evidence that the reaction proceeds chaotically in time over a range of oxygen concentration; a hysteresis characteristic was also evident. The results were obtained with oxygen concentration from 0% to 7% (by volume) and gas temperature of 231°C.

References

1. NICOLIS, G. & J. PORTNOW. 1973. Chem. Rev. **73**: 365–384.
2. SCHMITZ, R. A. 1975. Adv. Chem. Ser. **135**: 156–211.
3. SHEINTUCH, M. & R. A. SCHMITZ. 1977. Catal. Rev. **15**: 107–172.
4. MAY, R. M. 1974. Science. **186**: 645–647.
5. MAY, R. M. 1975. J. Theor. Biol. **51**: 511–524.
6. LI, T. Y. & J. A. YORKE. 1975. Amer. Math. Mon. **82**: 985–993.
7. MAY, R. M. & G. F. OSTER. 1976. Am. Nat. **110**: 573–599.
8. MAY, R. M. 1976. Nature. **261**: 459–467.
9. RÖSSLER, O. E. 1976. Z. Naturforsch. **31a**: 259–264.
10. RÖSSLER, O. E. 1976. Z. Naturforsch. **31a**: 1168–1172.
11. RÖSSLER, O. E. 1977. Bull. Math. Biol. **39**: 275–289.
12. MAY, R. M. & W. J. LEONARD. 1975. Siam J. Appl. Math. **29**: 243–253.
13. SCHMITZ, R. A., K. R. GRAZIANI & J. L. HUDSON. 1977. J. Chem. Phys. **67**: 3040–3044.
14. OLSEN, L. F. & H. DEGN. 1977. Nature. **267**: 177–178.
15. RÖSSLER, O. E. 1978. Nature **271**:89–90.

BIFURCATIONS TO PERIODIC, QUASIPERIODIC, AND CHAOTIC REGIMES IN ROTATING AND CONVECTING FLUIDS*

P. R. Fenstermacher and H. L. Swinney†

Physics Department
City College of CUNY
New York, New York 10031

S. V. Benson and J. P. Gollub

Physics Department
Haverford College
Haverfold, Pennsylvania 19041

INTRODUCTION

We have performed a comparative experimental study of the bifurcations in circular Couette flow and Rayleigh-Bénard convection, two systems which long served as classical prototypes for experimental and theoretical investigations of hydrodynamic stability.[1,2] In circular Couette flow in its simplest form a fluid is contained between concentric cylinders with the inner cylinder rotating, and in Rayleigh-Bénard convection the fluid is contained between horizontal thermally conducting plates heated from below. The bifurcation parameter for the Couette flow system can be taken as the Reynolds number R, which describes the distance away from equilibrium and is proportional to the angular velocity of the inner cylinder. Similarly, for the Rayleigh-Bérnard problem the distance away from equilibrium is given by a dimensionless number R, the Rayleigh number, which is proportional to the difference between the temperatures of the two horizontal plates.

At small R the flow in the Couette cell is purely azimuthal, but when the Reynolds number exceeds a critical value R_c the azimuthal circular Couette flow is no longer stable and there is a bifurcation to a flow with a horizontal toroidal vortex pattern superimposed on the azimuthal flow. This bifurcation was predicted and observed by Taylor in 1923 in work that stands as a classic study of hydrodynamic stability.[3] In 1916 Rayleigh showed, in the pioneering theoretical paper on the convective instability, that above R_c the pure conduction state is unstable to horizontal disturbances, and the system bifurcates to a new state consisting of parallel convection rolls.[4]

As R is increased further both the circular Couette flow and Rayleigh-Bénard systems bifurcate from the time-independent vortex patterns to time-dependent flows. These secondary bifurfactions have been studied theoretically and experimentally for the past few years; however, there has been little detailed quantita-

*This work was supported by Grants DMR 76-11033, ENG 76-19810, and ENG 76-82511 from the National Science Foundation.
†Present address: Department of Physics, University of Texas, Austin, Texas 78712.

tive work extending beyond the secondary bifurcations, which are already quite difficult to treat mathematically.

We have studied the time-dependent rotating and convecting fluids using laser Doppler velocimetry, a technique that determines the local fluid velocity without perturbing the flow. In this collaborative study, measurements on circular Couette flow were made at City College[5,6] and on Rayleigh-Bénard convection at Haverford College.[7] The fluid velocity has been measured at well-defined points in the samples, and measurements in sequential time intervals were recorded in a computer. These velocities were then Fourier transformed to obtain velocity power spectra. As an aid in the interpretation of the velocity power spectra we have also made photographs of the flow patterns.

Our velocity spectra for the two systems reveal several distinct dynamical regimes containing one or more sharp frequency components, accompanied by broadband noise which grows with increasing R. At large R, the sharp components disappear from the spectrum; the resulting aperiodic flow might be termed weakly turbulent.

We shall now describe the experiments on circular Couette flow and Rayleigh-Bénard convection, and then we shall compare the dynamical behavior of the two systems.

CIRCULAR COUETTE FLOW

Background

Flow between concentric cylinders is parameterized by the boundary conditions, the Reynolds number, and the spatial state of the flow. The boundary conditions are specified by: the ratio of the radii of the inner and outer cylinders, r_i/r_o; the ratio of the rotation frequencies of the outer and inner cylinders, Ω_o/Ω_i; the ratio of the fluid height to the gap between the cylinders, $h/(r_o - r_i)$; and the constraints on the upper and lower horizontal fluid surfaces. Most studies (including ours) have been concerned with radii ratios near unity and with the outer cylinder at rest. The Reynolds number of circular Couette flow with the outer cylinder at rest can be defined as $R = \Omega_i r_i (r_o - r_i)/\nu$, where ν is the kinematic viscosity.

As stated previously, the flow is purely azimuthal for small R, but above a critical Reynolds number R_c, which depends on r_i/r_o, there is a bifurcation to a flow with horizontal vortices, as predicted by Taylor[3] and pictured in FIGURE 1. (Visualization of the flow is achieved by seeding the fluid with small flat flakes which align with the flow.[4]) Taylor vortex flow has been studied experimentally by many workers, including particularly Donnelly *et al.*, Koschmieder, Synder, and Gollub and Freilich[8-12]; Davey and others have extended the theory to more general cases not treated by Taylor.[13]

The next bifurcation, from Taylor vortex flow to a time-dependent flow with transverse waves superimposed on the horizontal vortices has been observed in several experiments, and this bifurcation has been considered theoretically for the small gap limit by Davey, DiPrima, and Stuart and by Eagles.[14,15] Some careful measurements of the onset of time-dependent Taylor vortex flow have been reported by Donnelly and coworkers, who used a sensitive ion technique that they developed.[8,9] The only detailed previous study of wavy vortex flow extending well beyond the onset of the wavy vortices is the photographic investigation of Coles.[16] As the Reynolds number was increased he observed several bifurcations to spatial states with different numbers of axial vortices and azimuthal waves. He also found that at a given Reynolds number several (as many as 26) distinct spatial states were accessible, depending on the way in which that particular Reynolds number was approached. Our system has the same ratio of radii as that of Coles, $\frac{7}{8}$, while our height to gap ratio, 20, is somewhat smaller than his, 28.

Rather than study the bifurcations between different spatial states, we have chosen to study in detail the dynamics of a single spatial state, one with 17 axial vortices and 4 transverse waves. (We have also made some measurements on the

FIGURE 1. Photograph of Taylor vortex flow, $R/R_c = 1.1$. The cell dimensions are given in the text. The verticals bars are external fiducials marking 10° intervals.

state with 15 cells and 4 vortices, with similar results.) Once the 17-vortex state is produced, it is stable throughout the Reynolds number range studied, $5.4 < R/R_c < 45$, provided that R is not increased or decreased too rapidly. Below $R/R_c = 5.4$ the 17-cell, 4-vortex state loses stability to the 17-cell, 5-vortex state as R/R_c is decreased through 5.4; above $R/R_c = 22.4$, the waves are absent.

We now describe the instrumentation and the results of the circular Couette flow measurements.

Instrumentation

The dimensions of the Couette flow cell are $r_i = 2.224$ cm, $r_o = 2.536$ cm, and $h = 6.25$ cm. The lower horizontal surface of the fluid is rigid and the upper surface is free. The cell is temperature-controlled to $\pm 0.05°$C. The inner cylinder

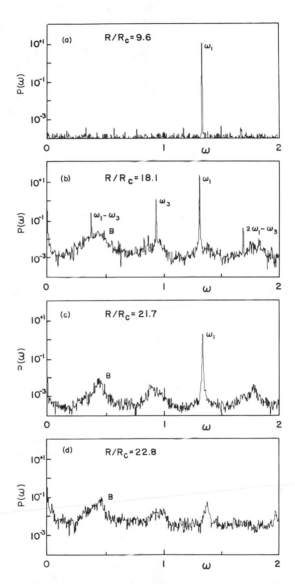

FIGURE 2. Power spectra of the radial component of the velocity in time-dependent circular Couette flow at different Reynolds numbers. The frequencies are expressed in units of the inner cylinder frequency; thus ω is dimensionless and is given by $\omega \equiv \omega(\text{rad/s})/\omega_{\text{cyl}} \cdot (\text{rad/s})$. The spectra are normalized so that $\int_0^{\omega_{\text{max}}} P(\omega)d\omega = \langle(\Delta V_r)^2\rangle$, where $P(\omega)$ has units cm^2/s.

frequency has an rms fluctuation of less than 0.3%. The fluid is water, seeded with 0.48 μm diameter polystyrene spheres for the laser Doppler velocimetry studies and with Kalliroscope suspension for the photographic studies.[5,6] The sample volume for the laser Doppler measurements has a linear dimension of 0.13 mm.

The fluid velocity is determined by measuring the Doppler shift of scattered laser light. The laser Doppler velocimetry technique is now widely used in fluid flow measurements, and our optical and electronic instrumentation has been described in previous reports.[5,6] The radial component of the velocity, measured midway between the inner and outer cylinders, is recorded in a computer for 8192 successive time intervals. Then the velocity power spectrum and velocity autocorrelation function are computed and plotted, and another program calculates the positions and linewidths of the spectral lines.

Experimental Results

The distinct dynamical regimes observed in our studies of Taylor vortex flow are illustrated by the velocity power spectra in FIGURE 2, obtained at R/R_c = 9.6, 18.1, 21.7, and 22.8, and FIGURE 3 shows photographs of the flow at the same Reynolds numbers.

Periodic Flow. The time-independent Taylor vortex flow exists over a small range in Reynolds number, $1 < R/R_c < 1.2$. Beyond the bifurcation to the wavy vortex state the power spectra contain a single fundamental frequency, which we have called ω_1, and its harmonics. This frequency, which can be seen as a narrow, intense component in the spectrum in FIGURE 2a, is the frequency of the azimuthal waves passing the point of observation.

Quasiperiodic Flow. A previously unobserved spectral component appears reproducibly (and nonhysteretically) at R/R_c = 10.1 ± 0.2, and grows in amplitude as R is increased. This component, called ω_3 for historical reasons, can be seen in FIGURE 2b. (A previously reported component, ω_2, appears to be a transient component and not a property of the steady flow, even though it has been observed on some occasions to persist for more than an hour.) The measured widths of ω_1 and ω_3 are instrumentally limited even in our highest resolution spectra, where $\Delta\omega$ = 0.0007 (half-width at half-maximum).

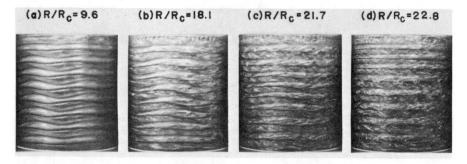

FIGURE 3. Photographs of the time-dependent circular Couette flow at the same Reynolds numbers as the velocity power spectra in FIGURE 2.

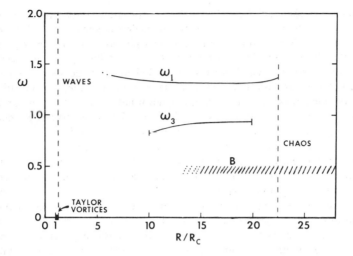

FIGURE 4. The fundamental frequency components observed in the power spectra of the radial component of the velocity in circular Couette flow. The time-independent Taylor vortex extends over a relatively small range in Reynolds number, from $R/R_c = 1$ to 1.2. The spectral measurements reported here were made on the state with 17 axial vortices and, where the waves exist, 4 azimuthal waves.

As R/R_c approaches 19.8 the amplitude of ω_3 decreases, and above $R/R_c = 19.8 \pm 0.4$ it is absent from the spectra, as FIGURE 2c illustrates. In the regime where both discrete peaks are present, their ratio ω_3/ω_1 increases monotonically from 0.62 at $R/R_c = 10$ to 0.71 at $R/R_c = 19.8$. Thus the flow in this range is apparently quasiperiodic, neglecting a weak broadband component to be discussed below.

Chaotic Flow. In addition to the sharp peaks, there is a weak broadband component B whose onset is more difficult to determine accurately. In a few runs B has been observed at Reynolds numbers as low as $R/R_c \simeq 12$. It is clear, however, that the sharp peaks and B coexist over a finite range in R/R_c.

As R/R_c is increased from 19.8 to 22.4 ± 0.4, the amplitude of ω_1 decreases to zero, leaving only the broad component B and its harmonics. Thus $R/R_c = 22.4$ marks the completion of a transition from a nearly periodic flow to a qualitatively different flow, one characterized by a continuous spectrum or, equivalently, a decaying velocity autocorrelation function. Spectra have been recorded up to $R/R_c = 45$ and no further bifurcations have been observed. FIGURE 4 summarizes the major spectral features observed in time-dependent Couette flow.

RAYLEIGH-BÉNARD CONVECTION

Background

To determine what features of the transition to aperiodic flow are similar in different hydrodynamic systems, we undertook, at Haverford College, experiments on time-dependent Rayleigh-Bénard convection similar to those just described on Couette flow. The state of the Rayleigh-Bénard system[17] depends on the boundary conditions and two dimensionless parameters, the Rayleigh number $R = g\alpha d^3 \Delta T / \kappa \nu$ and the Prandtl number $P = \nu / \kappa$, where g is the gravitational acceleration, α is the thermal expansion coefficient, d is the separation between the horizontal plates with temperature difference ΔT, κ is the thermal diffusivity, and ν is the kinematic viscosity. The Rayleigh number indicates how far the system is from thermal equilibrium, while the Prandtl number gives the relative effectiveness of the transport of momentum and heat in the fluid.

Much of the literature on this system is concerned with the primary bifurcation and its nonlinear development above R_c. In this regard, the nonlinear theory of Schlüter, Lortz, and Busse and the quantitative experiments of Bergé and Dubois are especially noteworthy.[18-20] The next bifurcation has been investigated by Clever and Busse,[21] who superimposed infinitesmal disturbances on the parallel convection rolls. Whereas the primary bifurcation is independent of Prandtl number, the next one is predicted to be strongly Prandtl number dependent. For large P, the secondary bifurcation is to a time-independent state in which the fields vary in all three directions. However, for $P \lesssim 5$, the next bifurcation is to a time-dependent state in which transverse traveling waves distort the basic convection rolls. The properties of this time-dependent state were predicted only near its onset. Its evolution at higher R, and the eventual transition to aperiodic convection have not been treated theoretically. Although there is no realistic quantitative theory of the onset of aperiodic convection, models showing aperiodic behavior can be constructed by Fourier transforming the equations of motion and performing severe truncations. The three mode model of Lorenz is well known, and a more complex model with 39 modes was more recently constructed by McLaughlin and Martin.[22,23] Numerical simulation of time-dependent convection in air has been undertaken by Lipps[24] who presented examples of both time-periodic and aperiodic flows in a relatively small volume with periodic boundary conditions. The numerical solutions, however, did not extend to large enough times to permit sharp distinctions to be drawn between periodic and aperiodic flows.

There have been only a few experiments on time-dependent convection in which quantitative measurements of the local velocity or temperature were made. Willis and Deardorff[25-27] used resistive thermometry to determine the space and time dependence of the local temperature field in several different fluids, and observed the oscillatory instability described above (in air). Bergé and Dubois have recently measured the local velocity field in time-dependent high P convection,[28] and observed an erratic oscillatory state. Neither of these investigations has permitted clear distinctions to be made between periodic and aperiodic regimes. Very recently, Olson and Rosenberger[29] have observed a clearly periodic regime at $P = 0.7$ in a tall cylindrical geometry, which is rather different from those normally employed.

Ahlers[30,31] has shown that extremely high precision heat flux measurements can be made in convecting liquid helium ($P \simeq 1$). He was able to study time-dependent states, since exceedingly small fluctuations in the heat flux could be detected and the thermal response time of the container was negligible. He observed a sharp onset of aperiodicity which was not preceded by periodic states. Very recent results (G. Ahlers and R. P. Behringer, private communication) indicate that the heat flux does have a periodic regime for a convection cell of small aspect ratio.

Finally, we mention several photographic studies that have provided important information about the onset of time-dependent phenomena. Busse and Whitehead[32] have shown that at a high Prandtl number, time dependence can occur by means of an oscillatory instability of a complex three-dimensional flow which is similar to the low Prandtl number instability discussed above. Krishnamurti[33,34] has presented a useful phase diagram summarizing experimental data on the various steady and time-dependent regimes as a function of R and P.

Although substantial effort has been devoted to the problem of time-dependent

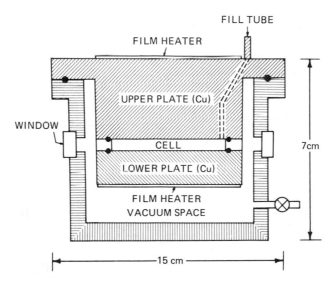

FIGURE 5. Schematic diagram of the convection cell, showing the vacuum space, heat shield with optical windows, and resistive heaters.

convection, it is still not known in what circumstances a periodic or quasiperiodic regime precedes the aperiodic or turbulent flow. In addition, it is not known whether the aperiodicity can be described by a model consisting of a small number of interacting modes. Our investigation, utilizing high resolution spectral measurements of the local velocity field, is aimed at these questions. Preliminary results are described below.

Instrumentation

In order to explore a range of Prandtl numbers with a single system, we chose to use water at elevated temperatures. By varying the mean cell temperature be-

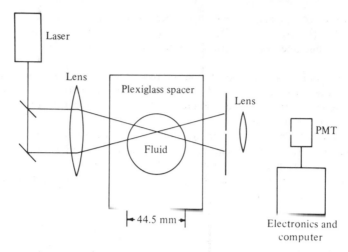

FIGURE 6. Schematic diagram of the optical system used for velocity measurements in the convection experiment. The plexiglass spacer, which is inside an evacuated heat shield (not shown), is square in order to minimize the extent to which it acts as a cylindrical lens.

tween 10°C and 90°C, P could be easily varied between roughly 2 and 9. The cell (FIGURE 5) was of cylindrical cross section, and its interior radius and height were 22.25 mm and 7.09 mm respectively, for an aspect ratio of 3.14. It was constructed of massive copper plates separated by a plexiglass spacer, suspended in a vacuum to reduce extraneous heat leaks, and surrounded by a copper heat shield controlled near the mean working temperature. The two plates, temperature-controlled by AC bridges, had a temperature difference constant to within 0.2%. We estimate that the spatial uniformity of the temperature over each of the plates was about 1% of ΔT.

Optical access was available through windows in the heat shield, and measurements of one horizontal velocity component (V_y) were obtained by observing forward scattering from crossed laser beams (FIGURE 6). The outside dimensions of the plexiglass spacer were rectangular rather than circular in order to minimize the extent to which the cell acts as a cylindrical lens. The electronics were similar to those used in the Couette flow experiments, and will not be described here.

Streak photographs of the flow were made to complement the velocity measurements. Photographs were obtained by illuminating a narrow slice of the fluid with a horizontal curtain of laser light propagating in the y direction, and then viewing scattered light propagating in the x direction (or vice versa). Because of the copper plates, only such a side view can be obtained. For photography the fluid was seeded with 2.02 μm diameter particles, whereas the velocity measurements utilized smaller 0.369 μm particles.

Experimental Results

The observations presented here are limited to $P = 2.5$, but experiments at other values are in progress. We begin by showing streak photographs for comparison with the velocity measurements to be presented subsequently. For R/R_c not too far above unity, a stationary pattern of circular convective rolls is generally expected, given the experiments of Koschmieder[10] and the calculations of Charlson and Sani.[35] However, the time-dependent regime of primary interest in this experiment begins at about $R/R_c = 15$. There is no reason to expect the flow to remain axisymmetric at such high Rayleigh numbers, even in the absence of asymmetries in the boundary conditions. Photographs taken in our cell at these high Rayleigh numbers show a spatial structure of the convection pattern that is generally not axisymmetric. For example, FIGURE 7 shows two orthogonal views of the convection pattern at $R/R_c = 20$. If the pattern were axisymmetric, both views would look the same. In fact the convection pattern is probably closer to parallel rolls than to the circular rings that would be obtained near R_c. While asymmetric boundary conditions can produce similar effects, we do not believe that is the reason here.

The onset of time dependence does not drastically alter the overall shape of the flow. Even at $R/R_c = 28$, where the flow is not only time-dependent, but also aperiodic, the roll structure is preserved. However, there is an irregular transverse oscillation of the rolls, as can be seen from the sequence of photographs in FIGURE 8, which are 1-sec exposures taken at 5-sec intervals. The aperiodic motion which is the subject of this paper is only weakly turbulent in the usual usage of this term.

Measurements of the time-dependent velocity field at a representative point within the fluid are shown in FIGURE 9 for four values of R/R_c. In each case, the velocity was sampled 8192 times at 2-sec intervals. Only a short portion of each velocity record is shown together with the velocity power spectrum. Time dependence first appears around $R/R_c = 17$, and is dominated in the interval $17 < R/R_c < 25$ by a single sharp spectral peak we shall call f_1, along with harmonics of this basic frequency. This feature is seen most clearly in FIGURE 9b ($R/R_c = 20.6$), where the peak is about 2.5 orders of magnitude above the broadband

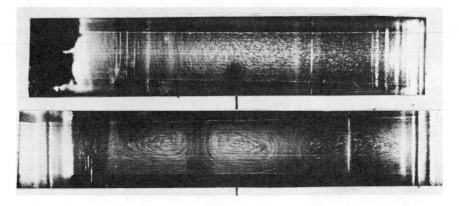

FIGURE 7. Two orthogonal views of the convection pattern at $R/R_c = 20$. The illuminated region in each case is a thin slice along a diameter of the cell. The edges are not visible due to the curvature of the cell walls. The difference between the two views indicates that the convection pattern is not axisymmetric.

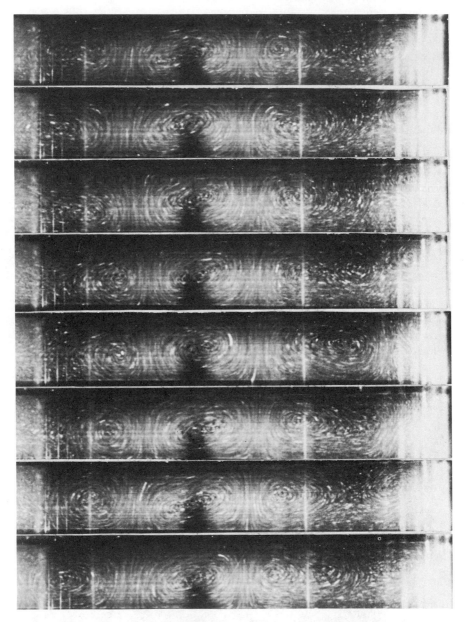

FIGURE 8. Time lapse sequence of photographs at 5-sec intervals when $R/R_c = 28$, showing irregular transverse oscillations of the convection rolls. Velocity measurements (FIGURE 9d) indicate that the fluid motion is actually aperiodic.

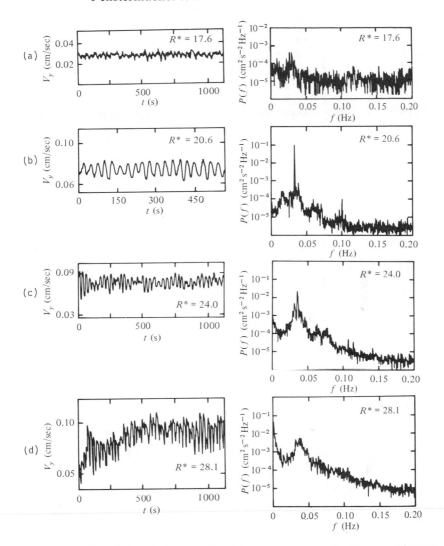

FIGURE 9. Portion of the velocity record and its power spectrum for the convection system when (a) $R/R_c = 17.6$, where the fluid is barely time-dependent; (b) $R/R_c = 20.6$, showing a strong periodic component f_1; (c) $R/R_c = 24.0$, where there is a substantial amount of broadband noise; (d) $R/R_c = 28.1$, where no periodic component is visible.

noise which surrounds it. This peak has a full width at half maximum which is less than 10^{-4} Hz, i.e., less than about 0.3% of the frequency of the peak. This linewidth, determined by the length of the data record, implies that the oscillation is coherent over hundreds of cycles. This is strong evidence for a strictly periodic state in time dependent convection. The small amount of broadband noise appears to be superimposed additively on the periodic component. Some of this broad-band noise usually accompanies even the first appearance of the periodic component, as in the spectrum at $R/R_c = 17.6$.

As R/R_c is increased beyond about 21, the amplitude of the periodic component declines, and the broadband noise increases in amplitude, until at $R/R_c = 28.1$, a periodic component is not visible above the noise (FIGURE 9d).

The integrated area and frequency of the periodic component f_1 as a function of R/R_c are shown in FIGURE 10a and FIGURE 10b, respectively. There are two branches in FIGURE 10b separated in frequency by about 10%. Sometimes the system will be found on one branch, and sometimes on the other. Transitions between the two branches seem to occur unpredictably. We do not understand

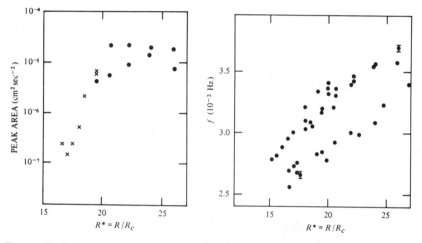

FIGURE 10. Integrated area of the periodic component f_1 in the convection experiment as a function of R/R_c. Because measurements could only be made where V_y did not change sign, it was necessary to use a different scattering volume for high and low values of R/R_c. This is indicated by the two symbols. However, peak areas at $R/R_c = 19.5$ for the two scattering volumes are approximately the same. (b) Frequency of the periodic component as a function of R/R_c, showing two distinct branches (see text).

the difference between these two closely spaced modes. The transverse oscillation may have a different wavelength in the two cases, or there may be two underlying spatial states of comparable stability.

The period of oscillation seen in these experiments is in semiquantitative agreement with that predicted by Clever and Busse.[21] However, no theory of the development of the broadband noise that overwhelms the periodic component presently exists. We conclude this section with the remark that our experiments on time-dependent convection are presently limited to one aspect ratio and one Prandtl number. It is possible that extensions of this work to other aspect ratios and Prandtl numbers will reveal qualitatively different phenomena.

COMPARISON OF THE TWO HYDRODYNAMIC SYSTEMS

The Couette flow and Rayleigh-Bénard convection experiments both show a sequence of bifurcations that culminates in an aperiodic ("weakly turbulent") flow. There are striking similarities in the behavior exhibited by these two systems, even though their hydrodynamic properties are rather different. Both systems show a bifurcation to a strictly periodic regime, a gradual growth of broadband noise, and a disappearance of the periodic component at high Reynolds or Rayleigh number. However, the disappearance of the periodic component in the convection experiments seems to be more gradual than in Couette flow.

No explicit solutions have been obtained that account for the observed behavior. The theory of differentiable dynamical systems may be able to account for the qualitative behavior, but at the present only the most general features are supported. For example, Ruelle and Takens[36] showed that chaotic behavior after only a few bifurcations is generic to a class of ordinary differential equations. They then suggest that such generic properties may apply to viscous hydrodynamics. Another approach is to construct finite dimensional models to describe the data, but some of the features of our results are not characteristic of models that have been studied to date.[22,23] The coexistence of sharp spectral peaks and broadband noise, which occurs in both Couette flow and Rayleigh-Bénard convection, is not contained in the current models. The continuous variation of the ratio ω_3/ω_1, or the absence of locking onto small integer ratios, is also uncharacteristic. Perhaps more sophisticated models including symmetry arguments will exhibit these features.

ACKNOWLEDGMENTS

We thank Dan Joseph for numerous stimulating discussions.

REFERENCES

1. CHANDRASEKHAR, S. 1961. Hydrodynamic and Hydromagnetic Stability, Oxford University Press, Oxford.
2. JOSEPH, D. D. 1976. Stability of Fluid Motions, Vols. 1 and 2. Springer. Berlin.
3. TAYLOR, G. I. 1923. Phil. Trans. R. Soc. (London) A 223: 289–343.
4. LORD RAYLEIGH. 1916. Phil. Mag. 32: 529–546.
5. GOLLUB, J. P. & H. L. SWINNEY. 1975. Phys. Rev. Lett. 35: 927–930.
6. SWINNEY, H. L., P. R. FENSTERMACHER & J. P. GOLLUB. 1977. *In* Synergetics, a Workshop. H. Haken, Ed. Springer. Berlin.
7. GOLLUB, J. P., S. L. HULBERT, G. M. DOLNY & H. L. SWINNEY. 1977. *In* Photon Correlation Spectroscopy and Velocimetry. H. Z. Cummins & E. R. Pike, Eds.: 425–439. Plenum. New York.
8. DONNELLY, R. J., K. W. SCHWARZ & P. H. ROBERTS. 1965. Proc. R. Soc. (London) A 283: 531–556.
9. DONNELLY, R. J. 1965. Proc. R. Soc. (London) A 283: 509–519.
10. KOSCHMIEDER, E. L. 1975. Adv. Chem. Phys. 32: 109–128.
11. SNYDER, H. A. 1970. Int. J. Non-Linear Mech. 5: 659–685.
12. GOLLUB, J. P. & M. H. FREILICH. 1976. Phys. Fluids 19: 618–626.

13. DAVEY, A. 1962. J. Fluid Mech. **14:** 336–368.
14. DAVEY, A., R. C. DIPRIMA & J. T. STUART. 1968. J. Fluid Mech. **31:** 17–52.
15. EAGLES, P. M. 1971. J. Fluid Mech. **49:** 529–550.
16. COLES. 1965. J. Fluid Mech. **21:** 385–425.
17. NORMAND, C. Y., Y. POMEAU & M. G. VERLARDE. 1977. Rev. Mod. Phys. **49:** 581–624.
18. SCHLÜTER, A., D. LORTZ & F. BUSSE. 1965. J. Fluid Mech. **23:** 129–144.
19. BERGÉ, P. 1975. *In* Fluctuations, Instabilities and Phase Transitions. T. Riste, Ed.: 323–352. Plenum. New York.
20. DUBOIS, M. 1976. J. Physique **37:** C1-137–C1-143.
21. CLEVER, R. M. & F. H. BUSSE. 1974. J. Fluid Mech. **65:** 625–645.
22. LORENZ, E. N. 1963. J. Atmos. Sci. **20:** 130–141.
23. MCLAUGHLIN, J. B. & P. C. MARTIN. 1975. Phys. Rev. A **12:** 186–203.
24. LIPPS, F. B. 1976. J. Fluid Mech. **75:** 113–148.
25. WILLIS, G. E. & J. W. DEARDORFF. 1965. Phys. Fluids **8:** 2225–2229.
26. WILLIS, G. E. & J. W. DEARDORFF. 1967. Phys. Fluids **10:** 931–937.
27. WILLIS, G. E. & J. W. DEARDORFF. 1970. J. Fluid Mech. **41:** 661 672.
28. BERGÉ, P. & M. DUBOIS. 1976. Optics Commun. **19:** 129–133.
29. OLSON, J. M. & F. ROSENBERGER. In press. Convective Instabilities in a Closed Vertical Cylinder Heated from Below, Part I: Monocomponent Gases.
30. AHLERS, G. 1974. Phys. Rev. Lett. **33:** 2285–2288.
31. AHLERS, G. 1975. *In* Fluctuations, Instabilities, and Phase Transitions. T. Riste, Ed.: 181–193. Plenum. New York.
32. BUSSE, F. H. & J. A. WHITEHEAD. 1974. J. Fluid Mech. **66:** 67–79.
33. KRISHNAMURTI, R. 1970. J. Fluid Mech. **42:** 309–320.
34. KRISHNAMURTI, R. 1973. J. Fluid Mech. **60:** 285–303.
35. CHARLSON, G. S. & R. L. SANI. 1975. J. Fluid Mech. **71:** 209–229.
36. RUELLE, D. & F. TAKENS. 1971. Commun. Math. Phys. **20:** 167–192.

DISCUSSION PAPER:
EXPERIMENTAL INVESTIGATIONS OF BEHAVIOR AND STABILITY PROPERTIES OF ATTRACTORS CORRESPONDING TO BURST PHENOMENA IN THE OPEN BELOUSOV REACTION

P. Graae Sørensen

Department of Chemistry
H. C. Ørsted Institute
University of Copenhagen
Universitetsparken 5
DK-2100 Copenhagen Ø, Denmark

INTRODUCTION

The Belousov reaction is a slow oxidation of certain organic compounds, e.g. $CH_2(COOH)_2$, by $KBrO_3$ in the presence of a catalyst, e.g. Ce^{IV}. In a closed system the oscillations proceed with decreasing amplitudes, and disappear after a time when the reactants are used up and the oxidation products have accumulated.

The loss of reactants can be compensated in a stirred tank reactor by an input flow of $KBrO_3$, $CH_2(COOH)_2$, and Ce^{IV}, and an output flow of the reaction mixture. Under these conditions true attractors of the system exist and are dependent on the experimentally controllable external flows. The attractors may be steady states, limit cycles, or attractors of more complex type.

This work describes a stirred tank reactor with instrumentation for investigating these phenomena, including computer-controlled burettes permitting the use of four independent time-dependent flows.

THEORY

A closed system of n chemical species A_ν, with concentrations c_ν is reacting by r different reactions with stoichiometric matrix N and reaction velocities $f(c)$. The differential equation is:

$$\frac{dc}{dt} = N f(c).$$

If the system exchanges chemical species with the surroundings by m flows with specific rates $j(t)$ (sec^{-1}), and concentrations C where $C_{\nu,\mu}(\nu = 1 \ldots n)$ is the concentration of species ν in flow μ, the differential equation is changed to:

$$\frac{dc}{dt} = N f(c) + C(t) \cdot j(t).$$

667

0077-8923/78/0316-0667 $1.75/1 © 1979, NYAS

The j_μ is negative for flows out of the system, and if

$$\sum_{\mu=1}^{m} j_\mu(t) = 0,$$

the volume is unchanged.

Necessary conditions for the applicability of this type of equation are that $C_\nu(t)$ be independent of the position in the vessel, and that the addition of reactants be done homogeneously through the whole volume. These conditions are in practice approximated by fast stirring of the fluid such that the mixing rate achieved by the turbulence is much greater than any of the reaction rates.

EXPERIMENTAL EQUIPMENT

The main part of the equipment is a stirred flow reactor with capability for simultaneous monitoring of the reactions by UV/visible light absorption and several types of electrochemical potentials (FIGURE 1).

The flow of the reactants is controlled by four pairs of two stepper motor-driven piston burettes. Solenoid-driven shift valves connect one of each pair of burettes to the reaction vessel and the other to a reactant reservoir, thus

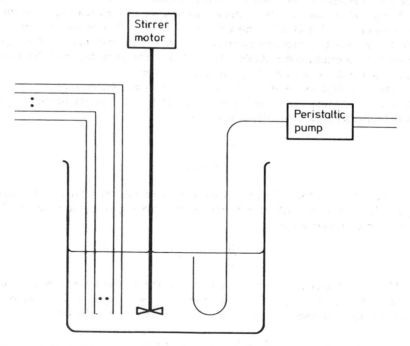

FIGURE 1. Reaction cuvette with feeding tubes, stirrer motor, and peristaltic suction device.

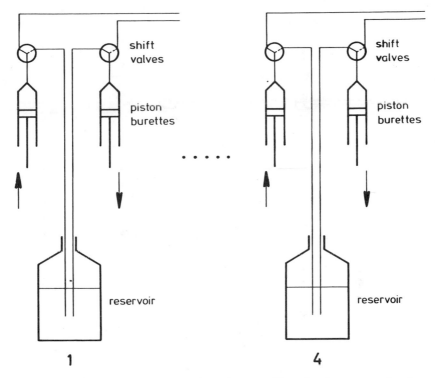

FIGURE 2. Liquid flow scheme for the piston burettes. When the pistons are going down, the burettes are connected to the reservoir. When the pistons are going up, the burettes are connected to the reaction vessel.

enabling one of the burettes to be filled from the reservoir while the other is feeding the reaction vessel (FIGURE 2).

During a run the volume in the reaction vessel is kept constant by sucking excess liquid from the vessel with a peristaltic pump. The pulses to the stepper motors and the currents to the shift valves are controlled by a computer, such that the pulse rates approximate the external input flow rates. $J_\mu(t) = V \cdot j_\mu(t)$ where V is the volume.

The software system is divided in two parts, one residing in a mcs8008 microcomputer and the other in a medium size multiprogrammed RC4000 computer (FIGURE 3). The mcs8008 receives a string of four bit words from the RC4000. For each interrupt from a programmable pulse generator, the burette that has a one in the dataword is moved one step. The mcs8008 counts the number of pulses sent to each burette. When one burette of a pair is empty, the following pulses are sent to the other burette, the shift valve of the first burette is changed to the reservoir position, and the first burette is filled to be ready next time the other burette is empty. In this way the flow can continue as long as necessary.

The bit strings for the four flows are computed by an algol program in the

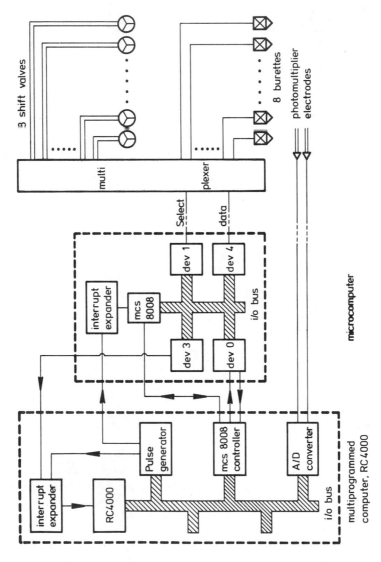

FIGURE 3. Signal connections between the multiprogrammed RC4000 computer, the mcs8008 microcomputer, the stepper motors, and the shift valve solenoids.

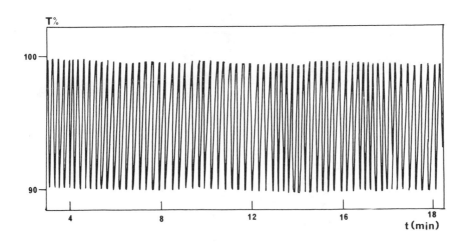

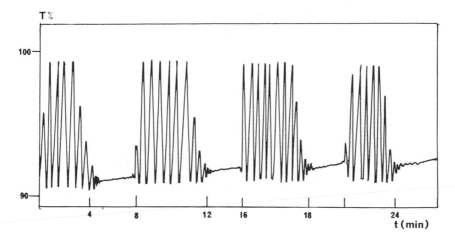

FIGURE 4. Percentage of light transmission T at 330 nm with flow compositions

A_1 = $KBrO_3$ C_{11} = 0.09 M
A_2 = $CH_2(COOH)_2$ C_{22} = 0.5 M
A_3 = $MnSO_4$ C_{33} = 0.0008 M
A_4 = H_2SO_4 C_{41} = C_{42} = C_{43} = 3 N

and flow rates

A: $j_1 = j_2 = j_3 = 0.00050$ sec^{-1}
B: $j_1 = j_2 = j_3 = 0.00072$ sec^{-1}

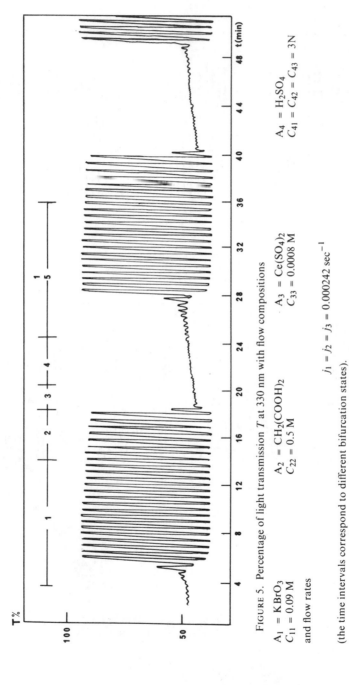

FIGURE 5. Percentage of light transmission T at 330 nm with flow compositions

A_1 = KBrO$_3$ A_2 = CH$_2$(COOH)$_2$ A_3 = Ce(SO$_4$)$_2$ A_4 = H$_2$SO$_4$
C_{11} = 0.09 M C_{22} = 0.5 M C_{33} = 0.0008 M C_{41} = C_{42} = C_{43} = 3N
and flow rates

$$j_1 = j_2 = j_3 = 0.000242 \ \text{sec}^{-1}$$

(the time intervals correspond to different bifurcation states).

RC4000 from mathematical expressions for the volume flows $J_\mu(t) = V \cdot j_\mu(t)$, the burette volumes, and the interrupt rate. The algol program runs under a time-shared operating system, and the mcs8008 contains a data-buffer to smooth the swapping-out periods in the operating system.[3]

The analog output from the photomultiplier and the electrodes is displayed on a recorder, but can also be sampled directly by the RC4000 and stored on the backing storage for later processing.

Thus the controllable parameters are the four different input flow rates $j_1(t) \ldots j_4(t)$. The output flow rate $j_5(t)$ is given by the condition

$$j_5(t) = -\sum_{\mu=1}^{4} j_\mu(t).$$

Also the concentrations of the reactants in the input flow can be varied. If the μth flow contains the species ν_1, the differential equation for ν_1 is:

$$\frac{dc_{\nu_1}}{dt} = \sum_{\rho} N_{\nu_1,\rho} f_\rho(\underline{c}) + C_{\nu_1,\mu} j_\mu(t) + C_{\nu_1} j_5(t)$$

EXPERIMENTS

If the Belousov reaction takes place in an open system with constant input and output flow rate, the system may show burst oscillations.[2] Examples are shown in FIGURE 4 and FIGURE 5. The flow parameters are indicated in the figures by species numbers (ν in A_ν) and flow numbers (μ in j_μ). The concentration in the input flow is given by indicating those values of elements in the C matrix that are different from zero.

That the burst phenomena are not caused by some special property of the cerium compounds is supported by FIGURE 4, where $MnSO_4$ has been substituted for $Ce(SO_4)_2$.

To get a survey of the dynamic features of the burst attractor, we have used the capability of the setup to generate the time-dependent input flows. In FIGURE 6 are shown the transmission curves of an experiment where the j_3 flow is periodic, with period P approximating the period of the damped oscillations for perturbations around the pseudosteady state in the nonoscillatory phase. The figures show forced oscillations around the pseudosteady state. After some time the system again jumps to the self-oscillatory state, and in this state the oscillating input flow has only a small effect.

CONCLUSION

The foregoing experiments suggest that the burst attractor is composed of two independent but interacting pseudoattractors, a pseudolimit cycle, and a pseudosteady state with the stability properties given in TABLE 1.

In stages 2 and 4 it is possible to perturb the system from the current stable state to the other stable state.

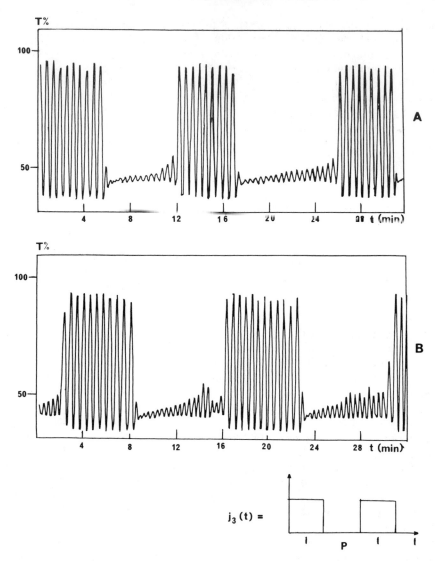

FIGURE 6. Percentage of light transmission T at 350 nm with flow compositions

$A_1 = KBrO_3$		$C_{11} = 0.09$ M	
$A_2 = CH_2(COOH)_2$		$C_{22} = 0.5$ M	
$A_3 = Ce(SO_4)_2$		$C_{33} = 0.0008$ M	
$A_4 = H_2SO_4$		$C_{41} = C_{42} = C_{43} = 3N$	

and flow rates

$$j_1 = j_2 = \overline{j_3(t)} = 0.000269 \text{ sec}^{-1}$$

($j_3(t)$ showing the time variation in flow 3)

$$A: P = 23.4 \text{ sec}$$
$$B: P = 27 \text{ sec}$$

TABLE I

STABILITY PROPERTIES OF PSEUDO-LIMIT-CYCLES AND PSEUDO-STEADY-STATES

Time Mark (FIGURE 5)	Pseudo-Limit-Cycle	Pseudo-Steady-State
1	Stable*	Unstable focus
2	Stable*	Stable focus
3	Unstable or bifurcating in other invariant subsets	Stable focus*
4	Stable	Stable focus*
5	Stable*	Unstable focus

*Current state.

The main advantage of the flow system with time-dependent flows is that the system can be varied in the neighborhood of critical phases and studied for long time periods, thus minimizing the risk of confusing statistical fluctuations with regular patterns.

REFERENCES

1. FIELD, R. J., E. KÖRÖS & R. M. NOYES. 1972. J. Am. Chem. Soc. **94:** 8649.
2. Faraday Symp. Chem. Soc. 1974. **9:** 88–92.
3. LINDGARD, A., P. GRAAE SØRENSEN & J. OXENBØLL. 1977. J. Phys. E (Sci. Instrum.) **10:** 264.

DYNASIM: EXPLORATORY RESEARCH IN BIFURCATIONS USING INTERACTIVE COMPUTER GRAPHICS

Ralph H. Abraham

Department of Mathematics
University of California
Santa Cruz, California 95064

HISTORICAL BACKGROUND

The crucial early experiments regarding bifurcation theory and applications as the experimental branch of differentiable dynamics may be described in three overlapping periods. The *period of direct observation* may be much older than we think, but let us say it begins with the musician, Chladni, contemporary of Beethoven, who observed bifurcations of thin plate vibrations. Much can still be learned from his work, painstakingly reproduced by Waller.[1] Analogous phenomena discovered in fluids by Faraday are still actively studied.[2-4] These experiments, so valuable because the medium is real, suffer from inflexibility—especially in choosing initial conditions.

The next wave of bifurcation experiments, which I shall call the *analog period*, begins with the triode oscillator. The pioneering works of van der Pol, with improvements by Hayashi, produced a flexible analog computer, and institutionalized the subharmonic bifurcations. These devices offer exceptional speed of convergence, but even with the recent development of modular electronics, only a limited class of dynamical systems are tractable.

The development of the early computing machines ushered in the *digital period*. Well-known numerical methods were implemented from the start, and graphical (CRT) output began to appear in the literature by 1962. The pioneer papers of Lorenz,[5] and Stein and Ulam,[6] are still studied. By 1967, the Association for Computing Machinery recognized this new field with a symposium entitled "Interactive Systems for Experimental Applied Mathematics."[7] Special systems for experimental math that have evolved since the Culler-Fried device of 1961 are fully described by Smith.[8] The current state of the art is now readily available in the form of a very large general purpose computer with BASIC or APL language and a color video graphics terminal. The currently available terminals of this type are listed in TABLE 1.

An equivalent, less expensive system would replace the large computer with a minicomputer, and a fast array processor. Such systems now exist at several institutions.

These devices are extremely flexible, accommodating a very wide class of dynamical systems, but suffer from the cost/resolution quandary: high resolution implies either a vast machine (high capital costs) or long run times (high operating costs.)

Our experiences over the past three years with forced oscillation machines in all three categories (direct observation of fluids, analog systems, and digital com-

676

0077-8923/78/0316-0676 $1.75/1 © 1979, NYAS

puter graphics) has resulted in the design of a fast, economical, special purpose digital computer graphic device, the *dynasim device,* which is described here.

CONVENTIONAL COMPUTER GRAPHIC TECHNIQUES

The dynasim device is designed around a new algorithm for dynamical systems, the *push-pull algorithm,* which is described in the next section. Here, for background, we describe the techniques now in use with general purpose computer graphic hardware to draw the phase portrait of a dynamical system.

To fix for once and for all the goal of these techniques in bifurcation research, we now specify the context. Let C and P be manifolds; $\mathfrak{X}(P)$, the space of smooth

TABLE 1

COLOR VIDEO GRAPHIC DISPLAY SYSTEMS

Manufacturer	Model	Resolution	Max. Depth
Aydin Controls	5212/5214A	240 × 256	12
Fort Washington, Pa.		512 × 640	12
DeAnza Systems	IEC 2212	256 × 256	12
Santa Clara, Cal.			
Genisco Computers	GCT 3000	256 × 256	16
Irvine, Cal.		512 × 512	16
		1024 × 1024	16
Grinnell Systems	GMR-26	256 × 256	32
Santa Clara, Cal.		512 × 512	8
Interpretation Systems	VDI	256 × 256	32
Overland Park, Kansas		512 × 512	8
Lexidata	200-D	256 × 256	3
Burlington, Mass.			
Ramtek	6000	256 × 512	3
Sunnyvale, Cal.	9000	256 × 256	24
		512 × 512	6

vector fields on P; and $\mu: C \rightarrow \mathfrak{X}(P)$ a family of vector fields, with smooth graph map

$$\Gamma_\mu: C \times P \rightarrow TP: (c, p) \mapsto \mu(c)(p).$$

We want to know for each control value, $c \in C$, the complete phase portrait of $X_c = \mu(c)$, a dynamical system (vector field) on the phase space, P. But machine computation can never reveal the asymptotic motions (ω-limit sets) of probability zero (separatrices, basic sets of hyperbolic type) so we ask only for the attractors, their basins, and (optionally) the phase foliation of each basin into isochrons (stable manifolds) of points of the attractor. As the control parameter, $c \in C$, can only be sampled at a finite set of points, our problem is just to *draw the ABP portrait* (attractors, basins, and phase foliations) of a fixed vector field. To be reasonable, let $P \subset R^n$ be an open box,

$$P = I_1 \times \cdots \times I_n, \quad I_\alpha = (a_\alpha, b_\alpha) \subset R.$$

We now describe the conventional algorithm, using color video graphic output from a general purpose computer, for the ABP portrait:

(1) Write a vector field, X, into the program.

(2) Define a finite set of initial points in P, such as a uniform grid, and a time duration, $t_0 > 0$.

(3) Draw the integral curve forward from each until $t = t_0$, using the Runge-Kutta or Adams-Moulton formula.

(4) Continue until $t = kt_0$ if necessary, until all ω-limit sets are identified in white (PUSH).

(5) By manual (interactive) graphic input, define a new grid of initial points near the attractors, variously colored,

(6) Integrate retrograde $(-X)$ until $t = kt_0$ (PULL).

(7) Observe the ABP portrait, with basins and isochrons (if any) in different colors.

(8) Repeat in a smaller domain if needed, for higher resolution.

Our experience with this algorithm, even for a single basin system such as the van der Pol equation and with a very convenient interactive program (ORBIT, written by R. Palais) indicates that most of a day is required for a single portrait. A dedicated computer, and fast array processor, would reduce this interval to perhaps an hour. Important discoveries revealed by such experiments are understandably rare. Subharmonic resonance, for example, might still be unobserved if analog devices were not available. Note that if $C \subset R^2$ (say force and amplitude) and $P \subset R^2 (X_c =$ forced Duffing or van der Pol equation) the flow

$$F: C \times P \times R \to P$$

was a domain of five-dimensions, so with a resolution of $w = 2^b$ bits for real numbers, $w^5 = 2^{5b}$ calculations are necessary. To shorten the time requires parallel processing (more calculations per unit time) or a faster processor (there is a limit) and increased device cost, either way. Doubling the resolution increases the cost by a factor of $2^5 = 32$, or lengthens the time of computation by the same factor.

This cost/resolution or speed/resolution problem is the main obstacle to effective experimental work in bifurcations. Yet many conjectures of bifurcation theory, and of differentiable dynamics, could be furthered by such experiments. See, for example, the problem set of Palis and Pugh.[9]

THE PUSH-PULL ALGORITHM

This technique, inspired by our studies of the video feedback phenomenon,[10] is based on the induced action of a map upon subsets or functions.

If $\varphi: P \to P$ and $f: P \to R$, let $\varphi^*(f) = f \cdot \varphi$ denote the *pull-back* of f by φ. Then $\varphi^*: R^P \to R^P$ is the induced action on real-valued functions on P, and likewise, $\varphi^*: 2^P \to 2^P: A \mapsto \varphi^{-1}[A]$ pulls back subsets. Considering only functions with discrete values, say z possible values, we have $\varphi^*: z^P \to z^P$, which pulls back partitions of P into z disjoint subsets. If $\varphi: P \to P$ is bijective, then we also have a push-forward action on functions, $\varphi_* = (\varphi^{-1})^*$. For example, $\varphi_*: 2^P \to 2^P: A \mapsto \varphi[A]$ takes disjoint sets into disjoint sets.

If $X \in \mathfrak{X}(P)$ is a vector field on P, choose a common (for all initial points) step-size, h, for numerical integration, fix $t = nh$ for some (large) integer n, and compute the flow diffeomorphism $\varphi_t \cong (\varphi_h)^n$ by the chosen formula, say Runge-Kutta. Rather than integrating separate initial points ad infinitum, our idea is to calculate the graph of φ_t at a fixed grid (as fine as memory resources allow) and store the graph as a look up table. Then this map, $\varphi = \varphi_t$, will be iterated (with its approximate, calculated inverse, $\psi \cong \varphi_{-t}$) as a cascade, without further numerical integrations, to draw the phase portrait, coarsely, of X. Dispensing with initial points, we choose a partition $f \in z^P$ of P into z disjoint sets. Typically, $z = 8$ or 16, in a practical trial. In fact, f is visualized (with a color video graphic terminal) as a colored map (partition) of P. Now push forward with φ, look at the image map $\varphi_* f$, and iterate. This is the *push cascade*. A few trials, with different partitions f, reveals the most probable attractors after n iterates, say. Then a new partition g is chosen by inspection, covering each attractor with a different color. Iteration of the *pull cascade*, $\psi_* g \cong \varphi^* g$, for n iterates (or more) colors the basins in different colors. This is the partition desired, as the goal of the ABP portrait. Finally, one attractor may be covered by small boxes of the various colors, in a new partition h. Iteration of the pull-back cascade, $\Psi_* h$, reveals (roughly) the phase foliation in the basin of the chosen attractor. This can be accomplished for all the basins, in a single step, yielding the ABP portrait.

Comparison of this algorithm with the conventional (orbit-by-orbit) technique described above shows that they are identical, except for the order of steps, and the more important distinction: repeated numerical integration (serial calculations) is replaced by storage of the graph (serial access to mass memory).

THE DYNASIM DEVICE

The push-pull algorithm sacrifices resolution in favor of speed. Implementation of the algorithm in conventional general purpose computers requires an enormous core, or cache memory, rarely available to mathematicians. Special purpose, fast, random-access memory (RAM) is available as a peripheral device for general purpose computers, but is very expensive at present. An ideal example is the Array Processor (of Floating Point Processors, Inc., Portland, Oregon.)

The idea of the dynasim device is to read the graph of the transformation, φ, always in the same serial order. Thus, serial access memory (SAM) can be used in place of RAM for the graph memory. Current SAM units (CCD) are much less expensive than RAM, and this inequality is expected to increase in the coming years. Even with the more expensive RAM storage, the dynasim device is substantially less expensive than general purpose hardware.

The dynasim device, illustrated in FIGURE 1, is a peripheral device for a general purpose computer, to which it may be serially interfaced, and made of existing

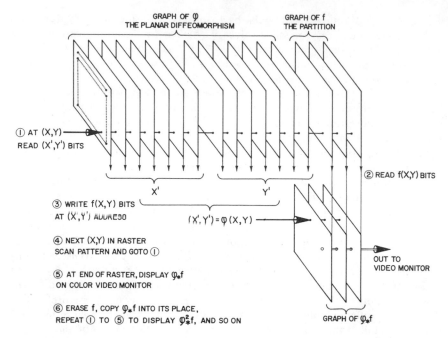

GRAPH OF φ
THE PLANAR DIFFEOMORPHISM

GRAPH OF f
THE PARTITION

① AT (X,Y)
READ (X',Y') BITS

② READ f(X,Y) BITS

X'

Y'

③ WRITE f(X,Y) BITS
AT (X',Y') ADDRESS

$(X',Y') = \varphi (X,Y)$

④ NEXT (X,Y) IN RASTER
SCAN PATTERN AND GOTO ①

OUT TO
VIDEO MONITOR

⑤ AT END OF RASTER, DISPLAY $\varphi_* f$
ON COLOR VIDEO MONITOR

⑥ ERASE f, COPY $\varphi_* f$ INTO ITS PLACE,
REPEAT ① TO ⑤ TO DISPLAY $\varphi_*^2 f$, AND SO ON

GRAPH OF $\varphi_* f$

FIGURE 1. Schematic of the dynasim device, 128 × 128 pixels by eight colors.

hardware modules of computer graphics technology. It contains a number of parallel planes of memory, each a bit matrix corresponding to the domain P. For simplicity, let P be a two-dimensional unit square, discretized into a $w \times w$ array. Then each memory plane in the dynasim device is a $w \times w$ array. Suppose $w = 2^x$. Then the graph of φ is stored in $2x$ planes. If $\varphi(i,j) = (i',j')$, $i,j,i',j' = 0, \ldots, 2^x - 1$, then x planes store the i' in binary, and the remaining x planes store j' in binary. An additional three planes store the partition (color) $f(i,j)$ if $z = 2^3$. Finally, three more planes record the push-forward, $\varphi_* f$, as it is computed, point-by-point. These last are read by a conventional color video graphic imaging device, creating a visible copy of $\varphi_* f$. In fact, the entire device can be built within a video graphic image device, if it has a large enough card cage to contain the memory planes, $2x + 6$ planes of $2^x \times 2^x$ bits. Commercially available devices are listed in TABLE 1, and the memory size and costs of the dynasim device, for various x values, are tabulated in TABLE 2.

To iterate φ_*, the f planes are erased, f and $\varphi_* f$ memories swapped (electronically), and another routine video raster scan proceeds. Each video frame (1/30th second) executes an iterate of the push (or pull) cascade.

In practice, the dynasim device would implement the push-pull algorithm as follows:

(1) The vector field is written into the numerical integration program in the core of the computer (CPU).

(2) The domain box is written in, and the dynasim resolution, $w \times w$. The program calculates the array of initial points (x_i, y_j).

(3) The CPU program calculates $\varphi_t(i,j) = (i',j')$ and $\varphi_{-t}(i,j) = (i'',j'')$, stores these graphs in arrays in its mass storage, and copies the graph of $\varphi = \varphi_t$ into the dynasim device.

(4) The operator indicates a partition f with a data tablet, light pen, or cursor. The CPU loads the partition into the dynasim device. At "PUSH", the device iterates the push cascade until "STOP." Repeat with another trial partition if necessary.

(5) The operator describes a new partition g (or h) to the CPU with interactive graphics input device.

(6) The CPU downloads g into the dynasim device, and the inverse map φ_{-t}. At "PULL," the device iterates the pull cascade until "STOP."

(7) Observe the AB (or ABP) portrait, with basins and isochrons (if any) in different colors.

(8) Repeat in smaller domain if needed, for higher resolution.

As in the conventional algorithm described previously, the output portrait can be stored in CPU mass storage, local floppy disc, videotape with verbal annotations, or color photograph. The time required for a successful portrait of a single vector field in two-dimensions can be estimated. Steps (3) and (6) require transfer of a large array, of $2^x \times 2^x \times 2x$ bits. If the CPU/Dynasim interface is serial, at 9600 BAUD, and $x = 8$, then about two minutes are required for downloading, in each step. The other steps take a few seconds, only.

And not only is this many times faster than the conventional algorithm for a single vector field, but if a bifurcation is being explored, the steps can be overlayed. Thus, the CPU integrates X_2, while the X_1 flow is being downloaded, and so on.

While speed has been obtained at the cost of resolution, the accumulation of round off errors will not indicate false attractors, but only miss some, and

TABLE 2

DYNASIM MEMORY REQUIREMENTS

EIGHT COLOR VERSION

2-Dimensional Resolution (pixels)	3-Dimensional Resolution (pixels)	Memory Size Per Plane (Cube) (kilobytes)*	Total Memory Cost	
			(RAM/RAM)†	(RAM/SAM)‡
128 × 128	25 × 25 × 25	2	\$ 2,000	\$ 1,020
180 × 180	32 × 32 × 32	4	4,200	2,100
256 × 256	40 × 40 × 40	8	8,800	4,320
360 × 360	50 × 50 × 50	16	18,400	8,880
512 × 512	64 × 64 × 64	32	30,400	18,240
720 × 720	80 × 80 × 80	64	80,000	37,440
1024 × 1024	100 × 100 × 100	128	166,400	76,800

*Kilobyte = 2^{13} = 8192 bits.
†At \$50/kilobyte.
‡At \$15/kilobyte.

misrepresent the shape of basins through the well studied phenomenon of *aliasing*. We now turn to this problem, common to all simulation schemes.

DECIDABLE STABILITY

Suppose that a dynasim device is available—or a general purpose computer fast enough to execute the push-pull algorithm—with an interactive color video graphic terminal for graphic input and output. If a vector field is entered, an ABP portrait is returned, showing calculated attractors, basins, and phase foliations. No matter how brightly colored, the picture is but a pale ghost of the time phase portrait. We have formula error and round-off error for the diffeomorphism φ. As the round-off error dominates in the context of the practical design characteristics described above, we may forget formula error altogether. The round-off error for φ is reducible by zooming in: a small piece of the domain is restudied, using all available memory. But this will still miss long thin basins of attraction. We will expect in practice:

(1) The discrete φ is many-to-one, not onto.
(2) The $\varphi * f$ partition gets errors on the boundaries of the colored regions.
(3) The image $(\varphi *^n f)$ shrinks with increasing iteration.
(4) The screen eventually goes black, except for twinkling attractors.

As not all the qualitative features of a dynamical system are decidable (by machine) we seek now *decidable versions of stability and bifurcation*.

Firstly, we take account of the fact that only relatively probable attractors are machine discoverable.

Let M be a manifold; $X \in X(M)$, a vector field on M with complete flow; and $\{A_i \mid i \in I\}$, the set of all topological attractors of X. Let B_i denote the maximal basin of A_i, $B = \cup\{B_i \mid i \in I\}$, and $S = M \setminus B$, the *separatrix* of X.

Now suppose M has a uniform topology, let S^ϵ be the ϵ neighborhood of S for $\epsilon > 0$, and $B^\epsilon = M \setminus S^\epsilon$. Then $B_i^\epsilon = B_i \cap B^\epsilon$ is the *ϵ-reduced basin* of A_i.

Given two vector fields, $X, Y \in X(M)$, and $\epsilon > 0$, we say X and Y are *ϵ-basin equivalent* if $X \mid B_X^\epsilon$ and $Y \mid B_Y^\epsilon$ are topologically equivalent.

Now this looks more machine approachable, as the fine structure at the separatrix (homoclinic cycles, etc.) has been pruned away. Further, many small basins of attraction (infinitely many, no doubt) have been thrown out with the ϵ-neighborhood of the separatrices. Yet many small basins remain (perhaps too small to be machine discoverable) that are nowhere near the separatrix. So now, we will throw these out as well.

Let M now have a probability measure μ, and for simplicity suppose that μ and the uniformity of M are both derived from a common metric (distance) function. Without assuming compactness, we may suppose that the volume of an ϵ-disk, $\mu[N_\epsilon(m)]$ for any $m \in M$, is bounded below, by $k\epsilon^d$, where $k > 0$ is a constant, and $d = \dim(M)$.

Now suppose, for the sake of discussion, that all basins are measurable. The probability of attractor A_i is then $p_i = \mu(B_i)$, and $p_i^\epsilon = \mu(B_i^\epsilon)$ is substantially smaller, perhaps even zero. Our discrete algorithm can discover an attractor only if its ϵ-reduced basin is sufficiently large. Thus, let

$$I^\epsilon = \{i \in I \mid p_i^\epsilon > k\epsilon^d\}$$

and $D^\epsilon \subset B^\epsilon$ be the union of the discoverable basins,

$$D^\epsilon = \cup \{B_i \mid i \in I^\epsilon\}.$$

At last, we say $X, Y \in X(M)$ are ϵ-decidably equivalent if $X \mid D_X^\epsilon$ and $Y \mid D_Y^\epsilon$ are topologically equivalent.

Likewise, $X \in X(M)$ is ϵ-decidably-stable if every Y sufficiently close to X is ϵ-decidably-equivalent, and a family of vector fields $\{X_\mu \in X(m) \mid \mu \in R^c\}$ has an ϵ-decidable bifurcation if not all its members are ϵ-decidably-equivalent.

Here, ϵ is not to be chosen arbitrarily small, but is a fixed fraction of the width of the domain, a characteristic of the simulation machine at hand.

CONCLUSION

We have described here a decidable version of the stability and bifurcation concepts of differentiable dynamics, an algorithm for exploring the decidable bifurcations in low dimensions using interactive computer graphics, and an inexpensive special purpose computer for implementation of the algorithm. The design characteristics of this system may now be correlated as follows: The cost of the device, C, is proportional to w^d, where w^d is the number of bits in the width of the domain, and d is the dimension. If $\epsilon - 1/w$, attractors of reduced probability ϵ^d can be resolved. Thus, the resolution is inversely proportional to the cost. The smaller ϵ, the more basins discovered, and the more bifurcations observed, especially, in the neighborhood of the separatrix (where the principal action is.)

We propose that dynasim devices are useful not only in bifurcations research, to develop theory, but also in applications, as fast, graphic presentation of the most probable equilibrium states is one of the goals of the qualitative theory.

REFERENCES

1. WALLER, M. D. 1961. Chladni Figures: A Study in Symmetry. G. Bell. London.
2. BROOK BENJAMIN, T. & F. URSELL. 1954. Proc. R. Soc. London A **225:** 505–517.
3. ABRAHAM, R. 1976. The macroscopy of resonance. *In* Structural Stability, The Theory of Catastrophes, and Applications in the Sciences. P. Hilton, Ed.: 1–9. Springer-Verlag. New York, N.Y.
4. MARSDEN, J. E. & M. MCCRACKEN. 1976. The Hopf Bifurcation and its Applications. Springer-Verlag. New York, N.Y.
5. LORENZ, E. N. 1962. The statistical predictions of solutions of dynamic equations. Proc. Internat. Symp. Numerical Weather Prediction, Tokyo: 629–635.
6. STEIN, P. R. & S. ULAM. 1964. Nonlinear transformation studies on electronic computers. Rozprawy Mat. **39:**66.
7. KLERER, M. & J. REINFELDS, Eds. 1968. Interactive Systems for Experimental Applied Mathematics. Academic Press. New York, N.Y.

8. SMITH, L. B. 1970. A survey of interactive graphical systems for mathematics. Computing Surveys **2**: 261–301.
9. PALIS, J. & C. PUGH. 1975. Fifty problems in dynamical systems. *In* Dynamical Systems—Warwick 1974. A. Manning, Ed.: 345–353. Springer-Verlag. New York, N.Y.
10. ABRAHAM, R. 1976. Simulation of cascades by video feedback. *In* Structural Stability, the Theory of Catastrophes, and Applications in the Sciences. P. Hilton, Ed.: 10–14. Springer-Verlag. New York, N.Y.

CLOSING REMARKS

Okan Gurel

IBM Corporation
White Plains, New York 10604

The Conference on Bifurcation Theory and Applications in Scientific Disciplines was one of those rare occasions where researchers from different disciplines focused their attention on a mathematical topic and discussed both common interests and basic differences. It is now time to proceed in our own directions with not only new ideas but also a warm feeling that our scholarly relationships with our colleagues are reinforced by stronger friendships. We all are grateful to The New York Academy of Sciences for making this possible.

This is also a special event for all of us to have the honor of Professor Eberhard Hopf's participation at this conference on bifurcations, a field in which he played a key role in bridging the past to the present. On the occasion of his 75th birth year, it is our privilege to dedicate the proceedings of this conference to Professor Eberhard Hopf, with many thanks for providing us with inspiration.

Although the hours were long, stretching from early mornings to late evenings, we felt exhausted but never tired. We covered the history, the present, and the future trends. We noticed that some concepts were clearly defined and analyzed, while there were others yet to be answered and even to be determined.

The sessions were bounded by the calendar only, otherwise overlapping at many instances with common characteristics. In the sessions on the theory of bifurcations, topological approaches, algebraic considerations, and analytical methods were reviewed, presented, and suggested as the theory stands today. At times new refinements on the old theoretical considerations were introduced. In the application areas, certain models analyzed by various research groups were presented, current shortcomings and difficulties were discussed. In addition, results obtained in some experimental studies were outlined. In chemistry, physics, and biology, as basic sciences, and in medicine and engineering as applied sciences, possible bifurcation analysis of known or recently constructed models were compared with each other. Deterministic and stochastic systems, continuous and discrete models, ordinary and partial differential equations, or functional equations were shown to play a fundamental role in analyzing actual systems by referring to bifurcations. A new dimension in application areas, problems of both ecology and economics were discussed as important topics related to interacting species.

While the debate on the precise definition of "bifurcation" was not quite settled in the minds of many researchers, other phenomena such as "chaos" and the creation of "strange attractors" were discussed in not one but many papers presented in diverse sessions. This will undoubtedly flourish in the days ahead. It is clear that the theory will be expanded in the directions incorporating these new objects; how they enter into the bifurcation analysis and how significant they are

0077-8923/78/0316-0685 $1.75/1 © 1979, NYAS

in understanding various phenomena observed in different disciplines experimentally will be discussed further.

We wish to express our gratitude to and would like to thank once again our sponsors for supporting our efforts and our colleagues for chairing, speaking, or participating by joining in discussions and interacting during the conference. This conference owes its success to each of you.

Author Index

(Boldface page numbers refer to an author's paper appearing in this volume. Lightface page numbers refer to reference lists where an author's work has been cited.)

Abraham, R. H., **676**, 683, 684
Adler, R., 120
Agmon, S., 277
Ahlers, G., 666
Akazawa, T., 637
Alekseev, V. M., 117
Alexander, J., 126, 406
Alexandroff, P., 38
Alling, D. W., 233
Al-Rawi, A., **590**, 603
Amundson, N. R., 330
Anderson, J. R., 235
Andronov, A. A., 4, 25, 622
Antman, S., 4, 570
Aris, R., 278, **314**, 331
Arnold, V. I., 406, 462, 622
Aronson, J. D., 506
Ashkenazi, M., 77
Asimov, D., 107
Athens, J. W., 235
Auchmuty, J. F. G., 62, 77, 250, **263**, 278, 313
Auslander, D., 529
Auslander, J., 107

Babloyantz, A. 313, 451
Baladi, J. V., 406
Balakhovskii, I. S., 295
Balasko, Y., 548
Ballow, M., 233
Balslev, I., 637
Bar-Eli, K., 294
Barkin, S., 294
Bass, G. E., 187
Bauer, L., 313
Bazley, N. W., 177
Beauchamp, J. J., 235
Beddington, J. R., 529
Bedeaux, D., 452
Belousov, B. P., 294, 637
Benjamin, T. B., 167
Benettin, G., 415
Benson, S. V., **638**
Benton, R., 234
Bergé, P., 666
Berger, J. M., 463
Berlinski, D., 570, 589
Bernard, P., 4
Berry, M. V., 570, 571

Betz, A., 637
Bhatia, N., 544
Bilous, O., 331
Bimpong-Bota, E. K., 356
Birkoff, G. D., 117, 392
Birman, J. L., 63
Bishop, R. E., 603
Bixon, M., 294
Blevins, R. D., 604
Blumenthal, R., 313
Boggs, D. R., 233
Boggs, S. S., 233
Boiteux, A., 212, 213
Borckmans, P., 262
Borner, G., 233
Bornmann, L., 294
Bowen, R., 126, 415
Bray, W. C., 294, 637
Brenig, L., 262
Breslow, R., 249
Broda, E., 249
Brolin, H., 85
Brook Benjamin, T., 683
Brown, H. W., 233
Bruice, T. C., 294
Buchner, M., 622
Buckingham, E., 187
Budiansky, B., 570
Budkin, A., 234
Buneman, O. P., 480
Bunting, P. S., 637
Burlington, H., 233
Busch, H., 451
Busse, F. H., 63, 666
Busse, H. G., 213, 294

Cairns, S. S., 117, 392
Canale, R. P., 489
Caplan, S. R., 313
Carandente, F., 233
Carberry, J. J., 331
Cassie, R. M., 505
Casting, C., 544
Cesari, L., 26, 42, 249
Champsaur, P., 543
Chanana, A. D., 233
Chance, B., 637
Chance, E. M., 249
Chandraskhar, S., 665

687

Subject Index

$\mathbf{A}$A (aplastic anemia), 222, 223, 230
Abelian group, 46
Accessibility to stability of weak areas, 106
Adaptability, 468–70
   operational definition of, 468–69
   structure of, 469–70
Adaptability theory, ecosystem stability and, 465–81
   bootstrap principle of theory, 470–72
   dynamical autonomy and role of unrepresented adaptabilities, 472–74
   formalism of hierarchical adaptability theory, 466–70
   functional significance of stability concepts, 475–79
Adaptation of phenotypes, 186
Aliasing phenomenon, 682
Alternative stable states
   ecological systems with, 517–18
   hysteresis and, 246
Analog period of bifurcation experiments, 676
Anemia, aplastic, 222, 223, 230
Anisotropic Kepler problem, homoclinic orbits in, 108
Anticlinal bifurcation points, 566, 575
Aplastic anemia (AA), 222, 223, 230
Apnea, 215, 217
Apneic patterns of respiration (periodic breathing), 215
Arrhythmias, 230–231
ASOP reaction (autocatalytic, second-order in products reaction), 236–38, 242, 244, 246
Asymmetry
   origin of molecular, 242–43
   See also Symmetry
Asymptotic measures of dynamic systems, 411–13
Attractors
   Axiom A, 126
   behavioral and stability properties of, corresponding to burst phenomena in open Belousov reaction, 667–75
   chaotic, 352
   elementary, 350
   fractals, fractal dimension and, 463–64
   Lorenz, 42, 121, 353, 370, 379, 393–99
   multiple dimension, 352
   multiply periodic, 352
   one-dimensional (limit cycles), 352
   periodic, 124
   perturbed, *see* Perturbed attractors
   strange, 84, 124, 379, 401, 463, 618

Autocatalysis, pattern formation by, 189–92
Autocatalytic, second-order in products reaction (ASOP reaction), 236–38, 242, 244, 246
Automorphism, shift, 110–13
   near homoclinic orbits, 111–13
Autonomy, dynamical, of ecosystems, 472–75
Axiom A, 122, 126, 404, 411
Axisymmetric domains of bifurcating waves, weakly coupled parabolic systems on, 264–67

$\mathbf{B}$anach space, 35, 64, 150, 427
Behavior surface
   diffusional, 354
   kinetic, 353
Belousov reaction, 667
Belousov-Zhabotinskii reaction (BZ reaction), 279–95, 332, 623
   behavioral and stability properties of attractors corresponding to burst phenomena in open, 667–76
   bistability in, 286–88
   echo waves in, 288–92
   mechanistic details about, 279–81
   oscillations in, 281–86
Bénard problem, 51–62
Bifurcating waves, 263–78
   bifurcating points and spectral theory and, 267–74
   bifurcation of waves, 274–76
   echo, 288–92
   electrochemical, 356
   Padé approximants and center, 355
   stability of, 276–77
   weakly coupled parabolic systems on axisymmetric domains of, 264–67
Bifurcation
   analysis of
      in complex systems, *see* Complex systems
      by Poincaré, 5–26
   cusp, 153
   described, 296
   diagrams of, 127, 296, 496
   flip, 80, 619
   in general bifurcation theorem, defined, 28
   generic, 85
   global, 86, 608, 621
   historical background of theory, 1–4
   imperfect, 127
   secondary, 64, 169, 297, 579
   subcritical, 55

695